KW-467-956

WITHDRAWN
FROM
UNIVERSITY OF PLYMOUTH
LIBRARY SERVICES

CASS LIBRARY OF INDUSTRIAL CLASSICS

No. 4

EVILS OF THE FACTORY SYSTEM

EVILS OF THE FACTORY SYSTEM

DEMONSTRATED BY

PARLIAMENTARY EVIDENCE

CHARLES WING

FRANK CASS & CO. LTD.

1967

Published by
FRANK CASS AND COMPANY LIMITED
67 Great Russell Street, London WC1

First edition	1837
New impression	1967

Printed in Great Britain by
Thomas Nelson (Printers) Ltd., London and Edinburgh

EVILS

OF

THE FACTORY SYSTEM

DEMONSTRATED BY

PARLIAMENTARY EVIDENCE.

BY

CHARLES WING,

SURGEON TO THE ROYAL METROPOLITAN HOSPITAL FOR CHILDREN, ETC.

" Ætas parentum, pejor avis, tulit
Nos nequiores, mox daturos
Progeniem vitiosiorem."— HORACE.

LONDON

SAUNDERS AND OTLEY, CONDUIT STREET.

1837.

TO

LORD ASHLEY, M.P.

My Lord,

The permission which you have granted me to dedicate my work to your Lordship, is an evidence of your readiness to encourage any effort, however humble, in furtherance of a common cause,—a cause which you have so ably and so perseveringly, I wish I could add, so successfully, advocated. But hostile interests have set themselves in array against justice; the voice of humanity has been drowned amidst the clamours of political economists, and a system which has disgraced us as a nation has been pronounced to be absolutely necessary to our commercial prosperity. Of late, however, the dogmas which, though supported only by sturdy assumptions, were at first received with implicit confidence, have been impugned by the researches of Mr. Cowell; the false alarm which was created by misrepresentation has gradually subsided, and a legislative measure, which, in one or other of its enactments, has given universal dissatisfaction, has, I trust, prepared the way for sounder and more efficient legislation. That the cause of which your Lordship has never despaired, and upon the success of which the welfare and happiness of so numerous and important a class of the community depend, may at length triumph, is the fervent wish and prayer of,

My Lord,

Your Lordship's most obedient
and humble servant,

CHARLES WING.

London, January 1837.

PREFACE.

I AM anxious to guard the reader from a misconception of my opinion as to combinations. They were formerly illegal; but as the law, which made them so, placed the employed at the mercy of their employers, it was repealed. This repeal of the law, though founded in justice, was not unattended with evil. Combination gives power; power is liable to abuse; and that the operatives have abused it, is asserted by their opponents, and cannot be denied by their friends. The only remedy for this abuse is, not to make combinations illegal, but unnecessary. By rendering combinations illegal, you stifle complaints; but the object of good government is, not to stifle complaints, but to redress grievances. By redressing grievances, you render combinations unnecessary. The office of a delegate becomes a sinecure; and as sinecures are not popular, the people will not support them long. The statesman who shall think that he has done enough by stifling complaints, will find his error in times of public commotion.

With regard to my motives for publishing, nobody will suspect the fame of authorship to be one; a work that consists of authorities, and is more a record of the opinions of others than a vehicle for the promulgation of my own, can confer no literary distinction. As little am I influenced by the hope of gain; for if the sale of my work cover the expenses of publication, it is more than I expect. But there is a motive by which I am influenced, and which I need not scruple to avow,—the ambition of being of some service to a numerous class of my fellow-men. It was this ambition that influenced the late Mr. Sadler, who fell a martyr to the cause which he so perseveringly advocated; it is this ambition which now actuates Lord Ashley, and that band of philanthropists who support the same principles, with regard to the Factory Question, in both Houses of Parliament.

The reader will observe that I have divided my work into two parts. My reason for so doing was to expedite the publication of it.

With regard to its imperfections, they arise partly from its very nature, partly from professional interruptions. When a work is written upon the spur of the occasion, it is necessarily limited by time; for if time were taken to make it as complete as the author might be able to make it, the season for publication would pass by; and a work, however complete, that should be published out of season, would be of no use. It is scarcely necessary to add, that my time has been further limited by professional avocations. I trust the candid reader will find that, notwithstanding imperfections, I have brought evidence enough before the public to justify the title which I have given to my work,—a work which I should have gladly left to others, possessing more leisure, had I heard of any others likely to undertake it. I have endeavoured to supply a desideratum; how I have succeeded the public is the only judge.

CONTENTS.

PART I.

APPENDIX TO PART I.

PART II.

EVILS OF THE FACTORY SYSTEM.

REASONS FOR PUBLICATION.

I. THE object of the present publication is to prevent, as far as bringing evidence within the reach of the public can do so, a partial return to the factory system, as it existed previously to the year 1833, since a partial return to that system is in itself an evil, and may be the prelude to a total return. An act was passed in that year, of which the operation was to be gradual, but its ultimate aim was to prevent, from the 1st of March, 1836, any child who had not completed his thirteenth year from working more than eight hours a day. This act contained many provisions which the manufacturers found exceedingly troublesome and vexatious, and which were consequently observed by the conscientious, and defied or evaded by the less scrupulous. Memorials were sent to the Board of Trade, not seeking the total repeal of the act, but of that part of it which afforded protection to children under thirteen years of age. At the suggestion of the memorialists, the president of the Board of Trade brought in a bill to repeal so much of the act as prohibits the working of children under thirteen years of age beyond eight hours a day, and leaving it in the power of children, from twelve and upwards, to work twelve hours a day. The sole object of bringing forward the bill is said to be to prevent 35,000 children being thrown out of employment, and this object is effected by withdrawing the protection of the act from all those children who are in the thirteenth year of their age, and suffering them to be worked twelve hours a day, exclusive of meal-times. The inconsistency of ministers, in bringing forward this bill, is obvious. They threw out Lord Ashley's ten-hour bill at the recommendation of their own commissioners, who gave it as the result of the evidence they had collected, that the labour of children ought to be restricted to eight hours, and that, therefore, a ten-hour bill would not afford them sufficient protection. And now these same ministers would drive back a large portion of these children to twelve hours, that is to say, to the labour of adults, for twelve hours is the usual period of adult labour, though, upon emergencies, it is protracted to thirteen or fourteen hours.

Quicquid delirant reges, plectuntur Achivi.

Ministers bring in an inefficient bill, and 35,000 children are to suffer for it. Two thirds, or perhaps three fourths, of the hands employed in mills, are children or young persons, and their labour is strictly connected with the labour of adults. Ministers, anxious to afford protection to the children, but reluctant that the benefit of this protection should extend to adults, attempted to legislate for the children only. Their own inspectors tell them that the main provisions of the bill, framed upon this principle, are impracticable. That, in many situations, relays of children, which this bill renders necessary, cannot be procured. Now, as there is a mutual dependence of the hands upon each other, if the children, who are employed principally by the spinner, are dismissed, his work ceases, and the mill is at a stand, so that if children are restricted to eight hours' labour, and relays cannot be procured, the labour of the adult will, in fact, be restricted to eight hours. If, on the other hand, relays can be procured, either the labour of the children must be less than eight hours, or the labour of the adult must be extended to sixteen hours. Though the inspectors call the main provisions of the act impracticable, they must be understood as limiting this assertion to peculiar localities, for they admit that they have found benevolent men who adhere to the very letter of the law. Men act from mixed motives; and though the epithet conscientious seems more applicable to men who observe a law that is easily evaded, and that many do evade, yet we can easily conceive that one great incentive to this conscientious conduct is a repugnance to overwork the children, whatever they themselves may suffer by the law, and therefore they are well entitled to the epithet benevolent. Ministers found themselves in a dilemma; either they must overwork the children, or underwork adults,—and they have got out of the dilemma by determining to overwork the children. In their alarm, they have thrown consistency overboard; and the very same men who declared even ten hours' labour too long for a child in his thirteenth year, would now expose him to be worked twelve hours. The case of the factory children is not a party question; for though conservatives have been among their most active champions, yet both whigs and radicals have fought in the same ranks; and perhaps the present ministry would not be found so lukewarm in their cause, if it were not for the formidable front presented by the manufacturers, some of whom have seats in parliament, and most of whom can make their influence felt at elections. To the credit of this country, I repeat, this has not been made a party question; and I can well conceive that there are many noble minds in the present administration who would gladly have their humane intentions strengthened by the force of public opinion from without, and the main object of this work is to effect that purpose, by affording the public that well-attested information which has not yet been laid before them, and which, without a publication of a similar kind to the present, would obviously not be within their reach.

To one of the documents I have published, namely, the evidence taken before Mr. Sadler's committee, it has been objected, that it is exparte evidence, and that it was not taken on oath. With regard to the first objection, the principal part of this evidence was brought forward at the expense of the operatives, and, with the exception of the clergy, the medical men, and a few others, the witnesses were operatives. They who had actually experienced the evils of the system bore witness to those evils; and if their evidence could have been disproved or contradicted by witnesses sent by the manufacturers, and was not so, it was not the fault of the operatives; they could hardly have been expected to be at the expense of bringing evidence against themselves. With regard to the second objection, that the witnesses were not sworn, I question whether this would much affect the evidence of the clergy, or of the medical men, and perhaps it would affect the evidence of the operatives much less than their opponents imagine, even if the operatives deserved the character which they give them; for that his word is as good as his oath may be said of a knave as well as of an honest man. I think that there is internal evidence that these operatives are honest men, for they assert nothing that is improbable, nothing that is at variance with the evidence of the clergy or of the medical men, nothing that is at variance with the evidence that had been repeatedly brought before the House upon former occasions. If they speak in a more indignant tone than their predecessors, it is because their predecessors had repeatedly spoken in vain. If they exaggerate, prove their exaggerations, but do not on that account entirely reject their evidence. Exaggeration is the natural language of the injured. Where did you ever meet with the man who felt himself deeply wronged, and who was at the same time prudishly solicitous lest a word should escape his lips that might be construed into exaggeration? We read, indeed, in ancient history, of a client of Demosthenes, who applied to him to undertake his cause against a person who had assaulted him, and that, in giving his account of the assault, he was not guilty of a single exaggeration, but stated the matter with the utmost coolness. "Tut, tut," said Demosthenes, "you been assaulted!—I cannot believe it." "What!" said the man, his eyes flashing with indignation, "have I not received these wounds?" "Now, indeed," replied the orator, "you speak like a man that has been injured." The operatives speak like men that have been injured, with considerable warmth, with something, perhaps, of exaggeration, and this is the utmost that can be alleged against them: not a single assertion, as to a matter of fact, has been disproved; not a single contradictory statement pointed out. They sometimes, in answer to the questions proposed to them, advance things as matters of opinion with regard to the probable effect of Mr. Sadler's bill; and here they evince great diversity of judgment, and frequently are at variance with the political economists. But is there not great diversity of judgment in the political economists themselves, and are they not frequently at variance with each other? I am far from joining in the ignorant clamour against political economy; I consider that clamour to be raised by persons who are utterly unacquainted with the science, and their evidence against it ought to have just the same weight as the evidence of the countryman against letters, who, when he saw his townsman carried to execution for forgery, exclaimed,—"Aye, aye! this comes of writing and reading!—thank God I can do neither!" But there are sciolists in political economy, as in other sciences, and to this branch of knowledge is emphatically applicable what Pope says of knowledge in general:

> A little knowledge is a dangerous thing;
> Drink deep, or taste not the Pierian spring.

There is no description of persons so apt to dogmatise as sciolists; and it is not those who have sounded the depths of political economy that have pounced upon the discrepancies of opinion existing among the operatives, evinced their liberality by descanting upon the freedom of labour, and driven our ministers from their consistency by exaggerating the danger of foreign competition.

Another objection against the evidence brought before Mr. Sadler's committee is, that it describes things that happened long ago, and that the state of the factory system is now very different from what it was formerly. With regard to this objection, I maintain that, though some parts of that evidence relate to things that happened long ago, the greater portion of it relates to things that are recent, and that he who wishes to form a correct estimate of the factory system must not confine his attention to its present improved state, but take both its present and past state into consideration. To what is this improvement to be attributed? To the perpetual agitation of the question, to the frequent strikes among the operatives, to legislative enactments, and to the voice of public opinion. It was the evidence brought before Mr. Sadler's committee that first awakened the public to the enormities of the system,—enormities which all the ingenuity of its advocates has not been able to gloss over, and to which all the exertions of its opponents have not been able effectually to put a stop, as is but too well proved by the numerous convictions of the masters for overworking the children. The present mitigation of these enormities, so far from being an argument for relaxing our efforts, is an incitement to perseverance; if those efforts had been made in vain, then there would be reason to despair; but they have already done much good, and, if persisted in, will make the factory system a blessing instead of a curse to the community.

There are two sufficient reasons for publishing the evidence of Mr. Sadler's committee, the first is, the praises it has received from the friends of the children; and the next, which is equally cogent, is, the abuse which has been accumulated upon it by their enemies. We do not scruple to call them enemies; for he who would subject a lad, in the thirteenth year of his age, to the labour of an adult, whatever he may babble about freedom of labour and foreign competition, is, in the strictest sense of the term, the children's enemy. With regard to the praises of this evidence, it will be sufficient to quote the opinion of Mr. Fielden, himself a manufacturer, and brought up from boyhood in a cotton-mill, of which his father was the proprietor:—

"In 1832, the late Mr. Sadler made great efforts in favour of the factory children. He brought a bill into Parliament to limit the hours of labour for all under eighteen years of age to fifty-eight hours in the week; and the provisions of this bill were to extend to woollen, flax, and silk, as well as cotton mills. On moving the second reading, on the 13th of March, he was met by strong opposition, and a cry for investigation. Unable to resist it, he acceded to a committee being appointed; and of that committee he became the chairman. It made no report to the House, excepting of the very *valuable evidence* that it had taken, so that we have to wade through an immense volume in order to arrive at conclusions upon all the points which it embraces: but no one can open this volume of evidence without perceiving that all the attempts that have been made by parliament are shamefully set at nought, that children are still worked beyond their strength, and that they cry aloud for protection. The most important evidence, however, taken before this committee, was that of the eminent surgeons and physicians of London. There were examined, amongst many others, Dr. Blundell, Sir A. Carlisle, Sir Benjamin Brodie, Dr. Roget, Dr. Farre, Sir G. Tuthill, Sir C. Bell, Mr. Green, Mr. Key, Mr. Travers, and Mr. Guthrie. I appeal to every man but the sordid, to read the evidence of these gentlemen. From some of them we have mild, to be sure, but serious, expressions of horror at our cruelty; by others we are openly denounced as the murderers of infants. From them all we have a lesson that we ought to attend to; and the parliament is distinctly told, that, if it is a part of its duty to make laws to protect men from the arm of the murderer, laws of the same protecting kind are necessary in the case of these children, where the murder is as certain as in any other instance, and more cruel, because the death is more lingering." (Fielden's Curse of the Factory System, pp. 13, 14.)

With regard to the abuse that has been accumulated upon the evidence of 1832, I might quote from our opponents passim, but I defer this part of my subject till I come to the examination of Dr. Ure's works.

As to the report of the government commissioners, since it has been acted upon by those whom we are obliged to consider as our opponents, of course they will raise no objections to it; indeed, it appears to be the report of men exercising the utmost, perhaps excessive, candour towards the manufacturers, and compelled, in spite of themselves, to give their verdict against the factory system by the sheer force of truth. For the information of those of my readers who may not have previously acquainted themselves with the subject, I hasten to do what has already been ably done by several of my predecessors, namely, to give a brief historical survey of the factory system, as far as regards the acts of the legislature.

HISTORICAL SKETCH OF THE FACTORY SYSTEM.

II. There is no class of persons to whom the celebrated lines of Virgil—

Sic vos non vobis vellera fertis oves, &c.

are more applicable than to inventors. It is not generally the inventor, but the man who makes his fortune by the invention, that carries away the honour. The invention of the mode of spinning by rollers, generally attributed to Sir Richard Arkwright, justly belongs, as is satisfactorily proved by Mr. Baines, in his History of the Cotton Manufacture, to John Wyatt, of Birmingham. The patent for the invention was taken out in the year 1738, in the name of Lewis Paul, a foreigner, with whom Mr. Wyatt was in partnership; and though the name of John Wyatt only appears as a witness, Mr. Baines has adduced sufficient evidence to prove that he was in fact the inventor. Sir Richard Arkwright took out his patent for a similar machine in 1769. The invention of Wyatt was identical in principle, though not in all its details, with the machine of Arkwright; but for the want of capital, and the inability of Paul, who was a mere adventurer, to fulfil his promises of pecuniary aid, and also, perhaps, from the want of that indefatigable perseverance in perfecting the details of the machinery, which distinguished Arkwright, Wyatt was not destined to reap the fruits of his invention. Another claimant to the invention is Thomas Highs, reed-maker, of Leigh; but his claim, though supported by Mr. Guest, in his History of the Cotton Manufacture, wants that which alone could substantiate it, namely, priority of publication; and is strongly controverted by Mr. M'Culloch, in his article on the "Rise, Progress, Present State, and Prospects of the British Cotton Manufacture," in the Edinburgh Review. James Hargreave, a weaver, of Stand-hill, near Blackburn, was the author of the spinning-jenny. The date of this invention is generally supposed to be 1767, two years before Arkwright obtained the patent for his water-frame, so called from its being turned by a water-wheel. The spinning-jenny differs so completely from the machines of Wyatt and of Arkwright, that there is no dispute as to the originality of the invention. Hargreaves was driven from his native county by the persecution of the spinners, who raised a mob, broke into his house, and destroyed his jenny. He went to Nottingham in 1768, and having entered into partnership with Mr. James, and erected a small mill in Hockley, he took out a patent for the jenny in 1770. These two inventions, that of the water-frame and the spinning-jenny, instead of coming into conflict, were brought into use together, and, by their combined operation, gave an impulse to the cotton manufacture which all the efforts of the workmen, and of individual manufacturers, who placed every obstruction in the way of the inventors, could not counteract. The carding machine, for which Lewis Paul obtained a patent in 1748, was introduced into Lancashire by a gentleman of the name of Morris, and afterwards adopted by Mr. Peel, who is said to have erected a machine with cylinders, by the aid of James Hargreaves, at Blackburn. The numerous machines which Arkwright brought out in a finished state completely changed the condition of the manufacturing population. The cottage of the workman could not afford space for the improved machines, and a greater power than that of the human arm was requisite to put them in motion. Mills were built, at first, one at Nottingham, which was driven by horses, and afterwards another, on a much larger scale, at Cromford, in Derbyshire, which was turned by a water-wheel. That the system, thus introduced, should be violently opposed by the workmen, is not to be wondered at, though it does not justify their violation of property in the destruction of machines. They were taken from their cottages, where they worked at their pleasure, with more or less intensity, and at a time when, in consequence of the demand for labour being greater than the supply, their wages were amply sufficient to maintain them, and placed in mills, where their labour was regulated by the machinery, and where sordid masters dictated what wages they chose, and what hours they chose; and the workman who should remonstrate was immediately thrown out of employment. The workmen were refractory; and as the employment of water for the harder part of the work, and the improvement of the machinery, made the masters almost independent of the adult operatives, children were employed in their stead. This demand for children was at first supplied by apprentices from the workhouses of London, Edinburgh, and other large cities. The overseers of the mills were paid by the master manufacturer according to the quantity of work performed, and consequently, it was their interest that the children should do as much work as possible. They did not neglect their interest, and accordingly the poor apprentices, having no natural protectors, were worked to the utmost, and treated with all the brutality of which the human heart, hardened by the desire of gain, is capable. Benevolent individuals became acquainted with the wrongs to which the children were subject, and made known their sufferings through the channel of the public journals so early as 1796. This first cry of the children was, at that time, unheeded, owing partly to the magnitude of the events which absorbed the public mind, partly to that apathy with regard to minor sufferings, which the contemplation of great atrocities is apt to create. Pity and terror had been too much exhausted by the excesses of the French revolution to be easily excited; and the appeals that were then made, not to the legislature, but to the humanity of the public, were, for a time, made in vain. The evils of the system went on increasing rather than diminishing, till at length the children met with a champion from a quarter whence it was least expected. Sir Robert Peel, himself a manufacturer, took up the question in 1802, and brought in a bill, which became an act of parliament, for the preservation of the health and morals of apprentices and others employed in cotton and other mills, and cotton and other factories. The provisions of this act were—

1. For the due ventilation and washing of the factories.

2. The proper clothing of the apprentices.

3. The limitation of their labour to twelve hours daily, and not permitting it at night.

4. Requiring each apprentice to be instructed, in some part of every

working day during the first four years of his apprenticeship, in reading, writing, and arithmetic.

5. The separation of the sexes.

6. Sunday instruction, and the attendance of the apprentices at divine service, and occasional examination by the rector, vicar, and curate, of the parish.

7. Authorizing the justices at quarter sessions to appoint visitors of such factories, with requisite powers.

The hardships of the system to which the apprentices were subject, may be collected from the fact, that a limitation of the hours of labour to twelve, exclusive of meals, and of the time appointed for instruction, was all the mitigation sought for. The act had no other effect than gradually to do away with the employment of apprentices. The manufacturers were at first compelled to employ apprentices, from the impossibility of procuring a sufficient number of children from the surrounding population; most of the mills were in thinly populated districts, their position being determined by the facility of employing water to work the machines. But the plan of employing apprentices was always troublesome to the master, he had to feed them, clothe them, lodge them, supply them with medical advice and religious instruction; and though the latter duty was, for the most part, neglected, the former could not be. He was at length relieved from this weight of responsibility. The improvement of the steam-engine by James Watt, so as to adapt it to the production of rotatory motion, and the working of machines, enabled the manufacturer to build his mills in the midst of the most dense populations; and the children of the neighbours were, on almost every account, preferable to apprentices from distant quarters, and particularly, because they were exempt from the restrictions of Sir Robert Peel's bill. The children in the immediate neighbourhood were consequently employed; and though the act of parliament, while recent, might have some little effect in mitigating the evils of the system, and the presence of their parents, or guardians, was some protection to the children, these salutary effects soon ceased, and the state of the children then employed became nearly as bad as that of the apprentices had been. This return to the evils of the former system took place gradually; and after the interval of thirteen years, Sir Robert Peel again became the champion of the children. In 1815 he made a motion for the appointment of a committee, to inquire into the expediency of applying the former act to children of every description. After the production of a vast mass of evidence in 1816, (extracts from which I have published) proving, beyond the possibility of doubt, the existence of the most scandalous abuses, and coinciding, in the main, with the evidence afterwards brought before Mr. Sadler's committee, in the year 1819, he succeeded in obtaining an act, by which no child under nine years of age should be allowed to work in a cotton factory, and no young person under sixteen be allowed to work more than twelve hours a day, exclusive of meal times. This miserable pittance of relief extended only to the children working in cotton-factories, though equal abuses existed in the woollen and linen factories. After the passing of this act, there were four others to render it valid, all of which were repealed by the act, called Sir John Hobhouse's act, by which it was unlawful to work any child in a factory who should be under eighteen years of age, more than sixty-nine hours a week. Like the act of Sir Robert Peel it was confined to the cotton factories; and the relief afforded, though somewhat more, was still very inadequate. On the 15th of December, 1832,

Mr. Sadler, at that time a member of the House of Commons, moved for leave to bring in a bill to regulate the labour of children in the mills and factories of the kingdom. Leave was given, and the bill was ordered to be brought in by Mr. Sadler, Lord Viscount Morpeth, and Sir Richard Vyvyan. On the 1st of February in the same year, Mr. Sadler presented a petition, signed by 9000 persons, operatives and others, employed in the woollen, flax, worsted, and other mills and factories in Leeds, praying for the restriction of the actual labour of children and young persons employed therein. In the discussion that took place upon this petition, the speakers were nearly unanimous in its favour; among those who have since a little changed their sentiments upon the subject was Mr. O'Connell. Colonel Torrens, it is true, suggested, that without the repeal of the corn laws, and without imposing the taxation, which he represented as grinding the poor to the earth, upon another class of the community, no substantial good could be done for the factory children; but he, at the same time, declared, that he did not rise to express his dissent from Mr. Sadler's measure. The petition was consequently ordered to be printed. This was only one of a series of petitions presented by Mr. Sadler, and numerously signed. Other gentlemen, also, presented petitions in favour of restrictions on the hours of labour, and petitions of an opposite description were also presented, though not half so many. Why they were not so numerous the master manufacturers can best explain; I am inclined to attribute it rather to want of willingness on the part of the operatives to sign the petitions drawn up by their masters, than to want of activity on the part of the masters. The operatives in full employment would naturally do much for their employers, but to sign petitions against themselves was a task, to which, at that time, there seems to have been some difficulty in bringing them. That it required no ordinary degree of courage to bear evidence against their masters is evident, from the two petitions brought forward by Mr. Sadler on the 30th of July, 1832, from persons deprived of their employment in consequence of their having given evidence before Mr. Sadler's committee. However cautious witnesses might be not to say anything that they could avoid saying against their masters, their caution did not prevent their being dismissed on their return to the factory. To illustrate the difficulty of getting evidence against the master manufacturers, I shall give some extracts from Mr. Sadler's statements upon the subject:—

"I hold in my hand two petitions from individuals of the humbler ranks of society. These petitioners complain of dismissal from their employment in consequence of having been summoned to give evidence before the committee on the bill for regulating the hours of labour of children employed in factories, and pray for some compensation. I want to know from honourable gentlemen who may laugh, what course these men are to take to obtain redress, who are summoned before Parliament to declare the truth? I ask, whether the summons is not imperative? I ask, whether it is fitting that these individuals should be exposed to suffering and ruin, without any indemnification being afforded them.

"The first petition I shall present—and it is, perhaps, the first only of a series on the same subject—is from a person named Charles Stewart. It was suggested that this person should give evidence before the committee. He was a reluctant witness, but, nevertheless, conducted himself in a most becoming manner; and though absent from his work only so long as was absolutely required, found himself, on his return home, supplanted in his situation. This has been the case, too, with several others. This petitioner casts himself upon the kindness of the House. He states that he was called before the committee of this House, and upon the authority of the House—a summons which he could not evade—but that, in consequence of having given evidence conscientiously, he is deprived of his bread, and therefore prays the House to grant him some compensation.

"I hold in my hand a similar petition of another individual, named Alexander Deans. I beg to say that this individual, without being a voluntary witness, was pointed out to me as a person whose evidence was desirable; and his name having been transmitted to me, which was without his knowledge, he was accordingly summoned. Having given his evidence, he returned home as expeditiously as he possibly could; and now finds himself dismissed from his employment.

"I have several similar representations to those made in these petitions regarding other parties brought before committees on this subject; and so convinced am I of the accuracy of the circumstances stated, that I have declined bringing forward many important witnesses, under the apprehension of their being ruined for giving evidence before the committee now sitting.

"I beg leave to say, that nothing could be more respectful than the manner in which the testimony of these witnesses was given in regard to their employers; and I may mention an instance, in particular, in which the witness said,—'With regard to my employer, he is one of the best and kindest masters known in the business, and I hope that my testimony will not influence his mind against me.' But it did so influence his master's mind, that he discharged him from his employ."—(*Mirror of Parliament for* 1832, vol. iv., p. 3462.)

Though the operative might not run the same risk in signing a petition in favour of the factory bill as in giving evidence in its favour, because his master could not so easily ascertain the fact of his having signed it; yet if the fact should by any means transpire, and come to the master's knowledge, we may infer that the action, being voluntary, would be visited with at least as heavy a penalty as that of an involuntary, and frequently reluctant, witness. I must therefore attribute the numerous petitions in favour of the bill to the general feeling with regard to its object, while I ascribe the petitions got up against the bill either to the wish of the operatives in full employment to curry favour with their masters, or to the fear of incurring their masters' displeasure.

On the 16th of March, 1832, Mr. Sadler moved the second reading of his bill for regulating the labour of children and young persons employed in the mills and factories of the kingdom, of whatever description. Mr. Sadler's speech, and the debate that took place upon the occasion, I have given at length. The bill was read a second time, and, on the motion of Mr. Sadler, a committee was appointed, consisting of some of the most distinguished members of the House, without reference to their opinions upon the bill. This committee made no report, but the evidence brought before them was printed, and spoke for itself. It coincided with the evidence brought before former committees, and fully made out the different statements in Mr. Sadler's affecting and convincing speech. But though the opponents of Mr. Sadler's bill could not maintain their opinions, they were determined to retain them; they accordingly objected to the evidence as ex parte, and not taken upon oath, and, owing to their pretended scruples upon a question that seemed to admit of no scruple whatever, a mass of evidence from workmen, clergymen, and medical men, was to be consigned to oblivion, and considered as if it had never been brought forward; and for the expenses to which the operatives had been put in making out the grievances under which they and their children laboured, they were refused compensation, upon the plea that the central committee of operatives was not a body recognised by the House, and that therefore no petition from them could be received. This evidence, in justice to the operatives, that they may not have spent their money for nothing, I have published; that of the operatives themselves and of the clergymen in an abridged form, that of the medical men at full length. We have given the evidence of the latter class at full length, instead of merely stating their opinions, that our opponents may be enabled to examine the reasons upon which their opinions are grounded, and not complain of being overwhelmed by the authority of great names. The public may now judge whether the evidence brought before Mr. Sadler's committee deserves the abuse so lavishly heaped upon it, or whether it does not bear the clearest marks of truth; whether it does not coincide with preceding evidence, and whether it has not been confirmed by subsequent inquiries.

In the session of 1833, Mr. Sadler was not a member of the House, and the children lost a faithful, persevering, and undaunted champion. Their friends were somewhat disheartened, for they thought, with regard to Mr. Sadler,

> Si Pergama dextrâ
> Defendi possent, certè hâc defensa fuissent.

But the cause of humanity was not suffered to languish. On the 5th of March, Lord Ashley moved for leave to bring in a bill to regulate the labour of children in the mills and factories of the United Kingdom. The motion was agreed to, and the bill ordered to be brought in by Lord Ashley, Mr. John Fielden, and Mr. Cutlar Fergusson. On the 17th of June, when Lord Ashley moved the order of the day for the second reading of his bill, the Chancellor of the Exchequer mentioned that the opinion of the commissioners, who had been appointed by government to examine into the state of the factory system, was, that with respect to children under fourteen years of age the bill did not go far enough, but that in other respects it went too far. The commissioners had not the evidence that was brought before them taken down at the time of its delivery, but had merely made abstracts of it, and required further time to make their report. The bill was read a second time; but, at the request of Mr. Poulett Thomson, Lord Ashley postponed the committal of the bill to the 28th of June, that the House might have adequate time for considering the report of the committee. When the 28th came it was further postponed to the 5th of July, upon which occasion, after a long discussion, the House resolved itself into committee upon the bill. On the 18th of July, the committee of the House began its labours on Lord Ashley's bill, and the second clause, which contained the principle of the bill, being rejected by a vast majority, there being only 93 for Lord Ashley's motion, and 238 against it, his lordship gave up the bill, and left the question entirely in the hands of his Majesty's ministers. On the 9th of August, the House having resolved itself into committee, the Chancellor of the Exchequer explained the character and provisions of the bill, the adoption of which he wished to press upon parliament. The three leading principles of the bill were:—first, the extension of Sir John Hobhouse's bill from cotton to all mills and manufactories, with certain exceptions with respect to silk-mills; second, that children under thirteen years of age are not to be allowed to work for more than eight hours a day, and that those under eighteen years of age are not to work more than sixty-nine hours in the week, the observance of which regulations are to be secured by the appointment of inspectors; and, third, the introduction of a general system of education for all the children employed in the manufactories.

On the 16th of August, 1833, the bill was read a third time, and passed; with some unimportant alterations, that were acquiesced in by the Commons, it passed the House of Lords, and received the royal assent on the 29th of the same month. This act I have published; its provisions were to come into operation gradually; and it was not till the 1st of March, 1836, that that part of it came into operation which prohibits all under thirteen years of age to work more than eight hours a day. On the 9th of May, a debate took place upon the second reading of Mr. P. Thomson's Factories' Act Amendment Bill, which debate I have published. The object of Mr. P. Thomson was to repeal so much of the former bill as prohibited all under thirteen years of age to work more than eight hours a day, and to permit them to be worked twelve. In 1833, in the course of debate, Mr. Thomson had complained that the bill which government brought forward was forced upon it, though he did not explain by what individual, or by what class of individuals, it was forced upon government. It was not forced upon government either by the late Mr. Sadler or by Lord Ashley, for they wished for a ten hours' bill, and the latter publicly washed his hands of the government measure; it was not forced upon government by the operatives, for they also wished for a ten hours' bill; it could not be forced upon government by their own commissioners. But, though it would have puzzled Mr. Thomson and his coadjutors to name the individual or party by whom they had been forced to bring in a bill upon the subject, yet they acted as most persons who act upon compulsion do, they brought in a bill which should give satisfaction to no party,—a bill, in short, that should speedily call for a repeal. You may lead a horse to the trough, but you cannot make him drink. You may, it seems, from Mr. Poulett Thomson's confession, force government to a measure, but they will take good care that it shall neither be sound in principle, nor lasting in its effects. To bring in a measure that should really mitigate the evils of the factory system, the heart must be in the cause; when it is not, the right arm is withered. Mr. Thomson gained his point by a majority of two only, there being 178 for the second reading of his bill, and 176 for Lord Ashley's amendment. Such a victory was little better than a defeat, since it was impossible for ministers, with any sense of decency, to persist in undoing what they themselves had done for the protection of the children, when backed by so small a majority.

Having thus given a sketch of the factory system, as far as regards legal enactments, I shall proceed to discuss two preliminary objections against legislative interference.

PRELIMINARY OBJECTIONS WITH REGARD TO THE FREEDOM OF LABOUR AND FOREIGN COMPETITION.

III. There are some who are ready to admit the evils of the factory system, but who contend that they cannot be remedied without interfering with the freedom of labour, and bringing our commerce into danger from foreign competition. It was urged as an objection against the ten-hour bill, that as the labour of the adult was linked to that of the children, if the children were restricted to ten hours' labour, the adult must also be restricted to the same time, and that this would be an infringement upon the freedom of labour. This, it is true, was not the only objection to the ten hours' bill, for ten hours were also declared to be too long a period for children to work; but it was an objection much insisted upon by the opposers of the bill. It was contended that the operatives themselves would probably be the first to call for the repeal of an act which, by restricting their hours of labour, would reduce their wages, and that to interfere between the master and the adult operative was against the first principles of justice and of political economy. It could not be contended that the operatives themselves looked upon this infringement on the freedom of their labour as an act of injustice, for they loudly called for it, but that they were led to do so by a fallacious hope that their wages would not be diminished by a rise in prices. Freedom of labour, according to Mr. Rickards, as quoted by Mr. Fielden, in the forty-first page of his work on the factory system, is explained to be the right which a man has to work himself to death if he pleases. The adult is a free agent, and consequently you cannot work him beyond his strength against his will. This is perfectly true; you cannot compel the will, for to compel it would be to destroy it; a man's actions are not his own unless they are voluntary. But when the alternatives proposed to the will are none of them eligible, abstractedly considered, the freedom of the will is of little benefit. The traveller who delivers his purse to the highwayman rather than run the risk of losing both his life and his purse too, is actuated by his will. When we talk of a man's acting against his will, we merely use popular language, and are conscious of its logical inaccuracy. When Shakspeare talks of poverty, and not the will, consenting, he speaks as a poet. Now, in consequence of the improvement of machinery, and the substitution of the labour of children for that of adults, the adult operative is at the mercy of his employer. If he exercise the freedom of his will, however conscientiously, in such a way as to displease his master, he is thrown out of employment. There are more than enough of others to supply his place. If he apply to the parish, relief will not be extended to a man who refuses work. He would gladly, therefore, give up that freedom of labour which leads him into such a miserable dilemma. But there is another sense in which the words freedom of labour may be understood, and he would gladly receive a little of this other commodity in exchange for what he gives up. Freedom is defined, by Sir James Mackintosh, in his Lectures on the Law of Nature and Nations, delivered in the hall of Lincoln's-inn, to be, security from wrong. Political freedom is security against wrong from our governors; civil freedom is security against wrong from our fellow-subjects; religious freedom is security against wrong from those in whose hands ecclesiastical power is lodged. Freedom of labour, according to this definition, is security against wrong from our employers, and we can possess it only by legislative interference. That freedom of labour with which the mill owner would on no account interfere, but allow the operative its unlimited possession, is attended with no blessing; it brings in its train bodily exhaustion, decline of health, moral deprivation, and abridgment of life. On the other hand, that freedom of labour which so great a majority of the operatives wish for, and which government alone can give, would remedy these evils; it would mitigate the irksomeness of toil, cheer the hours of confinement, and sweeten the intervals of leisure. It would restore vigour to the workman's limbs, and cheerfulness to his mind; it would do more to disperse that infidelity which is reported to prevail so extensively in the manufacturing districts than a host of polemics. The master and the operative would no longer frown defiance upon each other, but consider their interests as conjoined, and the shrill cries from the factories that have pierced the public ear for more than thirty years would be heard no more. "No such thing," say some of the mill-owners, backed by some political economists, "instead of these salutary effects, our commerce would be ruined, and the vast mass of individuals whom our manufactures have brought together would be thrown upon the community for support. Foreigners are every year advancing upon us with rapid strides, and if they once trip us up, we fall to rise no more." It is satisfactory to be able to say, that only some of our mill-owners are now of this opinion; they were at one time almost unanimous, but the obstructions which the government act has put in the way of the master manufacturers have made many a convert to the ten-hour bill. If foreigners are yearly advancing upon us in the race of competition, they must, after a time, overtake us; and, consequently, by allowing the masters to work the operatives as long as they choose, we are not preventing, but only delaying, the crisis which the master manufacturers apprehend will come, and the longer it is in coming the more tremendous the crash when it does come, since whatever number it might throw upon the community at present, it will throw a much larger number at any future period; and sound legislation looks to posterity, and will not remove a comparatively smaller load from the shoulders of the present generation, to impose a ruinously heavy load on those unborn. Besides, should foreign competition be more rapid in its advance upon us than is calculated upon, we ourselves might have to rue the effects of our own selfishness; but, after all, the danger of foreign competition destroying our commerce may not be either so immediate or so certain as it is represented to be. This destruction is talked of by some as if it were inevitable, and as if it were to come upon us like a thunder-bolt; but there are others who doubt whether it will come at all; there are others who have the hardihood to assert, that promoting a good understanding between masters and workmen is one way of averting it; and these opinions are held even by some of the masters themselves. We all know that prophecies with regard to the ruin of commerce are by no means infallible. When the bill for the abolition of slavery was discussed, it was confidently predicted that, if passed, the commerce of Liverpool would be utterly annihilated, and the handsome houses which adorn that fine town be converted into wretched fishing huts. Yet, notwithstanding these gloomy forebodings, there are at present no signs of this utter destruction of its commerce and dilapidation of its buildings, and perhaps the vaticinations of the political economists with regard to the ruin of our manufactures may proceed from false alarm; but our opponents will reply that these vaticinations are founded upon statistics. With regard to this very useful, though much-perverted science, I quite agree with Laplace. "Appliquons aux sciences politiques et morales la méthode fondée sur l'observation et sur le calcul, méthode qui nous a si bien servi dans les sciences naturelles."*—(*Laplace, Essai sur les Probabilités.*) But I would subjoin to this recommendation a very necessary caution of Quetelet's:—"Il serait à desirer que chaque fois qu'on emploie des nombres et qu'on en déduit des résultats, on êut soin de calculer les erreurs probables. Il ne suffit pas de posséder des materiaux, il faut encore en connaitre la valeur. Un des plus grands défauts des statistiques actuelles est de présenter indistinctement, sur la même ligne, tous les nombres que l'on a pu recueillis et de les faire concourir à un même

* Let us apply to political and moral sciences the method, founded on observation and arithmetical calculation, which has been of such essential service to us in the natural sciences.

résultat, sans tenir compte de leur importance ni de leur valeur probable. Cette confusion doit nécessairement apporter de grands obstacles à la marche des sciences et faire prévaloir pendant long-temps encore des erreurs dangereuses."*—(*Quetelet, sur L'Homme*, tom. ii., p. 295.) I will quote a celebrated writer of our own country, speaking of the proportion of crimes in the manufacturing and agricultural districts:—"In the course of my inquiries on this subject, I have met with two or three tables, laboriously compiled and quoted from many authorities, the names of which are respectable. I should have given them, for they proved what I do not doubt, viz., that a far greater number of crimes, more especially crimes against property, are perpetrated in the manufacturing than in the agricultural countries of Europe. But, strange to say, on referring to such official documents as were within my reach, though I came speedily to the same result as the tables in question, I hardly found one of their figures correspond with those before me. There is in general, I believe, much guess-work in the framing of those imposing-looking statistics, which are meant at once to settle the doubting reader."—(*Bulwer's France*, second series, p. 151.) What is here said of particular statistical tables is true of statistical tables in general; they must not only be thoroughly sifted, but the utmost caution must be used in deriving results from them. The disadvantages under which our country labours with regard to commerce are, the weight of taxation, which the national debt renders it very difficult materially to reduce, and, what is generally considered as one of the consequences of the weight of taxation, the corn laws. Some political economists do indeed tell the agriculturists that free trade in corn would promote their interests, as well as those of the manufacturers; but the agriculturists, seeing that political economists are divided in opinion on the subject, are *rather* sceptical upon that point, and would wish our legislators not to withdraw the protection they have hitherto afforded them till taxation is sufficiently reduced to make the experiment *safe*, since, should it fail, their condition will be most deplorable; for in the factories, so far from there being a demand for adult labour, there is an over-supply; and, indeed, if the contrary were the case, after being accustomed to handle the plough and the flail, the agricultural labourer is not very sanguine that he should give satisfaction to his employers as a spinner. We must therefore take it for granted, that the manufacturers will long have to contend with the above difficulties, the weight of taxation, and the corn laws; but, on the other hand, the advantages they possess are such as more than to counterbalance these disadvantages. I give them as stated by Mr. Baines, who, being a zealous advocate for the manufacturers, will not be suspected of over-rating them. The passages I shall extract are from his History of the Cotton Manufacture, published in 1835.

"No symptom has yet appeared to indicate a decline, or even a stagnation, in the cotton manufacture of England. Every year, with scarcely an exception, presents an increase in the raw material imported, and the manufactured goods exported." (p. 507.)

"With so many natural and acquired advantages, which, in their combi-

* It would be desirable, that whenever numbers are employed, and results deduced from them, their probable errors should be carefully considered. It is not sufficient to possess materials; we must also ascertain their value. One of the greatest faults of statistical tables, as at present drawn up, is, that they present upon the same line all the numbers which had been collected, and make them concur to the same result, without taking into account either their important or their probable value. This confusion necessarily occasions considerable obstacles to the progress of the sciences, and for a long time has been the cause of dangerous errors.

nation, are altogether unrivalled, and with an entire absence of any symptom of declension, there is good reason for believing that the cotton manufacture of this country will continue to flourish; and if it does not, as in the nature of things is impossible, still advance with the same giant strides as in the period that immediately followed the great mechanical inventions, we yet feel a confident expectation that its course will be steadily onward."

"In each of the countries mentioned as likely to compete successfully with England, there are circumstances unfavourable to such competition. In the United States, the high rate of profit, the expensiveness of machinery, and a rate of wages higher even than in England, will, for a long course of years, prevent the manufacturer from selling his goods so cheap as the English manufacturer; whilst the advantage of having the raw material produced within the boundaries of the republic is small, seeing that the cotton is not grown within many hundred miles of the manufacturing states." (p. 508.)

"If the English cotton manufacture is in little danger from that of the United States, it is certainly not in greater peril from the same manufacture in France. The French consume a somewhat larger quantity of cotton-wool than the Americans, and are indeed second only to England, though their production is only about one-fourth that of the English. In the silk manufacture, the French are unequalled, though our own country is pressing hard upon them in this respect; they are pre-eminent in taste and fancy, possess much ingenuity, and rank very high in chemical knowledge. But they labour under such serious disadvantages for conducting manufactures on the large scale, that there is not the least prospect of their ever successfully competing with this country in the manufacture of cotton.

"1st. The national character and habits of the French are unfavourable. Though they have an abundance of energy, they lack that close attention and persevering application which are indispensable to the attainment of the highest skill, and to regularity of operations in an extensive manufactory. The weavers, and even many of the spinners, cannot be induced to work the year round at their looms or mills, but, in the months of summer and vintage, turn to agricultural pursuits for relaxation; a practice which, however agreeable and healthful, is incompatible with high proficiency in any manual art, and most seriously interrupts the operations of the manufactory. It is the combination of perseverance with activity and intelligence, that makes the English artisan unrivalled.

"2nd. The political state of France is unfavourable. Wars, invasions, and revolutions, and the liability to their occurrence, have shaken credit, and prevented the manufacturing establishments from gaining that duration and firmness which are needful to the perfection of their arrangements, and to the full development of mercantile enterprise.

"3rd. France has natural disadvantages, especially in the comparative scarcity of fuel and iron. Coal is not largely found in that country, nor is it raised without considerable expense, and the supply of wood is inadequate to the wants of the manufacturer. The manufacturers of Paris use the coal brought from Mons, but it costs them ten times the price given for that article at Manchester. Iron is also far from abundant, and is therefore dear.

"4th. The artificial state into which French manufacturing industry has been brought, from being propped up on every side with protections, and therefore incapable of free movement, greatly aggravates the natural disadvantages of the country. Coal and iron might be imported far more cheaply than they can be raised in France, but duties, nearly prohibitory, are levied upon these articles when imported, to protect the domestic iron and coal proprietors. Of course, these duties fall directly upon machinery, which is in consequence double the price in France that it is in England. The protection of the proprietors of iron and coal mines renders it necessary to protect the makers of machinery; and the protection of the latter renders it indispensable to protect the cotton manufacturer. The system is a grand series of blunders, and all its parts must stand or fall together. So long as they stand, the body of the French nation will pay for it dearly, in the high price of their cotton and other goods; and if it should fall, their manufacturers will atone for an unfair monopoly by extensive ruin. The manufacturers have been seduced, by absurd legislation, into a false and dangerous position, where they enjoy no real advantage, and from whence they have no retreat. They have the monopoly of the home market, and of the French colonies, except in so far as the smuggler disturbs them; but they hold it under perpetual alarm, and on conditions which prevent them from ever enjoying an export trade of any moment.

" 5th. As an effect of the political and natural causes already mentioned, the manufacturing establishments in France are small; they are scattered in many parts of the country, in order to supply the wants of the inhabitants; and each spinner and manufacturer is obliged to make a variety of articles to suit his customers. It is a necessary consequence of this state of things, that the attention both of the manufacturer and of his workmen is divided among several kinds of work, and they are prevented from acquiring excellence in any; whereas the concentration of the manufactures in England, and the extent of their market, enables each to confine himself to one or to a few articles, which he brings to the highest perfection, as well as makes with the greatest economy of time and money.

" 6th. The defective roads and inland navigation of France render the carriage of raw materials and goods expensive.

" 7th. The duty on the importation of the raw material is 2 per cent. more in France than in England.

" 8th. Capital is much less plentiful in France, and fetches a higher interest.

These, with other minor causes, place the French cotton-spinner and manufacturer in so disadvantageous a position, when compared with the English, as to forbid all prospect of successful competition."—(pp. 512, 513, 514, 515.)

In the *Edinburgh Review*, for October, 1836, there is an article on the " Commerce and Manufacture of Switzerland" written on the data afforded by Dr. Bowring's Report. The reviewer quotes what Dr. Bowring says upon the state of that country, with regard to commerce, as a striking exemplification of the advantages of free trade.

" It could not but excite the attention of any reflecting person, that the manufactures of Switzerland, almost unobserved and altogether unprotected, had been gradually but triumphantly forcing their way into all the markets of the world, however remote or seemingly inaccessible. That such a result was not the consequence of geographical position is obvious, for Switzerland neither produces the raw material which she manufactures, nor, when manufactured, has she any port of outlet, except on the conditions which her maritime neighbours impose upon her."

That a nation adopting the principles of free trade, and surrounded by nations acting upon the prohibitory system, should, in spite of all natural disadvantages, exhibit an example of commercial prosperity beyond the most favoured nations, and force its way into all the markets of the world, however remote or seemingly inaccessible, does indeed seem rather paradoxical. But the reader will find in Dr. Bowring's speech in the parliamentary debate on the factories' act amendment bill, which I have published, that he talks of the smuggler being, under some circumstances, a national benefactor; and though the assertion was received by part of the House with an "Oh, oh!" it appears that the Doctor did not speak unadvisedly, for the reviewer intimates, at the conclusion of his review, that Swiss industry has enabled her commerce to compete successfully with that of foreign nations, by the aid and intervention of those practical advocates for free trade—the smugglers. Let it be recorded to the honour of these national benefactors, that they strenuously and indefatigably advocated the cause of free trade at a time when it was by no means popular, when political economists had not thought of it—when political economy, indeed, as a science, did not exist—and that they were at all times ready to risk both life and liberty rather than suffer free trade to languish.

But though I would not be behind-hand, even with Dr. Bowring, in doing justice to the smugglers, however we might differ as to what is due to them, I do not think that a nation, labouring under great natural disadvantages, and surrounded by nations acting upon the prohibitory system, is ever likely, merely by virtue of free trade and smuggling, to become a formidable rival to Great Britain. Lest my readers should suppose that I have not quoted the reviewer fairly, I subjoin his very words:—

"The Swiss Cantons have, from age to age, been visited by myriads of strangers, from the surpassing beauty and magnificence with which nature has invested their territory; but it is only of late years that they have become remarkable as being the best practical political economists in the world; and if it be true, as we believe it is, that their liberal policy has increased their wealth as much as it has redounded to their honour, we trust that, when again tempted, under feelings of mortification at the treatment they receive, or of annoyance at the high duties we still impose upon their productions, to retrace their steps, or to make common cause with their German neighbours, they will meet only to re-enact the memorable resolutions of 1833, above quoted (resolutions in favour of free trade); and they may be assured that, at no distant time, a British code of commerce will arise, which will compete in its liberal character with theirs, and enable them, *without the aid and intervention of the smuggler*, to reciprocate with us an honourable and remunerating commerce."

After all, it would be difficult to shew how an act of parliament, which, by regulating the moving power in each factory, should render evasion next to impossible; which, by equalizing the labour of children and adults, as far as time is concerned, should enable the parent and his children to go home together, and take their meals in common; which, after the first mortification of defeat should have subsided, would promote a good understanding between the employed and their employers; which would prevent frequent strikes, and the almost habitual violation of the law on the part of some of the master manufacturers; which would put an end to that perpetual struggle of the manufacturers with each other, of which the consequence is perpetual oscillations of fortune, would ruin our foreign commerce. At present many who are overworked at times, are at other times without employment, and all that is wished for is, so to diffuse labour that it should not be at one time excessive—at another deficient, or none at all. If labour were more equally spread, each could support his own share with health and comfort, and though on any one day he might not be worked so much as he is now, the amount of his year's work might be the same, and though his wages might not be so much for a single day, the amount of his wages during the year might not be diminished. His expenses would certainly be diminished, by himself and his children taking their meals in common, by improved health, which would not only curtail medical bills, but enable him and his children to work with fewer intermissions from sickness, and above all, by the introduction of domestic economy into his cottage. His work, too, under such circumstances would be better done, because he would work with greater alacrity. If a certain quantity of work is done in ten hours it does not follow that a proportionally greater quantity would be done in sixteen. The rule of three direct is not applicable to the case, for the workman cannot perform the same quantity of work, when exhausted, as he could in the same time when not exhausted. There are many of the master manufacturers who have tried the experiment of shorter hours, as far as home competition will allow them, and have not been ruined by the experiment, or made converts to the system of working operatives to the utmost for fear of foreign competition. I am far from certain that in introducing an efficient bill for restricting the hours of labour, we should violate a law of political economy, but I am quite certain that we should be obeying a higher law. Dr. Johnson, in answer to the argument against the abolition of slavery, that slaves were happier under their masters than they would be under their native chiefs, replied—"We have no right to make people happy against their will." Johnson's position might admit of discussion; but that we have no right to make people miserable with their will is incontrovertible, and justice, as loudly as humanity, calls for an effectual restriction of the hours of labour, both as regards children and as regards adults. I shall proceed to consider the physical evils of the system, which are, first, to be considered in the order of time, as being the primary source of intellectual and moral evils. But before I enter upon this branch of my subject, I must be allowed to quote some striking remarks of Mr. Rickards, in his Report, dated the 20th of August, 1834, with regard to foreign competition.

"I do not anticipate successful rivalry on the part of foreign competitors. I believe their efforts will be found unavailing, as long as tranquillity and satisfaction be maintained at home. The admirable tables of Mr. Cowell, published with the supplementary report of the factory commission, go a great way to establish this most interesting and important fact, which is, indeed, further corroborated by evidence on the records of Parliament. Although personal wages are much higher in this country, yet the quantity or value of the products yielded for a given sum is greater. In other words, the cost of production of our manufactures, taken in the aggregate, is less than in rival countries; whence we are enabled to undersell competitors, in a great majority of products, even in their own markets. A stronger proof can hardly be exhibited than the circumstance of our importing raw cotton from India, returning it in a manufactured state, and underselling the productions of the natives, where manufacturing labour may be rated at twopence-halfpenny or threepence per diem. The reason of this superiority is to be found in the great magnitude and judicious application of capital; in the wonderful ingenuity displayed in the construction of our machinery; and that ingenuity exercised in discovering constant improvements, which succeed each other with almost incredible rapidity; and in, what I believe has not hitherto been adverted to, the constantly improving skill of the operatives themselves, producing in a given time more and more, or better work. These combined circumstances, and particularly the latter, which is stimulated by increased remuneration for every increase of produce, have hitherto contributed to maintain that superiority which is our present national boast. Being the result also of the practical experience of vast multitudes of manufacturers, whose skill and ingenuity must be supposed to bear some ratio to their numbers, whose numbers, and consequent ingenuity, are almost daily increasing, and possessing natural and artificial advantages, either unknown or not yet existing in other countries, it is to me inconceivable that the puny efforts of foreign rivals can be a source of dread to us, whilst this our advancement in the race continues to be maintained, and our other accessory advantages to be unimpaired." (pp. 40, 41.)

PHYSICAL EVILS OF THE SYSTEM.

IV. In proof of the physical evils of the system, I would refer those who place the same confidence that I myself do in the evidence brought before Mr. Sadler's committee, to my abridgment of that evidence, and publication at full of the medical evidence given upon the occasion. To study the medical evidence thoroughly, is the best preparative for judging of the evidence that has since been elicited by the government commission. The evidence brought before the commissioners is so bulky, that to abridge it would take so much time, that the question might be finally settled before the abridgment should come out, and, in consequence, nobody would take the trouble to read it, or be at the expense of buying it. There is the first report from the commissioners, printed in 1833, a volume which it would require at least three months for one man to abridge; next comes the second report from the commissioners, printed in the same year, a modest volume enough, and which might be abridged in a few days; then comes the first supplementary report, printed in 1834, which might take a week; after that comes the second supplementary report, printed in 1834, and of about the same dimensions as the first report in 1833; to these must be added the reports of the inspectors. Of these reports in general it may be said, that they abundantly confirm the evidence brought before Mr. Sadler's committee; but the evidence brought before the commissioners is conflicting, and to deduce fair results from it, we must be well acquainted with the medical evidence à priori, and must also bear in mind the fact, that the immediate interest of the operatives who are employed, is to speak well of the system, and that those operatives who speak against it, do it at considerable risk. Some of the witnesses in favour of the system, in their zeal to please their employers, so widely overshot the mark, that if upon receiving their evidence the commissioners preserved their gravity, the utmost credit is due to them for command of countenance; and the common phrase, "as grave as a judge," ought to be converted into "as grave as a commissioner." For instance, a girl of fifteen gives the following evidence:—"Been here since I was quite a child; health very good; the girls generally have very good health. You see, Sir, we have plenty of air. There's one bad thing here (laughing), we have no over-hours. I've heard of your coming, and what it was about. I think it is a very good thing. I hope you wont make us work shorter hours, though." Girl, seventeen.—"Been standing twenty-four hours; has no pain in knees or ankles, else would not do it." Boy, eleven years old.—"Before coming to the mill, used to play about in the streets: like being in the mill as well as playing in the streets." As we have not the questions, which drew forth these answers, we cannot tell whether, when the girl of fifteen laughed, she was asked why she laughed; or the girl of seventeen asserted, that after standing twenty-four hours, she felt no pain in knees or ankles, or the boy of eleven asserted, that he liked being in the mill as well as playing in the streets, any surprise was expressed. Of the reports, I have published the first and second, the first was printed by order of the House, on the 28th of June, and the second on the 15th of July, 1833. The act of parliament, which embodied the recommendations of the central board of commissioners, was passed on the 29th of August, 1833. These two reports are themselves sufficient to prove the evils of the factory system, and were considered sufficient by the government who appointed the commission. From these reports I shall make extracts, and intersperse them with my own observations. The first report considers the evidence with regard to—

1. The regular hours of labour.
2. The time allowed for meals.
3. The extra hours of labour.
4. The age at which children begin to work.
5. The nature of their employment.
6. The state of the buildings in which that employment is carried on.
7. The treatment to which the children are subject.
8 The ultimate effects of their employment on their physical condition.

1. With regard to the regular hours of labour.—It appears from the evidence, that in Scotland, with the exception of two or three factories, in which they do not exceed from ten to eleven hours daily, in general they range from twelve to twelve hours and a half; and that in several districts they are not less than thirteen. In England, in the northern district, in a few factories, the regular hours of labour do not exceed eleven. In general, they are not less than twelve; occasionally they are thirteen. There are many places in the western district, in which they do not exceed ten. These hours are exclusive of meals. The hours, in short, vary considerably in different districts; and in some factories in the several districts there is no intermission of work, day or night. In such cases, two sets of workpeople are employed, each set commonly working twelve hours. Occasionally, there are three sets, and then each set works eight hours.

2. With regard to the time allowed for meals there are similar differences. In one or two factories in Scotland, the meal hours are one hour for breakfast, and one hour and a half for dinner. In a few others, three quarters of an hour is allowed for each of these meals; but in general the time allowed is half an hour for breakfast, and half an hour for dinner, with no stoppage for tea, or drinking, as it is termed. In the north-eastern district, the practice in some factories is to stop half an hour for breakfast, one hour for dinner, and half an hour for tea. Sometimes there is no stoppage, either for breakfast or tea, but only for dinner; in some factories for an hour, in others—and this is the more general rule—for half an hour. In some factories practices have been resorted to, to cheat the workpeople of a portion of their time for meals, by moving the hand of the clock at dinner-time. In the western district, in all the branches of manufacture, it is customary to allow an hour for breakfast, an hour for dinner, and half an hour for tea, though in some factories only half an hour for breakfast is allowed. I con-

clude that those persons, whether master manufacturers or managers, who allow the least time for meals, and resort to practices which are at variance with common honesty, have been driven into a state of nervous sensibility and moral apathy by the dread of foreign competition; and in order that they may be brought into a sounder state of mind, I would recommend to them the careful perusal of Mr. Cowell's Preface to his tables.

3. With regard to extra hours of labour, in many factories it is not an unusual practice for the workpeople to stop during a part of the dinner hour to clean the machinery; this sometimes occupies them half their dinner hour, at other times not more than ten minutes. The children commonly stop to clean their own work. In some factories, care is taken, on the part of the proprietors, to secure to the workpeople the whole of the time allotted to their meals; while in others, this time is infringed upon without scruple. Occasionally, but not often, the work continues without intermission during the whole of the meal hours, the engine never stopping, excepting about ten minutes to be oiled, and the workpeople "eating how they can." In order to regain the time lost by stoppages, whether from the breakage of machinery, from the want of water, or from holidays, it is the custom for the people to work sometimes half an hour, at other times an hour, and occasionally even as much as two hours daily, until the whole of the lost time be made up. When the children do not clean the machinery out of the hours allotted for their meals, they clean it at extra hours. In Scotland this appears to be the general practice. For additional labour to make up lost time from stoppages arising from any contingences, with scarcely a single exception, no additional wages are paid; and the workpeople, young and old, perform this labour with reluctance. On the other hand, when from any cause there is a press of work requiring extra hours of labour, for which extra wages are paid, there seems to be no limit to the period for which the people will continue at their employment, sometimes, indeed, reluctantly, but more often not only without grudging, but with thankfulness, looking upon the permission to do so as a privilege and boon. To the young persons, and especially to the young female workers, this extra labour is often extremely irksome and harassing, and the younger the age the more injurious the consequences.

I must inform the reader, en passant, that this privilege and boon, as the commissioners call it, *must* be accepted, or the children are turned out of employment. That they are not allowed to look at the gift-horse in the mouth is evident from the following witnesses:—

"Am twelve years old. Have been in the mill twelve months. Begin at six o'clock, and stop at half-past seven. Generally have about twelve hours and a half of it. Have worked over-hours for two or three weeks together. Worked breakfast-time and tea-time, and did not go away till eight.

"Do you work over-hours or not, just as you like?—No, them as works must work. I would rather stay and do it than anybody else should come in my place."

"Have worked here two years; am now fourteen; work sixteen hours and a half a day. I was badly, and asked to stop at eight one night lately, and I was told, if I went, I must not come back."

"I have worked till twelve at night last summer. We began at six in the morning, I told book-keeper I did not like to work so late; he said, I mote. We only get a penny an hour for over-time."

There are several of the children who profess themselves very fond of extra-work, and the reader, if he likes, may follow the example of the commissioners and believe them, provided he does not insist upon my doing so. Their evidence will be found in the commissioners' first report.

It appears that parents encourage their children to make extraordinary efforts, by leading them to consider the wages which they thus earn as peculiarly their own, although a cheat is often practised upon them with regard to their extra-wages. While all the witnesses agree in the statement, that whatever the child earns by its regular hours of labour is uniformly appropriated by the parent, it appears that a large portion of the additional wages earned by extra hours is also taken by the latter. I here insert an authentic document with regard to the violation of the act in overworking the children, which is one among many instances of the little scrupulosity with which some of the master-manufacturers break the law, and also of the tenderness of magistrates towards such delinquents.

Correspondence, relative to the Firm of Taylor, Ibbotson, and Co.

London, June 14th, 1836.

My Lord,—In obedience to your lordship's orders, I have the honour to inform your lordship, that the report of the firm of Taylor, Ibbotson, and Co. having overworked young boys between twelve and fifteen years of age is correct.

The district where the mill of these gentlemen is situated is near Dewsbury, in Yorkshire, a neighbourhood inadequately populated, and entirely employed in the manufacture of shoddy cloths and blankets. Shoddy cloth is made from old woollen rags, torn in pieces by a powerful machine, reduced as far as possible to their original form of wool, mixed with flock, and worked up again into cloth, for any purposes where coarse woollens are required.

In the tearing up of these rags a great quantity of dust is produced, so much so, that, generally, persons standing three yards apart cannot very clearly distinguish each other. The rooms are excessively small, and the machinery propelled with great velocity. These shoddy machines require great power to drive them, and as they have been added to the ordinary machinery of the mills in most instances after the power of the engines has been calculated, it has been found that the engines are unable to turn the machinery required for the day with the shoddy machines added to it. These shoddy machines had, therefore, in four mills, been worked in the night, after the other machinery had ceased.

My Lord, in the case of Taylor, Ibbotson, and Co., I took the evidence from the mouths of the boys themselves. They stated to me that they commenced working on Friday morning, the 27th of May last, at six, A. M., and that, with the exception of meal hours, and one hour at midnight extra, they did not cease working till four o'clock on Saturday evening, having been two days and a night thus engaged. Believing the case scarcely possible, I asked every boy the same questions, and from each received the same answers. I then went into the house to look at the time-book, and, in the presence of one of the masters, referred to the cruelty of the case, and stated that I should certainly punish it with all the severity in my power. Mr. Rayner, the certificating surgeon of Bastile, was with me at the time.

My Lord, by a clause in the factory act, it is enacted, that "not more than one penalty can be obtained from any one person for any one description of offence in any one day." Had I laid the information, therefore, on all the boys for the same offence, the penalty could only have been 20*l.*, a sum which appeared to me inadequate to the gross nature of the offence which had been committed; I therefore, my lord, varied the informations, so as to enable me to sue for 80*l.*, being four full penalties, could I have obtained them, laying the first for working two boys more than twelve hours on the Friday; the second, for working a boy before five o'clock on the Saturday morning: the third, for working two boys in the night time, between Friday and Saturday; and the fourth for keeping a false time-book, as to the hours of working. These views, however, my lord, were not responded to, I very much regret to say, by the magistrates before whom the case was heard. The parties were fined only 5*l.* on each information, and with an aggregate of penalties of 20*l.* escaped.

My Lord, to the informations which were thus laid, the defendants pleaded "guilty;" but in justification, stated, that "by bursting of the boiler, the power of the engine was insufficient to turn the whole of the machinery, and that had they not worked at night, the other hands would have been unemployed through the day, as the whole body of the operatives in each mill are dependent on each other for the necessary material." They stated also, that "they had given the boys four hours rest during the night-time," and that "this was their first offence." It is more than probable, my lord, that this was the first offence of Messrs. Taylor and Co., but from night-working having been practised by other persons, more than once, in the same neighbourhood, and from other circumstances, I have reason to believe that had not those parties been discovered and punished, such practices would have been continued. I trust your lordship will think me justified in the view I took of this affair, and that in thus endeavouring to meet the justice of the case, I was acting up to the spirit of your lordship's directions.

I have, &c.

(signed) ROBERT BAKER,
Superintendent of Factories.

The Right Hon. Lord John Russell, &c., &c.

4. With regard to the age at which children begin to work, it appears in evidence, that in some rare instances children begin to work in factories at five years old; it is not uncommon to find them there at six; many are under seven; still more under eight; but the greater number are nine; while some, but comparatively few, branches of manufacture do not admit of the employment of children under ten years of age.

The necessity of producing certificates of the ages of the children has given rise to much fraud on the part of parents, and of children under their influence, and to a very objectionable regulation of Mr. Horner, one of the inspectors, to obviate the commission of fraud, as far as it can be obviated. I here insert a letter of Mr. Fielder, addressed to the editors of the Champion, upon this subject.

Waterside, October 11th, 1836.

DEAR SIR,—As you have promised your readers to pay some attention to this question, I beg to furnish you with some information, which I have received from a surgeon, who has been appointed by Mr. Horner, the inspector of this district, to grant certificates uuder the factories' regulation act.

The act prohibits the employment of children under nine years of age, in any factories except silk-mills. Those between the ages of nine and thirteen must be certified by a surgeon or physician as being "of the ordinary strength and appearance of children of or exceeding the age of nine years," and the labour of such children is restricted to forty-eight hours a week; but when a child attains the age of thirteen years, it is lawful to employ such a child sixty-nine hours a week.

Mr. Poulett Thomson, in the last session of Parliament, tried to obtain an act to substitute twelve for thirteen years in this part of the act, by which children of twelve years of age, of whom, from the inspector's returns, there are 35,000, might lawfully have been employed sixty-nine hours a week, instead of forty-eight hours; but the right honourable gentleman did not succeed.

The seventeenth clause of the act makes it lawful for his Majesty, by warrant under his sign-manual, to appoint four persons to be inspectors of factories in places where the labour of children and young persons under eighteen years of age is employed; and the eighteenth clause authorizes the said inspectors "to make all such rules, regulations, and orders, as may be necessary for the *due execution of this act*, which rules, regulations, and orders, shall be binding on all persons subject to the provisions of this act."

The information I wish to communicate to you, and through the Champion, to your readers, consists of a rule, regulation, or order, made by Mr. Horner, for defining the age of children, and authorizing, by certificate, their employment for forty-eight hours, or sixty-nine hours a week, according to the height they measure, let their ages be what they may.

Mr. Horner's instructions to the certifying surgeon are, in substance, as under. No children can be certified to be of the ages mentioned below, that do not come up to the following standard:—

			Ft.	In.
A child of	9	years must measure	3	10
—	10	—	3	11½
—	11	—	4	1
—	12	—	4	2
—	13	—	4	3½

and not then, unless the children are of the ordinary strength and appearance of children of those ages. So that by this regulation, a child being only 9, 10, 11, or 12 years of age, but four feet three inches and a half in height (the standard for thirteen years) may be certified to be of the latter age, if its ordinary strength and appearance, in the opinion of the surgeon, correspond with those of children of that height. On the other hand, Mr. Horner directs that a child being fourteen years of age, but under the height of four feet three inches and a half, shall not be certified as being thirteen years of age, and cannot be worked longer than forty-eight hours a week. This constitutes the humane part of the regulation; more care appears to be taken of the child of fourteen, than of one of nine. In my case, however, the humanity of it does not appear, as I have not a single child of fourteen years that is below the standard.

The heights above given struck me as being a low standard for children of those ages, and having a boy of my own by me at the time, who was nine years old on the 8th March last, I immediately measured him, and found his height to be four feet three inches and a half; exactly the height Mr. Horner authorized to be certified as being thirteen years of age! I will not say what were my feelings, when I had measured my own boy, arising out of the reflection of what a vast number of little urchins, whose time of actual labour the law had restricted to forty-eight hours a week, would, by this regulation, be driven back again to sixty-nine hours a week.

Having, however, upwards of 100 children between the ages of nine and thirteen in my own employ, working forty-eight hours a week, I measured the height of every one of these children, also, to ascertain how they would agree with Mr. Horner's standard, and the following is the result:—

		The lowest.			The highest			Average.	
		ft.	in.		ft.	in.		ft.	in.
12 Boys, 9	years of age,	3	9½	...	4	3¾	...	4	1
6 Girls, 9	—	3	11½	...	4	3	...	4	1
17 Boys, 10	—	3	9	...	4	5	...	4	1¾
13 Girls, 10	—	4	0¾	...	4	6½	...	4	3
16 Boys, 11	—	3	11½	...	4	7½	...	4	3¾
11 Girls, 11	—	4	0¾	...	4	7½	...	4	3¼
14 Boys, 12	—	4	3	...	4	9	...	4	5½
14 Girls, 12	—	4	2½	...	4	8	...	4	5

In the 103 children, whose ages and average height is above given, I found that there are—

		ft.	in.		ft.	in.
3 of 9	years, measuring	4	3½	to	4	4½
12 of 10	—	4	3½	to	4	6½
16 of 11	—	4	3½	to	4	7½
26 of 12	—	4	3½	to	4	9

In all, fifty-seven children out of 103 in my service, between the ages of nine and thirteen, now enjoying the benefit of the act, will, by this humane regulation, pass as being thirteen years of age, and I may have the benefit of screwing four hours labour a day out of them more than was contemplated by the law-makers in St. Stephen's when they passed this act. And thus, what Poulett Thomson could not prevail on the legislature to do in the last session, has been far outdone by a superior law-maker, an inspector. Will the writer in the "Westminster Review" persist, after this, in saying that a central board in London is necessary to instruct the inspectors? I am, dear sirs, your obedient servant, JOHN FIELDEN.

It is of the utmost consequence that this rule, regulation, or order, of the inspector should be prohibited, for, if allowed, it will violate the spirit of the act, and its most important provision for the protection of the children will be rendered null and void by the perverted ingenuity of Mr. Horner. Quetelet, in his work, Sur L'Homme, has given a table of the ages and heights of children and adolescents, from birth to the twentieth year, measured without their shoes. They are calculated by the French metre, which is equivalent to 39,371 English lineal inches. He has also given Mr. Cowell's table of the average height of children employed at the factories at Manchester and Stockport, measured with their shoes on. Quetelet allows half or a third of an English inch for the soles of boys, and one-eighth or one-sixth for the soles of girls, and reduces Mr. Cowell's calculation to the French metre. I subjoin these tables for the reader who wishes to compare their calculations with those of Mr. Fielden and Mr. Horner. I have given that part only of Quetelet's table which relates to children from nine to eighteen inclusive.

Ages.		Garçons. m.		Filles. m.		Differen.
9	...	1·221	...	1·205	...	16
10	...	1·280	...	1·256	...	24
11	...	1·334	...	1·286	...	48
12	...	1·384	...	1·340	...	44
13	...	1·431	...	1·417	...	14
14	...	1·489	...	1·475	...	14
15	...	1·549	...	1·496	...	53
16	...	1·600	...	1·518	...	82
17	...	1·640	...	1·553	...	87
18	...	1·665	...	1·570	...	95

Taille moyenne des enfans des classes inferieures à Manchester et Stockport, travaillant dans les fabriques.

Ages.		Garçons. m.		Filles. m.
9	...	1·222	...	1·218
10	...	1·270	...	1·260
11	...	1·302	...	1·299
12	...	1·355	...	1·364
13	...	1·383	...	1·413
14	...	1·437	...	1·467
15	...	1·515	...	1·486
16	...	1·565	...	1·521
17	...	1·592	...	1·535
18	...	1·608	...	1·593

The average heights of factory children, from Mr. Cowell's table.

Ages.		Boys. in		Girls. in.
9	...	48·139	...	47·970
10	...	49·789	...	49·624
11	...	51·261	...	51·155
12	...	53·380	...	53·703
13	...	54·477	...	55·636
14	...	56·585	...	57·745
15	...	59·638	...	58·503
16	...	61·600	...	59·811
17	...	62·673	...	60·413
18	...	63·318	...	62·721

	ft.	in.	
Mr. Horner's average height, for a child of nine is . . .	3	10	whether boy or girl.
Mr. Fielden's	4	1	whether boy or girl.
Mr. Cowell's	4	$0\frac{139}{1000}$	or 3 $11\frac{970}{1000}$ if a girl.
For a child of twelve :			
Mr. Horner's	4	2	
Mr. Fielden's	4	$5\frac{1}{2}$	if a boy, 4 5 if a girl.
Mr. Cowell's . . .	4	$5\frac{380}{1000}$	if a boy, 4 $5\frac{703}{1000}$ if a girl.

Mr. Horner's standard is very unjust to tall children, who are often weaker than they appear to be, from the rapidity of their growth; and such is the degree of injustice done to them, as entirely to withdraw the protection of the act from them, unless they happen to be sickly in appearance.

5. With regard to the nature of factory employment, it appears generally not to be inconsistent with health, provided the hours are sufficiently limited; but the effectual limitation of the hours is the grand difficulty, a difficulty, in my opinion, not to be overcome, except by restricting the moving power, and equalizing, as far as time is concerned, the labour of children and adults. A very great and perpetually recurring difficulty will, in the long run, operate as an impossibility. The vigilance of inspectors is not a match for the money-getting spirit of the masters, and the stimulating effect of higher wages upon the workmen. The apparatus for carrying the present act into effect is palpably inadequate; the few inspectors and superintendents that are appointed would need the eyes of Argus, the hands of Briareus, and the seven-league boots of Jack the giant-killer, with his coat of invisibility, to discharge their duties effectually; and wherever the various clauses of the act have been rigidly observed, it has been owing either to the benevolence, or to the conscientiousness, of the masters. An act, to be effectual, should resemble the machinery, and be in some measure automative. The present act enlists so many interests against itself, that it has been, and will continue to be, defied or evaded. A ten-hour bill would enlist nearly all the operatives in its favour, for those of the operatives, who at present oppose it, do so, in most instances, from no other motive than that of keeping in favour with their masters, and full employment. As the masters derive very great advantages from employing children instead of adults, and as the labour of both classes is linked together, the adult ought surely to derive some advantage from a circumstance which has deprived so many of his class of bread, or made them dependent upon their children. It is stated, by witness after witness, that, from the nature of the employment, the factory workmen, at a period when others are in full capacity for labour, are incapable of any very gainful employment, from impaired eyesight and general debility. The system of relays acts against the adult, and aggravates the evils of which he has, year after year, been complaining. He dares not quit his occupation, not only from the dread of losing work altogether, but from the consciousness how little he is fitted for any other employment, which will support either himself or his family. It is a spurious kind of humanity which would protect the child and leave the adult unprotected, upon the futile plea that he is able to protect himself. How have those fared who have sought to protect themselves? A candid answer to this question would abundantly show the fallacy of the pretext, that they are able to protect themselves.

In the reports of some of the medical men, I am often astonished at the little dependence of their conclusions upon the premises. Sir David Barry, in his general report, would limit the protection, which he admits to be necessary, to infant workers and adolescents, and yet there are passages in his report which clearly show that the adults also need it.

"Although both the young and the *adult* mill-workers may command more abundant food, and better clothing, than their *unemployed* neighbours, there are causes to whose operation they are exposed, which, in a sanitory point of view, counterbalance the advantage alluded to.

"1. The first and most influential of all is the indispensable, undeviating necessity of forcing both their mental and bodily exertions to keep exact pace with the motions of machinery propelled by an unceasing, unvarying, power.

"2. The continuance of an erect posture for periods unnaturally prolonged and too quickly repeated.

"3. The privation of sleep.

"All the *adult* male spinners are pale and thin; they are subject to capricious appetite and dyspepsia.

"Both *adult* males and females, whose work obliges them to stand constantly, are more subject to varicose veins of the lower extremities, and to a larger and more dangerous extent, than ever I have witnessed even in foot soldiers. The females are more subject than males to evening swellings of the feet and ankles, but I know of nothing which unfits them from becoming prolific and healthy mothers, if married at a suitable age."

However prolific they may be as mothers, and prolific mothers, where the order of nature is reversed, and parents depend upon their children for support, are valuable commodities, it would be no recommendation to them as wives, that they are more subject to evening swellings of the feet and ankles than the male operatives, and more subject to varicose veins of the lower extremities, and to a larger and more dangerous extent than Sir David has ever witnessed even in foot soldiers.

"The *adults* who work in the preparing rooms of small mills, where there is much dust, are generally affected with cough, and a kind of mechanical asthma, or tightness of the chest.

"Those who apply the dressing paste to the yarn, the web-dressers, who work in the highest temperature, are constantly perspiring, and look pale and exhausted."

Now if the nature of their employment is such as to produce the effects described by Sir David Barry upon the adults, I think a little more time for sleep, recreation, and the improvement of their minds, than they at present enjoy, might be fairly recommended. Nor do I at all wonder that they are in a state of discontent, notwithstanding their being able to command more abundant food and better clothing than their *unemployed* neighbours, when I find, upon Sir David's admission, that these advantages are counterbalanced by the undeviating necessity of forcing both their mental and bodily exertion to keep pace with the motions of machinery propelled by an unceasing and unvarying power; by the continuance of an erect posture for periods unnaturally prolonged and too quickly repeated; and by the privation of sleep.

My only wonder is, how any operatives, young or old, can be found to praise the system, and I am morally certain that, though they praise it with their lips, they detest it in their hearts. They think, perhaps, that a reform will be brought about without their assistance, and that they shall enjoy the benefit of it, without incurring the displeasure of their masters, which they know, not indeed from their own experience, but, which is much better, from the experience of others, is no slight evil.

Dr. Loudon, in the remarks with which he concludes his medical report, speaks much to the same purport, with regard to the injurious effects which factory employment has upon adults, when excessively protracted:—

"I think it has been clearly proved, that children have been worked a most unreasonable and cruel length of time daily, and that even *adults* have been expected to do a certain quantity of labour, which scarcely *any human being* is able to endure. The result of this has been, that many have met with a premature death; many have been affected constitutionally for life; and the idea of posterity being injured from the shattered frames of the survivors, is, physiologically speaking, but too well founded. Independent of the accidents which have arisen from machinery, it is unquestionable that the existence of the local diseases alluded to by the medical gentlemen examined before the House of Commons in 1832, and by myself during the period of our commission, as resulting from labour in factories, is but too true."

Dr. Hawkins, in his report respecting the counties of Lancashire, Cheshire, and Derbyshire, would extend protection to all under eighteen.

"Although fully aware of the pecuniary evil which may possibly result both to the masters and to workpeople from a reduction of the hours of labour, I am compelled to decare my deliberate opinion, that no child should be employed in factory labour below the age of ten; that no individual under the age of eighteen should be engaged in it longer than ten hours daily; and that it is highly desirable to procure a still further diminution of the hours of labour for children below thirteen years of age.

"In order to ascertain the state of the health of the youthful factory classes, compared with youth in other conditions, I made a careful examination of the Bennet-street Sunday-school at Manchester, in which abundance of all trades exists. I accordingly took an account of 350 of both sexes not engaged in factories, and of 350 of both sexes engaged in factories. Of the former, several remain at home and do nothing, some are in service, some are dress-makers, some engaged in warehouses, and in shops. Their age varied from nine years to twenty, for the most part.

Of 350 not in factories—	But of 350 in factories—
21 had bad health.	73 had bad health.
88 had middling health.	133 had middling health.
241 had good health.	144 had good health.

"Again, at the St. Augustine's Sunday-school, at Manchester, I compared fifty boys engaged in factories, with fifty boys not in factories, some of whom lived at home doing nothing, while others were engaged in shops, and in various trades.

Of the 50 not in factories—	But of the 50 in factories—
1 had bad health.	13 had bad health.
18 had middling health.	19 had middling health.
31 had good health.	18 had good health.

"I believe that most travellers are struck by the lowness of stature, the leanness, and the paleness, which present themselves so commonly to the eye at Manchester, and, above all, among the factory classes. I have never been in any town in Great Britain, nor in Europe, in which degeneracy of form and colour from the national standard has been so obvious."

With the vast mass of evidence which we possess, as to the injurious effects of factory employment, not only upon children but upon adults, it is the height of injustice to tax the adult operatives with discontent, and with being under the influence of factious agitators, because they wish that their own hours of labour should be restricted to ten, exclusive of meals.

We must judge of the nature of any employment by its effects. Many employments may require considerable exertion of strength, and yet, from being less monotonous, from requiring less of continued attentiveness, and from being carried on by day-light, and in the open air, may be much less injurious than factory labour, or may be even conducive to health. But however light, however easy, however healthy, an employment may be, it may be so protracted as to become neither light, nor easy, nor healthy; and that this has been the case with the factory labour no one who reads the evidence brought before the several committees that have from time to time been appointed, can for a moment doubt. There appears such an evident bias in some, both of the commissioners and of the inspectors, to favour the master manufacturers, that I am scarcely doing the operatives justice in resting their cause upon their reports; and if time were at my command, I could produce such a case, from the evidence upon which these reports are founded, by diligently sifting it, and trying the conflicting evidence by the laws of probability, as not to leave the opponents of a ten-hour bill an inch to stand upon. But Time shakes his glass, and compels me to be content with doing much less than I am perfectly conscious that I could do, and that ought to be done.

In answer to Mr. Tufnell and others, who have expatiated so much on the lightness of factory labour, I must be allowed to quote a passage from Mr. Fielden's work on the factory system, a work as creditable to the talents as to the principles of its author. Nor is the author a mere theorist; I have inspected his factory, and I can safely say, that if all factories were equally well regulated, and all masters equally benevolent in their views, parliamentary interference would be unnecessary, and the expensive apparatus of commissioners, inspectors, and superintendents, might be dispensed with.

" But to return to the question, from which I have somewhat digressed, to that of the " light and easy" work performed by children in factories, and to the representations which constantly hold it up as a species of amusement rather than work; representations in which some of the commissioners of 1833 joined, and in which the inspectors seem to be unanimous, while some of them back up their notions by opinions professing to come from medical men, *whom they have employed to grant certificates* to children. Returning to this, I will shew, by minute calculations, what is the work performed *in mere walking*, by a factory child; and, after that, I want no philosopher of any description, nor even any medical man, to tell me whether or not it is more than a child ought to bear.

" This question was mooted at Manchester on the 1st of December last year, by certain delegates from the factory people, who were appointed from Bolton, Bury, Ashton, Oldham, Chorley, Preston, and Manchester, to meet a few members of parliament on that day at the last named place. One of these delegates gave a statement, with particulars, of a minute calculation of the number of miles which a child had to walk in a day, in following the spinning machine; it amounted to *twenty-five!* The statement excited great surprise. But this delegate was followed by another, who had also made calculations, and who has put them in print in the " Manchester Advertiser" of the 28th of December, 1835. He calculates that a child has to walk twenty-four miles in the day; and if the distance that it frequently has to walk, to and from home, be thrown in, it makes, not unfrequently, a distance of nearly *thirty miles.*

" Observing the impression that these statements made on the minds of my brother members of parliament, and being myself desirous of testing their accuracy, I resolved, on my return home, to make a calculation myself, by watching a child at work in the factory in which I am myself concerned.

To my own surprise, I found that the distance was not less than *twenty miles* in twelve hours; and therefore I can easily believe the statements of the delegates, seeing that the machinery in my own works is not driven at anything like the speed of that on which their calculations are founded.

"I will not go into minute details of my calculation, because I should be obliged to use terms that an ordinary reader would not understand; but I stood by a child with the clock before me, and thus found the number of times that she walked certain distances in a given time; I knew those distances, and, upon this, calculated the whole distance that she would walk in that day if she worked twelve hours. I was very careful to keep my calculation rather under the truth than over it; and therefore I can give implicit reliance to the representations of the delegates, believing that their statement as to the speed of the machinery in some factories is also strictly true."—The Curse of the Factory System, pp. 39, 40.

One of the commissioners, Mr. Stuart, has brought several grave charges against the Central Board. He accuses them of recommending the *relay system*, in direct opposition to the advice of several of the district commissioners; of having paid no attention to the recommendation of the district commissioners, relating to the *wet-spinning of flax*, and the *web-dressing* in power-loom-weaving factories,—employments stated to be attended with imminent danger to the health of the young people, and of having omitted all notice of the mass of evidence, both in Scotland and Lancashire, unfavourable to their views.

The writer of an article on the factories in the Westminster Review exculpates the Central Board by taking for granted the very question in dispute.

"The fact is, that not only have those answers not been suppressed, for it *appears* that they were all published in the form of a supplementary report as soon as the actuary, who had been specially entrusted with the care of forming statistical tables from the returns to the printed queries (to which the answers in question were made), could prepare the results; but that such of the answers as can be considered to bear at all upon the point are more decidedly in favour of the plan than the answers to the vivâ voce examinations which were published in the first report."

That the reader may judge for himself I have published the correspondence between Mr. Stuart and Mr. Wilson, secretary to the Central Board. It has certainly a very awkward appearance to make recommendations to Government, and *after* the Government has acted upon those recommendations, and settled the question by passing a law, to give the evidence upon which the recommendations were grounded. But whether this awkward appearance has resulted from delay on the part of the Central Board, or precipitation on the part of Government, is a different question. Government has certainly repented of its own act, and sought to repeal a most important clause in it, and probably this apparent inconsistency would have been avoided, if it had waited till the whole evidence on which the Central Board either founded, or ought to have founded, their recommendations, had been laid before it.

6. With regard to the state of the buildings in which factory labour is carried on, it appears generally from the reports of the commissioners, that the large factories, and those recently built, have a decided advantage over the old and small mills. Of the old and small mills the report is almost uniformly—"dirty; low-roofed; ill-ventilated; ill-drained; no conveniences

for washing or dressing; no contrivance for carrying off dust and other effluvia; machinery not boxed in; passages so narrow that they can hardly be defined; some of the flats so low that it is scarcely possible to stand upright in the centre of the rooms;" while the account of the recent structures and the large establishments in general is—"infinitely better managed in respect to ventilation; height of roofs; a freedom from danger to the workers near the machinery, by the greater width of the passages in the working-rooms, and by the more effectual boxing in of the machinery, than those on a small scale." The commissioners have given several examples of mills in which every advantage of this kind is combined in an almost perfect degree; but they admit that there are too many instances in which an utter disregard is shewn, not only to the convenience and comfort, but even to circumstances, which must influence, in no inconsiderable degree, the moral feelings and habits of the operatives. "But one water-closet for both sexes, which children, and men and women, use indiscriminately." "Privies situated in view; common to males and females; this, in his (witness's) opinion, has a tendency to destroy shame, and conduces to immorality." "Workers complain of smells from the water-closet." "Picking-rooms pretty well freed from dust by effective fanners; but there is considerable annoyance to the workers from the effects of the water-closets; the effluvia must be unpleasant in warm weather; it made the walking through the apartments to-day very disagreeable in several places." This is the account given by one of the commissioners of a factory stated to be remarkable as that at which the finest cotton is spun in Scotland, and as having the greatest number of spindles in Glasgow, about 43,000." There is one district in England in which the privies are in a condition no less disgusting and disgraceful.

7. With regard to the treatment to which the children are subject, it appears from the report of the commissioners, that in Scotland, and in the eastern district of England, where the harshest treatment of children has taken place, the greatest number of bad cases occur in the small obscure mills belonging to the smallest proprietors, and that the bad treatment is inflicted by violent and dissipated workmen, often the very men who raise the loudest outcry about the cruelties to which the children are subject in factories. A striking picture of a mill of this class, one of the very mills, indeed, in which various witnesses depose that the treatment is oftentimes harsh and brutal, is given by Mr. Stuart:—"It seemed more to resemble a receptacle of demons than the workhouse of industrious human beings."

There are many factories in which corporal punishment is strictly forbidden, and, as is proved by the testimony of all classes of witnesses, is never inflicted. By all classes of witnesses it is stated, that "strapping was more customary in former times than it is now," "that formerly there was a great deal of strapping, but there is very little now."

There is a curious incidental allusion in Mr. Stuart's report from Greenock:—"All the respectable witnesses throughout Scotland agree in declaring, that whatever may have happened in the beginning of the factory system, at a period when coercion was far more resorted to even in *public schools* than now, they are ignorant of any recent instance of punishment attended with severity, or with anything like unpleasant consequences."

Now before the precedent of public schools can be pleaded in excuse for the cruelties inflicted upon factory children, it ought to be shewn that the cases are parallel. If Dr. Hawtry, or Dr. Longley, in their zeal for learning, should confine boys to their books for fourteen, thirteen, twelve, or even ten hours a day, exclusive of meal-times; if they should make them, upon their return to school, work extra hours to regain the time lost by holidays, of which, including half-holidays, they have at least four months in the year; if they should occasionally work them thirty-six hours consecutively, or at least with very short intervals; if they should adopt the system of night-work; if they should apply the birch, not only for misconduct, but as a stimulus, when the poor lads should sink to sleep from sheer physical exhaustion, then I would readily allow their example to be pleaded as a precedent, though not as an excuse, for the cruelties exercised in the factories. But in our factories the children have been confined to their work as many hours during one day as the children at our public schools have during the whole week, and I would therefore advise the advocates of the factory system not to prejudice their cause by indiscreet allusions to our public schools, where time enough, and more than enough, is allowed for meals, and for all sorts of recreation; and where children are seldom punished, except when they richly deserve it.

In the report in which this allusion to our public schools occurs, the reporter steps out of his way to make an attack, and a most clumsy attack it is, on Lord Ashley. After mentioning the numerous factories where the proprietors have done all in their power to promote the education, the religious instruction, the morals, and general comforts of those employed by them, he adds, "Is it then fit, even supposing the chief enactments of the bill to be necessary, that the phraseology of Lord Ashley's bill, and its severe enactments by penalties, and their payment to common informers, should be applied to individuals who have acted, and are acting, in the most liberal, disinterested, and benevolent manner, or generally against the other proprietors of factories in the great manufacturing districts of Scotland." To this it may be replied, that the phraseology of Lord Ashley's bill, its severe enactments by penalties, and their payment to common informers, will less affect those who have acted, and are acting, in a liberal, disinterested, and benevolent manner, than those who are actuated by no other principle than that of getting money, and that one, though not the main, object of Lord Ashley's bill was, to protect the benevolent from the rivalry of sordid competitors. Whether common informers would not have been more successful in enforcing the enactments of the government bill than inspectors, is a question on which much may be said on both sides. Inspectors, even upon their own shewing, have not been so successful as to throw informers quite into the back ground. The machinery of common informers would, at any rate, have been less expensive. Each of the inspectors receives a salary of 1,000*l.* per annum; the informers' pay would depend upon the number of convictions, and it is thought that this circumstance would have given a keenness to their vigilance that could not easily have been evaded. Those who are paid by the job, provided it is not a government job, work harder than those who are paid by the year, though I am far from accusing the inspectors of inactivity. If I were requested to supply them with a motto, I should offer them this:—

"Magno conatu magnas nugas agunt."

It is said that many of the operatives live by agitation, and that a final settlement of the question is the last thing they wish for. The inspectors occupy too respectable a station in society to make a similar objection applicable to them; but their warmest admirers must admit, that having more to do than they can do, and 1,000*l.* per annum—not for doing it, or for making believe to do it, (they are too candid for that,) but for *** (I employ asterisks, till it shall be made apparent what the inspectors are paid for doing,)—they are not likely to hurry themselves, nor would they be at all sorry if the office of inspector, which they discharge in a manner so gentlemanly, so courtly, so conciliatory, so entirely to the satisfaction of the master-manufacturers, should be made perpetual. No wonder, then, that they are a little bitter against common informers, and a little nettled at Lord Ashley for wishing to introduce such sharp but coarse machinery to render his bill effectual.

The physical effects of factory labour on children, as stated by the commissioners, are immediate and remote: the immediate effects are, fatigue, sleepiness, and pain; the remote effects, such at least as are usually conceived to result from it, are, deterioration of the constitution, deformity, and disease. The Central Board have been much struck with the perfect uniformity of the answers returned to the district commissioners by the young workers in this country, in the largest and best regulated factories as well as in the smaller and less advantageously conducted. In fact, whether the factory be in the pure air of the country, or in the large town; under the best or the worst management; and whatever be the nature of the work, whether light or laborious; or the kind of treatment, whether considerate and gentle, or strict and harsh; the account of the child, when questioned as to its feelings of fatigue, is the same. Young persons of more advanced age, speaking of their own feelings when younger, give to the commissioners such representations as the following:—"Many a time has been so fatigued that she could hardly take off her clothes at night, or put them on in the morning; her mother would be raging at her, because, when she sat down, she could not get up again through the house." "Looks on the long hours as a great bondage." "Thinks they are no much better than the Israelites in Egypt, and their life is no pleasure to them." "When a child, was so tired that she could seldom eat her supper, and never awoke of herself."

The truth of the account given by the children of the fatigue they experience by the ordinary labour of the factory is confirmed by the testimony of their parents. "Her children come home so tired and worn out they can hardly eat their supper." "Has often seen his daughter come home in the evening so fatigued that she would go to bed supperless." "Has seen the young workers absolutely oppressed, and unable to sit down or rise up: this has happened to his own children."

These statements are confirmed by the evidence of the adult operatives. "The long hours exhaust the workers, especially the young ones, to such a degree that they can hardly walk home." "The young workers are absolutely oppressed, and so tired as to be unable to sit down or rise up." "The younger workers are so tired they often cannot raise their hands to their head." "All the children are very keen for shorter hours, thinking them now such a bondage that they might as well be in prison."

The depositions of the overlookers and managers are to the same effect. "Many a one I have had to rouse in the last hour, when the work is very slack, from fatigue." "The children were very much jaded, especially when we worked late at night." "Exhausted in body, and depressed in mind, by the length of the hours and the height of the temperature." "I found, when I was an overlooker, that after the children from eight to twelve years had worked eight, or nine, or ten hours, they were nearly ready to faint; some were asleep; some were only kept to work by being spoken to, or by a little chastisement, to make them jump up. I was sometimes obliged to chastise them, when they were almost fainting, and it hurt my feelings; then they would spring up and work pretty well for another hour; but the last two or three hours were my hardest work, for they then got exhausted." "I have never seen fathers carrying their children backwards nor forwards to the factories; but I have seen children apparently under nine, and from nine to twelve years of age, going to the factories at five in the morning, almost asleep in the streets."

"I have always found it more difficult to keep my piecers awake the last hours of a winter's evening. I have told the master, and I have been told by him that I did not half hide them. This was when they worked from six to eight." "I have seen them fall asleep, and they have been performing their work with their hands while they were asleep, after the billey had stopped, when their work was over. I have stopped and looked at them for two minutes, going through the motions of piecening fast asleep, when there was really no work to do, and they were really doing nothing. I believe, when we have been working long hours, that they have never been washed, but on a Saturday night, for weeks together." "Children at night are so fatigued, that they are asleep often as soon as they sit down, so that it is impossible to waken them to sense enough to wash themselves, or scarcely to eat a bit of supper, being so stupid in sleep. I experience it by my own child, and I did by myself, when a child; for once I fell asleep, even on my knees to pray on my bed-side, and slept a length of time, till the family came to bed."

Pains in the limbs, back, loins, and side, are frequent, but not as frequent as fatigue and drowsiness. Pain is seldom complained of when the labour did not commence until the age of nine, and was not immoderate. Girls suffer from pain more commonly than boys, and up to a more advanced age; though occasionally men, and not unfrequently young women, and women beyond the meridian of life, complain of pain, yet there is evidence that the youngest children are so distressed by pain of their feet, in consequence of the long standing, that they sometimes throw off their shoes, and so take cold." "Feet feel so sair that they make him greet." "Was quite well when she went to the mill, but the confinement brought on a complaint in the head, and her left side is now pained." "Many nights I do not get a wink of sleep for the pain." "At first suffered so much from the pain that he could hardly sleep, but it went off." "Knee failed from excessive labour; severe pains and aches would come on, particularly in the morning: it was better in the evening; felt no pains in any other parts. There were two or three complaining at the same time of their knees aching." "I have seen children under eighteen years of age before six at night, their legs have hurt them to that degree that they have many a time been crying." "Swelling of the feet is a still more frequent source of suffering." "Obliged to bathe her feet to subdue the swelling." "The long standing gives her swelled feet and ankles, and fatigues her so much that sometimes she does na ken how to get to her bed." "Night and morning her legs swell, and are often very painful." That this affection is common, is confirmed by the concurrent statements of parents, operatives, overlookers, and managers.

8. With regard to the ultimate effects of their employment on their physical condition, it appears that this excessive fatigue, privation of sleep,

pain in various parts of the body, and swelling of the feet, experienced by the young workers, coupled with the constant standing, the peculiar attitudes of the body, and the peculiar motions of the limbs, required in the labour of the factory, together with the elevated temperature, and the impure atmosphere in which that labour is often carried on, do sometimes ultimately terminate in the production of serious, permanent, and incurable disease. From cases detailed in the evidence, and the accuracy of which has been strictly investigated, the commissioners do not conceive it to be possible to arrive at any other conclusion.

9. With regard to accidents from machinery, there are factories in which everything is done that it seems practicable to do to reduce the danger from this source to the least possible amount, and with such success that no serious accident happens for years together. By the returns which the commissioners have received, however, it appears that there are factories, and that they are by no means few in number, nor confined to the smaller mills, in which serious accidents are continually occurring, and in which, notwithstanding, dangerous parts of the machinery are allowed to remain unfenced. It appears also in evidence, that cases frequently occur in which the work-people are abandoned from the moment that an accident occurs; their wages are stopped, no medical attendance is provided, and, whatever the extent of the injury, no compensation is afforded.

I here subjoin, from the medical reports by Sir David Barry, a list of the cases of mutilation and injury which have happened at spinning-mills in Arbroath.—Second Report from Commissioners, pp. 32, 33.

1. Jessie Smith, aged twenty-three. Thirteen years at mills; twelve years ago lost fore-arm within three inches of elbow, on left side, at Gordon's old mill; allowed full wages till able to work.

2. Janet Neish, aged fifty-six. Thirty years at mills; loss of arm about middle, of humerus, at Lumgair's mill, two years ago; allowed half pay for about four weeks only; wound at that time not healed, and still confined to her house; not able to work since then; supported by parish.

3. James Adam, aged fifty-eight. At mills for seventeen or eighteen years; previously a weaver; lost about half of fore-arm six weeks ago, at Gordon's new mill; still on full pay.

4. Anne Gandy, aged twenty-five. Went to mill at ten years of age; severe injury of three fingers, soft parts of arm, and elbow joint, producing complete inability to use the hand, and great weakness of whole limb, at Brottock mill, ten years ago; unable to work for twelve months; on full pay during that time.

5. Margaret Gandy, aged twenty-two. Went to mill eight years ago; loss of thumb, with severe injury of soft parts of arm, at Inch mill, ten weeks after commencing to work; allowed full pay till able to work.

6. Vannet Thomson, aged eleven. Went to mill three years ago; loss of thumb and two phalanges of fore-finger, at Campbell and Cargill's mill, two years ago; five months off work; allowed full pay till able to return.

7. David Salmond, aged twelve. Went to Inch mill in seventh or eighth year, and four weeks after lost four fingers of left hand; allowed half pay for four weeks after.

8. Katharine Parvis, aged thirty-six. Went to mill eleven years ago; loss of thumb and three fingers of left hand nine years ago, at Smart's mill; allowed full pay till able to work.

9. Robert Booth, aged fourteen. Went to mill four years ago; severe

injury of right fore-arm, producing contraction of fingers, and complete inability to use the hand, at Spink-street mill; allowed half pay for three months; still subject to uneasy feelings in the wound, and, since the accident, to dyspnæa and palpitation.

10. Agnes Landie, aged nineteen. Went to mill eleven years ago; loss of fore and middle finger of left hand, at Brottock mill, eight years ago; allowed full pay till able to work.

11. Alexander Beattie, aged thirteen. Says he has been seven years at different mills; loss of middle and ring fingers of right hand, at Inch mill, about a year ago; allowed full pay for six months, till able to work.

12. William Peter, aged sixteen. Nine years at mills; fracture of right thigh bone, with injury of knee, producing stiffness and shortening of the limb, at Pool mill, three years and half ago; allowed full pay till able to work.

13. Robert Lamb, aged twelve. Went to Gordon's new mill two years ago; six months after lost fore and middle fingers of right hand; full pay till able to work.

14. Isabel Matters, aged ten. One year at mill; lost middle finger of right hand, with subsequent necrosis of metacarpal bone, and injury of the other fingers, at Shark's mill, ten weeks ago; on full pay still; not yet able to work.

15. Elizabeth Mathew, aged eighteen. Went to mill six years ago; loss of two phalanges of ring finger, and complete contraction of little finger of left hand, at Inch mill, five years ago; allowed full pay till able to work.

16. Isabel Hanton, aged twenty-three, Went to mill fifteen years ago; loss of little finger, with injury of hand, and contraction of the fingers, at Gordon's new mill, seven years ago; full pay till able to work.

17. Aikman Davidson, aged fourteen. Worked in mills for six years; fore-fingers on both hands injured at Gordon's new mill.

18. John Ramsay, aged sixteen. Worked five years; lost fore-finger of right hand, at Gordon's new mill, five years ago.

19. Alexander Dean, aged thirteen. Went to mill three years ago; loss of thumb, middle, and ring fingers, with severe injury of fore-finger of left hand, at Gordon's new mill, two years ago; allowed full pay till able to work.

20. Jean M'Cabe, aged twenty-six. Attended mills for ten years; loss of right thumb, with injury of the muscles of the palm, producing contraction of the fingers, at Prestonholme mill, near Edinburgh, on the second day after beginning work; she was sent to the Edinburgh Infirmary, and remained for three months; was unable to work for some time after; had no pay or half pay, but received a gratuity of 1*l*.

Arbroath, 16th May, 1833.

We have carefully examined all the cases in the above list, and hereby attest its correctness.

John Traill, Surgeon.
William Traill, Surgeon.

TEN-HOUR BILL.

V. The commissioners object to the ten-hour bill, that it does not accomplish the object at which it purports to aim. Its professed object is the protection of children; but it does not protect children. Now all I contend for is, that it would protect children better than the present act. It would be better for children to work ten hours, and to have those ten hours diffused over a day, and to go home at the same time with their parents, and to take their meals in common with their parents, and to receive a day's wages for their work, than to work eight hours, and have those hours limited to half a day, and to go home by themselves, and to be absent from their parents half the day, and to receive half a day's wages for their work. As a ten-hour bill is loudly and generally called for by the operatives, and probably earnestly wished for by many of them who, through fear of their masters, observe a discreet silence, such a bill would be likely to be strictly observed, and, from its simplicity, its observance would be easily enforced. The present act, from its complexity, is difficult to enforce; this difficulty is admitted by the inspectors; by some its observance is said to be impracticable; by almost all it is complained of as exceedingly annoying in several of its enactments, and as the interests of parents, adult operatives, and masters, are against it, it is never likely to be observed except by the conscientious, and consequently it gives the unconscientious an undue advantage. If what are called relays of children are to be employed, I should object to eight hours for the children, as pressing too heavily upon the adults, who will have to work twice that time. I should prefer, upon the supposition of relays being employed, six hours for the children, and consequently twelve for the adults, and I advocate a ten-hour bill only in preference to that ineffective piece of legislation recommended by the commissioners, and adopted by government. By a ten-hour bill, the father who is in employment in the factories would be able to pay some attention to his family, and this would be no small advantage to the children. The children would receive a day's wages instead of half a day's, and this would be another advantage. Parents and children would take their meals together, and this would save considerable expense, as well as promote domestic comfort. Under the present act, the parent may be worked till, from physical exhaustion and mental depression, he seeks relief from stimulants, and when that is the case, so far from being in a state to take care of his family, he cannot take care of himself. The demoralizing effect of excessive toil is clearly proved; the adult who is overlaboured is not likely to employ wisely his intervals of leisure: relief from distressing feelings is what he seeks, and he is liable to be carried wherever the impulse of appetite may direct. It is impossible to read the evidence brought before the commissioners without being convinced that the adults require protection, and as they are, in a majority of instances, the natural protectors of the children, the children themselves cannot be effectually protected, to whatever degree their hours of labour may be diminished, unless the adults are protected. The adults are, in one sense, more at the mercy of the masters than the children, for the masters can manage with comparatively few adults, and have, therefore, multitudes to choose out of; but they can do nothing without an abundant supply of children and young persons. As the commissioners, in their report, lose no opportunity of attacking the adult operatives and their delegates, I may be allowed to say something, if not in their exculpation, at least in mitigation of judgment.

"Such acts of severity and cruelty towards children employed in factories as are still found of occasional occurrence, are, for the most part, chargeable neither on the masters nor on the overlookers, but on the spinners or the slubbers themselves."

To account for this, they adduce the evidence of Mr. Rowland Detrosier, a witness extensively acquainted with the cotton manufacture:—

"Is not correction sometimes substituted for fines?—Yes; but that takes place principally amongst a distinct class of children. It is necessary to premise, that the children employed in cotton factory labour are not all under the control of, or employed by, the proprietor. A very considerable number is employed and paid by the spinners and stretchers, where there are stretchers. There are what are called piecers and scavengers, the youngest children being employed in the latter capacity, and as they grow up, for a time, in the double capacity of scavengers and piecers. In coarse mills, that is, mills in which low numbers of yarn are spun, the wages of the scavengers are commonly from 1*s.* 6*d.* to 3*s.* 6*d.*, according to size and ability. The men do not practise the system of fining, generally speaking, and especially towards these children. The sum which they earn is so small, it would be considered by many a shame to make it less. They do not, however, scruple to give them a good bobbying, as it is called, that is, beating them with a rope thickened at one end, and perhaps a strap, or, in some few brutal instances, with the combined weapons of fist and foot."

"But this severity, you say, is practised towards the children who are employed by the men, and not employed by the masters?—Yes."

"And the men inflict the punishment?—Yes."

"Not the overlookers?—Not in these instances."

"But how do you reconcile your statement with the fact that the men have been the principal complainers of the cruelties practised towards the children, and also the parties who are most active in endeavouring to obtain for the children legislative protection?—My statement is also fact. I do not profess to reconcile the apparent inconsistency. The men are, in some measure, forced by circumstances into the practice of that severity of which I have spoken."

Now, supposing the greatest complainers to be the very persons who inflict the cruelties, I see nothing inconsistent in their wishing such an alteration in the system as should remove the temptation to inflict cruelties. If a spinner were sufficiently disinterested to allow the children to work no longer than they could work with ability, and thus render severity on his part unnecessary, would he obtain the approbation of his master? Would his master be satisfied with the diminished production, which would be the effect of the spinner's humanity?

The master may wash his hands of the cruelty inflicted by the spinner; but the children are sharp-sighted enough, though the commissioners are not, to trace the evil to the system, which system is authorized by the master.

But let us hear the commissioners.

"An apology is made for the conduct described in the foregoing statement, by supposing that the workmen are constrained to this severity by the pressure of the system upon themselves. This, however, is rebutted by the fact given in evidence, that in numerous well-regulated establishments the infliction of corporal punishment is effectively prohibited; and that in these the amount of production is apparently not less than in others where coercion is allowed on the part of the operatives."

This statement of the commissioners appears to me a much better apology for the conduct of the operatives than that suggested by Mr. Detrosier. It appears from the foregoing statement, that in some establishments coercion is allowed on the part of the operatives; now, in such mills, the severity of the spinners is clearly traceable to the masters; but in numerous well-regulated establishments the infliction of corporal punishment is effectively prohibited. Now in such mills, as no severity is practised, there is no accusation against the operatives. The masters can effectively prohibit corporal punishment if they will, and consequently, if they will not, they, and not their instruments, are most to blame.

"It is established, by a mass of concurrent testimony, such as rarely has been brought to bear on any point of inquiry, that the former proceedings of these men have, in hardly any case, been successful, so far as regards the object ostensibly aimed at by agitation. This uniform result affords a presumption, which is confirmed by as much of direct evidence as the nature of the case admits, that their leaders have other objects in view besides those to which their efforts are professedly directed. It appears that agitation is the trade by which they live, and that success in the attainment of the objects at which they profess to aim would involve the loss of their actual occupation, which consists in keeping up discontent at such a height as to secure distinction and profit for themselves at the expense of their fellows."

I think these observations a scandalous libel upon the delegates, and extremely unjust towards the operatives who employ them. The commissioners maintain that the factory system is much improved; that the cruelties, which were formerly frequent, are now rare; that the comforts, both of the children and of the adults, which were formerly neglected, are now, with some exceptions, attended to; and I think that, in common fairness, some portion of credit is due to the delegates, as having their share in producing this beneficial change. That their success has not been greater than it is, is not justly attributable to them: many causes have obstructed their complete success; but, I trust, have only retarded it. Little did they think, in 1832, when, at the expense of the operatives, a mass of evidence was produced, which, in conjunction with the opinions of the highest medical authorities, thoroughly awakened the public to the evils of the factory system, that the result would be a government commission. Justice was demanded, and had been demanded for upwards of thirty years; and it was not the supineness of the delegates, but the excessive scrupulosity of their opponents, that caused delay. The case was most urgent; but government, thinking to find a knot in a bull-rush, appointed a commission. It is no more a reproach to the delegates that they live by agitation, than it is to the commissioners, that they add to their incomes by commission, or to the inspectors, that they receive a thousand a year each for inspection. The question is, whether they do their duty in their vocation; and if tried by this rule, the delegates, though not quite so well paid as either commissioners or inspectors, will not shrink from the trial. What have the commissioners done? They have brought forward a cart-load of evidence confirmatory of what was brought forward before Mr. Sadler's committee, and before other committees that preceded Mr. Sadler's, and proved, what none, except those whose judgment is completely blinded by self-interest, can doubt, that the children require legislative protection. By their representations and recommendations they have induced government to get an act of parliament passed, giving satisfaction to no party, not even to its framers, since they have since sought to repeal one of its most important clauses. What have the inspectors done? They have demonstrated, that their inspection cannot prevent continual violations of the law; and that where it is observed, it is observed from conscientious motives, and not from dread of their vigilance.

The delegates have agitated the question from year to year, and, in spite of the most determined opposition, have forced it upon government; and it will not be their fault if justice be not dealt, ere long, both to the children and to the adult operatives. They have retained the confidence of the people for years, and will continue to agitate, to the satisfaction of their employers, and to the dissatisfaction of their opponents, till improved legislation shall throw them out of employment. I trust, when that time arrives, the operatives will shew their gratitude, and not suffer men, who have served them so well, either to starve or to be deprived of the honours they have so dearly earned, and to which they are so justly entitled.

The commissioners consider that the completion of their thirteenth year is the natural period at which young persons may be placed on the same footing as adults, as far as regards the disposal of their labour. The president of the Board of Trade would fain have anticipated this period by a year, and have made the completion of their twelfth year the period for commencing adult labour. But even at the latest of these periods growth is far too rapid to allow of the exhaustion occasioned by twelve hours' labour, without the most serious injury to the constitution. With reference to this point, I must quote the opinion of Mr. Wakley, M.P., stated in his "Voice from the Commons," p. 5:—

"In the supplementary report of the commission several tables are given, which have been compiled by an actuary, Dr. Mitchell, purporting to represent the average number of days of sickness which are incidental to factory labourers. But they present grossly deceptive results, as the sickness of those who have died, and of those who have been dismissed from ill-health from the mills, was excluded from the calculation! Yet the result, deduced from his insufficient and imperfect materials, is styled 'The average duration of sickness for every person employed' in the factories. Thus have loop-holes been created by the commission, through which the masters continually attempt to leap, in order to escape from the force of the arguments employed by the supporters of a short-time bill. The legislature itself has been deceived, the commissioners, in fact, having led a majority in parliament to consider that the sickness of factory operatives is below the alleged average. The factory commissioners committed, too, this gross and wilful error: they decided, in direct opposition to the best medical opinions, and in the teeth of all the observations of physiologists, that puberty is established at the thirteenth year, both in males and females; that the adult period then begins in boys and girls, intending to prove thereby that children are, from that time, capable of sustaining the labour of grown persons. This is a foul crime against nature and humanity. In the natural state of youth, from the twelfth to the fifteenth year of life, the mortality is less than at any other period; the functions are then in full operation to form the basis of manhood—to create materials for the waste of future exertion, not to be devoted to labour now. Growth should then be at its most rapid rate in the frame, both sexes, under a healthy condition, undergoing, at that time, a great change, which does not fully terminate before the close of the sixteenth year. Cultivation of the mind should then occupy the time of the individual, and a large degree of alternate freedom of body and rest of the muscles should be ensured to the youth; yet it is on this very period that the 'amended bill' threatens to encroach. The boy of twelve is to be reckoned an adult! The girl of a dozen summers is to be regarded as a woman! The few hours granted by the former bill (guarded enough in its enactments) are to be snatched from the time devoted for their instruction, recreation, and growth, and given to the factory slave master !"

A day's labour, that requires all the strength of an adult, must be destructive, either in its immediate or in its remote consequences, to a child of thirteen, yet this labour it was contemplated to force upon a child of twelve. The children's enemies were defeated, and, wonderful to relate, our foreign commerce still exists. Hard, indeed, would be the case of the poor mill-owner, if it were impossible for him to be just and merciful without being ruined; desperate would be the commercial prosperity of the country, if it depended upon the continued violation of physical and moral laws. I proceed to consider the intellectual and moral evils resulting from the factory system, which are, for the most part, the consequences of the physical evils, the existence of which is admitted by the commissioners in their reports, and incontrovertibly proved by the evidence on which their reports are grounded.

INTELLECTUAL AND MORAL EVILS.

VI. In considering the intellectual and moral evils arising from the factory system, we have another argument for a ten-hour bill. The children are obliged, by the present act, to bring a certificate of their having been at school two hours daily for six days in the preceding week, on the Monday morning, before entering upon their work, or the master cannot legally employ them. Now, I will venture to assert, that no compulsory education can counteract the demoralizing effect which is almost certain to follow from separating children from their parents for that lengthened period which the system of relays renders necessary. The precedents which the commissioners advance in favour of compulsory education do not, in the least, sanction the method enforced by the present act of parliament.

"It appears in evidence that, in other countries, care is taken to afford education to the people, and to prevent those who are engaged in labour at an early age from being deprived of this advantage."

"Mr. Peter Kennedy, (sworn) proprietor of a cotton-mill at Feldkirch, in the province of Tyrol, in the Austrian dominions :—What is the lowest age at which you employ children at present ?—The lowest age at which I employ them is from eleven to twelve years of age."

"Why do you not take any under that age ?—Because it is an inconvenience to lose their labour in the winter. By the law of the country, all children are obliged to go to school until they can read and write to the satisfaction of the priest. They usually go to school in the winter, when other employments fail; and it generally takes three winters for their education. The priest takes care of the education. If he found that a child was at the factory, or not sent to the school, he would remonstrate with the parent, and, on failure of sending the child to school, he would cite the parents before a judge, who would compel obedience by imprisonment, or very effectual means. It is to avoid these interruptions of the labour, by the education, that children above eleven or twelve years of age are taken."

"Then all you employ are usually educated ?—Yes ; I do not know that we have any workman who cannot read or write."—First Report, p. 69.

From this extract it appears that in the Austrian dominions, instead of compulsory education at the time that the children are daily working in the mill, the lowest age at which they are employed in the mill is from eleven to twelve years: and that they receive their education, as far as elementary instruction is concerned, before they enter the mill.

The other precedent is taken from the practice in the United States.

"Have you any national system of education?—We have public schools, supported partly by state funds and partly by bequests. All children have the privilege of attending."

"Do they, in point of fact, very generally attend in the manufacturing states?—They universally attend; and I think that information is more generally diffused through the villages, and the whole community of the New England States, than amongst any other community of which I have any knowledge."

"What is the general view taken of these schools by the manufacturers and persons of wealth in America?—From their experience they deem them of the greatest importance to the welfare of the state. They are encouraged by the state government, and all the leading persons of the state."

"How do the children whom you employ obtain education?—The manufacturers are always anxious that the children should absent themselves from the manufactory during two or three months of the year, to attend the schools. The manufacturers very frequently suggest to the parents the necessity of the children being taken to school. The sending the child to school is generally an inconvenience to the manufacturer."

"Is the inconvenience of the children going to the school such as to increase the cost of production?—I do not think it does increase the cost of production. The only inconvenience is in the trouble of getting other hands. We think the advantage of their being educated more than counterbalances that trouble."

"What is the nature of this national education?—It consists of reading, writing, arithmetic, grammar, and geography."—First Report, p. 70.

It is evident, from this extract, that if our manufacturers would follow the example set them in the United States, compulsory education would be unnecessary.

The difficulty of carrying on education, according to the plan of the present act of parliament, during the time that the children are daily employed in the mills, is sufficiently evident from the reports of all the inspectors; a difficulty so great, that in many localities it acts as an impossibility.

"In no part of my district have I found the enactments of the statute, on this head, complied with; and if I had insisted upon the literal fulfilment of the twenty-first clause, I am persuaded that the immediate consequence would have been, that all the children under eleven years of age would have been discharged."—Report of L. Horner, Esq., 21 July, 1834. pp. 9, 10.

Mr. Horner's district comprehends Scotland, Ireland, and the four northern counties of England.

"Every child is required by the act to produce, every Monday morning, a certificate of having attended school for two hours at least, on six days of the preceding week, on pain of dismissal from the mill. Very few mills are situated near a school which is open at such hours as many of the factory children could attend; that is, early in the morning and late in the afternoon. There is no obligation, on the part of the schoolmaster, to take the trouble of making out the certificates; and, in fact, some in Glasgow have already refused to do so."

"Few schools are open on Saturday after mid-day; and although attendance on a Sunday-school may be taken as one of the six, as these are usually taught by several voluntary teachers, and the children are mixed up together in great numbers, it would be almost impossible to obtain certificates of attendance; and if they could be had, each child must then have two certificates each week."

"In a mill, where the clerks have generally enough to do, (and in the small mills there are no clerks,) the examination of these certificates would be a very considerable labour; besides that, the children would be perpetually losing them, and the variety of excuses for non-attendance, which the mill-owner would have to investigate, would be endless. Then, if the children play truant, or are kept away from school for frivolous or insufficient reasons, the mill-owner, who is in no way to blame, must dismiss them, possibly to his great inconvenience and loss, by the stoppage of his machinery until new hands can be procured."—Report of L. Horner, Esq., 21 July, 1834. p. 12.

Mr. Howell, whose district comprehends several of the western counties of England, the south of Wales, and of Ireland, states the same difficulty.

"The act absolutely prohibits the master from employing any child who does not, every Monday, produce a schoolmaster's voucher that it has attended school for two hours at least in six days of the preceding week; and this, say the masters, places the employer at the mercy of every child in his factory; for that he has no control over the child after its hours of work, and that a truant disposition, employment for any domestic purpose by its parent, and other the like circumstances, will cause many schoolmaster's vouchers to be deficient on the Monday. The children cannot be employed; the spinners are short of their required number of assistants; and their labour, and their master's machinery, are rendered thereby proportionally unproductive, besides the consequent derangement and interruption caused in every subsequent department of the manufactory; and, it is added, that any uncertainty or irregularity in the supply of labour is even more prejudicial than a positive but defined limitation."—Report of T. J. Howell, Esq., 20 December, 1833.

"We anticipate," says Messrs. Fox—and their reasoning appears to me to apply to the west of England generally, where it is by no means unfrequent to find two, three, four, and even five mills, belonging to the same firm, but each distant from the other—"we anticipate serious difficulties in enforcing the attendance of the children at schools. In large towns, or in the instance of factories so extensive as to furnish a school within themselves, these difficulties may not be experienced; but where small factories are situated in secluded situations, insurmountable obstacles appear to be presented. It is true, the inspectors are authorized to establish schools; but as no funds for this purpose are placed at their disposal, we consider this provision to be almost nugatory. In our own case, the difficulty arises from the circumstance of the children not living exclusively in the neighbouring towns, but scattered over the surrounding villages and hamlets. We should feel pleasure in establishing schools on our own premises, and under our own control, were it not that our mills are distant from each other, and would require separate schools, involving an expense more serious than would perhaps be justified by the number of children employed in each establishment."—Report of T. J. Howell, Esq., 20 December, 1833. pp. 21, 22.

Mr. Rickards, whose district comprehends Yorkshire, Lancashire, Cheshire, Derbyshire, part of Staffordshire, and the North of Wales, speaks to the same effect.

" But the procuring of a sufficient number of young hands for the system of relay being impossible, they had no resource but to discharge all under eleven, twelve, and thirteen years of age, in six, eighteen, and thirty months after the passing of the act, and procuring older hands, or with the aid of improved machinery, to complete their daily twelve hours' work. This was, and still is, the unanimous voice of the whole country visited by me, with the exception of three mill-owners, Mr. Burley, of Manchester, Mr. Thomas Ashton, of Hyde, and Mr. Marshall, of Leeds. These three gentlemen, having large establishments, with schools attached to their mills, and everything conducted on a corresponding scale, propose to work out the schooling clauses according to the provisions of the act; but in regard to others who have no such schools attached to their buildings, or in the vicinity, or where no schools are, or can be established in villages and distant parts of the country, the difficulties, it is contended, will be insurmountable; whilst the restriction imposed by the act on masters not to take into their mills, on each successive Monday, children who could not produce the prescribed certificate of schooling in the preceding week, might subject them, they said, to the greatest inconvenience and injury, by the stoppage, for a time, of the work of the mill. Many children, it is thus apprehended, will be thrown on the parish for relief, with all the consequences of so deplorable an alternative."—Report of R. Rickards, Esq., 24 December, 1833.

" In these documents I am in hopes your lordship will see that the principal objects of the act have been duly attended to, save only the schooling clauses and the working of children under eleven years of age by relays, and there I must candidly confess that I despair of being able to reduce into practice, or of rendering in any way useful, except in a very few instances, as adverted to in a former report to your lordship, wherein the limitation of children's work to six or eight hours per diem, or the working of them by relays, are declared to be fraught with insuperable difficulties; whilst the utter impracticability of the schooling clauses being attended to by the masters of the mills, or being enforced by me, is so obvious, that should this matter be hereafter called in question, I trust I shall stand excused for dispensing with their observance, at all events in cases where proof to demonstration of this impracticability shall be found to exist. Where schools can be established, or a disposition is shown to do so, I invariably encourage it; but here at least the necessity for the schooling prescribed by the act is superseded, I regret to say, by the mill-owners having already very generally dismissed all their working hands under eleven years of age, and many of these cases, with pain I relate it, are truly distressing. Both masters and operatives have represented to me, in strong colours, the suffering thus endured, and, hard and heart-rending as these cases are, I have been obliged to answer that I had no power of relief."—Report of R. Rickards, Esq., 10 February, 1834, p. 29.

"These clauses, (those which relate to the attendance of children at school,) I am sorry to say, have, in operation, defeated the humane objects which the framers of the act had in view. The instances are extremely rare in this division, (I could not enumerate half a dozen,) where they are observed. They are so very objectionable to mill-owners generally, that to avoid what they deem the intolerable inconvenience and trouble of complying, they have discharged from their mills all children under eleven years of age. This has been carried in these districts to a distressing extent; and the children so discharged, instead of receiving the proposed education, get none at all; they are turned out on the wide world to seek other employment, or to wander about the streets of towns in idleness and vice. I entered on this subject to Viscount Melbourne, at considerable length, and submitted a plan for the education of the youth of both sexes, which, I am still confirmed in my opinion, would, if adopted, be attended with much benefit to the rising generation and to the public at large. It provides for the attendance of all young persons in mills under eighteen years of age at Sunday-schools; but interfering in no respect with the regular course of mill-labour, is consequently free from the objections which attach to the system prescribed in the present act."—Report of R. Rickards, Esq., 12 August, 1834, p. 39.

"When I first submitted this proposition to my Lord Melbourne, it was accompanied with another, in which masters and operatives would equally agree, that the employment of children under ten years of age in mills or factories, should be prohibited; that, on the contrary, up to that age, they should be furnished with the means of receiving a sound, moral, and religious education; and that every young person from ten to eighteen, after entering a mill, should be compelled regularly to attend a Sunday-school. It is only, as it appears to me, by a systematic education of this nature, long and uninterruptedly continued, that the vicious habits of this population can be corrected. Restrictive laws will avail us nothing. Let all the gin-shops and beer-houses of the country be hermetically sealed; as long as the propensity to drink shall be the ruling vice of the lower classes, drunkards will be sure to find liquor. The axe must be applied to the root. It is the vicious propensity that requires to be eradicated; the inward man, the heart, out of which are the issues of life, that needs correction. Education can alone accomplish it; and until the moral habits of the people be thus improved, drunkenness, with its concomitant evils, will prevail, and continue to be the characteristic reproach of our country."

"In my report of the 15th of April last, I explained the difficulties of conforming to the schooling clauses of the present act, and the distress occasioned to families by the consequent discharge of children under eleven years of age, to avert what the mill-owners consider an intolerable annoyance. In my present circuit I have found some instances where the retention of the younger hands was indispensable to the working of the mill, and where schooling was at the same time impracticable. I have found some others, where the children have been retained, worked according to the act, and have produced, every Monday morning, on returning to their work, schooling certificates; but the masters have at the same time assured me that they could not place much reliance on the authenticity of these tickets. There are other instances, but these are few, where the benevolence of mill-proprietors has induced them to provide schools in their own mills, and to furnish children with regular and good instruction at their own expense. Still, by far the greater number of the children under eleven years of age have been discharged, and I am informed by magistrates, that, since this act came into operation, applications for parish relief have increased, which the overseers ascribe to this cause. In short, the mill-owners of this division very generally declare, that when the clauses here adverted to come to be applied to children under twelve and thirteen years of age, they must either close their mills, or work them in open infraction of the provisions of the law; in other words, that the intentions of the legislature, in regard to the labour of children, which now work beneficially, will be wholly disregarded and defeated."—Report of R. Rickards, Esq., 12 August, 1834, p. 45.

Mr. Saunders, whose district comprehends the eastern, several of the midland, and all the southern counties of England, adverts to the same difficulties.

"The subject which seems surrounded by the greatest difficulties is, the education of the younger children."

"In my first visit to several towns and factories, I had reason to hope that a plan of forming schools for this particular purpose would be adopted, and that the younger children would not be discharged, but information since received assures me the various details and regulations required for the satisfactory formation of such establishments, and the many difficulties to which the manufacturer may be subject by the failure of his attempt to secure the regular attendance of the children at the school, and the consequent penalty on him should he employ them the subsequent week, will, in every case where it is practicable, oblige him to adopt the simpler course of employing none under the restricted age; and this system has already begun to be acted upon."—Report of R. J. Saunders, Esq., 28 December, 1833, p. 59.

With regard to the evils arising from the children not being under the protection of their parents, a statement was made by the Bishop of Exeter, in presenting several petitions for shorter hours from cotton-spinners, master-manufacturers, and from the delegates of the workmen, that in the town of Manchester no fewer than 8,000 children had been deserted by their parents, and found abandoned in the streets, within four years. Though there is an inaccuracy in the term 'deserted,' the statement will appear from the following document to be substantially correct.

"Return to an Address of the honourable the House of Commons, dated 7 September, 1835; for, A Return of the Number of Children taken up in the streets of Manchester, and the districts adjacent, (distinguishing the districts,) and deposited in the police offices to be owned, for the year between August 1832 and August 1833;—The same for the corresponding years between 1833 and 1834, and 1834 and 1835.

"Manchester, 30 September, 1835.

"Sir,—In compliance with Lord John Russell's direction, I inclose to you a Return, so far as it can be made, of the number of children taken up in the streets of Manchester within the period for which it is required, to be laid before Parliament.

"The Return from Manchester has been carefully extracted from the books of the police office, which is generally known as the proper place to which such children are to be conveyed by persons finding them in the streets, and to which parents always resort for information if any of their children are missing.

"In this manner all that have been thus brought to the office have been claimed, and restored to their homes.

"The township of Manchester, to which this Return relates, comprises the greater portion of the town. In the districts forming the suburbs of the town, I find that correct lists cannot be furnished; the practice having been, from time to time, to make minutes as to such children when brought to the office, but not to file or preserve the minutes.

"The only further return I can make is, therefore, the inclosed Return from Salford.

"The number will be found to be very great, and although the Return is from a district containing an immense population, it is doubtless greater than it would otherwise be, because the nature of the employment followed by most of the parents in the cotton-factories, prevents their attending personally to their small children during the day, and compels them to place the charge in the hands of their older children, or others, who are not sufficiently attentive to them.

"I have, &c.

"The Hon. F. Maule, &c. "J. Fred. Foster."

A Return of Lost Children registered at the Police Office, Manchester, in the years 1832, 1833, 1834, *and* 1835:—*viz.*

From the 1st August, 1831, to the 31st July, 1832	...	1,954
From the 1st August, 1832, to the 31st July, 1833	...	2,140
From the 1st August, 1833, to the 31st July, 1834	...	2,117
From the 1st August, 1834, to the 31st July, 1835	...	2,439
Total	...	8,650

J. S. Thomas, Deputy Constable.

An Account of Children Lost, brought to this Office, from May, 1835, *to September* 14, 1835; *also, Children Found.*

Lost.			Found.		
1835.	May	89	1835.	May	35
	June	105		June	29
	July	110		July	31
	August	132		August	32
	Sept., up to the 14th	35		Sept., up to the 14th	11
	Total ...	471		Total ...	138

Salford Police Office, Sept. 15, 1835.

If children were not admitted into mills without a certificate of their being able to read and write, compulsory education after their entrance would scarcely be necessary; and if ten hours were to be the common limit of actual labour both to themselves and to their parents, they would be more under the immediate protection of their natural guardians, and be able to attend Sunday-schools and places of public worship with satisfaction and advantage. Works teaching clearly those laws of political economy that regulate wages, and immediately concern both masters and operatives, might be put into their hands to be read by them during the intervals of leisure, and would interest, as well as improve them, more than Penny Magazines, treating of birds, beasts, fishes, insects, ancient ruins, and a thousand other matters, remote from their immediate business. What the labouring classes want is, not so much physical and antiquarian instruction, as sound religious, sound moral, and sound political knowledge; not one of these three branches of practical wisdom can be neglected with impunity—they are all intimately connected, all indispensable to guide men, whatever be their stations, in their duties as members of the community. Teach them as much more as you choose, but these things must be taught; you have learnt from experience that you can derive no security from their ignorance; afford them, therefore, every facility for learning what it is their duty and interest to know, and you may leave the rest to themselves.

I need scarcely advert to the national evils resulting from the factory system, as at present established; they are obvious to the most careless observer. That so numerous a class of the community should be suffered to remain in a state of discontent, after petitioning for relief from their grievances for upwards of thirty years; that they should witness the continued infraction by their masters of those laws which were meant for the protection, not of themselves, indeed, but of their children; that there should be no bond of union between them and their employers, but, except in comparatively few instances, mutual animosity; that from the want of religious, moral, and political knowledge, they should be left, as their opponents assert, the slaves of their fierce passions and ignorant prejudices; all these are fearful considerations, when joined to the reflection of their facilities for combination, and of the activity and intelligence that characterize many of their delegates. Well do they know their importance to our national resources, an importance likely to increase in proportion as our commercial intercourse with the world at large increases, and if they are not now heard, the time may come when they will make themselves felt. Many of the master-manufacturers have long been persuaded that the disaffection subsisting between them and the operatives should be put an end to; that the advantages of moderate labour should extend to all; and from the greater or less, if not utter, inefficiency of all former acts, and from the daily vexations which the present has caused, and is causing, I feel confident that a ten-hour bill, rendering by its simplicity observance easy, and evasion next to impossible, will at length be conceded by Government, as the only measure that can produce permanent tranquillity and general satisfaction. There are many of the master manufacturers who deserve the commendation which the Bishop of Exeter has justly bestowed upon Mr. Fielden; many who, whether radicals, whigs, or conservatives, are *worthy of respect and veneration;* many who will rejoice to see the consummation of the benevolent wishes which they have so long entertained; and perhaps the masters of a different description, after having indulged in a little sordid regret, may find that their interests and their duties are not so much opposed to each other as they now imagine.

DR. URE'S WORKS.

VII. Of all the writers in favour of the factory system, Dr. Ure is the most noted. In 1835 he published a work entitled "The Philosophy of Manufactures," and in the following year, a work called "The Cotton Manufacture of Great Britain," with an introductory view of its comparative state in foreign countries, drawn chiefly from personal survey. In the introductory view he states, as the result of his investigations, that in the different countries of Europe, and in America, the hours of labour are longer than in England. He does not advert to the disturbance in France from an attempt to introduce the English factory system, as stated in the evidence of Smith; upon which occasion the freedom of labour was obliged to be supported by calling in the military, and several lives were lost. He does not advert to the less intensity of labour in France,—hands leaving when they please for recreation. He does not advert to the holidays that are usually taken in France. He does not advert to the difficulty of getting piecers in America, except at a very high price, owing to the demand for agricultural labour, and the consequent ability of parents to maintain their children. These statements rest for the most part on the evidence of operatives, and to such evidence Dr. Ure turns a deaf ear, except when it happens to chime in with his own prejudices. His object is to excite an alarm with regard to foreign competition; and therefore the disadvantages under which foreigners labour are either slightly stated, or entirely overlooked. But that part of Dr. Ure's writings to which I would first request the attention of the reader is, the "Philosophy of Manufactures." The circumstances under which he undertook this work may be best explained in his own words.

"Having been employed in a public seminary for a quarter of a century, in expounding to practical men, as well as to youth, the applications of mechanical and chemical science to the arts, he felt it his duty, on being solicited from time to time by his pupils, now spread over the kingdom as proprietors and managers of factories, to prepare for publication a systematic account of their principles and processes. With this view he resolved to make afresh such a survey of some of the great manufacturing establishments, to which he had liberal access, as might qualify him to discharge the task in a creditable manner. This tour of verification would have been executed at a much earlier date, so as to have enabled him, ere now, to have redeemed his pledges, both publicly and privately given, but for an interruption of unexpected magnitude."

"The right honourable the Lords of the Committee of the Privy Council for Trade and Plantations requested him, about three years ago, to undertake a series of experiments on the refining of sugar, in order to ascertain the relation of the drawbacks on exportation of refined loaves to the duties paid upon the raw article. Under an impression that these researches might be set sufficiently in train, in the space of two or three months, to lead to the desired information, in the hands of experienced operatives, he undertook their arrangement, but encountered so many difficulties from the delicacy of the material operated upon, and other circumstances stated in his official report, printed by order of the House of Commons, that he did not get entirely extricated from them till nearly two years were expired, nor till he had suffered considerably from anxiety of mind and bodily fatigue.

"Being advised by his medical friends to try the effects of travelling, with light intellectual exercise, he left London in the latter end of last summer, and spent several months in wandering through the factory districts of Lancashire, Cheshire, Derbyshire, &c., with the happiest results to his health, having everywhere experienced the utmost kindness and liberality from the mill-proprietors."—Preface to the Philosophy of Manufactures, p. viii., ix.

Now we may collect from this statement that his investigations were principally among the great manufacturing establishments, in which the evils of the factory system appear in the most mitigated form; and the excessive kindness and liberality of the mill-proprietors, some of whom had been his pupils, would not have a tendency to make him lynx-eyed with regard to the evils of the system. Accordingly I find in his works such an account of it, that, if I did not think the Doctor's judgment to be biassed by his prepossessions, and if I could by any possibility keep out of consideration the length of the hours of labour, I should wish it perpetual. Skipping the Inferno, and the Purgatorio, he calls the attention of his readers to the Paradiso of the system, and the reader of Dante, who should confine his attention to the third great division of his work, would form as correct a notion of the whole as the reader of Dr. Ure's publications would of the factory system from the Doctor's account of it. To animadvert upon every position in his writings that requires animadversion would exceed my limits; I must be content with drawing the reader's attention to some of the most prominent.

"The sentimental fever excited by the craft of the operatives' union was inflamed into a delirious paroxysm by the partial, distorted, and fictitious evidence conjured up before the Committee of the House of Commons on factory employment, of which Mr. Sadler was the mover and chairman. It commenced its sittings on the 12th of April, 1832, and did not terminate them till the 7th of August following, when it published a mass of defamation against the cotton-mills, spread over upwards of 600 folio pages."—Philosophy of Manufactures, pp. 290, 291.

Be it recollected that this evidence, which Dr. Ure calls a mass of defamation, was called for by the opponents of Mr. Sadler's Bill, and was necessarily partial; for who but the parties that wished for the passing of that bill could be induced to give evidence against the master-manufacturers at their own peril? for coming up to London to give evidence was generally punished by the loss of their situations on their return. The operatives and their friends thought that the evidence which had been brought forward previously at different times was quite sufficient to shew the necessity of abridging the hours of labour. Their opponents were the conjurors who had conjured up, to use Dr. Ure's expression, this evidence before the Committee of the House of Commons. Right glad would they have been if they could have conjured it down again; but since this was beyond their art, their only resource was to characterize it as a mass of defamation. Mr. Tufnell, whom Dr. Ure styles a most able and candid observer, because his observations confirm his own opinions, found out that one of the eighty-nine who gave evidence was an atheist; but it never appears to have struck either Mr. Tufnell or Dr. Ure, that an atheist being found among the operatives may cut both ways. His individual testimony may be invalidated by his principles, but where did he imbibe those principles? Why, it appears that, between the age of eleven and twelve, he was sent out as an apprentice from St. James's, Westminster, to Douglas and Co., Holywell, in Flintshire, North Wales. Now, one of the charges against the factory system is, the little time allowed for religious instruction; and an atheist being bred up in factories does in some measure corroborate the charge. This charge has been brought against it by the clergy who live in manufacturing towns; among others, by Mr. Horsley, of the Episcopal Chapel at Dundee, son to the late Bishop Horsley; and if Mr. Horsley had wished to substantiate his accusation by a particular instance, he could scarcely have brought forward a more unexceptionable one than that of an atheist being bred up in factories.

The ministers of the day, not satisfied with the evidence brought before Mr. Sadler's committee, and terrified by the representation made to them that a ten-hour bill would enable foreigners to undersell us, appointed commissioners to enter upon the investigation of the subject de novo; and the result was that a ten-hour bill was pronounced an insufficient protection to the children, and the term of eight hours recommended for all children under fourteen.

"We have seen that the union of operative spinners had, at an early date, denounced their own occupation as being irksome, severe, and unwholesome in an unparalleled degree. Their object in making this misrepresentation was obviously to interest the community in their favour at the period of their lawless strike in the year 1818."—Philosophy of Manufactures, p. 298.

Now we think that the operatives are quite as good judges as their masters, or even as Dr. Ure, or any occasional visitor of the factories, as to whether their occupation is irksome, severe, and unwholesome; and, as during that strike they were receiving wages which, according to Dr. Ure, enabled them to maintain a stipendiary committee in affluence, and to pamper themselves into nervous ailments, by a diet too rich and exciting for their indoor occupations, that the irksomeness, severity, and unwholesomeness of their employment were among the causes that occasioned the strike. We should be careful how we impute motives to any individuals, much more to any class, that are not clearly deducible from their actions. How indignant would Dr. Ure himself be if any man should impute his three volumes on the factory system, abounding as they do with the most unqualified praise of the master manufacturers, and with the most rancorous abuse of the operatives, to the servile motive of currying favour with influential men, or even to the more commendable one of gratitude for the kindness and hospitality he has experienced from the great master manufacturers.

"Within a week after the factory commissioners arrived at Manchester, the operatives' union dramatized the miseries of the children in a public procession. They collected about 4,000 of the youngest, mustered them in tawdry array, and paraded them through the streets, heading the motley throng by themselves or agents, brandishing straps and bludgeons as emblems of their masters' tyranny, but really the instruments of their own wickedness, if punishments were ever inflicted by them with such weapons in their moments of ill-humour."—Philosophy of Manufactures, pp. 299, 300.

The masters pay the spinners according to the amount of work thrown off, and the spinners pay the children they employ out of their own wages; and as the beating is neither inflicted nor allowed by the master, he is, according to Dr. Ure, completely exculpated. But does not the beating result from the system adopted by the masters of paying the spinners by the piece, and leaving them to pay the children? The spinner does as much work as he can, both to give satisfaction to his employer and to obtain higher wages; but, to keep the children up to long protracted hours of labour, he is obliged to use coercion; and though Dr. Ure cannot, the children can, trace the treatment to which they are subjected to the master. The men are fined for bad work, and bad work must be the consequence of the children not being on the alert, and the children cannot be kept on the alert during the latter hours of the day without coercion; leave them to nature, and they will fall asleep. Besides, if there is any credit due to the testimony of the operatives, in some mills, masters have not only witnessed the chastisement of children, but taken a part in it, and men who treated the children with the most brutal cruelty, and were legally convicted of it, were yet retained in their employment. The spinners are paid in proportion to the work they perform; this operates as a premium upon cruelty, and there is no remedy to this evil but a limitation of the hours of labour enforced by legal enactments.

"I have visited many factories, both in Manchester and in the surrounding districts, during a period of several months, entering the spinning rooms unexpectedly, and often alone, at different times of the day, and I never saw a single instance of corporal chastisement inflicted on a child, nor indeed did I ever see children in ill-humour.

"It was delightful to observe the nimbleness with which they pieced the broken ends, as the mule-carriage began to recede from the fixed roller beam, and to see them at leisure, after a few seconds' exercise of their tiny fingers, to amuse themselves in any attitude they chose, till the stretch and winding on were once more completed. The work of these little elves seemed to resemble a sport in which habit gave them a pleasing dexterity. Conscious of their skill, they were delighted to shew it off to any stranger."—Philosophy of Manufactures, p. 301.

Unless the Doctor possessed the privileges and immunities of invisibility, we must still prefer the evidence of the operatives to his. His evident delight and admiration might stimulate the children while he was in the mill; but if he had prolonged his visit to an undue length, he would have found languor succeed what he calls the light play of their muscles, and if they had dared to address him, they would have said, "Alas! good Doctor, what is sport to you is death to us." The Doctor does not tell us whether, among the numerous attitudes which the children chose, that of sitting was one. As to his never having seen an instance of corporal chastisement, nor having found the children in ill-humour, perhaps he did not enter the spinning rooms quite so unexpectedly as he imagined, and perhaps the presence of one who so much admired the pleasing dexterity of their motions might put the children into good humour during his presence. Raumer, the historian, mentions his visiting one of the factories, and being received with great civility by the proprietor; but as the proprietor was expatiating upon the happiness of the children, one of them shook his head. It was about the time of dismissal; and Raumer, who did not, like Dr. Ure, take all that the masters said for gospel, followed the child into the street, and asked him what it was that made him shake his head. "I shook my head for myself," said the boy, "and not for others. I once lived in the country, and tended swine. I did not like the employment; but now I have tried the mill, not a day passes but what I wish myself back with the pigs again. With them I could shout and whistle, and do what I liked. Now, I am obliged to be silent, and attentive to what I am about, from morning till night. I can neither shout, nor whistle, nor run, nor sit, nor do a single thing that I like."

I cannot here forbear quoting some stanzas from a poem lately published, and called "A Voice from the Factories," as remarkable for their truth as for their poetical beauty.

Beyond all sorrow which the wanderer knows,
Is that these little pent-up wretches feel;
Where the air thick and close and stagnant grows,
And the low whirring of the incessant wheel
Dizzies the head, and makes the senses reel:

There, shut for ever from the gladdening sky,
Vice premature and care's corroding seal
Stamp on each sallow cheek their hateful die,
Line the smooth open brow, and sink the saddened eye.

For them the fervid summer only brings
A double curse of stifling withering heat;
For them no flowers spring up, no wild bird sings,
No moss-grown walks refresh their weary feet;—
No river's murmuring sound;—no wood-walk, sweet
With many a flower the learned slight and pass;—
Nor meadow, with pale cowslips thickly set
Amid the soft leaves of its tufted grass,—
Lure them a childish stock of treasures to amass.

Have we forgotten our own infancy,
That joys so simple are to them denied?—
Our boyhood's hopes—our wanderings far and free,
Where yellow gorse-bush left the common wide
And open to the breeze?—The active pride
Which made each obstacle a pleasure seem;
When, rashly glad, all danger we defied,
Dashed through the brook by twilight's fading gleam,
Or scorned the tottering plank, and leapt the narrow stream?

In lieu of this,—from short and bitter night,
Sullen and sad the infant labourer creeps;
He joys not in the glow of morning's light,
But with an idle yearning stands and weeps,
Envying the babe that in its cradle sleeps:
And ever as he slowly journeys on,
His listless tongue unbidden silence keeps;
His fellow-labourers (playmates hath he none)
Walk by, as sad as he, nor hail the morning sun.

(A Voice from the Factories, pp. 16, 17.)

But I must return to Dr. Ure:—

"As to exhaustion by the day's work, they evinced no trace of it on emerging from the mill in the evening, for they immediately began to skip about any neighbouring play-ground, and to commence their little amusements with the same alacrity as boys issuing from a school."—Philosophy of Manufactures, p. 301.

That play has irresistible attractions for children, however hard they have been worked, I readily admit; indeed it was asserted by the witnesses before Mr. Sadler's committee, that children employed all night in factory labour, would play during the day, and ramble about, to the curtailment of their due portion of rest. There is a passage in Dr. Darwin's Zoonomia, which Dr. Ure may probably recollect, in which he illustrates the effect of mental stimulus in overcoming fatigue by an anecdote. "A learned Doctor had been taking a long walk with his little son, and, towards the end of the walk, the little boy, thoroughly tired, asked his papa to carry him. The Doctor, either not liking so fatiguing an employment, or unwilling to compromise his dignity, said, 'Here, my boy, take my gold-headed cane, and ride on that.' The delighted child got astride the cane and gallopped away, no longer sensible of fatigue." I am not at all surprised that children should have been caught playing after the fatigues of the mill, though there is evidence enough to prove that they are frequently so little inclined to play, or even to eat, so thoroughly exhausted, in short, as to fall asleep over their suppers.

Let us hear the sequel of the Doctor's panegyric on the happiness of the children :—

"It is, moreover, my firm conviction, that if children are not ill-used by bad parents or guardians, but receive in food and raiment the full benefit of what they earn, they would thrive better when employed in our modern factories than if left at home in apartments too often ill-aired, damp, and cold."—Philosophy of Manufactures, p. 301.

What a notable discovery is this! or, rather, what a lame and impotent conclusion! The sum total of the boasted happiness of the children amounts to this,—that if well fed and well clothed, they would be better off in the factories than if left at home in ill-aired, damp, and cold apartments. But even this comparative state of happiness, or rather of mitigated suffering, is hypothetical, and depends upon their receiving in food and raiment the full benefit of what they earn; now as, in many instances, the parents are dependent upon their children for support, it is impossible for them to receive the full benefit of what they earn. The factory system, by employing children where adults were formerly employed, has inverted the natural order of things, and made the parents dependent upon their little ones for food and raiment; so that if the children are to receive the full benefit of their earnings, their parents must starve.

If this be the amount of all that Dr. Ure, the enthusiastic admirer, the undaunted champion, of the factory system, can say in its defence, with reference to the happiness of the children, we would advise him for the future to maintain a discreet silence, and if he has nothing better to say, to say nothing. Other topics are open to him. Let him expatiate upon the iniquities of the operatives; let him denounce their testimony as a mass of defamation; let him characterize the evidence of his medical brethren as *moonshine*, or *twaddle;* and when he has exhausted the resources of his vituperative rhetoric, let him try the laudatory strain, and celebrate the benevolence of those disinterested philanthropists by whom he has been so courteously received and so hospitably entertained; but let him not, if he has any regard for the cause he has undertaken, let him not touch upon so delicate a topic as that of the happiness of the children, for great as is the eloquence of Ure, the force of truth is greater.

"Though carding and spinning lie within the domain of automative science, yet slubbing is a handicraft operation, depending on the skill of the slubber, and participating therefore in all his irregularities. If he be a steady, temperate, man, he will conduct his business regularly, without needing to harass his juvenile assistants, who join together the series of card rolls, and thus feed his machine; but if he be addicted to liquor, and passionate, he has it in his power to exercise a fearful despotism over the young pieceners, in violation of the proprietor's benevolent regulations. This class of operatives, who, though inmates of factories, are not, properly speaking, factory workers, being independent of the moving power, have been the principal source of the obloquy so unsparingly cast on the cotton and other factories, in which no such capricious practices or cruelties exist. The wool slubber, when behindhand with his work, after a visit to the beer shop, resumes his task with violence, and drives his machine at a speed beyond the power of the pieceners to accompany; and if he finds them deficient in the least point, he does not hesitate to lift up the long wooden rod from his slubbing-frame, called a billy-roller, and beat them unmercifully. I rejoice to find that science now *promises* to rescue this branch of the business from handicraft caprice, and to place it, like the rest, under the safeguard of automatic mechanism."—Philosophy of Manufactures, pp. 8, 9.

The modicum of happiness which the children share is not, it seems, from this statement, without alloy; they are subject to the *fearful despotism* of the slubber, and the benevolent regulations of the proprietor avail them not. The comfort which the Doctor gives the children under these deplorable circumstances is in the form of a promise, and their emancipation is to spring from an improvement in mechanism, which shall throw the slubber out of employment. In the meantime, the poor things are to be subject to a *fearful despotism*, vainly attempting to keep up with the speed of the slubber's machine, and, upon the least deficiency, unmercifully beaten with that dreaded instrument of cruelty, the billy-roller.

In the course of his work, Dr. Ure again attacks the slubbers. His greatest enemy will not accuse him of being a sentimentalist; but his humanity never slumbers when slubbers are mentioned; they are the objects of his implacable disgust.

"It is given in evidence that the 'slubbers generally are a morose, ill-tempered set. Their pay depends on the children's work. The slubbers are often off drinking, and then they must work harder to get the cardings up. I have seen that often. Mr. Gamble (the proprietor of the factory) is one of the most humane men that ever lived, by all that I hear, and he will not allow the slubbers to touch the children on any pretence; if they will not attend to their work he turns them away.' It is, however, unhappily, of such consequence to poor parents to eke out their own earnings with that of their children, that they are often willing to wink at their ill-usage by the slubbers who hire them, and to stifle their complaints. These operatives are described as unamiable, refractory spirits, requiring overlookers of a determined temper to keep them in order, and they are often the fathers or guardians of their pieceners."—Ibid., pp. 180, 181.

In describing the slubbers, he seems to be describing animals of a distinct genus from their masters; they are generally morose, ill-tempered, unamiable, refractory, spirits. But it appears that there is a Prospero who can keep them under control, and make them do his biddings. This Prospero is Mr. Gamble, who will not suffer a child to be touched by a slubber upon any pretence. It follows, that if other proprietors possessed the active, resolute, humanity of Mr. Gamble, the slubbers would be taught forbearance. But the source of the evil is, their pay depends on the children's work; it is this, and not the inherent bad qualities of the slubber, as distinguished from other men, that makes him what he is. The master who would reform the slubber, must reform the system.

"When this new career commenced, about the year 1770, the annual consumption of cotton in British manufactures was under four millions of pounds weight, and that of the whole of Christendom was, probably, not more than ten millions. Last year the consumption in Great Britain and Ireland was about two hundred and seventy millions of pounds; and that of

Europe and the United States together, four hundred and eighty millions. This prodigious increase is, without doubt, almost entirely due to the factory system, founded and upreared by the intrepid native of Preston."—Philosophy of Manufactures, p. 17.

Upon the subject of exports there appears little reason for fear.* When we look upon a map of the world, and consider the small extent of our island, scarcely larger than a province of France, we are lost in astonishment, and can scarcely conceive to be possible what we know to be fact. Well may Mr. Rickards characterize the efforts of our rivals as *puny efforts.* I feel as proud as Dr. Ure can feel of our commercial prosperity; but my pride would be converted into shame and grief, if I thought that those who, whether children or adults, have been the main instruments of that prosperity, should continue a degraded class. I admire, with him, the wonder-working power of automative mechanism; but I sympathize with flesh and blood, and, tracing the vicious habits even of the slubber, bad as he is represented, to something more than a mere natural propensity to drunkenness and cruelty, I would rather reform him than throw him out of employment. By raising him in the social scale, I would teach him to respect himself; and he who respects himself can never be brutal towards others.

"If, then, this system be not merely an inevitable step in the social progression of the world, but the one which gives a commanding station and influence to the people who most resolutely take it, it does not become any man, far less a denizen of this favoured land, to vilify the author of a benefaction which, wisely administered, may become the best temporal gift of providence to the poor, a blessing destined to mitigate, and in some measure to repeal, the primeval curse pronounced on the labour of man, 'in the sweat of thy face shalt thou eat bread.' "—Ibid., p. 17.

There is no man who is better acquainted with the virtue of an *if,* than Dr. Ure; but as this benefaction has been hitherto, in far too many instances, unwisely administered, his hypothesis falls to the ground, and not only does the primeval curse remain unrepealed, but cruelly augmented by the avarice of man, and his consequent insensibility to the sufferings of his fellow men. The best temporal gift of Providence has been perverted, and that which, in its legitimate use, should have mitigated, has actually increased, the irksomeness of toil, as well as added to the hours of confinement.

"It is, in fact, the constant aim and tendency of every improvement in machinery to supersede human labour altogether, or to diminish its cost, by substituting the industry of women and children for that of men, or that of ordinary labourers for trained artisans."—Ibid., p. 23.

It is this tendency which has caused the adult operatives so much annoy; it is this tendency that first took him from his cottage, and placed him in a mill, and that has since, in many instances, transferred him from the mill to the workhouse.

"It is one of the most important truths resulting from the analysis of manufacturing industry, that unions are conspiracies of workmen against the interests of their order, and never fail to end in the suicide of the body corporate which forms them; an event the more speedy, the more coercive or

	£.		£.
* 1820.—French exports....	1,091,300	1820.—British exports ..	16,695,539
1830.—Ditto............	2,192,240	1830.—Ditto	31,810,474

I leave it to arithmeticians to calculate in what year of the world the French, at this rate of progression, will overtake us.

the better organized the union is. The very name of union makes capital restive, and puts ingenuity on the alert to defeat its objects."—Philosophy of Manufactures, p. 41.

Whether unions, however objectionable in many points, end in the suicide of the body corporate which forms them, remains to be proved; at present, I think, they are likely to obtain their object, because that object is a legitimate one. Capitalists, however keen-witted, have sometimes been the dupes of their own avarice; and the combinations of the workmen have proved a salutary check to the rash speculations into which an over-eager desire of gain has betrayed them. In confirmation of this assertion, I must be allowed to quote a passage from Dr. Combe's work on the "Constitution of Man considered in relation to External Objects," pp. 216, 217:—

"According to the principles of the present treatise, what are called by commercial men 'times of prosperity,' are seasons of the greatest infringement of the natural laws, and precursors of great calamities. Times are not reckoned prosperous unless all the industrious population is employed during *the whole day* (hours of eating and sleeping only excepted) in the production of *wealth*. This is a dedication of their whole lives to the service of the propensities, and must necessarily terminate in punishment, if the world is constituted on the principle of supremacy of the higher powers.

"This truth has already been illustrated more than once in the history of commerce. The following is a recent example:—

"By the combination laws, workmen were punishable for uniting to obtain a rise of wages, when an extraordinary demand occurred for their labour. These laws, being obviously unjust, were at length repealed. In the summer and autumn of 1825, however, commercial men conceived themselves to have reached the highest point of prosperity, and the demand for labour was unlimited. The operatives availed themselves of the opportunity to better their condition; formed extensive combinations; and because their demands were not complied with, struck work, and continued idle for months in succession. The master manufacturers clamoured against the new law, and complained that the country would be ruined if combinations were not again declared illegal, and suppressed by force. According to the principles expounded in this work, the just law must, from the first, have been the most beneficial for all parties affected by it; and the result amply confirmed this idea. Subsequent events proved that the extraordinary demand for labourers in 1825 was entirely factitious, fostered by an overwhelming issue of bank paper, much of which ultimately turned out to be worthless; in short, that, during the combinations, the master manufacturers were engaged in an extensive system of speculative over-production, and that the combinations of the workmen presented a natural check to this erroneous proceeding. The ruin that overtook the masters in 1826 arose from their having accumulated, under the influence of unbridled acquisitiveness, vast stores of commodities, which were not required by society; and to have compelled the labourers, by force, to manufacture more at their bidding, would obviously have been to aggravate the evil. It is a well-known fact, accordingly, that those masters, whose operatives most resolutely refused to work, and who, on this account, clamoured most loudly against the law, were the greatest gainers in the end. Their stocks of goods were sold off at high prices during the speculative period; and when the revulsion came, instead of being ruined by the fall of property, they were prepared, with their capitals at command, to avail themselves of the depreciation, to make new and highly profitable investments. Here again, therefore, we perceive the law of justice vindicating itself, and benefiting by its operation even those individuals who blindly denounced it as injurious to their interests. A practical faith in the doctrine that the world is arranged by the Creator in harmony with the moral sentiments and the intellect, would be of unspeakable advantage to both rulers and subjects; for they would then be able to pursue with greater confidence the course dictated by moral rectitude, convinced that the result would prove beneficial, even although, when they took the first step, they could not distinctly perceive by what means."—Combe's Constitution of Man, pp. 217, 218.

I have great satisfaction in quoting from Dr. Ure any passage that may allay the fears of those who labour under apprehension with regard to foreign competition; because, as in several parts of his works he tries to excite alarm on this point, he is a most unexceptionable witness to the needlessness of such alarm.

"American cotton-wool, which three years after the peace was 1*s.* 6*d.* a pound, is now 7*d.*, and other cottons in proportion. Now, in respect of a very common article, supposing the price of the cotton-wool 7*d.*, the price of the cloth would be from 1*s.* 1*d.* to 1*s.* 2½*d.* a pound. The throstle twist, which has been so largely exported of late years, is made with almost no manual labour; and though it be low priced, it still remunerates the skilful manufacturer of good capital. And notwithstanding the competition of the cotton trade in foreign countries, the export of twist from us to them has materially increased. This increase may be attributed to the very low price at which we can sell it. Another advantage we possess, is the cheapness as well as excellence of our machinery. A machine, which would cost only 30*l.* in Manchester, would cost 54*l.* in the United States; and a dozen of spindles, which would cost only 4*s.* in the former place, would cost 14*s.* 6*d.* in the latter."—Philosophy of Manufactures, pp. 71, 72.

Dr. Ure asserts, in several parts of his works, that the rate of wages on the continent, and in the United States, is lower than in Great Britain. With regard to this point, I must beg the reader's attention to the observations of Mr. Cowell, in his Preface to his tables, as he made it one of the most particular objects of his researches:—

"When I reflect that I was unable, with the utmost care, to form any general and well-defined conclusions concerning the rate of payment for work done in the cotton manufacture in England, or in any district, I cannot attribute much weight to the comparisons, which are so confidently instituted, respecting the comparative rate of wages in foreign and English cotton-working. In cotton-working, the rates alleged to prevail abroad (putting out of consideration that they are never stated with reference to hours, but merely relate to periods comparatively vague, such as a day, a week, a month, or a year,) convey no information of the quantity or quality of work turned off in a given time, to realize a given sum of money. They amount, in fact, to nothing more than this, viz., that since 10*d.* is less than 2*s.*, so the rate of wages in France, Belgium, Switzerland, and Germany, is lower than in England; and then, by confounding the 'rate of wages' with the 'payment of the work done,' it is asserted or implied, that the rate of payment for work done is likewise lower abroad than at home."

"This consequence would not be true, even if the generality of the evidence, as to wages or net earnings, approached to accuracy, (which no one can suppose very probable;) still less is it true, when no distinction is ever made in this kind of evidence between rate of wages and rate of payment for work done."

"A variety of evidence, supported by documents and letters, was tendered in Manchester, to prove that the rate of wages in cotton-working was nowhere so high as in England. I found myself incapable of attaching much weight to it, even as far as regarded the rate of wages, or that of the daily earnings of an individual, and none whatever as regards the rate of payment for work done. I allude to the evidence furnished by Messrs. Pooley, Birley, H. Hoole, Ashworth, and Robert Hyde Grey, particularly to the statistical documents furnished by the latter gentleman."

To return to Dr. Ure—I beg the reader's attention to the following passages:—

"The locks of flax are screwed into the holders by a boy, called the *screwer*, generally younger than his companion, the *machine-minder*; and his labour appeared, to a most intelligent factory commissioner, (John Elliot Drinkwater, Esq.) to be, beyond comparison, the most fatiguing that he had seen children subjected to, independently of the noxious atmosphere, loaded as it is with particles of flax, incessantly detached and scattered by the whirling of the machines. The *screwer* seems not to have, with the common heckling-machine, an instant's cessation from labour; bunch after bunch is thrown down before him to fix and unfix—actions performed by him with incredible rapidity. If he does not do his business properly, he mars the work of the machine-minder—generally, a *bigger boy*—and is apt to suffer the usual punishments inflicted by the stronger workpeople on the weaker, who happen to obstruct their labour and their gain, annoyance or oppression in some shape or other. If the difference of age and strength were the same between the machine-minder and the screwer as between the slubber and piecener in wool-spinning, there is little doubt that the ill treatment, which is in our factories now almost exclusively the reproach of the slubber, would find its parallel in a flax-mill."—Philosophy of Manufactures, p. 214.

"Though the boys soon become expert at this business, (the business of machine-minders in flax-mills,) a great many of them examined by the commissioners had been wounded, sometimes severely so, during their noviciate."—Ibid., p. 215.

These passages need no comment. I cannot imagine what induced Dr. Ure to characterize the evidence brought before Mr. Sadler's committee as a mass of defamation, when even from the Doctor's own works the evils of the factory system might be almost as fully demonstrated, by a diligent selection of passages, as from that evidence. These passages are, indeed, disjecta membra, much of admiration and applause comes between, but, if collected, they would be sufficient to prove the existence of the evils which his opponents impute to the system.

"In spinning the rovings of the flax-line into fine yarn, it is necessary to pass them through a trough of water, placed at the back of the spindles, in consequence of which a dewy spray is continually thrown off in the front of the frame from the yarn, as it is rapidly twirled by the flys of the spindles; and as another spinning-frame is placed at no great distance, the spinner is exposed to this small rain both in his front and rear, whereby he may, without certain precautions, have his clothes thoroughly soaked through in a few hours, especially if stationed between two frames set close to each other. The propriety, therefore, of separating each pair of frames to a suitable distance is obvious, and has been properly insisted on by many visitors, more especially the factory commissioners and inspectors. It is, in fact, an instructive instance of avarice over-reaching itself, by crowding machines into a space too small for them to be advantageously worked in; a circumstance scarcely ever observed by me in any of the numerous cotton-mills which I entered."

"The ill effects of this condensation are felt in two ways, first in the exposure of the hands to discomfort, and the danger of being laid hold of and injured by the revolving wheels; and secondly, in the pollution of the air by a crowd of persons breathing in a small space."—Philosophy of Manufactures, pp. 219, 220.

One might imagine one's self, from these passages, to be listening to one of Mr. Sadler's witnesses. By what precautions a spinner might be saved from being soaked through is not pointed out; it seems scarcely possible for a person placed between two showers to escape a soaking except by a precipitate retreat. "Si sic omnia dixisset," Dr. Ure might have ranked among the friends of the operatives.

"When the handicraftsman exchanges hard work with fluctuating employment and pay, for continuous labour of a lighter kind with steady wages, he must necessarily renounce his old prerogative of stopping when he pleases, because he would thereby throw the whole establishment into disorder. Of the amount of the injury resulting from the violation of the rules of automatic labour, he can hardly ever be a proper judge; just as mankind at large can never fully estimate the evils consequent upon an infraction of God's moral law."—Ibid., p. 279.

When the handicraftsman is deprived of his old prerogative by the laws of automatic labour, he thinks it but fair that the hours of labour should be fewer; he finds, from experience, that he can endure hard work, with the prerogative of stopping, better than continuous labour, however light. A diminution of the hours of confinement would violate no law of automatic labour; it is this which he seeks, and he thinks that his master, in pertinaciously refusing it, is guilty of infringing upon one of God's moral laws, indeed, upon one of the cardinal rules of the gospel, "Do unto others as ye would that they should do unto you." As mankind at large never fully estimate the evils consequent upon an infraction of God's moral law, let the master-manufacturers, without exception, imitate the example so nobly set them by some of their own class, and religiously observe the laws of humanity, leaving the consequences to Supreme Wisdom.

"During a disastrous turn-out in Manchester, in 1818, when fifteen thousand of the factory people refused employment for several months, paraded the streets, besieged such cotton-mills as continued to work in defiance of their commands, and threatened to destroy their industrious inmates, the committee of the Spinners' Union issued the following proclamation,—the fountain head of the torrents of calumny since so profusely discharged on our factory system:—

"'We believe there is no species of labour so fraught with the want of natural comforts as that the spinners have to contend with; deprived of fresh air, and subjected to long confinement in the impure atmosphere of crowded rooms, continually inhaling the particles of metallic or vegetable dust, his physical powers become debilitated, his animal strength dwindles away, and few survive the meridian of life, and the grave is often the welcome asylum of his woes. His children!—but let us draw a veil over the scene!—our streets exhibit their cadaverous and decrepit forms, and any attempt to describe them would be impossible. Let it not be understood that we attach blame to our employers, as applied to these calamities; they are perhaps inseparable from the very nature of the employment, and our masters may lament but cannot redress them.'

"We shall presently prove," says Ure, "that this picture is distorted in every respect."—Philosophy of Manufactures, pp. 287, 288.

The Doctor's proof that this picture is distorted, does not consist in a well supported negation of the existence of the evils imputed to the factory system, but in tracing them to the operatives. He does not attempt to get over the circumstance, that in some mills the operatives are effectually prevented from coercing the children; whence it follows, that where they are not prevented from coercing them, the proprietor has at least his share in that blame, which Dr. Ure would fix exclusively upon the operatives. He lays great stress upon a passage, which he has extracted from Mr. Tufnell's factory commission report: as, in the process of extraction, he has adroitly introduced some softening alterations in the wording of the passage, I shall give it from the report itself:—

"Now the inevitable conclusion from this evidence is, that the reason, which has been so prominently put before the public in favour of the ten-hour bill, is wholly groundless;—that children in cotton-mills are not injured by their labour, are not overworked. What other explanation can be given of the circumstance, that of the numerous witnesses examined not one rested the case for the bill on the ground of its being necessary for the protection of the children? This notion is wholly repudiated in the greatest manufacturing district in England! How else can we explain the fact, that persons of the greatest respectability are in the habit of sending their children to work the usual hours in cotton-mills? To take a single instance: Mr. Rowbotham, the superintendent of nearly 400 workmen in Mr. Birley's mills, a man of equal respectability with any London shopkeeper, has brought up all his children in cotton-factories, and three out of four of them in that department which is usually considered the most unhealthy of all, the card-room. Are we to suppose that Mr. Rowbotham, and the hundreds such as Mr. Rowbotham, are so devoid of parental affection as to wish to deform their children, and to subject them to all the miseries described in the report of the factory committee; or so unobservant, as not to have discovered, if such a discovery be possible, that twelve hours' labour in a cotton-mill injures their children. One of two things must be true; either the tales of the hardships of the factory children are unfounded, or the Lancashire people are very fools in intellect and demons in feeling. If those persons who are most acquainted with, and constantly reside in, the cotton districts, do not rest their advocacy for the bill on the plea of humanity, and if by their conduct they show that they disbelieve this plea, what evidence can surpass or even equal in force this argument? It overlays and supersedes all other evidence whatever, medical or other, that can be adduced. If it is contradicted by all the physicians in England, the physicians must be wrong; if returns of sick societies or mortality tables say otherwise, they must be false; all the testimony that can be produced from other sources cannot weaken this evidence, unless, indeed, it can be proved that the whole population of the cotton districts possess the heads of idiots or the hearts of brutes."—Supplementary Report, Part 1, D. 2, 196.

Now in this redoubted argument I confess that, for my own part, I can see little but confusion of ideas and self-sufficiency. The conclusion which Mr. Tufnell would draw, so far from being inevitable, depends upon an hypothetical assumption, that those persons who reside in the cotton districts do not rest their advocacy for the bill on the plea of humanity, and by their conduct shew that they disbelieve this plea. Now with regard to this point, it is surely a matter of opinion, and not an inevitable conclusion. The advocacy for the bill has at almost every public meeting rested on the plea of humanity, of humanity to adults as well as to children. Parents, by sending their children to the mills, and at the same time complaining of the hardships of the system, evince their necessitous condition, not inconsistency, still less hypocrisy. The physicians may be right, the returns of sick societies and mortality tables may be perfectly correct, and yet *the whole population of the cotton districts* may neither possess the heads of idiots, nor the hearts of brutes. Mr. Tufnell's conclusion is rash in the extreme, and, so far from deserving the name of argument, is scarcely entitled to be called a fallacy, as it has not enough plausibility to deceive any one, except, perhaps, Dr. Ure and its author. I shall, therefore, for the purpose of distinction, call it the Rhodian Leap, or Tufnell's Short Method with the Factory Question. With regard to Mr. Rowbotham, it is not necessary to suppose either that he wishes to deform his children, or to subject them to *all the miseries* described in the report of the factory committee; he may be quite as respectable as any London shopkeeper, and yet be under the necessity of placing his children in a factory; and the report to which Mr. Tufnell refers, proves, by direct evidence, that there are factories that are well regulated, and in which the humanity of the proprietors mitigates, as far as possible, the evils of the system. I have received a letter from a correspondent residing in Manchester, and personally known to me, by which it appears that four of the young Rowbothams left the factory in which their father placed them on the first opportunity, and that the fifth fell a victim to factory labour.

Manchester, Oct. 27th, 1836.

Dear Sir,—I duly received your favour of the 24th, and was glad to hear of your safe arrival. You will, I have no doubt, think I have been very negligent in not forwarding to you the information I promised. I found much difficulty in obtaining correctly the real position of Mr. Rowbotham's family. He had five sons, who are thus disposed of: one (I believe the eldest) is going about the town with an ass, hawking milk, and has, over and over again, declared, that if he had remained in the factory he must long before this have ceased to live. As a proof that he did not like the mill after he came to the use of reason, I need only assert that the difficulties he had to encounter in establishing his business were so great, that he was obliged actually to go almost without shoes to his feet, and all this he was willing to and actually did suffer, rather than return to the mill, where he could have had good wages, and what is termed a comfortable situation. The second son also found that the system was not so good as his father is represented to have made it appear by Ure and Tufnell, and he also left the mill as soon as he could, and is now travelling for some small-ware manufacturer, or something of that sort. Another of the sons is gone to Belgium, to superintend some branch in the cotton trade. The fourth is still in the employ of Mr. Birley, but not in the card-room, where his father said he had brought them all up, but in the warehouse, taking in the work as it comes from the reelers, where there is neither dust nor necessity for a heated room. The fifth is dead, and, if report be true, he fell a victim to his father's avarice, which was acknowledged by all who knew him to be very great. The report runs thus, (and I have no reason to doubt it, as I have it from good authority:) that he was taken ill in the factory; he asked leave from his father to go home, and was refused; the malady increased so rapidly without the advice and attention of a surgeon, that death was the result, and may have made the father often reflect on his own imprudence for not allowing him to go home when he first requested it. Whatever Dr. Ure may say, there is this fact, that they were all glad to get out of the mill on the first opportunity."

So much for the wholesale argument of Mr. Tufnell, so triumphantly quoted by Ure. There is a passage in Dr. Ure, which I extract, as putting in a striking point of view the repugnance that is felt even by the most necessitous towards factory labour.

"Mr. Strutt, learning that the inhabitants of a village a few miles from Belper, occupied chiefly by stocking-weavers, was in a distressed state, from the depreciation of their wages, invited a number of the most necessitous families to participate in the better wages and steadier employment of their great spinning-mills. Accordingly, they came with troops of children, and were delighted to get installed into such comfortable quarters. After a few weeks, however, their irregular habits of work began to break out, proving, both to their own conviction, and that of their patrons, their unfitness for power-going punctuality. They then renounced all further endeavours at learning the new business, and returned to their listless independence."—Philosophy of Manufactures, pp. 333, 334.

Neither comfortable quarters nor better wages could reconcile these necessitous stocking-weavers to power-going punctuality; the delight they at first felt was succeeded by weariness, and they preferred to factory labour, in its mildest form, (for Mr. Strutt is noted for his humanity, and for regulating his mills in the very best manner,) their former listless independence, with all its accompanying distress. Nor were they at all singular in their taste. It appears from another passage, that the lace-embroiderers hate factory labour quite as much as the stocking-weavers do.

"The embroidery of bobbin-net, called lace-running, also a non-factory household work, painfully illustrates our position. No less than 150,000 females, chiefly of very youthful ages, get their livelihood from this employment in Great Britain. The work is wholly domestic; and though requiring more skill and harder labour than any other branch of the lace business, it is the worst paid. Almost the youngest of them, says Mr. Power, (and they begin at the age of nine or ten,) is able to speak with regret of a better state of earnings, and a period of less necessity for constant labour. They begin early, and work late, and during this long daily period their bodies are constantly bent over the frame upon which the lace is extended, the head being usually kept within five or six inches of the frame, the edge of which presses against the lower part of the chest. One effect universally produced by this habit is short-sightedness, and often general weakness of the eyes; with consumptive tendency, distortion of the limbs, and general debility from the confinement and the posture."

"Aversion to the control and continuity of factory labour, and the pride of spurious gentility, or affectation of lady-rank, are among the reasons why young women so frequently sacrifice their comfort and health to lace-embroidery at home. One girl in her examination states, 'I like it better than the factory, though we can't get so much. We have our liberty at home, and get our meals comfortable, such as they are.'"—Ibid., p. 335, 336.

Now though the complaints of the operatives with regard to the irksomeness of factory labour are thus confirmed both by the stocking-weavers and lace-embroiderers, Dr. Ure will persist in pronouncing it not only easy but delightful, and asserting that the children are not only contented under it, but gay and sportive. The Edinburgh reviewer, to whom he refers in his recent publication on the cotton manufacture, controverts many of his positions, and therefore he is stigmatized by him as *the most venal parasite of absolutism;* independent men of the highest reputation in the medical profession give opinions diametrically opposite to his own, and therefore their evidence is characterized as *moonshine or twaddle.* I have not time to examine the numerous passages in Dr. Ure's work that make for the cause of the children and adult operatives, either directly or indirectly; but, if other evidence were wanting, I could almost venture to rest their cause upon the Doctor's reluctant admissions. I call those admissions reluctant where he admits a fact, but ascribes it to some far-fetched cause, as when he admits that the lace-embroiderers hate factory labour, but ascribes their hatred of it, not to the causes which they themselves assign, but to the pride of spurious gentility, or affectation of lady-rank.

The evidence with regard to the factory system is conflicting, and I am compelled to say, that Dr. Ure's summary method of arriving at a conclusion is, for the most part, to disregard what is said against it, and magnify whatever has been said, or can be said, in its favour. He lends an unwilling ear to the operatives when they complain, and characterizes their evidence as a torrent of defamation; but let them give testimony in favour of the system, and they become splendid exceptions to individuals in general of the class to which they belong; their evidence is the dictate of truth, and they themselves models of respectability.

I neither possess the patience of Job, nor even of a factory operative, one or other of which the Quarterly reviewer pronounces to be an indispensable requisite to the perusal of Dr. Ure's work,—but I have read enough of them to appreciate their exact value, all, in fact, that relates to the treatment, as well as to the condition, both of the children and of the adults; and I forbear to inflict further extracts upon my readers.

The aim of the present publication, whether it be productive or not of any improvement in the physical and social condition of the different parties in whose cause it was undertaken, is to procure an impartial consideration of their case, and such treatment, both of the children and of the adults that are subjected to factory labour, as justice, as well as humanity, demands.

It was obviously necessary, in the prosecution of this object, to prove that they were sufferers under the present laws, whether from their inherent defects, or from the facility of evading them; and as the remote consequence of their sufferings is to despoil them of their constitutions, the evidence of medical men was indispensable, to demonstrate the disproportionate resistance of the human frame to the destructive agency of the factory system. To those who may be disposed to view the professional evidence, which forms so large a portion of my work, as redundant, let it be observed that it divides itself into different periods of time, and that it derives much of its irresistible force from this very circumstance. The testimonies of so many individuals, all eminent for professional skill, given at separate times, confirming and illustrating the same important truths, may be considered as the radii of a circle converging to a common centre. At the latter part of the last century, Aikin, Hunter, and Percival, drew the attention of the public to the evils of infant labour. Men more eminent in their profession can scarcely be named; and I extract from Mr. Gaskell's work on the factory system the statements of Dr. Percival, made on the 15th of January, 1796, on the establishment of the Manchester Board of Health, that the reader may compare it with the statements of other medical men, equally distinguished in their profession, made about forty years afterwards.

"The Board have had their attention particularly directed to the large cotton-factories established in the neighbourhood and town of Manchester, and they feel it a duty incumbent on them to lay before the public the result of their inquiries :—1. It appears that the children, and others, who work in the large cotton-factories, are peculiarly disposed to be affected by the contagion of fever; and that when such infection is received, it is rapidly propagated, not only amongst those who are crowded together in the same apartments, but in the families and neighbourhoods to which they belong. 2. The large factories are generally injurious to those employed in them, even when no particular diseases prevail, from the close confinement which is enjoined, and from the debilitating effects of hot or impure air, and from the want of the active exercises which nature points out as essential in childhood and youth, to invigorate the system, and to fit our species for the employments and the duties of manhood. 3. The untimely labour of the night, and the protracted labour of the day, with respect to children, not only tends to diminish future expectations, as to the general sum of life and industry, by impairing the strength, and destroying the vital stamina, of the rising generation, but it too often gives encouragement to idleness, extravagance, and profligacy, in the parents, who, contrary to the law of nature, subsist by the oppression of their children. 4. It appears that the children employed in factories are generally debarred from all opportunities of education, and from moral and religious instruction."

Such is the dauntless evidence of a man living in the very heart of the factories, and as eminent for success in his profession as for professional talent. The evidence of one such man is worth that of a host of others; and if all who have equalled him, perhaps, in medical skill, had possessed the same degree of moral courage, the factory question would, I am confident, by this time have been satisfactorily settled. But every allowance must be made for medical men residing and practising in manufacturing districts; the ties of kindred, of family connexion, of social intercourse, and of private friendship, as well as a justifiable apprehension of offending those upon whom, in many instances, their success in life depends, present obstacles to the explicit declaration of the truth, which nothing but great professional eminence, independence of fortune, and, what is still more rare, a high degree of moral courage, stimulated by ardent and active humanity, can completely overcome. When all these circumstances are taken into consideration, we shall cease to wonder that the evidence of those medical men who reside in manufacturing districts is conflicting. It is rather matter of surprise that so many should be found who have fearlessly come forward to denounce the system and expose its evils, with so many and such cogent motives for silence. Much as I differ from Dr. Holme, of Manchester, in general opinion, with regard to the factory question, there is one point in which we entirely coincide,—namely, the necessity that all who approach the subject should come to it with unbiassed minds, and in his consequent recommendation of the subject to the consideration of visitors, rather than to that of his fellow-townsmen. But I cannot agree with him, that a day or two would be sufficient to sift the subject, though the master-manufacturers should afford—as Dr. Holme assured the meeting they would—every facility for instituting the inquiry. Even upon the supposition that the gentlemen who were congregated at Manchester last July, to celebrate the fourth anniversary meeting of the Provincial Medical Association, could have afforded the day or two, which their chairman, Dr. Holme, thought necessary for the investigation, their time would have been lost. Much courtesy, no doubt, would have been shewn on the part of the manufacturers, which would have been met by a correspondent feeling on the part of the medical men; a short, and therefore cursory and superficial examination would have been gone through, and the whole affair would have ended entirely to the satisfaction of the master-manufacturers and to the dissatisfaction of the public. The sagacity of the Provincial Medical Association must be something supernatural, if, during a day or two's residence at Manchester, they could decide a question upon which the greatest diversity of opinion has existed, and which has occupied the public mind for upwards of forty years. Dr. Holme's speech is recorded in the "Lancet," No. 19, which contains an account of the fourth anniversary meeting of the Provincial Medical Association, held at Manchester, on the 20th and 21st of July, 1836. To justify Mr. Sadler's assertion, that there was evidence enough before the public to demonstrate the evils of the factory system before the appointment of that committee which has been called by his name, because he was its chairman, I have made extracts from the evidence delivered in 1816, 1818, and 1819. I have given some of the strongest evidence in favour of the system, as well as some of the strongest evidence against it; and, in making these extracts, I have been limited by nothing but the indispensable necessity of keeping my book within reasonable compass. From the report of the committee, of which Sir Robert Peel was chairman, in 1816, I have given extracts from the evidence of the following medical men, whose names are too celebrated to require any epithet to be affixed to them—Dr. Baillie, Sir Astley (then Mr.) Cooper, Sir George Tuthill, Sir Anthony (then Mr.) Carlisle, Dr. Pemberton, and Sir Gilbert Blane; to which is added the no less important evidence of Mr. Archibald Buchanan, Mr. Robert Owen, Mr. Nathaniel Gould, and of Sir Robert Peel himself.

From the report of the committee of the House of Lords in 1818, of which Lord Kenyon was chairman, I have made extracts from the evidence of Dr. Holme, Mr. Whatton, surgeon, Dr. Hardie, Mr. Thomas Wilson, surgeon, Mr. W. J. Wilson, surgeon, Mr. Hamilton, surgeon, Dr. Carbutt, Mr. Ainsworth, surgeon, and Mr. Turner, house-surgeon and apothecary of the Manchester poor-house.

Of these, Dr. Holme was requested to examine into the health of the children by the proprietors of several large establishments; Mr. Whatton went to inspect the factories, in consequence of an order from the chairman of the committee of cotton-spinners; Dr. Hardie, in consequence of a similar application; Mr. Thomas Wilson, by the desire of Mr. William Ellis, the owner of a cotton-mill; Mr. W. J. Wilson, at the request of Mr. Douglas, chairman of the committee of cotton-spinners; Mr. Hamilton, at the request of Mr. Birley, manager; Dr. Carbutt was a physician to the Manchester Infirmary, as well as to other medical institutions in that town, and was employed by Mr. Holt, a factory owner, and by Mr. Douglas; Mr. Ainsworth was requested by the committee of the proprietors of spinning-mills to examine their mills; circumstances which it is necessary to bear in mind, in appreciating the value of their respective testimonies.

From the report of the committee of the House of Lords in 1819, of which also Lord Kenyon was chairman, I have made extracts from the evidence of Thomas Wilkinson, operative, Richard Coar, operative, John Farebrother, operative, Dr. Ward, Job Bottom, operative, Mr. Dadley, surgeon, Mr. Boutflower, surgeon, Mr. Simmons, surgeon, Dr. Jarrold, and Dr. Jones.

Of the medical men who were examined, Dr. Ward had resided at Manchester for upwards of forty years, and nearly thirty years as a medical man—his evidence is against the system; Mr. Dadley had practised in Manchester for fourteen years, in public institutions, and among the lower orders—his evidence is against the system; Mr. Boutflower had practised at Salford for twenty-six years—his evidence, also, is against the system; Mr. Simmons had practised at Manchester for thirty years—his evidence is against the system; Dr. Jarrold had practised fourteen years at Manchester, and four or five at Stockport—his evidence is against the system; Dr. Jones had practised for nearly five years at Chester, where he was physician to the infirmary, and various other public institutions—his evidence, also, is against the system; and I would particularly draw the reader's attention to one of his answers, as illustrating the very different results that two persons would come to, one of whom should inquire of the managers, and the other of the children themselves, without estimating the comparative probability of answers diametrically opposite:—

"Did you inquire whether the children preferred working by day or night?—I did; the managers gave me to understand, without inquiry, that they preferred working at night. That statement led me to inquire of the children themselves, which I did. Under these circumstances, going down on the morning following the night already spoken of, I met the night-workers returning to their homes; I inquired of them, and found that they all preferred working in the day, except one little girl."

It appears from a subsequent part of Dr. Jones's evidence, that *one little girl* had from three to four miles to come to her work, and that she preferred doing so in the afternoon, because it was too early for her to get up if she came in the morning to work in the day.

After having diligently perused the parliamentary documents relating to the factory system, as well as almost every publication of celebrity upon the subject, I determined to pay a visit to Lancashire, and judge for myself. I found, upon visiting the factories, many of my previous anticipations verified; in some of the mills the evils of the system were considerably mitigated by the humane regulations of the proprietor; in others, they appeared in a form but too palpable. During the time I stayed in the latter, I almost gasped for breath. But when I got out of the mills into the open air, my case was not always much mended. Some of the streets are justly described by Dr. Kay as almost pestilential, the atmosphere loaded with the smoke and exhalations of a large manufacturing city. I must refer my readers to two very able articles by Dr. Lyon, in the numbers of the "North of England Medical and Surgical Journal" for August and November, 1830, for a sketch of the medical topography and statistics of Manchester, or to the work of Dr. Kay on the Moral and Physical Condition of the Working Classes, published in 1832. The facility of procuring coal and water must have been the inducements to place the metropolis of the manufacturing districts, as it has been aptly called, in a locality possessing such an atmosphere as Manchester. I refer to the publications of Dr. Lyon and Dr. Kay, that the reader may not attribute my descriptions to the prejudices of a stranger; and I prefer, for the most part, making extracts from them, to giving the results of my own observations, with regard to the climate, police, &c., of this far-famed city. There are two points, however, with regard to which it would be unfair to withhold the tribute of praise—I mean the number of schools and of medical institutions in which it abounds. I subjoin a table of the day and evening schools in the borough of Manchester, from a report of the committee of the Manchester Statistical Society, printed in 1835.* It is true that, in this city, the education of the children of the lower classes is, for the most part, confined to reading, writing, and arithmetic, and that these are, in too many instances, very imperfectly taught; but I know of no town in England to which the same observation might not be applied, and therefore the inhabitants of Manchester are to be praised for having done much, rather than censured for not having done more. Education, as a means of forming the character, is more attended to on the continent, particularly in Prussia, and in several of the German States, than in our country. All children, of whatever class, are obliged, by law, to attend school between the ages of seven and fourteen; while in Manchester not quite two-thirds of those between the ages of five and fifteen receive even nominal instruction. I mention not this to disparage the exertions of many benevolent individuals, both master-manufacturers and others, but I would willingly draw their attention to this important conclusion,—if mental cultivation, coupled with moral and religious instruction, are judged so necessary under a despotic form of government, where the liberty of the subject is so effectually coerced by the iron hand of government, how much more necessary are they under our own more liberal constitution, where so much is left to the freedom of the will, that even the very appearance of coercing it is denounced as an infringement upon British liberty. It is of the utmost importance that that which is likely to be so powerful in its operation, should not be mischievous or uncertain in its effects. Combinations among the workmen were formerly illegal; but as the law which made them so had a tendency to drive them to secret combinations, or to place them, individually weak, at the mercy of their masters, it was repealed. A new instrument, of tremendous power, is put into their hands, to be dreaded in proportion to the ignorance of those who wield it. That it has too often been employed, not defensively, but vindictively, cannot be denied; and it is to be feared that this will never cease to be the case till the improved physical and moral condition of the operatives, whether children, adolescents, or adults, brought about not merely by legal enactments, but by the strenuous and persevering efforts of the master-manufacturers, shall disarm combinations of their malignancy, and convert the office of a delegate into a sinecure. Happy will it be for all parties, if the germ of a national education should be found in the manufacturing districts. To religious and moral instruction, together with the ordinary branches of education, should be added a competent knowledge of political economy, as far as concerns the regulation of wages, that they may not, by erroneous views upon this point, injure either their masters or themselves. The interests of both parties are so intimately combined, that they can never be separated, except

* See the printed Report, p. 33.

for a time; this is true, even if the term interests be taken in its most contracted sense. But there are many among the master-manufacturers of enlarged views, who have proved, by their actions, that justice and humanity are their predominating principles; some of them I know personally, others from the voice of fame; for the operatives are by no means backward in bestowing praise where they think it due.

The liberal spirit which the inhabitants of Manchester and its suburbs have displayed in providing medical institutions of every description for the relief of the sick and needy, is a subject of unqualified praise. It will not detract from this praise, that these charities have been sometimes misapplied, as the same observation is applicable to almost all charities. Dr. Kay, in his Letter to Mr. Murray, mentions, that in the year preceding July, 1831, in the township of Manchester alone, when the population of the township was 142,000, the number of patients received at the dispensaries and hospitals was 27,804. With regard to the utility of many of these institutions, the political economists may entertain doubts: the rapid increase in the claims of the Lying-in Hospital, from 833 in the year 1796 to 4903 in the year 1830, rather argues that the effect of that institution is not favourable to habits of prudence and forethought; but no doubt can be entertained that charity flourishes at Manchester, and as so many influential men there are master-manufacturers, this fact is highly creditable to them as a class.

The two Mechanics' Institutions, one of which was established in 1824, the other in 1829, of which the utility is generally acknowledged, are patronized by many of the manufacturers and merchants. The utility of these institutions is not only acknowledged by their patrons, but appreciated, according to a recent report, by many individuals, who attribute their advancement in life to the opportunity which they have afforded them of improving themselves.

Before I dismiss this part of my subject, I extract the following account from Dr. Kay's pamphlet of the number of patients relieved by the medical charities of Manchester and the adjacent townships, and of the expenses incurred in each institution, in 1832, besides the workhouse, where 1000 patients are treated.—See Dr. Kay's Letter to Mr. Murray, p. 15.

I proceed to extract some of the most important passages from Dr. Kay's Pamphlet on the Moral and Physical Condition of the Working Classes employed in the Cotton Manufacture in Manchester, only premising, that as far as my own observations extended, they fully confirm Dr. Kay's statements.

"The township of Manchester chiefly consists of dense masses of houses, inhabited by the population engaged in the great manufactories of the cotton trade.

"The rapid growth of the cotton manufacture has attracted hither operatives from every part of the kingdom, and Ireland has poured forth the most destitute of her hordes to supply the constantly increasing demand for labour. This immigration has been, in one important respect, a serious evil. The Irish have taught the labouring classes of this country a pernicious lesson. The system of cottier farming, the demoralization and barbarism of the people, and the general use of the potatoe as the chief article of food, have encouraged the population in Ireland more rapidly than the available means of subsistence have been increased. Debased alike by ignorance and pauperism, they have discovered, with the savage, what is the minimum of the means of life upon which existence may be prolonged.

" Instructed in the fatal secret of subsisting on what is barely necessary to life,—yielding partly to necessity, and partly to example,—the labouring classes have ceased to entertain a laudable pride in furnishing their houses, and in multiplying the decent comforts which minister to happiness. What is superfluous to the mere exigencies of nature is too often expended at the tavern; and for the provision of old age and infirmity, they too frequently trust either to charity, to the support of their children, or to the protection of the poor laws.

" When this example is considered in connexion with the unremitted labour of the whole population engaged in the various branches of the cotton manufacture, our wonder will be less excited by their fatal demoralization. Prolonged and exhausting labour, continued from day to day, and from year to year, is not calculated to develop the intellectual or moral faculties of man. The dull routine of ceaseless drudgery, in which the same mechanical process is incessantly repeated, resembles the torment of Sisyphus,—the toil, like the rock, recoils perpetually on the wearied operative. The mind gathers neither stores nor strength from the constant extension and retraction of the same muscles. The intellect slumbers in supine inertness; but the grosser parts of our nature attain a rank development. To condemn man to such severity of toil is, in some measure, to cultivate in him the habits of an animal. He becomes reckless. He disregards the distinguishing appetites and habits of his species. He neglects the comforts and delicacies of life. He lives in squalid wretchedness, on meagre food, and expends his superfluous gains in debauchery.

" The population employed in the cotton factories rises at five o'clock in the morning, works in the mills from six till eight o'clock, and returns home for half an hour or forty minutes to breakfast. This meal generally consists of tea or coffee, with a little bread. Oatmeal porridge is sometimes, but of late rarely, used, and chiefly by the men; but the stimulus of tea is preferred, and especially by the women. The tea is almost always of a bad, and sometimes of a deleterious quality; the infusion is weak, and little or no milk is added. The operatives return to the mills and workshops until twelve o'clock, when an hour is allowed for dinner. Amongst those who obtain the lower rates of wages, this meal generally consists of boiled potatoes. The mess of potatoes is put into one large dish; melted lard and butter are poured upon them, and a few pieces of fried fat bacon are sometimes mingled with them, and but seldom a little meat. Those who obtain better wages, or families whose aggregate income is larger, add a greater proportion of animal food to this meal, at least three times in the week; but the quantity consumed by the labouring population is not great. The family sits round the table, and each rapidly appropriates his portion on a plate, or they all plunge their spoons into the dish, and with an animal eagerness satisfy the cravings of their appetite. At the expiration of the hour, they are all again employed in the workshops or mills, where they continue until seven o'clock, or a later hour, when they generally again indulge in the use of tea, often mingled with spirits, accompanied by a little bread. Oatmeal or potatoes are, however, taken by some a second time in the evening.

" The population nourished on this aliment is crowded into one dense mass, in cottages separated by narrow, unpaved, and almost pestilential streets, in an atmosphere loaded with the smoke and exhalations of a large manufacturing city. The operatives are congregated in rooms and workshops during twelve hours in the day, in an enervating, heated atmosphere, which is frequently loaded with dust or filaments of cotton, or impure from constant respiration, or from other causes. They are engaged in an employment which absorbs their attention, and unremittingly employs their physical energies. They are drudges, who watch the movements, and assist the operations, of a mighty material force, which toils with an energy ever unconscious of fatigue. The persevering labour of the operative must rival the mathematical precision, the incessant motion, and the exhaustless power, of the machine.

" Hence, besides the negative results, the abstraction of moral and intellectual stimuli, the absence of variety, banishment from the grateful air and the cheering influences of light, the physical energies are impaired by toil and imperfect nutrition. The artisan too seldom possesses sufficient moral dignity, or intellectual or organic strength, to resist the seductions of appetite. His wife and children, subjected to the same process, have little power to cheer his remaining moments of leisure. Domestic economy is neglected; domestic comforts are too frequently unknown. A meal of coarse food is hastily prepared, and devoured with precipitation. Home has little other relation to him than that of shelter; few pleasures are there; it chiefly presents to him a sense of physical exhaustion, from which he is glad to escape. His house is ill furnished, uncleanly, often ill ventilated, perhaps damp; his food, from want of forethought and domestic economy, is meagre and innutritious; he generally becomes debilitated and hypochondriacal, and unless supported by principle, falls the victim of dissipation. In all these respects, it is grateful to add, that those among the operatives of the mills who are employed in *the process of spinning*, and especially of fine spinning, (who receive a high rate of wages, and who are elevated on account of their skill,) are more attentive to their domestic arrangements, have better furnished houses, are consequently more regular in their habits, and more observant of their duties, than those engaged in other branches of the manufacture.

" The other classes of artisans of whom we have spoken, are frequently subject to a disease in which the sensibility of the stomach and bowels is morbidly excited; the alvine secretions are deranged, and the appetite impaired. Whilst this state continues, the patient loses flesh, his features are sharpened, the skin becomes sallow, or of the yellow hue which is observed in those who have suffered from the influence of tropical climates; the strength fails, the capacities of physical enjoyment are destroyed, and the paroxysms of corporeal suffering are aggravated by deep mental depression. We cannot wonder that the wretched victim of this disease, invited by those haunts of misery and crime, the gin-shop and the tavern, as he passes to his daily labour, should endeavour to cheat his suffering of a few moments by the false excitement procured by ardent spirits; or that the exhausted artisan, driven by ennui and discomfort from his squalid home, should strive, in the delirious dreams of a continued debauch, to forget the remembrance of his reckless improvidence, of the destitution, hunger, and uninterrupted toil, which threaten to destroy the remaining energies of his enfeebled constitution.

" The following table, arranged by the committee of classification appointed by the special Board of Health, from the reports of inspectors of the various district boards of Manchester, shews the extent to which the imperfect state of the streets of Manchester may tend to promote demoralization and disease among the poor :—

Number of District.	Number of Streets inspected.	Number of Streets unpaved.	Number of Streets partially paved.	Number of Streets ill ventilated.	Number of Streets containing heaps of refuse, &c.
1	114	63	13	7	64
2	180	93	7	23	92
3	49	2	2	12	28
4	66	37	10	12	52
5	30	2	5	5	12
6	2	1	0	1	2
7	53	13	5	12	17
8	16	2	1	2	7
9	48	0	0	9	20
10	29	19	0	10	23
11	—	—	—	—	—
12	12	0	1	1	4
13	55	3	9	10	23
14	33	13	0	8	8
Total......	687	248	53	112	352

" A minute inspection of this table will render the extent of the evil affecting the poor more apparent. Those districts which are almost exclusively inhabited by the labouring population, are Nos. 1, 2, 3, 4, and 10. Nos. 13 and 14, and 7, also contain, besides the dwellings of the operatives, those of shopkeepers and tradesmen, and are traversed by many of the principal thoroughfares. No. 11 was not inspected, and Nos. 5, 6, 8, and 9, are the central districts, containing the chief streets, the most respectable shops, the dwellings of the more wealthy inhabitants, and the warehouses of merchants and manufacturers. Subtracting, therefore, from the various totals, those items in the reports which concern these divisions only, we discover in those districts which contain a large portion of poor—namely, in Nos. 1, 2, 3, 4, 7, 10, 13, and 14, that among 579 streets inspected, 243 were altogether unpaved, forty-six partially paved, ninety-three ill ventilated, and 307 contained heaps of refuse, deep ruts, stagnant pools, ordure, &c.; and in the districts which are almost exclusively inhabited by the poor—namely, Nos. 1, 2, 3, 4, and 10, among 438 streets inspected, 214 were altogether unpaved, thirty-two partially paved, sixty-three ill ventilated, and 259 contained heaps of refuse, deep ruts, stagnant pools, ordure, &c.

"The replies to the questions proposed in the second table, relating to houses, contain equally remarkable results, which have been carefully arranged by the classification committee of the special Board of Health, as follows:—

District.	Number of Houses inspected.	Number of Houses requiring white-washing.	Number of Houses requiring repair.	Number of Houses in which the soughs want repair.	Number of Houses damp.	Number of Houses ill ventilated.	Number of Houses wanting privies.
1	850	399	128	112	177	70	326
2	2489	898	282	145	497	109	755
3	213	145	104	41	61	52	96
4	650	279	106	105	134	69	250
5	413	176	82	70	101	11	66
6	12	3	5	5	- -	- -	5
7	343	76	59	57	86	21	79
8	132	35	30	39	48	22	20
9	128	34	32	24	39	19	25
10	370	195	53	123	54	2	232
11	—	—	—	—	—	—	—
12	113	33	23	27	24	16	52
13	757	218	44	108	146	54	177
14	481	74	13	83	68	7	138
Total......	6951	2565	960	939	1435	452	2221

"Some idea of the want of cleanliness prevalent in their habitations may be obtained from the report of the number requiring whitewashing; but this column fails to indicate their gross neglect of order and absolute filth. Much less can we obtain satisfactory statistical results concerning the want of furniture, especially of bedding, and of food, clothing, and fuel. In these respects the habitations of the Irish are most destitute; they can scarcely be said to be furnished; they contain one or two chairs, a mean table, the most scanty culinary apparatus, and one or two beds, loathsome with filth. A whole family is often accommodated on a single bed, and sometimes a heap of filthy straw and a covering of old sacking hide them in one undistinguished heap, debased alike by penury, want of economy, and dissolute habits. Frequently the inspectors found two or more families crowded into one small house, containing only two apartments, one in which they slept, and another in which they eat; and often more than one family lived in a damp cellar, containing only one room, in whose pestilential atmosphere from twelve to sixteen persons were crowded. To these fertile sources of disease were sometimes added the keeping of pigs and other animals in the house, with other nuisances of the most revolting character.

"With unfeigned regret we are constrained to add, that the standard of morality is exceedingly debased, and that religious observances are neglected amongst the operative population of Manchester. The bonds of domestic sympathy are too generally relaxed, and, as a consequence, the filial and paternal duties are uncultivated. The artisan has not time to cherish these feelings by the familiar and grateful arts which are their constant food, and without which nourishment they perish. An apathy benumbs his spirit. Too frequently the father, enjoying perfect health, and with ample opportunities of employment, is supported in idleness on the earnings of his oppressed children; and, on the other hand, when age and decrepitude cripple the energies of the parents, their adult children abandon them to the scanty maintenance derived from parochial relief.

"That religious observances are exceedingly neglected we have had constant opportunities of ascertaining, in the performance of our duty as physician to the Ardwick and Ancoats Dispensary, which frequently conducted us to the houses of the poor on Sunday. With rare exceptions, the adults of the vast population of 84,147 contained in districts Nos. 1, 2, 3, 4, spend Sunday either in supine sloth, in sensuality, or in listless inactivity. A certain portion only of the labouring classes enjoys even healthful recreation on that day, and a very small number frequent the places of worship.

"The early age at which girls are admitted into the factories prevents their acquiring much knowledge of domestic economy; and even supposing them to have had accidental opportunities of making this acquisition, the extent to which women are employed in the mills does not, even after marriage, permit the general application of its principles. The infant is the victim of the system; it has not lived long ere it is abandoned to the care of a hireling or a neighbour, while its mother pursues her accustomed toil. Sometimes a little girl has the charge of the child, or even of two or three collected from neighbouring houses. Thus abandoned to one whose sympathies are not interested in its welfare, or whose time is too often also occupied in household drudgery, the child is ill-fed, dirty, ill-clothed, exposed to cold and neglect; and, in consequence, more than one-half of the offspring of the poor (as may be proved by the bills of mortality of the town) die before they have completed their fifth year. The strongest survive; but the same causes which destroy the weakest, impair the vigour of the more robust; and hence the children of our manufacturing population are proverbially pale and sallow, though not generally emaciated, nor the subjects of disease. We cannot subscribe to those exaggerated and unscientific accounts of the physical ailments to which they are liable, which have been lately revived with an eagerness and haste equally unfriendly to taste and truth; but we are convinced that the operation of these causes, continuing unchecked through successive generations, would tend to depress the health of the people, and that consequent physical ills would accumulate in an unhappy progression.

"Before the age when, according to law, children can be admitted into the factories, they are permitted to run wild in the streets and courts of the town, their parents often being engaged in labour, and unable to instruct them. Five infant schools have been established in Manchester and the suburban townships, in which six hundred children (a miserable portion of those who are of age to learn) receive instruction.

"Frequent allusion has been made to the supposed rate of mortality in Manchester, as a standard by which the health of the manufacturing population may be ascertained. From the mortality of towns, however, their comparative health cannot be invariably deduced. There is a state of physical depression which does not terminate in fatal organic changes, which, however, converts existence into a prolonged disease, and is not only compatible with life, but is proverbially protracted to an advanced senility.

"*A Table, shewing the Mortality of every Year, from* 1821 *to* 1831 *inclusive, in the Town of Manchester.*

Year.	Interments of Churchmen.	Interments of Dissenters.	Total of Interments.	Population.	Rate of Mortality.
1821	1561	1726	3287	152,683	46.45
1822	1285	1044	2329	156,663	67.223
1823	1585	3230	4815	160,664	33.36
1824	1428	3219	4647	166,117	35.74
1825	1398	3530	4928	173,083	35.12
1826	1548	3804	5352	180,052	33.64
1827	1604	3235	4839	186,462	38.53
1828	1615	4106	5721	192,874	33.73
1829	1479	3719	5198	201,691	38.80
1830	1590	4383	5973	212,913	35.64
1831	- -	- -	6736	224,143	33.27

"Some error appears to have occurred in the returns of interments for the first two years, therefore, omitting them, the mean annual rate of interments, acting as a divisor on the mean numbers of the population from 1823 to 1831 inclusive, will give an approximation to the mean rate of mortality, or 188,666÷5356=35.22, the mean rate of the annual mortality of Manchester."

According to Dr. Young, the mean mortality of all England is one in forty-nine; but on the subject of the rate of mortality I must refer my readers to the posthumous letters of Mr. Sadler, published by the London Society for the Improvement of the Condition of Factory Children, and which I have obtained permission to republish in my present work—a permission for which I feel most grateful. The talent of the author is universally acknowledged, and his zeal in the cause induced him to spend more time and labour on the subject than can be expected from those who have merely taken it up as in duty bound.

I cannot agree with Dr. Kay, that when existence is converted into a prolonged disease, it is *proverbially* protracted to an advanced senility, even though there should be no fatal organic changes. That it *may* be so protracted, where wealth affords the means of resisting physical depression, and procures for the sufferer all that medical skill can prescribe, I readily allow; but the poor man, whose life is a disease from physical depression, and who drags on his existence, in the midst of toil and want, to a protracted senility, must, indeed, be an exception to the general rule. I have extracted largely from Dr. Kay, because his professional eminence and long residence in Manchester entitle his opinions to the deepest consideration. It is not the opinion of a partizan, and therefore will not be suspected of exaggeration; nor of a superficial observer, and therefore cannot be considered as the result of precipitate judgment; but of a man every way qualified to bear testimony to the evils which he so forcibly describes. When our opponents expatiate upon the pseudo humanity of those who presume to advocate the cause of the operatives, without having visited a single factory, the names of Aikin, of Percival, of Peel, of Kay, of Fielden, and of Brotherton, are exceedingly useful to stem the current of their declamation. I must be content with referring my readers to Dr. Lyon on the medical topography and statistics of Manchester, as I have only room for two extracts, one relating to the climate of that town, the other to the excessive multiplication of its dram-shops.

"The climate of Manchester may justly be called mild: the air is humid, and the quantity of rain falling in a year is rather above the common average of England; but deep falls of snow rarely occur, and any considerable fall is generally followed very soon by a thaw.

"The prevailing winds are the S.W., the W., and the S., which are generally loaded with moisture: strong west winds occasionally waft hither salt water from the Irish sea, distant thirty miles, windows exposed to the west having been repeatedly covered with a copious incrustation, proved by Mr. Dalton and others to be really sea salt. The easterly winds are next in frequency, and always produce more or less of a disagreeable feeling of coldness, and contraction of the cutaneous vessels.

"In the winter Manchester is subject to very dense fogs, deeply tinged with particles of soot, which are at all times disagreeably abundant in the town's atmosphere; there is little reason, however, for supposing the smoke to be really deleterious. At all seasons, the valley below the Crescent in Salford is apt to be covered with a collection of vapour, in the form either of mist or of fog.

"It is necessary to advert to an evil, of recent origin in this town, which has rapidly attained an enormous magnitude,—namely, the excessive multiplication of dram-shops, now almost universally attached to the public-houses, and frequented to an alarming extent, especially by the female part of the population, and even by children. Some of these baneful places of resort are scarcely ever closed; they stand open to receive the latest wanderer by night; and again, to tempt, with a treacherous warmth, the earliest of those who repair to their morning's work. If the practice be suffered to continue, it must inevitably debase both the moral and physical character of the people."

THE RESULTS OF PERSONAL OBSERVATION IN THE AUTUMN OF 1836.

Having deduced, from indisputable sources, the existence in Manchester of certain evils external to the mills, and which, from my own observations, I can corroborate, I proceed to record some further circumstances, to which I was an eye-witness, within the mills. In doing so I shall be induced to dwell, in some measure, upon subjects which I have previously noticed.

Having provided myself with a thermometer, I visited ———, a coarse mill, and found the operatives, both adults, and children of thirteen, worked the twelve hours, exclusive of meal times,—that is to say, from six o'clock in the morning to half-past seven at night, with half an hour for breakfast at eight, and one hour for dinner at twelve; about four another meal is taken, without stopping the machine; a copper of hot water on the outside supplied the power of having tea and coffee, which the operatives thus obtain; the different parties, as they come, unfold a paper containing tea or coffee, with a little sugar; the contents are placed in their pitchers, the water is poured upon the ingredients, they quickly depart to their different operations, and can scarcely be said to rest at this period of refreshment. In the card-room, a boy, who was in-kneed, sickly-looking, short, and of spare muscular development, and apparently about twelve, I found, to my astonishment, was seventeen years old. "Had worked five or six years in a mill. On a Monday morning was so stiff could hardly move, and at night was sadly tired; his limbs were straight when he entered the mill." This unfortunate boy was doomed to work the full time. The temperature of this room was sixty-seven. A boy belonging to this mill having told me he was thirteen, upon being closely interrogated, said, that he should be of that age the next month, he worked the full time; my belief is, he was barely twelve. In the dressing-room of this mill, there were nine adults, who looked of a factory cast. The temperature was ninety-eight degrees, and the overseer said, that when the gas was up, the increase of heat would be to 103 or to 105 degrees. The work necessarily required a high temperature. The general appearance of this mill was not favourable. The overseer, however, appeared to be a man of humane disposition.

A girl, belonging to ——— mill, told me she was not nine. She was working eight hours.

In ——— mill, twelve adult women were in the batting and picking room, which was dusty and unhealthy. I was informed that these operatives were rather disorderly, and fond of ardent spirits. The temperature was eighty-four, the gas being lighted at the time I was in the room. One person had worked in this mill thirty-five years; I did not meet with many such instances.

I perceived a girl with others coming out of a large mill, whose young and delicate appearance particularly struck me. Upon asking her age, she said she was thirteen. Not crediting her statement, I asked the age of her next sister, if she had one. A child standing next to her said, that she was her sister, and I learnt that her own age would be "fourteen in February;" they both worked full time.

This was a cotton and weaving mill, which I had not an opportunity of seeing internally. I was near when the children left it at twelve o'clock. The general character of this mill was, I understood, good; but the children looked ill, and several were lame. This lameness appeared to proceed from the flat state of the feet, in consequence of the plantar arch having been weakened. An anatomical account of this arch has been ably drawn up by Dr. Roget, in his Bridgewater Treatise, from which, as it has frequently been made a subject of inquiry, I shall introduce the following extract:— "The base on which the whole body is supported in an erect position, is constituted by the toes and by the heel, the bone of which projects backwards at right angles to the leg. Between these points the sole of the foot has a concavity in two directions, the one longitudinal, the other transverse, constituting a double arch. This construction, besides conferring strength and elasticity, provides room for the convenient passage of the tendons of the toes, which proceed downwards from the larger muscles of the leg, and also for the lodgment of the smaller muscles affixed to each individual joint, and for the protection of the various nerves and blood-vessels, distributed to all these parts. The concavity of the foot adapts it, also, to retain a firmer hold of inequalities of the ground on which we tread." The evil consequences, therefore, of reducing this arch are obvious, and excessive pressure is followed by pain and lameness. I was surprised to see great numbers of the children without shoes or stockings.

A girl belonging to ——— mill could not tell her age: "Mother knew how old she was." She said she worked the full time; if thirteen, she was of stinted growth.

A little boy who worked in ——— cotton-mill, said he had not turned nine, he worked eight hours. Another said, "he was going of eleven," but he worked the full time. A third said he was thirteen; in my opinion he was not, if I am to be guided by Mr. Horner's instructions relative to *ordinary growth* and *appearance*. He worked twelve hours. A great many children looked ill, and some were lame. I could multiply, from my own knowledge, similar instances to a wearisome extent. The reluctance of the children to state their age, and the prevarications by which, upon being closely interrogated, they evade giving a direct answer, is most distressing, and must proceed from instructions given them. That they are taught to disregard truth, not merely when simply interrogated, but even when upon oath, will appear from what occurred not many weeks since, an account of which I extract from one of the papers:—

"Charges were preferred at Manchester New Bailey, on the 28th ultimo, (the 28th of December, 1836,) by Mr. Heathcote, factory inspector, against Messrs. Longshaw and Co., cotton-spinners, Pendlebury, for overworking children in their factory. The children were brought up to give evidence, but they had been instructed by their mothers, immediately before going into court, to swear falsely; and, for more than an hour, each successive charge fell to the ground. The witnesses were then kept away from their mothers, and in nearly every subsequent case convictions were obtained, and fines, amounting to about 26*l.*, were imposed upon Messrs. Longshaw."

In one mill which I visited, there were about forty operatives in the card-room, which was very dusty, and oppressive to respiration. I found the effluvia particularly offensive. To describe the effluvia in some of the rooms is difficult, partaking, as it does, of the combined qualities of the friction of the machinery and of oil; the effect is of a faint and sickly nature. I had an opportunity of seeing spinal deformity, and, upon making the relative inquiry, I found that it was induced by the system, and I saw wounds caused by the machinery. The impairing effects of the former are innumerable, and the fatal influence which it has upon childbirth was treated of by the late Mr. James Lucas, of Leeds, so long ago as the year 1788. In speaking of cases where a curvature "in the pelvis" had been *gradually* contracted by weakening influences, Mr. Lucas said, "Some such unfortunate cases have fallen to my lot, where delivery could not be effected without operative means; the majority of such cases have been the wives of manufacturers, and consultations were previously obtained. It would be well if this *unhappy change* in the pelvis could be either prevented, or early discovered, its cause investigated, or a stop put to its progress."* With regard to the pathology of such cases, I hope that I have fully described it in two letters to the editor of the "Lancet," one contained in the number for October 22, 1836, and the other in that for December 24, of the same year. In alluding to the necessity for early attention to the slightest yielding of the spinal column, I particularly noticed, in the words of the following extract, that kind of deformity spoken of by Mr. Lucas:—

"Numerous are the individuals in whom the spine takes a stationary curve, and who yet enjoy a considerable portion of health, the only failing being confined to an unusual liability to fatigue, the primary invasion upon visceral space and visceral functions having been so gradual as to make no material inroad upon the animal economy beyond the resources of relief and accommodation which nature commonly displays when her footsteps are turned from their natural course. Such cases do not, however, exist, without one appalling exception to the period of safety, and that the best and fairest part of the creation is destined to experience. The list of female victims is by no means small who have lived under, to the world, an unknown curvature of the spine, the progressive states of which have been slow, passive, and free from any disturbance of the general health, until the great trial of

* Memoirs of the Medical Society of London, vol. ii.

childbirth. The fatal truth has then appeared, that the projecting vertebræ have so encroached upon the space within the inferior portion of the pelvis, that nature cannot accomplish her purposes. To prevent the impending doom, a severe operation has been resorted to, though scarcely ever leading to the preservation of either parent or child."

To put a stop to the progress of such cases, whether numerically or physically, which take their origin in the factories, we must put a timely stop to that amount of labour which the human frame has been again and again proved unable to sustain, and under which the spinal column, the mainstay of the vital organs of the chest and abdomen, has been so repeatedly known to give way, and with accumulated evils, to lose that rectilinear form which nature assigned to it. There has been abundant parliamentary evidence, both of a remote and recent date, with regard to spinal curvature arising from factory labour. I declare at the same time my firm persuasion, that the evils of the system have been so far mitigated, as to have materially lessened the number of such cases. The general temperature of the cotton-mills I found to be oppressively high to a stranger like myself. The silk-mills generally presented a more cleanly appearance; the children's employment in them is for the most part carried on in a standing position, the evils of which some writers have considered to be of a severer nature than the extensive walking in the cotton-mills. The children in some of the silk-mills looked more healthy than those in the cotton-factories. I visited a well-regulated silk-mill, in which the hours of labour were limited to ten. In this mill the children were decidedly of a more healthy appearance than those in the cotton-factories. Several of the silk-mills receive not only children, but very little children, within their closed doors of work; and I saw during my walks, at different stations which I took, several entering mills who looked wretchedly ill. The cotton-mills to which I have alluded worked the full time, and were in Manchester. I will now speak of others a few miles from that place. And first, of an extensive cotton-mill at Todmorden, which employs 840 hands. The labour is sixty-seven hours and a half a week, being an hour and a half less than most others. The females were particularly good-looking and handsome in this mill, and their complexion partook less of the waxy cast than I observed in any other mill. The appearance also of the whole mill, and its general order, were such, that I said to the owner, the member for Oldham, "I cannot help telling you, that I am disappointed. I came to see the evils of the factory-system, but in your establishment they are indeed difficult to discern." No children were employed under nine. A school is attached to the mill, and in addition to high-minded and benevolent management, are the inestimable blessings and agencies of country life, with a pure and bracing air out of the mills. But even these advantages are not sufficient to counteract the mischiefs of the system as carried on in some country mills, as I had afterwards an opportunity of witnessing; and I may here introduce a remark, as to the large being better conducted than the small mills. In a rural and healthy district like Todmorden, I was led to the conviction, that if the liberal management pursued in Mr. Fielden's mill were generally adopted, there would be few evils to complain of; but as Manchester, and other large manufacturing towns, cannot have the benefit of pure air, and must abound with certain agencies destructive of health, there is the greater necessity for reforming and removing remediable evils. I was particularly struck with the factory visage. Its pale and yellow hue frequently presented a striking and painful contrast to that of the agricultural labourer. Mr. Gaskell, page

231, says, "Few amongst the population can be said to enjoy perfect health; all are more or less ailing, and are deprived of every chance of restoration by removing themselves from the influence which is around and within them." After forcibly describing the effects of indigestion from innutritious, and badly-cooked, and ill-masticated food, he comes to the conclusion, that the consequences lead to a singularly miserable aspect, presented by many of the operatives, who exhibit an epitome of everything that melancholy can impress on the human face. The diet is frequently most objectionable for children. In several instances the children were taking their dinner out of tin cans, containing potatoes, with either liver or bacon. I did not see either bread or salt.

The next point to which I shall come, from personal observation, is, the spare form of the factory child. Between the weight of the factory and non-factory children, there is a considerable difference in favour of the latter, being, with regard to the boys, 3.5 lbs., though, with regard to the girls, only 0.3 lbs. We are told by one of the commissioners, "that factory-labour requires no muscular exertion whatever; consequently, many of the muscles are never developed, and the additional weight which their development would give to the body is lost. If their employment were more laborious than it is, they would doubtless weigh heavier." I feel compelled to differ from such conclusions; factory employment, with its concomitant evils, not only does not conduce to muscular development, but it prevents it. I cannot, consequently, concur in the opinion that an increase of labour would cause an increase of bulk and weight. The course of life pursued by young ladies in a boarding-school requires no muscular exertion; yet nature does her duty, and the proper growth of parts is established. The eye of a beholder sees not, in such cases, pale faces and squalid forms. I present the following table of Mr. Cowell on this subject, from the first report, 1833:—

Table, shewing the Comparative Weight of Factory and Non-Factory Children.

Age.	MALES.				FEMALES.			
	Employed in Factories.		Not employed in Factories.		Employed in Factories.		Not employed in Factories.	
	Number weighed.	Average weight.	Number weighed.	Average.	Number weighed.	Average.	Number weighed.	Average.
		lbs.		lbs.		lbs.		lbs.
9	17	51.76	41	53.26	30	51.13	43	50.44
10	48	57.00	28	60.28	41	54.80	38	54.44
11	53	61.84	25	58.36	53	59.69	29	61.13
12	42	65.97	20	67.25	80	66.08	27	66.07
13	45	72.11	22	75.36	63	73.25	18	72.72
14	61	77.09	16	78.68	80	83.41	16	83.43
15	54	88.35	24	86.83	81	87.86	13	93.61
16	52	98.00	13	110.30	83	96.22	6	91.16
17	26	104.46	20	117.80	75	100.21	9	102.44
18	22	106.13	14	126.30	65	106.35	2	122.00
	420		223		651		201	

Average of an equal number of males of all the above ages, from nine to seventeen inclusive:—

Boys employed in factories 75.175

Boys not employed in factories 78.680

Average of an equal number of females of the above ages, from nine to seventeen inclusive:—

Girls employed in factories 74.739

Girls not employed in factories 75.049

SAMUEL STANWAY.

I have now arrived at the third point—the stature of the operative; and I cannot come to any other opinion than that in the factory race it is diminished from the effect of the system. That embarrassment and perplexity, in the absence of a general system of registration, should attend the operation of the laws, which so greatly countenances the labour of children in the factories, no one can be surprised; and whilst I am disposed to admit this on the fullest basis of candour, I must be permitted to express my surprise also at the method adopted to give additional effect to the law, as it appears from the following document:—

"FACTORIES REGULATION ACT.

"*Circular Letter addressed to the several Mill Occupiers in the District of Leonard Horner, Esq.*

"Whitehall, 10th October, 1836.

"Sir,—When I received instructions from the secretary of state, last June, to take charge of the district which had been subject to the inspection of the late Mr. Rickards, it was my first intention to continue to act upon the regulations which he had issued, until, by experience, I should find alterations to be necessary. But it was represented to me that transgressions of the law had been *so frequent*, and *detection in many cases so difficult*, that it was necessary to make some additional regulations, and some alterations in those of Mr. Rickards, in order to check those persons who were disposed to *violate* the provisions of the act, and to render the conviction of *offenders more certain*. It was further represented, that frequent complaints had been made, by those who were obeying the law, of the advantages which less scrupulous mill owners were having over them, by working a greater number of hours than is allowed by the act, and by employing children under the restricted ages, for full time, with impunity. To check the latter offence, which was said to be of most frequent occurrence, I was recommended by the secretary of state, in consequence of representations made to him on the subject, to order the time book No. 3 of Mr. Rickards to be kept for all children under thirteen years of age.

"Such being the state of the district, I sent for Mr. Heathcote, the superintendent of the Lancashire division, who had had most experience under Mr. Rickards, and, with his assistance, I drew up the rules and regulations which I issued on the 6th of July. In these, I continued the time book No. 1, that had been ordered to be kept by Mr. Rickards; I substituted for his time book No. 2, which had been found insufficient for the object intended to be attained by it, his time book No. 3, for the purpose of registering the hours of work of the children, as advised by the secretary of state; and I continued his time book No. 3 for those mills which work more than twelve hours a day. I issued the regulations respecting these time books with reluctance, because I had not found it necessary to require such minute details to be recorded in my former district, because I saw that they must occasion great trouble to the mill owner, and that the book No. 2 must throw a serious obstacle in the way of the employment of children; but I felt at the same time convinced, from the extensive evasions said to be practised in my new district, of the necessity of an accurate account being kept of *the periods of the day* when the children and young persons subject to the restricted hours of work are employed, and I did not then see how that could be done by a shorter process than by the time books of my predecessor.

"My anticipations of the annoyance to the mill owner which these time books must occasion have been fully realized. I stated at different meetings which I attended, during my late circuit in my district, the reasons why they had been ordered; but that if any plans were suggested to me by which the necessary information could be given with less trouble, I should carefully consider them, and readily adopt any one which I saw would be efficient. I stated, at the same time, that I should myself try to discover some less objectionable mode of registration, as it is my earnest wish, and will at all times be my endeavour, to carry the law into execution with the least possible inconvenience to the mill owner and his workpeople.

"When I found the strong objections which were urged against these time books, and with so much reason, not only in my own district, but in that of my colleague, Mr. Howell, who had adopted them by direction of the secretary of state, and by my colleague, Mr. Saunders, who had considered them so objectionable as to ask and obtain leave of the secretary of state to suspend ordering them in his district until there had been time for further consideration, I suggested to Mr. Under Secretary Maule, as soon as I heard of the appointment of the new inspector, Mr. Stuart, that it would be better that these time books should not be ordered to be kept in Mr. Stuart's district, as they had been found so very troublesome, and as I then saw the possibility of attaining the object by a more simple method. Only one suggestion had been made to me in consequence of the invitation I had held out to the mill owners, and the proposal was not such as I could adopt; but a plan had occurred to myself, which appeared to me to remedy the evil so much complained of to a great extent—viz., by the substitution of a general declaration by the mill owner for the daily record, when the hours of work are regular, and requiring a special entry on those occasions only when there are deviaticns from the regular hours.

"In compliance with the 45th section of the act, which requires that the regulations of the four inspectors shall be 'as uniform as is expedient and practicable,' at our first statutory meeting since the law came into full operation, and which commenced on the 8th of September, we took into consideration the rules and regulations which had been issued in our several districts; and the secretary of state, with special reference to the time books, recommended us to prepare a code of rules and regulations which should be applicable, as nearly as possible, to all the factories in the United Kingdom. At that meeting I laid before my colleagues my scheme of a new form of time register, and Mr. Saunders explained the forms which had been adopted, at his recommendation, by a considerable number of mill owners in his district. In several particulars the general principle of these two plans were similar. The principle being also approved by Mr. Howell and Mr. Stuart, we proceeded to make some alterations and improvements in the plan I had proposed, and we finally agreed upon the forms which you will find explained in the paper I now inclose. That paper is a copy of the general rules and regulations agreed upon by all the inspectors; they have been approved of by the secretary of state, and are to be observed in all the factories of the United Kingdom subject to this act. Besides the alterations as regards the time books, you will find some other points of difference between my regulations of the 6th of July and those now sent, and which you are now to act upon; but they do not require any particular explanation. All former rules and regulations issued in my present district, which are in any respect at variance with those now sent, are hereby set aside, and are to be held invalid.

"The inspectors have prepared a REGISTER OF WORKERS, and TIME REGISTER, in such forms, and with such explanatory directions, as, it is hoped, will leave no one uncertain as to the way in which the entries in it are to be made. The record required to be kept of the hours of employment of the young persons between thirteen and eighteen years of age, of that of the children under thirteen, and the 'register of time lost and recovered,' are all contained in one book.

"To secure uniformity, and to reduce the expense, the inspectors have had the 'time register' and 'register of workers' printed and published by Shaw and Son, of 137, Fetter-lane, London, who print and publish the forms and books required by parochial unions under the new Poor Law Act, and who have correspondents for the sale of these in all parts of England. The price of the 'time register' is four shillings and sixpence; of the 'register of workers,' two shillings. The 'register of workers' and 'time register,' bound together, price four shillings and sixpence, are prepared for the use of mills where the total number of persons does not exceed two hundred. They can be procured from them by any bookseller; but to save time, they will send an immediate supply to their correspondents in Manchester and Leeds—viz., Messrs. Bancks and Co., Manchester; and Messrs. Baines and Newsome, Leeds.

"*You are hereby required to adopt this new form of register of workers and of time register on or before the first day of November next, and, until you have procured them, the time books and register of workers required by my regulations of the sixth of July must be continued, so that there may be no interval without a regular and distinct account of the hours of work of all the persons in your employment who are subject to restricted hours of labour.*

"You will be pleased to remove the sheet of my rules and regulations of the 6th of July, and put up that now sent; and I beg you to attend that it must be pasted upon *a moveable board*, so that it may be taken down to be consulted at any time, and not, as I have often seen it, fixed up or pasted upon a wall or partition. Additional copies of these rules and regulations, for hanging up in the factory, may be had, price two shillings and sixpence for six copies, of Mr. Hartnell, the government printer, Wine Office-court, Fleet-street, London, through the medium of any bookseller. In applying to Mr. Hartnell, he must know whether the copy wanted be for a cotton, woollen, or flax factory, or for a silk-mill, and for whose district, otherwise a wrong copy might be sent. An octavo edition, for convenience of reference, may also be had at the same place, price two shillings for six copies.

"These are the rules and regulations which, as I have said, are required by all the inspectors to be observed in their several districts: but other regulations and orders may be issued by them for special cases; such, for example, as that of a mill working more than twelve hours a day, professing to employ no person under eighteen years of age. If I find any mill occupier disobeying or evading the law, and that the present regulations are not sufficient, it will be my duty to require him to observe such other regulations, authorized by the act, as I may consider necessary to keep him in check, so as to prevent him from obtaining any unfair advantage over his neighbours, who are faithfully complying with those restrictions which the legislature has deemed it necessary to impose for the protection of the children and young persons, and the other provisions of the statute.

"I have recently sent a circular letter to the surgeons who are appointed to issue the certificates of age in my district, containing some fresh instructions for their guidance; and as it is material that you should be made acquainted with some of them, I subjoin an *extract* from that letter.

"I remain, &c., "LEONARD HORNER."

The extract alluded to at the conclusion of this document I now give in a note, which I think demands the most serious attention.*

Mr. Horner, in the foregoing document, fully admits the frequent transgressions of the law; and what is his remedy? A proposition to send the children of the relative *ordinary strength and appearance* to factory labour, this being a collateral guide with his standard of height, which, I repeat, is 3ft. 10in. for nine, and 4ft. 3½in. for thirteen years. I shall insert here upon this subject extracts of letters from Mr. Baker, of Leeds, and another from Mr. Harrison, of Preston.

Extract of a Letter from Robert Baker, Esq., Surgeon, Leeds, to Leonard Horner, Esq., dated 8th September, 1836.

"The difficulty of arriving at an exact age, for want of an efficient system of registration, has prevented me hitherto from obtaining the number of measurements of individuals which is desirable for an accurate calculation

* *Extract from a Letter addressed by Mr. Horner to the Surgeons, dated Whitehall, 20th Sept.*, 1836.

1. "Since the date of my circular letter of instructions of the 6th of July last, on the occasion of my succeeding the late Mr. Rickards, I have visited a considerable part of my district; and having found that many of the appointed surgeons have been proceeding upon an erroneous interpretation of the act, by inquiring into the actual ages of the children who come before them, and that many have been uncertain how to proceed in some matters of importance connected with the duty confided to them, it is necessary for me to point out, more distinctly than it appears to have yet been done, the principle which is adopted in the act for the determination of the question of age; and to call the attention of the surgeons to other circumstances, which will tend, I hope, to introduce a greater degree of accuracy in this fundamental part of the act than has hitherto prevailed.

2. "The declaration which the surgeon makes in his certificate has no *reference* to the *actual* age of the person named; it states that, in his opinion, the person who has appeared before him '*is of the ordinary strength and appearance*' of the particular age, which he inserts. The surgeon must form that opinion *from the physical condition of the person*, without regard to the actual age. He ought even to abstain from asking any question as to the age; for the probability is, that a true answer will not be given; and the inquiry is more calculated to mislead than to assist his judgment. The *bodily strength* usually met with at the particular age mentioned, with reference to the fitness of the child for undergoing factory labour, is the object sought after, and not the determination of actual age; therefore, although you may have positive knowledge of a child being thirteen years of age, if, by imperfect development or by ill health, it is not, in your opinion, of the ordinary strength and appearance of a child of more than twelve years, you must insert the word *twelve* in the certificate: *on the other hand, if you find a child, which you know with certainty to be not more than twelve years of age, with such an unusual degree of development as to be of the ordinary strength and appearance of thirteen, you will be justified in inserting the word* thirteen *in the certificate.*

3. "It is very important that you should bestow great pains to acquire as correct a judgment as possible as to what *is* the ordinary strength and appearance of children and young persons among the working classes, in the district where you reside, at different ages; and it is very desirable that all the appointed surgeons should follow some general principle in forming their opinion. It is not possible, it is true, to point out any one invariable guide in a matter of this sort, where so many circumstances must be taken into account; but stature appears, upon the whole, the indication which will be least liable to error, and it may be at all events safely taken as the best initiatory step in the inquiry; its value as evidence being subject to modification from other circumstances in the physical state of the person examined, such as breadth of chest, muscular strength, and general healthy condition.

4. "Mr. Harrison, of Preston, one of the surgeons appointed to grant certificates of age there, instituted, some time ago, a series of examinations, with a view to ascertain the average height of children and young persons employed in factories at different ages; the results he obtained are given in a letter from him to the late Mr. Rickards, which forms a part of the report of the latter to the secretary of state, printed by the House of Commons, in August, 1834. I had lately an opportunity of conversing with Mr. Harrison on this subject; and, at my request, he has revised his materials, and has communicated the results in a letter to me, of which I subjoin an extract. Mr. Baker, surgeon, of Leeds, the factory superintendent in that division of my district, has had his attention for some time specially directed to this subject, and he has sent me the results of his observations, which will be found in the subjoined extract from his letter to me. I have been anxious to find out some work of authority containing some observations upon the growth of the human body at different ages, in this country; but on consulting some eminent medical men in London, whose attention has been particularly directed to inquiries of this nature, I have been told that there is no book which contains such information, as regards the people of this country.

5. "Until some more precise and accurate data can be obtained, it will be advisable for you to take the observations of Mr. Harrison and Mr. Baker as a guide; and therefore, unless in cases of

on this interesting subject. Some experiments made by me two years ago are rendered less valuable on that account; but I send you some others, which have been lately completed, sufficient in number to enable you to form some idea by which a recommendation of a standard might be issued, both to the surgeons and overlookers, as a guide, to be taken along with other evidence of a more satisfactory nature:—

AGES.	Number of Males Examined.	Average Stature.	Number of Females Examined.	Average Stature.
9 to 10	74	3ft. 11 in.	70	4ft. 0¼ in.
10 — 11	40	4 2	28	4 1
11 — 12	20	4 2	34	4 2¼
12 — 13	20	4 4½	41	4 4¼
	154		173	

unusual development of muscular strength, no child that, without shoes, measures less than 3 feet 10 inches ought to be considered as having the appearance of *nine* years of age, and no child less than 4 feet 3½ inches ought to be considered as having the appearance of *thirteen* years of age; there must, besides, be no deficiency of bodily health and strength to justify a certificate of nine and thirteen years respectively being granted to children of the above-mentioned stature. For the intermediate ages, the following measurements may be adopted:—for 10 years, 3 feet 11½ inches; for 11 years, 4 feet 1 inch; and for 12 years, 4 feet 2 inches. These measurements may apply to both sexes, when taken, as is now proposed, as a general guide, and not as an invariable rule.

6. "In my circular letter of the 6th of July, I particularly pointed out the necessity of your guarding against the too common fraud of an older child being substituted, under a false name, for a younger child, in order to obtain a certificate for the latter. My observations in my new district, since the date of that letter, have convinced me that there can be no effectual check against that species of fraud, so long as certificates are granted at the houses of the surgeons. Except under very particular circumstances, you must grant certificates only at the mills. It will therefore be necessary for you to visit each mill in your district once a week, to have a particular day for each mill, and keep to that day as nearly as you can. I must leave the amount of your remuneration to be settled by a private arrangement between you and the mill-owner, with this only condition, that when the certificate is to be paid for by the workpeople, the charge is not to exceed sixpence for each certificate. In Manchester, Leeds, Halifax, and several other places, the following arrangement has been very generally entered into:—The mill-owner agrees to secure the surgeon the amount of the charge he usually makes for a visit in his private practice, and which, in Manchester, is half-a-crown. If the number of certificates, at the rate of sixpence each, be less than that sum, it must be made up; if greater, the surgeon has the benefit of the surplus.

7. "I authorize you to state to the mill-owners in your district that such as enter into an arrangement with an appointed surgeon to visit their mills once a week will be held by me free of blame should they employ children without a certificate in the interval between the weekly visits of the surgeon, because, by such an engagement, they will not only be free of all charge of disobeying the law from wilful negligence, but will best secure its observance in that particular. But no child must be allowed to work at all, in the interval, which is less in stature than 3 feet 10 inches; and no child must be employed full time in the interval, which is less than 4 feet 3½ inches. The mill-owners must, at the same time, be informed that these measurements are given merely as a guide to the overlooker in the meanwhile, and that it is very possible that, when the surgeon comes round, children of these heights may not be found upon examination to have 'the ordinary strength and appearance' of nine and thirteen years respectively. If the mill-owner decline to form such an engagement, he must be held responsible if he employs any child, for however short a time, without having first obtained a certificate of age.

8. "As a farther security, you will please to measure every child for whom you grant a certificate, and mark the height *on the counterpart* which you retain in your possession. This may be done very speedily by marking the side of a door or window of the room where you grant the certificates at the mill, with the feet and inches, from 3 feet to 5 feet, measured from the floor, placing the persons to be measured with their backs to that line, and laying a flat ruler or book upon their heads. By this precaution of marking the age in the counterpart, if a child of lower stature fraudulently get possession of a certificate, and if a judicial inquiry were to take place, you would be able to swear positively as to the actual height of the child which appeared before you, when the certificate in question was granted. By not granting certificates at your own house, you will be relieved from the importunate solicitations of parents and other persons interested in getting the child passed; but you must keep yourself equally secure from all interference with the free exercise of your judgment at the mills, by having no one near you at the time, except the children to be examined. If you grant a certificate under any particular circumstance at your own house, you must take an early opportunity of visiting the mill where the child is at work, to see that no fraud has been committed."

In all the above, the individuals were bordering on the highest age. The common standard of children of nine years of age is from 3ft. 9in. to 3ft. 10in.; and I am of opinion, that none under that ought to be admitted. I have used, for some time, 4ft. $3\frac{1}{2}$in. as the standard for thirteen years, which is a fair average of physical condition, as to stature, in an ordinarily healthy child, at that time of life. The great advantage I have found in the adoption of this standard has been, that it gives a degree of equality of advantage to all the mill owners. A child not 4ft. $3\frac{1}{2}$in. for one mill, cannot be so for any other; and thus all parties are made alike."

Extract of a Letter from James Harrison, Esq., Surgeon, Preston, to Leonard Horner, Esq., dated 16*th August*, 1836.

"I HAVE made the examination which you desired, the results of which you will find in the following table; the measurements were taken, of course, without shoes:—

AGES.	Number of Children Examined.	Average Stature.	Number of Males Examined.	Average Stature.	Number of Females Examined.	Average Stature.
11 to 12	244	4ft. $2\frac{3}{4}$in.	146	4ft. $2\frac{1}{2}$in.	98	4ft. $3\frac{1}{4}$in.
12 — 13	235	4 $4\frac{3}{4}$	99	4 $3\frac{1}{2}$	136	4 $5\frac{3}{4}$
13 — 14	193	4 $6\frac{1}{3}$	102	4 6	20	4 $7\frac{1}{4}$
14 — 15	257	4 $8\frac{3}{4}$	117	4 $8\frac{1}{4}$	140	4 9
15 — 16	188	4 $10\frac{2}{3}$	82	4 $10\frac{1}{2}$	106	4 $10\frac{3}{4}$
16 — 17	133	5 $0\frac{1}{4}$	43	5 $0\frac{1}{2}$	90	4 $11\frac{1}{2}$
17 — 18	159	5 0	47	5 0	112	5 0
	1409		636		702	

"This table has been formed chiefly, though not entirely, from the examinations, from which the one furnished to the late Mr. Rickards was calculated; and on comparing it with that table, you will find that there exists a slight, though not material, difference; it is, however, sufficient to shew the necessity of a more extended investigation, in order to arrive at a correct result.

"The investigations were made on the first coming into force of the Factories Regulation Act; and as the children and their parents, with few exceptions, did not know for what object the inquiries were made, nor how they would affect their interest, they had no inducement to give in false statements of age. The greatest doubt must exist with regard to the average of the children represented to be between the ages of eleven and twelve years, as eleven was, at the time of the examination, the lowest period at which children were allowed to work twelve hours a day; and the parents of such children as were under that age had a strong inducement to make false statements; but as the parties did not then know much of the provisions of the act, it may perhaps be regarded as an approximation to the truth. Above eleven, there was no conceivable motive for practising deception. In making any future examination, the difficulty of arriving at the real ages of the children under thirteen will, I conceive, be great. With respect to such children, the truth is very rarely told; and almost every conceivable expedient is adopted to produce an erroneous impression on the minds of the surgeons. Above the age of thirteen, much valuable information, that may be relied on, may still be collected by the certifying surgeons.

"It is singular, that from both the tables I have given, there seems to have been no increase of stature from the period between sixteen and seventeen, and that between seventeen and eighteen years. The most remarkable feature of the present table is, that the average height of the females examined, with two exceptions, exceeds that of the males. This is a result for which I was not prepared; and it induced me, at first, to think that I had made some miscalculations; but after repeated examinations, I found the fact to be so with respect to the children who had come under my notice. If any reliance may be placed upon my investigations, it would seem to indicate that between the ages of eleven and sixteen the growth of the female is more rapid than that of the male. I do not, however, think that my observation has been sufficiently extended to justify my drawing any general conclusions from it. It is a subject, however, of much interest to the physiologist; and if the surgeons who are connected with the Factories Regulation Act could be induced to extend the inquiry, some important results might be obtained.

"I am of opinion that, in the absence of better evidence, the table, which I have given above, might be rendered available for fixing a standard, by which the ages of children to be employed in factories might be conjectured. The statures which are there given are the average ones for the respective ages specified; and, being the average height, it is obvious that there may be expected to be as many children below as above. My plan in examining children, since the first averages were ascertained, has been, to regard the minimum height for a child of eleven years of age, to be 4ft. 1in.; of twelve years, 4ft. 2in.; and of thirteen, 4ft. 3in.; and not to give certificates of those ages to children who were under the minimum size fixed for their respective ages. By this means I have had an uniform mode of proceeding, which has, I believe, afforded pretty general satisfaction, and seems to me to have been in accordance with the spirit of the factory act."

Messrs. Baker and Harrison's averages, with Mr. Cowell's and Mons. Quetelet's, are all above Mr. Horner's. Surely this gentleman must either deem the factory population to be of stunted growth, or he does them great injustice in exacting labour by a rule so much at variance with the experience of others, and at periods of life which the spirit of the law does not sanction. If anything can prove how much nature is liable to be outraged when once we turn away from her legitimate footsteps, the statements and the practices to which this departure from them has led, must fully do so. The desire to obtain the benefits and profits of the labour of childhood is restrained within no ordinary bounds of feeling and of justice; and both parents and masters lay themselves open to the severest condemnation.* If this country will permit extended labour to be fastened upon the early and tender years of life, the legislature is bound to look out for the safest principle by which the imperfect growth and strength of childhood may be measured against twelve hours of exclusive labour, and the untiring power of the machinery. Does not the

* I have very recently received a communication from a most respectable authority, and to which I can at any time refer, as to a piece of most deceptive sagacity having been lately displayed by parents in cramming cotton into the stockings of their children, so that a fictitious height of one inch, or one and a quarter, has been obtained beyond the real height and laws of nature. Where will the evils of such a system end?

law regulate factory labour by actual age? From my own experience, I know how difficult it is to ascertain the ages of children, yet I would not put aside reference to age, which is the very rule and basis of the law; and for that, too, which begs the question—if I may be permitted the expression—by giving collateral judgment from the *ordinary strength* and *appearance* of so numerous a portion of the rising generation. This is a licence too fallacious to be entrusted to human nature. The sound and wholesome boundaries of the law would be again and again overstepped, and that high and enlightened spirit which so generally characterizes the enactments of the lawgivers of this kingdom would be repeatedly and cruelly broken in upon. The factory surgeon is told by Mr. Horner, that he should abstain from asking any question as to the age. With regard to this point, I beg the reader to peruse the *present act of parliament*, embodied in this work. Desirous to act as much as possible upon the late Mr. John Hunter's valuable precept, "Take nothing upon trust," I have examined several hundred children: the different results I now give. I took the height of between two and three hundred children, in an establishment in one of the most healthy districts, a few miles from London, where they were farmed. Their ages I found so little to be depended upon, that I could not make a just deduction from the investigation. The inferences would have been unfavourable to Mr. Horner's conclusion, as far as I could determine upon the averages. My visit was not made in vain, for I became possessed of several valuable particulars.

The following tables are taken from the Marylebone Workhouse; Christ's Chapel National School, Maida Hill; and St. Andrew's Parochial School, Hatton Garden:—

General Abstract of the Children at Marylebone Workhouse; Christ's Chapel National School, Maida Hill; and Saint Andrew's Parochial School.

Ages.	No.	BOYS.			No.	GIRLS.		
		Greatest Height.	Least Height.	Average Height.		Greatest Height.	Least Height.	Average Height.
		ft. in.	ft. in.	ft. in.		ft. in.	ft. in.	ft. in.
8 to 9	39	4 1½	3 5¾	3 9½	49	4 1	3 5	3 9
9 to 10	51	4 5	3 5¾	3 11	51	4 3¾	3 6¾	3 10¾
10 to 11	81	4 5¼	3 7¼	4 0¼	55	4 5¼	3 7¼	3 11¼
11 to 12	44	4 6	3 10¼	4 1¾	53	4 7¾	3 9¾	4 2¼
12 to 13	62	4 8½	3 9¾	4 3¼	31	4 9¼	3 11¾	4 4
13 to 14	66	4 10½	3 11	4 4¼	38	5 0	3 11¾	4 6¾
	343				277			

Half an inch was allowed for the measure of the shoes.

Abstract of the Heights of Children at Marylebone Workhouse; Christ's Chapel National School, Maida Hill; and Saint Andrew's Parochial School.

BOYS.

Age.	Place.	Greatest Height.	Least Height.	Average Height.	No.
		ft. in.	ft. in.	ft. in.	
8 to 9	Marylebone Workhouse	3 11¼	3 5¾	3 7¾	9
	National School	4 1½	3 7½	3 10	22
	Saint Andrew's School	4 0¾	3 7½	3 10¼	8
	The three Schools together	4 1½	3 5¾	3 9½	39
9 to 10	Marylebone Workhouse	4 1½	3 6¼	3 9½	19
	National School	4 5½	3 10½	4 1	16
	Saint Andrew's School	4 2½	3 5¾	3 11½	16
	The three Schools together	4 5	3 5¾	3 11	51
10 to 11	Marylebone Workhouse	4 2	3 7¼	3 11¼	31
	National School	4 5¼	3 7¾	4 0¾	21
	Saint Andrew's School	4 5¼	3 9	4 1	29
	The three Schools together	4 5¼	3 7¼	4 0½	81
11 to 12	Marylebone Workhouse	4 3½	3 11	4 1	17
	National School	4 3½	3 10¼	4 1½	8
	Saint Andrew's School	4 6	3 11½	4 2¼	19
	The three Schools together	4 6	3 10¼	4 1¾	44
12 to 13	Marylebone Workhouse	4 5½	3 9¾	4 1½	26
	National School	4 11	3 10	4 4½	12
	Saint Andrew's School	4 8½	3 11½	4 4½	24
	The three Schools together	4 11	3 9¾	4 3¼	62
13 to 14	Marylebone Workhouse	4 7¾	3 11	4 3½	39
	National School	4 10½	4 0	4 5¾	10
	St. Andrew's School	4 9	4 1¾	4 5¼	17
	The three Schools together	4 10½	3 11	4 4½	66

GIRLS.

Age.	Place.	Greatest Height.	Least Height.	Average Height.	No.
8 to 9	Marylebone Workhouse	3 9½	3 5	3 7	15
	National School	4 1	3 7½	3 10	24
	Saint Andrew's School	4 0½	3 6½	3 9	10
	The three Schools together	4 1	3 5	3 9	49
9 to 10	Marylebone Workhouse	4 1¾	3 6¾	3 9¼	19
	National School	4 3¾	3 8¾	4 0	19
	Saint Andrew's School	4 2¼	3 8½	3 11½	13
	The three Schools together	4 3¾	3 6¾	3 10¾	51
10 to 11	Marylebone Workhouse	4 1	3 7¾	3 10¼	25
	National School	4 5¼	3 11	4 1½	14
	Saint Andrew's School	4 1½	3 7¼	3 10½	16
	The three Schools together	4 5¼	3 7¼	3 11¼	55
11 to 12	Marylebone Workhouse	4 5½	3 10¼	4 1¾	16
	National School	4 7¾	4 0	4 3½	14
	Saint Andrew's School	4 5½	3 9¾	4 2	23
	The three Schools together	4 7¾	3 9¾	4 2¼	53
12 to 13	Marylebone Workhouse	4 4¼	3 11¾	4 1¾	12
	National School	4 9¼	4 2¼	4 6¾	8
	Saint Andrew's School	4 6¾	4 2	4 4½	11
	The three Schools together	4 9¼	3 11¾	4 4	31
13 to 14	Marylebone Workhouse	4 10	4 3½	4 6¼	9
	National School	4 5¼	3 11¾	4 3½	3
	Saint Andrew's School	5 0	4 3½	4 7½	26
	The three Schools together	5 0	3 11¾	4 6¾	38

Christ's Chapel National School, hours ... { Morning, 9 till 12.
Afternoon, 2 till 5, from Lady-day to Michaelmas.
Afternoon, 2 till 4, from Michaelmas to Lady-day. }

The school room of the boys is 14ft. 3in. high, 40ft. long, 25ft. wide, and contains daily 160 boys. The girls' room is rather larger.—On Sunday they attend divine service three times.

Saint Andrew's Parochial School, hours ... { Morning, 8 to 12, Afternoon, 2 to 5 } from Lady-day to Michaelmas.
{ Morning, 9 to 12, Afternoon, 2 to 4 } from Michaelmas to Lady-day.

The school room of the boys is 18ft. 9in. high, 40ft. long, and 24ft. 8in. wide. The girls' room the same.—On Sundays they attend divine service at their parish church, morning, afternoon, and evening; they have a fortnight's holiday at Christmas and Midsummer, a week at Easter, and a few half-holidays.

The System pursued with the Boys and Girls in reference to the Diet, School Hours, and Hours of Labour, in the Workhouse of Saint Marylebone.

The hours of rising .... { Summer, 6 o'clock. Winter, 7 o'clock. }
The hours of going to bed......... { Summer, 8 o'clock. Winter, 7 o'clock. }

This applies to both boys and girls.

Days.	Breakfast.	Dinner.	Supper.
Sunday	milk porridge, (about $\frac{3}{4}$ of a pint,) 4 oz. bread.	6 oz. cooked meat, vegetables, 4 oz. bread.	broth, 4 oz. bread.
Monday.........	ditto	pease soup, 4 oz. bread.	2 oz. cheese,* 4 oz. bread.
Tuesday.........	ditto	Irish stew,	broth, 4 oz. bread.
Wednesday ...	ditto	pease soup, 4 oz. bread.	4 oz. bread, and treacle.
Thursday	ditto	6 oz. cooked meat, vegetables, 4 oz. bread.	broth, 4 oz. bread.
Friday	ditto	pease soup, 4 oz. bread.	2 oz. cheese,* 4 oz. bread.
Saturday	ditto	suet dumplings, (about 1lb.,) and potatoes.	4 oz. bread, and treacle.

The boys and girls attend in their separate schools from nine to twelve in the morning, and from two till five in the afternoon, with the exception of Wednesdays and Saturdays, being half holidays; the other portions of the day are devoted to meals and recreation. They usually walk out into the fields once a week, and in fine weather oftener.

The boys above ten years of age attend in the workshops alternate days, where they are instructed in the different trades, such as tailors, shoemakers, rope makers, &c. The maximum of labour is six hours per day. They are instructed by persons employed from out of the house, and are not permitted to have any communication with the adult paupers.

The elder girls are trained up in the house to do the usual domestic work, and are also instructed in plain needle-work, knitting, &c., by which means they become well qualified to take places as servants at an early age.

Corporeal punishment is almost entirely done away with; the increased vigilance of the school-master and mistress will, it is hoped, in a short time, render punishment of any description of rare occurrence.

* Girls, 4 oz. bread and treacle.

During the summer months a properly qualified person, from the Military College, Chelsea, is employed to give a lesson, once a week, in the drill and gymnastic exercises.

There is an Infant School for children under seven years of age: they are kept apart from the others, and have a mistress and other assistants to look after them.

The first of the tables contains an average of the whole, and this taken from children in the *metropolis.* The measurements are higher than Mr. Horner's standard. To the second table I attach some consequence, from its shewing the relative heights between the children of the national schools, who have much freedom of action, and the children of the Mary-le-bone workhouse. The growth of those in the latter is less than that of those in the former, which is the more extraordinary from the healthy appearance of the juvenile inmates of that establishment, which struck me so forcibly, that I entered upon my minutes that three-fourths had a healthy colour in their cheeks. The arrangements throughout reflect high credit upon its management. The children exhibited great agility and discipline, which I had an opportunity of witnessing in their gymnastic exercises. Here indeed was muscular development in the absence of labour, from the adoption of humane and well-combined regulations. Bulk and apparent strength were the consequences. In point of *height* there was a difference, with one exception, in the girls from thirteen to fourteen years of age, against this establishment, which, though well-conducted, necessarily exercises a restraint over the daily proceedings and actions of the children, confining them so much to a given spot. Growth and freedom of range are greatly allied. The temperature of the two schools, and of the school-room of the work-house, I found to range between fifty-seven and sixty-two. I have subjoined to the tables the hours employed in education, with other material circumstances, that in the removal of such children to factory-life, a due estimate may be made of the favourable and unfavourable states attendant upon a sudden and extensive change :—

I here subjoin a passage from one of Mr. Horner's letters, which he addressed to the different district surgeons :—

(Par. 3.) " Some surgeons have laid great stress upon the development of the *teeth* as a safe guide; and if the object were the ascertaining of the *actual age of the child,* such a test would perhaps be less liable to error than that of height; but as an evidence of bodily strength, it is obviously not to be depended upon."—(Further paragraphs I give below.*)

What a field for argument and inquiry does this single sentence open!

* (Par. 4.) " And that M. Quetelet, of Brussels, is the only writer of authority on the subject. I have consulted the accounts of the researches of that author, which are contained in two essays by him, in the Memoirs of the Royal Academy of Brussels for 1831 and 1833; but however curious and interesting the observations are in a physiological point of view, I have not found anything bearing sufficiently on the present object to serve as a practical guide. The average height given by M. Quetelet for the different ages are *higher* than those obtained by Mr. Harrison and Mr. Baker; but his observations were not confined to the children of the working classes, nor is it stated from what number of observations the averages were obtained. M. Quetelet makes the following remarks upon the great want of observations upon this subject:—'To shew how little advancement has been made in the study of the progressive development of the human frame, if it were required to establish the age of an individual by the combined consideration of his physical qualities, we should not be able to find any scientific rules to guide our determination, but should be obliged to have recourse to the most unsatisfactory empiricism.

(Par. 5.) "It is frequently asserted that there is a great difference between the ordinary strength and appearance of children at the same period of life in towns and in the country, and that the children in woollen-factories are stouter and better grown than those in cotton-factories. If that were really the case, a different standard would be required; but until the fact be determined by

With Mr. Horner's opinion before me, I sought for evidence upon this point from several distinguished practitioners in that department of the profession, and amongst others, from Mr. Bell, of Broad Street; Mr. Waite, of Old Burlington Street; Mr. Henderson, of Charlotte Street, Fitzroy Square; Mr. Clark and Mr. Saunders, both of Argyle Street:—

New Broad-street, Dec. 28, 1836.

Dear Sir,—I have always felt too deeply interested in the cause which you have so humanely espoused, not to be anxious to render it any aid within my humble power; and I should long since have answered your letter, had I not thought that, by a recurrence to some notes which I have from time to time made on the subject of it, I might have been able to shew how much or how little information may be gained as to the age of children by an examination of their teeth during the period of their change. But as I have not been able to lay my hand upon those memoranda, and as the time when most of my little friends pay me their periodical visit was at hand, I thought it might be of some avail were I to observe all that came under my notice during the present season of their holidays. The result of my recollection of past observations, and of my recent examination, is, certainly, far from satisfactory, as it convinces me that only a very vague and uncertain judgment can be formed upon such observations. I have occasionally seen children who had shed several of the temporary teeth before they were five years old—often before six; and I have this day seen two children, of the

actual observation, on an extensive scale, one rule must be followed for all. It is very much to be desired that the different surgeons should institute an inquiry as to the stature of children in their several districts, in order to obtain some sure guide for the better execution of the provisions of this act. To render the results truly valuable, it is important to attend to the following particulars:—To confine the observations to the children of the working classes; to measure those only whose real ages can be ascertained with tolerable certainty; to distinguish the males and females; to exclude those who are not in an ordinary state of health; and to distinguish the measurements by differences of half-years.

"I inclose two papers, ruled in a convenient form, for recording the observations; and if you should undertake the inquiry in your neighbourhood, I shall be happy to supply you with additional copies of the form, as soon as those now sent are filled up. The larger the number of individual measurements, so much the better. If I should receive a tolerably extensive series of observations from the medical gentlemen in my district, I may be able to construct a table upon which considerable reliance may be placed. The value of such a table would not be confined to the immediate object of this inquiry, but would supply information of great interest in physiology and medical statistics.

(Par. 9.) "In case of any fraud being practised, it is as well you should know the law on the subject; and I, therefore, now send you a copy of the 28th section of the factory regulation act. 'If any person shall give, sign, or countersign, indorse, or in any manner give currency to any false certificate, knowing the same to be untrue, or if any person shall forge any certificate, or shall forge any signature, or any endorsement to any certificate, or shall knowingly or willingly give false testimony upon any point material to any certificate of any inspector or schoolmaster, such persons shall be deemed guilty of a misdemeanor, and shall, on conviction thereof before any inspector or justice, be liable to be imprisoned for any period not exceeding two months in the house of correction, in the county town or place where such offence was committed.'

"If any instance of fraud should come to your knowledge, you would do well to lay an information against the offender before the magistrate. There is no necessity to wait before doing so for the coming of the inspector and superintendent.

"I request the favour of you to send me, as early as you conveniently can, a list of all the factories situated in your district, distinguishing those mill-occupiers who have made an engagement with you to visit their mills once a week, and those who have declined to enter into that engagement. I shall be obliged if you will state whether it be a cotton, woollen, flax, or silk factory, and, if in the country, the parish or place where it is situated. I beg you also to give me your Christian name in full, and your exact address.

"I remain, sir, your obedient servant,

Leonard Horner."

age of seven, who had acquired the four permanent incisores in the upper as well as in the lower jaw. On the other hand, I have sometimes observed the change only just commencing at the age of eight, or eight and a half; and, in one instance, in a child of the age of eleven, although many of the temporary teeth had fallen, not one of the permanent ones had come through the gum, excepting the first molares. After some time, however, most of them made their appearance.

Upon the whole, I should say that the statement which I have given in my work on the teeth is correct, as to the average of the cases that have come under my notice, and which I now subjoin; but the exceptions are so numerous as to render it, I fear, but a fallacious ground to depend upon. With the most sincere wishes for the full effect of your humane endeavours for the relief so many little sufferers,

I remain, yours, &c. THOMAS BELL.

P.S. I beg to add, that, as a general rule, an early change of the teeth appears to me to be indicative of precocious physical powers; but this rule, too, has numerous exceptions.

EXTRACT.—" The change of the temporary for the permanent teeth commences, in the majority of instances, at about seven years of age, though I have occasionally known it to occur as early as five, and as late as eight years and a half. The first permanent molares usually pierce the gum before the loss of the temporary central incisores; and their appearance may be considered as indicative of the approaching change. The following are about the medium periods at which the different permanent teeth are generally cut; but so irregular are they in this respect, that comparative little dependence can be placed on such a statement. Those of the lower jaw are here indicated, and they most commonly precede the upper by about two or three months:—

Anterior molares	6½ years.
Central incisores	7
Lateral incisores	8
Anterior bicuspides	9
Posterior bicuspides	10
Cuspidati	11—12
Second molares	12—13
Third molares, or dentes sapientiæ	17—19."

Anat. Physiol. &c. of the Teeth, p. 80.

2, Old Burlington-street, 7th January, 1837.

MY DEAR SIR,—In pursuance with your request, I have carefully examined the subject you led my attention to, not wholly without hesitation. My inquiries have been made through unexplored channels, and I have been wholly unassisted; for although, at your suggestion, I have communicated with some of the first in the profession, I have not been furnished with any facilitating information worth reporting to you.

Mr. Cartwright and myself have discussed the subject in person. We agree in many points. He is not as sanguine as myself; but I must take this opportunity of returning him my sincere thanks for the handsome and obliging way in which he came forward. Mr. Dumergue appears to have considered the question; he withholds his opinion, in consequence, as he says, of being about "to trouble the world" with a work on the subject. I feel a strong assurance that, when his work appears, his experience will fully support my statements. I have confirmed my opinion, by statistical information, that the ninth and thirteenth year, generally speaking, ought to have some peculiarities in the appearance of the teeth; and I can assert with confidence, that where an advanced or retarded development of the teeth exists to any great extent, it is caused by certain peculiarities of constitution, which physiologists ought readily to explain.

The following cases, which I have selected indiscriminately from my notes, will bear me out in the correctness of a table, which I add, vindicating the general appearance of the teeth at the ninth year:—

NATIONAL SCHOOL BOYS, NINE YEARS OLD.

J. Sharp, 4 molars, 2 incisors
J. Cavan, ditto
J. Foester, ditto
J. Vincent, 8¾yrs., ditto
G. Berry, ditto
J. Hill, ditto

R. Parkinson, 4 molars, 2 incisors
J. Dickey, 8 bicuspids, 4 molars, 4 incisors
— Jeffery, 7 incisors, 4 molars
G. Hasell, 4 incisors, 4 molars
C. Brown, 1 upper incisor, 2 lower
— Hickson, 5 incisors, 4 molars

NATIONAL SCHOOL BOYS, EIGHT YEARS OLD.

J. Baron, 6 incisors, 4 molars
— Viant, 8 incisors, 4 molars
— Marsden, 8¾yrs., 8 incisors, 4 molars
J. Barter, 2 lower incisors
H. Smith, 6 incisors, 4 molars
— Bridgeman, 8¾yrs., 8 incisors, 4 molars

J. Kemp, 8½yrs., 7 incisors, 4 molars
— Bridgeman, 8¾yrs., 8 incisors, 4 molars
J. Kemp, 8½yrs., 7 incisors, 4 molars
W. Cotton, 5 incisors, 4 molars
H. Topliffe, 8½yrs., 3 lower incisors, 2 upper

You will see that the eight central permanent teeth are, in most cases, formed in the mouth at the ninth year. The case of J. Dickey is somewhat curious—he has eight bicuspids formed, and only four incisors. It is a common occurrence to find the size and progress of certain teeth retarded or enlarged *at the expense of others* which are of a diminutive growth. In my notes I find this boy marked phlegmatic and strumous; and I must mention, that it is in the mouths of children of this temperament that preternatural growths are mostly seen. In the cases of the ages of 8, 8½, and 8¾, the teeth are progressing towards this condition, as marked in this table:—

$$3\left\{\begin{matrix} 0\ 0\ 0 & 2\ 1\ 1\ 2 & 0\ 0\ 0 \\ 0\ 0\ 0 & 2\ 1\ 1\ 2 & 0\ 0\ 0 \end{matrix}\right\}3$$

1 1 1 1 The four central incisors
2 2 2 2 The four lateral incisors.

0 0 0
0 0 0
0 0 0
0 0 0 } The twelve primary molars.

3 3 A permanent molar tooth, common to each jaw at each side.

NATIONAL SCHOOL, GENERALLY NINE YEARS OLD.

H. Nottage, 8 incisors, 4 molars
E. White, ditto
D. De Lisle, ditto
H. Garrett, ditto
A. Harrington, 8 incisors, 4 molars
E. Hambey, 8 incisors, 4 molars
C. Bedwin, 2 lower incisors, 2 upper
Cath. Adhers, 6 incisors, 4 molars
C. Brett, 9 large incisors, 4 molars
— Pooley, 10yrs, 2 lower, 2 upper incisors.
— Pooley, 7½yrs., no teeth shed

These children were principally examined at an establishment about ten miles from London; they were for the most part healthy looking, and subject generally to the same causes which influence growth and health.

We have greater difficulty in discovering the thirteenth year by the appearances of the teeth than the ninth year; this is partly owing to the number of teeth which have come forward, but principally, and, as Mr. Cartwright well observed, by the undetermined period at which the eye tooth comes through the gum. Statistics, however, clearly shew, that at the thirteenth year the number of teeth should be twenty-eight. I add a few cases, which I have taken indiscriminately as I have found them :—

WORKHOUSE GIRLS, THIRTEEN YEARS.

E. Mayo, all changed, 8 molars
M. A. Norris, all changed, 8 molars
L. Claws, all changed, 8 molars
E. Hazel, 12½yrs., 8 molars
J. Thorogood, all changed, 7 molars
E. Bradford, 12yrs., all changed, 4 molars
M. Poulton, 12yrs., one lower bicuspid on the right not changed
W. Penfold, 12½yrs., 8 incisors, 2 lower canines, 3 bicuspids, 4 molars
A. Thorogood, 12yrs., 6 bicuspids, 8 molars
M. Hodges, 12½yrs., 2 lower canines, 1 upper coming, 1 primary molar not shed
M. Salf, 2 bicuspids not shed, 4 molars
C. Collier, 13½yrs., all shed, 8 molars

Great facility exists in detecting the age of the horse, in consequence of the space between the incisors and the molars. In man, the teeth are jumbled together, there is no decided mark, and references must continually be made to many points which can only be known to a physiologist of great and scientific attainments. In all cases, those laws of the animal economy must be taken into consideration which influence growth and development. From the eye tooth great difficulties must often be expected; the least thickening of the jaw, or a too close approximation of the other teeth, will considerably retard its appearance; other circumstances connected with the shape of the jaw will also keep it back till a late period in life; and as a most curious coincidence, and one well worth relating, I mention that of a lady, who at the age of sixty-one cut an eye tooth. In instances also where a primary tooth is large, it frequently remains unshed till a late period of life; this is usually connected with peculiar habits, or hereditary tendencies. Retarded developments of the teeth depend on causes of a twofold nature—these are proximate and remote. The proximate are those connected with contiguous teeth, or with the shape, size, or thickness of the jaw; the remote are, debility, the negligence of fever, the confluent small pox, hereditary tendencies, impaired functions, imperfect nutrition, bad diet, climate, &c., &c. Your connexion with the metropolitan hospital for children will doubtless have shewn you how frequently fevers produced by dentition terminate in diseases of the brain. In all the severe cases, whenever life is spared, you will bear me out in my statement, that it is spared at the expense of development; so also you will have observed the debilitating effects which a bad attack of the confluent small pox often leaves behind.

As infantine fevers are connected generally with dentition, and as this process influences in a most extraordinary degree almost all formations, it is requisite that I call your attention to the manner in which disease of the membranes and substances of the brain becomes established. Irritation is first sent up by means of branches of the par trigeminum to the pons varolii of the brain, a turgescence of the neighbouring vessels takes place, and all those nerves arising contiguously soon become involved in the general turgescence, thus lesion gets established, and paralysis may fall on muscles supplied by branches of these nerves. The hydrocephalic appearance of a child subject to attacks of this kind renders him an object of peculiar pity. If such a child of nine years of age were taken for a child of thirteen, subjected to severe labour, or kept depressed in other ways as to his spirits, the grave would soon take him, or the workhouse receive him as an idiot. I made out the following tables, in conjunction with the late Mr. Rice, extracted, according to my papers, from upwards of twenty thousand cases, occurring during a long process of time at your institution: the records from which we extracted them are still in your possession.

ONE HUNDRED PHYSICIAN'S CASES.

ANALYSIS OF DISEASES.

Affections of the head	3	Gastritis	1
— of the bowels	3	Herpes	10
— of the lungs	1	Hydrociphalus	2
Catarrh post rubeola	1	Ictanus	1
Cephalitis	8	Marasmus	1
Cholera	2	Mesenteritis Chronica	1
Cynanche	1	Meneugitis	2
Croup	1	Pulmonitis	2
Dentitis	5	Pneumonæa	8
Dyspepsia	3	Pertussis	3
Febris Catarrhatis	2	Phthisis Incipiens	2
Febris bilio remittens	1	Paralysma Abdominalis	1
Gastrica	1	Swelled Face	3
Dentitionatis	6	Scrofula	8
cum eruptione	2	Tinia Capitis	10
Verminosa	1	Ulcers and Abscess	2
Mesenterica	2		

Out of these cases you will select eight of cephalitis, several of febris dentitionalis, dyspepsia, affections of the head, bowels, with cynanche and eruptive and other fevers, influenced in a great measure by teething.

On early developments I need not say much; they are connected with strength and powers of the frame. All that is requisite to the early formation of teeth is, a constant supply of blood to those parts having to form them. It little signifies for the actual formation of the teeth of what quality the blood is; an apparatus is at work, mechanically wrought as the weaver's loom, which only requires supply, and this supply must be of proper material, or the work, when completed, will fall short of perfection. In children of a scrofulous or strumous diathesis, or in those where a quickened action of the heart exists, there is frequently an early formation of teeth in the gum, but these children cut their teeth late, owing to the constitution not possessing the sufficient power to expel them from the gum. This is noticed by Underwood, when treating on weak and rickety children. The component parts of their teeth are not such as render them capable of the offices which nature intended them to perform. In children, decayed teeth indicate peculiarities of constitution, which a physiologist who has paid attention to the point knows well how to appreciate; he sees them often the forerunners of hip-joint and spinal diseases. In the mouths of persons possessing extreme sensitiveness of mind as of body decays invariably exist; and in old age, the breaking up of the mouth is too often in accordance with the breaking up of the constitution.

It is children, subject to decays and to a peculiar sensitiveness of the teeth, who are the greatest sufferers through life, both in mind and body. I might enlarge greatly on the influence which hereditary tendencies have on the teeth, but merely select the cases of the two Pooley's, at Christ's National School; the eldest, at ten years of age, had only cut two upper and two lower incisors; and the youngest, at nearly eight years of age, had changed no teeth: they were children of an unhealthy appearance; their parents are both short and unhealthy.

Climate and diet have much to do with general development. I saw seven young ladies from India a few days since; the two eldest at eleven years of age had twenty-eight teeth: this is only amongst other precocious developments which take place under an oriental sun. You must see the fallacy of all doctrines giving to the teeth no share in those sympathies which link them so closely to the economy of the frame.

Amongst other collateral evidences of the ages of children, I am certain that the voice, properly analysed, with due regard to its perfection, roundness, elasticity, and changes, might, to an accustomed ear, be of essential use in deciding the age during the early periods of growth and development. Your principal aim being to make correct returns of the ages of nine and thirteen, I am fully warranted in telling you that the teeth, with some few exceptions, would decide the point. You may recollect when some short boys, whose ages were far beyond their appearance, were shewn me at Christ's National School, I made some remarks as to their ages, which I found on inquiry to be correct; I must, however, admit several cases where I was much in error. I fear I may have gone too much into detail; further research into this interesting subject would lead to many valuable deductions of high interest to science. Any direct legislation by means of the teeth would be extremely difficult. I would in no instance set down a tall boy who had not changed his eight incisor teeth beyond nine years of age. No doubt can exist but that such a lad is less capable of fatigue than one of shorter but of more compact growth.

From statistical knowledge which I have lately acquired, and from my own professional experience, I am quite justified in asserting that the teeth afford good collateral evidences of the ages of children. As the interests of humanity are involved in this interesting question, and as science is never so well employed as in its promotion, it will afford me great pleasure if my mite of exertion can at all assist your praiseworthy efforts towards ameliorating the condition of the factory children.

Believe me, my dear Sir, with sentiments of the highest esteem, yours most faithfully,

G. Waite.

Charlotte Street, Fitzroy Square, Nov. 25.

Sir,—I have much pleasure in answering your note, and in reply to your question "Whether, in the absence of direct information from registration of the age, any dependence is to be placed upon a knowledge of the teeth, as a criterion of the age of a child?" &c., I beg to observe, I think my experience would enable me to determine with tolerable accuracy the age of a *healthy subject*, from between six and seven up to between thirteen and fourteen years old, by the appearance of the teeth and their progress towards development.

I have the honour, &c. J. P. Henderson.

5, Sackville Street, Saturday, 7th January.

Sir,—In reply to your inquiry as to the possibility of judging correctly of the age of children by their teeth; I beg to state, as far as my own observation goes, that a young person of healthy constitution may be considered to be thirteen years of age when all the primary teeth have been shed in a natural manner. In the majority of cases, I would not expect the eye teeth of the upper jaw (the last to come before the age of eighteen) to have descended sufficiently into their places although decidedly and largely in view. It seems a more difficult task to judge correctly whether a person has attained the age of nine years. In the majority of cases, I would consider a child to be nine years old when the four incisors, or front teeth of both jaws, were in view, even if the two upper lateral incisors had not descended sufficiently into their places.

I need not remind you that in cutting the teeth there is considerable diversity as to time in different children, even of the same family, varying from a month or two to a year or two; but the difference seems to depend a good deal on constitution; and this difference I apprehend to exist to a greater extent in high and pampered, than in humble life.

It seems to me that a very equitable standard might be obtained easily in the following manner, much more certain than the stature of children,—viz: an examination of the mouths of the children in the tenth and thirteenth years of age, at charity schools, on an extensive scale. I would willingly lend my assistance on any Sunday or Sundays for so humane a purpose.

For the guidance of others who might incline to investigate the subject, I would refer them to a map of the mouth in one of the plates of my little treatise, lately published by Longman and Co. It is quite new, and easily understood by professional and non-professional persons. It would be necessary to bear in mind that the human teeth consist of two sets, the first, twenty in number; the second, thirty-two; that the first set of twenty is usually fully developed at the age of two years, and are shed between the seventh and thirteenth years, when a permanent one succeeds to each. They consist of six single front teeth in each jaw, and four double teeth, or grinders, two in each side of the jaw. As only twenty teeth are changed by nature, it will be proper to observe that the twelve additional ones required to complete the second set never come but once; and that the first four permanent teeth usually cut the gum at six years of age, making the complement at that period twenty-four teeth. These four permanent teeth are the first of three series of large grinders: the original grinders of a child are succeeded by smaller ones, called the bicuspides, being the fourth and fifth teeth in both sides from the centre of the mouth. At twelve, the second series of large grinders make their appearance, being the sixth tooth everywhere from the centre of both jaws, so that the complement at this age is twenty-eight teeth, to which no more are added till eighteen. The upper eye teeth, as was already stated, are the last of the twenty-eight that belong to this period of life to take their proper places. In the map referred to, all that is here intended may be seen at a glance.

If the above description be not satisfactory to you, I shall be most happy to follow out any hint you may give on the subject.

I remain, Sir, very faithfully yours, J. P. Clark.

The following I was favoured with by Mr. Saunders, of Argyle-street, and is an extract from a lecture, lately delivered by that gentleman at Saint Thomas's Hospital, on the anatomy and diseases of the teeth :—

"Thus, then, the teeth appear to possess an economy of their own, and to be, to a great extent, uninfluenced by those affections and states of constitution in which the other parts of the system, more or less, participate, and with which, according to the degree of vitality which they possess, they are found to sympathize. Even in those more violent commotions of the system resulting from the catalogue of infantile diseases, the progress of the formation of the teeth does not appear to be retarded; there may be, and frequently is, an unequal and scanty deposition of enamel, and the bony portion of the tooth may be delicate, and the tooth may consequently be predisposed to those morbid affections to which these organs are liable, but the formative process is all the while in operation; it is never entirely suspended. Thus there is far less irregularity in the development of the teeth than of any other part. The growth may be checked, the signs of puberty may be retarded or may appear prematurely, but the teeth, as if belonging to another and a separate economy, are uninfluenced by any of these modifying circumstances, at least to anything like the same extent, and become developed according to a law of their own. We frequently see children who are the subjects of rickets or other congenital morbid influences by which the general growth has been retarded, having a very protuberant mouth, and teeth apparently too large for the maxilla: this will be found, however, on a closer inspection, to arise, not from the unusual size of the teeth, nor from their premature development, but from a want of normal growth and expansion of the containing and contiguous parts, the growth of the latter has been retarded, while that of the former has proceeded with its usual regularity. On account of this great regularity and independence of modifying circumstances, which are even greater among the lower animals, the teeth have long been regarded as the most certain, and indeed the only criterion by which their age may be ascertained; and it has been lately proposed to apply this test to the factory children; and the idea, I confess, appears to me to be a singularly happy one. It is difficult to conceive of a more accessible and ready means of arriving at a knowledge of the child's age up to the period of fourteen, at which time the process of changing the teeth is accomplished, and the child has acquired its second set, with the exception of the dentes sapientiæ, which occupy the next septenary period for their development, and while it is the most accessible evidence, it is also, as we have seen, the most certain and the least liable to fallacy. It will not be denied that a certain degree of latitude must be allowed for the time of their appearance, yet this, when once compared with the uncertainty which pertains to the other parts of the system, is so trifling, that it scarcely deserves to be mentioned."

The opinions of such eminent men upon this novel subject of inquiry will doubtless be read with much interest. Mr. Waite most handsomely undertook to obtain the sentiments of two very eminent gentlemen, Mr. Dumergue and Mr. Cartwright; their answer Mr. Waite has embodied in his own valuable communication. I have examined some hundreds of children, with reference to this inquiry, and if I may be allowed to offer an opinion, I must express my concurrence in the views of these professional gentlemen, who judge favourably of the teeth as a practical and very valuable *auxiliary* in determining the ages of children.

I could not help observing the pallid cheeks of young girls about the age of puberty, who were evidently suffering from oppressive toil. If human nature, after the first exposure to infantile or juvenile labour, were to cry aloud for a mercy-seat and a resting-place, would not that cry be made, and should it not be heard, at that delicate period?—a period, when advancing growth and increasing power are struggling to establish the constitution of a fellow-creature, and upon which is staked the physical character of the future man to accomplish the purposes, and to fulfil the duties, of an earthly destiny. More deeply still are our most elevated sympathies demanded, when man's helpmate, woman, is the subject of such a trial-season in factory life.

It is, indeed, to be regretted that the factory commissioners decided, "in opposition to the best medical opinions, that puberty is established at the thirteenth year." The ordinary strength and appearance may be enjoined as a test, but our inspectors, in their capacity of *law-makers*, fly from one criterion to another, from age to height, and from height to ordinary strength and appearance, as yet undetermined as to any fixed course of proceeding. The very able author of the 'Voice from the Commons' asks, "How many persons are to be thus directly, deeply, and cruelly injured? Mr. Stanway has calculated that of every 212,800 persons habitually working in the cotton-factories, more than 83,000 (nearly half of whom are *females*) are under eighteen years. Of these 83,000, about 44,000 (19,000 of them females) are *under* fourteen years of age. Of these 44,000, there are 4200 children who are not more than nine years old; and some hundreds of the factory labourers are not more than seven, while, even at the tender age of five, there were many little slaves in the mills."

Very recently, Mr. P. Thomson lent his powerful assistance to deprive children of a year's protection—"that a boy of twelve might be reckoned as an adult, and that *a girl of a dozen summers** might be regarded as a woman." Both in health and disease, notwithstanding the disputes upon the late Mr. John Hunter's theory, nature, in her general ways, applies herself to the accomplishment of one great leading principle. The farmer looks not for milk and flesh at the same time from his cows. "The dainty smells of flowers are out of those plants whose leaves smell not; as violets, roses, wall-flowers, gilly-flowers, pinks, woodbines, vine-flowers, apple blooms, lime-tree blooms, bean blooms, &c. The cause is, for that where there is heat and strength enough in the plant to make the leaves odorate, there the smell of the flower is rather evanid and weaker than that of the leaves; as it is in rosemary flowers, lavender flowers, and sweet-briar roses."—(See the valuable edition of Bacon's Works, by Basil Montague, Esq., vol. ii.)

* "At thirteen years of age, there are nearly 10,000 children who are directly affected by Mr. P. Thomson's bill, in every 213,000 persons."—Voice from the Commons.

I may plead the aptness of the following for its present repetition:—

"In the *natural state* of youth, from the twelfth to the fifteenth year of life, mortality is less than at any other period. The functions are then in full operation to form the basis of manhood, to create materials for the *waste* of *future* exertion, not to be *devoted to labour now.* Growth should then be at its most rapid rate in the frame in both sexes, under a healthy condition, undergoing, at the same time, a great change, which does not fully terminate before the close of the sixteenth year." It is evident that nature cannot supply sufficient energies for the purposes of inordinate labour, and for sound growth also; she cannot effectually comply with both demands, and under repeated endeavours to accomplish the two, she either perishes, or exhibits a powerless and degenerate form. Skill may augment the mechanical speed of iron and steel, those in authority may sanction and decree a course of labour beyond the power of the young frame to meet, yet they cannot command—they cannot quicken—the proper agencies of organization; neither can they spur on animate against inanimate matter in the competition of labour, without committing a series of outrages at which nature will revolt, and from which the law will be made to shrink. No surprise will be felt that a national evil should befal us under a system which has been described "as more hostile to human life than the most depopulating ravages of war or pestilence, or the most sanguinary forms of superstition."

With such an increasing proportion of factory population, we may anticipate in a few years a dreadful reduction in the physical power and construction of our armies. What did that experienced general, Sir John Elley, say, on the 9th of March, upon Mr. P. Thompson's motion?

"In Leeds, we are told, that amongst the manufacturing classes, there is even a redundancy of health. I have not been lately to Leeds, but when I last visited that town, so far from witnessing the plethoric health which I have heard described to night, I saw nothing of it; on the contrary, I saw nothing but want and misery." Sir John goes on to say, "If I wanted to fill up my companies, I should not from choice go to the manufacturing districts to look for recruits.* I have been on many recruiting parties;

* Evidence of LIEUTENANT THOMAS, 43d Light Infantry.—(From 1st Rep. Fac. Com. Examinations taken by Mr. Tufnell, p. 59.)

How many years have you been in the recruiting service?—Thirteen months, and during that time 175 recruits have been raised by the parties under my command; not all in this town, not above one-third; the remainder in the adjacent towns. Of the above number about thirty-one were weavers or spinners, and of the whole number twenty-one were born in this town, six of whom were weavers or spinners.

Do these factory workmen in general make strong soldiers?—No, much less so than other classes; in my opinion, men of agricultural counties are most eligible for the army, being in general most robust and healthy.

Of the recruits which you have collected in this town, do you perceive much difference in strength between those who work in factories and those who work at other trades?—I conceive that those who work in factories are of a less robust description.

And altogether worse, both as to health and physical appearance?—I cannot speak so positively as to their health, but inferior to their physical appearance.

Do you think that these two classes of town workmen are equal in point of courage?—It is a point that I could not speak to directly, as I have never been employed on active service.

Are their characters equal?—I have not sufficient experience to speak positively to that point; but, as a matter of opinion, I should prefer the character of those not employed in factories to factory men. My regiment has recruited fifty or sixty men from factory districts, chiefly from Bolton. These men improved very much in the course of six months.

(The witness subsequently sent in the following paper, containing the ages, heights, birthplaces, and trades, of 175 recruits raised in the Manchester subdivision, from March 1st, 1832, to May 13th, 1833: one half were raised in Manchester, and the remainder in the neighbouring towns.)

RECRUITS RAISED IN THE MANCHESTER SUBDIVISION.

Age.	Height. FT. IN.	Where born.	Trade.	Age.	Height. FT. IN.	Where born.	Trade.
17	5 8	Manchester -	Weaver.	18	5 11	Northwich -	Labourer.
18	5 7	Preston - -	Corder.	21	5 6	Liverpool - -	Painter.
20	5 6½	Manchester -	Dyer.	18	5 6	Macclesfield -	Gunsmith.
21	5 7	Bolton - -	Crofter.	19	5 6	Manchester -	Weaver.
17	5 7½	Blackburn -	Skinner.	18	5 6	London - -	Coachmaker.
23	5 6½	Leeds - -	Weaver.	19	5 6	Bilston - -	Butcher.
19	5 7	Derby - -	Labourer.	22	5 8	Rochdale - -	Labourer.
22	5 7	Longford - -	Armourer.	21	5 9	Manchester -	Ditto.
20	5 6¼	Altringham -	Labourer.	19	5 8	Bolton - -	Weaver.
18	5 6½	Rochdale - -	Weaver.	18	5 6	Blackburn -	Ditto.
17	5 7	Manchester -	Fustian-cutter.	22	5 7	Ashton - -	Skinner.
18	5 6½	Stockport -	Weaver.	20	5 8	Huddersfield -	Cloth-dresser.
18	5 6½	Oldham - -	Skinner.	22	5 10	Druncliffe -	Clerk.
18	5 7	Bolton - -	Draper.	20	5 9	Manchester -	Joiner.
24	5 9½	Rochdale - -	Weaver.	18	5 6	Hadleigh - -	Shoemaker.
18	5 6¼	Belturbet - -	Shoemaker.	21	5 9	Bolton - -	Weaver.
21	5 6½	Westminster -	Butcher.	17	5 7	Trowbridge -	Fustian-shearer.
18	5 7	London - -	Sadler.	21	5 9½	Clonmel - -	Plasterer.
20	5 6¼	Leeds - -	Joiner.	18	5 8	Chorley - -	Weaver.
20	5 6¼	Leeds - -	Ditto.	22	5 8	Chester - -	Plasterer.
17	5 7½	Middleton -	Weaver.	19	5 8	Middlewich -	Paviour.
18	5 6½	Leeds - -	Dyer.	18	5 6	Blackburn -	Chair-maker.
23	5 9	Wigan - -	Spinner.	18	5 6	Halifax - -	Cotton-weaver.
22	5 6	Burnley - -	Printer.	17	5 6	Shrewsbury -	Labourer.
17	5 7	Bolton - -	Piecer.	19	5 9	Chesterfield -	Ditto.
18	5 7	Macclesfield -	Printer.	18	5 8	Bolton - -	Bricklayer.
18	5 9	Manchester -	Ditto.	19	5 7½	Halifax - -	Wool-comber.
18	5 7	Tadcaster -	Weaver.	18	5 6	Manchester -	Labourer.
20	5 6	Worsley - -	Collier.	20	5 7	Ditto - -	Weaver.
18	5 7	Manchester -	Labourer.	22	5 6½	Eccles - -	Slater.
19	5 6	Dublin - -	Shoemaker.	18	5 6½	Dunfermline -	Labourer.
21	5 8	Plymouth -	Labourer.	20	5 8	Manchester -	Ditto.
23	5 6	Keighley - -	Woolsorter.	18	5 9	Middleton -	Weaver.
18	5 7	Blackburn -	Baker.	22	5 10	Bolton - -	Wheelwright.
22	5 6	Huddersfield -	Cropper.	20	5 8½	Nottingham -	Stocking-weaver.
22	5 7½	Preston - -	Plasterer.	19	5 7	Worksop - -	Labourer.
19	5 6	Sheffield - -	Grinder.	20	5 7	Lane End -	Grocer.
21	5 6	Hindley - -	Labourer.	19	5 6	Manchester -	Tailor.
18	5 6	Durham - -	Ditto.	23	5 8	Middleton -	Weaver.
18	5 6	Warrington -	Ditto.	18	5 8½	Stone - -	Shoemaker.
19	5 6	Liverpool -	Baker.	18	5 8½	Manchester -	Shoeing-smith.
18	5 6	Prestbury -	Spinner.	18	5 6½	Worsley - -	Weaver.
20	5 7	Salford - -	Cropper.	22	5 6½	Hilton - -	Apothecary.
23	5 8	London - -	Labourer.	16	5 7	Manchester -	Collier.
18	5 7	Macclesfield -	Silk-weaver.	18	5 7	Stockport -	Spindle-grinder.
19	5 8	Leek - -	Silk-twister.	20	5 8	Bolton - -	Weaver.
17	5 6	Ashton - -	Labourer.	18	5 8	Sandbach -	Shoemaker.
18	5 7	Glossop - -	Blacksmith.	21	5 8½	Middleton -	Labourer.
19	5 6	Glossop - -	Apothecary.	21	5 8½	Bolton - -	Ditto.
24	5 7½	Leeds - -	Shoemaker.	19	5 8½	Dewsbury -	Ditto.
23	5 7¼	Sittingbourne -	Groom.	18	5 7	Letterkenny -	Weaver.
21	5 7¾	Sleaford - -	Baker.	19	5 7	Dorchester -	Groom.
18	5 7	Manchester -	Printer.	22	5 10	Chatham - -	Plasterer.
18	5 7	Bradford - -	Stonemason.	19	5 8½	Northwich -	Skinner.
18	5 6½	Duckinfield -	Boiler-maker.	18	5 6½	Huddersfield -	Weaver.
18	5 6	Burnley - -	Servant.	21	5 7	Warrington -	Shoemaker.
17	5 6	Paisley - -	Dyer.	18	5 6½	Stockport -	Collier.
26	5 6	Preston - -	Painter.	20	5 7½	Manchester -	Spinner.
22	5 6½	Knaresborough	Tin-plate maker.	21	5 6½	Ditto - -	Clerk.
17	5 6	Huddersfield -	Moulder.	18	5 6	Middlewich -	Weaver.
18	5 6½	Birmingham -	Silversmith.	20	5 6¼	Bolton - -	Collier.
18	5 6	Blackburn -	Sawyer.	19	5 6½	Little Hutton -	Piecer.
18	5 8	Prosperous -	Weaver.	18	5 6½	Manchester -	Tailor.
17	5 6½	Perth - -	Calico-printer.	19	5 8	Ditto - -	Ditto.
18	5 7	Bolton - -	Mechanic.	18	5 7½	Coventry - -	Spinner.
19	5 7½	Duckinfield -	Weaver.	18	5 8	Rochdale - -	Cordwainer.
19	5 7½	Criddleton -	Ditto.	19	5 6½	Macclesfield -	Calico-printer.
18	5 6	Preston - -	Calico-printer.	19	5 6½	Chapel-le-Frith	Labourer.
21	5 6	Keighley - -	Wool-comber.	18	5 6	Brixworth -	Ditto.
21	5 6	Shrewsbury -	Joiner.	20	5 8	Northwich -	Tallow-chandler.
18	5 6½	Manchester -	Weaver.	19	5 6¼	Manchester -	Tailor.
22	5 7	Bolton - -	Ditto.	18	5 6¼	Halifax - -	Brickmaker.
18	5 6	Littleboro' -	Woollen-miller.	18	5 8	Ditto - -	Labourer.
19	6 0	Colne - -	Reed-maker.	18	6 0	Lancaster -	Stonemason.
20	5 6½	Rochdale - -	Labourer.	20	5 7	Ditto - -	Collier.
18	5 7½	Liverpool - -	Servant.	19	5 8	Ditto - -	Ditto.
19	5 7½	Wednesbury -	Ditto.	20	5 8	Ditto - -	Labourer.
18	5 6	Macclesfield -	Piecer.	20	5 6¼	York - -	Cordwainer.
19	5 8	Christchurch -	Labourer.	21	5 8	Chester - -	Clerk.
18	5 6	Kensington -	Ditto.	20	5 8	Derby - -	Collier.
18	5 8½	Manchester -	Cabinetmaker.	20	5 6½	Warwick - -	Dyer.
18	5 8	Oldham - -	Shoemaker.	18	5 8	Chester - -	Labourer.
17	5 6	Keighley - -	Labourer.	18	5 9	Lancaster -	Shoemaker.
18	5 10	Ashton - -	Clerk.	21	6 0½	Chester - -	Filer.
18	5 7	Okeham - -	Painter.	18	5 6¼	Rochdale - -	Weaver.
18	5 8	Manchester -	Hatter.	19	5 6	Ditto - -	Cotton-carder.
20	5 10	Altringham -	Farmer.	18	5 7¼	Ditto - -	Weaver.
18	5 7	Manchester -	Weaver.				

and when on emergencies we sometimes went to the manufacturing towns and populous districts, at periods when we expected to find many unemployed, though many came and expressed their willingness, I have found them for the greater part not possessed of sufficient physical power, and I am sorry to add, deficient also in mental endowments. Compare them with the agriculturists, and those whose avocations lead them to live in the open air, and there is the greatest difference imaginable, both corporeal and intellectual." Great difficulties attended the duties of surgeons in former times relative to recruits, which I fear are not diminished now. I refer to a note,* which, whether applicable or not to the present day, I do not know.

MORTALITY.

Upon this subject, of so much importance in all its bearings, the most conflicting evidence has prevailed. Unfortunately, statements have been sent forth which have no foundation in truth; and by which it has been sought to influence the public mind to the belief that the factory system is not more fatal to human life than other callings and avocations, whether of an agricultural community or otherwise. It was published in a census of Manchester for 1821, that the annual mortality of that place was one in seventy-four. Subsequently, Mr. Roberton has stated it to be one in forty-five, and Dr. Kay one in thirty-five; whereas we find it to be in the agricultural county of Lincoln one in sixty-two.† Mr. Gaskell having observed, that Dr. Hawkins had published Mr. Roberton's testimony as to fifty-four dying in Manchester in every hundred under five years of age, asserts, without giving the names, that this is contradicted by two surgeons at Bolton, and *partially* by a physician at Stayley Bridge. In the Voice from the Commons, which should be read throughout, there is a just complaint made in speaking of the chief duties of the commission, that "no statistical evidence of the *whole* of the factory children exists as to the fruit of its labours, to be compared with the ill health of other classes of children. The functionaries are certainly either very ignorant of statistical calculation on a large scale, or very neglectful of their business. They should have prosecuted an examination of the population of the manufacturing and other districts. They should have distinguished the actually sick from the healthy, at different ages. They did neither; in short, their neglect of certain means of ascertaining the amount of ill health in the entire neighbouring population was complete."

Dr. Hawkins, in his medical statistics, asserts, that we shall find the rate of mortality in counties dependent upon the number of towns they contain. There being several in Hampshire, the mortality is one in fifty-eight; in Sussex, where they are less numerous, there is only one in seventy-two; in Cornwall, for the same reason, one in seventy-one. Returns from large manufacturing towns or districts should be taken with great caution; for the same indefatigable author remarks, that the small proportion is not always real, because the constant influx of *adults* is likely to render the number of deaths less considerable than that which would occur in a stationary

* "If a man," says Hamilton's Regimental Surgeon, vol. i., "be rejected as improper, the officer who enlisted him is offended; indeed, in all probability, he is a great loser, for if a recruit be not approved of, all that was given him is lost to the officer; the surgeon accepts or signs his name to the sound list: he is blamed afterwards by the regiment, as well as by the commanding officer, when it is discovered the man is always in the sick reports, and really unfit for service."

† Vide tables at the end of the volume, given by Dr. Hawkins from Mr. Rickman.

population of all ages. Mr. Thackrah, whose residence was at Leeds, and who was well acquainted with the subject, after admitting the wonders which art and science have effected in this country, says, " But let us examine what are the physical and moral effects, if those millions who spend their lives in manufactories, or live by trade, civic arts, and professions, enjoy that vigour of body which is ever a direct good, and whether they attain the age of agricultural labourers." Mr. Thackrah, after some remarks from a table taken in the census of 1821, with reference to the three ridings of Yorkshire, thus concludes :—" It is therefore evident that the duration of human life is considerably less in the West Riding than in other parts of Yorkshire." The valuable work upon Trades and Professions, from which the foregoing passages are taken, is worthy of general perusal. I further state, from the same source, that in a comparison between Leeds and an agricultural district, Pickering Lythe, there was an excess of 321 deaths in the borough of Leeds. And Mr. Thackrah observes, " the destruction of 450 persons, year by year, in the borough of Leeds, cannot be considered by any benevolent mind as an insignificant affair."

Nor is it in Leeds only that inquiry produces so painful a result. Leaving out of the question London and sea-ports, we might prove that Sheffield, Manchester, Birmingham, and, in fact, all our great manufacturing towns, exhibit an equal or greater excess of mortality, and an excess *increasing* with the magnitude of the population. If we should suppose that fifty thousand persons die annually in Great Britain, from the effect of manufactories, civic states, and the intemperance connected with these states and occupations, our estimate, I am convinced, would be considerably below the truth. Such is Mr. Thackrah's opinion. I have given the evidence of Dr. Baillie and others, and before I enter upon the subject of remedies, so important do I consider the *recent* testimony given by the medical men on Mr. Sadler's Bill, that I shall now offer the principal heads thereof. There were not less than twenty-one members of the faculty examined upon that occasion.

Exercise and rest for meals — Bell, Carlisle, Hodgkin, Key, Malyn, Roget, Sharp, Smith, Thackrah, Travers, Tuthill.
Deformity — Blundell, Carlisle, Elliotson, Guthrie, Hodgkin, Key, Lutener, Morgan, Smith, Tuthill, Young.
Education after bodily exertion — Bell, Blizard, Roget, Thackrah, Travers.
Stunted growth — Blizard, Blundell, Brodie, Carlisle, Guthrie, Malyn, Roget, Smith, Travers, Young.
Gas-light — Bell, Guthrie, Malyn, Morgan, Tuthill.
Hours of Labour — Bell, Blizard, Blundell, Brodie, Elliotson, Farre, Green, Guthrie, Key, Lutener, Malyn, Morgan, Roget, Sharp, Smith, Thackrah, Travers.
Bodily exertion checking growth — Bell, Brodie.
Osseous state — Bell, Blizard, Key, Smith.
Health — Blizard.
Night-work — Bell, Blizard, Blundell, Carlisle, Elliotson, Farre, Hodgkin, Key, Lutener, Morgan, Roget, Tuthill.
Scrofula — Bell, Blundell, Elliotson, Malyn, Sharp, Smith, Young.
Long-continued labour — Blizard, Morgan.
Puberty — Bell, Blizard, Blundell, Carlisle, Elliotson, Green, Hodgkin, Malyn, Roget, Travers, Young.
Legislative interference — Blizard, Blundell, Green, Guthrie.
Female v. *Male power* — Green, Hodgkin, Key, Morgan, Roget, Travers, Young.
Pulmonary consumption — Blizard, Sharp, Smith.
Remission of the hours of labour — Blundell, Farre, Key, Smith.
Varicose veins — Smith.
Numbers of accidents — Bell, Hodgkin, Lutener, Malyn.
Ventilation — Brodie, Smith.
Temperature — Blundell, Carlisle, Elliotson, Farre, Hodgkin, Key, Morgan, Young.
Age of admission, &c. — Blundell, Carlisle, Farre, Hodgkin, Malyn, Morgan.

Illegitimate children — Blundell, Brodie, Carlisle, Malyn, Hodgkin.
Exhaustion and flagellation — Brodie, Key, Roget, Travers.
Sunday-school after week's work pernicious — Blundell, Brodie, Green.
Flat-footed children — Brodie, Hodgkin.
Factory visage — Green, Key, Smith, Tuthill.
Injury to succeeding generations — Brodie, Green, Sharp.
Abridgment of life — Hodgkin, Sharp.
Working at meal times — Key.
Labour "light and easy" — Carlisle, Green, Guthrie, Hodgkin, Key, Malyn, Roget, Smith, Travers, Tuthill, Young.
Necessary time for meals — Elliotson, Farre, Malyn, Roget.
Effects of the factory system—more illness in the factories — Blundell, Farre, Guthrie, Hodgkin, Key, Lutener, Smith, Thackrah, Travers, Tuthill, Sharp.
Wages ill employed — Thackrah.
Remedies — Farre.
Agricultural population more healthy — Smith.
Mortality — Blundell, Elliotson, Guthrie, Hodgkin, Key, Lutener, Malyn, Morgan, Sharp, Smith, Thackrah, Tuthill, Young.
Muscular exertion and erect position — Key, Malyn, Brodie, Roget, Sharp, Smith, Travers, Tuthill.
Flax and flues — Blundell, Malyn, Roget, Travers.
Transitions of temperature — Tuthill.
Difference of health in agricultural and in manufacturing life — Sharp, Thackrah.
Maturity — Smith.
Morality — Smith, Young.
Weight of children — Morgan.
Contrast between slavery in the West Indies and factory labour — Dr. Farre.

I readily terminate the painful task of exposing, from indubitable sources, the existence of evils which I have not designed to magnify, and of which no other country can present a similar record, either as to number or their astounding nature. My labours would be closed with something like satisfaction, if I could persuade myself that some obvious remedies would be adopted, in consequence of a corresponding feeling on the part of the government and the mill-owners. The remedies are of an external and internal order. In the first place I will speak of those which are external to the mills, particularly in Manchester. I have shown, from Dr. Kay and other gentlemen, that great improvements may be obtained in the construction of the houses, and at very little additional expense. Sewers should be regularly made, and the streets formed into passable roads, instead of being left in an impassable state of mud and dirt. The laws of nature, decency, and health, should be more strictly attended to. The invigorating influences of public baths should be resorted to, according to the custom of continental nations, amongst *all classes* of the community, even of those who have not the various and baneful effects of a confined factory life to call forth such a salutary practice. To the credit of Mr. Wood and some other masters, baths have been established upon their own premises for the use of the operatives. It was frequently mentioned to me, when in Lancashire, by mill-owners, that a system of education* of the children should precede their entering upon their duties in the mills. This plan would leave a child some time for the restoring effects of air and exercise after the labours of the day,

* Mr. Horner, whose ingenuity has been applied in many instances to make the Act of Parliament run smooth, in his rules and regulations, has left it open to the master-manufacturers completely to evade the intention of the twenty-first clause, part ii. p. 436. That clause requires that no child under a certain age shall work the usual hours appointed for that age without a schoolmaster's voucher, certifying that such child has, for two hours at least for six out of seven days of the preceding week, attended school. Mr. Horner, by construing two hours at least for six days out of seven into "twelve hours weekly," has given parents the option to thrust as many of these hours as are convenient into the Sunday, and thus making it no Sabbath to the little factory labourers under thirteen.

and would allow him the advantages and blessings of domestic guardianship. Gin-shops should be discouraged. I need not dwell here upon their pernicious tendency in every way. The order of nature is inverted, and the child supplies to the parent the means of enjoying his midnight orgies in these incentives to wasteful and pernicious expenditure. We must anticipate national degeneracy when a people fly more to fluids than to solids to satisfy the cravings of nature. If the poor factory children could have a warm meal of oatmeal-gruel and milk, with even one-third or one-fourth of the latter, with a proper supply of bread, night and morning, as part of their diet, they would receive a more healthy supply of blood than that which they now obtain from coffee and tea, such as they are.

As to the remedies connected with the *internal* management of the mills, a restriction upon the moving power is a paramount one. To understand the present picture of factory life, as to domestic comforts, or rather as to domestic misery and wretchedness, I refer the reader to Dr. Hawkins' admirable report. That much has been done with regard to ventilation in some mills, and principally by employing the lately invented fans, is generally admitted. There is a most extraordinary statement in one part of the published evidence, that the operatives in one mill, in which the improved method of ventilation was introduced, petitioned the masters either to remove the fan, or proportion their wages to the increase in their appetite. This of itself shews the advantage of improved ventilation. On this subject I have received the following letter from Dr. Reid, whose late successful experiment in the House of Commons I had much satisfaction in witnessing:—

Edinburgh, November 21st, 1836.

Dear Sir,—I have to acknowledge your letter of the 13th instant, and would have replied at an earlier period, had not the engagements which crowded upon me on my return, after a long absence, prevented me from obtaining an earlier opportunity. I have no hesitation in affirming, that a minute attention to many factories has convinced me that an effective system of ventilation might be introduced in innumerable situations where it has not hitherto been an object of special attention, and that the extent of human suffering which could in this manner be relieved, far exceeds anything which has ever been imagined by those who have not studied the effect of the long-continued operation of a vitiated atmosphere upon the human system, whether in the private abodes of the poor, or in the oppressive air which they breathe in many manufacturing establishments.

My time, at this moment, does not permit me to add more, than that I consider a current of air properly attempered and diffused in the apartment to be ventilated, and impelled towards a chimney containing a fire at the centre of the base, as the most simple and economical method of maintaining a pure atmosphere. Fanners may also be used with advantage where the proper machinery is available.

Wishing you every success in the important cause to which you have directed your attention. I remain, dear sir, sincerely yours,

D. B. Reid.

The following extract is from Mr. Inman's highly-interesting work, entitled, 'A Report of the Committee of the House of Commons on Ventilation, Warming, and the Transmission of Sound:'—" It is absolutely essential that the halls in which the legislature assembles for hours together should be rendered wholesome. Is there a member who has not experienced the noxious effect of the present vitiated atmosphere on mind and body? Instead of being able to devote his best energies to the calm deliberation and discussion of the public weal, his mental powers are exhausted, and his bodily feelings wearied; impatience and irritation succeed, and important measures are hurried over or left to be settled by those whose condition equally fits them for the arena of the amphitheatre or of the forum." Fully as I concur in the foregoing observations, I would offer this passage as an additional inducement for the legislature to afford as great a proportion of relief as can be expected to the relative situation of those employed in factory-life. In this work I find a note that death in an hospital before ventilation was one in six; after ventilation, one in twenty; as stated by Dr. Joseph Clarke, Edinburgh Meeting, British Association. I am glad to find that in the newly-erected factories the rooms are considerably higher; a most necessary improvement. From all I have read, seen, and heard upon the subject, I am confident that there can be no effectual remedy for the evils of the factory system which does not make a ten-hour bill its basis. The present act aims at the exclusive protection of the children, and on that very account it has failed to protect them. But even if it had succeeded, as it offers no relief to the adult operative, but rather aggravates his case, it can never restore tranquillity to the manufacturing districts. That the adult operative requires protection, and that it would be for the advantage of the children that he should be protected, I have given satisfactory proof; satisfactory, not indeed to those who think indifferent persons more unexceptionable witnesses to the existence of grievances than those who actually suffer them, nor to those who would suspend their judgment upon the subject till there should be no conflicting evidence, but to those who are content with such evidence as the nature of the case admits.

It is surely enough that medical men, eminent in their profession, and practising in the manufacturing districts, together with master manufacturers, whose interests are involved in the system, have forcibly denounced its evils. It is highly improbable that such persons should either have been deceived themselves or should have attempted to deceive others; and as probability is the guide of human affairs, we must consider the evils of the system as proved. That they should so long have been suffered to continue without a comprehensive and effectual remedy is matter of deep regret. It cannot be defended upon any principle of policy; since it cannot contribute to the stability of a state that so large a portion of the population should remain in a state of discontent, and the infraction of the law should become a habit. It cannot be defended upon any principle of justice; for justice requires that speedy and effectual protection from wrong should be extended to all classes of the community. It cannot be defended upon any principle of religion; for religion teaches us to consider every man as a brother, and the concerns of all mankind as our own. Religion enlarges our views beyond the narrow sphere of self into universal sympathy. Let not, then, our legislators be deceived by the flattering accounts of the system so industriously circulated by interested individuals—let them not be deterred by the clamour of political economists, nor by the sneers of men who, judging, perhaps, from their own hearts, are incredulous as to the existence of philanthropy in the hearts of others, from doing, speedily and effectually, what policy dictates—what justice demands—and what religion enforces by the strongest obligations.

EXTRACTS

FROM

THE EVIDENCE

OF

DR. BAILLIE, SIR ASTLEY (then Mr.) COOPER, SIR G. TUTHILL, SIR ANTHONY (then Mr.) CARLISLE, DR. PEMBERTON, SIR GILBERT BLANE, MR. ARCH. BUCHANAN, MR. ROB. OWEN, MR. NATH. GOULD, and SIR ROBERT PEEL,

(From the Report of the Committee, of which Sir Robert Peel was chairman, in 1816.

MATTHEW BAILLIE, M.D., called in, and examined, 29th April, 1816.

M. Baillie, M.D.

* 1. At what age may children, without endangering their health, be admitted into factories, to be regularly employed thirteen hours a day, allowing them one hour and a half to go and return from meals, and one hour for instruction?—I should say, that there was no age, no time of life whatever, where that kind of labour could be made compatible, in most constitutions, with the full maintenance of health.

2. Do you think that children from seven to ten years of age could be employed more than ten hours per day without injury to their constitution?—I think not; and if it was left to me to determine, I should say that they ought to be employed fewer hours for the full maintenance of health.

3. What do you consider to be the effect upon the development and growth of the bodies of children from six to ten years of age, of so many hours confinement per day?—I cannot say much from experience, not having attended children that have been labouring in manufactories; but I can say what appears to me to be likely to arise out of so much labour, from general principles of the animal economy. I should say, in the first place, that the growth of those children would be stunted; that they would not arrive so rapidly at their full growth; that they would not have the same degree of general strength; that it is probable their digestion would not be so vigorous as in children who are more in the open air, and less confined to labour; that they would probably be more liable to glandular swellings than children who are bred differently: and I think it likewise probable that, in particular manufactories at least, they would be more likely to be affected with diseases of the lungs.

4. How many hours would you recommend children of that tender age to be employed?—I should say that at that age, probably, for the first year, they should not be employed more than four or five hours a day; and for the two succeeding years they might be employed six or seven hours a day; that afterwards they might be employed ten hours a day; and beyond that, in my opinion, there ought to be no increase of labour.

5. In speaking of the injury to young persons arising from labour, do you mean to speak of labour which requires great bodily exertion?—I did not suppose that children at so early an age were employed in great bodily exer-

* As only extracts from the evidence are given, it is scarcely necessary to remind the reader that the figures are placed merely for convenience of reference, and not as denoting the order in which the questions were put.

M. Baillie, M.D.

tion, but I meant any bodily exertion in which they were confined in a given space, and their minds not allowed to wander into the various channels of thought, and their limbs allowed the sort of irregular exercise which takes place in children who are living in the usual manner.

6. Is not the state of maturity of children very different in those brought up in the country to those brought up in town?—With regard to children who are brought up in the country, they are more vigorous, and I have no doubt, in many instances, their progress towards maturity may be more rapid than in children who are reared in a large town.

7. What is the state of heat, as ascertained by a thermometer, in which children might work without injury?—I should say, that the temperature which is upon the whole most favourable, is about sixty degrees of heat.

8. Would children of the age of ten be employed in the ordinary business of the country?—No; but they would be doing a good deal of work of various kinds, as going of errands, or weeding, and a thousand employments which I cannot at present call up to my mind.

9. That answer seems to refer more particularly to children in the country; as the manufactories are generally in towns, it does not apply to them; therefore the committee wish to know whether you conceive, if children at an early period of life were prevented by act of parliament from working in factories, their situation would be better than it is?—I conceive it would be more favourable to health to be at large, although they might sometimes be not well nourished; and although sometimes they would be in hot rooms, they would have a great deal more time in which they could be playing about, and using their faculties of observation.

10. In a factory consisting of 875 persons, the annual deaths in which were not more than from two to five, should you conceive that the employment was inconsistent with the health of the people employed?—I should say it does not appear from that statement to have been inconsistent; I conceive a great many of those children might not have been in vigorous health, not in the same health in which they would otherwise be, and yet not be attacked with diseases which would occasion death.

11. Are you of opinion that the proportion between the cubic feet of air in a room, and the number of the persons employed in it, has an important effect upon the health of the people?—Certainly.

12. Then would you be of opinion, that if proper means of ventilation were attended to, a building containing a greater proportion of cubic feet of air to the number of persons employed in it would be more healthy than one containing a smaller proportion of air?—If the employments were the same, undoubtedly.

ASTLEY COOPER, Esq., called in, and examined, 29th April, 1816.

Astley Cooper, Esq.

1. At what age may children, without endangering their health, be admitted to close labour for thirteen hours per day?—I think at no age.

2. Do you think that children of from seven to ten years of age can be employed more than ten hours per day without injury to their health?—I think not.

3. What do you consider to be the effect upon the development and growth of the bodies of children from six to ten years of age, if so many hours confinement are imposed upon them?—The result of confinement, commonly, is not only to stunt the growth, but to produce deformity; and to that point I can answer, from a good deal of experience, that deformity is a common consequence of considerable confinement.

4. Which do you think the most prejudicial to children, an occupation which makes the children always in motion, or the occupations which are

sedentary?—That depends on the time that they are employed; if employed for a short time, it would be best they should be in exercise; but if for a great number of hours, it would be rather better for them that they should not be in motion. Astley Cooper, Esq.

5. Suppose a factory should be so free from mortality for a few years with the employment already stated, may not the constitution of the children be injured by that confinement, though death has not ensued?—Yes, I conceive that diseases may be founded in children by that employment, which may not shew themselves in so rapid a manner as to cause dissolution for a year or two, and yet which have their origin in that employment.

GEORGE LEMAN TUTHILL, M.D., called in, and examined, 30th April, 1816.

G. L. Tuthill, M.D.

1. Supposing children employed in manufactories from twelve to fourteen hours per day at the early ages of from seven to ten, do you conceive that they would sustain no injury in their constitutions?—I think they would sustain injury; but I am not in the habit of attending any manufactory.

2. What is your opinion as to the effect of confining children to such occupations in close rooms, from six years old to ten or twelve, for thirteen, fourteen, or fifteen hours?—I think it would be prejudicial.

3. Would it be desirable, in your opinion, that children so confined during the week should be called to religious or moral instruction on the Sabbath-day?—As far as their health is concerned, I should think that they had better be employed in the open air in exercise.

4. And take their education in the week time?—And take their education in the week time.

5. If the number of children be not great, and they should be confined only two or three hours on the Sunday, would that materially affect their health?—I should think not.

6. Do you think the proportion between the cubic feet of air in a room, and the number of persons employed in it, has an important influence on their health?—Very great.

7. Allowing the manufactories to be regulated in the best manner you have ever met with or heard of, do you think that the number of children brought up in them from an early age, and employed in the occupations there usually followed, for ten or twelve hours per day, would generally be found, when arrived at the age of maturity, to be possessed of as much bodily strength and soundness of constitution as the children brought up in the manner in which those of the peasantry of this country usually are?—Certainly not.

Mr. ANTHONY CARLISLE, called in, and examined, 30th April, 1816.

Mr. A. Carlisle.

1. Have you, in the course of your practice, had occasion to attend to the progress of the health and strength of children from seven years old upwards? —In the course of my professional life I have seen and attended several manufactories as a very young man; since I have been in London I have been twenty-three years a surgeon to the Westminster Hospital, where the consequences of confinement, and other unhappy modes of treatment, have brought upon children disease, and they have consequently come into such situations as have placed them under my care.

Mr. A. Carlisle.

2. Supposing that children, at that early age, were confined in manufactories thirteen hours per day, do you conceive that it would be attended with any serious consequences to those children?—From my experience, I believe that children cannot, with safety to their health, be confined thirteen hours a day in a close room; but, with submission, I must state that the difference upon different children will be very considerable, in proportion to their different constitutions: to some it will be fatal; as, for example, the children of scrofulous parents; others might resist a considerable portion of unwholesome influence with more impunity. I apprehend, that the earlier children are confined and subject to unwholesome regimen, the more deleterious it is to such persons throughout life afterwards; I think I have seen that to be the consequence in my experience.

3. What is your opinion as to the effects of confining children to such occupations when they are only six or seven years old, for fourteen, fifteen, or sixteen hours in the day?—The evil consequences will be in proportion to the youth of the person, his delicacy, or otherwise, the natural constitution, the length of time he is confined, and the confinement of the air; a considerable difference will result from there being a free circulation of air, or being confined in close rooms; a very considerable difference.

4. When so confined during the week, what is your opinion of the propriety of inducing children to attend schools for many hours on the Sabbath? —If that is a physical question, and not a moral one (which latter I beg to decline answering)—if it is a physical question, then the transfer from a sedentary and confined life in a manufactory to that of a school would certainly add to the unhealthiness of the party as contrasted with free exercise in the open air.

5. Has not the proportion between the cubic feet of air in a room, and the number of persons employed in it, an important influence on their health?—Certainly, both temporarily and permanently.

6. What, in your opinion, would be the effect of employing children from six to twelve years of age twelve or fourteen hours a day in the winter, in a room the temperature of which would be from seventy to eighty, and afterwards their being exposed to the air?—I should think it would be very deleterious to them.

7. Do you think it is necessary for the general health and growth of children that they should have hours for recreation and exercise in the open air?—I think it is absolutely necessary for them.

8. Even supposing that from the circumstance of well-aired rooms, and a sufficiency of good food, no particular ill effect should immediately appear upon the children, do you think that any considerable number of children of all ages, from six years to twelve, could be confined to such employments as have been spoken of, during twelve hours in the day, without its producing an injurious effect upon their constitutions?—I think they cannot be so employed without injury to their constitutions.

CHRISTOPHER PEMBERTON, M.D., called in, and examined, 29th April, 1816.

C. Pemberton, M.D.

1. At what age may children, without endangering their health, be admitted into close rooms to be regularly employed from thirteen hours upwards?—My opinion coincides with the opinions already given to the committee, that at no age they should be employed thirteen hours in the twenty-four.

2. What do you consider to be the proper time for the employment of

children under ten years of age, without endangering their health?—Under ten years, I should think, nine hours in the day. C. Pemberton, M.D.

3. What is your opinion of the employment of children under eight years of age?—I should think, under eight five or six hours a day.

4. What do you consider to be the effect upon the development and growth upon the bodies of children from six to ten years of age, if employed from twelve to fourteen hours a day?—I think it certainly would injure the growth and the health of children of that age to be employod fourteen hours out of the twenty-four.

5. The committee wish to know, whether in a factory where the number of persons employed in 1813 was 879, the deaths were only two; and in the same factory in 1812, the number being 891, there were only two deaths among them; and in 1811, the number employed being 873, and the deaths only three, you would conceive that such facts would be evidence of the healthiness of the employment in that manufactory?—I think there may be great fallacy in that calculation; before I could give an opinion on that, I should wish to be informed of the age of those persons, whether it included the whole establishment, or only the children.

6. The numbers were in this proportion: 22 males and 37 females under ten years of age; 98 males and 218 females under eighteen, and of course the rest must be above?—It must be recollected, that under eighteen years is the most healthy time of life; it is very difficult to give an opinion on this question. From these data, if those three deaths arose from anything that can be attributed to the manufactory, they are three too many; if, on the contrary, they are the mere casuality of small-pox or fever, or anything that might happen out of the manufactory, it is a small proportion: but no opinion ought to be given, that should have any weight, till it be ascertained what was the nature of the complaint of which they died.

7. Is no inference to be drawn from the fact of there being only two or three deaths in a year out of eight or nine hundred people?—I do presume that that calculation includes only the children employed, because this is a question respecting the children only; it very likely may appear, on inquiry, that these three deaths were among the aged persons, and then you would nearly come to my conclusion: between the age of six and eighteen none of our public schools average so much.

Sir GILBERT BLANE, Bart., M.D., called in, and examined, 30th April, 1816.

1. If children at a very early age were to be confined from fourteen to sixteen hours each day, though an immediate indisposition might not take place, do not you conceive that they would be prevented acquiring their full growth and strength when arrived at maturity?—I am assuredly of that opinion; for the consequence of confinement is to produce two or three diseases, which are particularly produced by confinement and want of exercise, I mean rickets and mesenteric obstructions, weakness of body, and imbecility of mind. Sir G. Blane, Bart., M.D.

2. Those disorders are particularly to be apprehended in children from seven to ten years of age?—Certainly, that is the period of life; I would say from birth to ten.

3. Have you had any opportunity of seeing the condition of children in any of the manufacturing towns, either lately or at an earlier period of life?—I was about twenty years ago in Buckinghamshire, and saw the lace-manufactories, which evidently have an injurious effect on the health of children.

Sir G. Blane, Bart., M.D.

4. You conceive the employment of children under ten years of age, from twelve to fourteen hours a day, is particularly likely to produce those complaints, and generally to be disadvantageous to their condition ?—Most assuredly; that great length of time in the day I should conceive to be highly injurious.

5. To what number of hours do you mean to refer when you speak of that great length of confinement ?—I think anything above ten hours at that early age,

6. Do you happen to know what number of children in the lace manufactories were employed ?—I did not enter into the detail; I was merely on a visit.

7. Did you observe from their appearance that they had sallow countenances and other indications of ill health?—I certainly did.

8. What was the apparent age of them ?—From seven or eight to fifteen.

9. Did it occur to you to make any remark upon what you conceived to be the cause of that sickly appearance of the children ?—I should say it was confinement in bad air ?

10. Are you acquainted with the general state of the stature of the people in that country ?—They are not the most robust, according to my observation.

11. Have not the improvements of ventilation introduced into the navy had a most beneficial effect in promoting the health of people in his Majesty's navy ?—The effect would almost appear incredible; cleanliness and ventilation go beyond calculation; they would hardly appear credible to this committee.

12. Would not an attention to the same principles in the manufactories be likely to be attended with the same beneficial consequences ?—There is not a doubt of it, to counteract the effect of the confinement and want of exercise.

13. From your great experience and practice, the committee would wish to ask you what is your opinion as to the feelings of affection and tenderness on the part of the lower classes towards their offspring ?—I have seen a great deal of all orders, and I never saw a want of affection on the part of mothers; I have seen hard-hearted fathers.

14. Is the proportion between the cubic feet of air in a room, and the number of persons employed in it, of great importance to their health ?—Very great; that is a subject I have particularly studied.

15. In rooms properly ventilated, and where the quantity of respirable air allowed to each person is 1,440 cubic feet, do you think that employment is likely to be prejudical to such persons ?—There is ample space for pure air there; in a hospital there is 700 feet to a patient, and we consider that a safe and proper space, still more so where they are in health and walk about. In a hospital well ventilated, we find 700 cubic feet is a safe and proper space for each patient.

16. Is it important to the health of children and others, that the temperature of the rooms in which they are employed in winter should be comfortable, and as nearly uniform as is consistent with proper ventilation ?—There is no doubt of it; I think comfortable and salutary to be one and the same thing; nature points out what is salutary.

17. Your attention seems to have been particularly called to the proportion of deaths in different places in this country; do you conceive that, in a factory where, in 1811, the number being 873, the deaths in that year being only three, in 1812, the number being 891, the deaths only two, and in 1813, the number being 879, and the deaths only two, such facts to be an indication of the healthiness of the employment in such factory ?—It is an indication of the greatest possible health; but it so far exceeds the common course of nature, that if I had it not from such respectable authority I should greatly doubt it.

18. Would you be surprised at the statement, if you were informed that when children are ill, and likely to die, they are removed from the manufactories?—That alters the case totally. Sir G. Blane, Bart. M.D.

19. Would your surprise of the small mortality cease, if you were informed that no persons are employed under nine years of age, only fifty-nine of the number under ten at the larger factory, and perhaps not forty out of the number above forty years of age, and the factory situated in the healthful country of Ayr, with which you are acquainted?—That renders it somewhat less marvellous.

20. Have you the means of informing the committee what the general mortality is in healthy districts in this country upon healthy persons between the age of ten and eighteen?—I had lately occasion to make inquiry about that. From some calculations I have made, I found that the mortality in England, between twenty and forty, was about one in eighty.

Mr. ARCHIBALD BUCHANAN, called in, and examined,
25th April, 1816.

1. What is your employment?—I am employed in the management of the cotton-mills in Scotland, the property of Messrs. James Finlay and Company, merchants, in Glasgow; of which company I am also a partner. Mr. A. Buchanan.

2. Have you been some time employed in that pursuit?—I have.

3. How many years?—Since I first learned the cotton-spinning business, about thirty-three years.

4. What number of persons are employed in their different works?—I can only speak to the works under my particular management, the returns from the others have not yet come to town; there are 875 employed at the Catrine works.

5. How many of those are under ten years of age?—Twenty-two males, and thirty-seven females.

6. What is the youngest labourer that you employ?—I cannot answer that question; I suppose the youngest may be eight or nine: we have no wish to employ them under ten years of age.

7. Wishing not to employ any under ten, what circumstances have led you to employ any under that age?—The circumstances, generally, of the condition of their parents; people with large families, who find great relief from having a child or two put in at an earlier age.

8. In that number of persons employed, how many are males, and how many females?—227 males, and 648 females.

9. How many of those are under eighteen years of age?—From ten to eighteen; 98 males and 218 females.

10. To that number is to be added the 59 who are under ten years of age?—Of course.

11. Of that number of 875 persons, how many are there who cannot read?—There are eleven males and twenty-six females.

12. How many who cannot write?—660, I think.

13. What are your hours of work?—Our working hours are twelve hours in the day.

14. At what time do they begin in the morning?—They begin at six o'clock in the morning, they stop at half-past seven at night, and they are allowed half an hour to breakfast, and an hour to dinner.

15. What has been the state of the health of those children, particularly those under ten years of age?—Generally very good; much the same as those children in the neighbourhood who are not employed in work.

Mr. A. Buchanan.

16. You have not observed that the twelve hours work has interfered with the health of the children?—I have not.

17. Suppose that the children were taken at six years of age, do you think they would be able to work that number of hours without great indisposition?—I should think they would.

18. That it would not injure their health?—I have seen many instances of children that were taken in even as young as six, whose health did not appear at all to suffer; on the contrary, when they got to greater maturity, they appeared as healthy, stout, people as any in the country.

19. Not crippled in their growth?—No.

20. Have you known weavers employed very long hours; have you generally understood in the country that they work fifteen and sixteen hours at particular times?—I have heard some of the weavers say so.

21. Do you know how many people were sick on any particular day?—Upon the 18th of this month there were sixteen: there is, however, a note here, which I will beg leave to read:—"Of the regular hands, there are six off work, from indisposition; but besides those, there are ten invalids and *superannuated* people, who have spent a great part of their lives at the works, who receive from the Company a weekly pension in proportion to their circumstances."

22. Do you know what the deaths were in the year preceding the last?—In 1814, the people employed were 926; the deaths were seven; in 1813, the number employed was 879; the deaths were two: in 1812, the number employed was 891; the deaths were two: in 1811, the number employed was 873; the deaths were three.

Mr. ROBERT OWEN, called in, and examined, 26th and 29th April, 1816.

Mr. R. Owen.

1. What is your situation in life?—I am principal proprietor and sole acting partner of the establishment at New Lanark, in Scotland.

2. How many persons, young and old, are immediately supported by the New Lanark manufactory and establishment?—About 2300: upon the first of January last the numbers were 2297, I believe.

3. At what age do you take children into your mills?—At ten and upwards.

4. What are your regular hours of labour per day, exclusive of meal times?—Ten hours and three quarters.

5. What time do you allow for meals?—Three quarters of an hour for dinner, and half an hour for breakfast.

6. Then your full time of work per day is twelve hours, out of which time you allow the mills to cease work for an hour and a quarter?—Yes.

7. Why do you not employ children at an earlier age?—Because I consider it would be injurious to the children, and not beneficial to the proprietors.

8. What reason have you to suppose it is injurious to the children to be employed in regular manufactories at an earlier age?—The evidence of very strong facts.

9. What are those facts?—Seventeen years ago, a number of individuals, with myself, purchased the New Lanark establishment from the late Mr. Dale, of Glasgow: at that period I found there were 500 children, who had been taken from poor-houses, chiefly in Edinburgh, and those children were generally from the age of five and six, to seven and eight; they were so taken because Mr. Dale could not, I learned afterwards, obtain them at a more advanced period of life; if he did not take them at those ages, he could not obtain them at all. The hours of work at that time were thirteen, inclusive

of meal times, and an hour and a half was allowed for meals. I very soon discovered, that although those children were extremely well fed, well clothed, well lodged, and very great care taken of them when out of the mills, their growth and their minds were materially injured by being employed at those ages within the cotton-mills for eleven hours and a half per day. It is true that those children, in consequence of being so well fed, and clothed, and lodged, looked fresh, and, to a superficial observer, healthy in their countenances; yet their limbs were very generally deformed, their growth was stunted, and although one of the best schoolmasters upon the old plan was engaged to instruct those children regularly every night, in general they made but a very slow progress, even in learning the common alphabet. Those appearances strongly impressed themselves upon my mind to proceed solely from the number of hours they were employed in those mills during the day, because in every other respect they were as well taken care of, and as well looked after, as any children could be. Those were some, and perhaps they may be considered by the committee sufficient, facts to induce me to suppose that the children were injured by being taken into the mills at this early age, and employed for so many hours; therefore, as soon as I had it in my power, I adopted regulations to put an end to a system which appeared to me to be so injurious. Mr. R. Owen.

10. In consequence, then, of your conviction that children are injured by being employed the usual daily hours in manufactories, when under ten years of age, you have for some time refused to receive children into your works till they are ten years of age?—Yes.

11. Do you think the age of ten to be the best period for the admission of children into full and constant employment for ten or eleven hours per day, within woollen, cotton, or other mills or manufactories?—I do not.

12. What other period would you recommend for their admission to full work?—Twelve years.

13. How then would you employ them from ten to the age of twelve?—For the two years preceding to be partially instructed; to be instructed one half the day, and the other half to be initiated into the manufactories, by parties employing two sets of children in the day, on the same principle that two sets of children were employed when proprietors thought it their interest to work day and night.

14. What time would you recommend?—About ten hours of actual employment, or at the most ten hours and a half.

15. Do you think if such an arrangement were made in regard to the number of hours, the manufacturers would suffer any loss in consequence?—My conviction is that no party would suffer in consequence of it.

16. Either with reference to the home or the foreign trade?—Either with reference to the home or the foreign trade.

17. What benefits do you contemplate from this limitation of time?—A very considerable improvement in the health of the operatives, both young and old; a very considerable improvement in the instruction of the rising generation, and a very considerable diminution in the poor rates of the country.

18. Do you give instruction to any part of your population?—Yes.

19. What part?—To the children from three years old upwards, and to every other part of the population that choose to receive it.

20. What employment could be found for the children of the poor, in those situations, till ten years of age?—It does not appear to me that it is necessary for children to be employed, under ten years of age, in any regular work.

21. If you did not employ them in any regular work, what would you do with them?—Instruct them, and give them exercise.

Mr. R. Owen.

22. Would not there be a danger of their acquiring, by that time, vicious habits, for want of regular occupation?—My own experience leads me to say, that I have found quite the reverse, that their habits have been good in proportion to the extent of their instruction.

23. You know a good deal of Manchester?—I do know something of it.

24. You have been, in fact, a resident in Manchester?—Yes; previous to the purchase of the Lanark establishment, I had the superintendence and management, some time as a manager and some time as a partner, of cotton mills in the neighbourhood of Manchester.

25. Do you think that the schools established upon Bell's and Lancaster's, and other plans, at Manchester, would be sufficient for the instruction of the children there, till the time that they are regularly employed in the manufactories?—When I was in Manchester last year there was more school-room than children to fill it, and upon inquiring the reason, I found that there were such strong inducements held out, from the different manufactories in the town and neighbourhood, to the parents, to send the children early to work, that it counterbalanced any inclination such people had to send them to school; therefore the schools were not filled, nor nearly so.

26. Do you conceive, if children were not to be introduced into these factories till twelve years old, they would be as well fed and supported as they are now?—I recollect the period well when there were not any manufactories in several parts of the country (I speak of England and Wales); and the children, as far as I recollect, of the poor, were then as well fed, as well clothed, and, as far as my memory serves me, looked as well as now, and few or none of them were employed regularly until they were twelve, thirteen, and fourteen years of age.

27. Did you see any apparent deficiency in the faculties of those children?—I saw a very striking deficiency in the habits of those children; it was visible in their countenances.

28. Could those effects have escaped the attention of an intelligent man like Mr. Dale; and if he had observed them, would not his humanity have suggested a remedy for those evils?—There were no other means by which Mr. Dale could carry forward his manufactory; and it required some time before Mr. Dale could be aware of the injurious effects of such a system on those children; and, moreover, he was not resident at these works; he lived twenty-seven miles distant, and was seldom there more than one day in three or four months.

29. Was he himself sensible of those effects being produced by the labour of the children?—I had much communication with Mr. Dale soon after I purchased these works, and told him my sentiments, and my determination to alter the system; and he seemed to think it very right it should be altered as soon as it could be effected in practice.

30. Your work is now three-quarters of an hour less per day than Mr. Dale's was?—Yes.

31. Then the committee is to understand that three-quarters of an hour more in a day occasions the children being stunted and dwarfish, and that three-quarters of an hour less in a day prevents those effects being produced?—Pardon me; I stated that those children were employed at the early ages of five, six, and seven, and that now we do not take any children into the manufactory under ten years of age, which I apprehend will make a very considerable difference, combined with the diminution of hours.

32. Then the age affects your judgment on this subject?—Very materially.

33. You do not then suppose, that if children of the age of ten years worked three-quarters of an hour more each day, the effect would be, that their growth would be stunted, and they would be dwarfish; and that if they did not work that three-quarters of an hour each day, they would not be

stunted nor dwarfish?—No; but I apprehend their health would be considerably better by being worked three-quarters of an hour a day less. Mr. R. Owen.

34. What was the state of the health at that time of those children?—I state distinctly their health to be very much short of children in a sound state of body.

35. From your knowledge, what is the state of the health of the children employed in the different works?—In those manufactories in which the time of work is not very long, where the children are not taken in at a very early age, and where the proprietors are men of enlarged minds, and possessing some degree of benevolence, to induce them to look after the welfare of those from whom they derive their support and their wealth, the children are perhaps better off than they are in many other situations.

36. Do you conceive that it is not injurious to the manufacturer to hazard, by overwork, the health of the people he employs?—If those persons were purchased by the manufacturer, I should say decisively yes; but as they are not purchased by the manufacturer, and the country must bear all the loss of their strength and their energy, it does not appear, at first sight, to be the interest of the manufacturer to do so.

37. The committee wish next to ask, whether, in reducing your hours of work, you also reduced your price of labour?—No, I did not.

38. Is the state of the works of the cotton-manufactories in England such as to make it any difficulty for them to obtain children at this moment? —When there are so many children offered from five and six years of age to ten, it is very likely there may be many applications, and those applications unsuccessful; but, I conceive, if the children were prevented from working till they were ten years of age, there would be a sufficient number for the demand, and not more.

39. Are there any children employed under seven years old?—Yes.

40. Where?—In Leeds, in Stockport, in Macclesfield, and in almost all the manufacturing districts.

41. There are children employed under seven years of age?—Yes.

42. Will you name the factories?—If the committee will have the goodness to lend me their aid, I will name the parties who will give direct evidence to those facts.

43. Then you only know it from hearsay?—I know it from the persons who have been employed; from the manufacturers themselves, and those of the highest respectability.

44. What you are now speaking is from hearsay, and not your own knowledge?—It is from the direct information of the most respectable manufacturers in the kingdom.

45. You do not know it of your own knowledge?—I have not been in those mills from morning till night; I have not been to investigate in person the details of other mills; but I saw a large mill dismissed at noon in Leeds, about ten days ago, and from the appearance of the children I should conclude that many of them were under seven years of age, some of them under six.

46. Supposing the children to be prevented, under ten years of age, from working, what security do you suppose there is that their parents would send them to school?—The security is one which I took the liberty of suggesting; and if I may be permitted, as it is a very short statement, I will read it: That the children of either sex shall not be admitted into any manufactory, after a time to be named, until they can read and write in a useful manner, understand the first four rules of arithmetic; and the girls be likewise competent to sew their common garments and clothes.

47. You have stated that the children, when you came to the works, though apparently fresh and healthy, were really deformed and stunted in

Mr. R. Owen. their growth in consequence of being overworked and received at too early an age; will you have the goodness to state what motives induced you to lengthen the time of working, as you say you did?—I did not say I found the children in that crippled state when I first got there, but afterwards I found that to be the case. The reason the hours were increased was, that other mills over the kingdom at that time employed their children a greater length of time; and the parties with whom I was connected wished the mills to be employed that time.

48. Did you continue employing the children this lengthened time after you had discovèred that, in consequence of being overworked and received at too early an age, they had become deformed and stunted?—I employed them as short a time as possible under the circumstances.

49. Then your observation upon the improvement of the children, from the shorter number of hours work, has only been made since last January?—Yes, from the shortest hours. That reminds me of one of the parts of the evidence which was rather defective: I was permitted to state on Friday some of the principal advantages that I found to arise from this limitation of hours; and it is a very important circumstance, which I should have pleasure to impress on the minds of the committee and the gentlemen present, that, previous to the first of January last, although we had an evening school open for many years, the utmost number of children that usually attended, when the hours of employment were eleven and three-quarters, were about 100 per night on the average, very frequently less: but so soon as a reduction was made of another hour, down to ten hours and three-quarters per day, the number immediately increased; and I believe, in less than a week, there was a regular attendance of those persons previously employed in the mills in the day, of 350, 360, 370, and, for a considerable time past, 390.

50. Do you say that from an accurate investigation of the fact, or upon conjecture?—From pretty accurate knowledge of the fact, the cause of which I will state; in a year or less after I had been at Lanark I found the number of illegitimate children to be considerable, and they increased for two or three years. I found it necessary to devise some means by which a stop should be put to a practice so injurious to the inhabitants and to the neighbourhood; and I made a regulation by which each young person, both male and female, connected with the fact, was obliged to pay a regular weekly sum to a poor's fund established in the place; the men two shillings a week, and the females one shilling a week: and this I found to be so effectual, that in two or three years afterwards, the instances became very rare indeed; and I have among my papers the name of every individual for the last six years who has had an illegitimate child.

Mr. NATHANIEL GOULD, called in, and examined, 12th, 13th, and 14th June, 1816.

Mr. N. Gould. 1. What is your line of life?—That of a merchant,

2. In what line as a merchant?—That chiefly of preparing goods and shipping them abroad; sending them to different parts.

3. You ship as well as prepare?—Both.

4. How long have you been resident at Manchester?—Upwards of five-and-thirty years.

5. Have you turned your attention to the state of the poor children at Manchester?—I have.

6. Can you form any estimate of the proportion of children and adults employed in the factories at Manchester?—Upon that I have no certain ground

to go; I have endeavoured to inform myself as well as I could upon it. In some factories there are nearly two under eighteen years of age to one above; in some others there may not be much more probably than half, but I think that is very rarely the case. From all I have learned, if I were to form a conjecture, I should say that I conceived three-fifths of them to be under eighteen years of age. Mr. N. Gould.

7. Do you conceive that the number not employed in factories, in Manchester, as much exceeds the number employed in factories, as the number in the Sunday-schools, not employed in factories, exceeds the number in those schools who are employed in factories?—By no means.

8. Do you conceive, then, that the children employed in factories do not attend Sunday-schools in a due proportion to those who are not so employed? —I do; I think that is exceedingly the case; the disproportion is, I think, exceedingly great.

9. Have you, from your knowledge and observation, been able to trace that disproportion to any particular causes?—There are two great causes; the one is, I think, that the parents of the children that do not work in factories are more attentive to their children than those are whose children work in factories; I am speaking generally on this. The other great cause, in my mind, is, that the children having been confined so very much in the factories all the week, have less inclination to go to the Sunday-schools, as well as their parents to send them; I think it is natural for many parents to wish their children to be as much as may be in the open air on the Sabbath-day, in order that they may get some strength against they begin work the next week, and recruit themselves a little.

10. Did you make any observation, when at the Sunday-schools, on the health of the children?—I did.

11. Did you take any means to ascertain the relative healthiness of those working in factories, and those who do not work in factories?—I did.

12. What means did you take?—On the 12th of May I accompanied Mr. Whitelegg to two large schools in the neighbourhood of several factories, in which were present 1,327 scholars, of whom 174 were too young to be useful in factories; of those who were above that very tender age, we were induced, from their not looking well, to ask 213 what was their employment in the week-days; of those so questioned, 172 replied they worked in factories; the rest of those so questioned, in number 41, were found to have other employments, or to go to a week-day school. On the 19th of May, Mr. John Allen and myself visited seven other schools, to six of which we were accompanied by Mr. Todd; in the seven schools were present 3,882 scholars, of whom 458 were too young to be useful in factories; of those who were not so young, we asked, on account of their sickly appearance, 759, what was their employment in the week-days; of those so questioned, 464 replied that they worked in factories; the rest of those so questioned, in number 295, proved to be otherwise employed, or to go to day-schools.

13. Are you Treasurer of the House of Recovery in Manchester?—I am.

14. Have you any means of knowing whether the patients who come from cotton-factories are numerous?—I have.

15. Are they numerous?—The matron lately told me, that she believed the greater part of the patients admitted have been employed in factories.

16. Did she, from her answer, lead you to believe that a great many of the young persons admitted into the House of Recovery come from factories? —She did.

17. Did she say that she inquired into the facts of the persons who were brought into the fever-wards?—She certainly did; she inquired from themselves.

18. She stated to you that it was her practice, when people came into the fever-wards, to learn what their trade was?—She did.

Mr. N. Gould. 19. And she assured you, she had learned from the people who came into the fever-wards generally, that they were employed in factories?—She did; I will read what I have written: she said they often came in very dirty, and that when they had become convalescent she had found, from conversation with them, that a great proportion of them are very ignorant, and cannot read.

20. Has it fallen under your observation to know whether masters are in the habit of taking care of those members of factories who become ill?—I have generally understood that they did not.

21. Have you any particular grounds for that opinion?—My opinion is grounded upon a knowledge that when children are ill and cannot attend factories, other children are taken in their stead.

22. Do you know whether there is a provision made for them by their masters or not, in case of their being sick?—I always understood it to be a generally understood fact that there is not; I know no more.

23. Have you had any opportunity of learning the sentiments of the gentlemen of the faculty, in consequence of the patients coming from factories?—I do not know that I have heard them make any observation.

24. Not respecting the effect of factories upon their health?—I have heard some of them express themselves very strongly indeed on the effect of the cotton-factories.

25. Name the medical gentlemen you have conversed with upon the subject?—I have conversed with several medical gentlemen, and I should be very glad to mention their names if I had their permission. I know, in general, the medical gentlemen have strongly expressed themselves; but they would rather be excused having their names made known.

26. Will you state the substance of those conversations, stating at the same time the names of the medical gentlemen, and the time when it took place? —I have seen Mr. Bill, surgeon, who says, he has long been seriously impressed with a sense of the ill effect of spinning-factories, both as to health and morals, and that the impression increases as he grows older. In speaking of the early age at which the children attend them, the protracted hours of labour, the heat and closeness of the rooms, he called it a miserable business. He says that these works crowd the Infirmary with patients; they promote scrofula; that if there be but a grain of it in the constitution, they are sure to bring it to maturity; that they also bring on consumptions; and that many die early in life in consequence of attending them. And I have the opinion of other medical men that have lived very long in Manchester, who corroborates this gentleman.

27. Are there any other gentlemen whose names you can mention with propriety?—I should be very glad to do it, but I am afraid it might injure them. I ought to observe, that Mr. Bill has been surgeon to our Infirmary, perhaps for twenty years, and he declined several years ago: he is a man of very good fortune now. I have also conversed with Mr. Boutflower, a surgeon, living at Salford, who says he considers cotton-factories a great national evil as they are conducted; that the attending them checks the growth of young people, causes much disease, much deformity, particularly in the legs and knees; makes a short-lived, puny, race; promotes scrofulous complaints, which prevail very much in Manchester. I have a note that my brother put into my hands as I was getting into the carriage to come here, which he received from the same gentleman on the subject of scrofulous complaints, which is as follows: "That of seven or eight thousand patients who are annually admitted to the Manchester Infirmary, one half of the surgical complaints are scrofulous. A difference very striking occurs at Liverpool: in a report of 15,000, only 152 are marked as scrofulous."

28. Can you assert, from your own knowledge, that scrofulous affections are more common among persons who are employed in cotton-mills than

among persons not so employed?—I have no doubt of the fact, from the opinions I have heard of medical men. Mr. N. Gould.

29. Do you know yourself the fact that scrofulous affections are more common amongst persons in Manchester, who are employed in cotton-mills, than among those who are not?—I certainly have observed it at the schools more in the children that are employed in factories than in others; and I have often been very much hurt at the sight, and have felt for the poor children exceedingly, being aware of their situation.

30. If scrofulous affections were very general amongst children employed in cotton-factories, must not the fact be known to the proprietors and managers of those factories?—I should think it would.

31. Do you think that you have friends and acquaintance who are managers or proprietors of factories who could give you evidence on this subject, and upon whose declarations upon the subject you could confide?—I am inclined to think that the owners of factories often know very little of the children employed in the factories.

32. Are not some of the proprietors of cotton-factories in Manchester among the most useful, benevolent, and respectable inhabitants of the town?—I know many very respectable men, indeed, cotton-spinners; some that I have long known intimately, and whom I am sorry to differ from in opinion upon this occasion.

33. Do not you conceive that if the proprietors of cotton-factories had regular returns of the health of the persons they employ, that fact alone would give them the means of judging of the effect of the employment on the health of the people better than persons who have not access to such facts?—Doubtless, if those returns were correct.

34. Is there any particular reason to suppose that those returns would be incorrect?—I think there is reason to suppose it; the people giving them are so used to observe children whose labour is excessive in factories, that I am inclined to think they are insensible to their state frequently.

Sir ROBERT PEEL, Bart., a Member of the Committee, examined in his place, 21st May, 1816.

Sir ROBERT PEEL.—I beg leave to hand in a paper as the substance of what I know respecting the state and management of cotton-manufactories within the scope of my acquaintance, which is not less than five-and-forty years. Sir R. Peel, Bart.

It was read, as follows:—

"The bill of last year, which I had the honour to bring into the House of Commons, respecting young persons in factories, was so far matured as to enable all persons interested to form an opinion of the extent to which they would be affected by the measure.

"But as the session was far advanced, and little opportunity afforded of knowing the sentiments of persons so interested, I thought it most advisable, at that time, not to press the completion of the bill.

"The pottery and many other trades in which young persons were employed, were never intended to be comprehended in the measure, and the benefit we have received from our late discussions will induce me (with the approbation of the committee) to confine the operations of the proposed bill to the three great manufactures of the United Kingdom, cotton, woollen, and flax, all of which are carried on to considerable extent in large buildings, dependent on machinery, and give employment to the children of the poor in great numbers. Having also learnt that the factories of persons of the greatest

Sir R. Peel, Bart.

respectability are the most distinguished for moderation in the hours of exacted labour, I flatter myself that the measure in question will not make any material alteration with respect to them. Mr. Arkwright was the inventor of machinery, of great national importance, which was employed at a time when steam-power was little known in large buildings, which were erected in situations commanding considerable water-power, but generally in country-places remote from inhabitants: to work these machines, the surplus population of large towns was sought after, and many thousand of parish children were supplied from London, Birmingham, and other populous districts.

"The house in which I have a concern gave employment, at one time, to near one thousand children of this description. Having other pursuits, it was not often in my power to visit the factories, but whenever such visits were made, I was struck with the uniform appearance of bad health, and, in many cases, stinted growth of the children; the hours of labour were regulated by the interest of the overseer, whose remuneration depending on the quantity of work done, he was often induced to make the poor children work excessive hours, and to stop their complaints by trifling bribes. Finding our own factories under such management, and learning that the like practices prevailed in other parts of the kingdom, where similar machinery was in use, the children being much overworked, and often little or no regard paid to cleanliness and ventilation in the buildings; having the assistance of Dr. Percival and other eminent medical gentlemen of Manchester, together with some distinguished characters both in and out of Parliament, I brought in a bill, in the forty-second year of the King, for the regulation of factories containing such parish apprentices. The hours of work allowed by that bill being fewer in number than those formerly practised, a visible improvement in the health and general appearance of the children soon became evident, and since the complete operation of the Act, contagious disorders have rarely occurred.

"Diffident of my own abilities to originate legislative measures, I should have contented myself with the one alluded to, had I not perceived that, owing to the present use of steam-power in factories, the forty-second of the King is likely to become a dead letter. Large buildings are now erected, not only as formerly, on the banks of streams, but in the midst of populous towns, and instead of parish apprentices being sought after, the children of the surrounding poor are preferred, whose masters being free from the operation of the former Act of Parliament, are subjected to no limitation of time in the prosecution of their business, though children are frequently admitted there to work thirteen to fourteen hours per day, at the tender age of seven years, and even, in some cases, still younger. I need not ask the committee to give an opinion of the consequence of such a baneful practice upon the health and well-being of these little creatures, particularly after having heard the sentiments of those eminent medical men who have been examined before us; but I most anxiously press upon the committee, that unless some parliamentary interference takes place, the benefits of the apprentice-bill will soon be entirely lost, the practice of employing parish apprentices will cease, their places will be wholly supplied by other children, between whom and their masters no permanent contract is likely to exist, and for whose good treatment there will not be the slightest security. Such indiscriminate and unlimited employment of the poor, consisting of a great proportion of the inhabitants of trading districts, will be attended with effects to the rising generation so serious and alarming, that I cannot contemplate them without dismay; and thus that great effort of British ingenuity, whereby the machinery of our manufactures has been brought to such perfection, instead of being a blessing to the nation will be converted into the bitterest curse.

"Gentlemen, if parish apprentices were formerly deemed worthy of the care of Parliament, I trust you will not withhold from the unprotected

Sir R. Peel, Bart.

children of the present day an equal measure of mercy, as they have no masters who are obliged to support them in sickness or during unfavourable periods of trade.

"Intended alterations to be made in the bill of last year:—

"Children not to be admitted into factories under nine years of age.

"Hours of attendance per day, thirteen; allowing one hour and a half for meals and recreation.

"Children from the age of nine to sixteen to be subject to the proposed bill.

"The magistrates, on complaint being made, to be empowered to appoint inspectors."

1. The committee wish to know what have been the changes that have taken place in the manufactories within your memory?—I have witnessed a very great change since my knowledge of the factories; they were at first badly managed, and subject to many inconveniences, not only to the masters, but to the children in them; and new buildings were made without any regard to the health of the persons employed in those buildings; there was no regard paid to the time of working, but every man regulated his own mill by his own views of interest: with those complaints existing, I was induced, assisted by the first gentlemen in the county, to bring in a bill, about fourteen years ago, in order to prevent those abuses; the House was so convinced of the necessity of the measure, that there was no difficulty in passing the bill. I did it not so much for the benefit of others, but finding that my own mills were mismanaged, and that, with my other pursuits, I had it not in my power to put them under a proper regulation, I conceived I could not do better for the children than to bring forward that bill; and it was soon perceivable that there was a very great difference in favour of children protected by that bill; but as I have stated, in the paper I have delivered in, the advantages of that bill are not likely to continue much longer, in consequence of the change in the trade; therefore I took the liberty of recommending a measure, last year, adapted to the circumstances of the times.

2. The committee are to understand that great improvements have been made in the ventilation of the rooms, and the cleanliness of the children, within the last forty years?—I wish I could say within the last forty years. I have had experience in the business, but little regard was paid to the ventilation of the rooms before the passing of that bill.

3. Are the committee to understand the first improvements that took place in the ventilation of cotton-mills originated in consequence of your apprentice-bill?—I am persuaded that was the case, because my correspondence with the trading world being very extensive, and from my knowledge of what passed in the county of Lancaster, and other trading counties, I was satisfied that ventilation could not have taken place, because there were no factories at that time but such as were worked by water-power; the steam factories, to any extent, had not then an existence; and I believe nine-tenths of the Manchester buildings at that time had no existence.

4. Are the committee to understand that you consider the improvements were in consequence of the apprentice bill?—No improvements took place till the passing of that bill, and great improvements have taken place since; I do not pretend to say that, in consequence of general inquiries, improvements would not have taken place if no such bill had passed.

5. Then you do not know that the improvements that have been made since that period were in consequence of the bill that passed?—I believe manufactories had not been improved much before the year 1800 or 1801, and at that time it was rumoured that the bill was going to take place; many meetings were held in consequence of it, and I believe my honourable friend who sits near me was one of those laudably employed to improve the con-

Sir R. Peel, Bart.

dition of the children employed in the manufactories. I see his name classed with a very respectable number of Manchester gentlemen, who lamented the abominable state of the mills at that day; and that might be one reason why my attention was more particularly called to bring forward a measure of this sort.

6. Were epidemic complaints frequent in the first years of your knowledge of cotton-mills?—In the early part of my acquaintance with business, after cotton-mills became established, those complaints were very prevalent, and many lives lost.

7. You have referred to the proceedings of a committee appointed in Manchester, in 1796, to superintend the health of the poor in the towns of Manchester and Salford, and the neighbourhood; and that committee having investigated the state of facts and suggested various modes of improvement, do you think those suggestions were perfectly useless, and that no improvements did actually take place till your bill was passed in 1802?—It would be presumption in me to suppose that no improvement had taken place after the investigation by some of the best men in that part of the country; but I have no recollection that anything was done by the Manchester Board of Health to induce the country to alter a system very ill calculated to benefit the children. With the permission of the committee, I will read the heads of resolutions for the consideration of the Manchester Board of Health, by Dr. Percival, January 25th, 1796. "It has already been stated that the objects of the present institution are to prevent the generation of diseases; to obviate the spreading of them by contagion; and to shorten the duration of those which exist, by affording the necessary aids and comforts to the sick. In the prosecution of this interesting undertaking, the board have had their attention particularly directed to the large cotton-factories established in the town and neighbourhood of Manchester; and they feel it a duty incumbent on them to lay before the public the result of their inquiries: 1. It appears that the children and others who work in the large cotton-factories are peculiarly disposed to be affected by the contagion of fever, and that when such infection is received it is rapidly propagated, not only amongst those who are crowded together in the same apartments, but in the families and neighbourhoods to which they belong. 2. The large factories are generally injurious to the constitution of those employed in them, even where no particular diseases prevail, from the close confinement which is enjoined, from the debilitating effects of hot or impure air, and from the want of the active exercises which nature points out as essential in childhood and youth, to invigorate the system, and to fit our species for the employments and for the duties of manhood. 3. The untimely labour of the night, and the protracted labour of the day, with respect to children, not only tends to diminish future expectations as to the general sum of life and industry, by impairing the strength and destroying the vital stamina of the rising generation, but it too often gives encouragement to idleness, extravagance, and profligacy in the parents, who, contrary to the order of nature, subsist by the oppression of their offspring. 4. It appears that the children employed in factories are generally debarred from all opportunities of education, and from moral or religious instruction. 5. From the excellent regulations which subsist in several cotton-factories, it appears that many of these evils may, in a considerable degree, be obviated; we are therefore warranted by experience, and are assured we shall have the support of the liberal proprietors of these factories, in proposing an application for parliamentary aid (if other methods appear not likely to effect the purpose) to establish a general system of laws for the wise, humane, and equal government of all such works."

EXTRACTS

FROM

THE EVIDENCE

OF

Dr. HOLME, Mr. WHATTON, Dr. HARDIE, Mr. T. WILSON, Mr. W. J. WILSON, Mr. HAMILTON, Dr. CARBUTT, Mr. AINSWORTH, and Mr. TURNER,

(From the Report of the Committee of the House of Lords in 1818, *of which Lord Kenyon was chairman.)*

EDWARD HOLME, M.D., called in, and examined, 22nd May, 1818.

1. You are a physician, at Manchester?—I am. E. Holme, M.D.

2. You are not a petitioner on the subject of this bill?—I am not.

3. How long have you practised as a physician at Manchester?—Four-and-twenty years.

4. Have you, in Manchester, occasion to visit any public establishments?—I am physician to the principal medical establishments. The medical establishments with which I am connected, and have been for twenty-four years, are, the Manchester Infirmary, Dispensary, Lunatic Hospital and Asylum, and the House of Recovery; the latter only since the year 1796, which I believe was the year in which it was instituted.

5. Has that practice given you opportunities of observing the state of the children who are ordinarily employed in the cotton-factories?—It has.

6. Did you visit them by appointment at any particular period, or at whatever time best suited your own convenience?—I believe that in two instances I visited them at a time agreed upon by the medical gentlemen employed along with me; in another instance the time was altered at my request.

7. Were those times appointed with the owners of the manufactories, or with any other medical gentleman?—I am not able to say whether the medical gentlemen had the appointment, or whether the appointments rested with the manufacturers; I rather think that the latter was the case, for this reason; it was necessary, for the purpose of examination, that a list should be made out of all the children and others employed in the manufactories for the last twelve months. Perhaps your lordships would have no objection to one of those lists being produced, which would explain the nature of the examination; or I will attempt a verbal description. Lists were made of every person who had been employed in the factories for the last twelve months; three of those lists, in which I was concerned, I have brought along with me.

8. In what state of health did you find the persons employed?—They were in good health generally. I can give the particulars, if desired, of Mr. Pooley's factory, with the descriptions annexed to them. He employs 401 persons; and, of the persons examined, two-and-twenty were found to be of delicate appearance, two were entered as sickly, three in bad health, and one subject to convulsions; eight cases of scrofula: in good health, 363. The cases of scrofula amounted to about one in fifty. Under the name of scrofula was included every instance of glandular swelling observed about the neck,

E. Holme, M.D.

though no inflammation had at any time taken place. Of distortion or deformity no instance whatever occurred.

9. Did it, upon the investigation, appear that the persons employed in cotton-factories were in a worse state of health than other descriptions of labourers?—The fever was supposed to prevail most in those parts of the town where a double set of workmen were employed, and to be propagated chiefly by the children of one set sleeping on the beds which had been recently occupied by the other; but it did not appear, on investigation, that the night-workers were less healthy than the day-workers.

10. Am I to understand you, that from the investigation that took place in 1796 you were induced to form rather a favourable opinion of the health of persons employed in cotton-factories?—That was the first occasion on which my attention was particularly drawn to the subject.

11. Have you had any occasion to change that opinion since?—None whatever. I, at that time, in common with, I believe, the generality of my colleagues, thought that the spinning-factories burdened our medical charities with a disproportionate number of patients; this led me to investigate, and I was undeceived. The proportion, if I may speak concerning what has passed many years ago, was not essentially different from what it is at present.

12. The proportions you have given in detail indicate a comparatively favourable state of health of the persons engaged in those manufactories?—They do.

13. To what cause do you attribute their superior advantage in point of health?—I do not know that I have stated any superior advantage in point of health; I have mentioned that they are as healthy as any other part of the working classes of the community: I state that as my opinion.

14. Has your attention been turned to the effect upon the children of the number of hours in which they are employed in these cotton-manufactories?—It has not; I can only speak as to the health of the children, particularly the health of the children in the factories which I have inspected.

15. If children were overworked for a long continuance, would it, in your opinion as a medical man, affect their health so as to become visible in some way?—Unquestionably; if a child was overworked a single day, it would incapacitate him in a great measure for performing his work the next day; and if the practice was continued for a longer period, it would in a certain time destroy his powers altogether.

16. Then you are to be understood, that, from the general health among the children in the cotton-factories, you should form an opinion that they were not worked beyond their physical powers?—Certainly not.

17. The result of your observation did not indicate any check of growth arising from their employment?—It did not.

18. Did you observe distortion of the lower extremities?—In Mr. Pooley's and Mr. Pollard's factories we met with none; in Mr. Murray's factory the instances of distortion were thirteen.

19. Did you make any inquiry or at all investigate the subject of complaints of the females employed?—No particular inquiries were directed to that, except as to what may be considered as the critical period of the life of every female, namely, that on arriving at maturity.

20. In what state of health did the females appear to you to be?—The state of health, according to the information we received, as to the manner in which that change had been effected, shewed it had been effected in the same course as among other females.

21. Was that the result of the inquiries made from the females themselves?—Yes; the mode in which we pursued the inquiry was, generally, to select a grown person of some years, and to desire her to make the inquiry of the young women.

E. Holme, M.D.

22. Looking at the individuals who stated their ages to you, was their appearance such as to indicate an effect upon them such as you might expect generally?—There was nothing that would not correspond with the ages of the persons, except that which I have before stated, the shortness of many of the children in Mr. Murray's factory; and it was only in that instance that we observed that. The children in the other factories appeared to me of quite a corresponding appearance to their age.

23. Are you not aware of certain occupations in which the infirmities and diseases of old age occur earlier than in others, in consequence of the insalubrity of those occupations?—I can scarcely say.

24. You are aware that many persons at a very early age are employed by calico-printers?—Certainly.

25. Is there not a very unhealthy atmosphere generated by the substances employed in that business?—I cannot undertake to say that there is.

26. There are considerable fumes arise from the process of boiling the different ingredients which are used, are there not?—Undoubtedly; but I am not prepared to say whether the effect of them is injurious or not, or how far even the temperature at which they are employed is injurious; it far exceeds that of any other temperature I am acquainted with under which any employment is carried on; I refer to the calico-printers and the dyers. The department to which I particularly allude is the drying stone, the heat of which is that of the point at which water boils, 212°.

27. Are children employed in that for a continuity?—I apprehend that it is impossible that they should; they are certainly occasionally employed in it, either children or grown persons; I do not suppose they could bear the heat for any considerable length of time.

28. Are children of a puny description, and whose constitutions are unequal to other employments, sent to the cotton-factories on account of the easiness of the employment in those factories?—I believe that to be the case, particularly with respect to Mr. Murray's factory; that was the only factory where I made the remark that many of the children were undersized. I wish to explain the reason why I say so: that it appeared decidedly, on our questioning the person or persons under whom the children worked, that those children had been selected because they were unfit for any other employment.

29. You mean selected for the cotton-manufactory?—Yes.

30. When you say selected, I suppose you mean by their parents?—No.

31. Do you mean that the manufacturers chose puny children?—The persons who work under the manufacturers; here are persons who work under the manufacturers, who can correct me if I am wrong. A person, I imagine, in a cotton-factory, from the manner in which our examination was conducted, has under him a certain number of children; he engages the children; and we found that, for instance, a person who brought up some children to be examined by us, on asking him why they were brought into the manufactory at all, he said, that they were the children of poor persons, who could not find any other employment. On mentioning the impropriety of this to Mr. Murray, the owner of the factory, he attributed blame to the man for engaging children of that description. That is all I know of the nature of the employment of those children.

32. Did the man make no explanation of the reasons why he employed them?—None, but the poverty of their parents.

33. Was anything said upon the subject of their being puny children, and unfit for other employments; that they could not get employment in any other trade?—If I may refer to the examination of Mr. Murray's factory, I can state the particular frame, in page 6, which is signed by me. 322, Thomas Walker is mentioned to be aged fourteen: the description we have annexed to him is—delicate, scrofulous, and two years ago laboured under mesenteric affection. 324 is Mary Lynch, of the age of ten, daughter of a poor weaver,

E. Holme, M.D.

and ill fed; ought not to be employed. 328, Michael Lynch, her brother. 325, James Constantine, aged twelve, delicate, son of a poor weaver. 326, Henry Carroll, delicate, son of a poor weaver. I only mention these instances; the children appeared so unlike the rest in this party, that we were led to make inquiry, and the answers given to us were uniformly those stated, that the parents were infirm and poor, and there was no means of keeping the family from the parish but by employing the children.

34. In one of your early answers on this subject, you have stated that one of the reasons assigned was, that they were too puny to get employment elsewhere in anything but the cotton-manufactories?—Yes; I fancy the substance of what I said was that.

Cross-examined.

35. You have stated, as part of your evidence, already, that you are not much conversant with cotton-factories?—That I have not particularly attended to the structure or management of cotton-factories.

36. You are not, then, enabled to speak of the manner in which cotton-factories are worked?—I am not.

37. You have mentioned that you have examined the state of three factories at Manchester, Mr. Murray's, Mr. Pooley's, and Mr. Pollard's?—Yes.

38. Who applied to you to examine Mr. Pooley's factory, for instance?—Mr. Pooley himself.

39. Did he go with you?—The children were brought out and examined in a large room annexed to the factory; a room chosen by myself and the gentlemen who attended along with me.

40. Did you go into the factory?—We went into the factory after the examination was concluded.

41. Had you ever been in the factory before any application was made to you upon this subject, as a stranger, who wished to see how cotton-factories were conducted?—Into Mr. Pooley's, never.

42. When you have been in a cotton-factory, how have you found the children engaged, sitting or standing?—Upon my word I can scarcely say.

43. Perhaps you may have been in there only once, that your attention was not drawn to it?—My attention, perhaps, was more bestowed upon the machinery than the children employed.

44. I might have mistaken you; and if I have, you will have the goodness to set me right; I have taken down, as part of your evidence, this; that you, either when you examined the children, or upon other occasions, found no difference in the health of those children that work during the night from that of the children that work during the day?—That question alludes to what took place so far back as the year 1796, since which night-working has been discontinued in our neighbourhood, except perhaps in a single instance.

45. Allow me, as a professional gentleman and a physician, to put this single question to you as a man of science; would it make any difference in the health of a child, in your judgment, whether it was employed the twelve hours of the night, or the twelve hours of the day?—From theory I should be extremely sorry to answer that; I should say, it must be answered by the fact; and the fact I believe to be as it was stated.

46. In your judgment, as a medical man, would there be any difference in the health of a child that was kept working during the twelve hours of the night, from the health of a child that was kept working during the twelve hours of the day?—I can form no opinion upon that, except what is deduced from facts; and I believe the fact to be, that there was no difference whatever.

47. Have the goodness, as a medical gentleman, to tell me, setting aside any question of fact, with reference to one, two, or twenty individuals, what is your judgment upon that subject as a physician?—My judgment, as a physician, is the inference I draw from facts; I know no other mode of deciding

upon the case, but attending to the generality of facts which have come under my knowledge.

E. Holme, M.D.

48. Then, independent of that, you can form no opinion as deduced from the general structure of a child's frame?—None whatever.

49. If you had the election, would you permit a child of eight years old, for instance, to be kept standing twelve hours during the twenty-four?—I presume my examination is merely medical; I speak merely as to the health of the children; I apprehend I am not come here to answer what I would do if I had children of my own.

50. I was not guilty of such a piece of impertinence as to ask you what you would do with children of your own. If you had the election, would you permit a child of eight years old to be kept standing twelve hours during the twenty-four?—Certainly, if I had my election, I would not.

51. Should you not think, generally speaking, that keeping a child of eight years old standing twelve hours in the day would be injurious to its health?—I conceive I have answered that also; if the children employed in the cotton-factories do stand eight hours, my evidence would go to prove that that is not inconsistent with their health; I believe that it is not.

52. Should you not think a child of the age of eight years being kept standing twelve hours in the day injurious to its health?—I should be able to form no opinion whatever upon the subject, except I knew how it turned out in practice.

53. Suppose I were to ask you, whether you thought it injurious to a child to be kept standing three-and-twenty hours out of the four-and-twenty, should you not think it must necessarily be injurious to the health, without any fact to rest upon, as a simple proposition put to a gentleman in the medical profession?—Before I answered that question I should wish to have an examination to see how the case stood. If there were such an extravagant thing to take place, and it should appear that the person was not injured by having stood three-and-twenty hours, I should then say it was not inconsistent with the health of the person so employed.

54. You, as a medical man, then, can form no opinion, independent of facts, as to the number of hours that a child might or might not be employed, that would or would not be injurious to his health?—I cannot.

55. You do not happen to know, from your own personal knowledge, what time is allowed to children in cotton-factories for their meals?—I do not.

56. Would it be injurious to a child, in your judgment as a medical man, if at the time he got his meals he was still kept engaged in the employment he was about?—Those are questions which I find a great difficulty in answering.

57. I will put it in a more medical form. Is it not detrimental to the process of digestion, that while a person is eating he should at the same time be working?—I cannot conceive how many actions may be carried on at the same time; there are certain actions carried on in the process of digestion, which, I conceive, do not impede the process of digestion.

58. Have you read Dr. Baillie's examination before the Committee of the House of Commons?—I have.

59. And that of the other medical gentlemen?—Some of them; it is some time ago.

60. Do you remember the examination of Mr. Astley Cooper?—I do not remember his.

61. Dr. Pemberton?—I do not recollect his.

62. I again ask you, as a medical gentleman acquainted with this subject, whether, supposing a person, during the time he was eating his meals, was employed in manual labour, is it your judgment that the food will be as nutritious to him as it would be if he were unemployed?—I should imagine that the food would be equally nutritious to him if he did the manual labour of handling his knife and fork.

E. Holme, M.D.

63. Who applied to you to undertake the office of examining all these children?—Mr. Murray, Mr. Pooley, and Mr. Pollard.

64. You had not been acquainted with them at all before?—Mr. Pollard I never saw before, nor Mr. Murray; I had a slight acquaintance in Mr. Pooley's family. I had been the regular attendant for several years whenever medical attendance has been necessary.

65. You did not see the children at work?—We did not; the children were brought into a large room at our request.

66. Not at any of the factories?—In Mr. Pooley's, after the examination was closed, we saw them at work.

67. Did you not know, at the time, that one important question would be, the heat of the factory in which those children were kept at work?—I conceived I had nothing to do with that; that the whole I had to speak to was, the health of the children at the time I saw them.

68. Your attention was solely directed to the individual health of the children brought before you?—Yes.

69. In one of the factories you state that you found the children generally under-sized?—No; many of the children; that was Mr. Murray's factory.

70. In the children of the other factories there was nothing particular?—No.

71. Did the children in Mr. Murray's factory appear to you to be sallow at all?—They had not the bloom of the children who live in the mountains of Wales and Westmoreland.

72. Were they sallow?—Not particularly sallow; like all the rest of the children in large towns, in such a town as Manchester.

73. You said, in some particular instances you had reason to think the children that were too puny to work in other employments were employed in the cotton-factories?—I do not think the word "puny" was my word, but "delicate."

74. Is it, then, your opinion, that children that are of too delicate a texture to be engaged in other factories, may be employed in cotton-factories?—It is.

75. Supposing I put this question to you, for that will be followed up by other evidence, that children in cotton-factories were employed twelve, thirteen, fourteen, fifteen hours out of the twenty-four, should you think that conducive to the health of a delicate child?—One part you must gain from another set of witnesses. My conclusion would be this: the children I saw were all in health; if they were employed during those ten, twelve, or fourteen hours, and had the appearance of health, I should still say it was not injurious to their health; beyond that I cannot say; I am unwilling to give any speculative opinion.

76. I had supposed that to a gentleman of science I could put questions which might be answered independent of facts; am I to understand you can answer that only as your mind arrives at any conclusion through the medium of facts?—That is my answer.

Re-examined.

77. As you doubted whether a child could work for twenty-three hours without suffering, would you extend your doubts to twenty-four hours?—That was put as an extreme case; my answer only went to this effect, that it was not in my power to assign any limit.

78. You said that the population of Manchester was about 120,000, and that the number of cotton-spinners was 20,000?—Yes.

79. Do not parents pay money to clubs, for the benefit of themselves and their children when ill?—I know nothing about that.

80. Do not persons who are disabled in factories go into other trades, and therefore when ill, though that illness may have been owing to their being employed in factories, yet still are they denominated from the trades in which

E. Holme, M.D.

they then work?—I know of no persons disabled in factories who have gone to other trades.

81. You said that the health of the children was equal to that of the working poor in general?—That is my belief.

82. You stated that you examined the factories of Mr. Murray, Mr. Pooley, and Mr. Pollard?—I did.

83. You state that you do not know the hours of work in Mr. Murray's factory?—Not in any of those.

84. You do not know whether the temperature was from 80 to 89°?—I do not know that fact.

85. You state that you inspected Mr. Pooley's factory?—I did.

86. Are you aware that he ordered preparations to be made before your visit?—Except the preparation of the list of names, I am not aware of any.

87. You did not hear that people were to appear clean?—I did not.

88. You were not aware that the temperature of the room was lowered previous to your visit?—I was not; I have stated that the temperature formed no part of my inquiry.

89. You are not aware that the speed of the machinery was slackened?—I cannot speak to that.

90. Were you aware that the hours of work were from fourteen to fifteen hours a day?—I am not aware of that fact, either one way or the other.

91. Is it your opinion, as a medical man, that recreation and exercise in the open air are necessary for growing children?—I cannot certainly give an opinion upon that; I certainly must look upon exercise and recreation in the open air as connected with the health of children.

92. Is it possible that children engaged for fourteen or fifteen hours in work can obtain this recreation and exercise?—I am of opinion that they cannot.

93. Was Mr. Murray acquainted with the day in which you intended to visit his factory, or not?—Mr. Murray was acquainted with the day we were to come.

94. You have stated that Mr. Murray's factory, when you saw it, contained 919 persons?—It did.

95. Did you state that there were very few children among those 919 persons?—The number of children under nine years was comparatively few, but I cannot, at this time, form an idea of the comparative number of the children and grown persons; the children, I believe, were the most numerous.

96. Cannot you say whether there were an hundred?—If your lordships will allow me, I will state why I speak with less certainty upon this subject. We divided ourselves into two bodies; the examination of the adults fell under Dr. Hardie and another gentleman, and most of the children came to my share, so that as to the precise proportion I cannot speak.

97. The whole number you saw were 919?—Yes.

98. Were you aware that there were 1,200 belonging to that factory?—There were 919 in the factory, all the others were employed out of the factory.

99. Do you conceive every one of those persons on the list was in the factory at the time you saw it?—They passed in review before us.

100. Do you believe none were absent from sickness?—They were brought before us in rotation: the persons who were sick are specified.

101. You have stated that the healthiest season observed among the poor in Manchester, was at that period when there was a great cessation of the manufactory work?—That was in the year 1793, the year before I was Physician to the Infirmary.

102. And you conceive that may be referred to the diminution in the power of purchasing liquors?—That has been the general opinion entertained by myself and my medical friends in Manchester.

E. Holme, M.D.

103. Must not you also attribute that to the diminution of the time they were employed in hot rooms?—I think not; the improvement in the health was general; the consumption of liquors was as great out of the town as it was within the town; wages were so low, that with respect to the time employed out of the factories, a greater number of hours labour must have gone to the support of a family than did in better times, when the head of the family could afford to spend a day or two in the week at the public house.

104. You say, in Mr. Murray's factory there were a number of children who were delicate, and who were admitted there because the work was light; what evidence have you that the children were delicate when admitted?—I have no evidence that they were delicate when admitted but the reason assigned by their employer, that they were unfit for any other employment; those were children who could not be employed in weaving, or anything which required active exertion.

105. In a year when the wages are very low, do the poor of Manchester labour more or less than in years wherein the wages are very high?—I believe that when wages are very low there is more labour performed in certain kinds of work; I believe I might say generally.

106. The committee are informed you wish to make some explanation of your evidence?—On my first examination, at the conclusion of my answer to the question, "Were those times appointed with the owners of the manufactories, or with any other medical gentlemen?" I stated, "Lists were made of every person who had been employed in the factories for the last twelve months; three of those lists in which I was concerned I have along with me." I wish to state, that whereas I have made use of the words, "for the last twelve months," I wish to be substituted the following—"at the time of our examination." With reference to a part of my examination of yesterday, I am anxious to explain, that the reason assigned by the man who had under his care the children I specified by Numbers 322, 324, 328, 325, and 326, for employing those children, was, that their parents were poor, and they were children who, from their delicate state of health, were not fit for any other employment, and could not be employed in any other manufactory.

Cross-examined.

107. How old did those children appear to be?—I am not aware that they appeared to be different from the ages assigned to them in the list.

108. Those children being delicate in the way you have described, is it your opinion that the cotton-factory was a proper place for such children to work in?—I have stated, with respect to one of the children, that she was unfit for any employment; the others are stated to be delicate or sickly. If I had thought the same of them, I should have made use of the same expression.

109. I ask you, explicitly, whether you think a cotton-factory is a proper place for delicate and sickly children to be employed in?—I do not know how far any employment may be proper for delicate or sickly children; all I mean to assert is, that children that are unfit for other employments may find employment in cotton-factories without detriment to their health.

110. Am I to understand by that, that there are employments more detrimental to the health of children than the cotton-factory?—I have not stated that the cotton-factory is detrimental to the health of children; more detrimental must mean something worse than that.

111. Do you conceive the cotton-factory detrimental, or not, to the health of sickly children?—I believe not more detrimental than any employment.

112. Generally speaking, with reference to a child of delicate health, and not as a measure of comparison, would you think a cotton-factory a prudent place to put a child of delicate health?—I feel a great deal of difficulty in answering that question, from the term made use of, "delicate."

E. Holme, M.D.

Re-examined.

113. If I understand you correctly, there are different degrees of delicacy in the children whom you have designated by the general term, delicate?—There are.

114. And that there are persons to whom that term might be appropriate and applicable, who do not indicate that state of health to which an employment in a cotton-factory is prejudicial?—Certainly.

115. From your observation in the course of examining the cotton-manufactories, is not the employment of the children more of the nature of occupation that requires attention than of labour which can fatigue the children?—I believe that to be the nature of the case; but I have stated, in former instances, that my knowledge of what takes place in the interior of a cotton-factory is limited.

116. Though you think that the nature of the case for a short time, would you think it of that nature for twelve or fourteen hours?—I should endeavour to ascertain how the fact stood; the presumption certainly would be against it; but if I found that the children were healthy I should carry my inquiries no further.

Mr. WILLIAM ROBERT WHATTON, called in, and examined, 25th May, 1818.

Mr. W. R. Whatton.

1. What is your profession?—I am a surgeon.

2. How long have you been a surgeon in Manchester?—Three years.

3. Have you had, generally, an opportunity of observing whether the work of the children tends to produce bad health among them?—No; I am not aware that that is the case.

4. Whose factory was that you inspected minutely?—A factory belonging to Peter Appleton and Company.

5. What was the occasion of your inspecting the spinning factories?—An order from the chairman of the committee of cotton-spinners.

6. At what time of the day was it you inspected those forty factories generally?—At all times, from five in the morning till late in the evening.

7. Did you observe any particular symptoms of disease or languor about the children?—No; I cannot say that I did.

8. What was the highest temperature you observed?—In one sole case there was a temperature of from 78° to 80°.

9. What was the lowest you observed?—Sixty.

10. What was the general temperature you observed, the average?—The general temperature appeared to be from 60° to 75° or 76°. Taking the whole number of factories together, I should put the average at about 70° or 72°.

11. Do you know whether the scrofula, or any particular disease, prevails more generally in factories than in other mechanical employments?—I do not think it does.

12. I will put the case of a tailor, whose sons are apprentices, and at work on the shop-board cross-legged the whole day; do you think that occupation, compared with that of a spinner in a factory, has the advantage of the spinner?—No; I think the spinner has the advantage.

Cross-examined.

13. You have stated to their lordships that the labour you saw was very moderate; what do you mean by the term labour?—The work that they did in the factories.

14. Then you confine that expression to the mere manual operation they were about?—No; to the general appearance of the work which they did.

Mr. W. R. Whatton.

15. Should you think it great labour for a child to be employed twelve hours or not, on what you saw them doing the short space of time you were there?—After a certain age I should not.

16. Should you not think that if a child of eight years old was kept standing twelve hours in the day in the work you saw them about then, it would be injurious to their health?—To a child of eight years I should.

17. If a child was of a delicate constitution, you would probably think it was too long to keep him at work?—The labour is so moderate it can scarcely be called labour at all; and under those circumstances I should not think there would be any injury from it.

18. What do you think, as a medical gentleman, of keeping a child standing, doing nothing, for twelve hours, where there was no labour?—I think it would be a very hard task.

19. Supposing he had the day to rest in, to go to bed and sleep, should you not think it injurious to the health of a child, having an opportunity of rest during the day, a child of eight or ten years old, to be kept at work during the night?—I should think it was.

20. What should you think a fair time for a child of eight years of age being employed in a cotton-factory; supposing you were asked what would be a proper number of hours to employ a child of eight years old in such a work, what should you say?—I cannot say.

21. Can you not give a judgment, as a man of science, knowing the nature of children's complaints?—No; I cannot form any idea.

22. You would say there must be some limit to the time they must be employed during the twenty-four hours?—Yes, I should.

23. What is that limit in your opinion?—I cannot state what would be the proper point for a child of that age.

24. Perhaps you would think a child of eight years of age ought not to be employed at all?—No, I should not give that opinion.

25. What would be the number of hours that such a child, generally speaking, ought to be employed?—My experience does not furnish me with an answer to that. I do not think I have seen any at work so young.

26. Take the case of a boy of ten years of age, a child of ordinary health; how long ought he to be employed in this species of labour, consistently with attention to his general health?—I cannot give a decided answer as to the number of hours he ought to be employed.

27. Nor a boy of twelve?—After twelve years of age, a person employed in a cotton-mill cannot be better employed.

28. How long should you say it would be safe for him to be employed?—I should say twelve hours; taking the practice generally, that seems to be the time; from twelve to thirteen hours I should say.

29. Do you happen to know whether or not any particular effect is produced upon a child's frame that is kept standing longer than his strength will permit, or rather, than he ought to be subjected to?—I am not aware of any effect.

30. You have no reason to know whether, in point of fact, if a child is kept standing longer than his strength will permit, it will produce a diseased affection of the sinews of the knee?—I have never seen that.

31. What is your judgment, if a child is kept standing longer than his strength will permit, what would be the effect upon his lower extremities, or would it produce any effect, in your judgment as a medical man?—Yes, I think it might.

32. What effect?—That it would weaken him.

33. Would it not produce an enlargement of the joints of the knee?—No, I am not aware that it would.

34. Would it not, in your judgment, produce an unevenness of the length

of thigh, and occasion a ricketty appearance?—I am not aware; I am sure I cannot say. Mr. W. R. Whatton.

35. Do not you think it would?—No; I think not, at the age I have stated.

36. Eight or ten?—Ten or twelve.

37. I will take the instance of a young person of eight years of age; would not the instance of a young person of eight years old, kept standing for twelve hours during the day, be likely to produce a ricketty appearance?—No; I should think not.

38. Would it not be likely to produce a diseased affection of some sort or other of the lower extremities?—It would, perhaps, produce a curvature of the extremities standing for twelve hours at eight years of age.

39. Did you go over the same factories the first time you went?—The first time I went by order of the proprietors was to Peter Appleton and Company's factory; that was a little before the month of May, I think.

40. Did your first examination of factories arise out of an application of the proprietors of factories?—The examination I allude to was by order of the proprietors of the concern; but I had been in factories before, from my own motives of curiosity.

Re-examined.

41. You have been asked whether a child being kept twelve hours to labour is not injurious to health; do you know of any species of labour to which an application of the same time would be less injurious to health than in a cotton-factory?—No; I do not know any.

42. What should you think of a member of your own profession confined in an apothecary's shop for twelve or fourteen hours in the day?—I should think it rather harder work than the factory.

43. The general purport of the evidence you have given has been to shew that other trades were more unhealthy; are the committee to collect, from what you say, that cotton-spinning is not injurious to health when carried on for more than twelve hours by children under sixteen years of age?—It does not appear to me, so far as my observations have gone, to have been injurious.

HENRY HARDIE, M.D., called in, and examined, 25th & 26th May, 1818.

1. You are a physician at Manchester?—I am.

2. How long have you practised as a physician at Manchester?—Seven years. H. Hardie, M.D.

3. Are you a physician to the Infirmary, and other public charities there?—I am.

4. Have you taken any general survey lately, and with reference to this inquiry, of spinning-factories in Manchester?—I have.

5. Did that arise from any particular application to yourself, or from any general application to the medical men in Manchester, to make a survey and communicate their observations?—I had an application made to myself by the chairman of the committee of cotton-spinners.

6. What observation did you make with regard to the state of health of the adults and children employed in those seven factories you have mentioned?—That it was generally good.

7. What was the state of ventilation of the several factories you visited?—It was good.

8. Have you any result of those several documents?—The total number, in Appleton's factory, was four labouring under scrofula, and three slightly distorted.

9. Does that include persons of all ages?—Yes.

H. Hardie, M.D.

10. Have you the ages of those specifically in those situations?—They will be found in the general table. In Simpson's, the total number of hands was 267; out of that number there was one scrofulous case, and one case of distortion. In Murray's factory, 919 persons, fifteen were scrofulous, and eight distorted. In Benjamin and William Sandford's, 139 total; no scrofulous case; three cases of distortion. In Thomas Houldsworth's the total number was 777; there were ten cases of scrofula, and eleven cases of distortion. In Jonathan Pollard's the total was 450; eight cases of scrofula; no cases of distortion. In Birley and Hornby's the total was 1020; nineteen scrofulous cases; eight cases of distortion. The total of all those factories was 3841; scrofulous cases in that total 57; and cases of distortion 34.

11. Have you found that persons who have grown up in the business of cotton-spinners are generally less healthy than those who have followed other occupations, either of agriculture or manufacture?—No; I have not found them worse looking.

12. Have you found that the infirmities of age came on sooner than with persons in general?—No; I have not.

13. Have you had an opportunity of comparing the general average of mortality and disease in the large towns of Manchester and Stockport, and the general mortality of the country?—No; I have not.

14. In the survey you have taken, does it appear to you that there is anything in the nature of the occupation, or the duration of it, as it is, in fact, carried on, that is injurious to the health of the persons employed in it?—I do not believe that there is anything in the employment prejudicial.

Cross-examined.

15. I observe that, in the two dirty and ill-ventilated factories, the temperature was at 80, and the hours per week eighty also?—In the first mentioned, that was the case; and in the second, it was only 68 degrees, and the hours of labour seventy-five.

16. Do not you regard both the temperature of 80, under those circumstances, as well as the hours of eighty per week, extremely injurious to infant subjects, if they were all in those states?—I do not think that a temperature of 80 is injurious, provided the air be pure; but it was not so in these two instances.

17. As to the number of hours, eighty per week, are not they injurious to infant subjects; you have singled out that with some degree of reprobation?—I believe you will find I did not single out that; I have taken the account in all the factories.

18. Is that or not an injurious degree of employment?—I cannot speak to that; I did not examine minutely the state of their health. I do not know what age is meant by infant subjects.

19. From six to twelve?—Perhaps for children of six years old it might be; at the age at which they generally work at factories, I should not suppose it was.

20. At what age do you think it would be perfectly safe to the constitution of an infant, working in the temperature of 80, to work eighty hours per week?—I have no fact to guide me in replying.

21. You do not feel capable of answering that question?—No; I do not.

22. You state yourself to have been in the habit of visiting factories; do you mean by that previous to this special visitation of forty generally and seven specifically?—Yes.

23. For how long before this?—Since I came to Manchester; within these seven years.

24. Some of them so far back as two or three years?—Yes, several. I did not at that time visit the factories professionally, but from motives of curiosity; principally with strangers coming to Manchester.

25. Forty you have visited by special desire?—Yes. H. Hardie, M.D.

26. By the desire of whom?—Of the chairman of the committee of mill owners, Mr. Douglas.

27. When was that desire conveyed to you?—I do not recollect the date, but it must have been in this month.

28. Some time this month?—Yes.

29. I speak now of those you visited generally; what proportion of young children might there be from eight years of age?—I think the number altogether was 13,000, and there were in that number about 200 under nine years of age.

30. Whether any were absent from the forty from illness of any description you cannot say?—No; I cannot.

31. Were any of the gentlemen, or their overlookers, aware that they were about to be visited?—I do not believe they were.

32. They were aware that a general order had gone out?—I cannot say.

33. Have you any doubt about it?—I believe they had not; for I met with considerable difficulty in gaining admission to several of the factories among the forty.

34. Did you find in the rooms any considerable degree of effluvia, arising from the hours employed, or the breath of so many persons together?—There were only two factories that were at all disagreeable out of the forty.

35. Out of the forty you found only two offensive, as to effluvia; all the others perfectly well ventilated, and no dust flying about, that occurred to your observation?—None.

36. How many rooms, in each of those factories, might you have taken the trouble to visit, in this rapid course?—We generally went into the middle room, equally distant from the top and bottom of the building, and sometimes, in going up, looked into the rooms we passed.

37. Do you mean that you fairly went into the rooms and inspected them?—Yes.

38. Then you inspected more than the middle rooms?—Yes, in some; but that was not general.

39. When you confined your observations to the middle room, were the children brought down to you to enable you to ascertain their healthy appearance?—No.

39.* If you visited only the middle room of the factory, how were you able to ascertain that the health of those children was good?—I could ascertain that only by their appearance.

40. If you visited only the middle room of the greater part of those factories, how could you ascertain that the appearance of the whole of those children was healthy?—I took that as the average of the health; I was not aware that the middle room was likely to be different from the others.

41. You went into the middle room and judged of the health of the whole factory from the appearance of the middle room?—Yes.

42. Had you any means of judging whether the children in that room had had their course directed by the overseer, or whether they worked habitually there?—I believe we gave them no time to make that arrangement; we rushed up stairs to that room, and afterwards sent from that room for the overseer to come to us.

43. How many hours in the day do you think children, from six years of age to twelve, may be employed in a temperature of 80, at an employment which requires them to stand much the greater part of the time, consistently with safety to their constitution?—I cannot answer that question; I have no fact to direct me to any conclusion.

44. You mean you can give no general opinion, unless you have the fact of their personal appearance stated as the consequence?—Certainly.

H. Hardie, M.D.

45. If a person were about to institute a factory, and wished humanely to be satisfied upon that subject, and asked you, as a medical man, how many hours a day you thought that they might be employed with safety; under those circumstances, what answer should you give?—I should say I could not tell; and for the same reason that I could not answer the last question.

46. Is a daily employment of thirteen or fourteen hours a day, in an erect position, and in a temperature of 80, consistent with safety to the constitutions of children from six years of age to sixteen?—I cannot answer that question, because I know of no employment wherein the children, and persons between those two ages, are in an erect position the whole of the day.

47. Supposing the fact to be, that, while occupied, they are standing or upon their legs during that period, can you make any other answer?—I cannot.

48. You were accompanied in this survey of forty factories by whom?—By Mr. Whatton.

49. Did Mr. Whatton, like yourself, confine his observations principally to the middle room?—He did; he accompanied me through the rooms that I entered; he was always with me.

50. By whose desire did you visit particularly the seven factories which you have enumerated?—By the desire of the several proprietors.

51. When did they desire you so to do?—At different times.

52. When they desired you to visit their respective factories, did they consult your convenience as to when your leisure would admit of your going?—I do not know that. I was sent for to visit the factories, and I went when I was sent for; but I do not know that they consulted my convenience in the time they appointed for the visitation.

53. Did you understand that a certain degree of high temperature is essential to the manufacture itself?—I have heard so out of factories, but never upon visiting them; I made an inspection of Mr. Houldsworth's factory, who spins the finest quality of twist spun in the country, and which is used in manufacturing the Nottingham lace, and his factory was found to be at 74°; and from all the inquiries I made of the people who worked in the several rooms where those fine numbers were spun, the heat was seldom or never above 74; and when it was, that extra degree of heat was not necessary.

54. Do you think there is any material difference in the salubrity of employing children to do night-work or day-work?—I am not aware that there is any night-work carried on in the factories in Manchester.

55. Supposing that one set of children are employed continually to do night-work, and another set employed to do day-work, as a medical man, do you think there would be any material difference in the effect upon their health respectively, whether the whole of their employment were by night or by day?—I should wish to be regulated in the opinion I give by facts, and I have no fact to go upon, and therefore cannot give an opinion.

Re-examined.

56. Who made the appointment for the time of visiting?—Myself.

57. Did you give previous notice of the time when you should attend for that purpose?—I did not.

58. In the visit of the forty factories, you stated that you generally went into the middle room, and sometimes into the others?—Yes.

59. Something has been said about dust and flue; are you of opinion that the flue and waste of cotton can be inhaled into the lungs so as to be injurious?—No, I am not.

60. Are you aware of any instances in which it has been done to the prejudice of the health?—I am not.

61. You are not aware that the flue they inhale might be injurious?—No, I am not.

H. Hardie, M.D.

62. Do you not think that the finer parts of the cotton and flue, for instance, if inhaled, would be injurious to the health of a young person?—I do not, from what I have seen, suppose the flue, when inhaled, would be injurious to a young person.

63. Do you think that if, day after day, the flue of cotton was taken into the lungs, and inhaled with the breath, that would be prejudicial to the health?—I do not; because the daily expectoration throws off the cotton; there is no accumulation takes place in the lungs.

64. Then if it were not for that throwing off, by the means of expectoration, you are of opinion that it would be injurious?—Yes, I am; it would eventually fill the lungs up.

65. Even with the assistance of that relief, such as it is, is it your opinion that a boy of ten years old, could, day after day, be inhaling this matter, and not receive any injury to his health?—Yes; I speak from what I have seen.

66. You have told the committee there were two factories very dirty, in which the people, according to your observation, were less healthy; name the persons to whom those factories belonged?—To Benjamin Potter and John Birch, junior, of Manchester.

67. Do you know whether the magistrates, under the power given to them, have visited those two factories?—I do not think, from the difficulty I had of entering one of the factories, that of John Birch, junior, that the magistrates could find their way there; there was no outer door to the place, it was bricked up.

68. Did you state your opinion to those gentlemen?—I stated it to the overlookers; I did not see the proprietors.

69. Did you make any inquiry whether these factories had been visited as the law directs?—No, I did not; but I concluded from the appearance, that the law had not been enforced as to whitewashing and cleansing.

70. There was no appearance of its having been lately whitewashed?—No, there was not.

71. Did you, on examining the several cases of distortion, perceive anything which led you to conclude that that distortion was occasioned by the employment in the cotton-works?—No, I did not.

72. Did you select the seven factories which you have visited particularly?—No, I did not.

73. Who determined as to the seven which were to be visited particularly, out of the forty-seven which were to be visited?—The several proprietors.

Mr. THOMAS WILSON called in, and examined, 26th May, 1818.

Mr. T. Wilson.

1. What are you?—A surgeon and apothecary.

2. How long have you been practising as a surgeon and apothecary?—Twelve years and a half.

3. Where is the usual place of your residence?—A place called Bingley, in the West Riding of the county of York.

4. Are you employed by the proprietors of the cotton-mills there?—A part of them.

5. Have you been called in to attend their people in case of sickness?—Frequently.

6. Have you lately made any examination of the cotton-mills at Bingley, or near it?—Yes.

7. Have the goodness to state when?—On the 15th of the present month.

8. What induced you to make an examination at that time?—I was desired by Mr. William Ellis to examine into the state of health of the people.

9. Is Mr. Ellis an owner of any cotton-mill?—One.

Mr. T. Wilson. 10. What is the name of the cotton-mill of which Mr. Ellis is owner?—Castlefield.

11. In consequence of the expression of that wish, what did you do?—I went throughout the mill, and examined the people employed.

12. In what state of health did you find them?—They were all following their respective employments; there were none absent from sickness.

13. Not one out of the ninety-eight?—No.

14. When you say the number is ninety-eight, what are your means of information as to numbers?—I went through the mill, and I counted the number myself.

15. And to the best of your observation and recollection, the number was ninety-eight?—Yes.

16. Have you a memorandum that the number was ninety-eight?—Yes, I have, in my hand.

17. How many children were there under nine years of age?—Two.

18. And, according to that information, you learnt there were two children under nine years of age in Castlefield mill?—Yes, only two.

19. What was the general appearance as to health?—The general appearance was healthy.

20. Some not so healthy as others?—Some looked rather delicate.

21. Did you see any deformity among them?—Only one was deformed.

22. How many persons did you find employed at Messrs. Knight's?—One hundred-and-twenty.

23. Did you take the same means of ascertaining the number there as at the former mill?—Yes.

24. How many were there under nine years of age?—Six.

25. In your judgment, as a medical man, is scrofula a frequent disease among the labourers in cotton-factories?—No; it is very little seen.

Cross-examined.

26. Scrofula arises from cold and damp principally, does it not?—There are several causes.

27. Those factories were principally in villages, were they not?—They were in the neighbourhood where I exercise my profession.

28. Were they in villages?—Not in villages; they were rather detached from the village I practise in.

29. And they were not in any large town?—No; not in any town.

30. There were 570 persons in the whole, you examined?—Yes.

31. And only one ill; and that one got well in the extraordinary manner which you have described, either charmed or frightened at the sight of the doctor?—She was almost well.

32. You would consider that an instance of remarkable good health, that out of 570 manufacturing people there should be only an instance of one person ill?—It was certainly something extraordinary at the time.

33. Did you ever meet with such an instance before in your life, in such a number of manufacturing people, (570 persons of different ages,) as only one to be ill?—I never had occasion to make a similar survey.

34. Who applied to you to visit those different factories; was it one person, or the different masters of the different factories?—One person, Mr. Ellis.

35. How long might it occupy you?—About ten hours and a half, the whole.

36. Were the factories close to each other?—At a little distance.

37. It was rather a rapid survey, I should think, you must have taken of the different factories, to have seen 570 people in ten hours; did you ever examine the health of so many patients in so short a time before; you had 570 persons to examine as to their state of health, and to go from factory to

factory to do it?—No; I never examined so many in the same time before. Mr. T. Wilson.

38. Then your examination must have been rather cursory and slight?—It was rather slight.

39. It was very slight, was it not?—No.

40. The question refers to six or seven factories to be visited, containing 570 persons, and the whole passed through in ten hours, was not your examination very slight indeed?—I could not exactly examine them so minutely from the time as perhaps I was requested to do.

41. Or as, perhaps, you might think necessary to form a decisive opinion as to the state of their health?—The appearance of all of them was good.

42. Am I to understand that the judgment you have given of their health was from their appearance, or inquiry of persons brought before you?—The method I took to examine was, that the foreman of each department went with me, and I asked him questions. There were eight, or ten, or twenty, or thirty, altogether; I asked them whether they had any swellings, any ulcerations, any stiff joints; or whether they could eat well and sleep well; or whether there was anything detrimental, or which prevented their following their employments with ease and comfort. Those I put to them individually.

43. Was that the way in which you made the inquiry through the different factories?—Yes.

44. And there was one that you found ill?—Only one.

45. You perhaps cannot tell me, either one way or another, whether any persons had been removed from the factory before you came there?—I cannot speak to that.

46. You found fifteen under the age of nine?—Yes.

47. You cannot therefore, probably, tell me whether any more persons of that age had been removed from the factory, previous to your coming there?—I cannot speak to that.

48. Should you not think that a lad of the age of fifteen years was sufficiently employed, if he was kept at work twelve hours out of the twenty-four, anywhere?—I never heard them complain of being overworked.

49. The question to you, as a medical man, was, whether you should not think twelve hours a sufficient time for a lad of fifteen years of age to be employed during the day in any light work, as light as you please?—Yes.

50. Should you not think a greater number of hours than that, even though the employment was of the most moderate description, if it occupied the attention solely, injurious, generally speaking, to the health of the person?—Yes.

51. Is it not, in your judgment as a medical man, necessary that young persons should have a little recreation or amusement during the day; is it not contributory to their general health?—I do not see it necessary.

52. Your opinion, as a medical man, is, that a boy of fifteen years old might be kept under a constant course of attention, day after day throughout the year, with the intermission of Sunday, without injury to his health?—Yes.

53. Should you think it would be a beneficial thing for his health, if he were kept fifteen hours out of the twenty-four employed, without amusement, or recreation, or intermission?—No.

54. Then, in your judgment, twelve hours is the extent at which, in prudence, you would think a person of fifteen ought to be so employed?—Yes.

55. Would you not allow, out of those twelve hours, an hour for his dinner?—No.

56. You would take the twelve hours, exclusive of the hour for dinner?—Yes.

Mr. T. Wilson. 57. Probably, exclusive of any time for his breakfast?—Yes.

58. What is the earliest time, in your judgment, supposing he should be employed twelve hours, neither speaking of a very robust nor a very delicate boy, that a lad of fifteen years ought to begin his employment in winter?—About six.

59. About six in winter?—Yes.

60. An earlier hour, probably, in summer?—Yes, half-past five.

61. How long have you practised as a medical gentleman?—Twelve years and a half.

62. In your judgment, as a medical man, is it not injurious to the health of a young person, of ten years, to keep him at work during the night, even though he has rest during the day?—I have never found it the case; at least I have not noticed it.

63. Perhaps no instance has come under your observation in which the thing has happened?—No.

64. But the thing not having happened, as a matter of science, what is your judgment of the point; is it not a detrimental thing to the health of a child to be kept at work during the hours of night?—More so than during the day.

65. Have you the least doubt of it?—Not the least.

66. Is it your opinion that a boy, of ten years old, could be kept in a room, heated to 76, thirteen hours and a half; and that then he can pass into an atmosphere as cold as 40, without injury to his health, day after day?—I have never found that detrimental.

67. What is your opinion, as a medical man, upon the subject?—It might be injurious.

68. Considering the situation in which you stand as a witness, have you the slightest doubt of it?—I think not.

69. Should you not think it a dangerous thing to a young person to be from day to day inhaling the finer particles of the filaments of cotton?—No.

70. You think it would not be injurious to the lungs at all, to be receiving, day after day, those particles of cotton?—No.

71. Do you think it would produce no effect at all upon the lungs of a young person?—I think not; or very little.

72. Be so good as to state how the constitution would be safe under such circumstances, from receiving those things into the lungs?—Expectoration is occasioned, which brings it back again.

73. Is not a constant state of expectoration injurious to health?—No.

74. Would not a constant state of expectoration be injurious to the health of a very young person?—Not a slight expectoration.

75. Have you ever been present with a young person when he has been working in the carding-room, and imbibing those particles?—Yes, frequently.

76. Are those persons unhealthy at all?—No; merely delicate, like some others in the factory.

77. To what do you attribute their delicacy; to natural ill health, or that brought on by their employment?—Their natural appearance.

78. You do not refer that to the employment they have been engaged in?—No.

79. Am I to understand you that that would have been their appearance if they had not worked in a cotton-factory?—Yes.

80. Even though they had been employed in agriculture?—Yes.

81. Be so good as to state upon what grounds you form that opinion?—It is the natural appearance of the children, of course. I see children look very pale that are employed in agricultural pursuits as well as in factories.

82. And as many?—I think equally.

83. It is your opinion, as a medical gentleman, that children employed in

the open air, and a free circulation of air, would look not a whit better than the children confined in those cotton-factories, even though it is twelve or thirteen hours a-day?—I think so. Mr. T. Wilson.

84. What species of yarn was worked in those factories you visited, where the 570 persons were; was it very fine?—I cannot answer that question.

Re-examined.

85. What number of hours in general do the children work in those factories you visited?—About twelve hours; from that to twelve and a half.

86. Supposing any of those children to have been ill, would they or not have been dismissed from those factories?—No.

87. You know the state of the health of those children; are there not usually more ill in those factories than there were at this particular time at which you visited?—I think in general there are.

88. You said that expectoration, if slight, is not injurious?—Yes.

89. Do you not think the slighter the better, and better still if not at all? —A degree of it is in that case beneficial.

90. You say that those children, confined for the time they are, are as healthy as other children; that they are equally healthy with agricultural children; do you mean to state that air and exercise is not requisite for the health of children?—The children are equally healthy in those factories as in other employments.

91. You state that you have examined factories containing 570 children; have you been in the habit of examining other children to that amount in the neighbourhood in which you live, and what was the comparative health of those children?—The state of health of children employed in cotton-factories is equal to that of children in any other capacity.

92. Have you been in the habit of seeing as many children employed in agricultural pursuits, or in no pursuit, but with their parents?—Not at one time.

93. You have at different times?—Yes, I have.

94. And you think that the one were as healthy as the other?—Yes.

95. Have those children employed in factories air and exercise?—They have sufficient.

96. When they work twelve hours in the day, have they exercise in the open air?—They have as much as is necessary.

97. Do not you think that the children have air and exercise enough in the factories?—Sufficiently.

98. You have said, in the course of your examination of those 570 you saw, that they were healthier than at other periods?—Yes.

99. Is not the month of May esteemed, among medical men, a healthy month?—Yes, I think it is in the place in which I reside.

100. Have you many patients in the course of the year out of those factories?—Very few.

101. Have you any idea how many, upon the average, you have out of those factories?—I have never taken notice of that particularly.

102. Do you suppose you have fifty?—I collected the statement from the manager of the mills, and the last twelve months there were only twenty-one absent.

103. Do you attend them all, or is there any other medical practitioner? —There is another medical man, but he is aged and infirm, and confines his practice to a few families.

104. Do you suppose that is as great a number as you attended in the course of the year out of the factories?—Twenty-one were absent last year from sickness; that was the information I collected from the owner of the mills.

Mr. T. Wilson.

105. Do you suppose that was as many patients as you had last year, of persons belonging to the factories?—I cannot exactly state.

106. Were you called in to more patients than twenty-one, of people belonging to the factories?—I cannot answer that question.

107. What was the employment of those children who you say were resting themselves; were they piecers, or what?—I really do not know their employment; I did not notice their particular employment.

108. Cannot you pretty nearly state the average of children you attend from those factories in the course of a year?—I cannot give an answer to that precisely; not more than thirty or forty, I should think.

109. And at this time you visited them, there happened to be only one ill?—Only one.

110. Are you in the habit of attending most of the factories?—The greatest part of the factories.

111. How long did you see any of them resting?—I cannot tell that, for I just passed gently through the mill.

112. Is it in the winter months that the children of poor people are more ill than in summer?—I think there is very little difference.

113. Can you speak to the average number of deaths out of those factories?—I cannot; but the number is very small.

Mr. WILLIAM JAMES WILSON, called in, and examined, 29th May, 1818.

Mr. W. J. Wilson.

1. Did you, about the 18th of May last, visit any cotton-factories in Manchester?—Yes.

2. How many did you visit?—Sixteen.

3. What induced you to go to them at that time?—I was requested by Mr. Douglas.

4. Who is Mr. Douglas?—He is a merchant in Manchester, and chairman of the committee of cotton-spinners.

5. What is that committee appointed for; what committee do you mean?—I understand it is a committee for the investigation of the business in regard to cotton-factories; I know nothing of it prior to that.

6. At the request of Mr. Douglas, the chairman of that committee, you visited sixteen cotton-factories?—Yes.

7. All in Manchester?—Yes, all in Manchester, or in the immediate neighbourhood.

8. Go on to the next mill, and state how many there were under nine years of age?—In the Commercial Mill there were three children under nine years of age; in Messrs. Smith and Guest's there were none under that age; in Messrs. Lloyd and Company's, none; in Gatley's Mill (occupied severally by Messrs. Stone, Stubbs, and Mayers, and William Lewis), in Stone's part of the mill, none under nine; in Stubbs and Mayers, one; in William Lewis's room, none.

9. Can you state how many persons there were altogether, those under nine as well as above, in the first factory?—In Samuel and Henry Marsland's, total number, 42.

10. How did the general health of those you saw appear?—Very good; the general appearance.

11. What measure did you take to get information as to their general health?—We merely went round the room, and took a general survey; if anything particular occurred to us, we had that individual down to examine him; if there was anything in the general appearance, or any thing that induced us to make particular inquiry, that individual was examined; otherwise it was a mere general survey.

Mr. W. J. Wilson.

12. Did you observe any appearances of deformity or lameness?—I have no instance, in my minutes, of any the first day.

13. What were the number of hands in the first mill you visited on the 19th?—In Buchan and Shaw's, the total number of hands, 92; in William Mitchell, Esq. and Company's, 540; in Taylor, Weston, and Company's, 300; in Brothertons, Harvey, and Company's, 140; in Slator's, 220; in Richards's, 120; in Hughes's, 170; in Marriott's, 420; in Brown's 50; in Smith's, 110; in Parry and Seton's, 175.

14. How many under the age of nine did you find in those factories respectively?—In Buchan and Shaw's, one child under nine years of age; in Mitchell's, six; in Taylor's, four; in Brothertons', none; in Slator's none; in Richards's, two; in Hughes's, five; in Marriott's, twelve; in Brown's, none; in Smith's, five; in Parry's, six.

15. Do you find lameness frequently affect them?—Not very frequently.

16. Out of the 428 children whom you have stated to be in those schools, how many cases of distortion were there?—Eight.

17. Boys or girls, or both?—Both.

18. Were they slight or otherwise?—Very slight.

19. Do you know from what they had arisen?—I cannot tell; some of them were ricketty cases, and others from disease.

20. Did you, as a medical man, see anything from which you could infer that that distortion had arisen from their employment in the cotton-factories? I think not.

21. Did you see any distortion in the other children in the Sunday-school? —That was not noticed at all.

22. Did you ever petition in favour of this bill?—I did.

23. You seem to have altered your opinion since you petitioned in favour of it?—Certainly.

Cross-examined.

24. You were formerly a petitioner in support of this bill?—Yes.

25. At that time you "feelingly deplored the sufferings of those who were thus employed?"—Certainly.

26. You have had very good reasons, I dare say: one ground of this lamentation was, "the protracted, unreasonable, and destructive extent of employ in point of time;" that was another of your reasons, was it not?—I cannot say that I read the petition; the petition certainly was not read by me prior to signing it; the whole of it was not read.

27. Lest I should be mistaken, what is your Christian name?—William James.

28. In what part of Manchester do you reside?—In Spring Gardens.

29. Are you in the habit of signing petitions that neither meet your eye nor your ear?—No; it was the first petition I ever signed in my life.

30. Did this petition either meet your eye or your ear; the substance of it?—I understood that this was a petition to abridge the labour of children in factories; that they were very much overworked; that they were subject to a vast number of diseases in consequence of that; and, as a friend to humanity, I certainly signed this petition.

31. You say your inducement for signing the petition was a belief that the children were overworked; how many hours did you then suppose they were worked?—Fifteen hours a day.

32. Allowing how much for meals?—That is a thing I have not thought about at all; the sum total of labour I took to be fifteen hours a day.

33. Of sheer labour?—Yes.

34. Without any allowance for meals?—Yes.

35. Could a gentleman of education, like yourself, suppose that they were

Mr. W. J. Wilson.

worked in any factory incessantly fifteen hours, without any time of refreshment?—No, certainly not for fifteen hours, incessantly labouring without food.

36. How many hours did you apprehend that they were labouring?—I cannot tell.

37. Upon no better premises you signed the petition?—Certainly.

38. Have the goodness to look at that, and see whether it is a copy of the petition to which you put your signature?—It very probably is; but I cannot say that I have to this day read it completely through.

39. Was it a petition to the Lords or the Commons?—I think to the Commons.

40. Cast your eye over it (*handing a printed copy to the witness*), and state whether that is, to the best of your belief, a copy of that petition you signed?—The witness inspects the paper, and says, I believe it is; I cannot tell exactly.

41. Is it your present opinion that there is no lengthened duration of employment, among the children of the factories, that is highly prejudicial?—I have no evidence that there is.

42. How long after your signing this petition was it, that you were applied to to visit those sixteen factories?—I cannot say the time; two months or six weeks, probably; I am not positive; I may be wrong for three or four weeks.

43. By whom were you applied to to visit those sixteen factories?—By Mr. Douglas.

44. Mr. Douglas, the chairman of the cotton-spinners?—Yes.

45. You did not consider it as what, in our profession, we call a retainer on the other side?—No.

46. But you was desired merely to visit those sixteen factories?—I will state my reasons why I did. I was told by Mr. Douglas, and two or three others, that I had signed a petition which contained a mis-statement of facts; that the statement contained in that paper was not correct, and that I, along with several others, had signed my name to a paper that was likely to do them injury, or at least that was not pleasant to their feelings; he therefore repeatedly requested that I would examine into those things for myself, and make a report to the committee, and after repeated solicitation I did it; I was requested four different times before I would comply with the wishes of those gentlemen.

47. How many hours would you recommend children from six to sixteen to be employed in an occupation which should induce delicacy and paleness of appearance?—I can give no opinion upon that whatever.

48. Would you think thirteen hours too long to be consistent with safety to their constitutions in a temperature of 75°?—I cannot tell.

49. In making this research, which you did in order that you might be enabled to judge for yourself, you did not even contemplate the question, how long children of those tender ages might be safely and constantly daily employed in such a factory at so high a temperature?—I only go to the facts as I found the children; it would have been impossible to have done that, from the great variety of the constitution, and a variety of other circumstances; it would have taken a man a year to have said how long each particular child might labour; I cannot draw a general line.

Re-examined.

50. You have been speaking of the actual hours of work; have you made allowance for the time of dinner and breakfast; do you include or exclude them?—The hours of labour are hours of actual labour; those I have put down are the hours of actual labour.

51. Without regard to the hours of refreshment?—Yes.

52. You stated, that when you signed the petition in favour of the bill, you

did not read the whole, but a part; do you think, on reading the petition, you should be able to point out the part you did read?—I should not. Mr. W. J. Wilson.

53. With regard to the number of hours, it appears that the greatest number of hours is $76\frac{1}{2}$?—Yes.

54. The smallest number 72?—Yes.

55. The average number is 74, or somewhat more?—Yes.

Mr. GAVIN HAMILTON, called in, and examined, 29th May, 1818.

Mr. G. Hamilton.

1. I believe you are a surgeon?—Yes.

2. How long have you practised in your profession at Manchester?—Eight and twenty years.

3. Did you sign a petition to the House of Commons in favour of this bill?—I did.

4. In consequence of the impression made by your name and others being to that petition, were you requested by any of the cotton-spinners to investigate the facts?—I was requested by Mr. Birley's manager to inspect the mills; and he said, as I had signed the petition, he would rather have me, on that account, to look to the state of the people.

5. Did it appear to you that the children looked more pale and unhealthy than children of the labouring class do in a great town?—They certainly looked pallid, but not more so than children in a great town; they did uot look like children employed in agriculture; but they made no complaints.

6. Did they exhibit any symptom of lassitude, and being exhausted by the species of labour?—I did not find it so.

7. If you had known as much on the subject as you do now, would you have signed that petition?—I certainly should not.

8. Did you read the petition which you signed?—I did not read it altogether; I signed it merely upon the idea that the employment of children from the early age of eight, for such a length of time, was an improper thing; and it was stated to me to be fourteen hours.

9. Without reading the petition particularly, was it explained to you that the object of it was an abridgment of the labour of infant children?—Yes.

10. Did you examine Messrs. Ewart's also?—I did; I have the report of that in my hand.

11. In what state was that factory?—In an admirable state indeed.

12. Did you observe any who, according to inquiries you made, had been many years working there?—A great number.

13. Who had been in that employment from their infancy?—There are some noticed in Mr. Wilson's report, of which I have a copy, who have been twenty-eight years in the works.

Cross-examined.

14. You have been asked whether you were one of those who signed the petition in favour of the bill, and you have stated that you were one of those who signed a petition in favour of the bill?—Yes.

15. Do you happen to know whether application has been made to all the medical gentlemen who signed that petition to correct the fallacious impression they had formed?—I do not know.

16. You were desired, however?—I was requested, in consequence of having signed that petition, to investigate.

17. Do you happen to know whether any gentleman besides yourself, and Mr. Wilson, who has been just examined, who did sign that petition, have come up to town for the purpose of giving evidence?—I do not know of any other, besides Mr. Wilson and myself, who did come up.

18. As far as you know, there are no others but yourself and Mr. Wilson

Mr. G. Hamilton. who signed the petition, who have come up to town to state the facts?—As far as I know.

19. It is detrimental to children's health, is it not, to be employed in a high state of temperature of atmosphere, and then exposed to the vicissitude of cold?—I should not conceive that could produce their death.

20. Is it not injurious to their health, in your opinion as a medical man?—At a very early age I conceive it would be.

21. Upon what early age do you put your finger, when you say that would be injurious?—I can hardly answer that question, for they seem to enjoy such good health, many of them, that I cannot say when it would be.

22. Would it not be injurious to a young person of twelve years, to be kept during twelve hours in the temperature of 75, and then exposed to the cold of 40; for instance, should you not, as a medical man, think it would be likely to be injurious to the health of such a person?—I should think so, certainly.

23. Is not the age of fourteen what you would call a delicate time of life, both for male and female children, when the constitution undergoes a considerable change?—It certainly is, in the female sex, a delicate period; but I did not find it had that effect upon them which I expected.

24. Is it not also a time of life when a male child would be more subject to complaints than he would at a period just antecedent to it, or when his constitution became more fixed?—I really do not know but that a boy at fourteen years of age is in as healthy a state as at any age.

25. In the case of a female child of fourteen, should you not think it would be injurious to be up at five o'clock in the morning working, and kept till seven o'clock in the evening?—I certainly would not from choice employ them; but I did not find that they suffered so much from it as I expected.

26. In your judgment, did they suffer from it at all?—I really can hardly say that they did suffer from it at all.

27. In your judgment, is not eleven hours in the day as much as a child of ten or eleven years ought to be employed in any occupation, adding thereto half-an-hour for breakfast and an hour for dinner?—I conceive that would be quite sufficient for that age.

28. Had you ever been in a factory, having lived in Manchester eight-and-twenty years, before?—Often.

29. Then you had some personal knowledge of the work to which children are put in factories?—Yes, I had.

30. Did you sign the petition solely on the representations of Mr. Gould, or were not you partly induced to do so from what you knew?—I certainly was induced to do it from what I was informed, and from what I conceive to be wrong, employing children so early for so long a period; I should have thought it a wrong thing to employ any children of that age for so long a time.

31. At what age?—Of the age of ten or twelve, or even fourteen, or any age, for fourteen hours.

32. Thirteen hours?—Thirteen hours is less; there is an allowance for dinner.

33. What do you say to thirteen hours?—Thirteen hours is too much.

34. Twelve hours?—They might work twelve.

35. To children of fourteen?—I should think it too much even for a child of fourteen, twelve hours.

36. What do you think of eleven hours?—They might work eleven hours.

Re-examined.

37. Do you think, as a medical man, that if they could be supported at the public expense, and allowed to play half the day, that would be better for their health, children under ten years of age?—They might be better; and would look better and stouter, I dare say.

38. Though the employment of the weavers is more laborious and more un-

healthy, do you mean to say that the employment of the cotton-spinners is not laborious and unhealthy, when carried on to a certain length of time?—It is certainly a laborious employment. Mr. G. Hamilton.

39. You signed this petition from a general impression?—Yes, from a general impression.

40. After your minute inspection of those factories, if the petition had been tendered to you, would you have signed it?—I would not.

41. Are the children at present employed fewer hours than twelve?—I cannot speak to that.

42. You know that they are employed the full twelve?—I had been told they were employed fourteen.

43. Then with this knowledge that they were employed fourteen hours a day, and with your opinion that twelve hours was more than enough, how could you, in consequence of your last examination, alter your opinion?—I am not conscious that I have altered my opinion.

44. You signed a petition, and now you say, from the late information you have obtained, you would not sign it?—I certainly would not.

45. On what do you found your present opinion?—On facts; finding them so perfectly well, comparatively speaking.

46. The opinion in your mind, at the time you signed that petition, proceeded from theory?—It did.

47. Then, after you had examined this and seen the facts, you changed your opinion?—I certainly changed my opinion.

48. How could it proceed from theory, when it must have proceeded from a knowledge of the illnesses of children, with which you, as a medical man, having practised twenty-eight years in Manchester, could not but be conversant?—Upon my examining the factories, I found so exceedingly little disease among them, that I was most decidedly of opinion that it hardly existed.

49. From the time you studied medicine first, and began to practise, have you not been constantly in the habit of correcting the theoretic opinions you have entertained, by practice?—Certainly.

50. Do you think a medical man who did not, could be trusted by his patients?—No.

51. Would not you correct the opinion of one day by the facts you experienced the next?—I have been correcting them all my life.

52. What interval of time was there between your petitioning and saying the children were suffering too much, and the time when you say they were not suffering from working too much?—A month.

53. In the interval you had examined the factories?—Yes.

54. And in the preceding time you had had many of the children under your care?—Yes, some of them; but children are liable to diseases independent of factories.

EDWARD CARBUTT, M.D., called in, and examined, 29th and 30th May, 1818.

1. Are you one of the physicians at the Manchester Infirmary?—Yes. E. Carbutt, M.D.

2. And to any other medical establishment in Manchester?—To the Dispensary, the Lunatic Hospital and Asylum, and the House of Recovery.

3. Have you in April last examined any cotton-factories in Manchester?—Either in April or May.

4. Be good enough to specify those factories you examined in Manchester?—I examined the Ancoat's cotton-twist company's; Mr. David Holt's; Messrs. Duckworth and Co.'s; Messrs. Greenways and Co.'s; those are all that I examined with minuteness.

5. Has the result of your examination in those factories enabled you to form any opinion as to the health of the persons employed in those factories

E. Carbutt, M.D.

compared with other persons in a similar situation of life, obliged to earn their bread by their labour?—It is rather difficult to give a decisive answer to a question of that kind.

6. Did they appear to be more unhealthy than similar classes of society?—They did not.

7. How long had any of the people you examined worked in factories?—In some instances I think thirty-five years. I recollect one in particular who had worked thirty-five years.

8. From his appearance, and the appearance of any others who had worked considerable periods in the factories, were you led to suppose that they were at all unfavourable to health?—There was one who had worked thirty-five years, who was so remarkably stout, that the surgeon and myself detained him particularly, to inquire whether he had worked so very long or not. When we first saw him we took him to be a blacksmith, he was so very stout.

9. What age was he?—He was of the age of forty-six years.

10. He must have worked in the mills pretty nearly from their origin?—I apprehend so.

11. The paper you are about to deliver in will give the particulars and ages; what did you observe in the children under the age of eight or nine?—The impression upon my mind is, that they were exceedingly few; but in those examinations I did not draw out, in the abstract, any account of the comparative ages.

12. Children of that age appeared to you to be but very few?—They did.

13. Did you ask any questions of any of those children, so as to enable you to ascertain the circumstances under which they came into the mill?—Not particularly of the children, but in many instances of the parents.

14. What accounts did you receive from the parents, whether they had solicited the employment of their children as matter of favour to themselves and to their families?—I can hardly give an answer to that.

15. Did you make any inquiry as to the employment of those children, whether they had any little indulgences as to time?—In some instances it occurred to me to make such inquiries.

16. Did they appear to you to flag at the close of the day, or did they appear jaded, as if they had been overworked?—They appeared in very good spirits and quite playful.

17. Should you, from their appearance, and what you observed, draw any conclusion that they were not overworked; that they had not been kept at work a longer period than they could work with ease?—I should draw the conclusion that they had not been overworked; and I have frequently turned the attention of persons not acquainted with factories to the circumstance of children, after leaving the factories, going home engaged in plays requiring a good deal of exertion.

18. Be so good as to state the highest temperature you found, and what was the usual temperature?—I think the highest was 76°; a very usual temperature was 74°; but in many instances it was much lower than that.

19. Did it appear to you to be of a light or laborious kind?—Of the lightest kind that can be given to a person to perform.

20. Did the children appear to go through it as a matter of labour or drudgery, or otherwise?—They appeared very attentive to it.

21. Were they at the same time playful and in good spirits, or did they appear dull and depressed by their work?—In some instances I observed them what I may call arch about it; what I mean by that is, that they were performing something like tricks while they were about it.

22. That they were amusing themselves like children while they were performing this occupation?—Yes.

23. Have you had an opportunity of comparing, in any large Sunday-

schools, the appearance of children employed in those factories, and children employed in other occupations?—I examined two Sunday-schools.

E. Carbutt, M.D.

24. Was the appearance of the children engaged in the cotton manufactories at all unfavourable as compared with the others?—Generally not unfavourable.

25. Am I to understand that their appearance would not lead you to the conclusion of its being an unhealthy employment?—Their appearance would not lead me to such a conclusion.

26. Has any application been made to you to sign the petition in favour of the bill?—Yes, there was one application.

27. By whom was that application made?—By Mr. Gould and Mr. Shelmerdine.

28. About what time was that application made?—I cannot answer precisely, not having kept any memorandum, but I presume that it was the commencement of April.

29. Did you decline signing that petition?—I did.

30. There are two mills at Darwen which you examined?—Yes.

31. What was the temperature in those mills?—The highest in both mills was 67 of Fahrenheit, the lowest was 61; in the open air the thermometer stood at 54; they spin forties.

32. Is the temperature you have stated at the different mills so high as in your judgment to produce any unfavourable effects upon the health of the children employed?—In my judgment the highest temperature I observed was not unfavourable to health; that of course is merely opinion.

33. Did you make any inquiries, when you were ascertaining the temperature in each of those mills, as to whether it was different from the general temperature of the mills?—In many cases they gave me the information without my making the inquiry.

[Dr. C. having stated that he had visited the Ancoat's cotton twist company's mill, was asked—]

34. What was the temperature in that mill?—On the day I went without being expected, the highest was 74; two or three days after, when I was expected, the highest was 76.

35. Do you know what numbers they work there?—I believe they spin all twist there, I rather think fine; judging from the appearance of things, there was less dirt than at some others.

36. Seventy-six was the highest temperature you found there?—Yes.

37. Did you find 76 at any other mill you visited?—I think not.

38. You have told me, that weavers who found themselves unequal to their employ, stated to you that they had resorted to the spinning mills to employ themselves?—In some instances.

39. Have you any reason to suppose that delicate children are sent to the cotton mills from the same reason, not being equal to more severe labour?—I do not know anything about that subject.

40. You have stated the case of an old man who had been employed thirty-five years, and appeared to you to be particularly healthy; did you make any inquiry of him as to the effect of the dust of the mill upon his lungs?—No, not upon that point; I had a great deal of conversation with him.

41. Did the result of that inquiry lead you to suppose that the dust or fluke of cotton was prejudicial to their health?—The answers which I received to the numerous cross-questions that I put to one individual upon that subject were such as rather surprised me, for I found by his account that it produced no effect upon his breathing, or upon the secretion of mucus in the air-tubes of the lungs; I could not learn that he received any injury at any time from the dust flying about in the card-room in which he worked.

42. Was the person of whom you made those inquiries employed in the room where there is the greatest quantity of dust?—He was employed in the card-room of a mill which spins low numbers, where there is of course most dust.

E. Carbutt, M.D.

43. Did that circumstance induce you to make the inquiries?—I can hardly recollect what led me to make the inquiries at the time; but I know he was in the card-room.

44. Were you one of a meeting who went to examine the Sunday-schools in Manchester, or thereabouts?—Yes; I was at one meeting.

45. Was it a large meeting?—A very large meeting.

46. Consisting of persons of what description?—Consisting of physicians, surgeons, apothecaries, and cotton-spinners.

47. About what time might they meet?—I think it was the 19th of April, if that day was Sunday.

48. In the present year?—Yes.

49. Did you proceed to examine any Sunday-schools after you had met together?—The purpose of our meeting was to proceed to examine Sunday-schools, and I did along with others so proceed.

50. Were the others medical men?—There were with me one surgeon and three or four other gentlemen; three of them cotton-spinners; one neither a medical gentleman nor a cotton-spinner.

51. How was this meeting assembled?—I received a letter from Mr. Douglas, the chairman of the committee of cotton-spinners, on the Friday preceding, requesting me to attend professionally for the purpose.

52. Did you make any observation upon the appearance of those engaged in cotton factories, and those engaged in other occupations?—I made a good deal of observation upon them.

53. What was your observation as to the appearance of health?—The appearance seemed to vary very little indeed.

54. If any body had asked you which of the two looked most healthy, what should you have said?—I should have said that I thought there was no difference.

55. Did any medical man, to your knowledge, decline attending that meeting at which you were present?—I know that one did, and I believe that more did.

56. Have you made any observations on the persons working in Mr. Holt's factory?—Yes; I made a very particular examination of the persons working in that factory.

57. In consequence of what did you go?—In consequence of the application of Mr. Holt, the owner of the factory.

58. Was that in Manchester?—It is in the township of Chorlton, in the vicinity of Manchester.

59. What number of persons were working in that factory when you went there?—234, of all ages.

60. Have you any memorandums of those above or below nine years of age?—I have none myself; I did not abstract that part of my memorandum. When I say none myself, I allude to an abstract which Mr. Holt himself made, and which I presume he is prepared to prove.

61. What was the general appearance of health among the persons there?—The general appearance was very good.

62. What was the general appearance among the younger persons working there, who appeared to be under sixteen or seventeen?—Not at all unfavourable, generally speaking.

63. Did you observe any appearance of any one disease prevailing more than another, such as consumption or scrofula?—I find, from a list I have got in my hand, that chronic head-ache was the most prevalent.

64. Among persons of what age, according to your judgment?—As far as my memory bears me out, I think they were very much among young women about the age of puberty.

65. Did you go to Mr. Greenway's factory?—I did go to Mr. Greenway's factory.

E. Carbutt, M.D.

66. How many hands were there at work in that mill?—337.

67. Of all ages?—Yes, of all ages.

68. Have you any note of the proportions of persons of different ages?—I have not.

69. Where is Messrs. Greenway's factory situated?—Very near to Mr. Holt's.

70. What was your observation as to the general appearance of their health?—I have a memorandum here: in very good health, 47; in good health, 221; delicate, and with slight ailments, 65; sick at home, 4.

71. Did you take that note at the time?—I abstracted it the same evening from memorandums I took on that occasion.

72. Have you any memorandum of any particular diseases to which the persons working in that manufactory were liable?—I have got memorandums of the cases of all the sixty-five whom I have stated to be delicate and to have slight ailments.

73. Are they different ailments, or do they come under two or three heads?—They come under twelve heads, besides the delicate ones.

74. Mention the heads, and how many persons under each?—Delicate, without any particular ailment, two.

75. By delicate, do you mean fit or unfit for work?—I should say fit for work; by delicate, I do not mean having any complaint, but being of a delicate frame.

76. Naturally of a delicate frame?—I mean of a delicate appearance, without our being able to detect disease; coughs, colds, catarrhs, and hoarsenesses, twenty; asthma, three; dyspeptic cases, thirteen; slight colic, two; chronic head-aches, six; rheumatism, seven; having pains in the legs, two; dysuria, —that is, difficulty with regard to urine, one; scrofulous, four; epileptic, three; ruptured from infancy, one; slight distortion, one; sick at home, four; but that is not in the number of the sixty-five. I have got a memorandum that among the healthy cases there were three slight distortions.

77. In those dyspeptic cases were they slight, such as you have mentioned before, or such as to disqualify them for working?—I apprehend slight dyspepsy is a case we cannot judge of, except from the account of the person.

78. You simply mean indigestion?—Yes.

79. You spoke of three persons having an asthma; did that incapacitate them for labour?—No, certainly not; I apprehend it was occasional.

80. Does your paper enable you to state whether the greater number indisposed were males or females?—It does not.

81. Have you a recollection upon that subject?—I have not particularly.

82. Did the place appear to you to be hot, so as in your judgment to make it unhealthy for a person to remain there several hours in the day?—I think it was not hotter than is consistent with health.

83. What would you say upon that point as to Mr. Holt's factory?—I should give the same answer to that.

Cross-examined.

84. Were you much acquainted with factories previously to these appointed visitations?—I never went into any, except occasionally, to shew a stranger the things that were worth seeing in the town.

85. During your previous visits it was merely to see the process of the manufactory?—Just so.

86. On those occasions you did not make any inquiries either as to age, health, or temperature?—Not the least.

87. When were you applied to to visit the factories you have described?—As nearly as my recollection serves me, I was first applied to on the 15th of April.

E. Carbutt, M.D.

88. At the Ancoat's cotton-twist manufactory, Mr. Windsor being engaged at the time, you took the house apothecary with you?—I did.

89. You understood that the inspection you were to make would be for the purpose of enabling you to give evidence?—I cannot say that, at the time I examined the Ancoat's twist mill, I had any particular idea as to the object or result.

90. In examining the others it was your general impression?—In the course of conversation with the owners of the mills, it certainly did arise in my mind, that the probability was, that I should have to come to London to give evidence.

91. How soon after the first application to you was it that you visited those mills you have enumerated?—The first application to visit the Ancoat's cotton-twist mill was on the Wednesday morning, and I think the examination took place on the Friday and Saturday.

92. How many days after that application did you visit the others?—The application to visit Mr. Holt's mill was on the following Sunday; the examination took place on the Tuesday.

93. All within a few days of your application?—Yes, of those within Manchester.

94. And those without Manchester also?—No, those were at a distance of more than a month.

95. There was a committee of masters sitting in Manchester at the time, to get evidence to oppose this bill?—So I understand.

96. You were applied to professionally?—Yes, I was.

97. And paid professionally, as any other gentleman would be?—At the time I examined the Ancoat's cotton-twist mill I had no expectation of payment; I have not yet received any, but it is probable I shall; when examining the others I perceived there was an intention of paying us professionally.

98. You mentioned yesterday that one of those mills you took by surprise; the other, of course, you had previously arranged with your colleague, and with the overseer, when you should visit?—Just so.

99. Did it not occur to you to inquire particularly as to the number between the ages of nine and sixteen?—We had an account before us of the ages of all the persons, but I did not abstract that part, not considering it within my province to prove that part of the business.

100. You are not capable of stating how many you found between nine and sixteen?—Not at the first mills, nor indeed at any of the mills.

101. Were much the greatest proportion of those you found between those ages?—There were very few appeared to be under nine; there did appear to be a good many between nine and sixteen, but there were also a great many above sixteen.

102. Do you think the greater proportion appeared to you to be under sixteen?—I am not able to answer that question.

103. Of those between nine and sixteen, were the much greater proportion females?—That question I am not able to answer.

104. In what room were those children examined; were they brought down into the middle room, or how?—In one of the factories, the Ancoat's cotton-twist factory, we went into every room successively; in the other factories we had them in a room separate from the workpeople; sometimes the overseers were present, and sometimes they were not.

105. Generally speaking, they were brought down into a room to you?—In all the factories, except the cotton-twist factory.

106. Do you mean that all the persons, adults and children, were so brought down?—I do.

106.* When you have been speaking of those distinctions of "very good,"

E. Carbutt, M.D.

and "good health," and so on, you apply that to the whole mass indiscriminately, without any distinction made?—Just so.

107. You knew nothing of their respective ages, but the report of the overlooker?—In almost every instance I asked the person who came before me what his or her age was.

108. Did you make any inquiry how long they had been respectively there?—I made inquiries how long they had respectively worked in the cotton factories.

109. The forty delicate ones for instance; did you make out how long they had been so employed?—Of course I knew how long they were so employed, because I had a statement to that effect lying before me; which, in every individual instance, as far as I recollect, I verified, by asking them how long they had worked.

110. Between them and the very healthy children, did you ascertain any distinction in time as to their respective employment?—I found that those who had worked the longest were the healthiest—the longest with regard to number of years.

111. Did you at all inquire how many might have gone away or been dismissed within the last year or two?—I did not make that inquiry.

112. How many hours did the children you examined work?—Of those in Manchester I cannot speak precisely, not having inquired; but it occurred to me to hear such remarks as led me to believe that none of them worked more than seventy-two hours in a week.

113. Generally speaking, you found the highest temperature 76?—The highest I can recollect was 76.

114. Do you recollect what it was in the open air about that period?—At the time of examining the Manchester factories I did not try the open air; but I apprehend, speaking generally, it was somewhere between 50 and 60.

115. Did you ascertain which of the factories you visited spun fine and which coarse?—In the case of the mills at Oldham and Bury, I did take a few memorandums of that kind.

116. In those at Manchester did it occur to you to ascertain whether they were persons who spun fine or coarse?—It did occur to me to hear that in the course of conversation, but I did not consider it one of those things that required to be either put down or recollected.

117. Do you recollect what they spun?—Some spun fine, and some spun coarse.

118. Are you not aware that to spin fine requires a higher temperature?—So I have heard.

119. Mr. Holt's factory and Mr. Greenway's you visited afterwards, on special application?—Mr. Holt's was on application by himself; Mr. Greenway's in consequence of the overseer's coming to me while I was in other factories, and saying they wished to have theirs examined.

120. During your taking this trouble for those gentlemen, you breakfasted and dined with them; for instance, Mr. Holt?—Yes; I am of the same religious profession with Mr. Holt, consequently intimate with him.

121. At Mr. Holt's the chronic head aches, among young women, were the most numerous, and next the dyspeptic—some indigestions and other disorders; and, after dining with Mr. Holt, you doubted whether any of them were ill or not?—That may be put down as your question, but it will not stand as my answer.

122. Am I to understand you doubted, after all, whether they were ill or not?—I did not say "after all;" if a precise question is put, I will endeavour to give a precise answer.

123. To whom did you apply the expression, that you doubted whether they were ill or not?—I do not recollect having used the expression.

E. Carbutt, M.D.

124. To whom did you mean to apply the expression, that you doubted whether they were ill or not?—I meant this; that through the whole of the cases, which we put as slight ailments, there was very little appearance of ailment among them; but that, by a number of pumping questions being put to them by my colleague, many of them were induced to say, "Yes, they had sometimes a head ache, or sometimes a cough;" when otherwise, if I had put a simple question, "Have you any ailment?" they would have said "No."

125. Suppose a person were about to institute a cotton factory, of the description such as you have visited, and were to ask you, for humanity sake, to advise how many hours children from six to sixteen should be employed, in an erect position; how many would you recommend, as consistent with safety to their constitution?—I should not recommend any particular number at all.

126. Supposing the temperature to be about 80?—I should not give any opinion upon the subject.

127. You would decline giving an opinion?—I should tell him it was a question which it was totally out of my power to answer; because the limit or distance between the minimum and maximum of work, as to every human being, proceeds by such imperceptible degrees, that it would be out of my power to say, "here you must stop."

128. Supposing he should say, how many hours should he employ children between six and nine years of age, what should you say?—I should give him the same answer, that there could not be, in a matter of that sort, any precise rule laid down. In all cases I should say, that the maximum of hours of employment could not be defined by any man; that if he pretended to define it he must be attempting that which he must be conscious he was unable to perform.

129. Do you think that children from six to twelve years of age being employed from thirteen to fifteen hours in a cotton factory, in an erect position, and in a temperature of about 80, is consistent with safety to their constitutions?—Not having examined children under those circumstances, I am totally unable to give an answer to the question.

130. After the examination you have made into those factories and those children, you say you are utterly unable to answer that question?—I say, that after the examination I have made, I am totally unable to answer the question which you put; which is one that differs entirely from what has occurred during my examinations.

131. Not having witnessed, according to your ideas, that particular fact, am I to understand you are incapable, as a medical man, of giving any general answer to the general question, whether or not children, such as I have described, might be safely employed from thirteen to fifteen hours?—Certainly; it is entirely out of my province or out of my power to decide what children can do, except in cases where I have seen them do it, and found them injured or not injured.

132. I am to understand you can give no answer to the general question, how far it would be safe for the constitutions of children, from six to twelve years of age, to be employed from thirteen to fifteen hours in a cotton factory in an erect position?—I have never examined children under those circumstances.

133. Are you capable, from your general studies, and your general reading and observation, to give a general opinion upon that question or not?—I am not; if I were to say "No," doubts would occur in my mind whether they could not very well bear that number of hours and that temperature; if I said "Yes," I should be totally committing myself, from the circumstance of my never having seen children who have done that work in that temperature.

134. Then you are incapable of giving a general opinion?—Certainly, for the reasons I have assigned.

135. You recollect being applied to to sign a petition in support of the bill; at what period was that?—I did not keep a memorandum of it, but I suppose it was either at the close of March or the beginning of April.

136. Before you were applied to to visit the factories?—Some weeks before.

137. When you said you could not conscientiously sign that petition, did you mean that you had not at that time any general knowledge of the facts, of the nature of factories; were you induced to make that answer from your general judgment of the nature of factories?—I made that answer for various reasons.

138. Was it one that you did not feel yourself informed as to the general nature of factories?—I was not then informed as to the general nature of factories; that was one of my reasons.

139. That you could not conscientiously sign the petition?—That was one of the reasons why I could not.

Re-examined.

140. As to all the factories that you visited, have not all your visits been since this bill has been pending, and at the desire of the owners or their agents?—All the factories that I have visited have been visited since this bill was pending; the Manchester factories were visited at the desire of the owners or their agents; the Oldham, Bury, and Darwen mills, were visited without any desire of the kind.

141. From any party whatever?—From any party connected with those particular mills.

142. At whose desire did you visit those country mills?—At the desire of Mr. Cririe, solicitor in Manchester.

143. Solicitor to some of the parties?—I presume he is.

144. Do you happen to know to whom he is solicitor?—I believe he is solicitor to the Committee of Cotton Spinners; he is partner with Mr. Eccles.

145. Does not the large number of persons seriously ill prove the general unhealthiness of the employment?—I think it proves nothing, until we have first ascertained the proportion of persons engaged in those employments.

146. You gave seventy-nine the proportion of cotton spinners, out of 325?—I did.

147. Are not the home patients in general those who are most seriously ill?—They are.

148. Does not then this large number of persons seriously ill, belonging to the cotton factories, prove demonstrably the general unhealthiness of that employment?—I am of opinion that it proves nothing, unless we first ascertain what is the proportion of persons in different employments.

149. You speak of the infirmary, the dispensary, and the fever wards; do you mean that the infirmary includes the dispensary and fever wards?—I mean that the infirmary includes the dispensary, but not the fever wards.

150. When you spoke of the proportion of 79 out of 325, taking the whole number, or of 79 to 246 as the differences, you meant to include the fever wards, as well as the dispensary?—No; I mean to include that description of patient whom we call the home patients.

151. The fever ward is excluded from that computation?—Yes, the home patients are dispensary patients.

152. The committee are to understand, then, when you speak of the infirmary you include the dispensary?—Yes.

153. How far does the operation of the infirmary extend; is it confined to

E. Carbutt, M.D.

the town of Manchester, or does it take in patients from other parts?—As far as the operation of the infirmary, strictly speaking, extends, it takes in a very considerable district, as far as concerns those patients who are provided with beds in the house; but with regard to patients who are seen at their own houses, it only takes in the absolute towns of Manchester and Salford.

154. As far as the operation of that part which is composed of the home patients goes, have you ever considered what proportion the number of persons employed in the spinning factories bears to all other persons in that district where the whole operation of the infirmary extends?—I have no knowledge whatever upon that subject.

155. Would it be one-fourth or one-third?—I have not the slightest information upon that subject.

156. You stated that the patients from the cotton-factories have a proportion to all other patients as 79 to 246?—That is of one description of patients.

157. That is one to three and one-eighth?—I presume it is.

158. The patients from the spinning cotton factories are in the number of as one to three and one-eighth; is not that a very large proportion?—It depends upon the number of persons employed in each occupation.

159. Does not that proportion incontestably prove, that the employ in the spinning cotton factory is more prejudicial than any other employment in Manchester?—I think it does not incontestably prove that; for this reason, that I think it very possible, that if I had authentic documents to point out to me the number of persons employed in spinning factories compared to persons in other occupations, I might then think it a small proportion.

160. You stated that you observed that those persons were the healthiest in the factories who had worked there the longest?—I said that those who struck me as remarkably healthy, were generally persons who had worked the longest; I intended to express myself to that effect.

161. You did not ascribe this healthiness of the persons who had worked there the longest, to the circumstance of their having worked there the longest; that the longer they worked the healthier they became?—I think it is very possible that persons may become so habituated to a particular kind of work, that they will have their health better after they have worked longer in it than when they commenced it.

162. Do you not think it may be ascribed to the circumstance, that so many have died in the seasoning, and that only the robust ones have held it out?—I do not think that; for this reason, that I have been informed, in a manner to induce me to give credit to it, that the deaths amounted to very few indeed.

163. When they become sickly in those manufactories, are not they removed, or do they continue to work there?—I presume that no man continues to keep a servant after he is unable to do the work which he has for him.

164. Then you cannot argue from the number of deaths in the factories themselves?—I presume I cannot.

165. The calculation you are giving at present is only founded upon your own patients?—Just so.

166. In the course of your practice at Manchester, have you had frequent opportunities of visiting weavers as patients?—I have.

167. Have you formed any opinion with regard to the comparative wholesomeness of the weaving business and the cotton spinning?—From seeing the habitations in which they perform the work, from hearing them at work at very late hours of the night as I go along the streets, and knowing the low price of their wages, I have formed an opinion that the employment of weavers is very unfavourable to persons employed in it.

E. Carbutt, M.D.

168. Is it more unfavourable than that of cotton-spinning?—I have no doubt it is, from the closeness and dampness of the rooms.

169. Do you think it is possible that children can inhale the dust and the fluke of cotton in those mills?—They certainly do inhale it.

170. They inhale it into their lungs?—Into the air pipes of their lungs.

171. Would you, as a medical man, prescribe emetics to get rid of that?—Certainly not.

172. Have you any idea of such a thing being done?—Certainly not, for that reason.

173. You have seen those children work?—Yes, I have.

174. Did it not strike you more in the nature of an occupation that requires attention, than of labour that could fatigue?—Certainly the occupation of a spinner requires attention, but is accompanied with very little fatigue; the piecers are under nearly the same circumstances as the spinners on that point; the scavengers seem to require very little either of attention or exertion.

175. If that be the nature of the occupation, when carried on for eight, nine, or ten hours, would it not become of a different description if carried on for twelve hours or more?—That would depend upon the strength of the person who was employed.

176. In the factories you visited, did you inquire what number of persons had died, and what number had been obliged to withdraw themselves from the factory within a definite time, in consequence of being unable to go on with the work?—I have not made any inquiry of that kind.

177. Is not the occupation the same the tenth hour as it is the first?—I will not say.

178. But the effect you would think different?—I am not prepared to coincide with that remark.

179. Can you state that, in point of fact, weavers are more unhealthy than cotton-spinners?—I cannot state it from any positive facts.

MR. JAMES AINSWORTH, called in, and examined, 30th May, 1818.

Mr. J. Ainsworth.

1. You are a surgeon at Manchester?—I am.

2. Are you surgeon to the infirmary and to the workhouse at Manchester?—I am.

3. How many years have you acted in that capacity at those places?—I lived six years at the infirmary as an apprentice subsequent to that period; I have been about twelve years a surgeon.

4. How many years at the workhouse?—Eight of the latter years I have been also surgeon to the workhouse.

5. Were you requested by the committee of proprietors of spinning-mills to examine, in company with other gentlemen, the mills at Manchester and in the neighbourhood?—I was.

6. What was the result of this investigation?—The result of this investigation was, that the hands whom we examined, who were employed in that mill, were as healthy as could be expected in any class of society who were obliged to work at a manufacturing employment.

7. Can you explain what is meant by numbers?—I suppose so many hanks to the pound; but I am not sure.

8. What is your deduction, as a medical man, from those facts?—That the cotton-factories are not more detrimental to the health of children employed than other trades are.

Cross-examined.

Mr. J. Ainsworth.

9. The seven mills you visited generally, are the country mills at and about Oldham?—Yes.

10. Those you describe; the hours of their work seem to run from about sixty-five to seventy per week?—I see there is one at seventy-two.

11. Who desired you to visit those country mills?—Mr. Douglas, the chairman of the committee.

12. Did you inquire the number between nine and sixteen years of age?—No.

13. They passed before you in review?—No: we went into the rooms separately, and looked round and observed their manner and general appearance. We did not examine any except those who looked unhealthy; and if we saw any who looked very small and unhealthy we made particular inquiries.

14. Where you saw a child who looked very unhealthy or small you made specific inquiries?—Yes; not so particularly as when we examined them individually in the mills at Manchester.

15. Was any particular time named for you and those gentlemen coming?—We sent word up on the preceding day, and requested them to get the lists ready, as it would facilitate our investigations most materially to prepare lists of all the hands which were present there, and that we would come; and we fixed the time ourselves when we would come up. They allowed us to fix our own day; but, in order to make it more convenient for ourselves, we sent word to them to prepare lists immediately.

16. Against a certain day that they were to prepare lists?—No; we could not fix that without seeing each other.

17. When was this application made to you from the owners?—I cannot speak to that; it was a very few weeks ago.

18. In the course of the month of May?—I think rather before that.

19. Had you a note from each separate owner?—Yes; several days intervened between our examination of one mill and another.

20. Within a few days after those notes you visited them accordingly?—Yes.

21. All this was comprised in a list made out by whom?—Prepared by somebody in the factories, not by ourselves.

22. Is it not quite a medical maxim, that the standing position for a length of time exhausts human strength in a greater degree than any other position of the human body?—If people stand too long, always standing, of course it would be; but as to any definite time I cannot answer, in the situation in which I am placed, never having seen the experiment tried; I can only speak from theory, and that sort of speculation, into which I am very desirous not to enter.

23. Is it not regarded among medical men, that the erect position, standing, say twelve hours, would exhaust the human body infinitely more than a reclining position during the same period?—Indeed I could not answer that question; I do believe that if a person were kept constantly in a recumbent position, as much weakness might probably ensue as if he stood a great many hours.

24. What is your general opinion in regard to my question?—I have never made the experiment; I have no facts to go by; and it would be giving a speculative opinion, from which I beg to be excused, in the situation in which I stand.

25. Am I to understand that you, as a medical man, can give no opinion whether it would not be more exhausting to the human body to keep in an erect position for twelve hours than in a reclining position?—I have no facts to lead me to conclude.

Mr. J. Ainsworth.

26. Supposing we were to increase the quantity half, and make it eighteen hours, could you then give an answer?—I have never seen that tried, nor has it come within my knowledge, and therefore I cannot say.

27. What do you say to thirteen hours: is that speculation?—I am really unwilling to answer these questions; I do not know the definite line to draw, and therefore am unwilling to answer.

28. You have given yourself the trouble of coming a long way to give information to the committee?—I have spoken to facts, and by those facts must be guided.

29. Can a child of six to twelve years of age be employed from thirteen to fifteen hours daily in a temperature of 80 degrees, and in an erect position, consistently with safety to its constitution?—I never saw an instance of the kind as a fact brought before me, and therefore cannot say; I am not aware if such an instance ever has occurred to me.

30. I am supposing such to be the fact, and ask you your opinion upon it?—Then I must meet that with a supposition which I wish to avoid. I have no fact; my experience does not enable me to answer that question.

31. You are incapable of answering the questions, not having before you the fact of a child so situate?—I have no facts, and must therefore beg leave to decline giving an opinion.

32. You are equally incapable, whether the question be thirteen, or fourteen, or fifteen hours?—There must be a limit; but with that limit I am unacquainted.

Re-examined.

33. Did you inquire what number of the children were absent from ill health?—We inquired whether those were all the hands they had; they told us, generally, they were. We inquired whether they ailed anything, and so on.

34. Did you inquire how many were absent from ill health?—When we went into the rooms, we inquired of the men, Are any of you absent?

35. And you inquired the number?—Yes; we inquired (not the number particularly), but—Are any of you absent? This was a very general inquiry.

36. Did you inquire what number had died off from the factories during any certain time?—Not in those instances.

37. As far as you have seen the limits of time, the works never injured the health of the children?—I never perceived it from the appearance on the individual examination of those persons.

38. Could you answer that question without knowing the number of persons who were ill, or who had died off, or were absent from other causes?—I speak only from what I have seen.

39. What hours do they work at those factories you visited in Manchester?—I did not inquire; I thought it best to tell what I saw, and to know nothing more about it; and I was very sorry I had known so much about Mr. Pollard's mill before.

Mr. THOMAS TURNER, called in, and examined, 1st June, 1818.

Mr. T. Turner.

1. You are the house surgeon and apothecary of the Manchester poorhouse?—I am.

2. You have never signed any petition on the subject of this bill?—I have not.

3. How long have you been in that situation?—Rather more than a twelvemonth.

Mr. T. Turner.

4. Your practice is exclusively confined to the business of the parish?—It is.

5. During your experience, have you had an opportunity of forming an estimate of the comparative state of health of the children employed in cotton-factories and in other occupations?—I have.

6. What is the result of your observation and experience upon that subject?—The result of my observation is, that persons working in factories enjoy a much better state of health than weavers, and as good a state as any class of workpeople.

7. Did you, in the course of the last month, inspect the mills of Messrs. M'Connell and Kennedy in Manchester?—I did.

8. How many persons did you find to be employed in that factory?—1125.

9. What appeared to be the general state of health of those you saw?—Their general appearance was good and healthy.

10. How many persons did you note down as being healthy, having no distortion whatever?—1037.

11. How many labouring under scrofula?—Twelve.

12. Coughs and difficulty of breathing?—Twenty-one.

13. Persons in-knee'd and otherwise distorted?—Eighteen.

14. Did you inquire the cause of those distortions from them?—We did; and the causes were noted in a book.

15. Cases of indigestion, and persons of unhealthy appearance and delicate, how many?—Thirty-seven.

16. Did you observe any material difference in appearance between the boys and the girls?—We did.

17. In what respect?—The boys, certainly, did not look so healthy as the girls, which I should attribute to their negligence of dress and cleanliness.

Cross-examined.

18. Have you been much conversant with the diseases of children?—I have had to do with the diseases of children, as constituting part of the practice of a surgeon.

19. Do you think it would benefit a child's health of eight years old to be kept twelve hours upon his legs?—Really I am not prepared to answer that question.

20. Is your medical skill so limited that you can form no opinion, whether it would or would not be injurious?—I conceive that would be quite a matter of opinion.

21. I ask your opinion?—As I have no facts to go by, I do not feel prepared to answer the question.

22. I am going to put an extreme case. Supposing you were asked, whether a man could take a pint of laudanum; do you think it would kill him?—Then I should know from observation and facts that it would kill him.

23. From the quantity?—Yes, from the quantity.

24. There is a time beyond which you would not, without knowing any precise fact, keep a young child standing upon his legs; as for instance, you would have no doubt that twenty-three hours would be too long?—None whatever.

25. Then there is a limit?—There is a limit, no doubt; but I consider it difficult to define the line between that which would be salutary, and that which would injure the constitution.

26. I guard the question; I only ask for the best of your judgment, in a case in which you have not had the benefit of facts to proceed upon; what is your opinion?—I really cannot give an opinion.

27. Should you think a child of eight years old being kept fourteen hours upon its legs without any intermission, that that would or would not be dan-

gerous, if he was kept standing the whole time?—I should think it might be fatiguing; whether the health would be materially injured by it, I am not prepared to say.

28. You can form no opinion whether a child of eight years of age being kept standing fourteen hours, without intermission, would be injurious to his health or not?—I have no facts to guide me.

29. I ask you, as a medical man, whether you can form an opinion, either one way or another, that it would or would not be injurious to a child's health?—I am not prepared to answer.

30. What is your opinion?—I should think you would wish me to have some ground; I have no ground for that opinion, and therefore do not wish to form it.

31. But from your knowledge of a child's structure?—I have no knowledge to guide me.

32. You do not know enough of a child's structure and constitution at eight years of age to guide you?—I do not know the nature of the effect of that upon a child; I know the physical strength of a child.

33. Do you not think it would be too much for the physical strength of a child to be kept fourteen hours a day upon its legs?—I am not prepared to answer to the fact.

34. I ask not to the facts, but to your opinion; I ask of a medical gentleman, a man who professes medical science, and would wish to be thought so, what is his opinion?—You would not wish me, or any other man, to advance an opinion, without any facts to found that opinion upon.

35. If you tell me, as a medical gentleman, that you can form no opinion at all, that you are not competent to form an opinion at all upon the subject, I am satisfied?—I am not competent, from not being in possession of facts.

36. Has it ever happened to you to be called upon to attend any persons who have been employed in beating feathers?—No.

37. Did you never hear that it is a well known fact, in medical science, that the persons employed in beating feathers were peculiarly subject to pulmonary complaints?—No.

38. Should you not expect that the persons employed in beating cotton, from which a great quantity of deleterious dust and dirt results, would be affected by it?—I have no reason to think so.

EXTRACTS

FROM

THE EVIDENCE

OF

THOS. WILKINSON, RICHD. COAR, JOHN FAREBROTHER, and JOB BOTTOM, (operatives;) DR. WARD, MR. DADLEY, MR. BOUTFLOWER, MR. SIMMONS, DR. JARROLD, & DR. JONES,

(From the Report of the Committee of the House of Lords in 1819, *of which Lord Kenyon was chairman.)*

THOMAS WILKINSON, called in, and examined, 8th March, 1819.

T. Wilkinson.

1. What age are you ?—Thirty-four.

2. Where do you live ?—At Bolton in the Moors.

3. Have you been employed in any cotton-factories ?—Yes.

4. In what cotton-factories ?—I was brought up in Mr. Thomas Ainsworth's till I was twenty-two years of age, and I have been in Bolton ever since, and worked there for different masters, Mr. Holyroode, and several others in Bolton.

5. Are you now employed in any ?—Yes.

6. In what ?—Mr. Crook's, in Bolton.

7. In what capacity ?—As a cotton-spinner.

8. Have you ever been employed in any other capacity than a cotton-spinner ?—Yes.

9. As what ?—As an overlooker and carder.

10. At what age were you when you first entered ?—Ten.

11. What are the hours of working in the factories you are acquainted with ?—From six to seven in the summer, and from seven to eight in the winter.

12. What time is allowed for dinner ?—An hour.

13. Is any other time allowed for meals ?—No.

14. Are the children ever obliged to be at the factory before or after the common hours of work ?—Yes.

15. For what purpose ?—They are generally called there by the journeymen spinners to meet them, to assist them in slipping the cops, assisting to set the rovings, and assisting to oil and clean before they begin.

16. Is no time allowed to get their breakfasts or their afternoon refreshment ?—No.

17. The mill continues going ?—Yes, except some accident happens.

18. Do the children often seem tired when they leave off at night ?—Sometimes they do.

19. Are they ever beaten to make them work ?—I have seen hundreds beat, to keep them awake, and drive them on.

20. Do you ever find them sleepy of a morning ?—No.

21. Are they ever forced to be beat to keep them to their work before dinner?—I have seen them beat in the forenoon, certainly; but that has been more from the cause of fatigue and neglect than sleepiness. T. Wilkinson.

22. Who have you seen beat them on this account?—I have seen the journeyman spinner, and the master, and the overlooker.

23. Have you often seen them beat in this manner?—Very often.

24. If they choose it, have they an opportunity of going home to get their dinner?—No.

25. Are they confined in the factory?—They are.

26. Suppose they choose to go away to get their dinner, what would be the consequence?—They must come no more there.

27. Is that the practice in the factory where you belong?—Yes, except they get leave; which they may get through a particular cause.

28. Is there any difference in the working hours of Saturday from other days in the week?—Yes.

29. What difference?—From six to four in the summer, and from seven to half-past four in the winter.

30. What do they do on that day with respect to getting their dinner?—They clean constantly on Saturday at their dinner hour.

31. Are the children detained after the usual hour on Saturdays to clean the machines?—They are.

32. Do the numbers from 40 to 110 require any greater heat than the common state of the air?—Yes.

33. What heat does number sixty require?—We always consider the warmer the cotton is spun through the roller, the better for the work.

34. What is the general state of the health of the people employed in factories, is it as good as in other trades?—No, it is not.

35. Do children under ten years of age employed in factories in general grow up healthy?—They do not.

36. Do they often become crippled and deformed?—I have known of many becoming crippled and deformed.

37. Do many who enter before the age of ten continue to work in the factory till the age of forty?—Very few—very few.

38. Are they often obliged to leave from ill health?—Yes.

39. Do you often feel much fatigue from your working?—Very often, I can scarcely get home of a night.

40. How many hours a day do you think people under sixteen can work as piecers or spinners without injuring their health?—I cannot speak to that; it would depend upon circumstances how long they could work without injuring their health; they could not work very long.

41. How many hours is it your opinion they could, without prejudice to their health, work, if they were under the age of sixteen?—Eight or ten hours, I would never allow any more, in my opinion.

42. Do you think it would be well if they were allowed to leave the factory to get their breakfast?—It would be a very great advantage.

43. From your own experience when you were young, did you feel there was any difference in the refreshment you received from eating your meals out of the factory, from what you received from eating them in?—When I first began to work in a factory, I had the privilege of going home to my meals for the first two or three years; and after that, I began to eat my meals in the factory, and many a time I have sent my meals home, because I could not eat them.

44. Have you been subject to any particular complaint or weakness?—I am subject to bodily weakness, generally speaking; I feel myself very tired on many days.

T. Wilkinson.

45. To what do you attribute that?—To the heat and long confinement in the factory.

46. Do you think that would have been the case if the hours of work had been fewer?—No.

47. How many hours a day do you think you could work without injury to your health?—I cannot fix any precise time; I am of opinion that eight hours is long enough for any person to work in a factory to be constantly employed.

48. Would you like to work eight hours instead of the time you have been in the habit of working, if your pay was in that proportion?—I should have no objection.

49. On the supposition of the hours being shortened, would labourers in general be content to receive the same price for the produce of their labour in proportion to the quantity?—All that I have ever conversed with upon the subject are.

50. Have you ever known any persons turned away from the factories on account of sickness or old age?—I have, both.

51. Is it a common thing to turn them off if they are sick?—If they are likely to continue sick; for a day or two they will not.

52. In the factory to which you belong is there often a change of the hands?—Yes; they change very often.

53. Is there often a change of hands on account of any of them being sick?—Yes; many leave when they fall sick, and never come more.

54. About what age do they often leave the factories?—I have known them leave at all ages.

55. How many are there in your factory above the age of sixty?—None.

56. How many above the age of fifty?—None.

57. Above forty?—I cannot speak exactly; I believe there are two or three above forty.

58. About what age is it they seem to be most equal to do the work of spinners?—From twenty to thirty they are the strongest, speaking generally; there are some at eighteen that do pretty well.

59. Have they ever the power of sitting down to rest themselves?—Never, when the engine was at work.

60. Within what period of time is it that those piecers are called upon to do any work?—They have not half a minute to spare at the longest.

61. On the average, how many pieces have they to join at each time?—I have known them join from one to twelve, owing to the badness of the spinning; I am now speaking of mule spinning; there is another sort of spinning called water spinning, and throstle spinning.

62. Do you mean to say that the work of the piecers in itself is hard work?—Yes, to follow it up; they have to reach over the mule, as it comes down, further than they can reach conveniently.

63. You have several times seen children beaten for neglect of work?—I have, and drowsiness and sleepiness.

64. What temperature do you hold to be necessary for spinning 100's?—I never practised the thermometer before that day, but I thought when it was at about eighty-two or eighty-four it was nearest the heat the factory ought to be kept at.

65. Do you not know that you can spin 100's with the heat of seventy?—Not so well.

RICHARD COAR, called in, and examined, 19th March, 1819.

Richard Coar.

1. How are you employed?—In cotton spinning.
2. How old are you?—Nearly thirty.
3. Where do you live?—Walton.
4. What made you quit the cotton factory for those six months?—I was so little able to manage them my master put me to the print shops.
5. Till what hour did they use to work on Saturday before the last shortening?—In summer they used to start at five in the morning, and the engine be engaged to five in the afternoon, and they then had to clean it afterwards.
6. Were the children then allowed to go home on the Saturday to get their dinner?—No.
7. Were they kept at work, without intermission, from five in the morning till five in the evening?—Yes; without breakfast and dinner.
8. Were they then forced to get their breakfast and dinner while at work?—Yes.
9. What numbers do you spin in that factory?—At present 50's.
10. Are the children's meals often covered with dust?—Yes, if they be in the mill a short time.
11. Do the children generally eat their meals well?—No, they very seldom eat them well when they are brought to the factory to them.
12. Are they beaten more at any one time of the day than another?—I have noticed that they have been beaten at the latter end of the day; and more before the hours were shortened than what they are now; they stand their work better now than they could then.
13. Is the heat so great as to occasion the spinners to throw off any part of their clothes?—Yes.
14. How are they dressed when they work?—In general their trowsers and shirt, and sometimes perhaps a waistcoat.
15. Are the children equally thinly clad?—The children are thinly clad.

JOHN FAREBROTHER, called in, and examined, 23rd March, 1819.

John Farebrother.

1. Are you a working spinner?—Yes.
2. How old are you?—Thirty-one.
3. Where do you live?—At Bolton.
4. Have you been employed in many cotton factories?—Yes.
5. In whose factory are you employed now?—Messrs. Ormerod and Company.
6. How old were you when you first went in?—Between five and six.
7. When the children have to eat their meals in the factory, do they generally finish it?—No.
8. Do they leave much of it?—Yes; I have seen it all left many times.
9. Is that owing to its being covered with dust?—Yes.
10. Have you seen it often left in the factory where you work at Bolton?—Yes.
11. How do they get their breakfast and afternoon meal?—As they can catch it; when the machinery is moving they eat it as they are piecing.
12. Are the factories at which you have worked lately kept much hotter than those where you used to work when a boy?—Yes, they are warmer in general; they are kept regularly near one heat, more near than they were before; before where I have worked, we were sometimes at a very great heat, before I came to Bolton, and sometimes very cold; that was at Penwortham, near Preston.

John Farebrother.

13. Have you noticed the effect working in a factory has on a child coming in healthy?—Yes; I have seen the children change in their colour, and in their eating their meat; my own children I have taken notice of.

14. Have you found their health get much worse in consequence?—Yes.

15. How soon do you begin to see a difference in a child's health?—I have seen a difference in one week.

16. Have they often been forced to be away from their work from being ill?—They have not been much off their work, but they have been poorly, and refused their meals; we are obliged to keep them to their work, it puts us so about to change our piecers; it is a great disadvantage to the workman if he cannot get a piecer.

17. Should you be willing, for the sake of your children, to be content with less wages, if you could have the hours shortened?—I should.

18. Do you remember the factory in which you are now employed being inspected by any magistrate?—Yes, I saw Major Watkins come in and try the heat of the mill.

19. How long did he stay in it?—About a quarter of an hour.

20. How did he take the heat?—With a thermometer.

21. Is there one hung up in your mill?—No, he brought one in.

22. When was it?—The 15th of February.

23. Was the factory at its usual heat then?—No, it was colder.

24. Did other spinners say it was colder as well as yourself?—Yes.

25. Before Watkins came in?—Yes; they complained that day because it did not spin so well as it should have done, not being so warm.

26. Could you account for that?—It is generally colder on a Monday, because the steam is kept out on the Sunday; the steam is out after four o'clock on Saturday.

27. Were the children as ill used while you were overlooker, as when you were working as a boy?—We were worse used; the mill was worked by steam; and my master had another to join him, and the hours were lengthened; they were more ill used than they were before.

28. Were they ill used, besides the number of hours being lengthened?—They have been very much beaten for not being there in time of a morning; they began at five o'clock.

29. Who used to beat them?—When I was an overlooker I used to beat them myself, and sometimes the master.

30. Was it your duty to beat them, if you could not get them to do their work without?—Yes.

31. Did your master tell you to do so?—Yes, and he has used bad language to me for not doing it.

32. Were you forced to see a certain quantity of work produced?—Yes, I was.

33. Which master was this?—Mr. Luke Taylor; I have seen him with a horsewhip under his coat waiting at the top of the place, and when the children have come up, he has lashed them all the way into the mill if they were too late; and the children had half a mile to come, and be at the mill at five o'clock.

34. How long ago was this?—Between fourteen and fifteen years; I was only young myself.

35. The mill to which your children went worked longer hours than you used to work when a boy?—Yes.

36. And for that reason you considered them worse used than you had been yourself?—Yes.

37. Could not the provisions taken into the mill be covered from dust if they took pains?—They have not an opportunity of uncovering it while they eat it at every stretch.

38. It would be an additional inconvenience if it was covered with any-

thing?—Yes; they have not an opportunity of taking the covering off every time the wheel comes out. John Farebrother.

39. How did you happen to take the heat of which you have given the committee an account?—I saw Mr. Watkins come in with a thermometer, and I thought the mill was cooler the following day, and I borrowed a thermometer.

40. At the time you took the heat, had you any idea of coming up to London to be examined?—I had not the least.

41. Did anybody help you to take the heat?—Yes, there was a man or two looked at it at the same time.

42. Does your master know of your coming to be examined?—He does; I gave him to understand it.

43. What passed upon that subject?—He asked me what I had got to say; I said I could not tell what I should be asked.

44. Is that all?—He asked me who was to pay me, and such like; he did not say anything else; my master said I must have my work when I came back.

45. Did your master tell you anything as to what he wished you to say, or what he wished you not to say?—Not anything more but what I have stated.

46. Have you related all that passed between your master and you?—Yes, I have.

MICHAEL WARD, M.D., called in, and examined, 25th March, 1819.

1. The committee are informed you wish to make some addition to your evidence?—There are some points I wish to speak more fully to than I had an opportunity of doing yesterday. Some stress seemed to be laid upon my having resigned the office of surgeon to the infirmary in the year 1805; but I wish to state I have continued to reside in Manchester; and, indeed, I have resided there upwards of forty years, including the whole time of my residing there; and within a few months of thirty years as a medical man; consequently, if I had not wished to make observations with regard to the cotton-factories, I could not have avoided it, from the frequent opportunities I have had of seeing the people coming out from the factories in crowds, and going into the factories; and occasionally attend-them as patients. Michael Ward, M.D.

2. Give to the committee such further information as you feel yourself able to do?—The result of my observation has been, and which has been particularly assisted by my visit to the three factories the last summer, that if three persons, whose names I have mentioned, Dr. Clough, of Preston, and Mr. Barker, of Manchester, and myself, if we could not remain ten minutes in the factory without gasping for breath, and without being induced involuntarily to run out of the factory with all the swiftness we could make use of; it certainly involves the question, how it is possible for those who are doomed to remain there twelve or fifteen hours to endure it; the temperature of the air is not only very much preternaturally heated, but it is rendered impure of a night by gas or candles; one or the other must be had; and, therefore, if we take into account the heated temperature of the air, and the contamination of the air, it is a matter of astonishment to my mind, how the work-people can bear the confinement for so great a length of time.

3. Do you mean to tell the committee that the condition in which you found the factories was still worse than you had expected?—It certainly was in those three factories. I was asked also, whether I had been written to,

Michael Ward, M.D.

to examine the cotton factories, by the spinning committee; I replied, I had not; I could, I dare say, have got admission to them; but my reason was, from every thing I had heard, I had reason to believe, that if I had requested to examine them, preparations might have been made, which would in a great measure have defeated the object I had in view; I, therefore, did not attempt to visit them, except those three; and my doing that occurred in consequence of meeting the gentlemen I have named, who requested me to accompany them, which I did; in one of which I have stated, that myself and Dr. Clough, who is a very robust man, and Mr. Barker also, could not bear the heat of the temperature for ten minutes without the greatest inconvenience.

4. Where do they live?—Dr. Clough lives at Preston, and Mr. Barker at Manchester.

5. Do they both practice as medical men?—Yes.

6. In your practice have you been as much in the habit of attending other descriptions of workpeople as those employed in factories?—Certainly; all descriptions of workpeople are equally eligible to be admitted into the infirmary.

7. What was your opinion of the relative state of the health between cotton-factory children and children in other employments?—I have no hesitation in saying, that the state of the health of the cotton-factory children is much worse than that of children employed in other manufactories.

8. That has been the impression upon your mind in consequence of your practice?—Yes, it has.

9. And that has been pretty extensive?—Yes, it must have been in consequence of being connected with one of the largest provincial charitable institutions in the kingdom.

10. Have you any further information to give the committee?—Though I have kept no register of the cases that have come under my observation; as far as my ability goes, I have formed in my own mind an opinion as to the result of those observations; which is, that the cotton-factories are highly unfavourable, both to the health and morals of those employed in them, generally speaking; and though the largest factories, and those which are the best managed, are much less injurious than the three I visited last summer, yet it is impossible for them to be otherwise than extremely unhealthy, on account of the heated temperature of the air required in all of them, and the very great quantity of dust which necessarily mixes with the atmosphere; and the contamination from either candles or gas; I conceive that the general opinion which has been stated with regard to cotton-factories, is a perfectly correct one, that they are really nurseries of disease and vice, that is, speaking generally upon the subject.

11. You have stated before that you have observed that children in the factories are liable to particular accidents?—It was last summer when Dr. Winstanley and Mr. Goold, along with myself, visited some of the schools, the Lever-street school, and the school in George-Lee-street; and in the course of my visit, I mentioned to Dr. Winstanley, that when I was surgeon, and he physician to the infirmary, accidents were very often admitted to the infirmary, through the children's hands and arms having been caught in the machinery; in many instances the integuments, and the muscles, and the skin stripped off down to the bone, and in some instances a finger or two might be lost; I wished to know whether he could inform me whether those accidents were still as frequent as they were at the time I alluded to; he said, that not being in his department, he could not say; I then suggested, that the probable means of obtaining the information might be for the master of the school to request, that all those children who had been hurt in factories, should go to a particular part of the room; the number of

children at that time in the school, who were employed in factories, was 106, and to my very great astonishment, when the number was counted of the children who had received injuries from the machinery in the cotton-factories, they amounted to very nearly one half out of the 106; there were forty-seven injured in this kind of way, so that had there been six more, the number would have been just half; I have mentioned, in the statement I formerly gave, that however the machinery might be improved, it required still further improvement; no doubt other children employed in other branches of manufacture are liable to similar accidents, in the callendering business for instance. Michael Ward, M.D.

12. During your early practice, did you often see children who had suffered these kind of misfortunes?—Yes, very often; I have seen accidents of a similar kind.

13. Do you wish the committee to believe, from your experience as a physician and surgeon, that forty-seven out of every 106 children employed in cotton-factories, receive serious injury from the machinery?—That is a latitude I can by no means speak to; I can speak to this fact, that the man called out that such children that had been injured in factories, should go to a certain part of the room, and they did so, to such a number as astonished me; there were forty-seven who had been injured more or less; we examined their arms and hands; some were injured as high as the elbow; one boy had lost the use of his right hand.

14. Were all those forty-seven serious injuries?—Some more so than others, but all of them had left marks of violence.

15. The committee wish again to ask you, whether from your experience as a medical man in Manchester, you believe that that proportion of children receive serious injuries?—I conceive we are entitled to draw that conclusion as far as this instance goes, as this was not a case picked out for the purpose, but happened incidentally; I conceive it is very strong as far as a single fact can go to lead to such a conclusion.

JOB BOTTOM, called in, and examined, 26th March, 1819.

Was not the breakfast which the children left that which they would have eaten if they had not been overworked; why do you suppose they left it?—They were at work, and had not time to eat it; it stands 'till it is cold; there is the flue in the mill; and they cannot eat it; their appetite gets bad. Job Bottom.

Mr. HENRY DADLEY, called in, and examined, 26th March, 1819.

1. You are a surgeon?—Yes. Mr. H. Dadley.
2. Where do you live?—In Manchester.
3. How long have you practised in Manchester?—Fourteen years.
4. Have you considerable practice?—I have been part of the time practising in public institutions, and I have a good deal of practice among the lower orders.
5. Has your practice led you to attend much to the health of persons employed in cotton-factories, and particularly children?—Yes, it has.
6. Are you of opinion that the employment, as now carried on in cotton-factories, is consistent with the preservation of the health of children?—I think it is injurious to the health.

Mr. H. Dadley.

7. Have you had any practice to enable you to say that you think that employment tends to shorten the lives of the persons employed in it?—It is my opinion that it does.

8. Does it appear to you much mischief would be likely to be obviated, if the hours were shortened?—It would do a great deal of good.

9. Have you had any opportunity of observing the difference between the health of children employed in cotton-factories, and the health of other children?—The opportunities afforded me at the Sunday-schools, and at other times.

10. What Sunday-schools did you visit, and when?—The Sunday-school in Jersey-street I visited on the 9th of March, 1817.

11. Give any information respecting it that occurs to you?—I observed the children employed in factories were much more sickly than those who were not; that was the general tendency of my observation, that their limbs were in some measure affected.

12. Were those children separated from the rest that you might observe them?—The manner in which the observation was made was this, we requested those children employed in factories to stand up.

13. What proportion of those children were cotton-factory children?—There were 96 boys, and 70 girls; the number of the children present was 200 boys, and 155 girls.

14. Did they look much more unhealthy than other children?—Yes; so much so, that I thought in a little time I had scarcely any occasion to ask; I could say, You work in a factory, do you not?

15. You found very little difficulty in discriminating?—Yes, very little.

16. Out of those 355 children in the Sunday-school, did you really distinguish any working in cotton-factories before they were pointed out?—Yes.

17. You were able to say, this child or that child is belonging to a factory, before they were separated?—I could say so mentally, and I found I was correct afterwards.

18. You mentioned there were 166 out of cotton-factories; are you not certain that there were many working in woollen-factories and other employments?—I believe there were.

19. You have said that your opinion is, that children who have been employed in cotton factories have a greater disposition to disease than others?—I think they acquire that disposition from being employed in cotton-factories.

20. Is that founded upon any facts?—Inasmuch as the employment tends to debilitate the constitution; and, in the course of my practice, I have generally been able to trace the origin of the disease to cotton-factories, at least, the patients themselves have referred the origin to it.

Mr. JOHN BOUTFLOWER, called in, and examined, 26th March, 1819.

Mr. J. Boutflower.

1. You are a surgeon, living at Manchester?—In Salford.

2. How long have you practised there as a surgeon?—Twenty-six years.

3. What is the general state of the health of children employed in factories?—The general state is, that they are very sickly; at least, they have the appearance of sickness in a very great degree; they are very much emaciated and reduced in their persons, and they are generally debilitated.

4. Do you think the employment in cotton-factories tends to shorten the

lives of those employed?—I think that those who are employed before the ages of nine or ten must suffer very severely from excessive employments; that after that period, perhaps, the injury may not be so severe to their constitutions. Mr. J. Boutflower.

5. Do you consider it would depend a great deal upon the number of hours they were employed?—The number of hours, the high temperature, the animal effluvia, the flue of the cotton, and other causes.

6. You mean to say, that from your observation of the children in the Sunday-schools, you had the best proof you could have, and so sufficient, you required no other, of the unhealthiness of the children working in cotton-factories?—Yes.

7. Did you examine into the complaints of the 152 children?—We investigated their circumstances very minutely, and, on inquiry, we found many of them were subject to coughs and hoarsenesses, and had distortions of their limbs, and had pains in their limbs, their legs, and thighs; and some few of them had glandular swellings, and tendency to scrofula; but those were but few when I made the observation.

8. Do you found that opinion upon any fact?—Upon the appearance of the children; they seem to have that appearance which constitutes a predisposition to disease; they are emaciated, and have wheezings and coughs, and some are narrow-chested; and, in short, any person of common understanding would say, they were not healthy children; I do not think it would require a medical man to make that observation.

Mr. WILLIAM SIMMONS, called in, and examined, 26th March, 1819.

1. You are a surgeon?—I am. Mr. W. Simmons.

2. Where do you live?—At Manchester.

3. How long have you been employed as a medical practitioner at Manchester?—I have lived there thirty years this summer.

4. And had considerable practice there?—Yes, extensive.

5. Are you of opinion that the employment as now carried on in the cotton-factories, is consistent with a due regard to the health of the children employed in them?—I am not of that opinion.

6. Your recollection enables you to represent to the committee, that those complaints are to be found in a very large proportion amongst those employed in cotton-factories?—They all, with very few exceptions, look very much enfeebled by it; the local complaints are occasionly scrofulous affections, and occasionally affections of the joints, chronic abscesses, sore eyes, glandular enlargements, and cutaneous affections, I have noticed to exist among them; and I have explained it by over exertion under unfavourable circumstances, that the system has been debilitated, which has given rise to these complaints; sometimes they are affected in one manner, since and sometimes in another.

7. Do they look much worse than the general population of Manchester?—I have pointed them out in the street, in the company of medical men, this question was discussed.

8. Is it your decided opinion they look most unhealthy?—It is.

9. Do you think eleven hours a day is quite as long as is safe for the health of any under sixteen?—I do.

10. As you have been surgeon to the Infirmary so many years, is there any other information you can give upon that point?—I think the hours are much too long, consistent with the health of the children, and I think too long with respect to adults; my knowledge is derived from my experience

Mr. W. Simmons.

at the Infirmary; they are a class of people I have no connexion with in any other way.

11. Do you wish the committee to understand, that you believe that 500 persons annually give up the work of a factory, or anything like that number, out of 1200?—I wish not to be understood in that light; that they do change, but what becomes of them I know not.

THOMAS JARROLD, M.D., called in, and examined, 26th and 29th March, 1819.

Thos. Jarrold. M.D.

1. You are a physician?—I am.

2. You live at Manchester?—Yes.

3. How long have you practised there?—Fourteen years at Manchester, and four or five at Stockport.

4. Are you of opinion that the health of cotton-spinners appears worse now than it did during the time you had the care of that factory at Stockport?—Very much so.

5. Does it appear to you now that there is any particular disease prevalent among those employed in factories?—I do not know that cotton-factories generate any specific disease; the children in the general now want appetite, and many of them appear in a broken state of health, and old age seems to be premature.

6. Do you attribute that to the long hours they work, and the temperature in which they are employed?—Yes; to the long hours principally.

7. Have you noticed the general state of the temperature?—It is hot, and somewhat oppressive to persons unaccustomed to it.

8. What do you think is likely to be the effect on young children being on their legs fourteen or fifteen hours a day, with very little interruption?—I could not have believed they would have sustained it; the effects I have noticed are deformity, but to what extent, compared with the whole number of children, I have not been able to ascertain, nor have I been able to ascertain how many are deformed in consequence of being in factories; but in the course of my practice perhaps I may have desired ten or twelve to relinquish the factory in consequence of deformity, who have recovered afterwards; the ankle joints, with a very considerable number, are rendered stiff, and with some almost obliterated, so as to render them splay-footed, the girls especially; I believe there are few factories without instances of this kind.

9. You have given in a statement as to the comparative ages to which spinners and people in other employments generally live?—I made some remarks upon that subject at my last examination; I fully confirm those remarks; they were made upon a large scale; I think I saw about 7000 children at the various Sunday-schools already specified, within a month of the present time; from the remarks I then made, if I state them without any fractional parts, I should say, that of one hundred children working in factories, the fathers of thirty would be dead; of four hundred children not working in factories, thirty would be dead; if I was to give a particular instance, I would take St. Clement's school; there I saw nine hundred and fifty children; there were one hundred and ten whose fathers worked in factories, of the children also working in factories; and I always remarked, that of the children who worked in factories and their fathers who worked in factories, of those the fathers of twenty-nine were dead; of the remaining eight hundred and odd, the number of fathers who were dead, and illegitimate, was, I think, ninety-one; there were one hundred and twenty in the whole. This I give from recollection, and I believe it is correct.

10. You mean the committee to understand, that taking the average ages of the school children, you found that many were fatherless, whose fathers had been employed in cotton-factories?—Yes. Thos. Jarrold, M.D.

LLEWELLYN JONES, M.D., called in, and examined, 29th March, 1819.

L. Jones, M.D.

1. You are a physician at Chester?—Yes.
2. How long have you practised as a physician?—Nearly five years.
3. Are you connected with any public institution there?—I am physician to the Infirmary at Chester, and various other public institutions there.
4. Where else have you practised?—At Holywell.
5. During the time you practised there, had you any opportunity of becoming acquainted with the state of the children employed in the cotton-factories?—I had.
6. Have you had any recent opportunity of making any observation with respect to the cotton-factories at Holywell?—About a fortnight ago I went over for the purpose of looking into them.
7. Are you acquainted with the temperature of the factories?—I prepared myself to observe the temperature accurately; I found some of the rooms of a higher temperature than others; one about 83, and others about 79, &c.
8. State any of the temperatures?—The first was 83, about noon, in the highest room of the factory; the temperature of the room immediately below was about 79, or upwards; the temperature of the room below was 79; and the different rooms varied from 79 to 63 in the four factories; there are not more than four.
9. Are you speaking of the temperature of the spinning-rooms?—Yes, the mule spinning-room was 83; the room first examined.
10. Do you know whether the Holywell factories are carried on in the night?—In two of the factories they work at night in water-spinning.
11. Did you inquire whether the children preferred working by day or night?—I did; the managers gave me to understand, without inquiry, that they preferred working at night; that statement led me to inquire of the children themselves, which I did; under these circumstances, going down on the morning following the night already spoken of, I met the night-workers returning to their homes, I inquired of them, and found that they all preferred working in the day, except one little girl.
12. From your observation of those factories, do you think that they are able to attend sufficiently to the ventilation to preserve the health of the children?—I am convinced that the ventilation is not sufficiently attended to, with a due regard to the children's health; and I received it in a statement from the manager, that full ventilation could not be admitted in any other than warm and calm weather; even in warm weather gusts of wind will prevent the proper progress of their work.
13. What general effect would be produced on their health from such causes?—The organs of digestion I should expect, and indeed found, to be materially impaired; the children were undergrown, and their appearance, in many cases, pallid and meagre, and their external appearance, to the practised eye, bespoke the fact of their having been subjected to some general causes of disease.
14. Have you observed an equal number of old people among cotton-spinners as other people?—No; I have looked to that fact, and I have observed there were but few old people, and, on inquiry, I found some of them had been absent for years as soldiers or sailors. There were, nevertheless, some old persons who had worked there from an early age.

L. Jones, M.D.

15. Have you been much employed in cases of midwifery ?—At that time I practised midwifery.

16. Have you reason to believe that the girls and young women have been much injured by the cotton-factory employment ?—I cannot give a decided answer to that question ; but I will state the fact, that during the short period of my practice at Holywell—viz., from eight to ten years—I met with more cases requiring the aid of instruments, that circumstance shewing them to be bad ones, than a gentleman of great practice in Birmingham, to whom I was previously a pupil, had met with in the whole course of his life.

17. You have mentioned as the cause of health three things, food, air, and exercise; with respect to food, you have given your opinion ; do you think that the air and exercise, as the children now have it, is or is not injurious to their health ?—I believe it is injurious.

ANSWERS of the principal PHYSICIANS, SURGEONS, and GENERAL PRACTITIONERS of the LANCASHIRE, CHESHIRE, and DERBYSHIRE District, to the Circular on the proper Limit of Labour for Children employed in Factories.

Question 17.—*What do you conceive ought to be the limit of factory labour for children from the age of nine to thirteen, and for the youth of both sexes from the age of thirteen to eighteen?*

I.—First Class of Answers: That children below twelve years ought not to work at all in Factories.

a.—I consider that children under twelve years of age ought not to be employed in factories. Under that age the osseous system is not sufficiently perfected to enable them to bear, without injury, any labour that is not easy and of short duration. Moreover, air and exercise are especially required at that period for the due assimilation of food, and for the full development of the frame.

Between the twelfth and eighteenth year it is desirable that factory labour should be limited to eleven hours; but if this limitation is to be attended by a decrease of pecuniary means, I would rather the time of labour were extended to twelve hours than see the factory operatives in the miserable condition of some of our poor artizans.

The object of shortening the duration of labour is to afford opportunities for the due growth of the body and the proper cultivation of the mind. These opportunities, however, would be in vain offered, if the wages were so reduced as to deprive the youth of both sexes of good nourishment and the means of procuring instruction.

II.—Second Class: That children should only work eight hours.

a.—From nine to thirteen years, eight hours labour; from thirteen to eighteen years, ten hours labour.

β.—I think that eight hours ought to be the extreme limit for children from nine to thirteen; and ten hours for youth of both sexes from thirteen to eighteen.

γ.—If it were a question of choice, I should think that children from nine to twelve years of age might perform with advantage the light labour of the factory for the space of eight hours a day.

Twelve hours a day is as long as it can be desirable for persons of any age to be actually employed.

δ.—I think no child between nine and thirteen ought to be allowed to labour more than one third of its time,—that is, eight hours; from thirteen to eighteen, perhaps ten.

III.—Third Class: That children should only work eight hours and a half.

a.—From nine to thirteen years no child ought to be worked more than ten hours, breakfast and dinner included; from thirteen to eighteen, twelve hours, breakfast and dinner included.

β.—From nine to thirteen, ten hours at the most, including meal-times; from thirteen to eighteen, twelve hours, exclusive of meal-times.

Protection against the employment of children under nine is required on account of the avarice of the parents, as well as for other reasons.

The hours for work should change with the seasons, so as to avoid the use of gas-lighting, which of course causes the rooms to be hotter, and the atmosphere more impure.

γ.—From nine to thirteen, ten hours; from thirteen to eighteen, twelve hours; the time for meals being included in these hours.

δ.—From nine to thirteen, perhaps ten hours; from thirteen to eighteen, perhaps twelve, though I have little knowledge of these matters; these hours include meals.

ε.—From nine to thirteen years, from eight to nine hours; from thirteen to eighteen, from ten to eleven hours.

IV.—Fourth Class: That children should not work more than ten hours a day.

a.—Certainly not more than ten hours in either case. Children exist when worked longer, but factory employment is so monotonous that the mind gets no relief; only to work and to sleep, as is the case now, is destructive of all moral principle.

β.—For all ages included in this query, I do not hesitate to offer the opinion that ten hours should be the limit of actual exertion daily; considering the nature of employment in factories, of course the younger and more delicate the individual, the shorter should be the term of active employment.

γ.—I cannot say; but perhaps ten hours for the first class, and twelve for the second, ought to be the limit, although I must confess that I have never seen any harm to arise from the first class working twelve hours; the labour is so light, it is more like play.

δ.—I cannot answer the first part of this query to my own satisfaction, from want of sufficient data. In some factories, properly constructed, in healthy localities and well-regulated, a child from nine to thirteen, in my opinion, would suffer less from eleven hours labour than in other unhealthy-situated and ill-regulated establishments, where the hours of labour did not exceed ten. I am moreover inclined to believe, that the limit of eleven hours for the youth of both sexes, from the age of thirteen to eighteen, under the favourable circumstance of well-regulated factories, would not be prejudicial to health.

ε.—I conceive that ten hours actual occupation in the duties of the factory is as much as a child from nine to thirteen years of age can bear with impunity. Children from thirteen to eighteen may probably be engaged for eleven hours, but every hour's labour beyond that period is undertaken at the expense of the constitutional stamina, and must lead to consequences the most deplorable.

ζ.—I am of opinion that ten hours a day from the age of nine to thirteen, and twelve hours a day from thirteen and upwards, ought to be the utmost limit.

V.—Fifth Class: That children should only work ten hours and a half daily, and that not before ten years.

a.—Children, I am inclined to think, ought not to be employed in factories earlier than their tenth year, otherwise they are not likely to become, under any circumstances whatever, educated and intelligent. The adults and the children ought, for many reasons, to work the same length of time daily; the adults have many duties to perform for themselves and families at home which the children have not; and besides, the children's work is lighter than that of the adults in proportion to their smaller powers. I am satisfied that the greatest temporal evil which could happen to this community would be the want of employment; as the next greatest evil, however, I should consider the permanent establishment of such hours of daily labour as must in the nature of things depress the energies of the operative, and render their moral and intellectual improvement all but impossible. Hence, as soon as, on economic grounds, the present hours of factory labour can be safely abridged, I am of opinion, considering there are so few holidays, that ten hours and a half of actual labour daily (the weekly market-day excepted) would be long enough.

VI.—Sixth Class: That children should work eleven hours daily.

a.—I have seen instances of children being apprenticed to Messrs. Greg at nine; not, however, many; from ten to twelve is the usual age, and the ninety apprenticed children are varying from ten to eighteen or nineteen years of age. The hours these are employed at the factory are, as I believe, sixty-nine hours per week, or the time which the act allows. By the answer to question 4, it would appear that this has not occasioned any extraordinary mortality; and I may further add, that not only are mortal diseases not frequent, but the health of the children is generally good. It is, however, right to add, that the children are well fed, well clothed, and comfortably lodged; how far, under different and less favourable circumstances, there might be a different result, I have not the means of knowing from my experience. It is very probable that if the factory labour of children from nine to thirteen was reduced an hour each day, or from sixty-nine to sixty-three hours, there might be some advantage to them from such reduction, though I have not had occasion to remark that the children from eleven to thirteen were not as well able to bear the labour as them from thirteen to eighteen.

β.—I am of opinion that the present time (limited by law to sixty-nine hours) is too long for any of the persons here described; at the same time I am aware that much greater evils exist in the unwholesome diet, insufficient clothing and bedding, dirty apartments, and immoral habits, than what arises from factory labour. I conceive the time ought to be limited to sixty-three hours per week, and it would be advisable to take this time from the morning work.

VII.—Seventh Class: The possibility of injury is admitted, but no remedy is proposed.

a.—Children from the age of nine to thirteen, of a weak or delicate habit, may have difficulty in standing the regular hours of factory labour, but strong and healthy children seem to suffer no inconvenience.

β.—I am unable to state; standing so many hours seems to be mainly injurious, and has partly a mechanical effect in producing disease, although the disease is generally preceded and accompanied by a general weakness and a deranged digestion.

VIII.—Eighth Class: Twelve hours labour is not considered injurious.

a.—As I have not been able to discover that the factory children have sustained any injury under the regulations that have existed hitherto, and seeing also the very advanced age to which persons have lived that have worked thirty, forty, and in one case nearly sixty years in cotton-factories, I should be disposed to discountenance any further restrictions in the factory regulation.

β.—I think twelve hours, with the usual intervals for food, for healthy children under the age of puberty.

IX.—Ninth Class: Children below eleven should not be employed at all in factories, and ten hours should be the limit for youth.

a.—I question very much the propriety of allowing children of the age of nine years, or under eleven, to work at all in factories; and I am certainly disposed to believe that more than twelve hours (including the intervals for meals), or ten hours actual labour, is inconsistent with the physical and moral health of our youthful population.

X.—Tenth Class: That the hours of labour for the youth below eighteen should be reduced to ten, ten and a half, or eleven hours daily.

Under this class fifteen medical gentlemen gave evidence.

APPENDIX.

The following Tables, I, II, and III, are extracted from the "Factories Inquiry.—Supplementary Report from Commissioners.—Part I.—(Ordered, by the House of Commons, to be printed, 25 March, 1834.)"

TABLE I.

AGES OF PEOPLE EMPLOYED IN COTTON FACTORIES IN LANCASHIRE AND GLASGOW.

FROM DR. MITCHELL'S REPORT.

[*The present Tables have been Abridged into periods of Five Years each.*]

LANCASHIRE.				
Age.	MALES.		FEMALES.	
	Number employed.	Average Weekly Wages.	Number employed.	Average Weekly Wages.
		s. *d.*		*s.* *d.*
Below 11	246	2 3½	155	2 4¾
From 11 to 16	1169	4 1¾	1123	4 3
— 16 to 21	736	10 2½	1240	7 3½
— 21 to 26	612	17 2½	780	8 5
— 26 to 31	355	20 4½	295	8 7¾
— 31 to 36	215	22 8½	100	8 9½
— 36 to 41	168	21 7¼	81	9 8¼
— 41 to 46	98	20 3½	38	9 3½
— 46 to 51	88	16 7¼	23	8 10
— 51 to 56	41	16 4	4	8 4½
— 56 to 61	28	13 6½	3	6 4
— 61 to 66	8	13 7	1	6 0
— 66 to 71	4	10 10	1	6 0
— 71 to 76	1	18 0	—	—
— 76 to 81	1	8 8	—	—
	3770		3844	
GLASGOW.				
Below 11	283	1 11¾	256	1 10¼
From 11 to 16	1519	4 7	2162	3 8¾
— 16 to 21	881	9 7	2452	6 2
— 21 to 26	541	18 6	1252	7 2¼
— 26 to 31	358	19 11¼	674	7 1
— 31 to 36	331	20 9	255	7 4½
— 36 to 41	279	19 8½	218	6 7¾
— 41 to 46	159	19 6	92	6 6
— 46 to 51	117	19 2	41	6 10
— 51 to 56	69	17 9¾	18	6 1½
— 56 to 61	45	16 1¼	16	6 0
— 61 to 66	17	17 7	7	5 5
— 66 to 71	15	15 9½	2	4 0
— 71 to 76	11	10 11	—	—
— 76 to 81	5	9 6	—	—
— 81 to 86	—	—	—	—
— 86 to 91	1	8 0	—	—
	4631		7445	

TABLE II.

WAGES PAID IN FORTY-THREE OF THE PRINCIPAL MILLS IN MANCHESTER.

FROM MR. TUFNELL'S REPORT.

PERHAPS the best test that can be produced of the flourishing condition of the workmen engaged in the cotton-trade is the wages they receive, which are so large as would appear almost incredible to those accustomed to regard the scanty earnings of the agricultural labourers. The average of the wages paid to all the persons employed in Messrs. Lees' mills, at Gorton, amounts to 12*s.* weekly per head; and as the establishment numbers 711 workmen from nine years of age upwards, we may imagine the wealth that this factory must diffuse in the neighbourhood. The average of the men's wages employed at Mr. Ashton's mills at Hyde is 21*s.* a week, while those employed out of the factory only receive 14*s.* But the following table of the wages paid in forty-three of the principal mills in Manchester will give an accurate idea of the earnings obtained at different ages.

SCHEDULE of the Number of Persons of various Ages, distinguishing Males and Females, employed in Forty-three Cotton Mills in Manchester, the average clear Weekly Earnings of each Age and Sex, the Per-centage which each Age and Sex bears to the whole Number employed, and the Per-centage of the Total of each Age relatively to the Gross Total employed.

Ages.	Number of Males.	Average clear Wages per Week.		Per-centage of Numbers.	Number of Females.	Average clear Wages per Week.		Per-centage of Numbers.	Number of each Age.	Per-centage of each Age.
		s.	*d.*			*s.*	*d.*			
From 9 to 10	498	2	9¾	2⅞	290	2	11½	1⅝	788	4·58
10 — 12	819	3	8	4¾	538	3	9½	3⅛	1,357	7·87
12 — 14	1,021	5	0¼	5⅞	761	4	10½	4⅜	1,782	10·34
14 — 16	853	6	5½	4⅞	797	6	4¾	4⅝	1,650	9·57
16 — 18	708	8	2½	4⅛	1,068	8	0½	6¼	1,776	10·30
18 — 21	758	10	4	4⅜	1,582	8	11	9⅛	2,340	13·58
21 and upwards	3,632	22	5¾	21	3,910	9	6½	22¾	7,542	43·76
	8,289	- -	-	- -	8,946	- -	-	- -	17,235	—

TABLE III.

AGES OF ONE THOUSAND PERSONS BURIED IN LIVERPOOL, MANCHESTER, & LONDON, IN ONE YEAR.

FROM APPENDIX TO MEDICAL REPORTS BY DR. HAWKINS.

THE following Table contains a classification of the ages of one thousand persons buried in Manchester, Liverpool, and London, in one year. The Manchester list is taken from the register of the collegiate church for 1830 (the latest we could obtain); and those of Liverpool and London from their respective bills of mortality for 1831. In all cases fractions are omitted.

Age.	Manchester.	Liverpool.	London.
Under 2 years - -	424	362	308
2 years and under 5 years	117	162	104
5 ——— 10 —	46	46	40
10 ——— 20 —	35	39	36
20 ——— 30 —	63	72	65
30 ——— 40 —	74	68	77
40 ——— 50 —	60	59	85
50 ——— 60 —	62	54	85
60 ——— 70 —	52	61	88
70 ——— 80 —	43	45	70
80 and upwards - -	20	27	36

TABLE IV.

AGE, SEX, AND NUMBER OF FACTORY ARTISANS.

DISTRICT.	AGES OF OPERATIVES.						TOTAL OPERATIVES.	
	Under 11.		From 11 to 18.		Above 18.			
	Male.	Female.	Male.	Female.	Male.	Female.	Male.	Female.
Scotland	285	343	6,629	14,902	8,904	19,113	15,818	40,358
North of England	39	34	542	1,021	685	1,261	1,266	2,316
Westmoreland, including Kendall	27	23	146	196	203	115	376	334
Lancashire	1,109	1,086	27,898	31,271	36,789	37,063	65,796	69,420
West Riding of Yorkshire	1,093	856	14,981	17,631	16,419	12,276	32,493	30,763
Cheshire	879	1,008	7,537	9,698	11,849	12,513	20,265	22,219
Flintshire	—	—	202	241	250	458	452	699
Derbyshire	28	28	938	1,287	2,825	2,863	2,841	2,885
Staffordshire	101	150	370	695	521	617	992	1,462
Leicestershire, &c.	1,055	1,563	2,938	6,951	3,274	8,516	7,267	17,030
West of England and Wales	181	193	3,646	3,009	5,017	4,905	—	—
South of Ireland	6	11	483	832	1,153	1,062	1,642	1,907
North of Ditto	8	13	893	2,088	970	2,050	1,871	4,151
	4,811	5,308	67,203	89,822	88,859	102,812	151,079	193,544

TABLE V.

[REFERRED TO IN PAGE LXIX.]

SUMMARY OF THE DAY AND EVENING SCHOOLS, AND OF THE CHILDREN RECEIVING EDUCATION, IN THE BOROUGH OF MANCHESTER,

EXCLUSIVE OF SUNDAY SCHOOLS, AND OF EVENING SCHOOLS ATTACHED TO THEM.

	Number of Schools.	Number of Teachers.	Number of Boys.	Number of Girls.	Total Number of Scholars.	Number who attend Sunday Schools.	ESTABLISHED.			
							In or before 1820.	1820 to 1830.	1830 and after.	Not ascertained.
Dame Schools	230	234	1,939	2,783	4,722	2,795	34	58	138	—
Common Boys' 116 and Girls' 63 Schools	179	210	4,107	2,795	6,902	3,757	27	56	94	2
Superior Private and Boarding Schools (Boys' Schools 36, Girls' Schools 78)	114	188	1,355	1,579	2,934	67	21	47	42	4
	523	632	7,401	7,157	14,558	6,619	82	161	274	6
Infant Schools	5	8	345	304	649	425	—	3	2	—
Endowed and Charity Schools	21	37	2,481	973	3,454	2,173	13	1	6	1
Total of Day Schools	549	677	10,227	8,434	18,661	9,217	95	165	282	7
Evening Schools	86	93	1,012	446	1,458	794	12	26	47	1
Total of Day and Evening Schools	635	770	11,239	8,880	20,119	10,011	107	191	329	8

PART II.

ABRIDGMENT OF THE EVIDENCE OF OPERATIVES,

CLERGYMEN, AND OTHERS,

WITH THE

EVIDENCE AT FULL OF THE MEDICAL MEN,

BEFORE MR. SADLER'S COMMITTEE IN 1832,

ALPHABETICALLY ARRANGED,

&c. &c. &c.

ABRIDGMENT OF EVIDENCE,

&c. &c.

ABERDEEN, CHARLES, age about 53,—examined 7th July, 1832,—a card-grinder in a cotton factory at Salford, in Manchester; apprenticed, when about 12, by the parish of St. James, Westminster, to Douglas and Co., of Hollywell, Flintshire; employed in different factories ever since; discharged by Messrs. Lambert, Hoole, and Jackson, on the 20th of April, for announcing his determination to support Mr. Sadler's Bill, and for refusing to sign a petition against it.

C. Aberdeen.

1. What was the nature of your employment?—I worked in a card-room, when first I commenced working in a factorv, spreading cotton.

2. Is it a very dusty apartment of the mill?—Very dusty; but it is superseded by machines; there is no spreading now by boys.

3. But still are there not various apartments of the cotton-mill now where there are many flues, and much dust?—Yes, men that are more lusty than myself, I have seen die daily for want of breath; because they were not allowed to let the fresh air in, and the foul air out.

4. Why so?—They consider that it damages the work; and that by not admitting so much air in the room, it makes a smaller surface on the flies of cotton; and that if they let too much air in, it becomes ouzy.

5. You are aware that it has been frequently asserted that the work-people in the mill have an objection to work in a tolerably cool and ventilated air?—I never heard an objection stated to let the foul air out and the fresh in; but a cry and craving for it.

6. What were the hours of labour in the first mill you were in?—From six in the morning, to seven in the evening: carding went on during the day; it was only spinning that went on in the night, while I was an apprentice.

7. What time had you for refreshment?—A whole hour for dinner, none for breakfast, or anything else.

8. During the hour that the moving power was suspended, had you to clean the machinery?—In the dinner hour, I, for one, used to have to clean and oil the machinery, and I could do that in half an hour, and eat my morsel afterwards.

9. Was it the common practice to employ the children in that interval to clean the machinery?—Not the children generally; but the scavengers for the mills were obliged to stop; they were the smallest of the children.

10. Does the business of the scavengers demand constant attention, and to be in perpetual motion, and to assume a variety of attitudes, so as to accomodate their business in cleaning the machinery to its motions?—Yes, to go under the machine, while it is going, in all attitudes, and in a most deplorable

C. Aberdeen. dress; perhaps a mantle made of the coarse stuff in which the cotton is brought, called the bagging.

11. Is it a dangerous employment in point of exposing persons to accidents?—Very dangerous when first they come, but by constant application they become used to it.

12. Do you think that the people who worked at night were less healthy than those who worked in the day?—I do.

13. Would the people have preferred to work by day, if they had had their choice?—They would have preferred to work in the day.

14. Do you think that the children who worked through the night took the rest in the day-time that they ought to have done?—I do not think they did.

15. That they were tempted, in point of fact, to play and move about in the day-time, instead of going to bed?—Yes; and in such weather as this, to go a blackberrying, and so on.

16. So that night-work left them without a proper degree of rest, and consequently deprived them of health?—Yes.

17. Could a hand choose whether he would be a day-worker or a night-worker?—If the hand, a male or female, would not come in the night, they would not give them a place in the day; and it has been rather compulsory to make them go to night-work.

18. So as to keep up their stock of night-labourers from those who have been employed by day?—Yes; it has been known that they have discharged persons who have refused to go to night-work often.

19. Are the hours longer or shorter at present, than when you were apprentice to a cotton-mill?—Much the same; especially at the place where I was last discharged. The master that I was last discharged from, had observed the Act of Parliament more than any master that I ever knew; indeed, it was framed, and hung up at the bottom of the factory stairs.

20. You say that the time of labour which is required from the children in those mills is much the same as when you first entered upon that employment; will you now inform the committee, whether the labour itself has increased, or otherwise?—The labour has increased more than twofold.

21. Explain in what way; do you merely mean that a double quantity is thrown off by some superiority in the machinery, or that a greater degree of exertion is demanded from the hands, and to the extent you mention?—The one is consequent upon the other; if the machine is speeded, it will turn off a double quantity; and it requires a double exertion and labour from the child, or from any person that is attending it.

22. Do you think there is double the quantity of labour required from the children that there used to be?—I am confident of it; since I have been working at the firm of Lambert, Hoole, and Jackson, I have done twice the quantity of work that I used to do, for less wages. The exertion of the body is required to follow up the speed of the machine.

23. Has this increased labour any visible effect upon the appearance of the children?—It has, indeed, a remarkable effect; it causes a paleness and a wanness; a factory-child may be known easily from another child that does not work in a factory.

24. Do you think it interferes with their growth, as well as with their health?—I do.

25. Has it had the effect of shortening their lives, do you suppose?—I am beyond supposing it.

26. Are you, then, confident as to that important and distressing fact?—Yes, I am confident of it from what experience I have had; and I think I have had a good deal.

27. What grounds have you for thinking so?—I have seen many instances,

but cannot state particularly: I have seen men and women that have worked in a factory all their lives, like myself, and that get married; and I have seen the race become diminutive and small; I have myself had seven children, not one of which survived six weeks; my wife is an emaciated person, like myself, a little woman, and she worked during her childhood, younger than myself, in a factory. C. Aberdeen.

28. What is the common age to which those that have been accustomed from early youth to work in factories survive, according to the best of your observation?—I have known very few that have exceeded me in age. I think that most of them die under forty.

29. Of course, if the period of their death is so much anticipated, a great deal of sickness must prevail before that event takes place?—I suppose there is not a week but what there are persons that are sick, who work in a factory; sometimes there may be ten; sometimes a dozen; sometimes half a dozen.

30. In consequence of their labour?—Not altogether in consequence of their labour, but for want of fresh air.

31. So that you consider that the hardship of the children and young persons confined to labour in factories does not altogether rest upon the circumstance of their being kept too long hours at their labour, but also has reference to the heated and unwholesome atmosphere which they have to breathe while at their work?—Yes; the friction of the brass, and the iron, and the oil, and the necessaries being in the same room; this all has a tendency to make them look ill.

32. Adverting to the trade, generally speaking, have you heard it as a usual remark and serious complaint, among the hands employed in factories, that their hours of labour were too long for them to endure with any comfort or safety?—I have heard it repeatedly said so by many.

33. What else have you to say with reference to the system?—I have something else to relate respecting the overlookers. They are men that are well paid, and are a great check to an advance of wages; I have known overlookers get 30*s.* a week, and 20*l.* a quarter bounty-money; according to the quantity of work that is thrown off, they get the bounty-money; but it is not half so much as they used to have.

34. Have the children any additional wages in proportion to the quantity of work done by their overlookers?—No; this all goes to the overlookers.

35. So that it only operates as an infliction of cruelty upon them?—Yes; those that do the most labour are the worst paid.

36. You have already stated your impressions as to the effect of the factory system, as now pursued, in reference to the health of those who are employed in it; will you state to this committee, whether it has not also a very pernicious consequence in regard to their education; and, first, have they a sufficient opportunity of attending night-schools?—I think they have not.

37. If after those hours of confinement and of labour they were to attend night-schools generally, do you think they are in a proper state, either of body or mind, adequately to avail themselves of the opportunity that might be afforded them under such circumstances?—I do not think they are.

38. Do you think that Sunday-schools are, in themselves, sufficient to obviate the great and manifest evils that must result from a total want of education?—By no means; the young persons, after they have been laboured during the whole of the week, are disinclined to attend Sunday-schools.

39. Will you state what, in the mill in which you were employed, according to your observation for the considerable number of years during which you have been engaged in it, is the actual state of morals, as resulting from excessive labour and want of education?—The morals of the children are in a bad state there; if their parents, and the Sunday-schools combined,

C. Aberdeen. were to use all their power to teach them morality, the superabundant hours and extreme debaucheries that are practised in factories would entirely choke it.

40. Do you say that from your own knowledge and belief?—From my own knowledge and belief; both debaucheries in words and in actions.

ADAMSON, JAMES, age 48,—examined 30th June, 1832,—overseer and manager at Arbroath, about eighteen miles from Dundee, upon the coast.

Mr. Jas. Adamson.

1. Have you any experience in mills and factories?—Yes, I have occasionally been engaged as an overseer, but rather more as a manager.

2. What were the hours of labour at the first mill in which you were employed?—Twelve were the stated hours, but then we made up all lost time.

3. Twelve hours independently of the time for refreshment?—Yes; we had one half hour for breakfast and another half hour for dinner, and the actual going hours were twelve; and if we lost time by holydays, or by any thing going wrong about the machinery, which stopped the mill in ordinary working hours, then we had to make it up.

4. Do you find that the children and young persons more particularly are fatigued with that length of labour?—I have seen its effects upon the young, and upon the old; I have found them upon myself.

5. The fatigue is more than the constitution can well bear?—Yes.

6. What were the hours at the next mill in which you were employed?—Twelve hours, the stated hours.

7. Still with the addition of the hours for meals, and having to make up lost time?—Yes.

8. At how early an age have you known children to labour in the mills?—I have known my own family labour before they were eight years of age, between seven and eight.

9. You have, as a parent, found that the hours of labour were too long for your children to endure, consistently with their health?—I have seen them, when they returned at night, so very tired, that if their meal was not ready when they came into the house, they were so sound asleep that we found it difficult to awake them to take their meal, and go to bed.

10. Have you observed in the mills in which you have been, that any considerable number of the hands have been absent from actual sickness, in consequence of the length of their labour?—We always found that we had occasionally a want of hands on account of sickness; and some part of the day it has been difficult to keep the work going for want of hands, on account of sickness.

11. Is it usual for you to beat the children up to that length of labour, so as to keep them attentive and vigilant at their work?—Yes; I must confess that I have beaten them a little myself.

12. It becomes necessary, when you have to exact that length of labour from children and young persons, to keep them up to it by downright chastisement?—Yes, I have found that to be the case.

13. And that not from any fault of the children, but from their inability to attend to the business with sufficient activity and success?—Yes.

14. Are you well acquainted with the operatives of Arbroath?—Yes, I am pretty well acquainted with them.

15. Are they favourable to the proposed limitation of the hours of labour of their children?—They seem very anxious for a reduction of the hours of labour.

Mr. Jas. Adamson.

16. How many mills are there in Arbroath of the nature to which you have alluded,—that is, flax or tow mills?—I think there are thirteen or fourteen different mills.

17. Have all those mills schools attached to them?—No, there are none that have schools, with the exception of one. There is a Mr. Gordon, that pays a teacher for attending at the mill two hours after labour: he has a school-room in one of his mills for that purpose.

18. Then supposing that the hours of labour, including the hour for meals, extend to thirteen hours and a half a day, and that there are two hours more for children to attend the night-school, making fifteen hours and a half a day, besides the time that must be taken in going to and from their homes, is there not a very little time for either rest or recreation, so as to preserve health, in the case of those children?—I think the time is too short for even the necessary rest, without any relaxation.

19. Can you state the general impression that prevails as to the health of the children so employed in the town of Arbroath; state your own impression, whether it is favourable, in the first instance?—I certainly consider that it is against the health of the children; and I have a certificate from the medical gentlemen of Arbroath to the same effect.

[*The witness delivered in the same, which was read, as follows:*]

"Arbroath, May 21st, 1832.

"We, the undersigned, medical practitioners in Arbroath, have no hesitation in stating it as our decided opinion, that the employment of children and young persons in the confined and impure atmosphere of spinning-mills, during the present long hours, must be, and from our observation and experience actually is, highly prejudical to health, soundness of constitution, and longevity. We need scarcely add further, that it is very unfavourable to morality, and leaves scarcely any time for mental culture.

"*Wm. Traill*, Surgeon.
John Traill, Surgeon.
Alexr. Mitchell, Surgeon.
Wm. J. Thomson, Surgeon.
Chas. Ginslay, Surgeon.
Robert W. Bruce, M.D."

20. Will you state your impression as to the effect that it has upon the character and the morals of the rising generation, those long hours of labour, interfering as they must do with instruction, both domestic and public?—My opinion is, that the length of the hours of labour puts out of their power their moral improvement, on account of their want of education, and I have a certificate from the ministers of religion to that effect.

[*The witness delivered in the same, which was read, as follows:*]

"Arbroath, 2d May, 1832.

"The undersigned have no hesitation in offering it as their decided opinion, that the present extended hours of labour in the flax-mills of this place have a most pernicious effect, both in a physical and in a moral point of view, upon the young persons employed in them. They have uniformly observed, that such young persons want the healthful aspect of children not similarly confined; that, in respect of education, they are far behind what used to be the average advancement of the same class of children in this part of the country, and that their moral and religious condition is such as was to be expected in the case of persons who are removed from school to the unwhole-

Mr. Jas. Adamson.

some air and dangerous companionship of a manufactory, at the age which is most available for the formation either of virtuous or of vicious habits. Although the practice of employing a great number of children at the mills is, in this place, only of recent date, yet enough has already appeared to prove that the tendency of such an unnatural system is to effect a rapid and certain deterioration of the race. And there is nothing of which the undersigned are more firmly persuaded than this, that if something be not speedily done to enable parents to resist the temptation, or dispense with the necessity of sending their children to work in mills for a longer period than is consistent with the preservation of their health and the improvement of their minds, Parliament will have, at no distant day, to legislate for a population tenfold more ignorant, improvident, pauperized, and immoral, than the present.

" *Thomas Doig*, Assistant Minister of Arbroath.
J. M'Culloch, Minister of St. Vigean's Chapel of Ease.
J. J. M'Farland, Minister of the Abbey Chapel.
William Henderson, Minister of the Episcopal Chapel.
P. Davidson, Minister of the Second Secession Congregation.
John Ramsay, Minister of the Independent Chapel.
William Allan, Minister of the Relief Congregation.
Robert Nicholson, Minister of the Methodist Chapel.
Joseph Hay, Minister of the First Secession Congregation.
George M'Ash, Teacher, Arbroath Academy.
David Grant, ditto - - ditto.
John Straton, ditto - - ditto.
Alex. Webster, ditto - - ditto.
Walter Low, Private Teacher, Arbroath.
John Grant, Infant-school Teacher, Arbroath.
John Adam, Teacher, Arbroath.
John Lundie, ditto.
Dd. Littlejohns, ditto.
R. Naughton, ditto.
John Hastings, ditto.
John Hackney, ditto.
George Sheriffs, Preacher of the Gospel, Arbroath."

ALLETT, JOHN, age 53,—examined 21st May, 1832,—a blanket manufacturer, began to work in manufactories when 14; eight children living.

John Allett.

1. Will you state, upon your own knowledge, whether the hours of labour have not been considerably increased (that is, in brisk times) since you were acquainted with factories?—When I went at first to factories I was at work about eleven hours a day, but time has increased to fifteen, to sixteen, and sometimes to eighteen, and sometimes to higher, even to twenty-four hours.

2. The labour of children and young persons has been increased as you have now stated?—Yes.

3. Is the work done harder than formerly?—It is much harder, and if we worked only ten hours now it is harder, because our machines are faster speeded than they were when they first began.

4. How long have you known children labour in brisk times?—Fourteen or fifteen hours a day, just as the work was.

5. How did you perceive them to bear it, especially at the latter part of the day?—I have seen my own children, when I have been in that depart-

ment of business, spinning for Monday, Tuesday, and Wednesday, and they seemed to be quite lively; but towards the other end of the week, when they began to be fatigued, whilst at every interval we have been getting refreshment, they have been sitting down, and could not abide playing, as they could at the fore-end of the week. John Allett.

6. Are they not almost continually upon their feet?—Always upon their feet; there can be no rest at all.

7. Were they excessively sleepy?—Very sleepy; I have seen them sleeping while we were at our drinking, and when in the evening my youngest boy has said, "Father, what o'clock is it?" I have said perhaps, "It is seven o'clock;" "Oh! is it two hours to nine o'clock?" I cannot bear it; I have thought I had rather almost have seen them starve to death, than to be used in that manner. I have heard that child crying out, when getting within a few yards of the door, "Oh! mother, is my supper ready?" and I have seen him, when he has been taken from my back, fall asleep before he could get it.

8. When did that child first go to the mill?—Between six and seven years old.

9. How are they kept up when they begin to be fatigued by this intense and long-continued labour?—If it be not a tender parent that is over them, they are kept up to their work by something like the lash of a slave-driver; this I have frequently seen.

10. Is the chastisement generally at the latter end of the day?—Yes, generally so; I have seen it also in the morning, because they have had so little sleep that they were hardly awakened; and I have known more accidents happen at the fore-end of the day than at the latter part; I mean before breakfast time. I was an eye-witness of one in the same place that I worked at many years: a child was working wool, that is, to prepare the wool for the machine; but the strap caught him, as he was hardly awake, and it carried him into the machinery; and we found one limb in one place and one in another, and he was cut to bits almost; his whole body went in, and was mangled.

11. Did you ever see any accidents happen at the latter end of the day from fatigue?—No, I cannot say; I have seen a spinner strike a child with a roller; I took up the child, and thought he was killed, but he got better afterwards. That man could not get any one to work for him in the whole town, and he went to another place, and he had spun there but very little time, not more than a week or a fortnight, before he took the roller, which is not less than three yards long, and four inches round about, and struck another child, and in six days he died. There was a coroner's inquest on the body, and it was brought in "accidental death."

12. You say that your children worked fourteen or fifteen hours a day; did you ever ask the overlooker to keep them only to the short hours?—Yes; I have myself asked the master, and said that we could not bear it; and my master told me that if we could not bear it, there were others that could, and we might go about our business.

13. Do not those long hours of labour not only render it impossible for the children to attend a night-school, or the Sunday-school, as they ought to do, but also prevent parents from having the opportunity of being with, and properly instructing their children themselves, and training them up in habits of domestic industry and virtue?—Yes, I am sure they do.

BENNETT, THOMAS, age 48,—examined, 18th May, 1832,—a slubber, at Dewsbury, eight children.

Thos. Bennet. 1. What were the regular hours of work at Mr. Halliley's mill?—Our regular hours, when we were not so throng, were from six to seven.

2. And when you were the throngest, what were your hours?—From five to nine, and from five to ten, and from four to nine.

3. What intervals for meals had the children at that period?—Two hours: an hour for breakfast, and an hour for dinner.

4. Did they always allow two hours for meals at Mr. Halliley's?—Yes, it was allowed; but the children did not get it; for they had business to do at that time, such as fettling and cleaning the machinery.

5. How long a time together have you known those excessive hours to continue?—I have wrought so myself very nearly two years together.

6. Were your children working under you then?—Yes, two of them.

7. State the effect upon your children?—Of a morning, when they had to get up, they have been so fast asleep, that I have had to go up stairs, and lift them out of bed, and have heard their crying with the feelings of a parent; I have been much affected by it.

8. Were not they much fatigued at the termination of such a day's labour as that?—Yes: many a time I have seen their hands moving while they have been nodding almost asleep; they have been doing their business almost mechanically.

9. While they have been almost asleep, they have attempted to work?—Yes; and they have missed the carding, and spoiled the thread, and we have had to beat them for it.

10. Will you state what effect it had upon your children at the end of their day's work?—At the end of their day's work, when they have come home, instead of taking their victuals, they have dropped asleep with the victuals in their hand; and sometimes, when we have sent them to bed with a little bread or something to eat in their hand, I have found it in their bed the next morning.

11. Were your own children obliged to employ most of their time, at breakfast and at the drinking, in cleansing the machine, and in fettling the spindles?—I have seen at that mill, and I have experienced and mentioned it with grief, that the English children were enslaved worse than the Africans. Once, when Mr. Wood was saying to the carrier who brought his work in and out, "How long has that horse of mine been at work?" and the carrier told him the time, and he said, "Loose him directly, he has been in too long," I made this reply to him—"You have more mercy and pity for your horse than for your men."

12. Do the accidents principally occur at the latter end of those long days of labour?—Yes, I believe mostly so.

13. Do you know of any that have happened?—I know of one; it was at Mr. Wood's mill. Part of the machine caught a lass who had been drowsy and asleep, and the strap, which ran close by her, caught her at about the middle, and bore her to the ceiling, and down she came, and her neck appeared broken, and the slubber ran up to her, and pulled her neck, and I carried her to the doctor myself.

14. Did she get well?—Yes, she came about again.

15. What time was that?—In the evening.

16. Could you not have got other children to supply the place of your children occasionally?—No, it was forbidden; and if one neighbour wished to take another neighbour's children, unless they were out of work they would not come.

17. When you were working in the mill, were you bound, when required,

to work the long hours?—Yes, if I had not done it, my master would have got somebody else that would. Thos. Bennett.

18. And the parish officers would not have relieved you if you had left? —No; they would have said, "You refused to work."

19. You would then have been left to starve?—Yes.

20. Did you ever know a case in which that question has been tried in a court of justice?—No, but I have tried it myself in practice. I came to some distress, and I went to the parish, and the parish then relieved me, but I obtained relief with great trouble; I was told to go back to my work; I was nearly a fortnight away; my master sent me a letter to come to my work, and we agreed again.

BENTLEY, ELIZABETH, age 23,—examined, 4th June, 1832,—as doffer, began to work, when six years old, in a flax mill, at Leeds.

1. What were your hours of labour?—From five in the morning, till nine at night, when they were thronged. Eliz. Bentley.

2. For how long a time together have you worked that excessive length of time?—For about half a year.

3. What were your usual hours of labour, when you were not so thronged? —From six in the morning, till seven at night.

4. What time was allowed for your meals?—Forty minutes at noon.

5. Had you any time to get your breakfast, or drinking?—No, we got it as we could.

6. And when your work was bad, you had hardly any time to eat it at all?—No; we were obliged to leave it or to take it home, and when we did not take it, the overlooker took it, and gave it to his pigs.

7. Do you consider doffing a laborious employment?—Yes; when the frames are full, they have to stop the frames, and take the flyers off, and take the full bobbins off, and carry them to the roller, and then put empty ones on, and set the frames going again.

8. Does that keep you constantly on your feet?—Yes; there are so many frames, and they run so quick.

9. Suppose you flagged a little, or were too late, what would they do?—Strap us.

10. Girls as well as boys?—Yes.

11. Have you ever been strapped?—Yes, severely.

12. Were you strapped if you were too much fatigued to keep up with the machinery?—Yes; the overlooker I was under was a very severe man, and when we have been fatigued, and worn out, and had not baskets to put the bobbins in, we used to put them in the window bottoms, and that broke the panes sometimes, and I broke one one time, and the overlooker strapped me on the arm, and it rose a blister, and I ran home to my mother.

13. How long were you in your first situation?—Three or four years.

14. Where did you go to then?—To Benyon's factory.

15. What were you there?—A weigher in the card-room.

16. How long did you work there?—From half-past five, till eight at night.

17. The carding-room is more oppressive than the spinning department?—Yes, it is so dusty; they cannot see each other for dust.

18. Did working in the card-room affect your health?—Yes; it was so dusty, the dust got up my lungs, and the work was so hard; I was middling strong when I went there, but the work was so bad; I got so bad in health, that when I pulled the baskets down, I pulled my bones out of their places.

19. You are considerably deformed in your person in consequence of this labour?—Yes, I am.

Eliz. Bentley. 20. At what time did it come on?—I was about thirteen years old when it began coming, and it has got worse since; it is five years since my mother died, and my mother was never able to get me a pair of good stays to hold me up; and when my mother died, I had to do for myself, and got me a pair.

21. Were you straight till you were thirteen?—Yes, I was.

22. Have you been attended to by any medical gentleman at Leeds, or the neighbourhood?—Yes, I have been under Mr. Hares.

23. To what did he attribute it?—He said it was owing to hard labour, and working in the factories.

24. Where are you now?—In the poor-house.

25. Do any of your former employers come to see you?—No.

26. Did you ever receive anything from them when you became afflicted? —When I was at home, Mr. Walker made me a present of 1*s*. or 2*s*.; but since I have left my work and gone to the poor-house, they have not come nigh me.

27. You are supported by the parish?—Yes.

28. You are utterly incapable now of any exertion in the factories?—Yes.

29. You were very willing to have worked as long as you were able, from your earliest age?—Yes.

30. And to have supported your widowed mother as long as you could?—Yes.

BEST, MARK, age 56,—examined, 2nd June, 1832,—an overlooker in flax-mills.

Mark Best. 1. What were your hours of labour in Mr. Marshall's mill?—The regular hours were from six to seven.

2. How many hours a day have you worked there, when they were throng?—From five to eight, or nine sometimes.

3. Was it the same in the other mills in which you were employed?—Yes.

4. What time was allowed in getting their meals?—Forty minutes.

5. Was any time allowed to take your breakfast, or your drinking?—No.

6. Speaking of the long hours of labour, how were the children treated when they were kept at their work for such a time?—In those rooms I have been in, spinning-rooms, they have small boys and girls to doff the bobbins off, and those that are the last they beat with a strap to make them look sharp.

7. Have you reason to think that, in any of the mills, the masters or the managers were aware that the children were thus beaten and strapped?—Yes, they knew it very well; they encouraged them to do it. Mr. Stirk's was the last place I was at; and the young Mr. Stirk made a strap for me himself, and told me to use it freely, and make them look sharp.

8. Do you think that you could have got the quantity of work out of the children, for so great a number of hours, without that cruel treatment?—No, I dare say I should not; the speed of the machinery is calculated, and they know how much work it will do, and unless they are driven and flogged up, they cannot get the quantity of work they want from them.

9. Does the better machinery increase the fatigue and labour of those engaged in watching it, or does it lessen their fatigue?—It gives more fatigue than it used to do; there are frames invented within the last few years, that they call water-frames. They spin all wet; they are heated by steam, and the place where the girls are minding them is all full of steam.

10. Is not that much more fatiguing employment for children than any spinning previously known?—Yes; this fresh system, which is called fine spinning, is spun all wet and in steam; and the frames stand so close in some

places, and the water flies from one frame to another, so that they are wet through to the skin. Mark Best.

11. Are they not liable to be exceedingly injured in consequence of this new process of spinning?—In winter-time the clothes of those that have a long distance to go from their work will be frozen to their back, and quite stiff, before they get home.

12. Had you ever any visitors come to examine your mills?—Yes; many a time.

13. Were you in the habit of making any preparation previous to strangers coming to look at your work?—Yes, they used generally to come round half an hour before those gentlemen came, to tell us to clean and get our machines tidy against that time.

14. Did any of this strapping and cruelty go on when the visitors came to look at the mills?—No.

15. Supposing a stranger comes to see a mill, would it be possible to make a room appear less dusty at that moment than it usually is?—In those dusty places, when any person is coming round to look at them, they generally send some one to acquaint them, and get all cleaned up; and during the time of their cleaning, the machinery is standing; so that if any gentleman comes up the room is clear of dust at that time.

BINNS, STEPHEN, age 39,—examined 2nd June, 1832,—began when about seven years old, to work as a piecener in a cotton factory; afterwards employed as an overlooker in several factories.

1. What is the temperature of the rooms in which hot-water spinning is carried on?—It varies; at the factory where I was employed, it was about 80°. Stephen Binns.

2. Is there any reason why the windows should not be kept open?—Yes; because as soon as the windows are opened the yarn becomes injured, because the temperature of the room is lessened; it cools the water, and the hot water dissolves the gum, and assists the rollers in breaking the flax.

3. Is not the water kept continually hot by a fresh supply?—No, by steam.

4. What is the temperature of the water?—About 110°, sometimes about 120°.

5. Have the children to plunge their hands and arms into the water?—Yes; continually, almost.

6. Has not the heat of the rooms, and the water, and the steam, the effect of almost macerating their bodies?—Yes; and their clothes are, as it were, all steamed, partially wet.

7. Are not the children so wet as to be very much endangered in going out into the street after a few hours' labour, especially in winter time?—Yes, I should think they would be frozen; I never saw any frozen, on account of having stopped last to lock up, and I never made any inquiry.

8. Do you consider this new system of spinning much more detrimental to the health than the old system?—Yes, on account of the steam and the hot water.

9. As long as this system is continued, must not they suffer in the mill, both from the steam and by being wet?—Yes; but shortening the time will lessen the evil in a certain proportion.

10. What were the hours of labour at Mr. Stirk's factory?—Thirteen hours a day, actual labour.

11. Could you keep the children to their work for that length of time without chastisement?—No; it is impossible to get the quantity of work from them without.

Stephen Binns. 12. What were your intervals for meals and refreshment?—Only forty minutes for dinner.

13. Was the chastisement inflicted principally at the latter end of the day, when they became weary?—Principally about two or three, or four or five o'clock, and in some degree all the day at times, but more after dinner than any time else, because they are more fatigued then.

14. How many hours' work had they before dinner?—Six, and seven after.

15. Are the children kept closely confined in those mills?—Yes; the same as in all others.

16. Is there anything to prevent them going to the privy when they choose?—Never when I was there; I never would allow it.

17. Are you not aware that that is the practice in some mills?—Yes, it has been so at Marshall's, in the card-room, they were allowed to go out only three times a day; but in case they are urged to go, the overlooker would let them go extra.

18. Did you observe a difference in the appearance of the children towards the latter end of the day?—Yes; in all factories there is a difference; they are all tired; the last hour but one is the worst hour of the day, and they are going to see what o'clock it is about every five minutes. I have stood in the dark, and have not spoken a word, but have given a good lacing to every one that came; they knew what it was for; but the last hour of the day is the best; after they know it is the last hour they go on more comfortably.

19. Is there little occasion for beating in the last hour?—There is less.

20. Is it possible that the masters can be ignorant of the excessive fatigue of the children employed in the factories?—I cannot say; when they go into the rooms, they do not look as to the fatigue, but if every one is not doing their duty, there is a very sharp look-out after them; if they fall sick it is hardly thought twice about, but they are sent home directly.

21. Have the children in any of those mills any opportunity of resting when they are exceedingly fatigued?—It is not allowed in any factory, I believe, to have seats at all. Whenever I see a seat in our concern, or a log of wood, or anything to sit upon, I order it to be taken away immediately, because if I do not do so I am called to account for it.

22. So that the overlooker is instructed by the master not to allow them to rest at all?—He does not say they are not to rest, but if they be sitting down he finds fault.

23. Is any conversation allowed in any of the mills amongst the children?—No, they are not allowed to be together, much less to speak with each other.

24. Have you reason to think that, upon inspections of visitors, the sickly children are kept away?—I have heard say so, but I have never done it.

25. Then the difference made in a mill by preparation for visitors is, that matters are made more tidy and clean than usual?—Yes, and the frames seem to spin better, and the hands seem to have less to do, and the girls wash their faces, and comb their hair, and make themselves look better; and being in their Sunday clothes, they appear to be in more prosperous circumstances then they really are.

26. Have you known masters themselves to assist in the chastisement of the children?—Yes; and I have beat children in the presence of a master.

27. Would they be chastised for an accident?—They are beaten for nothing at all, in fact, sometimes. This requires a little explanation: masters now get men for overlookers at low wages, and they do not properly understand the business, and in consequence of that they beat the children, when perhaps it is their own fault; the machine may be wrong, and a man of some judgment would rectify the machine with a little trouble; but a man that does not understand it lays the blame upon the poor children, and they get beaten for it.

28. Do you not think that there is a great additional expense incurred by the parents of children in consequence of those over-hours, for these reasons, among others: first, from the additional subsistence necessary, as well as waste thereby occasioned; secondly, from the medical treatment they have to obtain for the children when they are ill; and thirdly, from the loss of wages when they are kept by such illness from their work?—Yes, there are many expenses that arise from over-labour. Stephen Binns.

29. As there are so many competitors for labour, do you not conceive, that although perhaps one or two individuals of a family might get rather less, yet that the whole amount earned by the family might be as much, in consequence of more of them being employed?—I consider that it would be a general good both to the master and to the work-people, and the trifle of reduction that will take place will not be worth talking about.

30. Suppose the case of a child working ten hours a day, and of another working twelve, do you think the child working ten hours could do as much in those ten hours as a child who works twelve hours could do in twelve? —They cannot do as much in ten hours, because the machine is running at a certain speed by the clock, and it can only be minded, and all the ends be kept up, so that it cannot be exceeded; but it will be done better.

31. Does not a great deal go to waste at the latter part of the day, which diminishes the work done, and its value?—Yes, and that diminishes the profit; every individual that starts to work for a certain number of hours, proportions his labour to the time they have to work; if they have to work twelve hours they do not work so well, and they are not so attentive as they would be if they were to give over in a shorter time; I particularly observed it at the last factory, where I had to be answerable for the waste, and I found that I lost more from waste in the last few hours of the night than in any other part of the day.

32. Then you think it is from a misunderstanding of their own interests that the masters object to shorten the hours?—I do.

33. You conceive that there is much loss, in consequence of the family not being able to take their meals together, when the food is taken off to the mills?—Yes; the parents have not it in their power to calculate to a nicety, or to know how to send enough, without sending too much.

34. Have you observed a difference in this respect where Mr. Hobhouse's Bill prevails, in Lancashire and Yorkshire, as to their meals?—Yes, that was the first thing that gave me the idea: when I went to Preston, I thought it a great pleasure to have to go home together. I had two sons there, and we used to breakfast, and what was left was left for the good of the family: if I was away from home, it got cold, and I could not get it comfortably. I consider that as much is wasted as would make up, in large families particularly, for a less number of hours. I have heard parents complain, and they have said, "I will thank you to advance my children's wages." I have said, "I could not do it, I could get so many at low wages; but if they improved in their work, I would see if I could not do something for them;" and they have said, "It is very hard for us, we have only so much of income, and we must, for decency's sake, bring almost all we have for our children, and we are left nearly without food ourselves."

BRADSHAW, BENJAMIN,—examined 25th May, 1832,—a cloth-dresser at Holbeck Moor, near Leeds.

1. Have some of your children worked in mills and factories?—Yes, a few have; not all of them. Mr. B. Bradshaw.

Mr. B. Bradshaw.

2. What have been the usual hours of working to which they were subjected?—They have varied according to times, according to trade; but I have known them for years together to go at five o'clock in the morning, and work till nine at night in the woollen department, and sometimes longer than that; but I have known that for years together. When my eldest boy went to work in the mill, he was a little turned seven years of age, and he got a blow in the mill with the billy-roller, upon his loins, so that he has never been able to retain his urine for any length of time since.

3. For what time was he working at that period?—He was working from five o'clock in the morning till eight at night.

4. Is it your impression, from having several children working in mills, that much of the cruelty of treatment that is inflicted upon them results from the over-labour which they endure?—Reason dictates that when the children go to the mill at five in the morning, and work to ten at night, they are almost stupid with labour: and I know it from observation, because when I was working for Mr. Rosin, in the room underneath where I worked, frequently from seven to eight o'clock at night, you might have heard the cries of children that would have touched a heart of stone.

5. What did they beat them principally with in that mill?—With a strap; a sort of leathern belt.

6. Is that capable of inflicting a very serious hurt: are the children much hurt, and sometimes injured by it?—Yes; and I have had my own children come home beat with those things so severely, that it was hardly possible to tell the original colour of their backs.

7. Did they beat girls as well as boys in that manner?—Yes; I have known a girl of mine severely beaten for going to the privy.

8. Do you happen to know whether, in some establishments, they are limited as to the number of times they may be absent for the purposes of nature?—Yes, I do, and if they exceed those times, they are severely beaten.

9. When you remonstrated regarding the severity used towards your daughter-in-law, was the overlooker discharged?—No, the overlooker was not discharged; but the Saturday night following my children were all discharged. What was the reason of it I do not know; I met the overlooker a short time afterwards, and he asked me whether I was going to send my children back again; I told him I should not; that we would trust in God. I told him, "If you will assure me that the Almighty is dead, then shall the children come back; but as long as the Lord liveth, I will trust in him, however poor, and never let them come back to work again in that situation."

10. Supposing they do not attend punctually in the morning, are they subject to fines or to beating?—Partly to both; they are subject to beating if they are less than five minutes too late; and if they be five minutes too late, they have a quarter of a day to work for nothing.

11. Are you speaking of all mills?—I am speaking of that mill in which my children have worked, Dorp's. They have not worked at any other mill.

12. Some observations have been made upon the circumstance of females in factories not having a greater proportion of illegitimate children than females otherwise employed; do you conceive, as far as your observation has extended, that the morality of females in factories is as high as that of females otherwise occupied?—I do not believe it is; as to their not having so many illegitimate children, the reason is plain enough; there are certain books, which have gone forth, to inform depraved persons of a way by which they may indulge their corrupt passions, and still avoid having illegitimate children.

13. Do you mean that certain books, the disgrace of the age, have been put forth and circulated among the females in factories, to the effect you state?—Yes.

14. Have the children any opportunity of going to night-schools?—No, they have not. Mr. B. Bradshaw.

15. Are the Sunday-schools available for the purpose of their education?—In consequence of their excessive fatigue even on the Sunday, after such a week's labour, they are not. They will lie a-bed till twelve o'clock at noon. And my own children, though I always endeavour to preserve economical and proper arrangements in my own family, have on Sunday morning begged to lie in bed an hour or two later, and I have been forced to grant it; common feeling dictated it to me.

16. You have stated, that the children got much fatigued generally towards the end of the day; have you observed, also, whether their fatigue increased towards the close of the week, on Friday and Saturday, for instance?—Yes, on Friday and Saturday you will see a good deal more punishment inflicted upon children than you will at the beginning of the week, their powers are so relaxed and exhausted.

17. So that you attribute to these long hours of labour much of the misery that the children endure, much of the immorality which marks their conduct and character, and much of the general demoralization and ignorance that you unhappily witness?—Yes; and comparing the children of manufacturing districts with those of agricultural districts, I have every reason to believe it.

18. Do you imagine that the fluctuations of trade oblige the manufacturers to work those long hours against their own will?—I do not believe that it is altogether that; but we are all subject to temptation. One man has led the van, and another has considered that it is necessary, or would be advantageous, to follow in his steps, and consequently the practice has been followed in regular succession. That which one man has begun, another has taken up, and thus the demoralizing system has gone on increasing to the present time, and it does most powerfully call upon the legislature to use its influence in this matter.

BROOK, DAVID, age 38,—examined 16th April, 1832,—a cloth-dresser, at Leeds.

1. What age were you when you first began to work in the factories?—I did not begin to work in the factories until I was twenty-eight; I was brought up a hand cloth-dresser. Mr. David Brook.

2. When you began to work in the factories, what were the hours of labour?—From six to seven.

3. What were your intervals for refreshment?—Two hours; half an hour for breakfast, an hour at dinner, and half an hour at drinking.

4. Then the usual day was eleven hours, exclusive of meals; how long did it become?—It became as long as from four in the morning to eight, nine, and ten o'clock at night.

5. And the children that had to attend in that establishment were also kept for that length of time?—Yes.

6. Do you suppose that as much work is done in a given length of time, when the hours are increased?—No, I know there is not: and I wish to state it as my opinion, that if the ten hour bill was passed, it would operate in favour of the masters, contrary to what they expect; they would get, proportionably, considerably more work done in ten hours than is now done in twelve or thirteen.

7. From your experience, is the work done as well, when the hours are thus lengthened, as if they were kept within moderate limits?—It is not. The reason is this: we are obliged to work by gas-light, when we work in the

Mr. David Brook.

winter season, a great part of the time, and it is inconvenient to work in the night; it is impossible that a man can perform as much, or do it as well. It has been frequently found in the night, when long labour is required, that the goods have not been so well finished, and more damage has been occasioned; the goods have not pleased so well as when we worked the regular time.

8. Will you give this committee the benefit of your experience as to the effect excessive labour has on the health, especially of the younger part of the labourers employed in these factories?—As far as my experience has gone, I have generally seen that the children were very unhealthy in appearance, and that it operated very much to the injury of their constitutions, and prevented their growth. I took a boy of mine, when he was at school, at the period that I was an overlooker, because he did not seem inclined to continue at school, on purpose to see if I could not excite him to learn a little; and I kept him three months at a brushing-mill; I found in consequence of it that his health was materially hurt, and that his growth appeared entirely to be stopped, though he took more support in that time than he had done before. I took him away at the end of the three months, and put him to school, to which he was then very glad to go, and in a short time he was again lively and vigorous, and appeared quite a different boy. Generally speaking, it is found very injurious, even to men, and is not at all preferred; men do not find it at all advantageous to work long hours.

9. Explain why?—The reason is this, that although they get more money in one week, they require extra support; and additional expenses are incurred by being frequently under the doctor's hand.

10. Is it your impression that, while the public would not suffer, the master would be a gainer, rather than otherwise, by moderating the hours of labour?—I gave it as my opinion, that it would be an advantage not only to the men, but to the masters also, that the hours of labour should be regulated; it would, in some measure, destroy undue competition for labour, and the wages would be more likely to rise than fall.

11. But supposing you did anticipate a fall in the wages of labour, speaking conscientiously, would you prefer a moderate degree of labour, consistent with your general health and strength, with lower wages, or a longer period of labour with a proportionate increase of wages?—I should prefer the short labour and decreased wages, to long hours and larger wages.

12. You say that a man who refused to work nine days a week, on wishing to send a substitute, was refused; what was the reason of that?—This was a particular case, in the same shop where I am at present. The man was working in the night, and not the day part of the work; he was not very well, and requested that another person might be put in his place; the overlooker gave consent when asked at first; but half an hour after the answer was, "You cannot be allowed that; if I have a horse in the team that cannot go with the others, he may go about his business.

13. When you were an overlooker, what was the practice as to correcting children?—In our branch the practice generally has been, that the man is expected to keep the boy in order himself; he is, in a measure, left to the mercy of the man; and some men will use him better, and some worse, according to the dispositions of the men; they will correct a boy very severely sometimes; but other men will not beat them, but will make complaints. Generally speaking, in our line, I have not found any such severe treatment; I believe there is in the flax-line very severe treatment.

14. Do you think it necessary that any bill should be introduced to regulate men's labour?—I do venture to think that there are cases where that might be done with advantage. If men were free agents, as it is generally supposed they are, there would be no necessity whatever to legislate for their protection in this way; if we could work less or more as we wished, or could

bear it; if we could remove from place to place, and, if one situation did not suit, could get another; if machinery did not naturally make against us in the way it does, so as to render us at the mercy of our employers, then there would be no apology for it; but at present I do not know whether it might not be safely extended even to men. Mr. David Brook.

15. The advantage you contemplate by this bill is a more convenient and profitable employment of the men, as much as a relief for the miseries of the children?—Yes; we anticipate that.

16. That is the general opinion of the large body of the people at Leeds? —Of the operatives.

17. They do not wish it merely for the children, but for their own benefit as well?—They will, I believe, be very glad if the bill should pass for the children alone; at the same time they do contemplate it may in a measure relieve them from the excessive labour they have been subjected to.

18. You said, when you were at Mr. Gott's, you worked six days and three nights; then you worked thirty-six hours at a stretch without sleep?— I cannot exactly state how we worked at that time; it is seven years ago: but we began on the Monday morning, and worked through the night until Tuesday morning at six o'clock; then we had some sleep; about two hours; and began again at eight, and went on to a late hour that night, perhaps ten, and then we went home to rest till six in the morning. The next night we worked through the night again. The way we worked was to make up three days extra.

BROWN, HANNAH, age 23; examined 13th June, 1832.

1. Where do you live?—At Bradford. H. Brown.

2. How early did you begin to work in mills?—At nine years old.

3. At whose mill did you work first?—At Mr. Thomas Ackroyd's.

4. Will you state the hours you had to work at William Ackroyd's mill, the brother of Thomas Ackroyd?—I began at six o'clock, and wrought till eight.

5. What time was allowed for your meals?—None at all.

6. When business was very brisk at that mill, how soon have you began work?—We began at six in the morning and worked till nine at night, and sometimes at half-past five, and wrought till nine.

7. Did you ever work longer?—Yes; I remember beginning at five o'clock and working till ten.

8. Had you any time allowed for any of your meals during that period?— No, none at all.

9. I shall not ask you whether you were fatigued under such a system of labour as that, but were you ever poorly?—Yes, I was often poorly.

10. Did it at all affect your limbs?—Yes; I felt a great deal of pain very often in my legs.

11. Did it begin to produce deformity in any of your limbs?—Yes; both my knees are rather turned in; not much.

12. Were any seats provided for you?—No; but when we had our work right, we had windows for the bobbin to be put in, and if there were no bobbins in them, and the work was right, we could sit down a few minutes in the room where I was. I was under a very good overlooker.

13. Was there much punishment in that mill?—No; I cannot say there was much punishment.

14. Was there punishment?—Yes.

15. Was Mr. Ackroyd at all aware that the children were occasionally chastised?—Yes.

H. Brown. 16. Has he ever chastised you in any way?—Yes; he has taken hold of my hair and my ear, and pulled me, and just given me a bit of a shock, more than once.

17. Did you ever see him adopt similar treatment towards any others?—Yes; I have seen him pull a relation of mine about by the hair of the head, just by the side of the head.

18. Do you mean that he dragged her?—Yes; about three or four yards.

19. You again repeat, that there was no time allowed at that mill for meals or refreshment?—No, there was not.

20. Whether the length of labour was the ordinary day, or extended to those extraordinary hours, there was no time given for rest or refreshment?—No, there was not.

BURNS, CHARLES; age 13; examined 1st June, 1832.

Charles Burns. 1. What were your hours of working at Mr. Hives's, of Leeds?—From half past five in the morning till eight at night.

2. Had you any time allowed you for your breakfast there?—No.

3. Nor for your drinking?—No.

4. How much time had you allowed you for your dinner?—Forty minutes.

5. Had you sometimes to clean the machinery at your dinner hour?—Yes; and had to wipe all the machines.

6. How long did that take you generally?—About a quarter of an hour, and sometimes twenty minutes.

7. Pray how often were you allowed to make water?—Three times a day.

8. And were you allowed to make water at any time that you wanted?—No; only when a boy came to tell you it was your turn, and whether we wanted or not, that was the only time allowed us; if we did not go when he came round, we could not go at all.

9. Could you hold your water all that time?—No; we were forced to let it go.

10. Did you then spoil or wet your clothes constantly?—Every noon and every night.

11. Did you ever hear of that hurting any body?—Yes; there was a boy died.

12. Did he go home ill with attempting to suppress his urine?—Yes; and after he had been at home a bit, he died.

13. Were you beaten at your work?—If we looked off our work, or spoke to one another, we were beaten.

14. If you had not gone so fast as the machine, should you have been beaten?—If we let the machine stop half a minute we should have been beaten.

15. When you retired for the purposes of nature, how long would they allow you to stop?—If we were longer than five minutes we got beaten; and if we stopped longer they would not let us go out another time, when it was our turn.

16. Was the mill very dusty?—Yes.

17. What effect had it upon your health?—The dust got down our throats, and when we went home at night and went to bed, we spit up blood.

18. Is it not likewise, in what is called hot-water spinning, extremely hot in these mills?—Yes, very hot.

19. Is not the place full of steam?—Yes, and the machinery throws off water perpetually; so that we are wet to the skin by the hot water. And in winter time as soon as we get home our clothes are quite stiff with the frost.

20. What did you get for your breakfast and drinking?—I had tea, sometimes coffee, and butter, and bread; and my tea, for fear of wanting to make water, I used to throw out of the window. Chas. Burns.

21. In either of the mills you were in, Mr. Marshall's or Mr. Hives's, were you allowed to sit down?—No.

22. Were you not allowed to sit down during the whole of the day?—If we did we should get beaten; we had nothing to sit on unless we sat upon the frame by getting upon it.

23. Is it a common thing for you children to be beaten in this sort of way?—Yes; there used to be screaming among the boys and the girls every time of the day, and they made black and blue marks on the shoulders.

24. Where was this?—At Mr. Hives's.

25. Are accidents often occurring at these mills?—Yes.

26. State any that occurred within your own knowledge?—I had a sister who worked at Marshall's, and she got killed there by accident.

27. Were you able to attend the night-school?—No.

28. Were you able to attend the Sunday-school?—I was not able to go; I should have been too late; I had to rest on the Sunday morning.

BYWATER, DAVID; age 17; examined 13th April, 1832.

1. What age were you when you began to work?—I believe I was twelve, not turned. D. Bywater.

2. Where did you work at first?—At Mr. Hobblethwaite's, at Leeds.

3. Where did you work next?—At Mr. Brown's.

4. How old were you then?—I believe I was near thirteen then.

5. At what age were you when you entered upon night work in the steaming department?—I was nearly fourteen.

6. Will you state to this committee the labour which you endured when you were put upon long hours, and the night work was added?—We started at one o'clock on Monday morning, and then we went on till five, and stopped for half an hour for refreshment; then we went on again till eight o'clock, at breakfast time; then we had half an hour; and then we went on till twelve o'clock, and had an hour for dinner; and then we went on again till five o'clock, and had half an hour for drinking; and then we started at half past five; and if we had a mind, we could stop at nine and have half an hour then; but we thought it would be best to have an hour and a half together, which we might have at half past eleven; so we went on from half past five, and stopped at half past eleven for refreshment for an hour and a half at midnight; then we went on from one till five again, and then we stopped for half an hour; then we went on again till breakfast time, when we had half an hour; and then we went on again till twelve o'clock, at dinner time, and then we had an hour: and then we stopped at five o'clock again on Tuesday afternoon for half an hour for drinking; then we went on till half past eleven, and then we gave over till five o'clock on Wednesday morning.

7. Did you go home then?—No; we slept in the mill.

8. How did you sleep in the mill?—We slept among the white pieces, baulks, as they call them.

9. Did you undress yourself when you slept?—Yes; we took all our clothes off, except our shirts, and got into the warmest part of the mill, and amongst the driest cloth we could.

10. When did you commence on Wednesday morning?—At five o'clock, and then we worked till eight o'clock, and then we had half an hour again;

D. Bywater. then we went on to dinner time, and had an hour at twelve o'clock; and then at one o'clock we went on again till five, and then we had half an hour; and then we went on till half past eleven again; and then we started again at one o'clock on Thursday morning, and went on till five o'clock; than we had half an hour, and then we went on till eight o'clock; we had half an hour for breakfast, and then we went on till twelve, and got our dinner; then at one o'clock we went on till five o'clock, and then we had half an hour; then we went on till half-past eleven, and then we gave over till five o'clock on Friday morning; then we started again at five o'clock, and went on till eight; then we went on till dinner-time, at twelve o'clock; then at one o'clock we went on till five; then we had half an hour, and then we went on till half-past eleven; then we started again at one o'clock on Saturday morning, and went on till five; then we had half an hour, and went on till eight; then we had half an hour for breakfast, and went on till twelve; then we had an hour for dinner, and then went on from one o'clock till seven, or eight, or nine o'clock; we had no drinking-time on Saturday afternoon; we could seldom get to give over on the Saturday afternoon as the other people did.

11. Do you mean that you, as a steamer, could not give over as the rest of the people of the mill did?—Yes.

12. Did you take your meals standing, or was there a table set out for you?—We put our baskets on the boxes.

13. Did you attend a Sunday-school under these circumstances?—No.

14. How did you spend your Sundays?—I used to sleep till seven o'clock on Sunday morning, and then we got up and went a walking.

15. Did you go to any place of worship?—Yes; I always used to go to a place of worship.

16. Did you keep awake?—Yes; when I was working those long times I used to go twice or three times to the church on Sundays.

17. Were you perfect in your limbs when you undertook that long and excessive labour?—Yes, I was.

18. Shew what effect it had upon your limbs?—It made me very crooked.

[*Here the witness shewed his knees and legs.*]

19. Are your thighs also bent?—Yes; the bone is quite bent.

20. How long was it after you had to endure this long labour before your limbs were in that way?—I was very soon told of it, before I found it out myself.

21. What did they tell you?—They told me I was getting very crooked in my knees; my mother found it out first. She said I should kill myself with working this long time.

22. If you had refused to work those long hours, and have wished to have worked a moderate length of time only, should you have been retained in your situation?—I should have had to go home; I should have been turned off directly.

23. Have you received an intimation as to what will be the consequences of your having given evidence?—I was sent for to the White Swan, in Leeds, and when I got there they questioned me about what time I worked, and I told them, and they told me that I was to stop there all night; and the next morning the overlooker sent my brother down; and when he came, he said that I was to go back, or else both he and I were to be turned away; and when he went back the overlooker told him, that if I came up to London here I should never have any employment any more, nor my brother neither; but when he came again at night the overlooker cooled over it, and he told him to be at work in good time in the morning, and he has told him since that I should not be employed any more. My brother said he could not

help it; that it was not wrong in him; but I expect the first time he does a job which does not please that he will turn him away directly; because, if they work in a family, and one does wrong, they must all go. D. Bywater.

CARPENTER, JAMES; age 41; examined 4th June, 1832.

J. Carpenter.

1. Where do you live?—At Leeds, at Benk.

2. Have you worked in mills most of your life?—Yes; I began working when I was about seven years of age.

3. What were the hours of your labour when you were busy?—In the commencement of my working in factories we worked from six to seven; that was at the commencement, but afterwards we increased.

4. To what length of time was your labour increased when you became busy?—We worked from sometimes five, and sometimes half-past five, to eight at night, or half-past eight.

5. Did you ever work later than half-past eight?—Sometimes till nine.

6. In what branch of the business?—In the card-room, as a rover.

7. What time had the children allowed for refreshment?—Forty minutes at dinner.

8. Was that all?—Yes; that was all.

9. Had you to stand the remainder of the time during those long hours?—I had to stand altogether; I had no sitting; it would have been a great easement to me had I been allowed to sit.

10. State the effect which the nature of the employment had upon you?—It caused great weakness and loss of appetite; I felt gradually decreased strength of body.

11. What effect had it upon your limbs?—By becoming weak, and having to stand such a length of time, my legs were not able to bear the weight of my body, and they became crooked, as they are at present.

12. At what time was it your legs began to get deformed, as they are at present?—I was just turned twelve years of age.

13. Did this deformity come upon you accompanied with great pain?—Yes; it was great pain indeed, and has been attended with pain ever since. It was in my ankles and my knees that I felt the most pain; I did not feel much elsewhere.

14. Was your work frequently interrupted by your becoming so poorly and so deformed?—Yes; I was frequently off my work for a week, or sometimes two weeks; and I have been off a month together, and sometimes a longer time than that.

15. Do the masters, when the children become ill and diseased by labouring at the mills, continue to pay them their wages?—No; when they leave off work they leave off wages.

16. Do they usually employ a medical man to recover them?—No; they have to find their own doctor, and their own medicine, when out of work.

17. So that the master then completely loses sight of his hands?—Yes; when they are off work, they are, he considers, entirely out of his care.

18. Therefore the master, however much disposed he may be to make a fair return of the deaths of the hands, would not be able to do so?—No; their calculation would be a very imperfect one.

19. Were not the children excessively sleepy towards the termination of the day?—Yes; very much so.

20. What means were taken to keep the children to their work?—They had various means; sometimes they would tap them over the head, or nip the nose, or give them a pinch of snuff, or throw water in their faces, or pull them off where they were, and jog them about, to keep them waking.

J. Carpenter.

21. You say the hours of labour have increased during your experience: has not that increase of labour been attended with a great increase of cruelty in the management of mills?—Yes, I can speak from experience as to that; because, when I first went to work in mills, we had not much strapping and flogging; that has increased as much as the machinery has increased; there is great severity used now, more than there used to be.

22. Does not the long labour, and consequent fatigue, expose the children to a greater number of accidents than would have been the case if they had been more moderately worked?—Yes; because, when they begin to tire, they feel careless; they do not care what becomes of them; and I have seen a deal of misfortune occur to them.

23. State some of them?—I was witness to a girl that lives at Leeds, called Harriet Wilson; she lost both her arms; she is a well-known character in Leeds; I have seen many others, but that was the most serious.

24. You have stated that children are no longer heeded after they have lost their health; what is the practice regarding those servants in the establishment, when in early life, like yourself, they have lost their health in that employment?—When a servant has been in any employment, and he begins to decline in health, and cannot get through so much work, then it is that the master thinks he may lower his wages; because he has given the best of his time and the best of his service to him, he thinks he may work for less then; that I know from my own experience.

25. You find from these long hours of labour, that, if there be additional wages, the sum received is more than counterbalanced by the early destruction of their constitution, and their incapacity to perform the labour required of them?—Yes, I do.

26. You consider yourself to have been brought to a premature old age by the labour you have been exposed to?—Yes; I ought to have been worth something now, whereas I am worth nothing almost; I am not of a very great age.

27. Does not this length of labour prevent children from having any opportunity for improvement, in night-schools for instance?—Yes, it does; they have not the chance to go to a night-school, when they work so late as eight o'clock; by the time they go home and clean themselves it is bed-time.

28. They have no opportunity of deriving instruction from their parents under such circumstances?—No; the time is little at noon; and at night, when they come home, they sit them down, and are asleep from fatigue.

29. Will you have the goodness to shew the gentlemen of the committee your limbs?

[*The witness exhibited his limbs to the committee.*]

30. You have said that you were perfectly straight, as well as of a sound and strong constitution?—I was as straight as any boy that ever walked on two legs, till I was turned twelve.

31 Have you seen any medical gentleman?—Yes; I have been under many, at times.

32. What did they attribute the state of your limbs to?—They said it was owing to much standing, and working long hours; that brought on weakness of body, and in consequence the legs were not able to support the body.

33. If you wished to work those moderate hours that your constitution would have borne, you would not have been kept at your employment?—No, I should have been sent away, and another hand would have been got in my place.

34. You have always been regular in your habits?—Yes, if you were to make inquiry you would find it was so.

COLTON, ROBERT; age 16;—examined 25th May, 1832.

1. Where do you live?—At Bradford. Robert Colton

2. At what age did you begin to work in a mill?—Seven years.

3. At whose mill?—Benjamin Farrar's.

4. What were your hours of work at that mill?—From six to seven.

5. With how much time for dinner?—Half an hour.

6. Had you any time allowed for breakfast, or what you call drinking? —No.

7. Then you had twelve hours and a half actual labour?—Yes.

8. How old were you when you first went to Mr. Varley's?—Eleven years old.

9. What were your hours in summer-time, when you were busy?—From half past three o'clock in the morning, to half past nine o'clock; we increased as the days increased, and left off when we could not see; so that in summer we started at half past three in the morning, and were at it till half past nine at night.

10. How much time had you for drinking?—Half an hour.

11. How much wages?—Three shillings as regular wages the first week, and the second week they gave me three shillings and ninepence, and I never had any more.

12. As a boy, did not you feel exceedingly fatigued at that labour?—Yes; when I was at my work I kept nodding, and could not keep my eyes open; and then the others pieced up, and I was obliged to lie down on the floor, and then the man came and beat me up with a strap.

13. How were you roused so early in the morning; did your parents get you up?—No; there were two lasses appointed by the overlooker to come and knock at the doors, and make us get up.

14. Did they go throughout the whole village of Stanningley, where the work children were?—Yes.

15. Do you consider piecening hard work?—It is not very hard work, but they keep us running up and down, and it is hard work by being so long at it; it fatigues.

16. What were your hours of work at Mr. Holmes's?—From six to seven, and half an hour for dinner; and very often from six to nine.

17. Had you any more wages when it was from six to nine?—I had four shillings and six-pence, and nine-pence a week for those additional hours.

18. Had you any time to go to a day school?—No.

19. Nor a night school?—No.

20. Could you attend a Sunday-school with any profit?—No; all day on Sundays we lay a bed, we were so tired.

21. Can you read?—Very little.

22. Can you write?—No, not at all.

23. Had you, in fact, any opportunity of learning to read or write?—No.

24. At the time when you were worked for such a number of hours, supposing you had wished to work the regular hours, would they have permitted you to remain in the place, abating your wages accordingly?—No, they would not.

COOKE, SAMUEL; age 14; examined 17th April, 1832.

1. At what age did you begin to work in the factories?—I began between seven and eight years old. Samuel Cooke

2. At whose mill did you begin?—Benjamin Woodhead's, at Holmfirth.

3. State at what time in the morning you went to the mill, and, first, in the summer?—Before six o'clock.

Saml. Cooke.

4. At what time did you give over at night?—At dark, between nine and ten.

5. In winter at what time did you begin?—We began betwixt seven and eight in the morning.

6. At what time did you give over?—At nine, when there was plenty of work; and when we were slack, we gave over at dark.

7. Had you any time given you for your breakfast?—No.

8. What time was allowed for getting your dinner?—Half an hour, or about half an hour.

9. Did you stop for your drinking?—No.

10. Then you say that you had only thirty minutes during the whole day?—Yes.

11. What were you?—I was a piecener then.

12. Then you did not sit during the whole of the day, except these thirty minutes?—No.

13. How did you contrive to get your breakfast and drinking?—We got a mouthful when we could.

14. Then you had fifteen hours' daily labour when you were seven and a half years old?—Yes.

15. How were you treated at these mills; were you chastised or beaten?—Yes, sometimes.

16. Tell the committee what you were beaten with?—I was beaten with a roller.

17. Have you ever been beaten severely with a billy-roller?—Yes; when I worked at William Woodhead's I had my head broke with it.

18. Do you mean till it bled?—Yes.

19. What age were you when you were thus beaten with a billy-roller?—Between ten and eleven.

20. Did your mother ever make any complaints as to your treatment?—Yes.

21. What was said to your mother when she complained?—The slubbers all said, that if they did not like it they might take us away.

22. In what situation was your father at the time your mother complained?—My father was lame, and he could not work then; he had hurt his shoulder.

23. Where had he got lamed?—At work, when he was turning, in the factory.

24. Was he disabled from going on with his work?—Yes.

25. And your father and mother were in very poor circumstances?—Yes.

26. Could they do without your little wages?—They could not do without our working.

27. What age was your father when he died?—Between forty-six and forty-seven.

28. How long had he been ill?—He had been ill two years before he died.

29. What did he say his illness was brought on by?—He said it was brought on by working.

COOPER, WILLIAM; age 28; examined 12th April, 1832.

Wm. Cooper.

1. What is your business?—I follow the cloth-dressing at present.

2. When did you first begin to work in mills or factories?—When I was about ten years of age.

3. With whom did you first work?—At Mr. Benyon's flax mills, in Meadow-lane, Leeds.

4. What were your usual hours of working?—We began at five, and gave over at nine.

5. At what distance did you live from the mill?—About a mile and a half. Wm. Cooper.

6. At what time had you to get up in the morning to attend to your labour?—I had to be up soon after four o'clock.

7. What intermissions had you for meals?—When we began at five in the morning we went on until noon, and then we had forty minutes for dinner.

8. Had you no time for breakfast?—No, we got it as we could, while we were working.

9. Had you any time for an afternoon refreshment, or what is called in Yorkshire your drinking?—No; when we began at noon we went on till night; there was only one stoppage, the forty minutes for dinner.

10. Was not your food frequently spoiled?—Yes, at times, with the dust; sometimes we could not eat it, when it had got a lot of dust on it.

11. During the forty minutes which you were allowed for dinner, had you ever to employ that time in your turn in cleaning the machinery?—At times we had to stop to clean the machinery, and then we got our dinner as well as we could; they paid us for that.

12. To keep you at your work for such a length of time, and especially towards the termination of such a day's labour as that, what means were taken to keep you awake and attentive?—They strapped us, at times, when we were not quite ready to be doffing the frame when it was full.

13. Were any of the female children strapped?—Yes; they were strapped in the same way as the lesser boys.

14. When did you go to Mr. James Brown's?—I should think I must have been about twenty years of age when I went there.

15. Were you a gigger and a boiler when you first went to Mr. Brown's. —I was a gigger when I first went to Mr. Brown's; I was a boiler a good while after.

16. State what was your usual work, when you were only a gigger?—When I was only a gigger, I went at five o'clock on a Monday morning, and had half an hour at breakfast, and an hour at dinner, and half an hour at drinking; then went on till nine on Monday evening, and stopped half an hour; then went on to twelve at midnight, and stopped an hour; then went on to half past four on Tuesday morning, and stopped half an hour; then went on again from five to eight, and stopped half an hour; then went on till twelve, and stopped an hour; then went on again from one to five, and stopped half an hour; then again went on to nine o'clock at night, when we went home.

17. What did you do on the Wednesday?—Went again at five o'clock in the morning.

18. What time did you close at night?—At nine.

19. What did you do on the Thursday?—Went again on Thursday morning at five, and returned at nine at night. On Friday morning we went at five; worked all Friday night, until Saturday evening at five, with the same time for meals as before.

20. When you became a boiler, will you state the number of hours you had to labour at the same mill?—When I was a boiler I began work at one o'clock on the Monday morning; went on till five, and stopped half-an-hour; then went on to eight, and stopped half an hour; then went on to twelve, and stopped an hour; then went on to five, and stopped half an hour; then went on to nine, and stopped half an hour; then went on to twelve, and stopped an hour; then began again, and went on to half past four on Tuesday morning, and stopped half an hour; then went on to eight, and stopped half an hour; then went on to twelve, and stopped an hour; then went on to five, and stopped half an hour; then went on to nine, and then gave over on the Tuesday night. On Wednesday morning we went at five, and stopped half an hour at breakfast; then went on to twelve, and stopped an hour; then

Wm. Cooper. went on to five, and stopped half an hour; then went on to nine, and then gave over. Thursday was the same as Wednesday. On Friday morning we went at five, and stopped half an hour at breakfast; then we went on to twelve, and stopped an hour; then we went on to five, and stopped half an hour; then we went on to nine, and stopped half an hour; then we went on to twelve at midnight, and stopped an hour; then we went on to half past four, and stopped half-an-hour; then we went on to eight, and stopped half an hour; then we went on to twelve, and stopped an hour; then we went on to five o'clock on Saturday night, and gave over.

21. Then in the whole week you had only four nights' rest, exclusive of Sunday night?—No.

22. And that rest was after nine o'clock and before five?—Yes.

23. As I calculate, you laboured as a boiler forty-four hours running, from Monday morning till Tuesday night, having ten intervals, amounting alogether to only six hours and a half, and never going to bed?—You cannot go to bed.

24. And thirty-six hours of labour from Friday morning till you were let loose on Saturday evening, including five hours and a half for meals?—Yes.

25. On Wednesday and Thursday you had from five till nine, sixteen hours of labour, including meals?—Yes.

26. Then on Monday and Friday nights you had no rest?—None.

27. What was the effect of this excessive labour upon you?—We all felt unwell, and were stiff, and could not make proper use of our limbs till we had worked a little, when it went off.

28. After working at a mill to this excess, how did you find your health at last?—I found it very bad indeed; I found illness coming on me a long time before I fell down.

29. Did you at length become so ill as to be unable to pursue your work? —I was obliged to give it up entirely.

30. How long were you ill?—For six months.

31. Who attended?—Mr. Metcalf and Mr. Freeman.

32. What were you told by your medical attendants was the reason of your illness?—Nothing but hard labour, and working long hours; and they gave me up, and said no good could be done for me,—that I must go into the country.

33. Has it been remarked that your excessive labour from early life has greatly diminished your growth?—A number of persons have said that such was the case, and that I was the same as if I had been made of iron or stone.

34. What height are you?—About five feet; it is excessive labour that has hindered my growth.

35. When you were somewhat recovered, did you apply for labour?—I applied for my work again, but the overlooker said I was not fit to work; he was sure of that, and he would not let me have it. I was then obliged to throw myself on the parish.

36. Have you subsisted on the parish ever since?—Yes.

37. How did you spend your Sundays?—I never liked to go to a place of worship, for I was in the habit of falling asleep; I always kept away from such places, and either stopped at home or took a walk to get some fresh air.

COULSON, SAMUEL; examined 4th June, 1832.

Sam. Coulson. 1. Where do you live?—At Stanningley, near Leeds.

2. What is your trade?—A tailor.

3. Have any of your family worked in a mill?—Yes; three daughters.

4. At what age did they begin to work?—The elder was going twelve, and the middlemost going eleven, and the youngest going eight. Sam. Coulson.

5. At what time in the morning, in the brisk time, did these girls go to the mills?—In the brisk time, for about six weeks, they have gone at three o'clock in the morning, and ended at ten, or nearly half-past, at night.

6. What sort of mills were those?—The worsted mills.

7. What intervals were allowed for rest or refreshment during those nineteen hours of labour?—Breakfast a quarter of an hour, and dinner half an hour, and drinking a quarter of an hour.

8. Was any of that time taken up in cleaning the machinery?—They generally had to do what they call dry down; sometimes this took the whole of the time at breakfast or drinking, and they were to get their dinner or breakfast as they could; if not, it was brought home.

9. Had you not great difficulty in awakening your children to this excessive labour?—Yes; in the early time we had them to take up asleep and shake them, when we got them on the floor to dress them, before we could get them off to their work; but not so in common hours.

10. What were the common hours?—Six o'clock at morning till half-past eight at night.

11. Supposing they had been a little too late, what would have been the consequence during the long hours?—They were quartered in the longest hours, the same as in the shortest time.

12. What do you mean by quartering?—A quarter was taken off.

13. If they had been how much too late?—Five minutes.

14. What was the length of time they could be in bed during those long hours?—It was near eleven o'clock before we could get them into bed after getting a little victuals, and then at morning my mistress used to stop up all night, for fear that we could not get them ready for the time; sometimes we have gone to bed, and one of us generally awoke.

15. What time did you get them up in the morning?—In general, I or my mistress got up at two o'clock to dress them.

16. So that they had not above four hours' sleep at this time?—No, they had not.

17. For how long together was it?—About six weeks it held; it was only done when the throng was very much on; it was not often that.

18. The common hours of labour were from six in the morning till half-past eight at night?—Yes.

19. With the same intervals for food?—Yes, just the same.

20. Were the children excessively fatigued by this labour?—Many times; we have cried often when we have given them the little victualling we had to give them; we had to shake them, and they have fallen asleep with the victuals in their mouths many a time.

21. Had any of them any accident in consequence of this labour?—Yes, my eldest daughter, when she went first there; she had been about five weeks, and used to fettle the frames when they were running, and my eldest girl agreed with one of the others to fettle hers that time, that she would do her work; while she was learning more about the work, the overlooker came by and said, "Ann, what are you doing there?" She said, "I am doing it for my companion, in order that I may know more about it." He said, "Let go; drop it this minute;" and the cog caught her fore-finger nail, and screwed it off below the knuckle, and she was five weeks in Leeds infirmary.

22. Has she lost her finger?—It is cut off at the second joint.

23. Were her wages paid during that time?—As soon as the accident happened the wages were totally stopped; indeed, I did not know which way to get her cured, and I do not know how it would have been cured but for the infirmary.

Sam. Coulson. 24. Have any of your children been strapped?—Yes, every one; with regard to my eldest daughter, I was up in Lancashire a fortnight, and when I got home I saw her shoulders, and I said, "Ann, what is the matter?" She said, "The overlooker has strapped me; and, but," she said, "do not go to the overlooker, for if you do we shall lose our work." I said I would not if she would tell me the truth as to what caused it. "Well," she said, "I will tell you, father. I was fettling the waste, and the girl I had learning had got so perfect she could keep the side up till I could fettle the waste; the overlooker came round, and said, 'What are you doing?' I said, 'I am fettling, while the other girl keeps the upper end up.'" He said, "Drop it this minute." My daughter said, "No, I must go on with this;" and, because she did not do it, he took a strap, and beat her between the shoulders. My wife was out at the time, and when she came in she said her back was beat nearly to a jelly; and the rest of the girls encouraged her to go to Mrs. Varley; and she went to her, and she rubbed it with a part of a glass of rum, and gave her an old silk handkerchief to cover the place with till it got well.

CRABTREE, MATTHEW; examined 18th May, 1832.

Mr. Mat. Crabtree. 1. What age are you?—Twenty-two.

2. What is your occupation?—A blanket manufacturer.

3. Have you ever been employed in a factory?—Yes.

4. At what age did you first go to work in one?—Eight.

5. How long did you continue in that occupation?—Four years.

6. Will you state the hours of labour, at the period when you first went to the factory, in ordinary times?—From six in the morning to eight at night.

7. With what intervals for refreshment and rest?—An hour at noon.

8. Then you had no resting time allowed in which to take your breakfast, or what is in Yorkshire called your drinking?—No.

9. When trade was brisk, what were your hours?—From five in the morning to nine in the evening.

10. How far did you live from the mill?—About two miles.

11. During those long hours of labour, could you be punctual; how did you awake?—I seldom did awake spontaneously; I was generally awakened, or lifted out of bed, sometimes asleep, by my parents.

12. Were you always in time?—No.

13. What was the consequence if you had been too late?—I was most commonly beaten.

14. In whose factory was this?—Messrs. Hague and Cook's, of Dewsbury.

15. Will you state the effect that those long hours had upon the state of your health and feelings?—I was, when working those long hours, commonly very much fatigued at night when I left my work; so much so, that I sometimes should have slept as I walked, if I had not stumbled and started awake again; and so sick that I could not eat, and what I did eat I vomited.

16. In what situation were you in that mill?—I was a piecener.

17. Will you state to this committee whether piecening is a very laborious employment for children or not?—It is a very laborious employment; pieceners are continually running to and fro, and on their feet the whole day.

18. Do you think, from your own experience, that the speed of the machine is so calculated as to demand the utmost exertions of a child, supposing the hours were moderate?—It is as much as they can do at the best; they are always upon the stretch, and it is commonly very difficult to keep up with their work.

19. State the condition of the children towards the latter part of the day,

who have thus to keep up with the machinery?—It is as much as they can do, when they are not very much fatigued, to keep up with their work, and towards the close of the day, when they come to be more fatigued, they cannot keep up with it very well; and the consequence is, that they are beaten to spur them on.

Mr. Mat. Crabtree.

20. Does beating, then, principally occur at the latter end of the day, when the children are exceedingly fatigued?—It does at the latter end of the day, and in the morning sometimes, when they are very drowsy, and have not got rid of the fatigue of the day before.

21. What were you beaten with principally?—A strap.

22. Anything else?—Yes, a stick sometimes: and there is a kind of roller, which runs on the top of the machine, called a billy, perhaps two or three yards in length, and perhaps an inch and a half or more in diameter; the circumference would be four or five inches; I cannot speak exactly.

23. Have you yourself been beaten, and have you seen other children struck severely with that roller?—I have been struck very severely with it myself, so much so as to knock me down, and I have seen other children have their heads broken with it.

24. Do you think that if the overlooker were naturally a humane person it would be still found necessary for him to beat the children, in order to keep up their attention and vigilance at the termination of those extraordinary days of labour?—Yes; the machine turns off a regular quantity of cardings, and of course they must keep as regularly to their work the whole of the day; they must keep with the machine; and therefore, however humane the slubber may be, as he must keep up with the machine or be found fault with, he spurs the children to keep up also, by various means; but that which he commonly resorts to, is to strap them when they become drowsy.

25. You took your food to the mill; was it in your mill, as is the case in cotton mills, much spoiled by being laid aside?—It was very frequently covered by flues from the wool; and in that case they had to be blown off with the mouth, and picked off with the fingers, before it could be eaten.

26. So that, not giving you a little leisure for eating your food, but obliging you to take it at the mill, spoiled your food, when you did get it?—Yes, very commonly.

27. What is the effect of this piecening upon the hands?—It makes them bleed; the skin is completely rubbed off, and in that case they bleed perhaps in a dozen parts.

28. Is the work done as well when you are so many hours engaged in it as it would be if you were at it a less time?—I believe it is not done so well in those long hours; towards the latter end of the day the children become completely bewildered, and know not what they are doing, so that they spoil their work without knowing it.

29. You seem to say that this beating is absolutely necessary, in order to keep the children up to their work; is it universal throughout all factories?—I have been in several factories, and I have witnessed the same cruelty in them all.

30. Could you attend an evening school during the time you were employed in the mill?—No, that was completely impossible.

31. Did you attend the Sunday school?—Not very frequently when I worked at the mill.

32. How then were you engaged during the Sunday?—I very often slept till it was too late for school-time, or for divine worship; and the rest of the day I spent in walking out and taking the fresh air.

33. How many grown-up females had you in the mill?—I cannot speak to the exact number that were grown up; perhaps there might be thirty-four or so that worked in the mill.

Mr. Mat. Crabtree.

34. How many of those had illegitimate children?—A great many of them; eighteen or nineteen of them, I think.

35. Did they generally marry the men by whom they had children?—No; it sometimes happens that young women have children by married men, and I have known an instance, a few weeks since, where one of the young women had a child by a married man.

36. Is it your opinion that those who have the charge of mills very often avail themselves of the opportunity they have to debauch the young women?—No, not generally; most of the improper conduct takes place among the younger part of those that work in the mill.

DANIEL, THOMAS; examined 22nd June, 1832.

Mr. T. Daniel.

1. Where do you reside?—In Manchester.

2. What is your business?—A cotton-spinner.

3. What hours of labour are generally observed in Manchester in the spinning department?—They generally adhere to Sir John Hobhouse's Bill, in point of the number of hours of working the engine; but the children are employed working considerably longer than the engine works.

4. State how much longer than the engine they work?—It is always considered that, as soon as the engine stops, every machine stops at the moment, and it is not proper to start the machinery till it is properly cleaned and oiled: there is no time allowed for oiling and cleaning whatever. There are some mills where they clean the machinery at the time it is going, but it is a great injury to the work to stop one portion of the machine, while the other is going, such as the spindles or rollers; the spindle is where the thread is twisted from, and stopping two or three of those while the others are going is a great injury to the work, and the masters do not allow it; neither do the spinners allow it, if they know it, but the children will sometimes do it.

5. So that the children work longer hours than the adults in the mills and factories, under those circumstances?—I should think that the children that are employed under me work at least three quarters of an hour a day longer than I do, and I am as strict in point of time as I possibly can be; in fact, some of my children have worked with me for a long time, which is a proof that I have treated them kindly.

6. How much should you calculate the confinement in those mills and factories to amount to at present?—The law is twelve hours, and I should think that they work not less than twelve hours and three quarters; that is, the children.

7. To which period must be added the time for meals and refreshment?—Certainly; the hour and a half which is allowed must be added to that, for they come to work at half past five o'clock in the morning, and work till seven in the evening.

8. In what temperature do those children usually work?—In the fine mills, I should think that the heat would average 83°; in some mills it is unquestionably higher; and in others the masters do not take so much notice of the heat; the men regulate it themselves.

9. Do not you consider that the working by gas, which must be necessarily used in those hours of labour for a considerable part of the year, is very injurious to the health and eye-sight of those so employed?—I do from my own experience think that it is injurious both to the health and eye-sight; for often after working by gas-light, when I have gone home and sat down to write, and have had to mend a pen, I could not see to point it.

10. So that the light that is used, and the heat of the room being raised

considerably by the gas-lights, is an additional reason why the length of the labour of those children should be moderated ?—I do consider so.

Mr. T. Daniel.

11. Will you state the average age of those you call scavengers ?—The average age of scavengers will not be more than ten years.

12. Describe to the committee the employment of those scavengers ?—Their work is, to keep the machines, while they are going, clean from all kinds of dust and dirt that may be flying about, and they are in all sorts of positions to come at them; I think that their bodily exertion is more than they are able to bear, for they are constantly kept in a state of activity.

13. State the effect that it has upon the children, according to your own observation ?—Those children, every moment that they have to spare, will be stretched all their length upon the floor in a state of perspiration, and we are obliged to keep them up to the work by using either a strap or some harsh language, and they are kept continually in a state of agitation ; I consider them to be constantly in a state of grief, though some of them cannot shed tears : their condition greatly depresses their spirits.

14. You consider, then, upon the whole, their state as one of extreme hardship and misery ?—So much so that I have made up my mind that my children shall never go into a factory, more especially as scavengers and pieceners.

15. It is superfluous to ask you whether, after a day's confinement similar to that which you have stated, there is any possibility of the children attending night schools, or at least of their attending them with any degree of advantage ?—Not any ; I have known instances where the parents have compelled the children to go to school; but they have been ready to turn round, and use almost any sort of language that they could lay their tongues to ; in some instances I have known parents to compel their children to go, but they have gone the other way ; sometimes I have known Sunday schools to appropriate certain nights in the week to instruct the scholars in writing and something of that sort, but they have been obliged to give it up.

16. Is not there a very perceptible difference in the quality of the work that is done when the hands are fresh, and when they are fatigued by the length of time which you have described as common in Manchester ?—Most decidedly the work is better; the hands follow it with a greater degree of vigour, and more pleasure, and especially the younger children; it is those hands that we principally depend upon for the quality of our work, such as pieceing; they are more cautious in pieceing the threads when they are not so fatigued; it is therefore a great deal better for a master, for every end that is not properly pieced has to be broken down again.

DAWSON, JOHN ; age 26 ; examined 8th June, 1832.

John Dawson.

1. Where do you reside ?—In Leeds.
2. Have you ever worked in any mills or factories ?—Yes, in a flax mill.
3. At what age did you commence ?—Between six and seven, I believe.
4. What were your hours of labour ?—In a general way, from six to half past seven.
5. What intervals for meals and refreshment ?—Only forty minutes at noon.
6. No time for breakfast or drinking ?—No.
7. When the trade was very brisk, to what extent were you worked ?—From half-past five to nine and half-past nine.
8. Only the same time allowed for dinner ?—That was all.
9. State the effect that that labour had upon you ?—It was very heavy work indeed ; carrying bobbins ; it caused the pain to come in my knees and

John Dawson. ankles, and it caused me to grow crooked, working the long hours that we had to work.

10. Shew the gentlemen of the committee your limbs.

[*The witness exhibited his limbs to the committee.*]

11. Were you originally quite straight?—Yes, as straight as could be, when I first went to the mills.

12. How long did you remain so before your limbs began to fail you; to what age?—About ten or eleven.

13. Had you any opportunity of going to school during the period you worked so long?—No, I had no opportunity, only on the Sunday; I wished to go to school, when I could; but I was obliged to stop at home on many a Sunday, being tired and fatigued; on the Sunday I could scarcely abide to walk to school.

14. Had you to be chastised to keep you active and vigilant during such a long day's labour?—Yes, the overlooker kept a strap with a strong handle, and sometimes they struck with the handle, and sometimes with the leather part; and if they had a bobbin in their hand they would fetch them a rap over the head with it, if they saw them standing and not looking at their work.

15. After you were no longer capable of working in a factory, what became of you?—I went to the workhouse.

16. At what age was it you were obliged to find an asylum in the workhouse?—I was nearly sixteen.

17. Do you think that those long hours are very prejudicial to the morals of the children engaged in those factories?—Yes, they are.

18. Do you conceive that the children were indecent in their conduct, and immoral in their behaviour generally, from being deprived of the opportunity of having education?—There is the most indecency going on in the mills and factories that ever I saw; I know it by what I have seen; and the girls are, many of them, I know, in the streets now; they would not stop any longer in the mills on account of the hard labour and long hours; I know it for a fact.

19. Were the girls beaten as well as the boys?—Yes.

10. And laboured for the length of time you have described?—Yes.

21. Would the parish have relieved those parents who had taken their children from this employment merely because they did not choose to labour so long?—They did not.

DEAN, ALEXANDER; age 26; examined 29th June, 1832.

Mr. Alex. Dean.

1. Where do you reside?—At Dundee.

2. What is your employment?—Overlooker, at present.

3. How many years have you been employed in the flax business?—About fourteen.

4. To what length of time do you think that your labour may have extended when you were a card-minder?—Not less than seventeen hours, to my recollection, exclusive of meals.

5. How were the children kept in the establishment, they having to labour to such an extent as you have described?—They were kept in a constantly standing posture; no leave was allowed for sitting.

6. Were they confined to that sort of work?—Yes, the doors were all locked, both with check and turnkey.

7. Did they sleep upon the premises?—Yes.

Mr. Alex. Dean.

8. Were the children and young persons kept to that work by chastisement?—Yes.

9. What were they beaten with generally?—Sometimes the master's son would take a strap; or if the master himself was about the work, and saw a small fault, he struck them both with his feet and hands.

10. Were you ever beaten?—Very often.

11. In what manner?—I was struck one time by the master on the head with his clenched fist, and kicked, when I was down, with his feet.

12. Did you see anybody else treated in that sort of way?—Yes; I saw one girl trailed by the hair of her head, and kicked by him when she was down till she roared "murder" several times.

13. Where were the hands that did not sleep in the mill sent at night?—The houses which the hands slept in were about fifty yards from the mill.

14. What were they called?—Bothies.

15. In any of those bothies were the boys and girls mixed indiscriminately at night?—Yes; I myself, with six boys, was in one apartment with oldish girls.

16. What were the ages of those boys so locked up?—From fourteen to sixteen.

17. And what were the ages of the females?—From twelve to fourteen.

18. What effect upon the children, the female children more especially, has this long standing to their labour?—The feet of the girls have swelled so that they have been ready to take off their shoes.

19. Does it occasion positive deformity sometimes?—Yes, very often; the girls become knock-knee'd and bow-legged.

20. Has it at all affected you?—Yes, I am very much knock-knee'd.

21. Have you seen one of the witnesses in waiting, of the name of Openshaw, a boy?—Yes.

22. Is there anybody that you have witnessed in your neighbourhood, that is as strikingly deformed as he is?—A great deal more so; one man that is working now at a mill near Brachin, about twenty miles from Dundee, and who is about thirty years of age; this man does not stand, with his deformity, above four feet six inches high, and had he grown to his proper height I think he would have been about five feet eight, or five feet nine; he has been in mills since he was five years old, and he is reduced to that state that he slides about upon a stool to do his work; and though he is about thirty years of age, he can now do no more than a girl's work.

23. What is that individual's name?—It is Saunders Crabb.

24. As he was one of the individuals who would have been examined before this committee, only that it has been represented that it is with great difficulty that he could be brought so far, will you have the goodness to give a description of him?—He has the form of the letter Z; his body is twisted in one direction, and his shoulders in another; the body is twisted backwards and forwards, and his legs are so bowed that he mostly sits upon his heels.

25. Is it known that that deformity has come upon him since he has worked in the mills?—Unquestionably; every one that knows him knows that.

26. Do you believe that he could not possibly have come up to be a witness upon this occasion?—No; to my certain knowledge he is not capable of crawling above half a mile; he is not capable of conducting himself without some person being with him to assist him; so therefore he could not attend.

27. To the best of your recollection, how long in any one day do you think you have been kept to work in the mill at Strathmartin?—I was possessed of a watch there for some days, and I recollect fifteen hours, exclusive of meals, was the time we worked.

28. Could you not, then, have made the same remarks the whole time you

Mr. Alex. Dean.

were there?—After the overlooker found I was possessed of a watch, I had lost the key, and he took the watch and broke it, and gave it me back, and said, I had no use for a watch, and chastised me for letting the hands know the time of day.

29. Were the hands principally young ones at the Duntruin mill?—Yes; there was a great number of them below twelve.

30. Where did they come from?—Some from the poor-houses in Edinburgh.

31. Were they sent young?—Yes; they came at six and seven years old.

32. And they were sent for a stated length of time?—Yes; I know some that were engaged for three and four years.

33. Did you ever hear of any one attempting to escape from that mill?—Yes; there were two girls that made their escape from the mill through the roof of the house, and left nearly all their clothes behind them.

34. What became of them?—They were not brought back during the time I was there.

35. Do you know any body that escaped, and was brought back again?—At the time I was in that mill, there was a young woman who had been kept seven months in the gaol at Dundee for deserting the mill, and she was brought back, after having been in the gaol for seven months, to make up for her lost time and the expenses incurred; one day I was alarmed by the cries of "murder" from the lowest flat, and when I went there she was lying on the floor, and the master had her by the hair of her head, and was kicking her on her face till the blood was running down.

36. Was that at Duntruin mill?—Yes.

37. How long ago?—About eleven years.

38. What was the consequence of that?—I understood it would break her engagement; and after the master had retired from the flat, I opened the door and let her out, and told her to run; and the master came back, and, missing her out of the apartment, began cursing and swearing at me for letting her out, and ordered me to run after her, which I refused to do. I stated that, owing to the ill-treatment she had received, I never would be the man that would run after her to bring her back to the torture, and therefore he and I separated.

39. Was she brought back?—No.

40. Was she in a situation to get any other employment?—No; she became a prostitute, and was tried at the circuit of Perth, and transported to Van Diemen's Land, for stealing.

41. Do you think that that severity of treatment has not unfrequently a similar effect, in driving females to improper courses?—Undoubtedly; because, from the way in which those children are brought up in mills, they have no time to get education; and when they become fourteen or fifteen years of age, and think of getting into service, no person will have them, because they know nothing; and even an operative will shrink at the idea of taking a girl out of the mills, because she knows nothing. I knew one person that had married a girl out of the mill, and at the time I knew him, they had three children, and he had to pay for the washing.

42. The wife not knowing even how to wash the clothes for the family?—No.

43. Could she then either make or mend them?—She could do neither; it was all paid for.

44. Did you know any individuals brought to trial for inflicting the extreme punishment you have described?—I heard of one; there was an overlooker in Mr. Edwards' mill at West End, Dundee, who was brought before the justice for licking a girl, and on being examined before the justice he was fined; but the master returned the fine back to the overlooker, and turned away this girl whom he had struck, and also her sister and two other girls,

who were connected with her. Mr. Edwards was questioned about it in the Advertiser Paper, and he refused to answer. The only reply he made was, that he could do anything he liked with his own; though four or five suffered by that transaction of taking the overseer to justice.

45. Statements have frequently been made to the House of Commons, of the number of persons employed in mills, and of the number of such persons that have died in the course of the year, and of the number of such persons who have been ill in the course of the year, and according to those returns it would appear that very few of the operatives have died, and that they have in general enjoyed better health than other classes of the community; will you have the goodness to tell the committee in what manner you suppose those returns can have been made up?—I will state that as briefly as I can. Those returns they cannot get out of their books, because the hands are always scored out of the books; I never knew one hand yet of whom it was registered in the book that she died; no one, whether master or overseer, can state the number of hands that have died in the course of a year. The way those returns are made is this: they have taken two weeks in the summer time, two of the best weeks, and two in the winter time, and they have calculated those up for the whole year. I see no other way in which they can have got up this statement.

46. Suppose that a certain number of individuals had been employed in the mill, say that the mill employs five hundred, if a considerable number of the operatives were taken ill and gave up their work, and went to their homes and died there, they would not be registered as having died in the employment of the mill?—I never knew an instance.

47. It has been stated, by preceding witnesses, that considerable preparation has been usually made in mills previously to the expected visits of strangers and others coming for the purpose of inspecting them; do you believe that anything of that kind occurs in your district?—It is always the case.

48. Would it occur in a more particular manner, do you think, if any attempts were made to inspect those mills, in order to make out a case in their favour?—Yes, undoubtedly; some of the hands would be changed out of the mills altogether, and they would not let you see them. It was in May, I think, I received an order, when the Board of Health was going about visiting the mills, and they wanted to see if the system would have any great effect upon this plague that was raging, and I got orders that day to get all my flats cleaned and my windows dusted; and that is always the case, even when our master comes to visit the mill; we are always told to getevery thing cleaned up, in order that it might not be seen so bad as it is; and every gentleman or lady that comes to our mills and flats is always kept ignorant by this blindfolding plan, and they never imagine that our case is so hard and so bad as it is.

49. Would the cruel beatings to which you have been alluding entirely cease during any such visits?—Undoubtedly; everything is tried, in a case of that kind, to make the thing appear more bright than it is.

50. Do you believe that the hands, especially the younger ones, are more liable to accidents after they get fatigued and drowsy, at the termination of such a day's labour as is usually endured in those mills, than at any other time of the day?—Yes.

51. Do you attribute the loss of limb and of life, in many instances in which accidents occur, to the excessive labour which those children are compelled to undergo?—Yes, I had one girl under my charge at Dundee; she could not see very well, and towards evening (it is not very long since this happened) she was drowsy and sleepy, and her thumb came in contact with the machinery. I was about three or four yards from her when I heard the snap, and by the time I came up to her, her thumb was away from her hand, and

Mr. Alex. Dean.

she held her hand out to me with her thumb wanting; it was the piece of rove that had gone round the metal roller, and this had caught her thumb; and her whole body would have to go round this metal roller, or the thumb be twisted off; so the thumb was taken right off, the same as by a cut with a razor.

DOWNE, JONATHAN; age 25; examined 6th June, 1832.

Jon. Downe.

1. Where do you reside?—Leeds.

2. Have you ever worked in any mills or factories?—Yes.

3. Where?—I first went to work at Mr. Marshall's, at Shrewsbury, when I was seven years old.

4. State the hours of labour in that mill?—The regular average was from half past five to seven, or from six to half past seven.

5. Were very severe methods adopted in order to compel you to work for that length of time?—Yes, very severe.

6. Describe them?—I have seen boys actually knocked down with a strap; they have been called from their work, flogged, and been knocked down on the floor by the blow of the strap, and when they have been on the floor, they have been beaten till they had risen, and when they have risen, they have been flogged to their work again.

7. Was that a common occurrence?—Yes.

8. You are the brother of Samuel Downe?—Yes.

9. He has stated to the committee an extraordinary instance of cruelty he suffered, in being bound to a pillar, and then flogged for a considerable length of time; have you reason to believe that it is true?—It is quite true; and I know many at Leeds now that have been bound to pillars in the same room.

10. He stated he was bound by leather straps to the pillar, while the overlooker flogged him to the degree he described to the committee; was that an uncommon occurrence?—No, quite common.

11. Do you mean to say it was common to strap the boys to a pillar, and beat them to the degree your brother was beaten?—Not to that extent; but they were often bound to the pillar and beaten; but my brother was beaten several times in one day.

12. Was any complaint made to Mr. Marshall?—I cannot say positively that there was; if they had complained to Mr. Marshall, they would have been discharged.

13. Do you mean to say, that a boy that was beaten and complained would have been turned away?—Yes; if not directly, he would have been soon afterwards.

14. Will you state whether the labour and severity that the females had to submit to in these mills has had the effect of driving many of them to improper courses?—Yes, I know many, many instances; I knew two young women, as nice young women as ever stepped, that were apprenticed to Mr. Marshall; they were taken formerly at Shrewsbury, when the mill started, and when there were not hands in the town to supply them; they were taken from different parishes; some came as far as from Hull, in Yorkshire; they were bound apprentice to Mr. Marshall and Mr. Benyon; and those two young women I knew in particular were apprentices to Mr. Marshall; they never worked anywhere else. They were discharged from there when they were twenty-four or twenty-five years of age, and Mr. Benyon and Mr. Marshall being then acquainted, the other would not employ them; they had not heart to go elsewhere to seek employment, and they turned common prostitutes.

15. Is it the usual practice to prepare mills previously to their being inspected for any purpose of a public nature?—Yes, frequently. Jon. Downe.

16. State an instance?—Only very lately there were two or three gentlemen in Yorkshire, who visited Mr. Holdsworth's silk-mill at the Bank; and I was in company with a person the very same day at a quarter past twelve o'clock, and the gentlemen that were there came in the morning; and when they went there the machinery was wholly cleaned, and the hands were sent home in the morning to dress themselves; and the engine was stopped forty minutes for them to dress themselves in their best attire; and, moreover, the engine was put to run slower, while the gentlemen went through the concern.

17. So that the engine was abated in its velocity, the apparent effect of which would be to persuade the visitors that less exertion and labour were exacted from the children than was actually the case?—Yes.

18. When you worked in mills, what methods were taken to rouse the children from drowsiness?—It is a very frequent thing at Mr. Marshall's, where the least children are employed (for there were plenty working at six years of age), it has been the regular practice, of late years, for Mr. Horseman to start the mill earlier in the morning than he formerly did; and, provided a child should be drowsy, the overlooker walks round the room with a stick in his hand, and he touches that child on the shoulder, and says, "Come here." In a corner of the room there is an iron cistern; it is filled with water, so that if any fire should occur in the room they could quench it with that water; he takes this boy, and takes him up by the legs, and dips him over head in the cistern, and sends him to his work for the remainder of the day; and that boy is to stand, dripping as he is, at his work; he has no chance of drying himself. Such, at least, was the case when I was there.

19. Are many crippled from standing?—Yes.

20. How are they affected?—It first begins with a pain in the ankle; after that they will ask the overlooker to let them sit down; sometimes the overlooker will allow them a few minutes to sit down, but they must not let the time-keeper see it; it depends upon the humanity of the overseer; then it goes on for a month or two months: then they begin to be weak in the knee; then they begin to be knock-knee'd; after that their feet turn out; it throws them knock-knee'd and splay-footed, and their ankles swell as big as my fists.

21. After they get to that state, can they go on with their work?—They may follow it for a few days, or sometimes for weeks, but then they are regularly off and on their work, and under the physician's care. When they are off work a few days, they consider they recover their strength a good deal; then they return to their work again, and after they have returned to their work a day or two, or sometimes the first day, they are as bad as before.

DOWNES, GEORGE; age 41; examined 7th July, 1832.

1. Where do you reside?—Ashton-under-Line. Geo. Downes.

2. In Lancashire?—Yes.

3. Have you been accustomed to work in mills and factories for any part of your life?—Yes, all my life.

4. What time did you first commence working?—A little before I was seven years of age.

5. In whose mill?—Mr. Robert Lees's, of Padfield, in Derbyshire.

6. What was your employment?—Back tenter of a rover.

7. How many hours of labour had you at that mill?—About fourteen, exclusive of meals.

Geo. Downes. 8. What time was allowed for meals then?—We had sometimes an hour for dinner; not always an hour.

9. Had you any time for getting breakfast, or what they call, in Lancashire, bagging?—No.

10. Had you not to be subjected to a considerable degree of cruelty, to keep you up to that degree of labour for that length of time?—Yes.

11. Describe the methods that were taken with you?—The rover that I worked for had a long heavy stick, with which he often struck me severely when he could reach me; when I got out of his way, the machine being between us, he would take off the top roller, partly composed of iron and partly of wood, which had at the end an iron pike to keep them asunder from each other; he would take them off, and throw them at me with all his force; the pikes often caught me on the head and elsewhere; he very frequently threw these rollers at me, till at one time he threw one of them with such a force, that, notwithstanding its catching one side of my head slightly, it bounced from the floor two yards high, and flew into the machinery standing near where I was, and damaged it exceedingly.

12. Have you observed, generally, that the girls are beaten as severely as the boys?—Yes; they do not make any exceptions in that respect.

13. Do you think that children could have been kept up to that degree of labour, if they had not been so chastised in the evening; could they have been kept awake?—No, they could not.

14. What have been your observations as to the effect of the length of labour required from the children in the different mills and factories in which you have been?—Generally speaking, the children have had ill health in consequence of it, and been very ignorant too, in consequence of not having time to attend schools.

DOWNE, SAMUEL; age 29; examined 4th June, 1832.

Saml. Downe. 1. Where do you live?—At Hunslet Car, near Leeds.

2. Are you a Leeds' man?—No, a native of Shrewsbury.

3. Have you been acquainted with factories?—From my youth.

4. At what time did you begin to work at one?—About ten years of age.

5. In whose mill did you work?—In Mr. Marshall's, at Shrewsbury.

6. What were the hours of labour in that mill when they were brisk?—When they were brisk we used generally to begin at five o'clock in the morning, and they ran on till eight at night; sometimes from half past five to eight, and sometimes nine.

7. Was very considerable severity used in that mill when you were there?—Yes.

8. Have you yourself been subjected to it?—Yes, I was strapped most severely till I could not bear to sit upon a chair without having pillows, and I was forced to lie upon my face in the night-time at one time, and through that I left; I was strapped both on my own legs, and then I was put upon a man's back, and then strapped and buckled with two straps to an iron pillar, and flogged, and all by one overlooker: after that he took a piece of tow, and twisted it in the shape of a cord, and put it in my mouth, and tied it behind my head.

9. He gagged you?—Yes; and then he ordered me to run round a part of the machinery, where he was overlooker, and he stood at one end, and every time I came there he struck me with a stick, which I believe was an ash plant, and which he generally carried in his hand, and sometimes he hit me, and sometimes he did not; and one of the men in the room came and begged me off, and that he would let me go, and not beat me any more, and consequently he did.

10. You have been beaten with extraordinary severity?—Yes; I was beaten so that I had not power to cry at all, or hardly to speak at one time. Saml. Downe.

11. What age were you at that time?—Between ten and eleven.

12. What had you done?—I believe that in the machinery I did not like the part he put me to, because I had never been in a mill where there was any machinery before in my life, and it was winter time, and we worked by gas-light, and I could not catch the revolutions of the machinery to take the tow out of the hackles; it requires some little practice, and I was timid at it, and pricked my hands very much with the hackles; I desired him to remove me to another part, which he did for some part of the day, and then sent me back again to that which we call doffing the hackles.

13. You say you were beat so that you could not even cry?—I cannot assign any other reason for it; it was not because I had not sufficient punishment; I did my endeavours, when he had used some mode of language that gave me to understand that he wanted me to cry, when he had flogged me on the man's back; I remember he repeated a verse about devils trembling, and said, "But this hardened wretch will not shed a tear." He was a member of a religious society, and I suppose that was the reason that made him use those words.

14. Was he discharged from that society?—Yes, I believe he was; my grandmother went to the class; it was held in the chapel, and he was discharged from it.

15. Was the overlooker that abused you to that extent discharged?—No; not from the mill.

16. Was this beating common in the mill?—He was a very severe man, and he beat many a one to a very great extent, more than any other man in the mill; he was remarkably well known for it; I went to a justice of the peace; it was late at night, and the justice said we must come in the morning; he said he had frequent complaints of him, and he would make an example of him.

16. Was he made on example of?—No; my father desired that we should not go on with it; I was brought up with my grandmother, and that caused her to take me to the justice. My father had been in the army, and he had not long gotten his discharge; and he had gotten work at Mr. Marshall's mill, and he was afraid, I suppose, if we went on with law, it would cause a difference between him and the other men, and he desired me to go no further with it, but I was not to work under that man any more.

DRAKE, JOSHUA; age 56; examined 13th April, 1832.

1. What is your business?—A woollen-weaver. Josh. Drake.

2. Where do you reside?—At Leeds.

3. How long have you resided there?—Twenty-eight years.

4. How soon was it before you became acquainted with the factory system?—The first thing after I was married, and had to work for my living, and provide for a wife and one child, was to go to Mr. Rawthorne's mill, at Cowper's Bridge; I had 12*s*. a week for carders-filling.

5. How old were you when you were married?—I believe, when I married, I was somewhere about twenty.

6. What comparison do you make between the wages you received and the wages given for the same employment at present?—I consider that the wages I received then were worth double the amount they are at present; I mean that 12*s*. would go as far as 24*s*. now in paying for rent and other things.

7. What did you pay for rent then?—I paid 30*s*.

8. What do you pay now?—Five pounds five shillings.

Josh. Drake. 9. Had you as good a house then as you have now?—I had one about three times as big.

10. How many hours did you work?—From six to seven.

11. What time had you for meals?—Two hours a day; we had half an hour for breakfast and drinking, and an hour to dinner.

12. Did any of your children go to a factory?—Yes; the first that I had went to Mr. Gott's.

13. How old was that child?—Between seven and eight; nearly eight years old.

14. A boy or a girl?—A girl.

15. In what situation?—As a piecener.

16. What were her wages?—At that time they went for one week to learn; they had 1*s.* 6*d.* the next week, and they rose 6*d.* a week till they got to 3*s.* and there they stood.

17. What were the hours of labour at that period?—From six to seven.

18. Were there any extra hours to those children at that time?—At that time they had not begun extra hours.

19. Then it is presumed there was no beating or chastising?—Not at that time, because Mr. Gott kept what is called a billy set of children; that is, three children more than was wanted; and if any one was ill, another was put in its place.

20. The children were rarely chastised?—By order of Mr. Gott they were only chastised with a ferule, if they would not obey; but no man was allowed by Mr. Gott to do more than use that; but he did not know the extent to which they carried even that sometimes.

21. Did you think it did any harm to your child's health, putting her in this mill?—No, I do not think it did her any harm; because when the child had been a few hours at the billy set, she went out to play.

22. When did the over-hours commence at that mill?—Before any over-hours commenced I had two if not three children at work at Mr. Gott's, and I never sent a child to work before they were seven years old.

23. Were there children sent there before they were seven years old?—Yes.

24. What were the extra hours?—At first commencing they were from five to eight.

25. Had the children to attend those hours as well as the adults?—Yes.

26. Did the health of the children suffer by those long hours?—Yes; I never could keep children working at those long hours without injuring their spirits and appearance; I always found that the children's colour faded, and their appetite began to fail them.

27. Where did your daughter work when she was a scribbler-filler?—At Messrs. Brown and Banker's mill.

28. What age was she then?—I think she might be between twelve and thirteen.

29. How many hours did she work?—She was then working from six to eight, or from five to nine, for they varied sometimes; there are no meal-times for the scribbler-fillers; they carry their victuals with them, and they eat them as they mind the machine.

30. Did you ever remonstrate against the long hours?—I went down and asked the overseer whether they could be modified; he said, that if I did not like to let my daughter go, there were others that wanted the job.

31. Was your daughter ever cruelly treated while she was employed?—She was once much beaten, and I went down and remonstrated; and at another time there was some dirt in the doffer, and he ordered her to clean it out; and she took hold of the crank to clean it out, and her hand was hurt, and she was taken to the Dispensary; and when she came home and got well I took her away.

32. During those long hours did you see the usual vigour and sprightliness of children in your children?—No, they were more like dogs dozing upon a warm hearthstone, than like children. Josh. Drake.

33. What is the length of your children's labour at present?—At present, for the last winter, they have only been working from daylight to a short time before dark, and I never saw them look healthier and sprightlier.

34. Had you any opportunity of sending your children to a day or a night school?—No, from their excessive employment I had not an opportunity, and I could not find in my heart, against their will and feelings, to send them to a sabbath-day school.

35. Why do you allow your children to go to work at those places where they are ill-treated or over-worked?—Necessity compels a man that has children to let them work.

36. Supposing there was a law passed to limit the hours of labour to eight hours a day, or something of that sort, of course you are aware that a manufacturer could not afford to pay them the same wages?—No, I do not suppose that they would, but at the same time I would rather have it, and I believe it would bring me into employ; and if I lost *5d.* a day from the children's work, and I got half-a-crown myself, it would be better.

37. How would it get you into employ?—By finding more employment at the machines, and work being more regularly spread abroad, and divided amongst the people at large. One man is now regularly turned off into the street, whilst another man is running day and night.

FIRTH, JOSEPH; age 35; examined 18th June, 1832.

1. Where do you reside?—At Keighley. Mr. Josh. Firth.

2. Have you had much experience in mills and factories?—Yes, I had eight years' experience in my boyhood.

3. At what time did you commence working in one?—At six years of age.

4. In whose factory did you commence working?—It was one James Cozens.

5. Was it a cotton-mill?—Yes.

6. What were your hours of work?—We began working at six and continued till half past seven; we had an hour at noon, if I recollect right.

7. And any time for breakfast or drinking?—No time at all; I never knew a mill that had any time for breakfast or drinking in our neighbourhood.

8. Did you feel it very distressing to be kept so long a time at your work?—Very distressing towards night, and in the morning also; I frequently used to fall asleep, both at night and morning.

9. Was your food brought to the mill to you?—It was; that is, breakfast and drinking; we used to go home to our dinner.

10. Is the appetite considerably affected by the labour that you have described?—It makes them squeamish and dainty about their meat.

11. What is the latest time that you worked in the evening?—When there used to be a great call for cotton yarn, we used to work till nine and ten at night; and the master would give orders to the overlooker to retain a number of us and lower the speed, so that we could keep all the spinning moving; the same factory is working fifteen hours a day.

12. Did you not become almost too drowsy and sleepy to pursue your work?—I should say that I slept naturally while I worked. I know well

Mr. Josh. Firth.

that at seven, or half past seven, I was in such a state of stupor when I pieced an end, that I have had the skin taken off my fingers and hands; being asleep, as one may say, the frames went against my fingers.

13. Did you not feel it, as a child, to be a positive act of cruelty, to be hurried to your labour so early?—Yes, excessively so; and when I talk to persons who have children going to factories, it makes my heart bleed; we cannot tell the tears which have been shed on those occasions; children are torn out of their bed in the morning; they have to encounter all weathers; they open the door and meet the snow-drift and the cold; and in these ways they have to undergo great suffering at a very tender age.

14. Is there, with those hours of labour which you have described, any time for education?—I will leave every man to judge for himself what time there is for that purpose; those have to be up at half past five that begin at six: those that are upon the best regulated system will have forty minutes or three quarters of an hour at noon; then they will give over at seven at night, or ten minutes after, and then by the time they get home it will be too late, half past seven; then they have to get their supper; and to get to the school would take them about half an hour more. It would be an impossibility to drive any learning into them after they got there.

15. Suppose they were to go to Sunday-schools, and it was the practice of Sunday-schools to instruct them in writing; is the Sunday, fatigued as these children are, available for the purposes of their education?—No; the children in general wish to escape from the school if possible, those that go to the factories. I went to a Sunday-school myself, and I know that we used to fall asleep.

16. Had you generally to be stimulated to your work in the evening?—Yes, in a general way in the evening; we had window bottoms, they were very convenient places to sit down upon, and we used to get upon them and fall asleep, and then the overlooker used to come and shake us by the ear, or give us a rap with the strap, and for a time we resorted to our work again to repair what was wrong, but if we could see an opportunity, we got to the window bottom again.

17. Is there any check over the overlookers by the proprietors of mills, as to the chastisement employed?—I should fancy not; or if there was a check, I should look upon it as checking them with one hand, and giving them the whip with the other.

18. There are many worsted-mills in Keighley, are there not?—Yes.

19. And the hours of their labour are unregulated by act of parliament?—Yes.

20. And therefore they continue their children at work for the length of time you have mentioned?—Yes, at different mills; but I would not cast reflection upon all of them; there are a good many well regulated mills, which work twelve hours and a quarter; the manufacturer that I work for, though he is disinclined to this bill, regulates his mill in that manner.

21. Who is that gentleman?—Mr Calvert.

22. Is it your opinion that twelve hours and a half would be safe for children to be worked in mills, under the best regulations?—No, it would not; I can judge of it by my own case; when I was a boy I had only twelve hours and a half, and I felt fatigued by it; and all natures are the same; and I had a stronger constitution, I believe, than some; my knees were giving way once, and my mother was cautioning me against it; I had an idea that I should not like to go up and down the streets in a deformed condition; and therefore I used constantly to make a spring up when I was working, to prevent my getting deformed.

23. Were you induced to do that from seeing many cases of deformity in

that town?—Yes, I had seen many cases of deformity, and I had a great fear of it; and therefore in my general walk I tried to counteract it as much as possible, but it is an impossibility. Mr. Josh. Firth.

FOX, BENJAMIN; age 56; examined 18th May, 1832.

1. Where do you live?—At Dewsbury. Benjamin Fox.
2. Have you worked in factories?—Yes, about forty-two years.
3. What were your regular hours of work when you first worked in a factory?—Twelve hours, including meal times.
4. How long were the meal times?—Half an hour at breakfast, and an hour at dinner.
5. So that in fact, you only worked ten and a half hours then?—No, when first I went.
6. How long did you work at Mr. Halliley's?—Thirteen hours, with meal times.
7. What hours did you work when you were busy?—From five to nine, and from five to ten, when we were very busy.
8. What effect has your working in mills had upon you?—It has had a great effect; I can scarcely walk; my knee is crooked and weak.
9. At what age did that come on?—It came on when I was twenty-five or twenty-six; it began to be bad, and it has been getting worse.
10. Were the children employed there compelled to work as long as yourself?—Yes, when I was at Mr. Hambrough's I had one of my own girls there, and she tumbled down, and, as we thought, knocked her neck out, and broke her elbow all to bits.
11. And she was hurt in her arm?—Her elbow-bone was smashed.
12. What was she doing when she stumbled down in this way?—When I left her, the child, being weary and tired, was rambling, and tumbled down.
13. Have you had any other children that have worked in factories besides her?—Yes, two.
14. Did they work with you?—Yes; I would not allow them to work with anybody else.
15. What was your reason?—I have seen some men beat the children so bad as to make the blood stream down their heads.
16. You found it necessary to beat your own children in order to keep up with the machines?—Yes.
17. But was it not principally at the latter end of the day that it became necessary to beat them more severely?—Towards night, when they began to be tired and weary, and to neglect their work.
18. What sort of treatment have you witnessed regarding children at the latter end of the day who have been over-worked?—I have seen the slubbers taking up the rollers and knocking them down; I have seen a good deal of that within the last forty years.
19. You yourself have become a cripple from over-labour in factories?—Yes.
20. Were you a healthy child when you first went?—Yes, as healthy as anybody.

FRASER, DANIEL; age 28; examined 2nd July, 1832.

1. Where do you reside?—I have been residing nearly five years in the vicinity of Huddersfield. Daniel Fraser.

Daniel Fraser. 2. Are you acquainted with any mills or factories in that neighbourhood?—Yes; I have been working in a spun-silk and throwing mill.

3. How long?—For upwards of four years and a half.

4. What was your business there?—I was a warper; I managed the warping department.

5. What were your hours of labour?—From six in the morning to eight o'clock at night.

6. What was your allowance of time for meals and refreshment?—I had two hours for myself; the children had half an hour for breakfast, and an hour for dinner, and they were kept during drinking-time; but they had time to take their refreshment though they were kept in; they were not allowed to go out.

7. At whose mill was this?—John Fisher and Co's.

8. How old were the children?—They were mostly children in the throwing department; I have known instances of them going under seven to about eighteen; and sometimes there are young females, who superintend and assist the overlookers, called tenders; they are young women generally, but all the rest are children.

9. Have you ever seen, personally, the effect of the labour upon girls?—Contrasted with other children, they appear pale, their eyes sunk, they are thinner in the flesh, and altogether they seem deteriorated.

10. Have you seen the effect of their labour upon them while they have been at it?—From the long confinement I have seen them faint away; I have seen instances of their fainting from the closeness of the room and imperfect ventilation, and from their excessive labour day after day; I concluded it was from that; and had they worked in the open air, or in a more ventilated apartment, or for shorter hours, I think it would not have occurred.

11. Does the owner of the works expect a given quantity of work done in a certain time?—I cannot accurately say how much, but I believe that unless there is a given quantity turned off, that is, a given quantity of twist with a given number of children, that the overlooker in question is not kept.

12. Is the calculation such, to the best of your knowledge, as to demand the constant activity of those young children?—I conclude, from facts, that the calculations are made so as to bring them up to the highest pitch of labour.

13. Are the hands occasionally worked by night in the silk-mill as well as by day?—Yes; they work night and day regularly.

14. Without distinction either of age or sex?—Yes; without any such distinction.

15. That is, two sets of hands perform the night and the day work?—Yes.

16. Have you reason to believe that the system at present pursued affects the morals of the rising generation?—I think that the factory system is a very immoral system, and one more so does not obtain anywhere.

17. Do you think that its effects are beginning to be very serious to the general interests of society and to the prospects of future generations?—Yes; and I think that the factory system, after it has a little more pervaded society, will be attended with very alarming effects. In the first place, it cuts up the root of education altogether, and the children are harshly treated generally, and they are abused; and when these children come to receive instruction in the Sunday-schools, they are told that they are to be just and good children, and humane in their conduct towards each other; and when they hear these precepts they do not believe them; their experience gives the lie to it; the treatment they experience is directly at variance with the precepts; and I conceive, as far as I can reason, that it is of no use to give children instruction, and to tell them to be just and humane, while their experience tends to make them throw it off.

18. Did you bear a summons to George Montagu, of Watford?—Yes, I did. Daniel Fraser.

19. To appear before this committee to give evidence touching an inquiry respecting a bill for limiting the hours of children employed in mills and factories?—Yes, I did.

20. Are you aware that an answer has been returned to that summons, stating that he is too unwell to attend?—Yes.

21. Do you know that, nevertheless, he worked on the day on which you delivered the summons to him?—Yes; and I found him coming out of the mill when I delivered the summons.

22. Were you desired to summons others also?—Yes, I was.

23. Have you not found much disinclination to attend as witnesses on the part of those that would have been brought before this committee had it not been for their expressing their apprehension, that if they appeared they should endanger their interests, and, perhaps, lose their situations?—Yes; I am sorry to see that this feeling is very deeply impressed upon the minds of the operatives in connexion with the factories; otherwise they said they would give evidence with cheerfulness; it is my deep impression, that though very favourable to the passing of a bill to limit the hours of labour to ten hours a day, they are afraid to express their sentiments upon this subject, lest it should affect their interests.

24. You have consequently been desired to make personal inquiries and observations yourself relative to the management of silk-mills in these parts of the country; have you done so?—I have.

25. What, speaking from your own knowledge, is the state of employment in the silk-mills which you have recently visited, in respect of the number of hours; state, also, whether they work several of them by night, and the general condition and appearance of the hands employed?—As far as I have had an opportunity of knowing, I find that there is very little difference between the number of hours that the children work in those various places which I have visited in Essex, Suffolk, and elsewhere, from those which I have already stated to the committee—viz., twelve hours actual labour per day. They work from half past five in the morning, at Braintree, in Essex, till half past six in the evening, and with the intermission for meals of one hour, half an hour for breakfast, and half an hour for dinner; and the children there are fined a penny per hour for being absent. I am quite confident that they would be glad of a regulation of the hours of labour, even if it was attended with diminished wages. One woman, in particular, said, "She would far rather have less wages than see her children so tired at night;" and another said, "That if a jolly girl goes into a mill, her countenance soon alters, and they become more unruly than they were before."

FRITH, PETER; age 17; examined 10th July, 1832.

1. What is your business at present?—Engineer. Peter Frith.

2. Where do you reside?—At Winsley.

3. Have you ever worked in any mills or factories?—Yes.

4. In whose mill?—Vernon Royle's.

5. At what age did you begin to work there?—At about nine years of age.

6. What were your hours of work in that mill?—From six o'clock in the morning to eight at night.

Peter Frith. 7. How were you and the rest of the children kept up to your work for that length of time; were you beaten to your work?—Yes, we were.

8. What were you beaten with?—Straps, and sometimes canes.

9. Are there many girls employed in silk-mills?—Yes.

10. Are they chastised also?—Yes, they chastise them; not the larger girls, but the smaller ones.

11. Are prizes given to the children in order to get the utmost degree of work out of them?—Yes; I earned a new pair of shoes.

12. Were you ever chastised yourself?—Yes.

13. For what?—For being about five minutes too late at dinner time, although my mother sent me on an errand; he met me in the room, and kicked me.

14. Will you state the effect of that abusive treatment upon you?—It broke my knee in three places, and I fell; and he laid upon me with the strap till I got up again.

15. Was your knee smashed?—Yes; I went to a doctor about it, and he said he never saw such a knee in his life.

16. And yet you were chastised when you were in that condition, and upon the ground?—Yes.

17. Who was the individual that so used you?—George Harpur.

18. Were you detained in the mill after that cruel treatment?—Yes.

19. How did you get home?—I put my hand on a boy's shoulder, and hopped all the way home.

20. Did the doctors that you had to your knee cost your parents a considerable sum of money?—Seventy pounds, independently of the loss of time.

21. Did the overlooker, Harpur, or any of the other employers, pay any thing towards that?—Two shillings and ninepence Harpur paid me.

22. Will you name the doctors that attended you upon that occasion?—Mr. Warburton, Mr. Barton, Mr. Taylor, and Mr. Harrison.

23. Did you suppose that you should have lost your limb?—Yes; Mr. Barton wanted me to have my leg taken off.

24. He is a surgeon of the Manchester Infirmary, is he not?—Yes.

25. Will you shew the committee your knee?

[*The witness did so, and the cap of the knee appeared to have been broken in several places, and the flesh lacerated.*]

26. Did your parents or yourself make any complaint to a court of justice respecting this?—Yes; they summoned him to a court of requests.

27. Was the violence that had been used towards you proved?—Yes; and Harpur was brought in guilty, and he was to pay two pounds.

28. How long had you been ill then?—I had been ill about six or seven months, and he was to pay two pounds, and they asked my father whether he would have his body or goods; so my father said he would have his body, for he had no goods; and then they asked him if he would have the money weekly, and he said, yes, he would have it weekly; and then George Harpur said he would pay nothing, he would go to Lancaster gaol before he paid anything; and Mr. Vernon Royle said, yes, and he would give him his place when he came back, and his father would keep him while he was there. It would have cost three pounds ten shillings to send him to Lancaster, and my father could not raise it; and he was not punished.

29. Did your parents ever apply to Mr. Royle on account of your having been thus maltreated, in order to obtain something from him?—Yes; and Mr. Royle abused my father very much, and he got nothing.

GOODYEAR, JOHN; age 44; examined 15th May, 1832.

Mr. J. Goodyear.

1. Where do you reside?—At Huddersfield.

2. Have you worked in factories?—Yes.

3. At what age did you begin to work?—When I was between ten and eleven.

4. At whose mill?—At Mr. Fryer's.

5. What were your hours of labour at that time?—Twelve hours per day.

6. With any allowance for meals?—We had twelve hours of actual labour, from six to seven, and one hour off for dinner.

7. You conceive that the labour of children and young persons in mills has considerably increased since you were first acquainted with that business?—Yes, it has.

8. And the hours of labour then being shorter, I may ask the question, whether you do not think the treatment of children was much less severe at that period than it is now?—As far as my own knowledge leads me, it was a great deal less severe than it is at present, and has been for some time back.

9. Is it a general habit in the woollen-mills to beat the children to make them work?—Yes, it is a very common thing to beat them, either with a stick, strap, or top roller; the top roller is almost always used for pieceners.

10. It is two or three yards long?—Yes; some are longer.

11. And two inches at least in diameter?—Some of them are, and some an inch and a half.

12. After you left Mr. Fryer's mill, where did you go then?—To Messrs. Clay and Earnshaw's.

13. What age were you?—I think I was turned twenty.

14. You say you worked twelve hours when you were at an early age, exclusive of the time allowed for refreshment?—Yes; that was actual labour.

15. Did it produce any effect on your limbs?—I always considered that was the cause of my growing crooked in this knee. I was very straight before.

16. At what age were you quite straight?—When I was between twelve and thirteen.

17. When did you first feel a pain in your knee?—When I was about thirteen years of age, or a little before.

18. You have said that since that period the labour of children, till quite recently, has become more and more severe?—Yes.

19. And their treatment less humane on the whole?—The chastisement I received in the first mill I went to was very trifling; indeed, I have seen children, I might say, one hundred times worse treated than ever I was in my early days.

20. As you have had considerable experience in that sort of employment, will you have the goodness to state the effect it produces on the health of the children?—By long standing, and being over-worked?

21. Yes?—It weakens them, and causes them to faint very frequently at nights, when they give up. As soon as they get home, I have seen them fall down in a chair, almost unable to stir.

22. Have you any children of your own?—I have two, who have worked three years.

23. Was that effect produced on them?—Yes; they have actually fainted, and I have had to catch my eldest daughter in my arms, to prevent her falling.

24. What effect had it on their appetite?—It impaired their appetite very much.

Mr. J. Goodyear.

25. And what effect upon their appearance?—It made them look very pale and very ill.

26. What effect do you think these long hours of labour produce on the moral conduct of those who endure it, generally speaking?—I believe, if the hours in factories were shortened, the moral conduct of children would be a great deal better than it is at present. They would have an opportunity of learning; and when young children have some learning, they become more moral. In general, children have been very illiterate who have worked in factories, and towards the latter part of the day they have kept talking all kinds of debauched talk to keep themselves in amusement till the time of giving up; I have observed that very frequently.

27. They have no time to learn any domestic duties?—None.

28. Neither yourself, as their father, nor their mother, had an opportunity of being with them sufficiently long to form their character, to improve their minds, or teach them their duties in future life?—It is impossible to do that from the length of time they had to work; they are up at five or six in the morning, and continue till twelve at noon, and have but one hour to dinner, and then remain till half past seven at night, when they are so tired that they go to bed as soon as possible.

29. Do you conceive it would be possible to regulate the hours of labour without regulating the engine?—I cannot say that it could be done so well, because the masters can evade the laws in so many different ways; but stop the moving power, and then they must stop the hands.

GUMMERSALL, BENJAMIN; age 16; examined 21st May, 1832.

Benjamin Gummersall.

1. Where do you reside?—In Bowling-lane, Bradford.

2. At what mill did you first work?—At Mr. Cozens's; I have worked nowhere else.

3. What were your hours of labour?—From six in the morning to seven, and half past seven, and eight at night.

4. What time was allowed you for dinner?—Half an hour at noon.

5. Any time for breakfast or drinking?—No.

6. At what age did you go into the mill?—About nine years old.

7. What sort of position do you stand in, in order to piece worsted goods? —If we are higher than the frames, we have to bend our bodies and our legs,—so.

[*Here witness shewed the position in which he worked.*]

8. Have you always to bend your body?—Yes, always.

9. Were you healthy and strong before you went to the mill?—Yes.

10. Could you walk well?—Yes, I could walk from Leeds to Bradford when I was eight years old.

11. How long did you work at that mill for those long hours before you found your limbs begin to fail?—About a year.

12. Have you ever been beaten?—Yes, till I was black and blue on my face, and have had my ears torn.

13. Were you generally beaten at the end of the day more than at any other time?—Yes, at the latter end, when we grew tired and fatigued.

14. Had you still to attend at the mill after your limbs began to fail?— Yes, I had.

15. How did you go on with your work when you became deformed?— Not so well as I could before.

16. Had you to stand thirteen or fourteen hours a day frequently?—Yes.

17. And to be constantly engaged as you have described?—Constantly. Benjamin Gummersall.
18. Will you have the goodness to shew the committee your limbs?

[*The witness did so, and they appeared to be excessively crooked.*]

19. Had you any opportunity of learning to write?—No.
20. Can you read?—Some little; but very poorly.
21. How are you supported?—I help my sister to knit heels to weave with.
22. Have you pains in your limbs now?—Yes, sometimes; when I stand I have very much pain.
23. When this came on, were you working long hours or short hours?—Long hours.
24. While you worked short hours, did you suffer any of this pain or inconvenience?—We never did work less than from six to seven.
25. When your legs got bad, did they shorten your hours?—No, I worked all the same hours.
26. What wages did you get a week?—I had 5*s*. 6*d*., and then I left and went again, and they gave me 4*s*. 6*d*.
27. Because you had got worse in your limbs?—Yes.
28. Though that had been in their service?—Yes.
29. Did you get 5*s*. 6*d*. from the time you were nine years old, when you first went into the mill?—I got 2*s*. the first week, when I was learning, and then I kept getting raised as I could get work.
30. Did you work as long for your 4*s*. 6*d*. as you did for your 5*s*. 6*d*.?—Yes.
31. How did you reach your work?—When I began to be deformed and crooked, they gave me a stool to sit on.
32. Can you stand at all without crutches?—Not without crutches or a stick, or something to lean against.

HALL, JOHN; examined 21st May, 1832.

1. Do you live at Bradford?—Yes. Mr. John Hall.
2. Are you the overlooker of Mr. John Wood, junior?—I am.
3. Will you have the goodness to state the present hours of labour in your factory?—Our present hours are from six to seven.
4. With what intervals for rest or refreshment?—Half an hour for breakfast, and forty minutes for dinner.
5. Is there very particular attention paid to the cleanliness and ventilation of your works?—Very particular attention.
6. Are the children employed in your establishment constantly visited by a medical gentleman?—They are.
7. Regularly?—Regularly every week; every Tuesday morning.
8. Who is the medical gentleman in Mr. Wood's employment?—Dr. Sharp.
9. Is he a man of great celebrity in that part?—Yes, and very much respected.
10. If any children are ill, does Mr. Wood send them to the dispensary, to be cured at the public cost, or is he at the expense of obtaining medical assistance for them?—We have authority from Mr. Wood to send them to Dr. Sharp ourselves; they all go to Dr. Sharp, and Mr. Wood pays the expense of medical attendance.
11. Has Mr. Wood sometimes sent the children he employs to Buxton, or

Mr. John Hall.

elsewhere, for the benefit of the baths?—I do not know where the place is that he has sent them to, but I know that he has sent them off many a time.

12. When they have become poorly?—Yes; he has sent them to watering places for the recovery of their health.

13. Have you any baths in the works?—We have cold and warm baths now on the premises, for the use of the persons employed in the manufactories.

14. For the children?—For the children, and for the men too; I have used them myself many a time.

15. Does Mr. Wood keep a larger number of children to do the same work than is usual in the worsted business?—Yes, he does.

16. Do you think he does that because he considers the labour of the children to be too severe?—Yes, I have no doubt of it.

17. Now, will you state to the committee whether, with all this care and attention, you find that the children can endure the labour you have been describing without injury?—No, I believe they do not endure it without injury.

18. Do you select healthy children, whom to take into your mills?—Yes.

19. Are you particular in so doing?—I am particular, and so are we all; we have often rejected children that we thought were not healthy looking, and taken others in preference. The book-keeper keeps a list of the names of persons that have applied, and we choose our hands from that list; and they send them down for the overlooker to look at them, to see if he likes them, and we choose the healthiest and the strongest, as far as we can judge.

20. When your fresh hands have been employed for some length of time, do you see any alteration in their appearance?—Yes, I think I can see a marked difference in the course of a month or five weeks; when they come into the mill they look rosy, and they are plump and fat, and generally lively and spirited; but in the course of a few weeks I can see a falling in their faces, and a paleness in their countenance, and they grow spiritless and languid.

21. Will you describe to the committee the position in which the children stand to piece in a worsted-mill, as it may serve to explain the number and severity of those cases of distortion which occur?—At the top of the spindle there is a fly goes across, and the child takes hold of the fly by the ball of his left hand, and he throws the left shoulder up and the right knee inward; he has the thread to get with the right hand, and he has to stoop his head down to see what he is doing; they throw the right knee inward in that way, and all the children I have seen, that I could judge, that are made cripples by the practice of piecening worsted, invariably bend in the right knee. I knew a family, the whole of whom were bent outwards as a family complaint, and one of those boys was sent to a worsted mill, and first he became straight in his right knee, and then he became crooked in it the other way.

22. Have you remarked that cases of deformity are very common in Bradford?—They are very common; I have the names and addresses of, I think, about two hundred families that I have visited myself, that have all deformed children, and I have taken particular care not to put one single individual down to whom it had happened by accident, but all whom I judge to have been thrown crooked by the practice of piecening, and of throwing up the left shoulder, and of bending the right knee.

HANNAM, JOHN; age 45; examined 18th June, 1832.

Mr. J. Hannam.

1. Where do you reside?—At Leeds at present.

2. Have you been engaged in mills and factories for the greater part of your life?—Yes.

3. From what age?—I started at about ten years of age; I was rather less than ten, I believe.

4. In whose mill did you commence your labour?—I was first in Mr. Blessard's cotton factory; he was the master then.

5. Where was it?—In the forest of Knaresborough; it goes by the name of Raikes's mill.

6. What were your hours of labour at that time?—We scarcely knew our hours; for, being in a country place, they worked us just as they thought proper; some days we worked till seven, and some till eight, and nine, and ten o'clock at night, just as it happened.

7. How long were you there?—I wrought there, off and on, many years; I should say pretty nearly twenty years.

8. What time had you allowed for your meals?—We had only half an hour for dinner; we had no time for breakfast or drinking.

9. Did that mill, so conducted, answer to the proprietor?—No, he stopped it, and sold it to Mr. Willet.

10. Was the same system continued by Mr. Willet?—The time was more regular then.

11. Was the time also long?—Yes, the hours were long; but after he got it, the hours were not so long, because he began to work day and night; when he got it, I began on night work; we worked from seven at night till six in the morning, and the others worked in the day, with half an hour for dinner; that was all the time they had for meals.

12. Did night labour answer to that individual?—No, he failed, and was made a bankrupt.

13. Since you have had much experience of night working, is it not true that children that have worked during the night have such a disposition to be up and play in the day-time, that, between their being obliged to work and their inclination to play, they hardly ever get sufficient rest?—It is the case.

14. Do not you think, from your experience, that the bill regulating the labour of children and young persons, which was enacted about the year 1820, was serviceable to a certain extent in mitigating the hours of labour?—Yes, in cotton factories; it only extended to them.

15. Do you know of your own knowledge the regulations of Mr. Holdforth's mill in Leeds at present, and for any length of time past?—Mr. Holdforth is not what I call regular in the working hours; I have lived there fifteen years, and sometimes he would work seven days, sometimes more, and sometimes less; in the last winter they have been working in the silk department eighteen hours for a time, and six hours off.

16. The same children working eighteen hours, and then six hours interval only?—Yes, some of my children go there.

17. For how many days together?—For a week together.

18. What ages are the children that work those eighteen hours?—Some ten and twelve, and some older and some younger.

19. And in what condition were they at the end of the week?—They were very much fatigued all the time through; it is too long for anybody to work; for children or anybody else.

20. Has that mill been recently inspected by any gentlemen?—Yes; there were some gentlemen came to look through it, and they were particular in cleaning; the children were sent home to dress themselves, and some of

J. Hannam. their mothers washed their troswers, which they were obliged to put on before they were dry.

21. Have you, as a general observation, seen that the health of the children has been very seriously affected in the flax factories?—Yes, particularly so in the card rooms; there are many that cannot stand them.

22. In what way have the children suffered in their health?—They are affected some one way and some another; they are delicate, and can scarcely take their victuals; it is as if something lay upon their stomachs, and their lungs caused them to be sick, so as not to take their victuals; and some have swellings in their glands.

23. The scrofula?—Yes.

24. Have you observed that consumptions frequently ensue?—I have seen that in part of them; but not so much of that complaint as the other.

25. Have you remarked whether the body attains its full growth in the instance of children who have to labour so early and so long in those mills?—We have always remarked that those who worked so long and were so ill treated were generally ill, especially when they worked so much at night; people were always saying, "How ill they look; how little they are!" And I have myself noticed, generally, that they do not reach their growth, and that they look ill, when they are in the habit of working such long hours.

26. Can you state to this committee whether you think it possible for children employed as you have described to obtain any suitable education?—No, they cannot get any education, and they are not fit for it if they had the opportunity; for when they go home after labour of this kind, they are either asleep or something like it directly. When they are working such long hours, we make it a rule, "Bed and work, work and bed;" for if they do not go to bed, we know that they will not be fit for work on the next day.

27. Do you think that they are capable of deriving as much improvement from the institution of Sunday-schools, as they would do were their hours diminished?—No, they want to lie in bed; I had sometimes a good deal to do with mine to get them up to go to Sunday-schools.

28. What do you think are the natural consequences of this undue labour, as undergone by young females?—According to my own simple ideas, I attribute great injury to the females from it, in different ways; some, before they will undergo this treatment, will turn out into the streets, and become prostitutes; and others, I can say it, and say truly, who have been as anxious for work as could possibly be, have been obliged to follow the same course, because they had no parents or friends to keep them, and could get no work.

29. With whom do you work at present?—Mr. Harris; we are not obliged to work there such long hours; Mr. Harris has often wanted the other employers not to work such long hours, especially when the trade was not likely to be so brisk, so as to keep the markets from being overstocked with yarns.

80. Have you had frequent intercourse with individuals in reference to the bill now under the consideration of the house?—Yes, I have been one of the head of them that support it.

31. You are chairman of the short-time committee at Leeds?—Yes.

32. Now state whether you could not multiply evidence to almost any extent, and from every part of that most populous district, so as to substantiate, by an almost infinite number of witnesses, the facts that have been already stated to this committee by those whom you have sent up for that purpose?—Yes, there might be witnesses to any amount; one thing, indeed, might prevent it, and that is, the fear of losing their situations.

33. Do not you consider that very long hours of labour occasion a great expense to families?—Yes.

34. In the first place, do they not sometimes lose much of their time by becoming sick and ill?—Yes. J. Hannam.

35. Have not the medical men to attend them frequently in consequence of those illnesses?—Yes; or they are obliged to go to dispensaries and infirmaries.

36. Has not a child, that might be employed in a mill, to stop at home in order to carry the meals of his brothers and sisters to the mill, they not being allowed to leave the mill to get their refreshments?—Yes; the parents are obliged to keep a child at home, or to do it themselves; and, besides that, it would be much cheaper for them to have their meals at home with their parents.

37. The long hours prevent their eating together, and destroy all domestic economy?—Yes, because they cannot carry just so much to the mill as they would do with at home; for instance, porridge and such things they would not only have at home more comfortably and better, but there would be less waste at home, because they cannot tell exactly how much to take to the mill; and then, if they have bread, there is a part wasted and thrown about; I have seen it many a time covered with dust and dirt, and they have laid it down, and it has got spoiled.

38. Do not persons thus over-laboured demand more support and sustentation than when they are worked only for a moderate length of time?—Yes, we always have thought that we had better work a shorter time, and have a meal a day less; and that we should be better, both ourselves and the children, who would then have a little time for recreation too, and other things; that has been our observation hundreds of times.

HANSON, JOHN; age 41; examined 2nd July, 1832.

1. Where do you reside?—I am residing at Huddersfield at present. Mr. John Hanson.

2. Have you had considerable experience in the manufactures of the West Riding of Yorkshire?—I was brought up in a domestic manufactory in the woollen business.

3. Do you think that the hours of labour, and the extension of the means of manufacturing, have exceeded the increase of the regular demand for manufactured goods?—Yes, I do.

4. And do you conceive that that has the direct tendency of lowering the remuneration of labour, and very much pressing down the condition of the operative classes?—Yes.

5. Speaking as to your personal experience, has any part of your family ever been employed in mills or factories?—Yes; I have two girls who have been employed in various factories in Huddersfield and its neighbourhood these four or five years.

6. What have been the usual hours of labour in those mills and factories, when trade has been brisk, and the demand for goods large?—They generally go between five and six o'clock in the morning, and come away at from eight to half past eight at night. I recollect that my children went at one time soon after five in the morning, and worked till about ten at night. I went when this commenced, to see what was the matter; it was a place where they did a little country work, and I found that they had more work than usual on hand, and that they had but very little time to get it done, and they exacted the labour from the operatives to that length of time; they worked from five or six in the morning to ten at night.

7. Speaking of the more moderate day's labour, do you conceive that the children and young persons are capable of sustaining it with any degree of

Mr. John Hanson.

safety to their health?—No, I believe not; but my children, though they were occasionally worked such long hours, have, owing to particular circumstances, which I did not regret, scarcely averaged much above half work in the different situations they have had.

8. Is it not your impression that the undue labour that is imposed upon the industrious classes at one period, has, of necessity, the effect of abridging the labour afforded to them at another?—Yes, it is; I have witnessed persons working there excessive hours, while others are standing idle in the streets, as it were, or running fruitlessly about, seeking work and unable to obtain it.

9. Will you give to the committee your opinion as to the nature of the factory system as at present pursued?—I think that the nature of machinery is to give out, by degrees, its productiveness to the public, and that there is a great struggle now going on between this tendency on the one hand, and the capitalists on the other; competition has now so far affected the profits of the present machinery, that the owners of it, in their wish to retain the benefits and profits to which they have been accustomed, resort to long hours, and the excessive labour of children.

10. You think, then, that those advantages which were formerly derived from the use of machinery by the capitalists of the country, namely, the advantages of rapid and sudden accumulation, are now sought in the undue labour of those that are employed?—Yes; and if I take a comparative view of machinery I think it will confirm this, for I see that almost every previous machine has, in its turn, been supplanted; and I recollect hearing a story of a person who had been working upon a forty-spindle jenny, and who was asked how times were; his answer was, "They are thirty-nine times worse than usual; for," said he, "they used to get their living by one spindle, and now I can scarcely get my living by forty.

11. Do you think that the operatives, under existing circumstances, can be, properly speaking, regarded as free agents: and more particularly the younger part of the operatives, whom it is the object of this bill to protect?—No, I cannot consider them so at all; for I think when circumstances are such that a man has no employment, and cannot get any, he is forced to submit to almost anything; for when there are too many labourers in the market, the master has only to say to those who are employed, "Do so and so," and they are obliged to do it, or be in the situation of those out of employment; and least of all are children free agents; they are obliged to submit to every regulation or imposition of the master.

12. Do not you think that the hours of labour which you have described as being general in regular times, are perfectly unfavourable to the obtaining of a sufficient degree of mental or moral improvement?—I can observe that my children, when they have been working those long hours, have no inclination to do anything in the house, to take any job off their mother's hands; they want to have rest, and to go to bed; they have to be, as it were, driven to anything that relates to domestic affairs.

13. You think it would be the interest of the employer, as well as of the employed, to mitigate the hours of labour?—Yes, taking them all as a body, I think it would; but we know that some of them have a greater power, and are willing to risk all chances to run the others out of existence, as it were, to monopolize the trade themselves. There are some establishments that have 136,000 spindles almost always kept in motion; and the utmost extent that I have been able to perceive that hand labour is capable of managing is one hundred spindles; therefore it is evident that, if we take a division on this scale, there is a monopoly of the labour of 1,360 persons. Besides, the persons who get possessed of these great establishments not only monopolize in that branch, but they have a great many other branches, which they accu-

mulate to themselves; power-looms, and other branches which ought to be kept perfectly distinct; so that they become overgrown capitalists, and by this means are occupying a station or position in the country which I think is not beneficial to society. Mr. John Hanson.

14. Did you ever know a period when the master manufacturer of the community acknowledged profits to be high?—Very seldom; they always kept that very dark; but I have perceived, even latterly, that some have risen rapidly to great eminence and wealth, and that is a sufficient evidence that there have been great profits.

HARGRAVE, ELDIN; age 14; examined 13th April, 1832.

1. You come from Leeds?—Yes. E. Hargrave.

2. What age were you when you were sent to the mill?—I was about eight years old.

3. To whose mill did you go?—I went to Messrs. Shaun and Driver's.

4. What were your hours of labour there?—I worked from six to seven.

5. You worked thirteen hours, with two hours for meals?—Yes.

6. What wages had you?—Three shillings a week.

7. Where did you go next?—I went to Lord and Robinson's.

8. What wages had you?—Half-a-crown a week.

9. How many hours did you work at Lord and Robinson's?—From five to nine.

10. What age were you, when you went to Mr. Brown's mill?—About ten years old.

11. In what situation were you there?—I was a sweeper and errand-boy.

12. What were your wages?—Three shillings.

13. What hours did you work?—I went from six to seven.

14. How long were you a sweeper?—About a year.

15. What situation did you get after that?—I was a brusher.

16. What wages did you get as a brusher?—I had 3*s.* 6*d.* a week; and I had sometimes 1*s.* for over hours, and sometimes 1*s.* 3*d.* I had three farthings an hour.

17. How long did you work when you worked over hours?—I went from five to ten.

18. What did you do after that?—I went to mind Lewises.

19. At the same mill?—Yes.

20. What had you for working at the Lewis?—I had 5*s.* a week, and over hours.

21. How much had you for over hours?—I had a penny an hour.

22. Will you describe the labour you had to do in attending to the Lewis?—I had a stool to stand on, and then I had to reach over as far as I could reach to put the list on.

23. In attending to this machine, are you not always upon the stretch, and upon the move?—Yes, always.

24. What effect had this long labour upon you?—I had a pain across my knee, and I got crooked.

25. Will you shew your limbs?

[*Here the witness exposed his legs and knees.*]

26. Were your knees ever straight at any time?—They were straight before I went to Mr. Brown's mill.

27. You say that you worked for seventeen hours a day all the year round; did you do that without interruption?—Yes.

E. Hargrave.

28. Could you attend any day or night school?—No.
29. Can you write?—No.
30. Can you read?—I can read a little in a spelling book.
31. Did you go to a Sunday school?—No, I had not clothes to go in.
32. Are you still working at Mr. Brown's?—I have got turned off for going to London.
33. Did they say anything?—They said that I was not to come any more again, if I went.

HARGRAVES, ALONZO; age 18; examined 15th May, 1832.

A. Hargraves.

1. Where do you reside?—At Little London, in Leeds.
2. What is your business?—A cloth-drawer, at present.
3. At what time did you begin to work?—When I was ten years old.
4. In what line?—Carping.
5. What were your hours of labour?—From six to seven, with the customary time for meals.
6. What were your wages a week then?—Half-a-crown.
7. How long did you work at that?—About a year.
8. With whom?—With Mr. Rushforth.
9. Where did you go then?—To Mr. Walker's.
10. What was your occupation there?—Priming.
11. What were your hours of work there?—From six to seven; sometimes longer.
12. What were your wages?—Three shillings a week.
13. Where did you go afterwards?—Messrs. Shaun and Company.
14. What were your hours of work there?—From six to seven; sometimes as late as to eight or nine.
15. What were your wages then?—Three shillings.
16. How long did you work there?—About one year.
17. Where did you go then?—To Mr. Brown's
18. What did you do there?—Mind the dry-beating gig.
19. State what your hours of work were then?—From five to nine; then we had only a quarter of an hour at breakfast, half an hour at dinner, and a quarter of an hour at drinking.
20. Did you occasionally work longer than that?—Yes; sometimes to ten.
21. Beginning at what time in the morning?—Five, and sometimes four.
22. Did you ever work all night?—About two nights a week.
23. All night?—Yes, for about two months together.
24. How did you employ your Sundays?—In sleep, in the forenoon.
25. That is, you were too much fatigued to attend the Sunday-school?—Yes.
26. What effect had this long labour upon you?—It made my knees ache and bend in.
27. How did you feel in your general health?—I felt sick and poorly: I had the head-ache.
28. You have given over this long labour?—Yes.
29. Have your limbs now recovered?—Yes, a great deal.
30. Just shew the gentlemen your legs?

[*The witness shewed his limbs.*]

31. Do you believe you would have been allowed to remain in your employment, if you had refused to work those hours which were demanded of you?—No.

HEBDEN, WILLIAM; age 19; examined 13th June, 1832.

1. Where do you live?—At Leeds. Wm. Hebden.
2. Are your father and mother living?—No; they are dead.
3. What time did you begin to work at a mill?—When I was six years old.
4. What sort of a mill?—A woollen-mill.
5. Whose?—Mr. John Good's, at Hunslett.
6. What were the hours of work?—We used to start at five, and work till nine at night.
7. What time had you for your dinner?—Half an hour.
8. What time for breakfast and drinking?—A quarter of an hour at each end of the day.
9. What wages had you?—Two shillings and sixpence.
10. Where did you remove to afterwards?—My father and mother removed to Leeds, and I had a brother working at Tetley, Tatham, and Walker's, and they got me work there.
11. How long did you stop there?—Three years and a half.
12. What were the hours of working there?—We used to start at half past five, and work till half past nine at night.
13. What time had you for dinner?—Forty minutes.
14. What time for breakfast and dinner?—Nothing for either.
15. What is Tetley, Tatham, and Walker's mill?—A flax-mill.
16. What were the wages there?—I had 3*s.* 6*d.* there.
17. What was your business?—A doffer.
18. Where did you go then?—To Mr. Hammond's flax-mill, at Leeds.
19. At about what age?—About ten years of age.
20. At what time did you begin at that mill in the morning?—We used to start at half past five, and work till eight at night.
21. What time was allowed for breakfast and dinner, and drinking?—Forty minutes a day was all that was allowed.
22. How were you kept up to your work during the latter part of the day?—The overlooker used to come with a strap, and give us a rap or two, or if they caught us asleep they would give us a pinch of snuff till we sneezed; they would give us a slap with a strap if we did not mind our work.
23. Was the strap an instrument capable of hurting you badly?—It was a heavy strap, with a small handle to it.
24. Where did they strike you with it?—Generally in the small of the back, and over the head.
25. Did they strike the young children as well as the older ones?—Yes.
26. And the females as well as the males?—Yes.
27. State the effect upon your health of those long hours of labour?—I was pretty fair in health, but happened with two or three misfortunes.
28. State, in the first place, the effect upon your health and limbs of those long hours of labour?—It produced a weakness in my knees; I was made crooked with standing the long hours.
29. Just shew the gentlemen your limbs.

[*The witness exhibited his limbs to the committee, which appeared exceedingly crooked.*]

30. Are you quite sure you were, as a child, perfectly straight and well-formed?—Yes.
31. How old were you before your limbs began to fail you?—About eight years and a half old.

Wm. Hebden. 32. Had you any other brother or sister working at the mill?—Yes; I had two sisters and a brother.

33. Have those long hours of labour had any effect upon the rest of your family?—Yes; upon one of my sisters.

34. Is she crippled?—She is nearly as bad as I am.

35. Was she originally perfectly straight and well formed?—Yes.

36. To what age did she continue to be perfectly well formed?—Till she was about nine years old.

37. How tall are you?—About four feet nine inches.

38. Are you quite certain that the deformity of your limbs was not consequent upon the accident you had?—No; it was not owing to that.

39. You were deformed, as you are now, before that?—Yes.

40. Were the children unhappy at the state in which they were?—Yes, they were.

41. Have you seen them crying at their work?—Yes.

42. Had you time to go to a day-school or night-school during this labour?—No.

43. Can you write?—No, not at all.

44. Had you to work by gas-light?—Yes.

45. What effect do you think that has upon the eyes?—It nearly made me blind; I was forced to go into the Infirmary; I was seven weeks there, and the doctors said, towards the latter end of the seven weeks, they did not expect they could cure me.

46. What do you do now?—I sell potatoes.

HEBERGAM, JOSEPH; age 17; examined 1st June, 1832.

J. Hebergam. 1. Where do you reside?—At North Great Huddersfield in Yorkshire.

2. Have you worked in factories?—Yes.

3. At what age did you commence?—Seven years of age.

4. At whose mill?—George Addison's, Bradley mill, near Huddersfield.

5. What was the employment?—Worsted-spinning.

6. What were your hours of labour at that mill?—From five in the morning till eight at night.

7. What intervals had you for refreshment?—Thirty minutes at noon.

8. Had you no time for breakfast or refreshment in the afternoon?—No, not one minute; we had to eat our meals as we could; standing or otherwise.

9. You had fourteen and a half hours of actual labour, at seven years of age?—Yes.

10. What wages had you at that time?—Two shillings and sixpence a week.

11. Did you not become very drowsy and sleepy towards the end of the day, and feel much fatigued?—Yes; that began about three o'clock; and grew worse and worse, and it came to be very bad towards six and seven.

12. What means were taken to keep you at your work so long?—There were three overlookers; there was a head overlooker, and there was one man kept to grease the machines, and there was one kept on purpose to strap.

13. Had you any brothers or sisters working in the mills?—I had at that time a brother and a sister; they called him John, and my sister Charlotte.

14. What ages were they, when they began working at the mills?—I cannot say how old my sister Charlotte was, but my brother John was seven.

14. Where is your brother John working now?—He died three years ago.

15. What age was he when he died?—Sixteen years and eight months.

16. To what was his death attributed by your mother and the medical attendants?—It was attributed to this, that he died from working such long hours, and that it had been brought on by the factory. They have to stop the flies with their knees, because they go so swift that they cannot stop them with their hands; he got a bruise on the shin by a spindle-board, and it went on to that degree that it burst; the surgeon cured that; then he was better; then he went to work again; but when he had worked about two months more his spine became affected, and he died. J. Hebergam.

17. Did his medical attendants state that that spinal affection was owing to his having been so over-laboured at the mill?—Yes.

18. How long was it before the labour took effect on your health?—Half a year.

19 Did it, at length, begin to affect your limbs?—When I had worked about half a year a weakness fell into my knees and ankles: it continued, and it has got worse and worse.

20. How far did you live from the mill?—A good mile.

21. Was it very painful for you to move?—Yes, in the morning I could scarcely walk, and my brother and sister used, out of kindness, to take me under each arm, and run with me to the mill, and my legs dragged on the ground; in consequence of the pain I could not walk.

22. Were you sometimes too late?—Yes; and if we were five minutes too late, the overlooker would take a strap, and beat us till we were black and blue.

23. Just shew the committee the situation in which your limbs are now.

[*The witness accordingly stood up and shewed his limbs.*]

24. Were you originally a stout and healthy boy?—Yes, I was as straight and healthy as any one when I was seven years and a quarter old.

25. Were there other children at the mill that were also made ill by this labour, and who became deformed?—Yes, there were some very often sick, and some were deformed; but the parents who were able to support their children took them away, in consequence of seeing that they would be deformed if they did not take them away.

26. Have any cases of accidents in mills or factories been brought into the Leeds Infirmary since you were there?—Yes; last Tuesday but one there was a boy brought into the Infirmary, about five or six o'clock in the evening, from a mill; he had got caught with the shaft, and he had both his thighs broke, and from his knee to his hip the flesh was ripped up as if it had been cut with a knife; his head was bruised, his eyes were nearly torn out, and his arms broken. His sister, who ran to pull him off, got both her arms broke, and her head bruised, and she is bruised all over her body. The boy died last Thursday night but one, about eight o'clock; I do not know whether the girl is dead, but she was not expected to live.

27. Did you not meet with an accident in the mill?—Yes, I had one of my arms broken.

28. What were you doing when that occurred?—I was working at what is called a brushing-mill; there is a pin they put into the roller to make it run round, and the pin caught my sleeve, and twisted my arm round and broke it, and another boy has had his arm broken in the same way.

29. Is there any way of avoiding such accidents?—Yes; at Mr. Brooke's mill they cannot break their arms by that part of the machine, owing to a different arrangement. There was a boy who, to fettle the machine, was kneeling down, and a strap caught him about his ankles and carried him round the wheel, and dashed his brains out on the floor.

30. Do you think these accidents usually happen at the latter end of the

J. Hebergam. day, when the children get tired?—Yes, that boy got killed at a quarter past seven at night.

31. Do you think it is possible for the children to be so active and nimble as to avoid the danger of the machinery, when they are so over-worked and fatigued?—No, because they are sleepy.

32. Have you found that in all those mills you have been required to labour longer than your strength could bear?—Yes, I have.

33. And have you found that, on the whole, you have been rendered ill, deformed, and miserable, by the factory system, as at present pursued?—Yes. Oh! if I had a thousand pounds, I would give them to have the use of my limbs again.

KENWORTHY, DANIEL; age 38; examined 15th May, 1832.

D. Kenworthy. 1. What is your present employment?—Woollen-weaving.

2. Are you the father of William Kenworthy?—Yes.

3. Was he a strong and healthy boy in his early youth?—He never had any sickness in his life till that weakness.

4. How long is it since he began to be deformed in his limbs and knees?—It is not past eighteen months, but I cannot speak exactly to the time; it is not more than that since he began to be so crooked.

5. Originally he was quite a straight and strong boy?—Yes; he was straight enough for any boy.

6. What do you consider was the cause of his deformity?—I think being over-worked, but what it is I cannot say.

7. What has been your employment?—When I began to work I was in a cotton factory.

8. What age were you?—I was turned six years of age.

9. What were the hours of labour at that period?—We began to work at six and worked till eight.

10. With what intervals?—We had one hour in the afternoon.

11. What was the general treatment at that period?—I never got any bad treatment from my employers.

12. Do you conceive that the usage of persons in mills has improved or got worse since you have had experience in them?—By all accounts it has got worse; but I have never seen any cruel treatment for my own part; I have heard tell of it.

13. What is your opinion of the effect of the long hours of labour early in life, commencing as you did as a cotton-spinner, and continuing in the different branches of business for so many years; what has been the effect on your own health?—I think it has done me a great deal of harm; I am very ill to do for many years.

14. What is the nature of your complaint?—I am troubled with an asthma.

15. You consider that a complaint people are frequently troubled with in cotton mills?—It is a very smothering, unhealthy job altogether.

16. Do you not conceive that your memory and your mental faculties have suffered by this constant confinement and labour?—I have a very poor memory to what I used to have; I am very much troubled with a pain in my head and in my back-bone.

17. You find yourself totally out of sorts?—I am scarcely able to do anything in the winter-time in consequence of the asthma.

KENWORTHY, WILLIAM; age 14; examined 15th May, 1832.

1. What is your business?—A scribbler-feeder. W. Kenworthy

2. When did you first begin to work in a factory?—When I was under seven and a half years old.

3. In whose factory did you then work?—Mr. Starkie's.

4. What were your hours of work?—I cannot say exactly what the hours of work were then; we began at six in the morning; I cannot tell at what time we left at night; I was too young then to recollect.

5. Where did you work next?—My father and mother then went to Leeds; I got jobs at different places, till I worked at Messrs. Riples and Ogle's.

6. State their business?—Woollen manufacturers.

7. How many hours did you work there?—We began at six and worked till seven; two hours for meals.

8. At what sort of work?—I was piecener.

9. How long did you work when you were busy?—When we were right busy, making seven days a week, we began at five and worked till eight; two hours out of it.

10. What were your wages at that time?—Three shillings and sixpence a week.

11. With over hours?—No; 4*s.* 1*d.* with over hours; we were paid for our over hours.

12. Where did you work when you left Messrs. Riples and Ogle's?—I went to Messrs. Starkie's, at Huddersfield.

13. Were you still a piecener?—Yes, till lately.

14. What were your hours at Mr. Starkie's mill?—We began at six and worked till half past eight.

15. What time had you for meals?—One hour and a half.

16. What effect had that long labour on your health and limbs?—It had no effect on me then.

17. How, then, did you become ill?—It was from the hard work I had.

18. Did you apply to any doctor?—The overlooker said I was to tell my father and mother that I had something the matter with me, I grew very crooked, and had better go to some doctor; so my father and mother took me to Dr. Day, but he did not seem to do me any good; and then they said I had better go to the Dispensary at Huddersfield, under Dr. Walker, and he did me a vast deal of good; he right cured me.

19. What did the doctor say was the matter with you?—He said I had an affection of the spine.

20. Will you just shew your legs?

[*Here the witness shewed his legs and knees, which appeared excessively deformed.*]

21. How long have your legs and knees been in that state?—I was as straight as ever I could be two years since.

22. What did the doctor state to be the cause of your becoming deformed?—He said it was hard work; it was being overworked.

KERSHAW, WILLIAM; age 41; examined 14th April, 1832.

1. What is your business?—A cloth and operative manufacturer. W. Kershaw.

2. Where do you reside at present?—At Gomersal.

W. Kershaw.

3. At what age did you begin to work?—From eight to nine.

4. As a piecener?—As a piecener.

5. At whose mill?—It was a company's mill; Joseph Kershaw and Joshua Wood were the masters.

6. What were your wages as a piecener?—Two shillings and sixpence a week.

7. What were the hours of work for children at that period?—Generally from six to seven, or half past seven, or from that to eight.

8. Had you any intermissions for meals?—Yes, half an hour for breakfast, an hour to dinner, and half an hour in the afternoon.

9. When you were a piecener, what was the general treatment at that time; were they much punished?—Yes; I have been ill-treated myself, and I have seen others that have been a great deal worse used.

10. How were you beaten?—There is a difference in the dispositions of the slubbers or persons under whom the child is placed; some have more humanity, and rather wish to encourage the children to attention than to punish them for negligence. I have been employed under both. Some of them who are kind have some rewards, such as some fruit, and say that those who have the fewest ends in a given time shall have this fruit; and others will keep beating the children, whether they are in fault or not. I have been beat with a billy-roller towards night, when I have been particularly drowsy, till I repeatedly vomited blood.

11. Did you go home in that condition?—Yes, I did, and I had never complained before; but this I could not hide, and I was obliged to tell the cause of it; and I entreated my mother not to make a complaint, lest I should be further beaten. The next morning after I went to work, she followed me, and came to the slubber that had used me in that way, and gave him a sharp lecture; and when she had done she retired into the engine-feeder's house, and left me to my work; and as soon as she was gone, he beat me severely again for telling, when one of the young men that served the carder went out and found my mother, and told her, and she came in again, and inquired of me what instrument it was I was beaten with, but I durst not do it; some of the by-standers pointed out the instrument, the billy-roller, and she seized it immediately, and beat it about the fellow's head, and gave him one or two black eyes.

12. How long ago was this?—About the year 1799.

13. Do you think the children are any better treated now?—I believe not; I have two children that actually work at the mill at present, and one that goes to learn—three girls; the oldest, when a piecener, has had to stop a day or two at home for three successive weeks together, on account of being beat upon the head; she is now turned fourteen years of age.

14. Was it in the former part of the day, or in the latter, when this beating usually took place?—I have seen it in all parts of the day; but it was more frequent at the latter part of the day, when a child has been labouring a long time.

KIRK, JAMES; age 17; examined 12th April, 1832.

James Kirk.

1. What business do you follow?—That of cloth-dressing.

2. In whose employment have you been?—Mr. Robinson's, first.

3. How old were you when you went to that mill?—Nine years of age.

4. What were the hours of work?—From six o'clock till nine.

5. What intervals had you for meals?—Two hours.

6. How long did you stop there?—I was there between five and six years.

7. What wages had you when you first went?—Three shillings a week at first. James Kirk.

8. How much was that increased to at last?—To 6*s*.

9. What was your last employment there?—The French machines.

10. How many hours a day were you employed at these machines?—From six o'clock till nine.

11. In what situation were you employed in Mr. James Brown's mill?—At the Lewises.

12. What age were you then?—Between fifteen and sixteen.

13. State the hours of labour you had to conform to at that mill?—At the Lewises we were working from five o'clock till nine, with two hours for meals.

14. How long did you continue at the Lewises?—For about a year.

15. What did you go to then?—The giggs.

16. What were the hours of labour at the giggs?—We began at five o'clock on Monday morning, and went on to Tuesday night at nine.

17. What age were you at that period?—About sixteen.

18. You began on Monday morning?—At five o'clock.

19. When did you rest?—At eight o'clock.

20. For how long?—For half an hour.

21. From half past eight to when did you work?—Till twelve.

22. How long did you rest then?—For an hour.

23. Was that for dinner?—That was for dinner.

24. Go on.—We then went on from one till five, and stopped half an hour; from half past five to nine, and stopped half an hour; from half past nine to twelve, and stopped an hour; from one to half past four, and stopped half an hour; from five to eight, and stopped half an hour; from half past eight to twelve, and stopped an hour; from one to five, and stopped half an hour; and from half past five to nine, and then we went off.

25. Then you worked for forty successive hours, including the intervals you have stated?—Yes.

26. What was your daily work on the Wednesday?—From five o'clock in the morning to nine o'clock at night.

27. With two hours' rest?—Yes.

28. And what was it on Thursday?—The same.

29. Then on Friday, will you state what your usual labour was?—We began at five o'clock on Friday, and went on till eight, stopped half an hour; from half past eight till twelve, stopped an hour; from one to half past four, stopped half an hour; from five till eight, stopped half an hour; from half past eight to twelve, stopped an hour; from one till five, and then went home.

30. What were your wages at this time?—Eight shillings a week.

31. Describe to the committee the sort of labour you had to undergo for this number of hours?—It was what we call gigging; turning over wet pieces of cloth.

32. Do you consider that very hard labour?—Yes.

33. For what length of time together was this excessive labour undergone by you?—About three or four months.

34. Tell the committee what effect it had on you?—I began to be very weak in my knees; one of my knees gave way.

35. What did you think this bending of your knees was owing to?—Owing to working such long hours.

36. Were you perfectly straight-limbed before?—I was.

37. Will you shew your limbs?

[*Here the witness shewed his knees and legs.*]

38. What did you do then?—I was so weak that I was forced to give over.

James Kirk. 39. To whom did you apply?—To the Leeds Infirmary.

40. What medical gentleman attended you in the Infirmary?—Mr. Samuel Smith.

41. What did he say your distortion was owing to?—To working such long hours.

42. Did your employers inquire after you, or pay any attention to you after you became thus weak and deformed?—No.

LONGSTON, WILLIAM; age 40; examined 6th July, 1832.

W. Longston. 1. Where do you reside?—At Stockport.

2. Have you had considerable experience in the spinning of cotton, and in other operations connected with that trade?—Some little in the spinning, but more in the previous preparations.

3. Have you known the operatives in the manufacture of cotton from your youth up?—Not without intervals; for about six years in my youth, and in maturer years as a manager.

4. Is it a subject, on the whole, to which you have paid much and serious attention, the management of the cotton factories of this country?—Yes, my attention has been very frequently called to it in disputed and difficult matters.

5. Is it your opinion, that personal punishment ought to await those who habitually and intentionally evade the law made to protect the infant and rising generation; I am meaning now to allude to the incarceration of offenders?—The question is, indeed, a delicate one to be answered; but in conscience I should think, that masters who set to their workmen the example of transgressing the law, and to diminish that respect that ought to be maintained in them for the laws, I think, and must say, that their criminality is augmented by that circumstance, and that (as you term it) personal punishment and (as you explained it) incarceration, I really think would be a punishment not more than adequate to the criminality.

6. Do the operatives connive at the violations of the laws which have been enacted for their own express protection?—Yes, almost universally.

7. State why and how they do so?—If they do not connive, they must be brought forward as witnesses against the masters; but, being brought forward as a witness against the master, if the person be a voluntary witness, it is ten to one if they can obtain employment anywhere in the district where they are known, and where their name could be sent, so that their persons could be in anywise identified; I think this explanation will shew and prove that employment would be exceedingly improbable for them anywhere, as employers would never deem themselves safe where they had in employ one who had been a voluntary witness against them.

8. You presume that there is a combination among the masters, to exclude from work those individuals, who give information upon the infraction of the law?—I am not aware that there is a combination, I have judged it rather to be a tacit consent, or a general practice, than a combination.

9. Have you observed any effects on the health of those employed in the cotton branches?—Those who have been brought up in factories are smaller than the average number of persons engaged in other employment.

10. Does it lead to frequent deformities, especially in the lower extremeties?—I have seen a few instances of it; I have frequently heard them complain of their feet and ankles; a person, about a fortnight before I came here, was brought to me, and I looked at his ankles; one ankle was exceedingly swollen, and I was informed that it was perpetual, that it had been so for

many years, and that it was with walking to and fro as a piecener, being perpetually on his legs. W. Longston

11. Does it not consist with the general remark of those engaged in that employment that it has a tendency to produce deformity, especially in the lower extremities of the body, and a general debility of the constitution?—It is universally admitted; I have never heard it questioned.

12. Will you state the moral effects that you conceive to result from this system, as far as you have observed it?—The want of time for educating children in the evening, and even if there were a little more time, it would not mend the matter much, unless they had considerably more time allowed, that they might not be so sleepy and so fatigued; they must be indisposed to learn anything when the spirits are sunk, and the languor of fatigue indisposes the mind; and hence in the mills an uneducated people grow up. By Sunday-schools they can all read, or nearly all, but in general they cannot read so that it can be of any kind of advantage to them. In conversation they understand little more than monosyllables; books are of no use to them; a common newspaper can be understood only by a few. From the want of that information which might be derived from books in the absence of other instruction, they are in a state of moral debasement.

M'NISH, JAMES; examined 15th June, 1832.

1. Where do you reside?—In Glasgow. Mr. J. M'Nish.

2. What is your trade?—An operative cotton-spinner.

3. Have any of your friends or relatives been in that pursuit?—Yes; my father has been in that line since he was seven years of age; he was manager of a factory twenty years.

4. So that you are pretty well acquainted with everything connected with the spinning branch of the cotton trade?—Yes; I believe I am.

5. Can you estimate the number of persons employed in the cotton branch alone?—We were expecting estimates of it up from our committee, but have been disappointed; however, I know that we can say there are about 11,000 or 12,000 employed in our department alone, but there are a number of power-loom factories, of which I cannot give any account.

6. What proportion of the spinners do you think are children or young persons?—There are 1,100 spinners, and each of those spinners employs three individuals under the age of eighteen, and there are a number of young persons employed in other parts of the works.

7. Making in the whole, what?—About 4,000 under eighteen, engaged in spinning alone.

8. A considerable proportion of those children being females?—There are more than two females to one male.

9. Are there any children under nine years of age?—Yes, I believe a good number; the mill-owners do not sanction the employment of those children, but I believe there is a good number.

10. State the hours of labour that the children are required to work in the mills?—Twelve hours a day, and nine on Saturday, are the hours stated in the bill; but I believe the children may be rated at half an hour a day more than that, as they return in the meal hour to clean the machinery.

11. In addition to those hours of labour, have the children to make up lost time, in consequence of any accident happening to the machinery, or interruption in other parts of the establishment?—Yes, they have to make up lost time; though they are confined during the time the repairs are going on, they have to make up the time at night.

Mr. J. M'Nish. 12. State whether the labour of those children for that length of time has not to be endured, especially in the fine-spinning department, in a very hot temperature?—The fine department requires a great heat. I have had the regulating of steam for three years, and, to do justice to the work, I had to keep it from 80 to 84 degrees of heat; but there are many of the works in Glasgow where the men have not the regulating of the heat, and where it is very much higher.

13. It has been occasionally stated that the work-people themselves, especially the children, were anxious to have this extraordinary heat kept up, and that it was difficult to induce them to have the windows kept open or the places ventilated; is that true or not?—It is not true; it is always a matter of anxiety with the hands to get the heat as low as possible; but, however, if the heat is too low, and the work not going well, it is then so severe upon the pieceners, that they request it to be a little warmer; they cannot keep it always exactly to answer the work.

14. What should you call too low?—Too cold.

15. Would it be below 60°?—Yes, that would be far too low. Without a greater heat, the fibres of the cotton will not open or draw; then it draws down the yarn and breaks it, and the little things have more work to do, in picking up those ends again.

16. Do you think that the children suffer very much by sudden transitions from heat to cold?—I believe they do, very much; if you view them coming out on a cold winter's night, or going in a cold winter's morning to the mill, you will perceive it.

17. Are the children very much emaciated and rendered tender and weakly in their constitutions by the exposure, through so long a day, to this heated and humid atmosphere?—Yes; a view of the working classes in a manufacturing town would convince any unprejudiced mind of that; and, besides that, they have often a cough upon them; even the young children, soon after they first go there, have often a cough.

18. Do you know, from your observation and experience, that individuals at forty years of age are discharged as not being able any longer to perform their work with sufficient activity?—Yes, I know it; we took an account through Renfrewshire and Lanarkshire, and out of 1600 men in the factories, there were not more than ten that were forty-five years of age, and they consisted of men who had been long in the employment of an individual master, and who, different to other masters, kept them on, though they were deficient as to quantity. When they come to the age of forty their eyesight fails, and their constitution is so debilitated that they cannot throw off the quantity that is required by their employers. A given quantity is required, and they are turned off, and young men employed; and there is always a redundancy of young men at the command of the employers.

19. So that the extravagant degree of labour exacted from them in early youth is balanced in a melancholy way by their being deprived of employment in after life?—Yes, there is many a man going about idle, but willing to work, and who would be able to work if he was in any other employment, but who is now forced to be a burthen upon his friends.

20. Have you made observations as to a larger quantity of work having been done, proportionably, when the number of hours have been reduced in Glasgow, in consequence of a temporary slackness in trade?—Yes, when our hours were reduced, we exerted ourselves better; we had more strength to do so when we were only working seven or eight hours a day; and I believe, on one occasion, we gained five or six shillings a week in wages, in proportion to the time, above what we could do when working the full twelve hours a day.

21. There is a difference in the quality of the goods manufactured, when

the hands are overlaboured?—Yes, there is one particular fault, in what we call bad pieceings, and those have to be broken, and the piecers have to follow up the machinery to break those bad ends; and when they do not follow up the machinery to break them, and those bad ends go on, they hurt the price of the yarn, and reduce its value in the market; and it is sometimes the cause of great complaint. I have known baskets of yarn, of two or three different hands, returned back upon the master spinner, owing to those bad ends, and those bad ends are through the neglect or weariness of the piecers in the winter season; this is more especially complained of during the long night work. Mr. J. M'Nish.

22. With regard to the necessity of chastising the children at the latter part of the day, as well as of other means to keep them awake, if you did not resort to those means, should you not fall short of the quantity of the work required of you, calculated on a pretty accurate knowledge of what the machinery will do?—Yes; and undoubtedly we are obliged to keep them at their work; for as much depends upon the exertions of these children as upon our own, in producing the quantity and the quality of work; and if we did not keep them to it, we should lose our employment, both from the badness of our work and the deficiency of the quantity.

23. Is it the spinner's interest to overlabour the children?—Yes, it is his interest to keep the children at their work, and throw off the greatest quantity he can produce, without injury to himself.

24. From your observation, should you say that the children become crippled in their limbs?—Yes, I have seen various instances of their being crippled, and their limbs growing crooked; and I have had an instance of a girl about sixteen that had actually to leave the factory from swelled legs and feet. Her parents were very poor, and I kept her at a very heavy loss to myself; but I was obliged to put her away at last for my own sake.

25. Is that a frequent occurrence?—I cannot speak generally. I know that the girls have frequently bad legs, with running sores; in fact, I may say so of factory people in general.

26. Have they sores in their legs?—Yes, running sores, arising from their long standing.

27. Do you consider that the reduction of the hours of labour would have any effect in raising your wages?—No, I do not think it would have any effect in raising the wages, but it would have the effect of preventing an unnecessary reduction. In times of stagnation, some sordid manufacturer comes forward, and offers the starving men a certain reduction of wages, which they are obliged to take, or starve; he then brings his yarn to market at a reduced price, and not only ruins the wages of the operatives, but occasions a great loss to the other mill owners themselves, by reducing the value of the stock they have on hand. He having got it done cheaper, and bringing it cheaper to market, the value of their stocks is reduced.

28. If a bill, regulating the hours of spinners, were carried into effect, there would be no particular advantage given to any individual or employer; they would all be then on a system of equality, in that respect, and could fairly compete with each other?—Yes, it is my opinion that it would be so.

29. Under such a regulation, could this country, in that particular branch of its industry, compete with its foreign rivals, and particularly with France? —Viewing, as I do just now, that we have not ten hours' labour per day, I do not see that, even if France and other countries were now equal to us in advantage, they could have any advantage over us. But allowing that we have now twelve hours, and two hours were to be reduced, still I do not think, from the knowledge that I have obtained of France, that France

Mr. J. M'Nish. (and France I allow to come nearer to us than any other nation) could reach us, or cause us any danger in the foreign market.

30. Have you had any opportunity of personal observation in that respect?—Yes; we went to France purposely to ascertain whether the French manufacturer could compete with the English manufacturer; we did not wish to hurt the interest of the employers, nor would we have advised this measure had we found, by actual inspection, that it would have hurt our masters' interests in the foreign market.

31. Have you any observations to make in reference to the hours of labour in the cotton-mills in France?—I have only this observation to make, that though I believe they have as long hours for work ostensibly as we have, from six to eight, they have not the same system of discipline that we have; the men can go and come when they please, and they do not attend those hours that we are obliged to do in Britain; Monday appears to be a day that they do not work at all; we got that from every individual connected with the management of any of those works, and even from some of themselves; and that they attended at their work when they pleased, and did as much work as they pleased.

32. So that that which is one of the main objects of the manufacturers in this country, and which seems so essential to the complete prosperity of their works, punctuality, is, comparatively speaking, but little attended to there?—Very little; and it undoubtedly must be a great loss to the master manufacturer, for the machinery is stopped when the spinner is absent, and day-wages are going on and overlookers' wages, and the waste of the engine-coal, and so on; therefore it must be a great loss to the master manufacturers there, when they do not get the same quantity produced from their machinery as our masters do.

33. Are the mills in France principally propelled by steam or by water?—There were a number about Rouen propelled by water, but the greater part of the mills that we saw were propelled by steam.

34. Will you have the goodness to state to the committee the observations you made regarding the price and quality of coal?—We averaged the price of coal in the various manufacturing districts at 36*s.* a ton; in some districts they were only 28*s.* and 32*s.*; in Paris (and there is a good deal of spinning around it) they were much higher.

34.* The committee wish to know whether you were careful to calculate properly the difference between the weights and measures in the two countries?—It was mostly from English overlookers, and managers of foundries, that we got the price of coal, and they gave them us in English weight; but the principal point that we depended upon was, reducing it down to the weight of a pound, and taking it upward.

35. Is the quality of the coal as good as that in this country for the purpose of raising steam?—No, their coal is not; but they use very little of it for any purposes of manufacture, and I believe the greater part of the coal comes from this country.

36. Of course, if the coal were imported from this country, it must be consumed at a much greater cost than here?—Yes, besides the inland carriage that they have.

37. Have you made any inquiries as to the price of the machinery necessary for spinning in France?—Yes; we were in the foundries, and got the price of the castings, and, to the best of our opinion, it was double the amount of what it was in our own country.

38. Speaking of the very best machinery you saw in France, and considering the disadvantages under which the French manufacturer labours, you consider that there is no possibility for him to compete with the English spinners?—No, there is no possibility of it; there are one or two reasons

that occurred to us, besides their machinery and fuel; the scattered condition of their manufactories gives rise to a heavy inland carriage, and there appears to be a complete want of that spirit of enterprise in the manufacturers which exists among us, besides a want of capital. Mr. J. M'Nish.

39. With regard to the price of the different numbers of cotton twists, when manufactured, was a corresponding number of the same article higher?—Fifty per cent. higher than ours, and of a quality not near so good; from all the information that we could get, both from the late weavers, by going with them to their warehouses, and from the overlookers of the fabrics, as they call them, we brought it, by our own calculations of the price of the yarn here, to this,—that there is full fifty per cent. difference in the fine numbers down to number forty; in the less fine numbers, there may be about from three to five per cent. less difference.

40. In the coarser numbers you conceive that from forty to fifty per cent. is the least difference between the prices of the English and French yarns?—Yes, we proved it in another way; we discovered that they used a great quantity of English yarn in France, although it was prohibited; and we also learned the way in which it was got into that country, and it proved to us that there must be this difference, or the smuggler could not carry on his business; he pays five per cent. at Brussels for the duty upon it, and at Brussels, or in Belgium, there is twenty per cent. insured against the loss of the yarn in crossing the frontiers; and I think the lace weavers calculated to us that there would be another five per cent. for the expenses, amounting in all to about thirty per cent.; now if the smuggler had not a heavy profit upon it he could not pay this and carry on his business. We likewise ascertained that the British yarn was so much better than the French in quality, that the lace manufacturers would rather have it at the same price as the French yarn, although they had to run the risk of losing the yarn, and of being fined in double the amount of the quantity that they had on hand.

41. Have you anything further to state, resulting from your observations in France, in reference to this subject?—No; but I would leave it to the committee whether they would allow me to put upon record the mode in which we got this knowledge; it may be of benefit in future to individuals going to France. We did not inform those people what we were, or what we wished to know, and we made an excuse that we were in search of a friend, got a great part of the English operatives together, and were introduced to the English overlookers, likewise, to ask if they had any knowledge of this lost friend, and thus got a knowledge of these matters. In other parts we altered our conduct a little, and appeared to wish to establish a factory of our own, inquiring from the makers of the machinery, and from gentlemen whom we considered were engaged in situations in factories; therefore we got both sides of the question, and could form our ideas.

42. Where do you think the French principally purchase their cotton?—I believe the great part of it comes from Liverpool; to the best of my knowledge it is there that they purchase. There may be one or two masters of extensive capital, we only, however, learnt of one that brought his cotton from America; that was Mr. Jolie, of St. Quentin.

43. After having stated the result of your observations regarding the introduction of British twists into France by the means you have described, it is almost superfluous to ask you, whether they could export any of their twists to any market to which the British manufacturer has established access?—No, I do not think that they produce within a tenth part of what would supply their own home market at this moment.

44. Something has been said regarding the competition which is to be feared from the American spinners, supposing that the hours of labour were

Mr. J. M'Nish. reduced in this country; have you had friends that have emigrated from your populous city to America that have given you any information as to this subject?—I have had a brother-in-law that went out last Easter, and in the first letter that he wrote to me he said that the machinery was so bad, and the method of conducting their mills so bad, that he could not stop at the spinning business, but had joined the plaistering, and had a dollar a day for it; which was more than he could have made at the spinning, owing to the badness of the machinery, having only one small mule at work; and there were about sixty that went out at the same time, all spinners, and out of that sixty, which were sent out at the expense of the association, there were not above six that continued at the spinning.

45. Are you aware that the price of labour is very much higher in America than it is in this country?—It must undoubtedly be higher when they can make such a price as a dollar a day on such small machines as they work.

46. If, therefore, they can make that price by producing so small a quantity of goods, it must of necessity follow that those goods are very much dearer than those produced in this country?—Yes, they make the same wages from producing, I believe, one half less quantity than we do; they are paid by the piece the same as we are, and I believe, from all the statements that have come in their letters, that they have double price for that quantity; the reason is, that they have to pay so much more for their piecers, because they cannot get children to stop with them without a great hire; parents there have something to keep their children upon, and will not put them into their mills; a single man going there will not be employed in the mills, but a man with a family they will employ; and in a very short time he becomes rich, and goes to agriculture: they are not allowed either to strike the children, or use the same discipline there as we have liberty to do here. I have in one letter an anecdote of an acquaintance that was taken before the justice of the peace for giving his little piecer a cut on the side of the head.

47. Speaking as to the healthiness of the employment; supposing an individual, wishing to ascertain that interesting fact, were to make inquiries in any mill or factory regarding the sickness that had occurred among the hands during a given time, and those that had died, belonging to the establishment, do you conceive that any such cursory inquiry as that could give the least idea of the actual health or longevity of the hands so employed?—No; I believe it would be impossible to ascertain that from any method whatever. When any of the workers lose their health, their parents endeavour to find some other employment, such as putting the girls to mantua-making, and the boys to some other work; and, therefore, though their health may have been ruined, they do not die connected with the mills, but with other trades; and, besides, the men are turned off at such an early period of life, forty years of age, that very few can be found who die as spinners.

OPENSHAW, ISAAC; age 22; examined 2nd July, 1832.

I. Openshaw. 1. Where do you live?—At Sharples, near Bolton.

2. At what age were you when you first began to work in a factory?—Nine years of age.

3. In whose factory did you begin to work?—Messrs. Taylor and Indle's, near Bolton.

4. What was your employment?—Piecening.

5. What were your wages after you had learnt?—Eighteen-pence a week.

6. At what time did you go in the morning?—About half past five, and we used to start at six. I. Openshaw.

7. At what time did you stop at night?—Half past seven was the regular time then.

8. Had you to work in a very hot room?—Yes.

9. How long did you stop at that work?—I have never been at any other work.

10. Will you shew the committee your limbs?

[*The witness exhibited the same, which appeared to be excessively deformed.*]

11. Were you originally quite straight?—Yes, I was as straight as any man here.

12. What age were you before that excessive degree of deformity began?—Seventeen. It is about five years ago since I began to be crooked.

13. What were you doing when you became crooked?—Piecening; I never worked at anything else than piecening.

14. How long did you stay working in the mill after your limbs began to be crooked?—It is eight weeks now since I left it.

15. State to the committee how you felt in your legs and knees when they were becoming so dreadfully crooked.—I felt myself quite tired, and stiff, and stark, till I could hardly stand, so I leant down to rest myself.

16. Were your legs swollen, as well as becoming crooked, then?—Yes, they swelled till I dropped down.

17. When you became so crooked, how could you get to your work?—I went a bit with a stick on the road, and afterwards they carried me by turns, three of them.

18. Did certain boys consent to carry you by turns to your work?—Yes, and back.

19. Had you to use a crutch as well as a stick?—Yes, always a crutch; a black-thorn crutch.

20. What did you earn at the time you fell ill?—Six shillings and sixpence.

21. What were the highest wages you ever received?—Six shillings and sixpence.

22. Have your employers made you any allowance since you have left?—The hands gathered me 1*l.* 8*s*.

23. You mean the workmen at the place?—Yes.

24. What for?—To buy an ass for me, to carry coals about.

25. Did you find you could do that?—No, I could not do with an ass; I could not follow it.

26. Has your master made you any allowance?—No.

27. Were you ever chastised in the mill?—Yes.

28. Did the overlookers, or others, beat you at your work?—No, my master beat me.

29. What did he beat you for generally?—Because I could not be sharp enough to do my work.

30. Were they continually in the habit of working more than twelve hours a day?—Yes.

31. Did you ever hear any of the grown up spinners complain of that?—Yes, that the hours were too long.

32. Do you know whether they ever complained to the neighbouring justices of their masters, because they broke the law?—No, if they had complained the master would have turned them off.

PATERSON, JAMES; age 28; examined 30th June, 1832.

J. Paterson.

1. Where do you reside?—At Dundee.

2. What is your business?—Mill overseer.

3. At what age did you go into a mill?—At ten years of age.

4. At whose mill did you begin to work?—Mr. Proctor's mill, Glames.

5. In what department?—In the carding room.

6. How many hours were you and the rest of the hands confined in that department for actual labour?—There were fourteen hours' actual work, and fifteen hours a day confinement, including meals.

7. What effect had that upon your health?—I was often very bad with shortness of breath and stoppage at the breast. I was obliged to leave the work in consequence.

8. How many years were you there?—Nearly three years.

9. Were other children similarly affected?—Yes, they were. I had a brother who was at that work too, and he was compelled to leave for bad health, and was laid up, and died of consumption. The doctor said it was occasioned by being confined at that work.

10. Were accidents frequent after the children became fatigued?—Yes, there were several accidents; they generally happened in the morning before breakfast, and some of them towards evening.

11. You mean to say, that before breakfast the children were still so drowsy as not to be capable of the vigilance and exertion that was required of them?—Yes, I myself happened to fall asleep while leaning on the canes. I happened to go forward, and fell amongst the wheels, and got my arm a little shattered by that means. I was three months off work at that time, I think.

12. When those accidents occur, do the employers usually pay you your wages while you are under medical treatment, or while you are using means for your recovery?—No, there are no wages given, to my knowledge, under those circumstances. I never received any.

13. How were you employed when at Messrs. James and William Browne's mill at Dundee?—I was employed as oiler there, in the night-time.

14. Was any time allowed for refreshment during the night?—None; we began at eight o'clock at night, and went on till six o'clock in the morning, without stopping at all.

15. According to your experience, is night-work, though it be much shorter, far more prejudicial to the constitution than day-work?—Much worse than day-work.

16. Did you leave that mill?—Yes, I left it in consequence of bad health.

17. What mill did you then remove to?—I went to Mr. Blyth's.

18. What age were you at that time?—I dare say I was between nineteen and twenty.

19. In what situation did you go?—As overlooker.

20. What were your hours of labour in that mill?—About fourteen hours, and about fifteen hours' confinement.

21. To whose mill did you go next?—I went to Mr. Braid's of Duntruin.

22. What were your hours of labour there?—We had no nominal hours there at all.

23. As there were no regulations at all, they laboured you as long as you could see, perhaps?—Yes, as long as we could see in summer time. There was nobody but the master, and the master's son, had a watch; and we did not know the time.

24. The operatives were not permitted to have a watch, were they?—No; there was one man that had a watch, I believe it was a friend that

gave it him, and it was taken from him, and given into the master's custody, because he had told the men the time of day. J. Paterson.

25. You had no clock at that mill?—None.

26. Were any of the children orphans?—Yes, most of them were orphans.

27. Had some of those orphans been in that mill for a considerable length of time?—Yes; I believe there were some from Edinburgh, that had been in it four or five years.

28. Were those children beaten, in order to compel them to work for that length of time?—Yes; whenever they refused to do anything, they were taken and beaten.

29. How were they kept in the premises under that method of treatment and labour?—They were all locked up.

30. While they were at work?—Yes; they were guarded up to their bothies to take their meals, and they were locked up in the bothies at night, and the master took the key away with him to his own bed-room; they were guarded to their work, and they were guarded back again, and they were guarded while they were taking their meat, and then they were locked up for rest. The windows of the bothies where they slept had all iron stanchions on the outside, so that they could not escape.

31. Were they allowed to go to a place of worship on a Sunday?—No, they were not.

32. Were the sexes always divided in those bothies?—No; there was one bothy for the boys, but that bothy did not hold them all, and there were some of them that were put into the other bothy along with the girls.

33. Of what ages were the boys that were put into the girls' bothy?—The boys might be, I should suppose, from ten to fourteen.

34. What were the ages of the girls?—Perhaps from twelve to eighteen.

35. Did the children and young persons attempt to escape from their labour and confinement, and, in fact, from the incarceration that you have described?—Yes; I have gone after them on horseback, and brought them back myself.

36. What was done with any of the hands that were brought back?—They were taken into the mill and got a severe beating with a strap. Sometimes the master kicked them on the floor, and struck them with both his hands and his feet.

37. Did you ever know any put in gaol?—I knew a woman put in gaol, and brought back after a twelvemonth, and worked for her meat; and she had to pay the expenses that were incurred.

38. For how long a period were the engagements made?—There were some engaged for two years, and some for three years; and some of those girls sent from Edinburgh, I heard them say, were engaged for five years.

39. This system gives a very great and a very improper control over those young women; have you, upon your own knowledge, reason to suspect that improper advantages are taken of that control, and of the defenceless situation of the female hands?—Yes, I have every reason to believe that such is the case, for the master of this same mill has a considerable number of females, that have been working under him, that have children by him.

40. Why did you leave that mill?—I left it because a woman had been severely beaten, and I opened the door, and let her out, and would not go after her; so I was ordered about my business myself.

41. Have you anything further to observe upon the subject of this inquiry?—I have nothing further to add; but only, that when I went to Dundee at the time I received my summons, I shewed it to our manager, and he said that Mr. Sadler had used them very badly, in not having given them more notice; and I asked him whether he had any fault with me, or whether he

J. Paterson. would give me work when I returned. He said he would tell me when I returned: and that he had no fault with me, only coming here to give evidence before the committee.

RASTRICK, WILLIAM; age 34; examined 23rd July, 1832.

W. Rastrick. 1. Where do you reside? At Watford.

2. Have you ever been in a silk mill?—Yes.

3. At what age did you go into one?—At eleven years of age.

4. How young have you known children go into silk mills?—I have known three at six; but very few at that age.

5. In whose mill did you work at that period? — Mr. Shute's, at Watford.

6. What were your hours of labour?—From six in the morning till seven at night.

7. What time had you allowed for breakfast, for dinner, and for tea?—Half an hour for breakfast, half an hour for dinner, but no tea time.

8. Was it not found necessary to beat children to keep them up to their employment?—Certainly.

9. Did the beating increase towards evening?—Their strength relaxes more towards the evening; they get tired, and they twist themselves about on their legs, and stand on the sides of their feet.

10. When you were employed as an overlooker, and had to superintend those children, was not the employer aware that you had to stimulate them up to labour by severity?—Certainly he was, and it is always considered indispensable.

11. Would he himself rather urge you to that course than to the contrary?—His object was, in every case, to get a certain quantity of work done; it must be done by some means or other; but when it was necessary for the overlooker to use severity, he had to bear the stigma of it, and not the master.

12. Did you not find it very irksome to your feelings, to have to take those means of urging the children to the work?—Extremely so; I have been compelled to urge them on to work when I knew they could not bear it; but I was obliged to make them strain every nerve to do the work, and I can say I have been disgusted with myself and with my situation; I felt myself degraded and reduced to the level of a slave-driver in such cases.

13. Is not tying the broken ends, or pieceing, an employment that requires great activity?—Yes.

14. Does not the material often cut the hands of those poor children?—Frequently; but some more than others. I have seen them stand at their work, with their hands cut, till the blood has been running down to the ends of their fingers.

15. Is there more work required of the children than there used to be when you first knew the business?—Yes; on account of the competition which exists between masters. One undersells the other; consequently the master endeavours to get an equal quantity of work done for less money.

RHODES, SAMUEL; age 19; examined 26th May, 1832.

S. Rhodes. 1. Where do you live?—At Keighley.

2. How soon did you begin to work at a mill?—I began when I was about six years and a half old, at Mitchell's mill—a worsted mill.

3. Your business was to piece up the ends ?—Yes. S. Rhodes.

4. What were your hours of work ?—We worked twelve hours; we began at six in the morning, and worked till seven in the evening.

5. What time had you allowed for meals ?—An hour for dinner, nothing for breakfast or drinking.

6. What wages had you at that mill ?—Eighteen pence a week.

7. How were you treated at Mitchell's mill by the overlooker ?—Very badly; sometimes he had a strap with nails in it.

8. Did Mr. Mitchell know anything about that ?—No; I do not believe he did.

9. He did not come into the mill much ?—No; I never saw him in the mill.

10. What were your hours of work, when at Bury Smith's ?—We generally began at six in the morning and wrought till eight at night.

11. How much time was allowed for your breakfast, dinner, and drinking ?—There was nothing allowed for breakfast, there was three quarters of an hour allowed for dinner, and nothing for drinking.

12. What age were you at that time ?—When first I began going there I was between eight and nine.

13. What effect had that labour upon your health, and upon your limbs ? I had very good health, but standing long hours, I began to be tired, and could not stand it; I got stiff in my limbs, and began growing deformed in my knees.

14. Will you shew the gentlemen of the committee your limbs ?

[*The Witness shewed his person, and he appeared to be very crooked and knock-knee'd.*]

15. How many boys were in that mill ? About thirty.

16. How many of those were deformed ?—Somewhere about eight of them were deformed in their legs and their knees.

17. Were you perfectly straight till you were between eight or ten years of age ?—I was perfectly straight between eight and nine years of age.

18. You stated that there were about eight of the thirty boys that worked with you in that mill deformed; have you remarked that other children, similarly employed at other mills, were also deformed ?—Yes; in Keighley, you may find wagon-loads.

19. Had you any time to go to a day or night school ?—No; I never went to any week-day school in my life—I have been to a Sunday-school.

20. You had no opportunity of learning to read or write ?—I had no opportunity of learning to write; I had a bit of opportunity of learning to read, on Sundays.

21. Did you feel yourself so fatigued, that when Sunday came you were disposed to spend it in a little ease, and occasional recreation ?—Yes; I was forced to go to school; our parents made us go, or else we should not.

ROWLAND, DAVID; age 38; examined 10th July, 1832.

1. Where do you reside ?—At Liverpool. Mr. D. Rowland.

2. What is your employment ?—I am agent to the Liverpool Bootle Waterworks Company, for the Shipping Department.

3. Where were you born ?—In Manchester.

D. Rowland. 4. At what age did you commence working in a cotton mill ?—Just when I had turned six.

5. What was the reason that you commenced working so early ; did your father and your mother send you ?—I was an orphan.

6. What employment had you in a mill in the first instance ?—That of a scavenger.

7. Will you explain the nature of the work that a scavenger has to do ?—The scavenger has to take the brush and sweep under the wheels, and to be under the direction of the spinners and the piecers generally.

8. Have they to put their bodies in constrained attitudes in order to get at their work ?—I frequently had to be under the wheels, and in consequence of the perpetual motion of the machinery, I was liable to accidents constantly. I was very frequently obliged to lie flat, to avoid being run over or caught.

9. How long did you continue at that employment ?—From a year and a half to two years.

10. What did you go to then ?—To be a piecer.

11. Did that employment require you to be upon your feet perpetually ?—It did.

12. You continued at that employment how long ?—I was a piecer till I was about fifteen or sixteen years of age.

13. What did you become then ?—A spinner.

14. So that, in point of fact, you have yourself gone through almost every department of the spinning-mill ?—I have.

15. What were your hours of labour, generally speaking, at that period ?—Fourteen ; in some cases, fifteen and sixteen hours a day.

16. How had you to be kept up to it ?—During the latter part of the day, I was severely beaten very frequently ; I felt an incapacity to keep up to that extent of labour which was required, and had to undergo severe flagellation.

17. Of course, in the time of labour to which you have been alluding, there were the usual intervals for meals and refreshment ?—Yes.

18. Will you state the effect that that degree of labour had upon your health ?—I never had good health after I went to the factory. At six years of age, when I was obliged to be put to the factory, I was particularly healthy—I was ruddy and strong ; I had not been long in the mill before my colour disappeared, and a state of debility came over me, and a wanness in my appearance.

19. To what age do those that are engaged in the cotton factories of the country, generally speaking, survive, according to your observation and impression ?—I think persons that have been brought up in the cotton factories, generally become inefficient for the labour to which they have been accustomed, soon after they have turned forty ; if they should survive to fifty or fifty-five, I think them extreme cases.

20. Certain of the medical gentlemen of Manchester, who have recently published on the subject of the factory system, have attributed the general prevalence of tippling, and especially in spirits, to the feeling of exhaustion and fatigue that is induced by the particular work under our consideration, in heated mills, and for an undue length of time ; do you believe, from your own observation and experience, and from the remarks you have been led to make upon that particular sort of labour, that such is, generally speaking, the result ?—From all that I have seen and known, and from my recent opportunities of information, I should be disposed, most decidedly, to concur in that opinion.

SADLER, JOSEPH; examined 16th June, 1832.

J. Sadler.

1. Do you live in Stockport?—Yes.

2. What has been your employment?—A dresser of cotton yarn.

3. In whose factory?—In Mr. Robinson's.

4. How many of the principal factories have you worked at?—Mr. Ratcliffe's, Mr. Smith's, and Mr. Robinson's.

5. How many hands do you think are employed in the spinning and weaving of cotton in the town and neighbourhood of Stockport?—At least 14,000.

6. What proportion of those are children?—I should think more than half.

7. Are a considerable majority of those children females?—Yes; more than half of the children employed in factories are females.

8. What temperature do you have generally in the factory?—It varies, but it is generally very high: as high as 80°, 90°, 100°, and 110°.

9. Are any children employed in that temperature?—Yes; there are children in those rooms. The warps are twisted and drawn in the dressing room, and there are children employed for those purposes.

10. What number of hours do the children work in the mills at Stockport?—The time is understood to be twelve, but it is generally, I think, thirteen, with the exception of a few mills, which adhere to the bill of Sir John Hobhouse.

11. Will you state your impression as to whether the hours of labour are not too many, even if Sir John Hobhouse's bill were rigidly observed and enforced?—I am decidedly of opinion that the hours of labour under Sir John Hobhouse's act are too many for children to endure; I believe that, in consequence of the excessive labour, their health is injured, that they are reduced to a state of debility and feebleness which is apparent to every person who observes them, and that it has a very bad effect in a moral point of view upon them.

12. How long have you been employed in a cotton mill?—With the exception of three years, I have been employed in a mill ever since I was nine years old.

13. Has your health suffered in consequence of that employment?—It has not suffered so materially as that of others, though, about 1821, I was affected with shortness of breath, and what I was afraid would prove an asthma, and had to take medical advice for some time, and to leave off my employment. However, I would take the liberty to observe to the committee, that I have been a person as careful of my health as most people; perhaps I am an exception to the general rule in cotton mills.

14. What were the hours during which you worked?—From half past five to half past eight at night.

15. That was continued for some years?—Yes, it was.

16. What intervals had you for meals? None for breakfast; generally three quarters of an hour for dinner.

17. Is the general impression, do you believe, warranted, that there are few who have been long employed in the factories of the country, as at present conducted, who survive the age of forty?—I know of very few instances of persons that have worked in mills the greater part of their life that have survived that age.

18. If, then, the persons so engaged are short-lived, is not the necessary consequence that of leaving upon the community a large portion of orphan children?—It is certainly a very natural consequence that there should be a considerable number of orphans; and indeed there is a considerable number.

Mr. J. Sadler.

19. To recur again to the labour of these children; do you conceive that it has a very visible effect upon them?—Yes; some of the strongest and most healthy, perhaps, may appear rather agile and lively; but take them in general—in fact, nearly the whole appear so fatigued that they are almost lifeless; and in place of manifesting that desire for recreation and sport which is prevalent in boys and girls generally who are employed in other branches of trade, or are not so confined, their only object is to get home as soon as they can, and get to bed.

20. Is not one of the results of machinery this, that the employer can and does calculate the exact quantity that a machine, at a given speed, can throw off?—Certainly; there is a given quantity required, and this quantity must be done, or the person must lose his employment.

21. Could the attention and the labour of the children be sufficiently kept up, so as to produce that quantity, without resorting to those extraordinary means of spurring them on to that labour, for such a length of time?—I believe that such is the intense application required at present, and the quantity expected, that it is impossible to make the children sufficiently attentive to their work, unless the stimulus of chastisement is used.

22. So that the employer must know the necessity of that case, and consequently is as responsible for that consequence as the individual who finds it necessary to resort to such means to obtain the work out of those children?—Certainly.

23. It has happened that certain mills have sent petitions signed by the operatives themselves against the limitation of the labour of the children and others so employed; can you account for the fact of such petitions having been presented to Parliament?—Yes, I can; and I will account for it simply by observing that the master's wishes are considered in the light of commands.

SMART, PETER; age 27; examined 23rd June, 1832.

Peter Smart.

1. Where do you reside?—At Dundee.

2. What is your business?—An overseer of a flax mill.

3. Have you worked in a mill from your youth?—Yes, since I was five years of age?

4. Were you hired for any length of time when you went?—Yes, my mother got 15*s*. for six years, I having my meat and clothes.

5. At whose mill?—Mr. Andrew Smith's, at Gateside, in Fifeshire.

6. What were your hours of labour in that mill?—In the summer season we were very scarce of water.

7. But when you had sufficient water, how long did you work?—We began at four o'clock in the morning, and worked till ten or eleven at night,—as long as we could stand upon our feet.

8. How were you kept to your work for that length of time; were you chastised?—Yes, very often, and very severely.

9. How long was this ago?—It is between twenty-one and twenty-two years since I first went.

10. Were you kept in the premises constantly?—Constantly.

11. Locked up?—Yes, locked up night and day; I never went home while I was at the mill.

12. Was it possible to keep up your activity for such a length of time as that?—No, it was impossible to do it; we often fell asleep?

13. Were not accidents then frequently occurring in that mill by over-fatigue?—Yes, I got my hands injured there by the machinery.

14. Have you lost any of your fingers?—Yes, I have lost one, and the other hand is very much injured. Peter Smart.

15. At what time of the night was it when your hands became thus injured?—Twilight; between seven and eight o'clock.

16. Did you attribute that accident to over-fatigue and drowsiness?—Yes, and to a want of knowledge of the machinery; I was only five years old when I went to the mill, and I did not know the use of the different parts of the machinery.

17. How old were you when you became an overseer?—Seventeen years of age.

18. At whose mill?—At Mr. Webster's, Batter's Den, within eleven miles of Dundee.

19. Did you inflict the same punishment that you yourself had experienced?—I went as an overseer; not as a slave, but as a slave-driver.

20. What were the hours of labour in that mill? — My master told me that I had to produce a certain quantity of yarn; the hours were, at that time, fourteen; I said that I was not able to produce that quantity; I told him if he took the time-piece out of the mill, I would produce that quantity; and, after that, I found no difficulty in producing the quantity.

21. Did not that almost compel you to use great severity to the hands then under you?—Yes; I was compelled often to beat them in order to get them to attend to their work, from their being over-wrought.

22. Did you find that the children were unable to pursue their labour properly to that extent? — Yes; they have been brought to that condition that I have gone and fetched up the doctor to them to see what was the matter with them, and to know whether they were able to rise, or not able to rise; we have had great difficulty in getting them up.

SMITH, THOMAS; age 27; examined 26th May, 1832.

1. Where do you reside?—In Keighley. Thos. Smith.
2. What height are you?—I believe, four feet eight inches.
3. Have you ever worked in mills or factories?—Yes.
4. How early did you commence?—When I was six years old.
5. In what sort of a mill?—A cotton mill.
6. How long did you stay there? — Only about a quarter of a year there; I was sent home; I was too little, I could not reach my work.
7. Where did you then go to?—To Bury Smith's.
8. What age were you then?—Perhaps nine.
9. What were your hours of work at Mr. Smith's? — Twelve hours and a quarter.
10. What time had you out of that for meals?—An hour.
11. Will you state what effect this excessively long labour had upon your limbs?—It made them very stiff constantly.
12. Shew the committee your limbs?

[*The witness shewed his person, and his knees appeared much distorted.*]

13. Were you originally perfectly straight-legged?—Yes; I was straight till I was nine or ten years old, and very active.
14. Did your long labour stop your growth, as well as cripple your limbs? —Yes.
15. Have you had any relations working at mills and factories?—Yes; I had a brother and a sister.

Thos. Smith.

16. What effect had this long labour upon them?—My brother was very ill, fatigued with it, I think, the same as me.

17. Did he become a cripple?—Yes; he is worse than me.

18. Was he, too, originally perfectly straight?—Yes.

19. Your sister also, you say, worked in mills; what effect had it upon her?—The same.

20. Was she perfectly straight at first?—Yes.

21. Did she become crooked?—Yes, she did.

22. Have the children any time to go to school on week days, during the day or night?—No.

23. Are they in a situation, as to health, to attend regularly the Sunday schools?—They are too tired; they take a liberty on Sunday mornings to lie in bed; they will not get up to go to school.

24. Are the children in danger from coming into contact with the wheels and straps of the machinery?—Yes; they are not cared of in any way.

25. Have you known any accident owing to that great neglect?—Yes; I had a very great accident myself; they thought I was dead: the overlooker went out, and we were playing, two or three of us, taking the advantage while he was out, and as I was going by the upright shaft, it caught me by my clothes, and took me round; my left arm was broken in three places, and my head was sadly damaged.

26. What age were you when you left the mill?—Turned of eighteen.

27. What did you become?—A weaver.

28. What can you earn a week, when you are in employ, in weaving?—About 6*s*.

29. Do you work at home?—Yes.

30. Do you prefer being a weaver to being at the mill?—Yes, a great deal.

31. What is your reason for so doing?—I have more relaxation; I can look about me, and go out and refresh myself a little.

32. To what do you attribute your crookedness and deformity?—It arises from standing so long.

33. Do any mill owners regulate the labour of their hands by what is called a speed-clock?—Yes; the engines have all speed-clocks.

34. Are you not aware that you had to work much longer than the regular or nominal hours by the timepieces?—Yes; if the engine lost speed, from ailing something, and not running so quick, we had to make it up.

35. Did it ever save you anything by running too glib?—We never gained five minutes a week by that.

36. Most of you were aware that you had to work longer than the hours nominally assigned you, in consequence of the labour being regulated by a speed-clock, instead of a timepiece?—Yes.

SMITH, WILLIAM; examined 13th June, 1832.

Mr. Wm. Smith.

1. Where do you live?—At Glasgow.

2. What is your business?—That of a cotton-spinner.

3. What were you before you became a cotton-spinner?—A piecer.

4. What are the usual hours of work in Glasgow in that particular branch?—Twelve hours.

5. Exclusive of the time allowed for refreshment and meals?—Yes.

6. Are those hours ever exceeded in consequence of breakages in the machinery of the mills, or from other causes; such as the necessity of occasionally cleaning the machinery?—The adult spinners work twelve

Mr. Wm. Smith.

hours a day, except when they have to make up time, which they often have to do; but the children may be said to work twelve hours and a half a day.

7. In consequence of the cleaning they have to do?—Yes; in the generality of the mills in Glasgow, and I believe in Scotland, they have to clean the machinery in the mill hours, which will take several minutes off each meal, amounting often to nearly half an hour a day; and, with the exception of a very few mills, there is no time allowed for that but at meals.

8. What is the average heat of the mills, speaking of those mills where the fine spinning is done?—The average heat is about 84°, or 86°, in the fine works.

9. Have the goodness to state whether that heat is necessary to the spinning of the yarn well, or whether it arises from keeping the rooms too close? —I think there is no necessity for any heat above 80° to spin any number that is spun in Scotland; at the same time there are a great number of mills in England and Scotland that rise as high as 90°.

10. Is the ventilation of the room in which spinning is carried on left to the overlooker and the hands, or is it regulated by order of the master?—It is in general left to the hands to regulate it as they think proper; but the fine numbers will not, when the wind is high, bear ventilation; the windows have then to be shut, or the wind blowing in upon the fine fabric would break it.

11. What proportion of the hands are grown up, and what proportion are children?—I should think two-thirds or three-fourths are children.

12. Do you conceive that the majority of those children are females?—Yes, I think there are two females to one male.

13. You have said that the children work as great or a greater number of hours than the spinners?—Yes, from a quarter to half an hour a day.

14. Do you conceive that their work is as hard, in proportion to their strength?—I think it is a great deal harder in proportion.

15. State your reasons for thinking so?—Those little children have to travel, in the course of working, considering the speed of the machinery, eight or ten miles a day, and are often stooping and creeping under the machinery to do their work.

16. Speaking of the hours of labour, how are those children to be kept vigilant for such a length of time, and attentive to their work?—The calculation of the working power of a cotton-mill is such, that they are obliged to attend to their duty; the spinner is the first hand in the employ, and it is known by calculation of the machinery how many hanks may be thrown each week, if he attends to his duty; if he does not do that, or near it, he loses his work; and in consequence of this, he is obliged to keep the piecers strictly to their duty; and if they are not able to do it, in consequence of being over-wrought, there is no other remedy but to use the strap. I have done it myself, though I have been grieved at it afterwards; but my employment depended upon keeping them to their work.

17. Does the punishment of the children depend a good deal upon the length of hours during which they are employed?—It must be so, on account of the child's getting tired in the after-part of the day, and not being able to fufil its duty.

18. Do you conceive there is sufficient time left after a day's labour like that, to obtain proper instruction, even if the children were not too much fatigued to avail themselves of it?—There are some parents most anxious to get their children educated, and they force them to go to school; but I have seen it to be of very little benefit; they have gone to school, but they generally went to sleep; those that are more lively, long for a little fresh air and amusement.

Mr. Wm. Smith.

19. Speaking of the effect of this excessive labour, state your impression of its effects upon the female character and conduct?—I consider that in many instances it greatly unfits a young woman for becoming a useful partner to a man, or for being an exemplary mother of children; their time is so long and incessantly occupied at the mill, that they have often neither leisure nor disposition to prepare, under parental instruction, for those duties that await them in after-life, if they become wives and mothers. I believe many of them have good inclinations, but the long hours they labour prevent them from availing themselves of the opportunity of improvement; and many of them are injured in their morals by the numerous bad habits and examples they see daily exhibited. Many of them from overlabour have abandoned their employment, and followed the habits of vice, which, if they had been moderately laboured, they never would have fallen into.

20. Are the effects of this long-continued labour speedily visible when children commence labour in these different establishments?—It is visible in a few weeks after they come to the cotton-mills; the rosy, blooming appearance they have generally soon fades; many of them look pretty well at times; but the proper time to notice those spectacles would be in the morning or at night; after work, there is a rosy hue, arising from the heat of the factory, and the work they follow; but in general their appearance is sallow, and their looks are quite different from those of children engaged in other occupations.

21. What knowledge have you of cotton-spinning in France, have you been there?—Yes.

22. Where?—At Rouen, St. Quentin, Lisle, and the neighbourhood of Paris.

23. Do you happen to know whether there would not be very great difficulty in introducing the discipline of an English mill into any part of France?—Yes; I was acquainted with a man who went to Rouen to superintend a factory there, and in consequence of his endeavouring to establish the English system, the workmen turned out to the amount of 4,000 or 5,000; the military were called in, and several lives were lost. I first heard of it from a Frenchman, who called upon me in Glasgow, and told me of it; and I inquired into it when I was in France, and found it to be the fact, that in consequence of endeavouring to establish the English mode of working, the hands struck, the military were called in, and many lives were lost; the military were kept in the town a great length of time.

24. Are the ancient holidays in France so rigidly observed as they were?—I believe the ancient Romish holidays, that were much attended to at one time, are mostly done away with, except eight or ten in the course of the year; but besides those, Monday may be called a holiday: there is no work done in the cotton factories in France on Monday worth speaking of.

25. You have already stated the disadvantages under which the French manufacturer must labour, by the objection to rigid confinement which the French labourers manifest upon all occasions; will you state any of the difficulties that the French cotton spinner has to encounter in competing with the English?—The first great cause of our superiority that appeared to my view of the subject was, that in the cotton trade Great Britain has the lead; and, while profits were high, she had established the manufactures to such an extent that no foreigners can now be induced to enter into the trade as it exists at present in Great Britain, as the profits are now so low. The next cause I found to be the greater expense in establishing a manufactory, from the price of the machinery; there are a few foundries near Lisle and St. Quentin; and we found the mouldings to be nearly a half higher than what they are in England; and the iron also is not so good; the coal, upon the average, in the different manufacturing towns we were in, was 36*s.* a ton, which is nearly six times as high as fuel in this country.

26. Considering the dispersed state of the population of America, and the rate of wages in that country, and the habits of the people, and the constant endeavours of all those who may have an opportunity of so doing, to locate themselves upon property which they may acquire and purchase in different parts of the Union, do you conceive that the cotton-spinning trade could exist at all in America if it had to encounter competition with that of this country? —No; I think a generation would pass away before it can compete with this country; and the land of the country will be more taken up before that arises. America will long continue an agricultural country. Mr. Wm. Smith.

STEWART, CHARLES; age 34; examined 28th June, 1832.

1. What business are you?—I am an overlooker of a flax-spinning mill. Mr. Chas. Stewart.

2. At whose mill?—Mr. William Boyack's, at Dundee.

3. At what age did you first go to a mill?—Ten years of age.

4. What were the hours of labour at Mr. William Boyack's?—Twelve hours and a half.

5. What time did you stop there?—I have been there, I think, nearly three years now; it is about two years and a half, or better, since I went there.

6. How long have you known the children work in the mill when the trade was brisk, or when stoppages had to be made up?—I have seen them work for a time for about twenty-eight hours; I remember them one time working that length.

7. Did they receive great additional wages on that account?—Yes.

8. Does the length of standing and of exertion tend to deform the limbs of the children so employed?—Yes, that is my opinion; I took an examination of those that were employed under me; there are fifty hands in the room altogether, old and young; and I found that out of that fifty there were nine who had entered the mill before they were nine years of age, who are now above thirteen years of age.

9. Having been at that employment then four years?—Yes; and out of those nine there were six who were splay-footed, and three who were not; the three who were not splay-footed were worse upon their legs than those who were; and one was most remarkably bow-legged; she informed me she was perfectly straight before she entered the mills; her name was Margaret Webster.

10. Do you conceive that the ankles are much affected by standing in mills?—Yes; they frequently swell; a number of them do so.

11. Is all this attended with considerable pain and difficulty of walking? —Yes; I have heard them frequently exclaim that they were hardly able to go home, those who were in that way.

12. Have you made any other examination?—I have examined those who had not entered the mills till after twelve years of age, and found that out of fifty there were fourteen of this class; two of them were splay-footed, and one with her ankle a little wrong; the others were all perfectly straight.

13. State whether the children attend the night-schools generally in Dundee?—In some of the mills they have no schools to attend, and in some they have.

14. In the mill in which you were had they any school?—They have; three nights a week they can attend an hour, and on Sunday evenings.

15. State your impression as to the capacity of children to learn as they otherwise would after they have endured this length of labour?—My own opinion is that they cannot learn, because I remember when I was a boy

Mr. Chas. Stewart.

myself, being at the school, I learned nothing under such circumstances; but I have likewise certificates from teachers, who state the same thing.

16. Will you be so good as to put those in?

[*The witness delivered in the same, which were read, as follow:*]

"Dundee, 14th April 1832.

"I hereby certify, that it is my most unqualified conviction, that the present system of employing children for so long a period each day, at spinning-mills, is in a high degree unfavourable to their moral and intellectual improvement. I have frequently observed the languid state of those children when in school; some of them, through the fatigue of the day, fall asleep when writing their copies; others when learning to read, &c.; it is therefore very difficult to communicate instruction to them.

"*Alexander Hutchinson*, Teacher of the Dens Mills School."

"I hereby certify, that I have taught a school for the last sixteen years in the immediate vicinity of a number of spinning-mills, and during that time I have frequently had numbers of young persons there employed attending my evening classes, of both sexes; and I feel no hesitation in saying, that I consider the long hours they are obliged to labour very injurious to their bodily health; and as for making any improvement in learning, it is nearly impossible, as they are generally so fatigued by the labour of the day as to fall asleep if not actually employed in receiving instruction. I have known instances of them being so exhausted as to hide themselves in the school, and fall asleep, and they were only discovered by their parents becoming alarmed at their absence, and, coming for the key, they have searched the school, and found them sound asleep.

"Blainshall-street School,
23d March 1832.

"*Andrew Stewart*, Teacher."

"Having had several years' experience in teaching an evening school, composed of young persons employed in spinning-mills, I feel no hesitation whatever in saying, that under the present system of conducting these works, it is nearly impossible for the young there employed to make any considerable improvement in learning during the short time allotted for that purpose in the evening; and I have invariably found them fitter subjects for repose than for mental cultivation. Nor is this the only evil attending the system; those who may be making some improvement are very frequently interrupted by working extra hours, making up lost time; and by the time they again return to school, they are nearly as bad as when they first came. I give this as my decided opinion of the system.

"Hawk-hill, Dundee,
23d March 1832.

"*Charles Edwards*, Teacher."

SWITHENBANK, WILLIAM; age 39; examined 17th April, 1832.

William Swithenbank.

1. What is your business?—A cloth-dresser.

2. Where do you reside?—Park-lane, Leeds.

3. At what age did you first begin to work in a factory?—Just turned eight years of age.

4. With whom did you work?—Mr. Gott.

5. What was your age when you were a gigger at Mr. Gott's?—I was thirty-one years of age.

6. As a gigger, and in that department of the mill, you have boys constantly attending upon the men ?—Yes. William Swithenbank.

7. What were your hours of labour, speaking of the excessive hours ?—We went to work on Monday morning at six o'clock, and remained till Tuesday night at eight o'clock.

8. What intervals were allowed for your food ?—We had half an hour at breakfast and an hour at dinner, half an hour at drinking, half an hour at nine o'clock, and half an hour at twelve at midnight, and the engine stopped at four until six.

9. Then you had two hours rest between four and six ?—Yes.

10. From thence you went on till when ?—Till eight in the evening.

11. Having of course half an hour for breakfast, half an hour for drinking, and an hour for dinner ?—Yes.

12. What time did you begin on Wednesday morning ?—Six o'clock.

13. What time did you conclude ?—On Thursday night.

14. You worked, with the same intervals as before, till Thursday night, at what o'clock ?—Eight o'clock on Thursday night.

15. Then on Friday what time did you begin ?—Six o'clock on Friday morning.

16. When did you end ?—Five o'clock on Saturday night.

17. How many nights' rest had you ?—Only two nights from Monday morning till Saturday night.

18. Had you children or young persons so occupied ?—Yes.

19. Boys from what age do you think ?—Boys from eleven years old up to fourteen.

20. For what length of time might you be working at that extravagant rate ?—About six or seven months together.

21. Did the boys ever suffer from fatigue ?—Yes; we have had them to seek when we wanted them; they would go into any part where they thought we could not find them.

22. Did you find them asleep, or merely hiding themselves ?—I have found them asleep.

23. Did this long labour have any effect upon your health ?—Yes; I was bad a long time.

24. So ill as to have advice ?—Yes; I was at home many weeks.

25. What did the medical men say your illness was owing to ?—It was the long hours; and I was to give up if I could, and I did give up when I was at home, but I began again when I went back.

26. You went to the mill at an early age ?—Yes, about eight.

27. Will you state to the committee whether you had any opportunity of going to a day or a night school ?—I had no opportunity.

28. Did you go to a Sunday-school, so as to learn the rudiments of a decent education ?—Sometimes I went to a Sunday-school, but being so close confined, we did not like to go; we were so long confined during all the week round, my father used to tell me we had better take a walk somewhere.

29. Can you read and write ?—I can do neither.

TURNER, JAMES; examined 20th June, 1832.

1. Where do you reside ?—At Manchester. James Turner.

2. What is your occupation ?—A cotton-yarn dresser.

3. What are the general hours of labour in that town and neighbourhood ?

James Turner. —The general hours are twelve; there is not above one or two exceptions in which they work more than twelve hours a day in Manchester.

4. What intervals are allowed for refreshment?—In most instances half an hour for breakfast and an hour for dinner. There are a few that let their work-people have a quarter of an hour for tea, but those work a quarter of an hour longer at night.

5. Are the children employed during any part of the dinner or breakfast hour in cleaning the machinery?—Yes, very often.

6. So that in fact the children, generally speaking, have a longer day's labour than the adults have?—In general.

7. Is it quite obvious to any one that those hours of labour are excessive, in regard to the power of enduring them, at least as far as the children are concerned?—Most certainly; when children have been employed those hours, I have seen them come home at night, and complain of being very tired; and in some instances they would scarcely attempt to go out to play, they would be so tired.

8. Does this excessive labour also interfere with the opportunity, which they would otherwise have, of obtaining the rudiments of a decent education? —It does. We have in our cotton factories a great many children that are very ignorant at twelve, fourteen, fifteen, and sixteen years of age, and many that, if they read or write a little, if they can write their own name, that is all they can do.

9. Can they avail themselves of the opportunity of going to Sunday-schools?—They go to Sunday-schools, some of them, but not many in proportion to what ought to go; having been confined so long in the week, they think they should have Sunday to themselves.

10. Is not the heat excessive in many of these mills and establishments to which you have referred?—In some parts of the place where the children work, I have known it as high as from 70° to 80°, and in some of the spinning rooms it runs as high as 84 degrees.

11. Does not that also necessarily tend to make that degree of labour more irksome and fatiguing to them?—Most certainly; it destroys the appetite; their appetites are not good, and consequently their bodies are weak, and then their toil becomes quite irksome and injurious.

12. Has not that sense of fatigue and exhaustion a tendency to drive them to the dram-shop when they are liberated from their labour?—It has, especially when they come to be fifteen, sixteen, or seventeen years of age; it is quite common for this faintness and exhaustion to drive them to the dram-shop; but the ignorance that I spoke of, I believe, has done still more in corrupting our young persons in the manufacturing districts, and in driving them to the dram-shop.

13. Will you give the committee some idea as to the ages of those employed in mills and factories, generally speaking?—According to the best observations I have made, I never can find that there is one in ten in a mill above forty years of age; they will not average more than one in ten; and at forty-five they will not average one in twenty.

14. In making just calculations on this subject, should you say that the labour of the adults, or of the children in mills and factories, is the more severe, having regard to the capacity of each class respectively?—Most certainly the labour of the children is more severe, and in general they are longer employed; the work of the children, in many instances, is reaching over to piece the threads that break; they have so many each to mind, and it is their work to keep them up; and they have only so much time, because the wheel, as we call it, of the machine is drawing out, and they have only so much time to piece those threads; consequently it keeps them very active, and they have to reach over while the wheel is coming out.

15. Is there not a particular description of children in those mills and factories called scavengers?—Yes. James Turner.

16. Will you describe their work?—The scavengers are the youngest and least of the children that come; it is the first thing that they are put to; their work is to sweep the floor under the spinning machine, and to wipe down the lint and waste that may get round on the carriages and about the spindles; to keep those clean, and the steps where the spindles run; and while this wheel is drawn out they run under, and if there is not very great care they are exposed to great danger; and children at nine or ten years of age do this work.

17. Do not you conceive that the danger is greatly increased, when they become more fatigued, towards the conclusion of a day's work, such as you have described?—From all the observations that I have made, I find that a great many of the accidents that children are subject to, of losing part of their limbs, and so on, happens in the latter part of the day, when they are tired out, and have not that activity about them which they have in the former part of the day.

18. When the hands, from age, are no longer retained at the mills and factories, they have to seek other employment?—In some instances, but that is not very common; for they seldom can get any other employment if they have been once employed in cotton factories, especially our young women. The boys are mostly kept on after they begin working in a factory till forty or forty-five years of age; at least that is the utmost length they are kept on; and then they are driven to selling sand, or gathering rags, or such pursuits as these. And the lowest classes that we have in Manchester, nine-tenths of them, are the refuse of cotton-mills, who have spent the best of their days in cotton factories.

19. Will you state whether there is as much hardship in the case of females in factories?—I am sure it is harder; in the first place, I never knew a master, in all my factory experience, however well he might think of a girl in his service in the mill, who would take her as a menial servant; and if they were to apply for situations as menial servants, there would be a disinclination to take them.

20. Do you mean as it respects their moral character?—I mean as being unfit, because they cannot do any of the work that it is necessary for a menial servant to do; they are ignorant of those things, and we say that if they are not fit for servants, they will make very poor wives for us working men, and these young women do in general make very poor wives; I have known thousands of instances where they could not mend a stocking, or do a little washing.

21. Do you think that there is, generally speaking, a suspicion of the morality, however unjust in many cases, of girls employed in factories?—There is a suspicion of their moral conduct; but the great aversion arises from their being unable to fufil those duties which I have mentioned.

URQUHART, WILLIAM; age 32; examined 28th June, 1832.

1. Where do you reside?—At Dundee. W. Urquhart.

2. What is your business?—An overlooker, in a flax mill.

3. At how early an age did you begin to work in a mill?—I cannot exactly say the period that I entered; it was between nine and ten, to the best of my recollection.

4. What were your hours of labour when you were overlooker at Messrs. Baxter and Brother's mill?—We had fifteen working hours, to the best of my recollection; including meals, sixteen in the mill.

W. Urquhart. 5. How late at night, and how soon in the morning, have you occasionally worked?—I have worked there from five o'clock in the morning. We were regularly called up at a quarter past four; there was a man who always went round to call us up, and he had a great number of streets to go through for that purpose. I was commonly called up at a quarter past four in the morning, and began work about five, and wrought till within a quarter or twenty minutes of nine at night.

6. Have you occasionally been kept longer at night than even that?—Yes; when I was there, the situation I held was the preparing the tow for spinning: there were not preparing machines enough to supply the spinning, and there were no additional hands kept for that, and they wrought the preparing engine longer. We commonly wrought up to from a quarter past one to two in the morning.

7. After having worked from what time?—At a quarter to five the mill started the preceding morning.

8. Speaking of the ordinary hours, did not you observe the time which you have stated, to be exceedingly distressing to the children?—Yes.

9. How did you contrive to keep them to their work?—I have had to take every means; I have had to take them by the nose, and to pour water upon them; sometimes one thing and sometimes another.

10. Were not they very liable to accidents in consequence of that labour?—Yes; and I saw one accident of a woman who entered the work in the morning about six o'clock, and was one hour too late; I considered, by the over-fatigue of the day, she had slept too long, and I was very hard upon her, as I was obliged to be, because the work was kept back because she had not attended; and before I had got to the other end of the flat she had lost her hand by the machinery.

11. What did you leave that mill for?—On account of losing my health.

12. Had you originally a robust constitution?—Pretty fair.

13. And did you find that that injured you?—Yes, it injured me for some nights; I lay for hours and did not sleep at all for the stoppage of my breathing.

14. To whose mill did you then go?—To Mr. Davis's.

15. Had you many children in that mill?—A great many young children.

16. How young?—Some of them would not exceed seven years, and some were even below it.

17. How long did you stay there?—About thirteen months, I think.

18. Where did you go then?—To Mr. Brown's mill, at Lochie, about two miles from Dundee.

19. What were the hours of labour?—They were twelve and a half when I entered there.

20. Including the hour for refreshment, you were confined for thirteen hours and a half?—Yes.

21. How did you keep them at their labour there?—I kept them to their labour there in different ways; sometimes by the "tawse;" and was often found fault with for not doing it enough.

22. Did you ever see any instances of gross cruelty under those circumstances?—Yes, I have seen the manager of the work whip the boys there; I witnessed a boy commit an offence; it went to the ears of the master and manager; the boy was taken out of the mill, and, to make an example of him, after having been strapped, and tied to a post, and stripped to the skin, about twelve boys were ordered to take the lash in succession, or for a certain number of lashes; I could not say how many were given by each.

23. Where did this happen?—At Walter Brown's, at Lochie.

24. When did this happen?—I could not exactly say the time, but it was three or four years ago.

25. Was the boy much hurt?—He was not greatly hurt; he was spared; for the boys had more feeling than those who had the charge of them. W. Urquhart.

26. What age was the boy?—The boy was, to my best recollection, nine or ten years of age.

27. Are the children fined as well as beaten?—Yes.

28. Who receives the fines?—The master of course got the children to pay the money to him, and I know no other purpose they went to but to his own pocket.

29. Is there a great deal of harsh treatment in those mills in Scotland?—In some of them more so than in others.

30. Do they often chastise the children with severity?—Yes; or the overlookers are found fault with if they do not chastise. I have often been found fault with myself, and have frequently thought that I should be very glad if I could get out of that labour; I have lost my situation from being, as they term it, too simple; but I know what would keep my situation; if I could use the strap a little more, or keep them under continual fear.

WILDMAN, ABRAHAM; age 28; examined 26th May, 1832.

1. Where do you live?—At Keighley. A. Wildman

2. Have you seen a good deal of the factory system in Keighley?—Yes, I have.

3. Have you had some of your own immediate relations engaged in it?—Yes, six in the whole.

4. Are any of them your sisters?—Four sisters.

5. Will you state where the eldest of them first commenced?—At Judson and Brother's, Castle mill.

6. What were the hours of work at that mill?—They were rather irregular at times; generally twelve hours a day; when they had a strong order on hand to finish, they got up in the morning by half past four o'clock.

7. To what time in the evening did they then work?—It has been eleven before they have got home, sometimes.

8. The same set?—Yes.

9. What intervals had they for meals?—Three quarters of an hour for dinner only.

10. Did they work that immoderate length of time for any considerable period together?—For four months together.

11. At what ages did your sisters begin to work?—My youngest sister began when she was seven years old; the others began about nine.

12. Were they all stout and healthy when they commenced?—Yes, a very healthy family they had been; but by working too long at that period two of them had the typhus fever.

13. Did the medical attendant assert that it was in consequence of over labour they had the typhus fever?—Yes; he said it was from early rising in the morning, and going into the cold air, and going out of the factory again at night into the cold air, at the end of the day.

14. Do you believe that circumstances occur in factories which render it very difficult for a person to escape the contagion of bad example, and resist the control exercised over them?—Yes; I know a factory where there are eighteen females employed, and out of those eighteen twelve have had illegitimate children; some as many as three a-piece.

15. From what you have yourself seen of factories, do you suppose that overlookers, and others concerned, sometimes avail themselves of their situation for very improper and immoral purposes, and corrupt the females under

A. Wildman. their charge?—Yes, I believe it, and I know facts to corroborate it; I know facts sufficient to prove that the system is bad in that respect.

16. In what respect do you think the limitation of hours of labour would remedy that evil?—It would be giving to the children a little time for improvement in their moral conduct, and they might, perhaps, receive religious instruction too; but where they are so long confined, they have not time to be instructed at all, and are subject to be led astray.

17. Do you think this long and exhausting labour has a great tendency to increase the crime of drunkenness?—Yes, I do think so, because there are young men that are very much addicted to drinking; when their bodies become exhausted, they go to the dram shop and get spirits, and they say that they do them good.

18. You have already stated the effect, in a moral point of view, which this system produces in your town and neighbourhood; have the goodness to state the result of your observations as to its effects upon the persons of those that have been subjected to it?—I have observed them in the Sunday-school, and at times in the street, living in the midst of them, that they have not that healthy appearance we see children generally have in the country; frequently without arms, without legs, and without fingers; and we can produce in Keighley 150 rickety, crooked-legged children, owing to their being overwrought.

WILSON, RICHARD; age 29; examined 25th May, 1832.

Rich. Wilson. 1. Where do you reside?—In Bradford.

2. Where were you first employed?—At Mr. Matthew Thompson's worsted mill.

3. At what age did you go?—At five years and upwards.

4. What were your hours of labour?—From six to seven, with half an hour for dinner.

5. What was your employment in that mill?—Piecening.

6. Does it require constant attention?—Yes, they must always be standing to it, walking backwards and forwards.

7. How did that length of labour suit you; did you feel much fatigued with it?—Yes, towards night. When I had been about two or three years at it, I began to be very much fatigued.

8. Had you any pains in your limbs?—Nowhere but in my knees.

9. What effect did it begin to have upon your knees?—First of all, in the morning, when I had to get up, the joints of my knees cracked; when I could scarcely get down stairs I slid down, or laid my hands on each side of the staircase, and so got down.

10. Did that labour produce the deformity in your knees?—Yes, they became crooked after a time; when I had been between three and four years at the mill.

11. You are sure you were perfectly straight originally?—Yes, I was as straight as anybody.

12. Are you labouring under that defect now?—No, I am straighter now.

13. What are you now?—I am a wool-comber.

14. Has any of your family besides yourself become deformed?—Yes; I had a brother who went to the mill when he was about eight years old, and he became so deformed that he had to be carried backwards and forwards. My parents were poor, and they could not maintain us except we went to mills.

15. Did your father use to carry him to the mill?—Yes, my father and myself used to carry him.

16. Was he straight before he entered on that employment?—Yes. Rich. Wilson.

17. Did it injure his health?—He never looked well after he went to the mill.

18. Is he living?—No, he is not; he died when about twenty-three years of age.

19. Have you any other brothers or sisters who have been in mills?—I have a sister now living upon parish pay, who is quite deformed; I have neither father nor mother now; my father died when I was about eight years of age; my father met with an accident in the mill about half past six one night, and he was dead before eight.

20. In what part of the mill did he meet with that accident?—He minded the wheel, and he was endeavouring to put a strap on the billy, and it caught him, and caused his death.

21. You say that your sister was very much deformed?—She was.

22. At what age did her deformity commence?—I think at ten or eleven years of age.

23. You are sure it was not in infancy?—No; she did not go to the mill till seven years of age.

24. How long had she been working at the mill before the deformity commenced?—She had been working there somewhere about three or four years; I cannot exactly say.

25. Can any of your family, who have been employed at mills, write?—There was not one of us that could read or write but this lame sister, who has learnt it since she has become a cripple.

MARSHALL, ELIZA; age 17; examined 26th May, 1832.

1. Where do you live?—At Leeds.

2. Were you born in Leeds?—No, I was born at Doncaster. Eliza Marshall

3. At what time did you remove to Leeds?—Nearly eight years ago.

4. When you were about nine years of age?—Yes.

5. Where did you get work at first?—At Mr. Marshall's, in Water-lane.

6. Was that a flax-mill?—Yes.

7. How happened you to leave that mill?—It was so dusty, it stuffed me so that I could scarcely speak.

8. Where did you go next?—To Mr. Warburton's, in Meadow-lane.

9. What is Mr. Warburton?—A worsted spinner.

10. What were your hours of work?—When I first went to the mill we worked from six in the morning till seven in the evening.

11. What time had you allowed for your dinner?—When first I went we had an hour, but we did not keep that long; we removed to Lady-lane, and then we had but half an hour.

12. When you removed to Lady-lane, how long were you required to work?—After a little time, in Lady-lane, we began at five in the morning, and worked till nine at night.

13. Did they allow you more time for dinner then?—No, we had half an hour for dinner then, and none for breakfast or tea.

14. How did you get your breakfast and drinking?—We got some little of it, and then went on with our work.

15. How old were you when you went to Mr. Warburton's?—Nine years old.

16. Were you not very much fatigued by that length of labour?—Yes.

17. Did they beat you?—When I was less they used to do it often.

Eliza Marshall

18. Did you not think this treatment very cruel?—I have cried many an hour in the factory.

19. Did the labour affect your limbs?—Yes, when we worked over-hours I was worse by a great deal; I had stuff to rub my knees; and I used to rub my joints a quarter of an hour, and sometimes an hour or two.

20. Were you straight before that?—Yes, I was straight before that; my master knows that well enough; and when I have asked for my wages, he said that I could not run about as I had been used to do.

21. Did he drop your wages in consequence?—No, but he would not raise my wages, as I hoped he would; I asked, "Could not I mind my work?" and he said, "Yes, but not so quick."

22. Are you crooked now?—Yes, I have an iron on my right leg; my knee is contracted.

23. Have you been to the Leeds Infirmary, to have, if possible, your limbs restored?—Yes, I was nearly twelve months an out-patient; and I rubbed my joints, and it did no good; and the last summer I went to the Relief, and that did me no good, and I was obliged to have a machine; and this last winter I have been in the Infirmary six weeks.

24. Under whom are you?—Mr. Charley.

25. They have put irons on your legs?—Yes, they cost 3*l*.

26. Have any of the surgeons in the Infirmary told you by what your deformity was occasioned?—Yes, one of them said it was by standing; the marrow is dried out of the bone, so that there is no natural strength in it.

27. You were quite straight till you had to labour so long in those mills?—Yes, I was as straight as any one.

CLASS II.

BULL, The Reverend GEORGE STRINGER; examined 5th July, 1832.

Rev. G. S. Bull.

1. Where do you reside?—At Bierley; I am incumbent of Bierley, near Bradford.

2. Have you, for some years, resided in a manufacturing district of Yorkshire?—I have resided about seven years in all in the manufacturing districts; but I am a native of an agricultural district; I am not a native of the manufacturing districts.

3. Of what county are you a native?—Essex.

4. Has your mind been directed to the consideration of the factory system as connected with the morals and education of the lower classes?—My mind was necessarily attracted to the system as soon as I became a resident clergyman in the manufacturing districts.

5. As a clergyman, what views have you been led to adopt, from your own observation, regarding that system as at present conducted?—I should say, in brief, that my views of the system are, that it is in general, as at present conducted, very injurious to the morals, and to the health, and to the social order and domestic comforts of the poor.

6. Will you speak as to its effects on the morals of the poor?—I beg to say, that my conclusions upon all those subjects are derived from my own observation; and that whatever I shall state before this committee I shall state as the result of that observation, and more especially as connected with my official station. With regard to the morals of the persons engaged in factories, my observation has led me to conclude, from the language which they are accustomed to use, from its profaneness and obscenity, from the indecency which I have witnessed in their conduct, from their general impudent bearing, and especially from witnessing these things, as I have done, in their returning from their work, when I have very frequently walked up with groups of them from the town of Bradford, within about a mile and a half of which I reside, I have been led to conclude, from my own observation upon their language, and their indecent conduct, that young persons become more corrupt in that employment than they do in any other. I have a great many young persons under my care; I have Sunday schools under my superintendence, containing 516 scholars; about one third of them are engaged in factories, and the other two-thirds are employed in collieries, and in assisting their parents at home in combing or in weaving; but I am led to conclude, from an observation of these several classes, that there is much more demoralization arising from the factory system than from any other system of employment for the children of the poor with which I am acquainted.

7. Will you state how you account for the melancholy result which you attribute to the factory system, as at present conducted?—I should say that

Rev. G. S. Bull.

it arises in a great measure from the nature of the employment, which brings together so many young persons, who are very frequently promiscuously mingled; I should say that my observations have led me to conclude, that it arises also from their going very young to this work, before any moral habits have been formed, before they have been placed under that kind of instruction which is calculated to form the character. I should also say, that they have had very little opportunity of learning good domestic habits under their parents, when they are sent so young as they are. The parental influence is very small indeed over them. They go very early in the morning; many of my little children (I call them mine) set off regularly at five o'clock, and do not return again till eight, almost the year round; perhaps in the course of the winter season some weeks of shorter employment may take place. They see, therefore, very little of their parents, and when they go to the mill their first impressions (and first impressions are generally the deepest) are of a very injurious kind. They meet there with their elders in age, and, I may also say, in vicious communications, and they very readily imbibe them. I conceive also, if the committee will permit me to state it, that the system of returning so late, especially those that live at any considerable distance from their work, is extremely injurious to their morals. I know very well the conversation that takes place amongst them as they go home, because I have often heard it myself; and I must say, that in the course of my life, though I have visited several of the sea-ports of this kingdom, and other places, where you might suppose very abandoned conversation to be heard in the streets, but (I am almost ashamed to say it) I have heard such obscene conversation from little factory children, as, I am sure, they could not know the meaning of themselves, and as has perfectly astonished me. In their returning home from their work, the elder of the young females especially, that are employed in the factories, are very frequently decoyed and seduced; assignations are made on the road home, and a great deal of evil of that description has come to my knowledge from the parents of the children and young persons, who have stated these things to me. I would also beg permission to say, that I esteem night-work to be a most fruitful source of immorality. I do not speak of my own knowledge, of course, but the parents of the young persons have reported to me most shameful scenes that have taken place during night-work; and overlookers of the mills, one or two respectable persons whom I have happened occasionally to converse with upon the subject, have told me very disgraceful things that occurred to their knowledge during night-work. I believe very frequently that scenes of great debauchery take place in the mills; and I do believe, that in many instances they are perfectly unknown to the masters of the mills. I would do them the justice to say, that I do not believe they would, generally, tolerate anything of the sort if they were acquainted with it.

8. But you consider that the confining the young persons of both sexes together during the whole night has not indeed the necessary but the too frequent tendency and effect of producing the immorality to which you are now alluding?—I have no doubt of it from my own observations and inquiries.

9. Will you proceed with any further remarks upon the subject you have to mention?—I should say that the system of coercion to which the children are exposed from an early age, is greatly calculated to blunt the moral principle. They are coerced by their parents to work; they are coerced by the overlookers when they get to their work, and some instances have been reported to me of the personal interference of the masters themselves in the way of punishment; but, at all events, the children have sense enough to know that the coercion of the overlooker is to be traced to the master; and

Rev. G. S. Bull.

it appears to me that this system of coercion from their parents, and from their overlookers and their masters, from so early an age, is calculated to produce very unfavourable impressions upon their minds, and to lower everything like good moral principle.

10. Will you have the kindness to explain what you particularly mean by the term "coercion?"—I mean that the parents oblige their children to go to the mill at a very early age. I have known instances myself, which the parents themselves have mentioned to me, where they have every morning pulled them out of bed for that purpose, and been obliged to shake them till they awoke.

11. Do you mean by coercion anything in the way of cruel treatment?—Yes; with regard to coercion from the overlooker, I refer to the use of the strap, which is very general in our neighbourhood. There are some one or two mills, I believe, I am not certain how many, but I know of one or two myself, from which the strap is banished, and it is found quite possible to control and direct the children in their work without it.

12. Do not you attribute much of the cruelty that occurs in mills to the overlabour of those children, and to the consequent inability which they, generally speaking, feel to perform the duties that are exacted of them for such an undue length of time?—I have no doubt, from my own observations, and not from any other source, that that is the case, because I have asked children myself sometimes whether they have got strapped, and they have said "yes," (they are ashamed to own it, they are not willing to own that that has taken place,) and I have inquired at what part of the day they get that punishment, and have found it to be towards the close of the day, when they get tired and fatigued, and cannot perform their work so well. I believe that this arises, in many cases, from the length of time which they have to stand. I should also say that it has been a very painful thing to my mind, as a minister of religion, to consider the manner in which we confine the children on the Sabbath-day, after the very close confinement of the week; they may think that our system on the Sabbath-day is a sort of justification of the system of the week-day; for we, while they are stowed up, if I may use the expression, in the mills during six days of the week, confine them in our crowded Sunday school-rooms on the Sabbath-day; and they are very little competent to attend to the instructions they receive, on account of the extreme fatigue to which they have been exposed during the previous days.

13. Do not you, as a clergyman, think that the treatment to which those children are subjected has the effect of generating in them a disinclination to learn their duties, and, in fine, a contempt for religion, generally speaking?—I do most decidedly think so; because, as soon as they learn to read the precepts of Christianity, they must see that, in many respects, the system of labour under which they are brought up is at variance with anything like the golden rule of doing to others as you would that they should do to you; and I do really believe, that a very great deal of the disregard of Christianity, or, in other words, the infidelity, which unhappily abounds in the manufacturing districts, may be traced to the system under which the population is found to exist, and under which they have to get their bread.

14. How many hours do the factories around you work?—The worsted-mills usually work in our neighbourhood twelve hours and a half; I mean actual labour.

15. To which length of labour is to be added, of course, the time necessary for taking their refreshment?—Yes, and also that of going and returning.

16. Do you consider that children, after working ten hours in the course of the day, are fit to go to an evening-school, and are able to pay that atten-

Rev. G. S. Bull.

tion which would enable them to obtain profitable instruction ?—I should say that some would still not be able to attend them ; but we are obliged by the opponents of the measure (for I do not hesitate to confess myself to be in favour of the measure) to be as moderate as possible in our request, or I should suggest a greater limitation than ten hours; but, as circumstances stand, I should of course think that ten hours would be far better for the moral improvement of the people than twelve hours and a half.

17. What further limitation would you give ?—I should be very well pleased with a limitation to eight hours, much more than ten ; but, under present circumstances, I and others are obliged to be satisfied with ten.

18. Allusion has been made to the moral evil resulting from this system, but you have not, perhaps, as a clergyman, left totally out of your consideration the physical evils and sufferings which are incident to it ?—I am necessarily conversant with the physical evils of the system, because (perhaps it does not become me to say it) I am a cottage clergyman, and therefore I must know something of the physical evils of any system of labour, as operating upon the lower classes.

19. Will you be so good as to state those evils, as far as they have come under your own personal observation ?—I have noticed the factory children, in a very considerable number of instances, to be very much stinted in their size, compared with other children ; I have noticed many deformities, particularly in the legs and ankles, and I would remark, respecting those deformities, that in general the children are extremely careful to conceal them ; I have understood they frequently reproach each other with them, and the females especially most studiously conceal them, and have, of course, a superior opportunity of concealing the deformities in the lower extremities ; but my Sunday-school female teachers, of whom I have inquired respecting the children in their classes, have informed me of many who were deformed, whom I should not have suspected of being so ; I should also observe that I have noticed the physical evils of the factory system in frequent consumptions, and consequent deaths, of many of the children of my own neighbourhood, whom I have attended myself on their dying beds ; I am in the habit of attending the sick at their own homes every month in the year, more or less ; there are several factory children now in my schools whom I do not expect to live the year out ; and some one or two, whom I left very poorly when I came from home, whom I never expect to see again.

20. You conceive that this undue length of labour has a fatal effect upon the health of the children ?—I have no hesitation in saying that it has. If the committee would permit me, I would also state that there is often not a little delusion in the minds of cursory observers with regard to estimating the health of factory children ; when such observers of the system go, for instance, to visit a mill, if they see a child of a florid complexion, and another child of a sallow complexion, they will conclude that the florid complexioned child is the most healthy one, but it very frequently happens that the contrary is the fact ; those who survive the severities of this system of excessive labour, are generally quite bleached by it, and have that appearance.

21. Speaking of the appearance of factory children, do you think visitors, generally, could take a just estimate of the effects of that system, especially considering the preparations which it has been asserted before this committee are made in anticipation of such visits ?—Very much depends upon the time when the visitors proceed to the mills. If visitors go to the mills on the Monday or Tuesday, and if they go in the morning of the day, they can have a very imperfect view of the physical effects which it inflicts upon the children. If they go to the mills when they have just got on their clean brats (or pinafores, as they are called in the south), when they have had

some little intermission on the Sabbath-day, or when they have only just come to their employment in the morning, or have been at it only for a few hours, they cannot at all estimate the intensity of the fatigue they have to undergo. They may see the clean brats, and the beautiful machinery, and admire it very much; but they cannot feel the aching ankles and the sinking sensations, which the little children have described to me when I asked them about it, especially occurring at the latter part of the day. I remember asking one very intelligent child who was in my school whether she felt fatigue at the latter part of the day, and I was very much struck with her answer, which was this: "At five o'clock, we often say, 'I wish it was seven.'"

Rev. G. S. Bull.

22. Will you please to proceed with what you have to state with regard to visitors?—I would merely say, that visitors to mills in general, in my opinion, can form a very imperfect judgment of the real state of the case; in many cases I know that preparations have been made, I have learnt it from persons engaged in the mills; and in some instances the sickly and the deformed have been sent out of the way, that they might not be seen.

23. Do you consider that the system is one which, without legislative interference, is likely to rectify itself; and have you made observations to this effect, that it is becoming ameliorated, or the reverse?—I cannot conceive that any permanent improvement of the system can take place without legislative interference. When any public inquiry is going on, such as the present, or when the public mind is in a great degree excited with reference to such a system, there will be many improvements of a temporary nature adopted, I must believe, from such influence, and these are in many cases intended to be merely temporary; but I conceive that those temporary improvements will wear away with the exciting causes of them, and that system, without restriction, is calculated to go on from bad to worse.

GORDON, the Rev. ABERCROMBIE LOCKHART; examined 8th June, 1832.

Rev. A. L. Gordon.

1. Are you a minister of the established church of Scotland?—Yes.
2. Residing where?—In Aberdeen.
3. You have the care of one of the parishes of that city?—Yes, the Grey-friars parish.
4. Have you remarked as to the length of the hours of labour in the manufactories of that city, that they interfere with the health, the education, and the morals of your parishioners and others?—With respect to health, I think that young persons employed at the age of eight, nine, or ten years, from six in the morning till eight at night, must be very much injured in this respect, even in the view of the most inexperienced person; but when I consider what medical men have said and written upon the subject, it is decisive upon that point. To myself, although unacquainted with the details submitted to medical men, it is quite apparent, from looking at these children, that they suffer from long hours of labour and confinement in those places; their wan and sickly appearance is sufficient to prove that, in my view.
5. You have the superintendence of a school in which the poorer classes of society are mainly taught?—Yes, the school is one established solely with the view to the young persons employed in the factories. There are other schools, particularly Sunday-schools, in the parish; but this is one I set up, with a view to these persons; they come in at eight o'clock; and there are 115, chiefly from the same factory.

Rev. A. L. Gordon.

6. Will you state the difficulty you have to encounter in consequence of these protracted hours of labour?—The obstacles may be summed up in a few words, and they are these: that the hours are so long, that in general they are unable to attend; and that to get a steady attendance on the part of those who do come is very difficult. At the same time, they are so exhausted, and the period is so short, that very little good is done.

7. What is the general impression of the ministers of religion in the large city of Aberdeen?—I have a document in my hand, signed by the whole of the clergy of Aberdeen, with two exceptions, and they were at the General Assembly at the time.

8. Have the goodness to read it.

"We, the undersigned ministers of the Gospel in the city of Aberdeen and its vicinity, do hereby express our approbation of the bill introduced into Parliament by M. T. Sadler, Esq. for ameliorating the condition of the working classes in manufacturing establishments, in as far as said bill proposes to prevent the employment in mills, or factories, of children under the age of nine years, and to limit the hours of actual labour for youth under eighteen years of age to ten hours a day, that is, allowing the usual and necessary two hours for refreshment and rest, from six o'clock A.M. till six o'clock P.M. We are induced thus to record our sentiments, because we cannot, as Christian clergymen, give sanction to a system such as now prevails; a system by which tender infants are subjected to labour beyond their strength, in a polluted atmosphere, and that too for a longer daily period than the adult felon, or the West India slave. We are further convinced, from our clerical experience, that the present long confinement of young persons in mills and factories, is prejudicial to their morals, inasmuch as religious instruction cannot be adequately obtained; to their mental culture, inasmuch as no regular system of education can be pursued; to their health, inasmuch as constitutional debility and disease are entailed. Given at Aberdeen, this 25th day of May, 1832.

James Kidd, D.D. Minister of Gilcomaton Chapel, &c.
James Foote, A.M. Minister of the East Parish of Aberdeen.
W. Skinner, D.D. Minister of St. Andrew's Chapel, King-street.
John Murray, A.M. Minister of North Parish, Aberdeen.
David Simpson, A.M. Minister of Trinity Chapel.
Alexander Thomson, Minister of George-street Chapel.
James Stirling, Minister of George-street Church.
Hugh M'Kenzie, Minister of Gaelic Chapel.
Samuel M'Millan, Minister of the Relief Church, St. Andrew's-street.
Thomas Cocking, Minister of Long Acre Chapel.
Gavin Parker, Minister of Union Terrace Chapel.
James Templeton, Minister of Belmont-street United Secession Church.
Charles Gordon, Roman-catholic Clergyman.
John Atken, Minister of Skere Terrace Chapel.
William Primrose, Minister, Nether Kirkgate Chapel.
W. Browning, Jun., Minister of St. Andrew's Chapel, King-street.
Walter Graham, Minister, late of Forgue.
P. Chryne, Minister of St. John's Chapel.
James Spence, A.M., Minister of Blackfriars-street Chapel.
John Brown, A.M. Cantab., Minister of St. Paul's Episcopal Chapel.
James Cordiner, A.M., Minister of St. Paul's Episcopal Chapel.
George Douglas, Minister of Long Acre Chapel.
O. Thomson, M.D. Minister of St. Clement's Parish.
Richard Penman, Minister of Frederick-street Chapel.
Henry Angus, Minister of the United Secession Church, St.Nicholas-lane.

Hugh Hart, Minister of the United Christian Church, Ship Row. Rev. A. L. Gordon.
Robert Smith, Minister of Old Macha.
Thomas Pottinger, Minister of John-street Chapel.
Abercrombie L. Gordon, Minister of Grey Friars Parish, Aberdeen.

OASTLER, RICHARD; examined 7th July, 1832.

1. Where do you reside?—At Fixby Hall, near Huddersfield. R. Oastler, Esq.

2. Has your mind been latterly directed to the consideration of the condition of the children and young persons engaged in the mills and factories of this country, with a view to affording them permanent legislative relief? —It has.

3. What was your inducement for directing your mind to these considerations?—The immediate circumstance which led my attention to the facts was a communication made to me by a very opulent spinner, that it was the regular custom to work children in factories thirteen hours a day, and only allow them half an hour for dinner; that that was the regular custom, and that in many factories they were worked considerably more. Being assured that this was true, I resolved, from that moment, that I would dedicate every power of body and mind to this object, until these poor children were relieved from that excessive labour; and from that moment, which was the 29th of September, 1830, I have never ceased to use every legal means which I had in my power to use, for the purpose of emancipating these innocent slaves. The very day on which the fact was communicated to me, I addressed a letter to the public in the "Leeds Mercury" upon the subject. I have since that had many opponents to contend against; but not one single fact which I have communicated has ever been contradicted, or ever can be. I have certainly been charged by the opponents of the measure, in general terms, with exaggerations, but on all occasions I have refrained from exposing the worst parts of the system, for they are so gross that I dare not publish them. The demoralizing effects of the system are as bad, I know it, as the demoralizing effects of slavery in the West Indies. I know that there are instances and scenes of the grossest prostitution amongst the poor creatures who are the victims of the system, and, in some cases, are the objects of the cruelty, and rapacity, and sensuality of their masters. These things I never dared to publish, but the cruelties which are inflicted personally upon the little children, not to mention the immensely long hours which they are subject to work, are such as, I am very sure, would disgrace a West Indian plantation. On one occasion I was very singularly placed; I was in the company of a West India slave-master, and three Bradford spinners; they brought the two systems into fair comparison, and the spinners were obliged to be silent when the slave-owner said, "Well, I have always thought myself disgraced by being the owner of slaves, but we never, in the West Indies, thought it was possible for any human being to be so cruel as to require a child of nine years old to work twelve hours and a half a day; and that, you acknowledge, is your regular practice." I have seen little boys and girls of ten years old, one I have in my eye particularly now, whose forehead has been cut open by the thong, whose cheeks and lips have been laid open, and whose back has been almost covered with black stripes; and the only crime that that little boy, who was ten years and three months old, had committed, was, that he retched three cardings, which are three pieces of woollen yarn, about three inches each long. The same boy told me that he had been frequently knocked down with the billy-roller, and that, on one occasion, he had been hung up by a

R. Oastler, Esq rope round the body, and almost frightened to death. But I am sure it is unnecessary for me to say anything more upon the bodily sufferings that these poor creatures are subject to. I have seen their bodies almost broken down, so that they could not walk without assistance, when they have been seventeen or eighteen years of age. I know many cases of poor young creatures, who have worked in factories, and who have been worn down by the system at the age of sixteen or seventeen, and who, after living all their lives in this slavery, are kept in poor houses, not by the masters for whom they have worked,—as would be the case, if they were negro slaves,—but by other people, who have reaped no advantage from their labour. It is almost the general system for the little children in the manufacturing villages to know nothing of their parents at all, excepting that in a morning very early, at five o'clock, very often before four, they are awakened by a human being that they are told is their father, and are pulled out of bed (I have heard many a score of them give an account of it) when they are almost asleep; and lesser children are absolutely carried on the backs of the older children asleep to the mill, and they see no more of their parents, generally speaking, till they go home at night, and are sent to bed. I think that the disaffected state of the working classes arises from no other circumstance than that complete inversion of the law of nature, making the little children into slaves, to work for their fathers and mothers, and leaving their fathers destitute in the streets, to mourn over their sorrows. The system is also manifestly unjust towards the proprietors of land, for when persons can get employment in these mills, they are tempted to bring their families from agricultural districts, and to dwell in the manufacturing villages. Those that come are perhaps strong and healthy at first, but are completely ruined by the over-working which they are subjected to; and the moment they become useless they are sent back again to the agriculturists for support. These cases are very common indeed.

4. Do you know instances in which parents live entirely on the earnings of their children?—Yes, I met with a case, a little while ago, of a man who lives a short distance from my house, and who said to me, "I hope you will get this Ten Hours' Bill passed; I have two children, one seven and the other thirteen, at work at the factories, and I have not had the least stroke for," I think he said, "the last thirteen months." He told me that they were earning seven or eight shillings a week; and he said, "That little girl has to go a mile and a half, very early, to her work, and she comes home at half-past eight, and all that I see of her is to call her up in the morning, and send her to bed, and it almost makes my heart break. We cannot get any work, and I know that I am living by the death of that child;" and he cried when he told me. In fact, they weep when they tell their tales, and the poor children weep too.

OSBURN, WILLIAM; examined 9th July, 1832.

W. Osburn, Esq.

1. Where do you reside?—At Leeds.

2. Have you been an overseer of the town of Leeds?—I have; I was an overseer from Easter 1830 to Easter 1831, at which time the overseers are changed.

3. Have you also been one of the trustees of the workhouse there?—Yes, I was a trustee of the workhouse at Leeds from May 1831 to last May.

4. Will you state upon what scale, or rule, you acted, in affording relief to families out of employment?—The scale of relief was one shilling and sixpence per week for children under ten years of age; no relief for the

W. Osburn, Esq.

parents, not any for the children above that age, except in case of sickness, or when they had been a very long time out of employment.

5. Do you think that, generally speaking, the hours of labour in the mills and factories of Leeds, and the neighbourhood, are excessive, at least with respect to the capability of the children and young persons employed?—Very excessive indeed. A case came under my observation only the other day, of a girl who was labouring in Mr. Hogg's factory, at Holbeck, nineteen hours a day.

6. What were the intervals for rest, or refreshment, during those nineteen hours?—I am not able exactly to speak of that; it came casually to my notice, from a complaint of the mother.

7. Supposing that the parents applying for relief for their children refused to allow them to labour in mills or factories, in consequence of their believing and knowing that such labour would be prejudicial to their health, and probably destructive of their lives; would they, in the meantime, have had any relief from the workhouse board, or from you, as overseer, merely on the ground that the children could not bear that labour?—Certainly not.

8. So that you would not relieve those children unless the labour had actually destroyed their health?—They are only relieved in case of positive sickness, a report of which, from one of the town's surgeons, would be required, who, having visited the applicants, should state to the board that they were actually ill, and incapable of working. If the persons applying are able to work in some degree, a mitigated relief is given to them, but still all paupers are expected to work to the extent of their capability of working.

9. So that the children of the poor, and their parents, have no alternative in such cases but submitting their children to this extravagant length of labour, or exposing them to absolute want and starvation, as the consequence of refusing so to be employed?—None whatever.

10. Have you attended at all to the education of the poor in that populous town?—I have paid a good deal of attention to it, both in the Sunday-schools and in the Lancasterian day-school; the latter is a large institution, always having, on the average, about 500 boys, or thereabouts.

11. Will you state what your impressions are as to the effect of these long hours of labour upon the education of the children of the poor?—I have very frequently boys of sixteen, and upwards, who come for a few weeks to my Sunday-school, and who are not able to read, in some instances, who do not even know their letters. They have worked all their lives in mills; and the invariable complaint is, that they have had no time to learn at all.

12. Do you think that those who work that length of time in the mills, and are sent to a Sunday-school, are, according to your observation and experience in those institutions, in a situation to derive the improvement which they would otherwise do from the opportunities afforded them?—I have the greatest difficulty in gaining the attention of children who have wrought in factories to the instruction I wish to convey to them; they are exceedingly dull and heavy. I ought to mention, that I have not many of such children in my school, not more than ten or twelve upon an average. In a place of worship, also, I find that they are much more inattentive and sleepy than the other children; the younger ones I have frequently directed the teachers to allow to sleep; they appeared so much fatigued, that I thought it was almost cruelty to prevent them from doing so.

13. Have you observed that deformity is very common among the children?—Very general.

14. Have you observed, also, whether there is any effect produced upon the children engaged in factories, as compared with the growth of children otherwise employed?—I have taken a good deal of pains in ascertaining that fact; and though I have not yet been able to make all the deductions

W. Osb Esq.

from the observations which I have made that I could wish, I am prepared to state that working in factories retards the growth of childrenupwards of a year, probably a year and a half.

15. That is, the children employed in factories will be only the same size as children a year and a half younger, who have been in other occupations ?—I believe that to be the case.

16. Have you observed whether there is the same degree of difference in point of muscular strength and bulk ?—I believe it to be the case, and in about the same proportion.

17. Is there a very perceptible difference in the appearance of those children ?—The difference in appearance is so perceptible that many other persons, as well as myself, are able to go into a Sunday-school, and select, at once, all the children that are employed in factories; and some persons, more conversant with the factory system than I am, persons employed themselves in factories, are able not only to do that, but also to point out the particular occupation in which each individual is engaged.

18. Do the children themselves seem to be sensible that their occupation is one involving a great degree of suffering, and are they anxious, do you think, to obtain other employments in preference, if the opportunity were afforded to them ?—I have always found that to be the feeling, both of the children, and of their parents.

19. Have you made any remarks elsewhere respecting the condition of the factory children, as compared with others ?—I believe the condition of factory children almost everywhere to be pretty much the same as I have found it at Leeds. I have observed the children in other manufacturing towns and districts; in the cotton-mills in the neighbourhood of Settle I took particular notice of them. They had the same unhealthy and squalid appearance there that they have in the close, confined atmosphere of Leeds, though the former is a very healthy country. I also observed but recently, in a mill in the neighbourhood of Oxley, Mr. Whittaker's, of Greenholm, I happened to be passing at the time that the bell rang for the children to go to dinner, and I must say I never saw more squalid and unhealthy-looking young people anywhere, or more of that peculiar deformity of the legs and feet which alters the gait so completely.

20. Have you not been engaged in assisting the operatives of the West Riding of Yorkshire to represent their case to this committee, and to Parliament, during the existing inquiry ?—I have; I was chairman of their central committee.

21. Will you state to this committee the fact whether you could not have multiplied the number of witnesses sent up, to almost any extent, if it had been deemed expedient or necessary to do so, in order to verify the facts which have been given before the public on this occasion ?—I have never conversed with any person employed in a factory (and I have conversed with many) who did not appear perfectly willing to corroborate those statements which have been made before this committee.

22. Do you know of any cases in which persons who have been examined as witnesses before this committee have been dismissed from their employment ?—The two Kinworthys, from Huddersfield, are both now dismissed from their employment; Goodyear, of Huddersfield, is also deprived of his work; a man of the name of Swithenbank, who came from Leeds, was dismissed from his employment immediately on his return; and another named Cooper, who had been promised employment the week that he returned from London, applied for it, and was refused, on the ground that he had come here and given evidence on the Ten Hours' Bill. This, also, was the case with Alonzo Hargreave, and I fear with several others.

SHARPE, GILLETT ; age 51 ; examined 6th June, 1832.

1. Where do you live?—At Keighley. G. Sharpe.

2. Are you overseer of the poor of Keighley?—I have been assistant overseer the last year.

3. You have had considerable opportunities of observing the customs and management of the mills and factories?—Yes.

4. What are the customary hours of labour?—With a few exceptions, the customary hours of labour are from six to seven.

5. With what intervals for rest or refreshment?—Some have an hour, some forty minutes.

6. At dinner?—Yes.

7. None for breakfast or drinking?—No.

8. Are those hours occasionally exceeded in brisk times?—Yes.

9. What do you think, from your own personal observation, has been the consequence of that length of labour endured by the children and young persons at Keighley; first, as it respects their health and their appearance? —I have observed that their health has been very much impaired by it in general, and that it is different in those places where they are not occupied in mills.

10. Do you believe that the cases of deformity in that town are exceedingly numerous?—Yes; I have had an opportunity of visiting other towns, and it is my opinion, according to my observation, that there is not another town worse, in proportion, to the size of it.

11. Have you always understood that that deformity is attributable to the too early and excessive labour of the children in the factories of that place?—Yes; I consider that labour from too early an age, and long standing, have been the causes of it.

12. Have you made any observation as to the moral effects of those long hours of labour?—Yes, I have.

13. What have they been?—I have made these observations: that by being confined so long in the factory, with so little time for relaxation or instruction, they have been prevented from attaining that knowledge which children ought to have in the morning of life; for the time they are employed in the week is so great that they have no opportunity, except on the Sabbath-day, of being instructed; and I have also remarked that when they have come to school on the Sabbath-day, they have been so fatigued, or so dull, in consequence of the confinement, that it has been very hard work to make any impression upon them, or to teach them.

14. Do the hours of labour during which the children are employed in Keighley leave any opportunity to the parents to supply the deficiency in their education by domestic instruction?—No, they do not; they are called to work by six o'clock in the morning, and they are confined to the factory, with the little intermission they have for dinner, when it is hard work for them to get back in time; and then, when they are released at night, they are only fit for bed.

15. Supposing that an individual had come to you to be maintained by the parish, stating that he found himself totally incapable of labouring for the length of time that he was required to do in the mill, but, having left that labour, had recovered his wonted strength and health; but knowing, at the same time, if he returned to it, he could retain neither, but should fail, and perhaps forfeit his life by it; would you, as the overseer, when you found a man under those circumstances, have relieved him?—I should have relieved him no longer than till he could get work.

16. Then, regarding the children, who should have been withdrawn from their employment by their parents, from seeing that they were totally inade-

G. Sharpe. quate to labour so long without ruining their health and crippling their limbs, would you have relieved those children if they had been withdrawn from any such plea of being overworked?—If they had been withdrawn upon the expectation of being injured, we should not have relieved them.

17. As overseer, had you many persons whom you used to relieve from being cripples, and in a state that they could not work, arising from the injury that you thought they had received in the factories?—We have had one we relieved that lost his arm; he had ten shillings a month; that was an accident: and there are others, by being crooked and deformed, and not able to get their living like other men, they are often upon the parish; they are not able to maintain themselves, and they are obliged to apply for relief.

18. What is the present population of Keighley?—Eleven thousand, and some odd.

19. You stated that working in a factory had a bad effect upon the morals of the people employed; do you know the number of illegitimate children in Keighley in the last year?—I cannot positively speak to the exact number, but I think there are now to those that are not out of their time, nearly fifty pounds a month paid in that way.

20. For the support of illegitimate children?—Yes.

21. How much is allowed a month for each; is it about 2*s.* a week a-piece?—It is very rarely there is an order made for 2*s.* now; it is 1*s.* 6*d.*

22. Do you know anything by report, or your own personal observation, as to the state of morality in mills?—I am informed that in a certain mill where there are, I believe, eighteen young women working, there are twelve out of the eighteen who have had illegitimate children; some one, some two, and some three each.

WHITEHEAD, ABRAHAM; examined 12th of April, 1832.

Mr. A. Whitehead.

1. What is your business?—A clothier.

2. Where do you reside?—At Scholes, near Holmfirth.

3. Is not that in the centre of very considerable woollen-mills?—Yes, for a space of three or four miles. I live nearly in the centre of thirty or forty woollen-mills.

4. Have you had constant opportunity of observing the manner in which these mills are regulated and conducted?—Yes.

5. At how early an age are children employed?—The youngest age at which children are employed is never under five, but some are employed between five and six, in woollen-mills, as pieceners.

6. How early have you observed these young children going to their work, speaking, for the present, of the summer-time?—In the summer-time, I have frequently seen them going to work between five and six in the morning, and I know the general practice is for them to go as early to all the mills, with one or two exceptions.

7. How late in the evening have you seen them to be at work, or remarked them returning to their homes?—I have seen them at work in the summer season between nine and ten in the evening; they continue to work as long as they can see, and they can see to work in these mills as long as you can see to read.

8. What intervals have these children allowed for their meals; for breakfast, for instance?—I have been in mills at all hours, and I never in my life saw the machinery stopped at breakfast time at any of the mills.

Mr. A. Whitehead.

9. How do they get their breakfast?—They get their breakfast as they can; they eat and work; there is generally a pot of water porridge, with a little treacle in it, placed at the end of the machine, and when they have exerted themselves to get a little forward with their work, they take a few spoonfuls for a minute or two, and then to work again, and continue to do so until they have finished their breakfast.

10. Is there any allowance of time for the afternoon refreshment, called drinking?—No allowance for drinking, more than breakfast.

11. How much at dinner time, as far as you have been able to judge, speaking now of the summer season?—In summer, some of the mills allow an hour for dinner.

12. And some less?—Some less; some forty minutes.

13. What has been the treatment which you have observed that these children have received at the mills, to keep them attentive for so many hours at such early ages?—They are generally cruelly treated; so cruelly treated, that they dare not hardly for their lives be too late at their work in a morning. When I have been at the mills in the winter season, when the children are at work in the evening, the very first thing they inquire is—"What o'clock is it?" If I should answer, "Seven," they say, "Only seven! it is a great while to ten, but we must not give up till ten o'clock, or past." They look so anxious to know what o'clock it is, that I am convinced the children are fatigued, and think that, even at seven, they have worked too long.

14. Do they frequently fall into errors and mistakes in pieceing, when thus fatigued?—Yes; the errors they make when thus fatigued are, that instead of placing the cording in this way [describing it], they are apt to place them obliquely, and that causes a flying, which makes bad yarn; and when the billy-spinner sees that, he takes his strap, or the billy-roller, and says—"Damn thee, close it, little devil, close it!" and they smite the child with the strap or the billy-roller.

15. You have noticed this in the after part of the day more particularly?—It is a very rare thing to go into a mill in the latter part of the day, particularly in winter, and not hear some of the children crying for being beaten for this very fault.

16. What moral effect do you think it has on the minds of the children who labour thus at this early period of life?—With regard to the morals of the children who work in mills, we cannot expect that they can be so strict as children who are generally under the care of their parents. I have seen a little boy, only this winter, who works at a mill, and who lives within two hundred or three hundred yards of my own door; he is not six years old, and I have seen him, when he had a few coppers in his pocket, go to a beer shop, call for a glass of ale, and drink as boldly as any full-grown man, cursing and swearing, and saying he should be a man as soon as some of them.

17. Is that an unusual case, or are there many such?—I do not know that there are many such boys, but the expressions of children in mills very much accord; you cannot go into a mill where even the most wealthy master clothier is called "Sir" or "Master;" they call them "Old Tom," or "Young Tom," &c.

18. Can children employed in this way obtain any instruction from day-schools?—There is no possibility of that; but since this Factory Bill has been agitated, when I have been at mills the children have gathered round me for a minute or two, as I have passed along, and have said—"When shall we have to work ten hours a day? Will you get the Ten Hours' Bill? We shall have a rare time then; surely somebody will set up a night school; I will learn to write, that I will."

DOCUMENTS laid before the committee by MR. CHARLES STEWART, 28th June, 1832, expressing the opinions of the ministers and teachers of schools, residing at Dundee, with regard to the factory system.

"Dundee, 20th March, 1832.

Documents.

"We, the undersigned, being ministers in Dundee, beg to state, as our decided opinion, that the long hours in which young persons are at present employed in our spinning-mills, are highly prejudicial to their health, their instruction, and their principles. By the present system, such persons have no time for amusement or relaxation of any kind, and too little for sleep. Any education which they receive is after about thirteen hours of labour, when they have more need of repose. As interested in, and in some respects accountable for, the instruction and religious improvement of the young, and anxious as we are for their comfort, we earnestly entreat the Legislature to adopt such a measure as will remove or mitigate the great evil of which we complain.

"*George Tod*, Minister.
David Russell, Minister.
Matthew Fraser, Minister.
Arch. M'Lauchlin, Minister.
P. Mainatt, Minister.
Alex. Duncan, Minister."

"Dundee, 13th April, 1832.

"An application having been made to me to give my opinion, in a few words, on the propriety of parliamentary interference to regulate the number of hours in which children shall be employed in mills and factories, I have no hesitation in saying, that if ever there was a subject which called more than another for legislative control, this appears to me to be that subject.

"The evils, physical and moral, which have resulted from the system which of late years has been pursued, can, I think, be denied by no one who has resided, as I have done, for five and twenty years in a large manufacturing city, and witnessed the lamentable change which has taken place in the appearance and habits of the rising generation; and it is the experience of these evils (for how else can it be accounted for?) which has led to the excitement which now prevails throughout the manufacturing districts on this subject, and to the numerous petitions thereon which have been laid on the tables both of the House of Commons and the House of Lords. With regard to the objections I have seen urged against the interference of Parliament in this matter, I confess they carry no weight with me. It is said that children in our cloth and linen manufactories (the manufactories which we have chiefly to do with in Dundee) are not harder wrought than those of a similar age in our potteries, glass, iron, tin works, &c. This may be very true for all I know to the contrary; but this appears to me to be an additional argument for the interference of the Legislature, because it shews that the abuse is not partial, but general; that it is not confined to one species of manufacture, but extends to all. Again, it is said that the child of the artizan is not more severely worked than the child of the peasant. This I deny. The labour of the latter is regulated by the light of day; and through a considerable portion of the year this precludes the possibility of his being employed so long as the former, who works by oil or gas-light for many hours after sunset, and for some previous to its rising. Besides, there is surely no parallel between the hardship of the two kinds of labour. The child of the peasant carries on his in the open air, and with scarcely any restraint over him; he is very much his own master as to the manner (I mean in what relates to the activity and intensity of his application to it) in which he shall perform his work. The child of the manufacturer is obliged to labour in the atmosphere of a crowded and frequently ill-ventilated apartment; an atmosphere impregnated with the effluvia, and clouded with the dust and miasma which arise from the article as it goes through its manufacturing progress, and heated also with the exhalations from breaths polluted with the frequent use of ardent spirits, and from the noxious vapours of gas and steam; and his occupation is carried on under the eye of a superintendent, who cannot admit of any further remission in his labour than what may be absolutely necessary to recruit his wasted strength and spirits. In a word, I have no doubt that the rate of mortality in persons of eight years old to twenty, will be found in the manufacturing class to be double that of persons of a similar period of life in the agricultural. The only objection against legislative interference in this business entitled to any consideration, is the plea, that Parliament has no right to meddle with the parent's authority over his child; that the parent is the best judge of what is most for his own and his offspring's benefit, of the capabilities of the latter, both of body and mind, and for how long a period they may be worked without injury to their health, and strength, &c. &c. This, as a first principle, is certainly true, but in the present case it must be received with considerable reservation; the fact is, (a fact which will not admit of denial,) that of late years the labourer, whether in trade, manufacture, or agriculture, in requital of his own labour receives a sum totally inadequate to maintain his family; he cannot earn for them, be his industry and sobriety what they may, a bare subsistence. To increase, therefore, the means of supplying to his family the absolute necessaries of life, he is compelled to put his children to work before their years and strength are fit for it. This, on his part, at first, is a measure of necessity, not of chance; but a little time, and the daily deterioration of his condition, reconcile him to it; the feelings of the parent become gradually deadened by want and privation, and he consents to the additional labour imposed on his children for the sake of the small additional gain which it brings him. Nor is this all; to recruit, as he imagines, the wasted strength of his child, he teaches him to have recourse to that stimulus which, in his own case, he has employed to steep his senses in forgetfulness of his former comforts and present misery; and thus our juvenile, as well as our aged labouring population, have become the victims of the constant use of ardent spirits. This is no exaggerated picture, as those can testify who reside in manufacturing towns, if they will speak the truth of the evils of the present system; a system which, together with the means of bettering the present miserable condition of the labouring poor in this once happy country, calls loudly for the consideration and interference of Parliament; and in which they would be better occupied than by deluding the people with visionary schemes, which will neither fill their bellies nor clothe their persons, but when they are consummated will be found to leave them just where they are.

" *H. Horsley*, Minister of St. Paul's, Dundee,
" Scotch Episcopal Congregation."

" Dundee, 20th March, 1832.

" I hereby certify, that it has been my opinion, most decidedly, for many years past, that the long protracted hours which the youth of our country are engaged in these mills, are highly injurious to their health and morals. I was at first impressed with a painful sensation, reflecting on the bodily fatigue to which many of the children are subjected in the very dawn of life; but the more I became conversant with this species of home-cherished slavery, the more convinced I am that it is pernicious to the morals of our

Documents. youth; because the young mind must and will have recreation and amusement, and if time for such purposes is denied on days that are lawful, then the Sabbath will be profaned to attain what is considered so necessary; and the pious parent, who desires devoutly to worship God in the place appropriated for that purpose, learns, with regret, that the youth he wished to accompany him has gone off with others, who, like himself, are freed from the yoke, to some recreative pursuit in the fields; or, if the youth be reluctantly drawn to accompany the parents, what is to be expected but, after the fatigues of the preceding week, instead of attending to and deriving benefit from the instructions communicated, he falls fast asleep; and the only consolation the parent has, is, that this youthful object of his anxious regard is not in worse society. I consider that this also is a prolific cause of crime, because numbers of uncultivated and uneducated youths, thus convening together, wantonly commit depredations, which, though at first but trivial in their nature, not unfrequently lead to acts of more awful consequence. As a minister of the gospel I feel deeply interested in the welfare and happiness of the community, and also that I am in some degree accountable for the education and religious improvement of the young; I am therefore convinced that a prudent restriction of these hours, such as would allow a moderate, reasonable period for recreation and rest, would be productive of much benefit, and would soon be extensively seen in the improvement effected on the morals of the rising generation. May I therefore earnestly entreat the legislator to adopt such a measure as his wisdom may suggest, and such as may tend to remove this growing evil, so severely felt in our country. " *Charles M'Alister*, Minister.

" *W. Johnston*, Minister."

" Dundee, 25th March, 1832.

" These certify, that I have, for the last nine months, taught the children employed in one of the spinning-mills of Dundee; that the children have been invariably so much fatigued by the labour of the day as to fall asleep almost immediately on their entering the school-room and taking their seats; that the master of the mill assisted me; that a considerable part of his time was occupied in keeping them awake, either by tickling their nostrils with a feather, or in making them stand a certain time on one of the forms, in order that they might be compelled to keep awake from fear of falling; that, from a long experience in teaching, I consider the present system of making the children work such a length of time, and then confining them to schools, is not only in the meantime prejudicial to their intellectual and moral improvement, but that it is also calculated, by enervating the faculties of the tender mind, to render them ever incapable of rising to any ordinary degree in the scale of intelligence; and that I consider the passing into a law the present measure before the Commons' House of Parliament, relative to lessening the hours of labour in our manufactories, is the only thing that will remedy the evil so justly complained of, that will give effect to teaching, by rendering the mind in some degree capable of receiving instruction, that will contribute to the happiness of the unfortunate creatures who at present labour under the grievance, and that cannot fail to promote the welfare and happiness of society at large. " *Andrew Ross*,

" Preacher of the Gospel, Dundee."

" Dundee, 30th March, 1832.

" Having been desired to state my opinion regarding the moral influence of the confinement of young people for so long periods in spinning-mills as are at present universally prevalent, I have no hesitation in declaring it to be my most unqualified conviction (from a considerably extensive observation of the condition of children employed in mills) that the present system is, in an extreme degree, unfavourable to their moral and intellectual improvement, whilst it is most prejudicial to the development of their physical energies. I consider the measure under the consideration of the Commons' House of Parliament, for abridging the hours of juvenile labour in our manufactories, to be loudly and imperatively called for; and holding those sentiments, I feel conscientiously bound to co-operate with those who, in the use of all lawful means, seek the interference of the Legislature for the accomplishment of an end so pregnant with benefit to so large and interesting a portion of the youth in the land.

Documents.

" *John Jaffray*,
" Assistant of St. Andrew's Chapel, Dundee."

" Dundee, 4th April, 1832.

" Having been desired to state my opinion respecting the many hours of confinement of the young people at work in the spinning-mills, in as far as I am able to judge, it is my humble opinion that the present system of long confinement at work is prejudicial to the health of the children, and tends so much to impair their energies as to render them unfit for intellectual and moral improvement. I cannot help thinking, therefore, that an abridgment of the hours of juvenile labour would be beneficial to the children employed at spinning-mills, and not unfavourable to the interests of their masters; and it is my earnest wish that some salutary legislative enactment may be made for the good of all concerned.

" *Alexander Peter*,
" One of the Ministers of Dundee."

" Dundee, 13th April, 1832.

" I do sincerely join in the above statement, and most earnestly desire a diminution of the time of labour.

" *William Reid*, Minister."

" Dundee, 20th March, 1832.

" I am free to state to all whom it may concern, that, upon my own observation, the present system of confining children to labour in the mills for thirteen or fourteen hours daily is a practice pregnant with a number of fearful evils to the children themselves, their parents, and the community. It not only injures the health of the children, but tends to rear them in ignorance and vice, of which ignorance is the prolific source. After the toil of a long day they are not in a state of either body or mind to acquire instruction; hence, at those works where evening schools are established, it is no uncommon thing to behold the little worn-out pupils locked in the arms of sleep, while the stronger and less wearied are acquiring knowledge. As Mr. Sadler's Factory Bill proposes to limit the hours of labour to ten per day, this arrangement would allow time for daily instruction; and as the acquisition of knowledge has a salutary influence on morals, it would tend to promote virtue; and on this account, if passed into a law, I shall regard it as a great boon conferred upon the community.

" *John Bowes*,
" Minister of the United Christian Church."

" So far as I have had an opportunity of observing, during a residence of nearly sixteen years in Dundee, I have no hesitation in declaring, that an abridgment of the hours of labour in public factories would be highly beneficial to all who are employed in the factories; it would allow more time both for diet and exercise, and afford greater facilities for religious and

Documents. moral improvement. It would be particularly favourable for these purposes to the young, especially to children of tender years, who cannot continue to labour during the long hours which are considered necessary at present without great disadvantage both to their health and to their education. I am happy to attest that I have witnessed with much pleasure, of late, the increasing efforts of the proprietors of public works here for the moral improvement of the youth under their care; but I feel confident that some legislative enactment of the nature proposed is absolutely necessary for the success of whatever plans may be adopted for this purpose. The abridgment in this case should be so extensive as to afford sufficient time both for healthful exercise and for instruction; while the rights of all should be so well secured as to prevent a return either to the present plan or to any modification of it which might disappoint the just and benevolent object for which such an abridgment is now sought.

" *Robert Arther*,
" Minister of the Associate Burgher Congregation.
" Dundee, 14th April, 1832."

" Dundee, 14th April, 1832.

" I hereby certify, that it is my most unqualified conviction that the present system of employing children for so long a period each day, at spinning-mills, is in a high degree unfavourable to their moral and intellectual improvement. I have frequently observed the languid state of those children when in school; some of them, through the fatigue of the day, fall asleep when writing their copies; others when learning to read, &c.; it is therefore very difficult to communicate instruction to them.

" *Alexander Hutchinson*, Teacher of the Dens Mills School."

" I hereby certify, that I have taught a school for the last sixteen years in the immediate vicinity of a number of spinning-mills, and during that time I have frequently had numbers of young persons there employed, attending my evening classes, of both sexes, and I feel no hesitation in saying, that I consider the long hours they are obliged to labour very injurious to their bodily health; and as for making any improvement in learning, it is nearly impossible, as they are generally so fatigued by the labour of the day as to fall asleep if not actually employed in receiving instruction. I have known instances of them being so exhausted as to hide themselves in the school, and fall asleep, and were only discovered by their parents becoming alarmed at their absence, and, coming for the key, they have searched the school, and found them sound asleep.

" Blinshall-street School, " *Andrew Stewart*, Teacher."
23d March, 1832."

" Having had several years' experience in teaching an evening school, composed of young persons employed in spinning-mills, I feel no hesitation whatever in saying, that under the present system of conducting these works, it is nearly impossible for the young there employed to make any considerable improvement in learning during the short time allotted for that purpose in the evening; and I have invariably found them fitter subjects for repose than for mental cultivation. Nor is this the only evil attending the system; those who may be making some improvement are very frequently interrupted by working extra hours, making up lost time; and by the time they again return to school, they are nearly as bad as when they first came. I give this as my decided opinion of the system.

" Hawk-hill, Dundee, " *Charles Edwards*, Teacher."
23d March, 1832."

CLASS III.

Sir CHARLES BELL, K.G.H., F.R.S., called in, and examined, 7th August, 1832.

Sir Charles Bell.

1. What is your profession?—A surgeon.

2. Are you an officer in any of the great medical establishments?—I am surgeon of Middlesex Hospital.

3. You have been in the habit of teaching your profession to medical students in the metropolis for a series of years?—I am a retired teacher of anatomy, and professor to the Royal College of Surgeons.

4. It is a universally received maxim in your profession, that exercise, confined within moderate limits, and with due intervals rest, is ordinarily necessary for the preservation of health?—Undoubtedly; that is an acknowledged principle in our profession.

5. Is it not also held by the profession as an undoubted principle, that undue labour, so as to produce great fatigue and weariness from its character or long continuance, and without due intervals for rest and refreshment, ordinarily considered, is prejudicial to life?—That corresponds with the suggestions of common sense.

6. Do you think that the customary day's labour in this and other countries, alluding mainly to the usual avocations of industrious life in agricultural and the mechanical arts, extending about twelve hours, including the usual intervals for meals, is, ordinarily speaking, as much as the human constitution can bear, with due attention to health?—Indeed I think so, sustained for any considerable time.

7. Assuming that the labour in the mills and factories of this country greatly exceeds that term, and extends sometimes to fourteen hours and upwards, and is endured in a confined situation, and often in an impure atmosphere, heated to a high temperature, can there be any doubt whatever that, generally speaking, labour of that description must be prejudicial to the human health?—I have no doubt of that; you present a very painful picture; and such a system must be attended with unhappy consequences.

8. Do you not conceive that undue labour, so long continued as to afford but short intervals of rest, whether for recreation or sleep, is yet more prejudicial to children and young persons than it would be to adults in the prime and vigour of life?—Certainly; all circumstances unfavourable to health will have a great influence on the constitution before it is settled, and before the strength is matured.

9. The labour in question to be endured for the length of time described often stretches into the night, more especially in the winter time; can you give any opinion whether labour is more insalubrious when undergone in the night-time than in the day?—That answer requires some detail; it implies want of comfort; it implies want of exercise, since they must sleep during the day; it more effectually, therefore, implies deficient exercise in the open air; in short, close apartments and confinement.

Sir Charles Bell.

10. Perhaps the working by artificial lights also might have a pernicious effect, inasmuch as the vital principle of the air would be, in some measure, consumed?—Additional light in an apartment has nearly the same effect as if so many more people were admitted into the apartment, exhausting the vital air.

11. Have you had any personal experience regarding the work and labour of children in mills and factories?—No.

12. Appealing, then, to the general principles of your profession, to the analogies which your extensive practice has afforded, and the studies you have pursued, have you any hesitation in tracing many injurious consequences to that system of labour, consequences which have been described to this committee at great length, as affecting the health and the limbs, and shortening the life, of those exposed to it?—I should answer in the affirmative, upon the acknowledged principles that such a state as that described would be very injurious to the constitution, and engender a variety of diseases; the great disease, emphatically using that word, is scrofula: wherever there is a want of exercise, deficient ventilation, depression of mind, and want of interest in the occupation, I should say, especially in young persons, scrofula, in its hundred forms, would be the consequence.

13. Does not that latent disorder affect the osseous part of the system as well as the glandular?—That is one of its effects, undoubtedly.

14. It has been alleged before this committee, that distressing cases of deformity are produced in the mills and factories, and that they take their rise in young persons sometimes as late as perhaps the 13th, 14th, 15th, or 16th year of their age; is not that an unusual occurrence, when unconnected with over-exertion in improper attitudes?—I shall divide the distortions into two kinds; distortion coming indirectly from constitutional defect or disease, and distortion arising from continued exertion of one kind, to which some artisans are liable; so that I should imagine the common distortion, the rickets, would make its appearance early in life; and that at a later period, deformities arise from the mechanical effort being continued in one mode, and without that variety which nature dictates.

15. Do you think that the female is as competent to long and continued exertions, and particularly in a standing position, as the male sex?—I rather think, where the work is light, that she is.

16. Do you think that there is any particular risk, at the age of puberty, from that description of labour long continued?—The period of puberty is a period of delicacy, and requires more particular attention.

17. Do you think that a child under nine years of age ought to be kept to the constant labour of a mill or factory?—It may be an alternative; on the general question, certainly not; but it may be a question of degree of suffering; the condition of the parents must be considered; the child may be starving.

18. But as a general question, you would think that long-continued labour under that age would be unfavourable to the health and future welfare of the child?—Very unfavourable.

19. From that period to the time of life at which it is generally supposed the osseous system may be about completed, is not ten hours' labour a day, to which must be added the time necessary for the taking of meals and refreshment, making therefore twelve hours a day, as much as can be endured, generally speaking, with impunity by those so occupied?—I should say yes; and more than that is a painful idea.

20. Are you of opinion, with the late Dr. Baillie, and many other eminent medical men who have appeared before preceding select committees of both Houses on this particular subject, that, under ordinary circumstances, and in a great plurality of cases, ten hours' labour a day is as much as can be

Sir Charles Bell.

safely imposed on human beings of either sex, and at any age?—I consider that if a person is interested in his subject, and the mind is carried on, he may labour as many hours, and often many more, with impunity; but where he is mechanically employed, without interest in his work, I think that is fully enough.

21. It is almost superfluous to ask whether the mind is capable of imbibing instruction when it is suffering under a considerable degree of fatigue?—Certainly, it is not; but to the contrary. Thus, a soldier cannot be kept too long on the parade without losing what he has gained in the exercise.

22. Alluding to one object which the framers of this bill have in view, namely, affording education to the lower orders of society by means of evening schools, it would be necessary to limit the hours of labour so as to render such instruction available to the lower classes of the poor?—If the child were to be instructed by its parent, that would be well; but as the question supposes a public school and a crowded room, the proposal would tend to aggravate its hard condition: that child is to be carried from a crowded manufactory into a hot and crowded school-room; it is adding to the hours of confinement.

23. You would judge, then, that to devote the only remaining day of the week, when the rest have been employed in factories, to the purposes of tuition, the question alluding to Sunday-schools, while the body continues to suffer under a considerable degree of fatigue, would add to the physical sufferings of those exposed to that system?—All the circumstances which we have considered disadvantageous to the constitution being incurred during the week-days, and repeated in this new form on the Sunday, of course they become more subject to the disorders proceeding from their mode of life.

24. Without alluding at all to the view of the question which a political economist might take, can there be a doubt on the mind of any medical man of knowledge and experience, that a remission of the hours of labour would be highly beneficial, in point of health and welfare, to the lower classes of society?—Certainly it would be highly advantageous to the health, and I should say, consequently, to the powers of the mind, to the improvement both of the mind and the body.

25. You have considered the subject of factory employment?—Only in one point of view; that in passing through England, and seeing the different hospitals in the manufacturing towns, I was very much struck with the nature and number of the accidents received.

26. You mean from machinery?—Yes.

27. It is stated, that those accidents mainly occur at the termination of the day's labour, and that they accumulate towards the conclusion of every week; would not that of itself be a strong indication that those engaged in this employment were generally over-worked?—It is the most pointed proof that can be given, I think.

28. Have you not been of opinion that some legislative means ought to be taken to decrease the number of those accidents, by inducing or obliging those masters who are careless in the fencing of those machines to adopt proper methods for protecting the persons of those so employed?—I did entertain such a notion some years ago; I mentioned it to the late Mr. Francis Horner, and he applied to some of his parliamentary friends; the answer that I then got was, it was not a practicable thing; but it is strongly in my mind that something ought to be done.

Sir WILLIAM BLIZARD, F.R.S., called in, and examined, 2nd Aug. 1832.

Sir Wm. Blizard.

1. What is your profession?—I am a surgeon; surgery is my profession.

2. Are you an officer in any of the great medical charities of this metropolis?—I am; and have been for upwards of fifty years surgeon to the London Hospital.

3. Have you been a medical or surgical lecturer?—Yes; for upwards of twenty years a lecturer on surgery, anatomy and physiology, in the London Hospital and the Royal College of Surgeons.

4. Your mind has consequently been directed, generally, to the causes and cure of those diseases to which the labouring classes of society are liable?—Undoubtedly.

5. May not the committee ask you, therefore, whether with moderate labour or employment proper intermissions for meals, refreshment and sleep, are not ordinarily necessary to the preservation of health?—Undoubtedly they are.

6. Are not those advantages still less to be dispensed with as regards children and young persons of either sex?—Unquestionably.

7. Do you not also hold it to be a maxim of the profession, that excessive labour, or labour so long protracted as to inflict great and continued fatigue of either body or mind, and endured without such due intermissions for meals, for sleep, and for refreshment, is inconsistent, generally speaking, with the preservation of health?—Certainly.

8. Would not such an excessive degree of labour, continued as described, without due intermissions, and inducing so much fatigue, be still more injurious to children and growing persons?—Undoubtedly.

9. In reference to adults, do you not think that the ordinary day's labour, whether in agricultural or handicraft pursuits, namely, twelve hours, with proper intervals of rest, is, generally speaking, as much as can be endured with impunity to the frame?—That is consonant with general observation.

10. Alluding to the labouring classes of society, are not such moderate intervals from their labour as are sufficient for taking their meals necessary to the preservation of their health, particularly with reference to the digestive organs?—Undoubtedly; with reference to those organs most important to health and life.

11. Supposing, then, that the labour of persons in mills and factories greatly exceeds the term in question, and extends to thirteen, fourteen, or fifteen hours out of the twenty-four, and sometimes to seventeen or eighteen hours, or more, should you have any doubt in saying that such a degree of labour must, in a great plurality of cases, be injurious to the health and the constitution?—Horribly so.

12. You of course mean to apply that expression to the extremity of the labour described to you?—Yes; and there is a question before, the spirit of which is involved in this; that which respects the relaxation and time required for taking food and exercise, and having relation to the digestive organs, and so on: all that is included then when you go to the maximum. I meant to reply generally, so that the application of my answer might be general, as I express it, and be taken in degree and ratio, as it would apply to the minimum and maximum respectively.

13. But you would think the average of such hours of labour would be an extravagant imposition upon the human frame?—Dreadful.

14. Then, when still further extended, so as to be continued for thirty or more successive hours, with inadequate intermissions, you would conceive it must lead to the most pernicious, and often fatal consequences?—I do.

15. You would perhaps scarcely believe, as your practice has been in this metropolis principally in the upper ranks of society, that children and young

Sir Wm. Blizard.

persons could be brought to endure that labour without the effects being immediately perceptible?—I should think it hardly possible.

16. Is it your opinion that labour pursued during the night is generally less congenial to the constitution than that undergone in the day-time?—It is contrary to the general course of nature; contrary to what nature seems to warrant.

17. Is not the employment in question, though it may, if contemplated for a moment, seem light or easy, yet when continued for such a length of time as to induce much fatigue of both mind and body, as it is asserted it does, likely to be more prejudical than even more strenuous labour pursued for a moderate length of time, and with due intervals for rest and refreshment?—I am clearly of that opinion; however light it may be, yet extended as it has been described, the consequence must be, in my opinion, as stated.

18. May the committee ask you, appealing now to the principles of your profession, whether it does not require some considerable degree of muscular exertion to maintain the erect position for a great length of time together?—No doubt of it; and it is a position which, if long maintained, is unfavourable in many respects, and leading to consequences very serious.

19. Then this labour having to be endured for that length of time in that position must, of course, render it more distressing to the feelings and exhausting to the animal frame?—I have no doubt of it at all.

20. If that description and degree of labour is to be undergone in a polluted atmosphere, rendered impure by the admixture of dust and flues constantly passing off from the material wrought, would that, in your opinion, constitute an additional hardship as it regards the hands so employed?—Oh, no doubt of it; I might enlarge upon this very much, but it is hardly necessary here or elsewhere: as to a polluted atmosphere, we know the deterioration that the health suffers from it. One should hardly imagine that what is separated from the material (though it will render the atmosphere unfit for respiratory purposes) could be called a thing that pollutes the air; but, when you come to the heated state of the atmosphere, the effect upon that atmosphere is very great from so much breathing; for, with regard to the office of the lungs, there is a certain quantity of pure air required; and if not supplied, there is something vitiated produced in the atmosphere that will pollute the whole frame. Now we know very well that heat does this; and that respiration does it by means of what is exhaled from the surface of the body and from the lungs: the surface of the body and the action of the lungs are continually affecting the atmosphere that is most suited to the existence of living persons, by exhalation and by absorption; and if there be a state of air not admitting of that which is exhaled from the lungs and the skin passing off, it will not admit of that which is of a vital character being absorbed; and the circumstances named are such as are most unfavourable and most pernicious to human existence, and to the health of any human being. With regard to that which is separated from the raw material, it is a foreign body, and, by being inhaled into the system, has a very material effect upon the small vessels of the lungs; it mingles with the mucus from the lining membrane of the lungs; and the air-pipes, which become smaller and smaller, until they terminate in what we call vesicles, have their functions impaired; for here you have these particles obstructing those tubuli, and rendering difficult the admission of that portion which should be received from the circumambient air.

21. Rendering the necessity for abridging the hours of labour in such an atmosphere still more imperative?—Most undoubtedly.

22. In alluding to the temperature in which this work is usually performed, should you not think it more healthy to pursue labour in an atmosphere at a given temperature, though high, but which was rendered so by natural causes,

Sir Wm Blizard.

than one artificially heated and defiled by the adscititious matter alluded to?—No doubt; for though, on the one hand, there is what might be abstractedly considered an evil; on the other, there is that which counteracts it, which is not the case where the atmosphere is heated by artificial means.

23. Then you would conceive, that children and young persons in this country, not being free agents, who are doomed to labour in an atmosphere like that, are entitled, upon every principle of justice and humanity, to protection equally to slaves in a tropical climate?—I have no doubt of it, both physically and morally.

24. To allude more particularly to children: taking into view all that has been described to you, that the length of their labour corresponds with that of all the rest of the hands in the factory, that it has to be pursued, therefore, for the same number of hours, sometimes in a heated temperature, and subject to other circumstances, to which allusion has been made; do you not think that, in regard of such children and young persons, a regulation of labour, so as to bring its duration within something like moderate limits, is highly necessary, and is imperative upon the legislature?—No doubt of it; it is a corollary of what you have stated.

25. It has been said by eminent men of your profession, who have had constant practical experience of the effects of the system as described to you, that to reconcile it to health would be to reverse all the principles of the profession; you have already stated your views upon this subject, and, from philosophical and general principles, you coincide with their declaration, deduced from actual experience?—Without a doubt.

26. During the course of this inquiry, and many preceding ones, it has been found that various afflicting diseases are attributed to this system of factory labour, as it has been described to you: may the committee ask you whether, reasoning upon the principles of your profession and the causation of disease, you do not think that that system would not only be very liable to cause certain diseases, but also to aggravate them, where there should be a constitutional tendency to them?—I have no doubt of it; long standing in one position has a very considerable influence on the circulating system; the veins become, as we denominate it, varicose or distended, and, of course, the return of the blood to the right side of the heart is not regular, nor in the right quantity or quality; and if it is either deficient in the one or the other, it is robbed of a certain degree of its stimulus, which is necessary, that all the other organs may be in a proper state. The heart is nothing more than a propelling organ, and the state of the circulation must become much affected by long standing.

27. That may be the cause which produces the languor of which these persons complain?—There is no doubt of it, and a thousand other distressing signs and consequences of debility.

28. Would ulcerated legs in many instances ensue?—Yes, and œdema, or a general swelling of the ankles.

29 Should you think that struma would be likely to be aggravated by this sort of labour?—Yes, and perhaps a strumous habit primarily induced by it. It is a disease of the lymphatic system, or lymphatic glands, and the various causes that have been expressed disturb the whole system and functions of the absorbents and their glands.

30. In alluding to the impure state of the atmosphere and to the irritating substances mixed up with it, should you conceive that pulmonary and bronchial affections might result from that sort of employment too long continued?—No doubt such a question involves in it the answer; when the structure of the lungs is attended to, how can it be otherwise? Pulmonary affections of every description may be induced by this disorganization of the pulmonary system.

Sir Wm. Blizard.

31. In allusion to the tender age at which this labour is often, and, in fact, mainly to be undergone, should you conceive that pernicious effects also might be produced by it upon the osseous system?—No doubt; the progress of the perfecting of the bony parts is gradual. In some parts there is no bone. At the extremities you have, not a union of the bones, but epiphyses; these are a long while before they are united to the bones themselves, and in all those the parts are imperfect; whence it is that you have distortions of the spine. The ribs and the cartilages are not formed with that relative perfection they ought to have for respirable purposes; hence you have such a train of complaints, and even distortion and disease of the spine itself.

32. Many distortions in the lower extremities have been exhibited to this committee, where the frame has been perfectly straight and well formed, even to so late a period as twelve, fourteen, or sixteen years of age; you would not confound those cases with ordinary affections of the limbs, or what are called rickets?—No; rickets have associated with them an imperfect formation of the bones, and an enlargement of the extremities of those bones; and, as the bones are not perfectly formed, they have ductility, so that they yield.

33. Is not rickets a disorder that is usually apparent in a much earlier period of life than those I have mentioned?—I can hardly say that; I am not quite certain upon that point; but what are called rickets are often ascribable to some or most of the causes you have stated.

34. So that the employment itself might induce that deformity of the skeleton that is sometimes called rickets, or at all events confounded with it?—Yes, unquestionably; I meant to convey that idea.

35. It has been stated, that the growth of children so circumstanced, compared with other children otherwise employed, is considerably stunted; should you be prepared for that effect also, as resulting from the system developed to you?—I have no doubt of it, both as to the extent of osseous growth and as to the condition of the bones, so far as they are formed.

36. Then it follows, that the muscular power and weight of the children so compared would exhibit similar variations, which also is found to be the case?—Undoubtedly there would not be an equal specific gravity in any of the parts; the bones would be lighter.

37. A paper has been ordered by this committee, and delivered to it, by which it appears that wherever this system prevails, it is accompanied by an extraordinary degree of mortality, especially in the earlier periods of life; taking the view you have done of the pernicious effects of labour so long pursued, and under the circumstances explained, you would be prepared for that result, namely, a greatly increased degree of mortality?—I do not know the fact; but, *à priori*, I should have no doubt of it, not the shadow of one.

35. It is a known fact, and often referred to as a sort of apology for this system, that it affords employment to females principally; would you conceive, arguing upon physiological principles, that the female is as well calculated to endure long and active labour as the male?—Certainly not; and universal observation would confirm that opinion.

36. Is it not necessary, with regard to any protection that the legislature may think it necessary to give to females so employed, that it should be extended at least to the period of puberty?—No doubt.

37. Is not the female constitution particularly liable to present and permanent injury, by undue exertion or improper treatment at that particular period?—No doubt of it; it is admitted that at an early period the bones are not permanently formed, and cannot resist pressure to the same degree as at mature age, and that is the state of young females; they are liable, particularly from the pressure of the thigh bones upon the lateral parts, to have the pelvis pressed inwards, which creates what is called distortion; and although

Sir Wm. Blizard.

distortion does not prevent procreation, yet it most likely will produce deadly consequences, either to the mother or the child, when the period of parturition arrives; it is a mechanical effect, produced by lateral pressure.

38. Do not you think that the labour as described to you, pursued often in a heated temperature, under the circumstances of sexual excitement, would be likely to anticipate the period of puberty?—It might, and I believe would; whatever affects one particular important organ tends to disorganize the whole frame; there is a dependence of one organ upon another, and particularly of certain organs, which have great influence with reference to all the other organs; these are the lungs and the liver. There is nothing in the world that is more readily affected than the circulation of the viscera by the causes mentioned, for its trunks go to form the *vena portæ* of the liver; this has reference to every organ in the cavity of the abdomen: then the other organ is the lungs. Now with regard to the liver, it is very important; it separates the bile; and if that is not properly separated, the blood retains it, and returns to the right side of the heart in a very imperfect state; and from thence it goes to the lungs. The heart is to be regarded as being no more than a projecting organ. The lungs and the liver and the surface of the body are the great organs; and every thing that has been stated goes to alter and deteriorate the healthy functions of those parts; it is not that organ or this organ; but there is that harmonizing movement in all the parts of the body, that when one of those organs is affected every organ is affected.

39. Should not you conceive that the languor and debility that this kind of labour is known to produce has a direct tendency to produce that tippling which is said to prevail so lamentably in these manufacturing districts?—I have not a doubt of it, for there is nothing in the world more inviting from their influence than spirits: spirits produce that which is most delusive; they produce a delightful sensation in the stomach; they tend to sooth and comfort, and to produce a favourable effect to the sense upon the whole system, but, ah! deadly is the consequence.

40. You think that degree of languor that seems to require a stimulus is likely to lead to tippling and all its injurious consequences?—Yes, beyond anything in this world; and when that habit is established, break it if you can.

41. Should you not conceive that the body is in a very unfit state to renew its daily exertions when it is not habitually refreshed by a necessary degree of repose?—Yes, no doubt; whoever has dwelt upon the intellectual functions must admit that at once.

42. And must not that labour which can alone be stimulated by the lash and cruel inflictions at the latter part of the day be peculiarly pernicious?—I have not a doubt of it.

43. Do you think that a child under nine years of age ought, under any circumstances, to be sent to the labour of the mill?—Certainly not.

44. Do you think that the limitation of the hours of labour to ten, which, including the meal-times, will make the day's work twelve hours, is an unreasonable limitation, or more than ought to be demanded in behalf of young persons between the ages of nine and eighteen?—Not more than that ought to be demanded; and it is consonant with general experience, like that of large seminaries and schools, where reason is the guide of humanity.

45. You are perhaps of opinion, with the late Dr. Baillie and other eminent medical men, that the usual duration of a day's labour, namely, ten hours of actual work, which, with the time necessary for refreshment, extends to about twelve hours, is quite enough, and as much as can be ordinarily endured at any age with impunity?—I heartily concur in that opinion with my late honoured friend; it is mine also.

46. One pernicious consequence of this promiscuous and long-continued

labour is stated to be gross immorality; in answer to that allegation, it is sometimes said that the number of illegitimate children is not greater among those employed under the system described; do you think that that would of itself be a proof of superior morality in that case than amongst an equal number of grown-up females in other pursuits?—No, certainly not; let any man reason upon that point, and he will decide otherwise. Sir Wm. Blizard.

47. You think that early and promiscuous profligacy is rarely accompanied by fertility?—Certainly.

48. Should you not consider it additionally injurious if, after those long days of labour, another hour or two were to be imposed upon the children by sending them to night-schools?—I should think it would have very little beneficial effect, but be demonstratively unfavourable.

49. Then supposing that the degree of labour leaves such fatigue and exhaustion as are now felt throughout the Sunday, often inducing the children to refrain from going to Sunday-schools, and rendering their attendance there very unprofitable, does not the institution of the Sunday-school increase the physical suffering of those children without those advantages being produced which would otherwise result?—Certainly; it increases their sufferings, and can produce very little benefit.

50. Can the youthful mind profit by instruction when the body is under a sense of great languor and fatigue?—No; every person's experience must have taught him that in some degree.

51. Then it is clear that it would be necessary to abate the extreme labour of those children and young persons if we mean to give them a decent and proper education?—Yes, no doubt. The mind, it is to be recollected, has reference to the brain and every nerve of the body, and every organ of the body has reference also to the brain; so that, morally speaking, and physically speaking, if you have reference to the brain and the functions of the brain, they cannot be properly performed under the influences named. The same thing may be said as to the nerves of the body, and the muscles, and every organ of sense; whence it is that every organ of sense and sight may become imperfect. There are two distinctive offices with relation to the brain; what has relation to the intellectual functions, and what has reference to the various organs of the body as to sensibility.

52. So that the employment that requires minute and incessant attention, and which makes constant demands upon the brain, would be most fatiguing?—Yes, certainly.

53. Would not, upon the whole, such a remission in the hours of the labour and toil of young persons as should afford them better opportunities of becoming educated, whether at home, or in evening or other schools, at the same time securing their health more effectually, be a great public as well as individual advantage in your opinion?—Very great; it would be to the personal comfort of the individual, and to the benefit of society at large.

JAMES BLUNDELL, Esq., M. D., called in, and examined, 27th July, 1832.

1. What is your profession?—Doctor of Medicine. J. Blundell, Esq., M. D.

2. Are you an officer in any of the great medical institutions of this metropolis?—I am a lecturer on physiology and midwifery in the school of Guy's Hospital.

3. Your mind has been necessarily directed to the consideration of the health and constitution of females and of children?—It has.

4. Is it not an universally received opinion among the medical profession that moderate exercise or employment, with due intermissions for refreshment,

J. Blundell, Esq., M.D.

recreation, and sleep, are ordinarily necessary for the preservation of health?—Certainly it is.

5. Are not those advantages still more necessary, and least of all to be dispensed with, in regard to children and young persons?—Certainly, in my opinion they are.

6. Is not it an equally received opinion with medical authorities, that excessive labour, or labour so long continued as to produce great and lasting fatigue to either body or mind, without having due intermission for meals or sleep, is inconsistent, generally speaking, with the maintenance of health?—Generally speaking, decidedly so; provided such labour is carried forward systematically for months together, and is not an occasional accident.

7. Would not such an excess of labour, then, habitually undergone, be still more injurious to children and young persons than to adults?—In my opinion it would, unless the adults were far advanced in years.

8. Is not the customary day's labour, whether in agricultural or handicraft employment, established as it seems by universal assent, in every age and country of the world, namely, twelve hours, inclusive of meals, as long, in your opinion, as the human frame, under ordinary circumstances, for a continuance, and generally speaking, is calculated to sustain with perfect impunity?—I do think that, for a continuance, it is as long a period as the human body will bear with perfect impunity, unless it be gifted with more than ordinary strength.

9. Are not intervals or cessations from labour, sufficiently long for taking necessary meals, essential to the health of the labouring classes, generally speaking?—I conceive that such intervals for such purposes are highly desirable.

10. Do you think that a cessation from labour at the time of taking meals ordinarily assists digestion, as is commonly supposed?—My opinion is, that where the digestion is very vigorous, it may go forward, though the meals are snatched up under circumstances unfavourable, perhaps, to their thorough manducation; I say my opinion is that this may happen, but nevertheless, unless the digestion is very powerful, I do think that the taking of meals week after week, and month after month, in a hurried way, without interruption, is an inpediment to their thorough digestion; and further, I imagine that physicians generally are of the same opinion.

11. Assuming, then, that the labour undergone in mills and factories almost always greatly exceeds the term already mentioned as that of an ordinary day's work, and sometimes extends to thirteen, fourteen, fifteen, and even eighteen or nineteen hours a day; do you doubt that such a system must, in many cases, prove highly prejudical to the health of those enduring it?—I should be exceedingly surprised if it did not.

12. You think, then, that the general result would be a derangement and interruption of health?—Decidedly so; with me the question does not admit of hesitation.

13. And that the exception would be in those cases where a moderate degree of health and strength could be preserved ultimately under such a system?—Giving my opinion on general principles, without experience in the matter, I should say that under this system the continuance of health and strength would be the exception, and not the general result.

14. Again to allude to that long term of daily labour, if it were to be undergone without any intermission, it would of course, agreeably to your opinions just expressed, be still more injurious?—Decidedly.

15. It seems almost unnecessary to press the question further, but supposing this labour to be continued to thirty or forty successive hours, twice or oftener a week, without any intermission for sleep, and sometimes so con-

J. Blundell, Esq., M.D.

tinued for months together; could such a system of labour by possibility be endured without, in most instances, very serious results?—All general principles, so far as I am acquainted with them, and all my experience of the powers of the body, are in favour of the negative; for these reasons, therefore, I would answer decidedly not.

16. Should you not, without the fullest evidence and assurance of the fact, believe it to be impossible that that extraordinary degree and description of labour could be endured for many months together, by children and young persons not yet arrived to maturity, without injury?—To convince me that such extraordinary labour could for months together be endured without injury by the class mentioned, I should certainly require very strong evidence; evidence unbiassed and cumulative, and of several consentient witnesses; and after all, I should feel a wish for the evidence of my own sight and touch.

17. Is it your opinion, that labour pursued in the night is less consistent with health than if undergone in the day-time?—Day-labour, I think, is more consistent with health than night-labour; generally speaking, the aptitudes of men are for day-labour. Many animals are by nature nocturnal, man is not; to them the star-light is, I presume, agreeable; but man finds it is a pleasant thing to behold the light of the sun.

18. In reference to the factory system, is not any employment, though it may be denominated light and easy, but being one of wearying uniformity, and inducing much fatigue of mind as well as of body, more exhausting and injurious than moderate locomotive exercise though of a more strenuous nature, undergone for a reasonable length of time and with due intermissions?—In my opinion it certainly is; and I would add, that it becomes so more especially if that wearisome exercise is to be undergone in a confined situation. Strenuous locomotive exercise in moderation is not injurious at all, but, on the contrary, healthful.

19. Is not the exertion necessary to sustain the erect position of the body for a great length of time more fatiguing than any other natural position ordinarily maintained, or moderate exertion equally and alternately exercising the various muscles of the body?—Decidedly it is; man, when he seeks repose, generally prefers the sedentary or the recumbent position, seldom standing.

20. According to physiological principles, might the excessive fatigue that a person endures who has to maintain an erect posture for a great length of time be explained and accounted for?—I think it might, inasmuch as the movements of the body in locomotive exercise tend effectually in man to help the circulation, and further, the standing position implies that the same muscles are kept continually in action, while in the locomotive movements there are alternate changes of the operative muscles.

21. Is there not alternate rest and exercise in the varied movements of the body, which is not the case in the standing position so fully?—Exactly so, alternate rest and exercise of different sets of muscles.

22. Again, in allusion to this labour, would it not be still more distressing if it had to be pursued in an atmosphere polluted with dust and flue perpetually evolved from the material manufactured?—Decidedly, in my opinion it would.

23. Would not labour so circumstanced and continued be still more injurious if undergone in an atmosphere artificially heated to from 70° to 80° and upwards, in order to facilitate the operations, and from which currents of fresh air must be excluded, as they would tend not only to lower the temperature but frequently to break the thread of the material spun?—My opnion is, that habit will do much to reconcile the body to the higher temperatures of the atmosphere; but I am farther of opinion that the tendency of those higher temperatures is to excite, exhaust, and weaken the body.

J. Blundell, Esq., M. D.

With respect to the effects of an atmosphere from which currents of air are so far excluded as to hinder the due changes by which it becomes fully refreshed and prepared for respiration, I am of opinion that such atmosphere must always be more or less noxious to the body, and more particularly that it tends to enfeeble the digestive organs and the nervous system : *cæteris paribus*, those species of animals whose structures are most completely ventilated seem capable of the most rapid movements, and are perhaps the most lively.

24. Does not the circumstance of the hands or operatives so employed, when they leave their work, pursued in this heated atmosphere, having to plunge into the open air, which in this climate is, for a considerable part of the year, cold and chilly, and often at the freezing point and under, most of them being very lightly and insufficiently clothed, greatly aggravate the injurious effects of the length of labour so pursued?—Habit is a second nature; and I am of opion that in the particular case proposed, habit will do much to reconcile the body to these sudden alternations of temperature; but I am further of opinion that where there is a tendency to inflammation of the lungs, these alternations are likely to induce it; and, moreover, I am of opinion that many individuals would suffer severely in the seasoning before the habit of sufferance was formed.

25. Is not, then, labour in an atmosphere thus impure and heated artificially more injurious to the constitution and exhausting to the strength than similar labour would be pursued in the open air, though at the same temperature, but being free, of course, from the impurities in question, and freshened by the natural atmospheric changes and influences?—Most decidedly.

26. Do not, then, that part of the British community, supposing them not to be free agents, compared with the Negroes of the Crown colonies and their children, equally require protection as to the duration of their labour?—Assuming that the children and their parents are not free agents, I presume that they do; but the question being rather political then medical, I do not feel qualified to decide. This, however, as a physician, I do say, that the labour before described to me is, in my opinion, more than children and young persons ought to undergo, and that to secure health a preventive of such excessive labour must be provided.

27. In addition to other effects it might be likely to produce, would not this labour, when pursued by night, and consequently by gas or other artificial light, tend, as the operatives allege that it does, to injure the sight?—I am not an oculist; I should, however, think that it had that tendency; but I am not prepared to say that where the retina is originally strong the conditions proposed would operate in a very powerful manner in weakening the vision; experience will best decide.

28. To apply, then, the preceding observations and the opinions that you have expressed to the immediate objects of this parliamentary inquiry, factory children, do you not conceive that the labour in question, continued for the length of time described, without sufficient and often without any intervals, even for meals, generally pursued in an erect, or at least a constrained position, and in a foul and polluted atmosphere, frequently heated to a very high temperature, and many times continued far into or during the whole of the night, can fail to be still more pernicious and destructive to children and young persons than it is to adults in the prime and vigour of life; and does not the condition of those children, in regard to either justice or humanity, demand legislation as imperatively as that of the adult felon or infant slave; can there be a reasonable doubt of the affirmative?—I should think decidedly that the conditions proposed could not fail to be more injurious to children than to adults, and though unversed in what relates to government, I should further think that the children, not being free agents, do require protection equally with the felon and the slave.

J. Blundell, Esq., M.D.

29. It has been stated by medical gentlemen, whose practice has necessarily made them conversant with the factory system and its effects, that such labour so early pursued is highly prejudicial. One of them, a witness before a preceding committee on the same subject, says, "Before one can at all reconcile one's ideas with the state of the factory being conducive to health, all past experience must be entirely reversed." Assuming the description of the factory system to be on the whole correct, do you, reasoning from the established maxims and principles of your profession, concur with that opinion, which was founded upon long and extensive experience?—I am decidedly of opinion that the factory system cannot be conducive to health.

30. Various results have been directly attributed to this system of early labour by many eminent persons of great experience, and resident in the manufacturing districts of the country; you have already stated that you are not practically acquainted with it; but what would be the consequences, reasoning from the principles of your profession, and from the analogies which your practice must have largely afforded, which you would expect that system in too many instances to produce?—Dyspeptic symptoms, and all their consequences; nervous diseases in greater or less degree; and, as the result of both, stunted growth; languors; lassitude; general debility; and a recourse to sensual stimulants, in order to rid the mind of its distressing feelings. These are the effects that I should expect.

31. Are not scrofulous constitutions likely to suffer severely under the factory system?—I conceive that they are; scrofulous constitutions are a sort of pageants fitter for show than for use.

32. You have already stated that pulmonary affections are likely to result from the vicissitudes of the temperature to which labourers of this kind are constantly exposed; will you state whether the nature of the employment, as described to you, would not tend also to produce such effects, now directing your attention particularly to the dust and the flue that must be constantly inhaled in many of the departments of those mills?—I am of opinion that an atmosphere loaded with dust and flue is likely to be injurious to the lungs. Much, however, will depend upon the quantity as well as the nature of those substances.

33. The dust is so prevalent from the flax spun in certain of the rooms, as almost to prevent persons from seeing each other at a few yards distance. Considering the material and the excess of the dust, should you doubt as to the pernicious effects of inhaling such an atmosphere as that?—As far as I can judge, without experience, I should have no doubt that inhaling such an atmosphere would be pernicious.

34. So that the circumstance of those engaged in such employments complaining constantly and almost generally of asthmatic and pulmonary affections, which may be classed together, would not be surprising to you as a medical man?—Certainly not; but the contrary would.

35. What would be the effects, commonly classed as surgical ones, which would probably be occasioned by the degree and description of labour that has been described to you?—Varicose veins of the legs; varicose ulcers of the legs; and œdema, especially in the women.

36. What do you think would be the effects upon the osseous system, as resulting from the same cause?—A rickety state of the joints, more especially the great joints of the lower limbs, curvatures of the spine, and contractions of the pelvis.

37. Can, then, the peculiarly distressing deformities seen by the committee in several witnesses, and which originated, as the parties themselves allege, from long standing at their work, which deformities commenced as late as nine or ten years of age, and in some cases as late as sixteen or seventeen, the body having been previously perfectly straight and well formed, which cases

J. Blundell, Esq., M.D.

the committee is assured are lamentably common in the factory districts, be attributed to that system of labour as now developed to you?—I am of opinion that they may be attributed to this cause with a very high degree of probability; and if the proportion of such cases is considerable, then I am further of opinion that they may be attributed to the factory system with certainty.

38. A witness before this committee has stated, that out of about 2000 children and young persons, upwards of or about two hundred were known to be deformed in a greater or less degree, and in many instances excessively so. He stated, also, that many of the cases of deformity among the females were of a nature not to be so easily detected, and consequently they have not been enumerated in forming that proportion; would you conceive that proportion which he stated to be the correct one in regard to the labourers in factories which he himself had examined and inspected, to be an excessive proportion?—Decidedly I should.

39. Then, in conformity with the opinion which you have just expressed, you would have no hesitation in ascribing it to the cause that the operatives themselves universally ascribe it to, namely, to the excessive labour and fatigue, and the long standing they endure in factories?—I should have no hesitation.

40. It has been stated as the result of very careful examinations, impartially conducted principally in Sunday-schools, that the stature of children who have worked in factories compared with those who have laboured at other employments, is, at corresponding ages, less by two inches; would a stunted growth be the probable consequence of such a system and degree of labour?—Even independently of experience in the matter, I should expect a growth more or less stunted as a consequence of such a system of labour; and with the preceding fact before me I can have no doubt that the factory system does produce such an effect.

41. The same examination has been made, regarding the comparative weight of those classes of children and young persons compared as before, and with corresponding results; should you be prepared to expect that system would diminish the weight and muscular power of those enduring it?—I should, certainly.

42. An official paper has been ordered by and delivered to this committee, which shews the great waste of human life, especially at its early stages, in the manufacturing or factory districts, as compared with the mortality at corresponding periods of life in other towns and places, and even in this great metropolis, commonly supposed to be unfavourable to life, especially in early youth; assuming the description of the factory system made to you to be substantially correct, should you be prepared to expect such a result from that system?—I should, certainly.

43. It is in evidence that a considerable majority of the children and young persons so employed are females; will you state whether the female sex is as well fitted to sustain long exertion, especially in a standing position, as the male, either in respect of the peculiar structure of certain parts of the skeleton or of muscular power?—Decidedly, females are not so well fitted to bear those exertions as the males; and the standing position long continued is, from the peculiarities of the womanly make, more especially injurious to them.

44. Then there seems to be some foundation in nature for the customary division of the pursuits of the industrious classes of society, which assigns the more active kind of labour to the male, and the more sedentary to the female?—I think there is such foundation in nature, and I think that the experience of this, and the instructive feelings of mankind resulting from this, have very principally led to the division of labour between the sexes; other

causes, however, have no doubt operated; among savages the woman is often the drudge. J. Blundell, Esq., M.D.

45. Is not it more than ordinarily necessary to give protection from excessive labour to females, especially when approaching the period of puberty?—I think that they do require protection at this age more than males.

46. Will you be so good as to favour the committee with the reasons for your opinion in that respect?—I am of opinion that long standing at this age has a tendency to produce contractions of the pelvis, and there can be no doubt that these contractions tend further to produce difficult and dangerous delivery; and further, it is certain that, in consequence of such contractions, it sometimes becomes necessary to destroy the offspring in order to save the mother.

47. Has not the labour in question, especially when pursued in a heated temperature, and under circumstances unfavourable to decency and morals, a direct tendency to anticipate that period, namely, the period of puberty?—I certainly think it has; and this for reasons which it may be unnecessary for me to particularize.

48. Is not the body in a very unfit state for renewed labour, especially that of children, when the fatigue of the preceding day is not removed by sufficient rest, and when therefore that labour has to be resumed in a state of comparative torpor?—Assuredly it is.

49. Is not the labour of children and young persons, when it is so long undergone as to have induced, as it commonly does towards night, great lassitude and drowsiness, and when it renders it necessary for those exacting its continuance to stimulate them by constant and severe beatings and punishments, highly injurious?—Assuredly it is injurious.

50. May not the numerous and afflicting accidents which occur more particularly at the end of the day, and are observed to increase towards the termination of the week, be fairly attributed to this over-fatigue, and lassitude, and stupor?—I think they may.

51. Do you think a child under nine years of age, of either sex, ought to be doomed to this description of labour?—Assuredly not.

52. Are you of opinion that young persons from nine to eighteen, or to about the period when the osseous system is arriving at its full development and strength, ought to labour more than twelve hours a day, including two hours for meals, refreshment, and rest?—I think that twelve hours a day, including two hours for meals, refreshment, and rest, are quite sufficient time for human beings to labour for a continuance.

53. Then you concur with the other eminent medical men of your profession, including Dr. Baillie, who have given evidence before preceding committees of both houses, as to that being a sufficient degree of human labour to be endured in a factory, especially by children and young persons?—Decidedly I do.

54. Would the proposed diminution in the hours of labour, if attended with a proportionate reduction of wages, have as injurious an effect, in your opinion, upon the health, as the present degree of labour, even with the higher wages that would otherwise be received?—Provided the reduced wages are sufficient to procure the necessaries of life, I conceive that such reduction, with an alleviation of the labour, would be an advantage to the health of the individual.

55. It is sometimes stated as a proof of the comparative degree of morality that exists in those large establishments where the sexes are promiscuously mixed, and where they work together for so many hours, and often far into the night, that a smaller proportion of females have illegitimate children than those in agricultural or in other employments; would you conceive that that

J. Blundell, Esq., M.D.

fact would sufficiently prove that their morals were in a better state?—Decidedly not.

56. Supposing that it is assumed that the state of morality in those establishments is lamentably defective, you would be prepared for a result like that in conformity with the principles of your profession?—I should, because ill-health and sexual profligacy tend to produce sterility; and where individuals are congregated as in the factories, I conceive that means preventive of impregnation are more likely to be generally known and practised by young persons; but it is to the first of these principles that I should look principally, in order to account for factory sterility.

57. But supposing that no such practices were resorted to, assuming that the morality of the factories was very much below that of the agricultural districts, would you not, agreeably to the principles of your profession, be prepared for an undue proportion of sterility, as compared with agricultural districts, where immorality was not carried to the same extent?—I should; in truth, under the circumstances, it appears to me that the sterility referred to rather tends to prove the insalubrity of the factory than the morals of the inmates.

58. You conceive, then, that early and constant profligacy is not conducive to prolificness?—I conceive it is not.

59. After a week's labour, such as that which has been described to you, should you not think that the confinement of the Sunday-school was, in reference to the health and comfort of the children, additionally injurious?—Decidedly I should.

60. Upon the whole, then, taking into consideration the undue length of labour endured by those children and young persons in factories, the circumstances under which they do ordinarily labour, and the whole of the case as now expounded to you, have you any doubt upon your mind that it would be highly beneficial to the individuals having to endure it if the legislature were to interfere, and to shorten and abridge that duration of labour?—I am decidedly of opinion that an abridgment of the factory labour is highly desirable in a medical point of view.

61. What effects is the factory system likely to produce upon the minds of those who are placed for a length of time under its operation?—I am of opinion that this system has a tendency to produce irritability of the nervous system, excitability of the feeling, and a certain busy play of the ideas when the mind is roused, together with that state of the mind generally which constitutes fretfulness and discontent; and I am further of opinion, that this system has a tendency to weaken the solid strength of the mind. In a word, if I may be allowed to use an expression which I have publicly employed some years ago, I look upon the factory towns as nurseries of feeble bodies and fretful minds.

BENJAMIN COLLINS BRODIE, Esq., F. R. S., called in, and examined, 1st August, 1832.

B. C. Brodie, Esq.

1. What is your profession?—A surgeon.

2. Are you an officer in any of the great medical institutions of this metropolis?—Surgeon of St. George's Hospital.

3. Have you given lectures in any institution?—I lectured formerly on anatomy, and for many years since on surgery.

4. Your attention has been directed, of course, to every branch of your profession?—Yes, I conceive so.

5. Your experience for many years is perfectly well known; are not due

exercise, and a sufficient degree of refreshment and sleep, in ordinary circumstances, necessary to the preservation of human life?—Certainly.

B. C. Brodie, Esq.

6. Is not that exercise connected with a considerable degree of what may be considered recreation in the open air and sufficient rest more especially necessary to children?—Certainly, this is necessary to adults, but more so to children.

7. Is not excessive labour, without due intermissions or sufficient rest, highly prejudicial to the human constitution?—Certainly; excessive labour, in fact, means that which is prejudicial.

8. Do you not think that the customary day's labour in various industrious pursuits, whether agricultural or mechanical, namely, twelve hours, with due intermission for refreshment, is, generally speaking, as much as the human constitution can endure with impunity?—As much as ordinary constitutions can endure with impunity; particularly strong constitutions may endure greater labour.

9. Assuming that the labour undergone in mills and factories generally much exceeds that term, and often extends to thirteen or fourteen, and sometimes even to eighteen or nineteen hours a day; do you think that ordinary constitutions can bear that degree of labour, of that description, without considerable injury?—I should think very few constitutions of adults can bear such a quantity of labour as is described with impunity, and much fewer still of children.

10. Would not that length of labour, if endured without any intermission whatever, be still more injurious?—It would be much more injurious if without intermission.

11. Do you think that labour which is pursued during the night is, for an equal length of time, more prejudicial than that pursued in the day?—It is often supposed so, but I am not able to make up my mind decisively upon that point.

12. If you were assured of this fact, that when labour is endured in the night by the children and young persons, the temptations to recreation and amusement, and other pursuits in the day, deprive them of sufficient rest, you would not, perhaps, have much difficulty in deciding that such a reversal of the order of nature would be prejudicial?—I should not. I observe that those who labour through the night, at least such among them as I have an opportunity of seeing, do not seem to get an adequate quantity of rest during the day, even though they have an opportunity; watchmen, for instance, I am surprised to find how little rest they commonly get in the day.

13. Have not you observed that that abridgment of what is thought necessary rest is prejudicial?—Certainly; first to the mind, then to the body.

14. Is not the employment, though it might be, and is denominated "light and easy" in itself, when continued for such a length of time as to become wearisome both to the mind and to the body, equally if not more prejudicial to the constitution than more strenuous exertion continued only for a moderate length of time, and with intermissions?—I think so.

14. Is not the maintenance of the erect position of the body fatiguing when continued for a great length of time?—It is more fatiguing than a recumbent or sitting posture, inasmuch as more muscular exertion is necessary to maintain it.

15. So that the labour described would be still more fatiguing as carried on to so great a length of time, if it had, generally speaking, to be performed in an erect position of the body?—Certainly.

16. In certain manufacturing establishments, there is a considerable deal of walking occasionally, united with labour, when, of course, they are still in an erect position; that, connected with the manual operations they have

B. C. Brodie, Esq.

to perform, would increase the degree of fatigue, would it not?—I do not know that the erect position, when the individual is partly walking and partly standing, is more prejudicial than merely standing: probably less so.

17. In those pursuits where the position is confined merely to standing, without any variation from that, you conceive that would produce much fatigue?—Yes, I think more fatigue than if some degree of walking were combined with it.

18. Would not the labour in question be still more injurious if it had to be pursued in an impure atmosphere, mixed up with the particles of dust and flue which are constantly evolved by the material prepared or manufactured?—Certainly, much more so; because the polluted atmosphere in itself, without the labour, is a great evil.

19. A polluted or impure atmosphere, if unaccompanied with exertion, would produce considerable languor and weakness perhaps?—Yes, certainly.

20. Would not that labour be even still more injurious if the atmosphere is at a high temperature, such as from 70° to 80°?—Undoubtedly.

21. Do you not conceive that considerable danger must accrue to the constitutions of children and young persons exposed to that sort of labour in the heated atmosphere described, where they have, for a considerable part of the year, to leave that work, and go by night into a cold and often a freezing atmosphere, not sufficiently clothed, and subject to all the vicissitudes of the weather?—That would add to the general ill effects of the labour in such an atmosphere.

22. You think that labour pursued in an atmosphere artificially heated as described, and subject to the other pernicious consequences incident to that system, would be still more pernicious than labour pursued in an atmosphere of the same temperature, but a natural one, and consequently subject to the freshening influences of those changes which a natural atmosphere experiences?—Certainly.

23. Presuming that a great proportion of the hands employed are, in no sense of the word, free agents, does it not become, in a medical point of view, and consulting the feelings and principles of humanity, as essential to protect those children and young persons so circumstanced, as the slaves in our West Indian colonies?—I should of course think that the children employed in labour, not being free agents, deserve protection as much as the West Indian negroes.

24. The inconvenience of working by night necessarily involves working by artificial and, generally speaking, gas lights, placed so as to be near the sight; would, in your opinion, that circumstance tend to injure the sight, as it is alleged by the operatives who have been before this committee that it does?—I have not made any particular observations on the effects of gas-lights; but I should suppose that working much in the glare of gas-light must be injurious to the eyes.

25. Numerous witnesses who have appeared before this committee, some of them medical ones, have attributed to the excessive degree of labour, which has been already partly described, very injurious effects of a description which may be denominated medical as well as surgical; what are the results which, reasoning from the principles of your profession, you imagine might ensue from labour so long continued, especially in an impure atmosphere, and by young persons and children, whose constitutions have not been sufficiently matured?—With persons who are disposed to pulmonary disease, whatever tends to debilitate the general system will cause the disease to become developed; scrofulous diseases of all kinds, I conclude, will shew themselves among children so circumstanced; scrofulous diseases generally appear in those who from any causes are in a state of

debility. I should suppose, in young children, rickets, leading to various kinds of deformity, would occur under such circumstances; perhaps, on consideration, I might mention others, but those are the principal points which occur to me at present. B. C. Brodie, Esq.

26. Many of the operatives, it is said, and especially the females, suffer from this labour other effects, namely, diseases in the legs?—Diseases in the legs generally, and especially varicose veins and ulcers of the legs, are more likely to occur in persons who are constantly in an erect posture. I observe in this town those who are a great deal in an erect posture, especially if they carry weights, become flat-footed, which is a very distressing complaint.

27. Do you observe the knees get affected?—Yes; I meant to include that in the distortions from rickets.

28. It has been stated by a preceding witness that out of about 2,000 children and young persons who have been carefully examined, about or nearly 200 were deformed, some of them very considerably, though it was considered many cases had escaped detection in the females which their dress would more easily hide; do you conceive that would be a great portion of deformity to befal persons under those circumstances?—Certainly, an immense proportion.

29. The result of accurate and impartial examination, conducted in Sunday-schools, has been that the growth of children who are employed in this long labour, and confined for such an undue length of time, is stunted; should you suppose that might be the result of the factory system as at present pursued, as it has been described to you?—I should not doubt it.

30. Similar examinations have been made regarding the difference in the weight of children so occupied; it has been found that those engaged in mills and factories weigh considerably lighter than others engaged in different pursuits, such as are ordinarily followed in the open air and under more favourable circumstances; should you be prepared to attribute that effect to the confinement in the factories?—I conclude that whatever prevents growth in particular children will cause a diminished weight in them as compared with other children.

31. Do you think that a deterioration of that description may become hereditary, especially if the causes of it are not removed, so as ultimately to impair the race and generation of human beings amongst us?—I should think it would become hereditary to a certain extent, as we find it to be in animals; but this is a difficult physiological question, and one requiring much consideration.

32. It is a known fact that a considerable majority of the persons employed in mills and factories are females; do you conceive that females are ordinarily as well adapted to endure long labour, and especially of the nature described, as the male sex?—After the age of puberty certainly not; but I suppose that the difference of the male and female constitution is not established till after puberty.

33. Do you conceive that it is more than ordinarily necessary to give protection to the female about the age of puberty?—Yes.

34. Should you not think that the labour which is often undergone in a heated atmosphere, and also amidst excitements of a peculiar nature, is calculated to anticipate that period, namely, the period of puberty?—I could not say without more particular experience, whether it would or not; most probably it would.

35. Should you think that the labour which has been described to you, inducing, as it does, considerable languor and a state of inanition, has a great tendency to induce those subject to it to the use of ardent spirits or other stimulants?—Yes, I should suppose it would; whatever produces a feeling of languor would be likely to lead those who experience it to the use of ardent spirits.

B. C. Brodie, Esq.

36. May not that labour be denominated excessive which leaves the body still greatly fatigued after an ordinary night's rest?—Certainly.

37. Is it not peculiarly injurious to continue that degree of labour, when that state of fatigue is constantly felt on commencing a succeeding day's exertion?—Undoubtedly.

38. Should you not think that the human body is in a great state of debility and exhaustion when it is necessary that it should be spurred on to its exertion by repeated punishments and flagellations?—Whenever that is required as a system the individuals employed must be in a state of exhaustion beyond that which a healthy degree of labour could produce.

39. Do you think that, upon any general system, a child under nine years of age ought to be subjected to the labour of a mill or a factory, that labour being such as has been described to you?—No, certainly not; children under nine years of age should be subjected to a very small portion of labour indeed.

40. What labour, in your opinion, speaking generally, is a child of nine or ten years old capable of enduring in the twenty-four hours in succes sive days throughout a year?—I should have thought four or five hours a day was quite sufficient for a child under nine years of age.

41. Are you of opinion that a young person from nine to eighteen, or about the period when the osseous system is arriving at its full development or strength, ought to labour beyond twelve hours a day, including the necessary intervals for taking meals?—I should think twelve too much for all, and indeed I think ten too much for children of ten years of age.

42. Are you of opinion with the late Dr. Baillie, and many other medical men of great eminence, who gave evidence before preceding committees of both houses on the subject of regulating the labour of children and young persons in factories, that ten hours a day, making the day's work to consist of twelve hours, is as long a time as labour ought in ordinary cases to be endured by any individuals?—Yes, I am of the same opinion with Dr. Baillie on that point; repeating that I think ten hours a day too much for children of ten or twelve years of age.

43. At what period, generally speaking, do you conceive a male is capable of undergoing ten hours' labour?—The period must differ in different individuals. I suppose that the changes which end in the complete formation of an adult are not to be considered as terminated until growth has terminated; probably, in general, growth is at an end at eighteen or nineteen years of age.

44. Do you conceive that while the body is rapidly attaining to its stature excessive labour can be endured with impunity?—Whenever growth is going on rapidly, labour ought to be very moderate; and when growth is going on slowly, a greater quantity of labour may be borne without bad effects. I am of opinion that boys of sixteen cannot generally go through the same labour as men can.

45. In allusion to the morals in many of the mills and factories where this degree of labour is habitually undergone, it is stated by many of the witnesses that they are in a most deplorable state; but that, notwithstanding this, it is said that not more illegitimate children are produced by females under those circumstances than by those engaged in agricultural and other pursuits; do you conceive that labour, interfering with the health of the females, diminishes the power of fecundity?—I should suppose that females under those circumstances would be less likely to bear children than females under other circumstances; and that, from this circumstance alone, the number of illegitimate children would be less.

46. So that that circumstance would not decide the point as to the comparative morality of the different classes?—No, I should suppose not.

47. In respect to the moral and mental improvement of the persons thus

employed, do you conceive that when they have been over-laboured during the week, and a considerable degree of fatigue is still felt on the Sunday, the institution of Sunday-schools can be rendered so beneficial as it might be were the hours of labour mitigated?—Certainly not. B. C. Brodie, Esq.

48. Do you conceive it possible, or at all events probable, that children subject to the labour you have heard described, can acquire any instruction whatever, either in the evenings of working-days or on the Sabbath?—I conceive children, under the circumstances which have been described, can obtain very little mental improvement in any way.

49. It is stated that the children and young persons feel a very great disinclination to attend Sunday-schools, and that they much wish to rest and to remain in bed all the principal part of Sunday; do you not conceive that the Sunday-schools, considered in connexion with that effect on the constitution of individuals so employed, add to the physical sufferings of those who have to resort to them?—Certainly, as far as their bodily health is concerned, the addition of the Sunday-school studies to the other labour must be an evil.

50. Is not it found necessary to put some limits, with regard to the preservation of health, to the study and mental exertions of the children and young persons in seminaries and academies?—Yes, in some girls' schools especially; the girls are engaged in study many more hours a day than is consistent with their health, and those girls frequently suffer in their bodies for the too great cultivation of their minds; they are especially liable to distortions of the spine.

51. So that those systems of education which require constant application for an undue length of time, leaving insufficient opportunities for relaxation and exercise, are prejudicial to the frame?—Yes.

52. Have you any doubt, then, that similar consequences would result from confinement carried on to an equal extent, and with an attention almost as constantly demanded in the case of children and young persons engaged in mills and factories?—No; I can have no doubt upon that subject.

53. It is wished by the supporters of the measure now under consideration of Parliament, namely, the remission of the hours of labour, which they seek in behalf of these children and young persons, that a part of the time should be appropriated for the purposes of education, especially in evening-schools; supposing, then, that the children would have to go from the factory to the evening-school, and be thus confined altogether, perhaps for as long a period as they now ordinarily are, would not the change in the occupation and the pursuit be beneficial to them in regard to their health?—I think, that if a child is to be occupied a certain number of hours in the day, he would suffer less if some of those hours were occupied by school studies, than if the whole of those hours were occupied in the factory.

Sir ANTHONY CARLISLE, F.R.S., called in, and examined, 31st July, 1832.

1. What is your profession?—I am a surgeon. Sir A. Carlisle.

2. Are you, or have you been, an officer in any of the great medical institutions of this metropolis?—I have been forty years principal surgeon in the Westminster Hospital.

3. Have you been a medical or surgical lecturer in any branch of your profession?—I have been a public lecturer in many situations.

4. Will you please to state them to this committee?—I have given lectures, especially upon surgery, for twelve years as a public teacher; I have given lectures upon comparative anatomy for eight or ten years; I have been the professor of anatomy to the Royal Academy for sixteen

Sir A. Carlisle. years; I have been professor of anatomy in the Royal College of Surgeons, and have also given the lectures appointed by the public trustees on the behalf of Mr. Hunter's Museum, and other public lectures endowed by the College of Surgeons.

5. Have you held any honorary appointment in the Royal College of Surgeons?—I have been the president of the college, and served the office of vice-president during three years.

6. Your mind consequently has been directed, for a considerable number of years past, to the causes of health, and those that originate disorders amongst mankind, and also, perhaps, in the animal creation, as you have alluded to the science of comparative anatomy?—My public duties and pursuits have called my attention, during the whole of my life, to those subjects, for I have always regarded preventive medicine, or the means of preserving and maintaining health, as the most important branch of the healing art, perhaps preferably so to the curing of diseases, the former being best understood.

7. Is it not a generally received maxim among your profession, that proper exercise, with due intermissions for refreshment and rest, and especially having a sufficient time for sleep, are, in ordinary cases, necessary to the preservation of health?—From my experience, from my studies and observations, I am satisfied, that man cannot be reared in health, nor maintained in health, without due attention to all those points, especially to exercise in the open air, to diet, to clothing, to recreations, and to sleep.

8. Are those several observances still more essential, and, least of all, to be dispensed with in reference to children and young persons?—Unquestionably so; because during the growth and formation of a young creature, its liability to deviate from the natural standard is much greater than in the adult subject, where the construction of the body is completed. I beg leave to say, that it is more liable to err in the development of its fabric while growing than after the growth has been finished. I think that fact is unquestionable with regard to men and animals. I have visited all the menageries in England, and some in other places, and I have attended to the history of training athletic persons, as a part of my duty at the Royal Academy, and, during the whole of my inquiries, I am quite satisfied, that unless the young creature is duly exercised, and not over-laboured, duly fed, and properly treated with regard to what have been formerly called the non-naturals, which are, in fact, the needful regulations of living creatures, unless all those things are attended to, there must be some deviation from health, and more so in young than in grown-up persons. If you will permit me, I will mention an instance. The great mass of the population of London have been brought up in the country. I would venture to say that half the shopkeepers and half the artisans in London have been born and bred in the country, and came to London at or after maturity to seek their fortunes, to obtain their livelihood, or to assist some of their relatives or connexions; such persons will go on for a considerable time without suffering much from the injurious effects of a confined city, or from an unhealthy occupation; but take an infant born and bred in London, and subject it to all the same vicissitudes, and that child will invariably become unhealthy.

9. Is it not an equally received opinion with medical authorities, that exercise or labour so long continued as to produce great fatigue of mind or body, without affording due intermissions for meals, recreation, and sleep, is inconsistent, generally speaking, with the maintenance of health?—I think every one of the points of that question may be answered in the affirmative. I can, from my own experience and knowledge, affirm that it is so.

10. Your affirmation in that respect is founded upon the principles of your profession, as well as upon your personal experience?—Certainly; from physiological inductions taken from the whole animal creation, man

being part of it, and subject to the same laws, or nearly the same laws, as the rest. Sir A. Carlisle.

11. Appealing to your knowledge of comparative anatomy, the principles to which you have alluded are universal throughout animal life?—Yes; hence all domesticated creatures that are kept in close confinement; all animals that are kept for labour, and that are worked at too early an age, or too severely, become deteriorated in form and vigour, and are more or less injured, so as to unfit them for the performance of their ordinary and habitual labours. It is, for instance, obviously so with the horse.

12. As, in consequence of the situations you have held, and the duties you have undertaken, you have, of course, directed your particular attention to the habits and capabilities of mankind, may I ask you whether, in your opinion, twelve hours a day labour, with due intermissions for meals and refreshment, being the customary time of human labour in the ordinary pursuits of life, and established as it should appear by the assent of all ages, and of all countries, is not, in your opinion, as long as can, for a continuance, be endured with impunity by the average of human beings?—I am quite satisfied, from my own experience, that is, from forty years of observation and practice in my profession, that vigorous health, and the ordinary duration of life, cannot be generally maintained under the circumstances of twelve hours' labour, day by day; it is incompatible with health, it is not to be done with impunity; that is my humble opinion.

13. So that you think that the general custom of society, which still further abridges the duration of labour in agricultural and mechanical pursuits during half the year, namely, in the winter months, is dictated by the nature and condition of human beings, as well as by the common usages of society?—I think it arises from the law of animal nature; and I suspect we never offend against those laws without being punished. I mean to say that in the winter season the whole animal creation requires greater rest than in the summer season, their physical capacities demand more repose in winter than in summer; and if, in an artificial state, men or animals are compelled to labour an equal number of hours in the winter season, as in the summer season, it would, in my opinion, be decidedly injurious to their health. If you will permit me, I will give a familiar illustration: the whole creation, man, animals, birds, fishes, and even insects, rise, if they be day creatures, with the rising of the sun, and go to rest with the setting sun, winter and summer. At Midsummer, the sun may rise in this climate before four o'clock, and it may set a little after eight; in the depth of winter it may rise at about eight, and set soon after four; still the creatures, whether quadrupeds, birds, or fishes, except the nocturnal ones, follow that rule; and even the nocturnal creatures do not wander all night, they only go out at twilight and early in the morning. During the stillness of midnight the whole creation is at rest; that is my own experience; even fishes, which I have included, obey this rule.

14. So that you conclude, labour undergone in the night is more pernicious to those enduring it than if it had to be performed in the day?—It is contrary to nature, independently of its being performed by artificial light, which necessarily deteriorates the atmosphere which the person breathes.

15. Alluding to the artificial lights by which those who labour long into the night, or during the whole of it, have to work, do you not conceive there may be some truth in what the operatives allege, namely, that their sight also is injured by having to work so long by artificial, and especially by gas-lights?—I have not any experience upon that subject, but I will take leave to say, that the diseases of the eye are so multifarious and so differing, that if even I had had experience, I should have hesitated in answering it. There are so many causes of dimness of sight and blindness, arising from

Sir A Carlisle. injuries to the organ of vision, from within and from without, and so many parts of its structure susceptible of injury, that I think the question cannot be answered generally.

16. Are not intervals or cessations from labour, sufficiently long for taking necessary meals, essential to the health of the labouring classes?—Unquestionably, labour could not be continued for ten hours with impunity without intervals. I would take leave to say that the brain is the great organ which regulates animal life; the cessation of the working of that organ is sleep; the working of that organ consists of two parts, the working of the attention of the mind, and the working of the muscles of the body, either of which equally exhaust. No man can continue the one for a great duration of time, without the other being impaired; for instance, no man can give the attention of his mind closely to any mental effort for a considerable nnmber of hours together, without finding, if he wants to perform some bodily action, that his muscular power is very much abated, so that he often cannot do it. On the contrary, no man can use a great deal of severe muscular exercise, particularly if it is constrained, and against his will, and which we term labour, without having a corresponding diminution of the power of giving the attention of his mind to any intellectual exertion.

17. I perceive you make a distinction between exercise and labour; will you please to explain that distinction?—The distinction which the best physiologists have made is this: exercise is that exertion of the muscles which is done with the free will or consent of the mind, or which is agreeable to it. Labour is that constrained exertion of the muscles which is done against the will and inclinations of the mind, and from necessity. I might appeal, for instance, to the poacher or the smuggler; the one may pursue game, and the other ride over a great tract of country, and carry extraordinary burthens; but performing those exertions with the will and consent of the mind, they will undergo exertions which they could not habitually sustain if they were compulsory or for hire; I would say, that in these voluntary efforts, and when the brain is in a state of hilarity, it gives out more of muscular volition than it does when labour is constrained.

18. To come, then, to the immediate subject of our inquiry: the employment of the factory is, according to your definition, labour; and assuming that it almost always exceeds the term you have mentioned as the extremity of the labour that can be endured without injury as an ordinary day's work, and that it generally exceeds thirteen hours a day, and is sometimes lengthened to fourteen, fifteen, or even sixteen or eighteen hours a day; do you entertain a doubt that such a system must, in a great many instances, prove ultimately prejudicial to the human constitution?—I think that such constrained labours must necessarily exhaust the constitution of the individual; and if they should be commenced in early life, they will so far impair the creature before it attains to manhood, that a manly frame, strength of body, and vigour of mind, will never be truly and properly developed; and the Roman saying, "Of a sound mind in a sound body," can never be realized in such a system.

19. Assuming, then, that the individual subject to this excessive degree of labour would be considerably deteriorated, physically and intellectually, do not you conceive that the offspring of such an individual, if subject to the same labour, under the same circumstances, would continue to deteriorate, so that ultimately the most prejudicial effect upon the physical and moral condition of future generations would be the inevitable consequence?—I do; but I think the subject admits of a delusive view in one respect. Children brought up from early life in warm rooms may enjoy an apparent degree of health until almost the age of maturity, but they never obtain vigorous health; for although they may not apparently perish in much

larger numbers, compared with those who are less protected against the inclemencies of the cold and wet weather, the effects of bad food, and so forth, still I would venture to add, that even supposing the mortality among the former were not so great as in the latter class, still they would be unfit to carry on a succeeding generation of healthy and vigorous human beings; they would be either nipped in the bud about the time of puberty, or they would beget weakly inferior children; for there is nothing more hereditary than family tendencies, particularly tendencies engendered by such habits as are hurtful to the first formations of animal structures. The same happens, again, in the animal creation: if you domesticate rabbits and keep them in close houses for a series of generations, they deviate into morbid varieties, and eventually cease to breed. Many instances of a similar kind occur in public menageries. If wild animals breed at all in that state of confinement, the next generation is a spurious variety; but, generally speaking, they cease to breed, the procreative power being one of the first animal faculties to cease with every diminution of animal health and vigour. Sir A. Carlisle.

20. Supposing it not to cease entirely, arguing from your views on that branch of physiology, you would naturally expect the human race to deteriorate if exposed to the same pernicious influences and subject to the excess of labour to which we have been adverting?—Unquestionably; I have not any hesitation upon that point; every succeeding generation would become worse until they stopped altogether. I would offer to the committee a matter in my own experience: the city of London would not maintain its population for fifty years if it was not refreshed by accessions from the country. I have had the curiosity to see if I could find a person of the fourth generation, by both the father and mother's side, in the city of London, and I have never been able to find such a person.

21. It is in evidence before this committee, that the term of labour at one stretch, if I may so express myself, is sometimes continued for twenty, thirty, and even forty hours, with but slight intermissions; there can be but little doubt on your mind, I presume, that the exceptions must be very rare indeed when, in such a state of things, irreparable mischief is not speedily done to the constitution?—Such a procedure must be, doubtless, soon destructive to the intellectual faculties and the bodily functions of the sufferers.

22. Should you not almost think it incredible, except you had full proof of the fact, that bodily labour could be pursued for any such length of time by children and young persons?—It must be very badly done; it must be done with many intervals of rest, taken by stealth. With submission to the committee, I will give you an example: there is no exertion of the muscles of our bodies that has not alternations of rest; those motions which seem to us to be continuous have their rest; the heart beats perhaps seventy times in a minute; but there are intervals between each of those beats which are perfect rests, perfect cessations, and one half of the heart takes its turn after the rest of the other half; so that muscular exertion is never continuous without the power of the brain, of the nature of which we know very little, being exhausted by it. I will give you two illustrations: a person will die from an extensive scald or a burn, say half the body shall be scalded by a person falling up to the middle into a boiling cauldron, or tumble in sideways, so that one half shall be scalded and the other not; the skin comes off; excessive pain ensues; it continues a certain number of hours; the person then becomes exhausted, and falls asleep; but it is the sleep of death; he wakes no more—it is the fatal symptom. Again, a man shall have a disease of the muscles, in which there is little or no respite of extreme muscular action. In a disease we call tetanus, or lock-jaw, the person dies generally within three days, apparently of muscular exertion, without a respite; simi-

Sir A. Carlisle. lar, also, is the case of canine madness. Now in the case of the burnt or scalded person, of tetanus, or of hydrophobia, the sufferers die from exhaustion of the sensorial powers, under the form of pain, in the former; and in the others, from exhaustion of the muscular power; for there is no lesion of the structure of the body; no loss of substance; no vital organ damaged; but they nevertheless die.

23. Is not that employment which has been sometimes denominated "light and easy," but being one of uniform and tedious, though not very strenuous exertion, but still inducing great fatigue of mind as well as body, more exhausting and injurious to the constitution than changeable locomotive exercise, when endured only for a moderate length of time, with due intermissions?—I am convinced of that; because the sensorial powers being exhausted by an uninterrupted succession of muscular actions, must necessarily weaken or exhaust the powers of life; and therefore such exertions cannot be followed without a violation of the laws of animal nature.

24. So that you would not consider it a sufficient apology for this exceedingly long-continued labour, if it were alleged truly, that, abstractedly considered, it demanded very little muscular exertion?—No; I would say, in such instances, it would still be a violation of the powers of nature.

25. In allusion to the position of the body which it is, generally speaking, necessary to maintain in mills and factories, namely, a standing and stretching position, is not that often more fatiguing to a person having to maintain it than moderate exercise of a more varied character?—The difference between standing and locomotion, with regard to the effect upon the living frame, will be this, that during the act of standing a great number of muscles are always in violent action to poise the frame of the body, for the trunk of the body rests upon two ball-and-socket joints, which are continually liable to slip backward, forward, or sideways; and unless the voluntary power and direction of the mind be unceasingly, although unconsciously, applied to the balancing of the body upon those two round joints, as well as upon the governing muscles of the shafts of the legs, the joints of the knees and ankles, the person must fall; for it requires, although insensibly, as I have said, to the person, continued direction of the mind over those large muscles, namely, all the muscles of the hips and spine, of the thighs and legs.

26. Involving, then, considerable exertion of the muscles in question?—The muscles must be incessantly exerted, or the skeleton would tumble down; it can only be maintained by the constantly and duly directed exertion of those muscles.

27. You have spoken of the needful alternations of muscular exertions so as to produce intervals of rest to each other, is not that one reason why the standing position is found to induce considerable fatigue?—In standing there is no respite for those muscles which maintain the erect figure; but if he walk, there is a respite for a great many of them.

28. Would not the exertion in question, endured in the position explained, be still more injurious if it had to be undergone in an impure air, impregnated with the flues and dust given out from the materials manufactured?—Certainly; if you will permit me, I would observe, that polluted air in a manufactory may be a contamination of the atmosphere from various sources; many persons breathing together pollute the air; the exhalations from many persons confined in the same space, and particularly if it is not ventilated; the burning of lights of any kind destroys that part of the air which is essential to respiration, the vital part of it; and any extraneous bodies floating in the air undoubtedly pass into the lungs, more or less. Many artificers' operations are injurious, because of the dust getting into the lungs. I will give you an instance, which I believe is unquestioned, but it

has never been sufficiently made known: the grinding of needle points is one of the most destructive employments in that sort of manufactory; in needle-pointing the workman stands at a wheel with a number of steel pins cut into lengths, and he rolls them between his fingers and thumb upon the face of the wheel; the vapour from the grinding of the steel to make the points gets into his lungs, and it produces what we call hepatized lungs—they become like liver; it stuffs them up by slow inflammation; and the life of the man seldom lasts above three years at that employment. He is tempted by high wages. Sir A. Carlisle.

29. Now the dust and steam in the preparing department of the flax-mill are sometimes so excessive as to render it difficult for persons to see each other distinctly at a very few yards distance; would not an atmosphere of that description, particularly alluding to the material of the dust, be highly prejudicial to the lungs and to the pulmonary system?—I should think so, but I have no experience upon that point.

30. Is not labour so circumstanced and continued, if undergone in an atmosphere heated to a high temperature, say from 70° to 80° and upwards, especially connected with the circumstance of the persons so employed being poorly clad, and having to pass suddenly into the open air at all seasons, after their long day's labour, often in a state of perspiration, likely to bring on serious disorders?—Yes; I can speak to that, because those sudden alternations of heat and cold, the going out of a very hot room into a damp cold air repeatedly, will inevitably produce slight inflammations of the lungs; those slight inflammations reiterated produce the groundwork of pulmonary consumptions, for all pulmonary consumptions are repetitions of little colds. The structure of the lungs, from this cause, becomes completely altered by those slight inflammations which disorganize the vascular tissue; when that has taken place, medicine is of very little use; but a sure mode of producing this malady is frequent alternations from hot rooms, with thin clothing, to cold damp air.

31. Rheumatism is a very general complaint amongst even the children and young persons so employed; should you be able to trace that disorder to the alternations to which you have been alluding?—I have a view of the causes of rheumatism of my own, and somewhat different from that, and also of gout, which I hold to be kindred complaints, arising principally from crudities of the stomach and bowels. A person may have a fit of rheumatism from a cold, but the raw materials, if I may so speak, were prepared in the stomach and bowels by improper aliment.

32. In allusion to an atmosphere thus impure and artificially heated, would you not say that labour endured therein is more exhausting to the strength than similar labour in the open air, though at the same temperature, but being free, of course, from the pernicious admixture alluded to, and being constantly freshened by the influences of currents of wind and other atmospheric influences?—I have no hesitation in saying that it would be better for a human being to labour at 90° out of doors, than at 85° in a close room. I think that a person often becomes very much fatigued from going to crowded rooms, such as play-houses and assembly-rooms, where a great number of people are confined together without a sufficient change of air, and that they consequently lose bodily vigour.

33. Is not a certain quantity of oxygen in the air necessary to respiration, and does it not consequently enable a person to endure fatigue?—Yes, certainly, if the other element of the air be also pure.

34. Do you not think that children and young persons, not being free agents, and having to labour in the manner alluded to, demand legislative protection, in point of humanity, equally with the slaves in the Crown Colonies of this country?—I am no judge of the slaves in the Crown

Sir A. Carlisle. Colonies; but I would say, in answer to the home question, they demand legislative protection for their own sakes, and for the sake of future generations of English labourers, because every succeeding generation will be progressively deteriorated, if you do not stop these sins against nature and humanity. Nature has been very wise in punishing all the offences we commit against her in our own persons.

35. Various affecting results have been attributed to this system of excessive and lengthened labour by different witnesses that have come before this committee, and by medical gentlemen who have witnessed the system and its effects; would you please to state to this committee what you conceive would be the probable results, in many instances, of such continuous and unwholesome labour?—I have already stated that, during the development of the body from the infantile state, the earlier it is opposed by labour and put into unwholesome circumstances, whether an impure atmosphere or an atmosphere of undue temperature, the more certainly will that child be injured in its general health, although it may not be so apparently for a few years, but ultimately its constitution will be utterly ruined. Comparing families brought up in rural situations with children brought up in factories, from seven or eight years of age to fifteen or sixteen, the mortality may be, for aught I know, less in the warm room of the manufactory for that period, but it does not at all alter the question of the greater mortality afterwards ensuing; that is the grand ethical question, and I am quite sure that the foundations for debility, decrepitude, and premature death, are to be found in these unnatural habits.

36. What are the effects which are usually denominated surgical, that you imagine might result from this system of labour, in addition to those to which you have already alluded?—In girls, affections of the spine would often occur, and a softening of the bones, which is vulgarly called the rickets. A softening of the bones, from deficient proportion of the earthy substance, may arise from bad air, from exhausting exertions, and from improper diet. This also takes place in many crowded boarding-schools in great cities: the girls have spine affections, or rickety curvatures of the spine, from being treated pretty much as they are at the manufactories, overworked and confined.

37. It has been stated by witnesses before this committee, that a very unusual number of cases of deformity are found among those who labour in factories, that have come on as late as at ten, twelve, fourteen, or sixteen years of age; should you have any hesitation in attributing those deformities, many of which have been seen by the committee, to the cause of factory labour?—I think it is more than probable, I would say certain, that the causes that have been related as to the confinement and labour of children in factories, are very likely to produce deformity, and more especially scrofulous or rickety deformities.

38. Should you suppose that long standing might have the effect of producing disease and ulcerated legs, especially in females, which it is alleged is a common result of the labour in question?—The circulation of the blood out of the lower limbs upwards into the body by the veins does not take place so freely in an erect posture if long continued, as if there were occasional rests, because the ascent of the blood is against gravitation; and this has the effect of distending the veins of the legs, and occasioning varixes.

39. Would that, if long continued, occasion ulcerated legs?—Yes; and ulcerated legs which would be very difficult to cure, arising from varicose veins: where the circulation of the blood is spoilt, it would be almost idle to talk of curing a diseased leg of a person in that disorganized state, and continuing that labour.

Sir A. Carlisle.

40. It has been asserted in this committee, by those who have made extensive and very careful examinations in Sunday-schools with reference to this subject, that the height of children of the same ages, comparing those that work in factories with those that are working in other employments, is considerably less; should you be prepared to say that a stunted growth must inevitably ensue from this undue labour so long continued?—I have no experience on these points; but I can tell the committee, that in this and in other countries agricultural persons are upon an average much taller than the persons born and bred in cities; the persons born and bred in cities throughout Europe are notoriously of less stature than those born in country places.

41. A considerable majority of those employed in mills and factories are females; do you conceive that the female sex is as well fitted to endure labour of the description alluded to as the male?—Oh, no, certainly not; they are by nature less muscular, and I would say there is less sensorial power about them, and less animal vigour.

42. Is it not especially necessary to give protection from excessive labour to females when approaching the age of puberty?—Quite important; if they are afterwards to become mothers it is quite essential.

43. Has not this description of labour, pursued in a heated atmosphere, connected as it is with other exciting causes, a tendency to anticipate that period, namely, puberty?—Yes; it has a tendency to give a precocity to the passions, certainly.

44. Do you not think that the labour and exhaustion of the mill is very likely to induce a habit of tippling?—A person reduced by muscular exhaustion, or depressed by over-exertion of mind, is induced to seek relief from an artificial stimulus, and, after finding its exhilarating effects once or twice, has recourse to it again; thus repeating the delusive remedy for muscular or intellectual exhaustion, he becomes an habitual drunkard; he finds that inebriation has that effect upon him, and he flies to it; half the drunkards you meet with, if you ask them how they became drunkards, will tell you that they begun the practice when they were fatigued in body or depressed in mind, or in a state of anxiety and distress; and the same thing may be said of persons in a better class, they begin the practice when labouring under some exhaustion of body or trouble of mind.

45. Do you think that a child under nine years of age ought to be doomed to habitual labour in the factory?—My own opinion is, as a matter of feeling, that to do so is to condemn and treat the child as a criminal; it is a punishment which inflicts upon it the ruin of its bodily and moral health, and renders it an inefficient member of the community, both as to itself and its progeny; it is, to my mind, an offence against nature, which, alas, is visited upon the innocent creature instead of its oppressor, by the loss of its health, or the premature destruction of its race.

46. Are you of opinion that a young person from nine to eighteen, or to about the period at which the osseous system becomes fully developed and perfected, ought to labour beyond the time in question, namely, twelve hours a day, including two hours for meals?—I think it quite incompatible with health, and I would venture to add, with moral propriety; for if the creature is to be a moral agent, and not a machine or mere animal, you must have some consideration for his moral character, as well as for his health.

47. So that you would consider it a great benefit to those individuals exposed to these tasks, with reference to their powers and capacities, if their labour were in some degree remitted, so as to give them an opportunity of obtaining some moral and mental culture?—Certainly, I think so; but it should not be done on the day of rest from labour, but it should be done at hours deducted from its hours of daily labour.

Sir A. Carlisle. 48. Then do you not think that the circumstance of over-labouring children and young persons so much as to render it actually impossible for them to receive any instruction on the week-days, and obliging them to have recourse to the Sunday as a day to be devoted to the obtaining the rudiments of education, is an additional hardship imposed on persons so circumstanced?—I do think so; and I would venture to say it is only changing the week-day labour of the body for the Sunday labour of the mind; they can take no pleasure in it, and gain little moral profit.

49. Considering that the Sunday-school is generally crowded, and that it is devoted to purposes requiring constant mental attention, do you not conceive that, when added to the labours of the week, the confinement of that day must produce an additional injury to health?—It is additional labour thrown upon them; I should say, let them go to church, let the church be well ventilated, and there, from a good scholar and divine, let them derive instruction, moral and religious. Let the little worn-out creatures have some little time for repose, for domestic enjoyment, and instruction, and for the exercise of the domestic and kindred affections. One unvarying round of attention is intolerable, and fit to be endured by neither man nor beast.

50. Do you think that the natural affections of uneducated persons in the lower ranks of life are less sincere and ardent than those of the affluent and educated orders of society?—I am of opinion that the instinctive and natural affections of the industrious classes of society are more pure, more sincere, and more active than among the educated classes; that is my humble opinion, and it is the result of long and varied experience; I have witnessed sacrifices on the part of people in the lowest condition of life, which I never saw among people educated artificially from the commencement of life; for the habits of simulation and dissimulation in which we are brought up, as a part of our manners, are totally different from the instinctive feelings and natural affections cherished among uneducated children and parents; the yearnings of those people after their progeny, and their filial affections, disparage the heartless manners and cold morals which too often prevail in fashionable life.

51. So that you conceive the over-labouring of those children must distress the parents, who have no sort of means or opportunity of resisting a system which operates upon their minds as a very great and grievous hardship and affliction?—I think the system is destructive of the natural and instinctive, parental and filial affections, that it too often weakens the natural ties and dissolves the most important duties of life.

52. It is sometimes alleged, in answer to these assertions and facts, which go to prove that this system of excessive labour is also an immoral system, that the number of the illegitimate children of females employed in mills and factories is fewer than among a similar number of females in other pursuits, should you consider that that was conclusive as to their superior state of morality?—Quite the contrary; premature sexual intercourse, or promiscuous sexual intercourse with very young females, almost invariably prevents their being prolific; this is notorious with regard to those unfortunate women who pass their lives in prostitution; they generally begin very young, and they often pass to the middle period of life without even once being pregnant; but it is a very curious fact with respect to those women who have been sent to Botany Bay, where they have been kept from sexual intercourse during a voyage extending to six or eight months, that they often become prolific afterwards, though they may have been for many years the lowest and most abandoned prostitutes.

53. On the whole, it is your opinion, then, that the limitation of the hours of labour would be most beneficial in preserving the health of the

children and young persons employed in mills and factories, and that it would be also most conducive to the present and future well-being of society?—I have no hesitation in saying so, not at all; I assert, it is beyond all doubt. Sir A. Carlisle.

54. But at the age of eighteen, taking into consideration the elasticity of the fibre in youth, do you not consider that a person of that age is as able to continue working as long as a person of more advanced years, provided the work done is not very difficult, but rather continuous than laborious?—I think that a person of eighteen years of age cannot endure so much bodily fatigue as a person of twenty-four or twenty-five: it may be for a spurt, as the horse-racers say, but I do not think they could do it continuously.

55. Not even light work?—No.

56. You have already stated you have been long a medical officer of one of the hospitals in this metropolis; have you been able to make any comparisons between the diseases and the mortality of that hospital compared with the diseases and mortality in the hospitals in the manufacturing or other districts of the country, so as to enable you to come to definite and practical conclusions upon this important subject?—I am sorry to say that there is not any established intercourse between the medical officers of the public hospitals of the kingdom, which I think there ought to be, for the advancement of the profession and for the public good. I beg leave to say, that even with respect to the public hospitals of this metropolis, the practice of the medical officers is unknown to each other; that there are no records published of the results of the practice of the public hospitals of this town, or of any of the public hospitals in the kingdom; and it is to be deeply lamented that there are no such authenticated records published. Many discoveries of the greatest value in the healing art have been elicited at county and provincial hospitals, and the whole profession have been generally unacquainted with them until eight or ten years afterwards, when the improver or discoverer thinks it convenient to publish them; so that we have no communion with each other for the advancement of the profession, or for comparing the practice of the several hospitals of the kingdom with each other. We know nothing of the respective mortality or of the results of the more dangerous surgical operations in the different hospitals in the kingdom; we know nothing of the different modes of treatment which may have been devised by skilful and experienced men; so that although there is hardly a hospital in the kingdom where there is not some mode of treating a disease, or symptom of a disease, or method of conducting a surgical operation, or some remedy either external or internal, which experience has shown to be superior to the common routine of practice, yet those things are unknown to the community of the profession, unknown except to a few. I have, in the capacity of president of the College of Surgeons, frequently sought to have tabulated registers sent to the college from all the different hospitals in the kingdom, to be by us published half-yearly; comprising authentic reports of all the operations and their results, and all the special modes of treatment of particular diseases or symptoms of diseases, and the applications made use of; and yet simple and manifestly useful as that will appear to every unprofessional man, it has not yet been carried into effect, although ten years ago I besought the profession to do so; the fact is, that great bodies move slowly in any but their established course, and we are, like all human beings who are concerned in a lucrative profession, so invaded by mutual jealousies, that what one proposes the other will not agree to; but in my humble opinion the profession ought to be compelled to do this most obvious public duty; and the governors and trustees of public hospitals in the kingdom ought to insist upon it. Now I submit that one great benefit to the public from such a measure

Sir A. Carlisle. would have been this, that the labours of this committee would have been very much abridged, for such tabulated reports would have shown you, in the manufacturing districts, in the county hospitals, and the London hospitals, all the results of our experience from one half year's end to another; and in order to be certain that there should be no undue representation to enhance the importance of an individual practitioner, or anything that might disparage him from the publication of unsuccessful enterprizes, I would not have the names of any of the practitioners attached to the cases, whether snccessful or otherwise, so that you would be sure to get at the whole truth: the worst of ordinary medical reports is, that instead of getting the truth, and the whole truth, you only get a part of it; you get only the successful cases, the unsuccessful cases being too often suppressed. Now, in the reports to be made by public hospitals, without annexing the names of the practitioners, you would get the whole truth, and that truth tabulated and printed, and sent round the country, the colonies, and all over Europe, at the expense of the College of Surgeons, whose resources would be competent to this important task.

57. Under those circumstances, which this committee regrets, you have not the power of presenting it with any comparative result as to the nature and the fatality of the diseases in the manufacturing districts compared with the metropolitan district?—I also lament that we do not possess such invaluable documents; for those comparative reports would have been upon this occasion, and upon all future similar occasions, so valuable and important.

JOHN ELLIOTTSON, Esq., M.D., F.R.S., called in, and examined, 2nd August, 1832.

J. Elliottson, Esq.

1. What is your profession?—I am a physician.

2. Are you a medical officer in any of the institutions of the metropolis?—Physician to St. Thomas's Hospital.

3. Have you been a medical teacher in any institutions?—In St. Thomas's Hospital; and I am at present also in the London University.

4. You have consequently directed your studies to every branch of your profession?—Yes, I have.

5. Is it not held by your profession that a moderate degree of exercise, with sufficient intervals for rest and recreation, is essential to the preservation of health in ordinary cases?—Certainly.

6. Do you not think that those advantages are more particularly necessary as regards children and young persons?—Certainly.

7. A longer period of rest and a longer period of intermission than would be required by healthy adults?—Certainly.

8. Do you conceive that excessive labour, without due intermissions, either for meals or sleep, must, generally speaking, be prejudicial to health?—Yes, always.

9. And more especially as it respects children and young persons?—Certainly.

10. Are you inclined to think that the usual term of daily labour, as established in agricultural and handicraft employments, namely, twelve hours a day, including the usual time for meals, is, on the whole, about as much as can be endured even by adults with impunity?—I should say certainly, the utmost for a constancy.

11. Do you think that the cessations of labour during the time of taking meals are essentially necessary to the health of the labouring classes?—Certainly.

J. Elliottson, Esq.

12. It appears, in evidence, that the labour endured in mills and factories greatly exceeds the term referred to, and extends, in many instances, to fourteen, sixteen, and even eighteen hours a day; should you have any hesitation in pronouncing that to be a degree of labour that cannot, generally speaking, be endured with impunity by the human frame?—No hesitation at all.

13. Then when that term is still further extended to thirty or forty hours, do you conceive that the effects must be eminently prejudicial?—Certainly.

14. When undergone without any proper intervals for rest or refreshment, do you conceive that such a system of labour as that must be destructive of health and tend to the diminution of human life?—Yes, certainly.

15. Is not labour frequently pursued in the night more prejudicial to the health than if it was followed in the day-time?—I have no experience upon that point; but I should presume it must be so from its being unnatural to labour when nature has produced darkness.

16. Have you not published a work upon physiology?—Yes, I have published an edition of a work, with considerable notes, in which this subject is mentioned.

17. In which you have illustrated that point?—Yes; it is ascertained by a very celebrated physiologist in Paris, an Englishman, Dr. Edwards, that if the light is excluded from tadpoles they continue to grow, but never become frogs.

18. So that you agree that light is a very powerful agent in the development of the capacities of the human body?—Yes; it is proved so by that animal; and since, Baron Humboldt, the traveller, found not a single instance of deformity among the millions of Caribs, Mexicans, Peruvians, and other savage individuals; this is ascribed by Dr. Edwards to their being continually exposed to the light; and he ascribes much of the sickliness of imprisoned persons, and scrofulous children living in close streets, to the want of light.

19. Supposing that the labour pursued during the night has the tendency to deprive the individual of sufficient rest in the day-time, lured, as he is, too often to pursue his recreations instead of taking necessary rest, that would be an additional reason for determining that night-labour is more prejudicial than daily toil?—Certainly.

20. Should you not conceive that labour, sometimes denominated "light and easy," considered in itself, and apparently demanding but little muscular exertion, but continued for so great a length of time as to produce much fatigue of mind and body, and ultimately great exhaustion, is more injurious than a still greater exertion endured for a less length of time, and with longer intermissions?—Certainly.

21. Would not that labour be still more prejudicial if it had to be pursued in an atmosphere polluted by dust and other effluvia from the machinery, and from the raw materials used in the work?—Yes, undoubtedly.

22. And most of all the labour, for the length of time mentioned, would be exhausting and pernicious if it had to be undergone in a temperature far beyond the natural temperature of this country, and rising to 80° and upwards? —Certainly.

23. Do not the circumstance of persons labouring in that atmosphere, and becoming exhausted in it, and often in a state of great perspiration, having, on leaving their work, to plunge into the open air, especially at those times of the year when, in this country, it is at a very low temperature, and being very insufficiently clothed, as many of them are, greatly aggravate the evils of such a state of confinement?—I should think so.

24. Do you not think that an atmosphere like that just mentioned, and polluted as described, is more pernicious to health than an atmosphere of

J. Elliottson, Esq.

equal temperature, but the heat of which proceeds from natural causes, and is freshened by various natural influences and changes?—Certainly.

25. Then, to refer to your observations already made to the main object of this Parliamentary inquiry, namely, that of children and young persons, would not labour, to the length described, and under the circumstances stated, be more injurious to young persons and children than to those who have arrived at their full strength and vigour?—Beyond a doubt it would.

26. You have not had personal experience in mills and factories?—I have not.

27. Nor had patients who have followed that pursuit?—No.

28. But, speaking from the principles of your profession, and the analogies your studies have afforded you, do you not conceive that many distressing diseases might be originated or aggravated by that degree of labour under the circumstances described?—There can be no doubt of it.

29. What do you conceive would be likely to be the result of such a system of labour?—Scrofulous diseases of every description, consumption, and deformities.

30. It is in evidence, that out of about 2,000 cases of individuals having worked at mills and factories, about 200 of them were considerably deformed, and some extremely so; should you consider that a disproportionate degree of deformity?—Yes, immensely so.

31. It has also been asserted before this committee, on very careful examination, that the stature of the children so employed, compared with the height of others differently occupied, is considerably less; would that be one of the results that might fairly be attributed to the factory system?—I should consider so.

32. The muscular power is stated to be less, and the weight to be equally diminished; perhaps that might be attributed to the circumstances alluded to?—No doubt at all of it.

33. Assuming that the degree of labour which has been described to you under the circumstances in which it is endured is correctly stated, should you not be prepared to believe that it might have a pernicious effect upon the duration of human life?—Most certainly.

34. It is known that a considerable majority of the individuals so employed are females; will you have the goodness to state your opinion as to the natural capability of the two sexes to bear labour; do you think that a female is as well calculated and fitted to bear the description of labour alluded to as a male?—I should think not.

35. This labour is ordinarily pursued in a standing position; will you state whether that position of the body does not produce considerable fatigue in the muscular system?—It does.

36. Then that circumstance, combined with the length of labour, would considerably aggravate the distress that would ensue from its endurance?—Yes.

37. Do you think it is necessary to give to females, especially when approaching the age of puberty, protection from the labour which has been alluded to?—I do.

38. Do not the languor and debility induced by excessive fatigue lead necessarily, according to the usual experience and observation of medical men, to the taking of artificial stimulants, and conduce to tippling and drinking?—Certainly.

39. Can you suppose the body to be in a very fit state for renewed exertion, when the fatigue of the preceding day's labour is not sensibly abated when the person having endured it rises on the ensuing morning?—Altogether unfit.

40. And that if towards the termination of the day it is the general practice to extort this labour by continual beating and punishment, and that it could

J. Elliottson, Esq.

not otherwise be obtained from the parties, such labour so extorted must be extremely distressing and prejudicial to the human body?—Yes, to the highest degree.

41. Do you conceive that a child under nine years of age ought to be put to the labour that has been described, for any length of time, in mills and factories?—No.

42. The answers with which you have already favoured the committee may render another question to the same purpose superfluous; but do you consider ten hours a day labour, exclusive of the time necessary for meals and refreshment, making altogether twelve hours, is as much as can be endured with impunity, generally speaking, by children and young persons under the age of eighteen?—Certainly.

43. Perhaps, then, you coincide with many other eminent men of your profession, that that term of labour is about as much as can be safely endured in the ordinary pursuits and occupations of life?—Yes, I certainly think so.

44. Do you think it is possible to put children to labour for the length of time stated, fourteen or sixteen hours, without, in the great plurality of instances, injuring most essentially the health and constitution of such children?—I should say it was impossible.

45. Do you conceive that the mind can be in a fit state for imbibing instruction after the bodily powers are in a great measure exhausted by a long day's labour?—By no means.

46. So that children having laboured to the utmost degree of their strength, and become much fatigued by their exertions, are not in a state of mind to derive sufficient improvement from evening or night-schools?—Certainly not.

47. The children and their parents who have come before this committee have stated that the fatigue of the preceding week's work has extended through the ensuing Sunday; do you conceive that in that state of body, the children are capable of deriving that advantage from Sunday-schools, or institutions of that sort, which they would otherwise be?—Certainly they are not.

48. Is it not a great increase of the physical sufferings of children so employed as not to have an opportunity of a moment's leisure during the week, that their only remaining day of rest should be devoted to another purpose, however laudable in itself, which utterly deprives them of any recreation during a course of years in the open air, or of any relief to their mental and bodily exertions?—Certainly.

49. Such a state of things, in general, you believe to be entirely inconsistent with the development of the mental and bodily faculties, and that it would deteriorate that class of society perpetually subjected to it?—Most materially.

50. You find that in schools and seminaries of learning, children and young persons cannot endure with impunity an unlimited number of hours devoted to their education?—Certainly not.

41. In those cases, hours of refreshment and recreation are found to be essentially necessary to health?—Yes.

41. Do you conceive such a remission of the hours of bodily labour as would give an opportunity of engrafting a system of education upon the factory system would be eminently beneficial, not only to the health, but to the moral and mental improvement of those concerned, and by consequence, a great advantage to the entire community?—Most certainly.

JOHN RICHARD FARRE, Esq., M.D., called in, and examined, 7th August, 1832.

J. R. Farre, Esq., M.D.

1. What is your profession?—That of a physician.

2. Have you had considerable experience in this city in attending to the diseases of the poor, whether in hospitals or in private practice?—The whole period of my medical study and experience in all the branches of the profession, for the purposes of a consulting physician, has extended to forty-two years.

3. Do you not regard it as an established principle in your profession, that moderate exercise, with due intervals for rest and for meals, are necessary for the preservation of health?—So essential that without it medical treatment is unavailing.

4. It is found, on the contrary, that excessive labour, exertion, or employment, so as to produce weariness of body and mind, and without due intervals for meals and refreshment, must in ordinary cases be extremely prejudicial to the human constitution?—Man can do no more than he is allowed or permitted to do by nature, and in attempting to transgress the bounds Providence has pointed out for him, he abridges his life in the exact proportion in which he transgresses the laws of nature and the Divine command.

5. You would think that one of those violations of the laws of nature is excessive labour, excessive both as to its length and the fatigue which it induces, and that it is more prejudicial to young persons than to those advanced to full maturity?—In exact proportion to their feebleness.

6. Do you conceive that the ordinary limits of a day's labour, as already settled by universal custom, whether in agricultural or mechanical pursuits, namely, about twelve hours a day, with due intermission, is, generally speaking, as much as the human constitution is capable of sustaining without injury?—It depends upon the kind and degree of exertion, for the human being is the creature of a day, and it is possible for the most athletic man under the highest conflicts of body and mind, and especially of both, to exhaust in one hour the whole of his nervous energy provided for that day, so as to be reduced, even in that short space of time, to a state of extreme torpor, confounded with apoplexy, resembling, and sometimes terminating in death.

7. Would labour, too long continued, of a less strenuous nature, but becoming very wearisome both to body and mind, be prejudicial if persevered in?—The injury is in proportion to the exhaustion of the sensorial power. Let me take the life of a day to make myself clearly understood. It consists of alternate action and repose, and repose is not sufficient without sleep. The alternation of the day and the night is a beautiful provision in the order of Providence for the healing of man, so that the night repairs the waste of the day; and he is thereby fitted for the labour of the ensuing day. If he attempt to live two days in one, or to give only one night and two days' labour, he abridges his life in the same, or rather in a greater proportion, for as his days are, so will his years be.

8. Upon the whole, then, recurring to the question, should not you think that a day's labour, consisting of twelve hours, pursued for a continuance, would, in ordinary cases, be as much as a human being could endure with impunity?—I think that twelve hours' labour is too much for a very large majority of human beings. If I am to state the precise quantity, in my experience, as tending to give the longest and the most vigorous life, I should take it, even in the adult, at eight hours' active exertion, eight hours' sleep, and eight hours allowed for recreation and meals. Those are the divisions of the day which would procure the happiest and the most vigorous life, and which would, I think, yield the greatest sum of labour; but the child requires

a greater proportion of sleep than the adult: for sleep is not simply repose, but it is a restorative process. J. R. Farre, Esq., M.D.

9. You are aware that a much greater duration of labour is required than that to which you have alluded as the most consistent with health; have you seen the effect of forced labour upon the human body?—In the course of my medical life I have seen it extensively; the first part of my medical education was conducted in the West Indies, where I had an opportunity of observing the effects of slavery, and the diseases produced by slavery.

10. Were you engaged in medical practice, as well as medical studies, in the West Indies?—I was.

11. Will you state to this committee whether, when human beings are regarded as mere property, it is not found necessary, with a view to the preservation of that property, to limit the duration of their labour, and more particularly the labour of young persons?—It has been found necessary to diminish that labour, especially in regard to the young, insomuch that the young are never accounted fit for labour; their employment is always of the lightest kind.

12. Until what age?—The exact time of employing them in the field varies with their growth. In the West Indies, human life is often precocious: in the island in which my observations were made, namely, the Island of Barbadoes, the oldest established colony, a colony in which there is the hardest work, for it is a hill country, and the whole production of that island is the effect of manual labour, as it is a rocky, and often an artificial soil, through the greater part of the island: it affords therefore a fair specimen of the effect of the fullest labour on the human constitution, and the returns of the increase in the slave population in that island would be a fair criterion to enable the committee to judge of the effect of that labour, and whether proper provisions were used to render it less injurious. I have reason to believe, it would be found that the population has been for a great many years considerably on the increase, notwithstanding that labour.

13. Have the regulations recommended by medical authorities in that island been adopted with a view to the preservation of the youth of that colony; and if so, what is their nature?—Each plantation has its regular medical superintendent; but there is a number of highly educated gentlemen in that island; and to my knowledge, an agricultural society was established about a quarter of a century ago, for the express purpose of improving the condition of the slaves, without reference at all to the proceedings on that subject which have occurred in this country. Medical gentlemen, I say, of the highest education were members of that society, and to my certain knowledge, took the two main questions of labour and feeding into their consideration, and communicated papers to that society for their guidance respecting the management of slaves, both old and young, as to labour and as to food, and as to the proper labour and rest; and I know that extraordinary care was taken, both in regard to regulating the labour of the young and in feediug them, to make them a vigorous race and fit for the work.

14. What were the regulations in respect of the labour of the children and young persons?—As far as I am acquainted with them, they consisted in not employing them in field labour, in digging or in carrying manure, but in exercising them in gathering the green crops for the stock. The plan of working them lightly in the open air and feeding them, not trusting to the food provided by the parents' care, but by the care of the master, had the most beneficial effect in improving their condition. When I contributed to superintend the negro population, I never knew the young overworked. I observed that their diseases were more frequently the effect of improper feeding.

15. Supposing that the employment of children in the factories of this country is spread over twelve or fourteen hours a day, and often with but very

J. R. Farre, Esq., M.D.

short intervals for the taking of their meals; is there anything equal to that sort of labour imposed upon the children of the slaves in that island?—Nothing of the kind; even the adult, in the most vigorous condition of body, is not subjected to labour of that duration.

16. Do you happen to know whether the owners of those slaves, alluding to the children, attempt to make a profit by their labour before they arrive at that period of life when they are capable of rendering it?—Never; I have always, as a medical observer, considered that their employment was used only as a training for health and for future occupation.

17. So that you consider that the limitation of the length and degree of the labour of the children and young persons in Barbadoes is eminently advantageous to the planter himself, with a view merely to his own interest and future advantage?—Certainly; it is necessary. In English factories, every thing which is valuable in manhood is sacrificed to an inferior advantage in childhood. You purchase your advantage at the price of infanticide; the profit thus gained is death to the child. Looking at its effects, I should suppose it was a system directly intended to diminish population.

18. What have been the provisions to guard against the ill effects of excessive labour?—The provisions consist in the proper management of the youth, in not sacrificing them, and in exercising them in the open air.

19. Do you consider that labour in the open air, at a given high temperature, is less pernicious than labour performed in a confined atmosphere artificially heated to the same degree of temperature?—There is only a certain portion of labour which any human being can bear in a day, but that labour is better accomplished in a pure atmosphere than in a confined and impure one, because air is vital food, and the relative value of a pure air to the blood is to the grosser aliments which supply blood, as one minute to fourteen days; and it admits of this demonstration, that if the human being, by malformation, disease, or the accidents of life, be deprived of the grosser aliments of bread, flesh, and water, he dies in a period of from eleven to nineteen days, and the ordinary period may be taken at fourteen; if he be *completely* deprived of air, he dies in one minute. The relative value therefore of food through the organs of supply, to sanguinification effected in the lungs by pure atmospheric air, is as the value of one minute to fourteen days, according to my estimate. And the deterioration of the atmosphere in which human beings are working produces a deleterious effect in proportion to that deterioration. Supposing that in a place where human beings are congregated, and the air, for want of ventilation, and from having been breathed, is impaired, the blood is immediately deteriorated; it passes from its red or vital state to black blood, and the vitality of the being is lowered; it gives a pallid or livid expression, according to its degree; it destroys the appetite, diminishes the action of the heart and arteries, lowers gradually the whole vigour of the system, and produces premature death. This is the inevitable effect of continued labour in a confined, heated, and foul atmosphere.

20. Are the children in the West Indies ever employed in night-work?—I never knew an instance of that.

21. None are employed in night-work but those who are called the strong gang?—It is only a privileged and higher class, often the most valuable persons on the property, who are employed during a part of the night in the manufacture of sugar in Barbadoes, as far as I recollect.

22. None but persons in the full enjoyment of strength are employed in night-work?—None but a select class of adults.

22. Should you conceive any child under nine years of age ought to be made habitually to labour in a mill or a factory; and do you think that

before that period a considerable number of hours of labour can be endured in such a situation with impunity?—I think that, to insure a vigorous manhood, the period of nine years of age is too early to commence labour. J. R. Farre, Esq., M.D.

23. Should you conceive that between that period and the time when the system has arrived at its growth and maturity, namely, about eighteen years of age, more than ten hours a day of labour could be imposed upon those children and young persons without producing, generally speaking, mischievous effects?—In considering the order of the human growth, the first seven years of the child properly constitute its infancy, which is closed by the change of teeth, in order to its future sustentation; the second seven years terminate in the development of the generic system, by which the being is to continue the species, and to prevent its deterioration. Now it is manifest, that if the child be deprived of its red blood and muscular fibre during these important periods of its growth, it cannot have a vigorous generic system; but at fourteen the changes are so delicate in relation to the provision for the future being, that if forced labour be carried to excess, even between fourteen and twenty-one, the value of manhood will be considerably diminished. I call the British system a forcing system, which departs from the truth of Nature, and from the revealed will of God. I have no hesitation in affirming, that there is not a due regard to the preservation of life, either in the British system of education or of labour generally, both as regards the child and the adult; and what I say of the adult applies still more strongly to the child.

24. Too great exertion is imposed on both mind and body?—Yes, certainly.

25. In what manner does this forcing system of too long continued exertion impair health, and occasion premature death?—The first effect of forced labour is to injure the organs of supply, commonly called the organs of digestion. The absorbent system is divided into two branches, one employed on the alimentary canal, to live on things without us; the other employed on the whole of the animal body; and there is a balance between these two functions, so that if the lacteal absorption be impaired by over-labour, the lymphatic system immediately begins to absorb the body; and this will explain why the body becomes not only pale, but thin. If the labour be pressed, the absorption of fluids is succeeded by the absorption of solids, the fibre is taken up, and thus the muscular fibre is diminished. If the labour be still further pressed, the earthy matter is absorbed, and the bones bend. Thus, by premature labour, the child is robbed of its blood, of its muscle, and of its bone; it is crippled for life, and, both in the male and in the female, the species is deteriorated.

26. It appears from an official document presented to this committee, that a greater proportion of morality exists wherever this system of long and irksome labour is allowed; would you be prepared for such results from the principles you have stated?—I think that the result is so inevitable, that I view it as a species of infanticide, and a very cruel, because lingering, species of infanticide, resulting from the over-extension of a principle in itself good, the principle of cheap production demanding over-labour; and that the only safeguard to the state consists in opposing this principle of political economy by the medical voice, whenever it trenches on vital economy.

27. You think that political economy, supposing it were made manifest that the system would produce national wealth, ought not to be suffered to interfere with vital economy?—It ought not to be suffered to trench on vital economy, because, if it does, it is guilty of homicide. I have no hesitation in affirming that the voice of the profession would maintain this truth, and never assent to life being balanced against wealth. That the life is more than the meat is a divine maxim, which we are bound to obey.

28. So that it is your settled conviction that what you have denominated,

J. R. Farre, Esq., M.D.

very properly, the forcing system, injures the present generation, and threatens the most injurious effects regarding the future ?—The vigour of the animal life depends upon the perfection of the blood, and the balance preserved between the pulmonary and aortic circulation ; but in the aortic circulation there is also a balance between the arterial and the venous systems, and the heart is the regulating organ of the whole. Now if the arterial circulation be too much exhausted, an accumulation necessarily takes place on the venous side, and the blood is deteriorated, and organic diseases, which abridge life, are produced. It would carry me too far to investigate this point ; I only state the fact which can be demonstrated ; but there is another and a higher effect ; for man is to be considered as something vastly better than an animal ; and the effect of diminishing the power of the heart and arteries by over-labour in a confined atmosphere is to deteriorate the blood, and to excite, in the *animal* part of the mind, gloomy and discontented trains of thought, which disturb and destroy human happiness, and lead to habits of over-stimulation. The reflecting or spiritual mind gradually becomes debased ; and unless education interpose to meet the difficulties of the case, the being is necessarily ruined, both for present and for future life.

29. Are there any countervailing means which suggest themselves to your mind, which would render this continued labour less injurious to present health and happiness, and more conducive to the prosperity of future life ?—Ventilation, exercise, and diminished exertion, are the most obvious means of meeting the difficulties of the case, joined to the change of ideas resulting from an education adapted to the spiritual nature of man.

30. So that you think it would be eminently beneficial if the hours of labour were somewhat abated during the day, so as to afford an opportunity for the change of mental exertion, and to direct that exertion to the purposes of education ?—It would be a positive gain to the child, which not only mercy, but justice demands ; and I think, also, it would be beneficial to the state.

31. It appears in evidence before this committee, that the labour in mills and factories undergone by children and young persons of both sexes, is rarely less than fourteen hours a day, including the time allowed for meals ; and that in many instances it greatly exceeds that term, and extends sometimes to eighteen or twenty hours, or upwards ; during your residence in the West Indies, was there any system of labour imposed upon any of the children and young persons at all equal to that ?—Never in the slightst degree approaching to it.

32. It is in evidence that the children and young persons employed in factories have often to be roused from their lethargy induced by exertions too long continued, and to be stimulated to their labour (their strength being exhausted) by constant whippings, beatings, or other means of a like nature ; are the children of the slaves in the West Indies more hardly worked or more cruelly treated than this state of things implies ?—In the colony of which I have spoken, I never saw nor heard of such inhumanity.

33. Assuming that the children of this country are not free agents, can you have any doubt whatever, the slightest hesitation upon your mind, that they demand protection equally with the child of the West Indian slave ?—I think the word demand is a very proper mode of putting the question ; for I myself consider that the nation is responsible for it , and, as a medical man. I assert that if you deem it a part of your duty to make laws against murder, I consider that legislation is equally necessary for the prevention of death in any mode in which it can be prematurely inflicted ; and certainly this must be viewed as a most cruel mode of inflicting it.

34. You have no hesitation, then, in saying, that whether considered as a medical or a political question, a remission of the hours of labour imposed

upon the children and young persons of this country would be essentially beneficial?—I view it not only as a benefit, but as a duty; and I would say, not only as a physician, a Christian, and a parent, but also from the common sympathies of a man, that you are bound to afford it. J. R. Farre. Esq., M.D.

JOSEPH HENRY GREEN, Esq., F. R. S., called in, and examined, 3rd August, 1832.

1. What is your profession?—A surgeon. J. H. Green, Esq.

2. Are you a medical officer in any of the great institutions of this metropolis?—I am one of the surgeons of St. Thomas's Hospital.

3. Have you any other public professional duties devolving upon you?—I am professor of surgery at King's College.

4. Do you lecture in your own hospital to the medical students there?—I give clinical lectures.

5. Your mind has necessarily been directed to the various branches of your profession?—It has.

6. Assuming that the labour undergone in mills and factories is continued for thirteen, fourteen, or fifteen hours a day or upwards, and sometimes without sufficient intervals for meals, and generally pursued in an erect, or at least in a constrained position of the body, often in an impure atmosphere, and frequently heated to a high temperature, and continued occasionally far into the night; what do you think would be the effect of such a system of unvarying employment or labour on the health and welfare of children and young persons up to the age of eighteen, or thereabouts, especially if persisted in for any length of time?—Although I cannot pretend to any practical experience in these matters, yet perhaps I may be able to contribute some information that my professional duties have enabled me to acquire, in aid of your benevolent purpose of duly limiting the hours of infantile and early labour. Confining, then, my observations principally to that period of life to which you have directed my attention, I would observe, that during childhood, and through the whole of that period which precedes and extends somewhat beyond the epoch of puberty, the powers of life are actively engaged in completing the *growth* of the body. There exists, at this period great difficulty in balancing the activities of life, in order to the due development and the building up of the various organs and structures, to the evolution and apportioning of their functions, and to the perfecting of the bodily frame; and it may therefore well happen, as we see even where the purposes of nature are aided by the tenderest care, that various forms of disorder and disease are produced or favoured by a too suddenly excited activity in one part or system, and a disproportionate elanguescence of life in another system. This may be assisted by original weakness of constitution, but it may be said to constitute the weakness of the living system in growing children, and determines their liability to the disorders incident to growth; and this liability is augmented in an especial degree, when the defective balance above mentioned is aided by the neglect or errors of nursing and physical education, and by want of protection from, or the wilful exposure to, a variety of external morbific causes of disease. It will be seen, then, that the period of growth is one of weakness, and characterized by a liability to disorder; but it will be well to consider briefly that condition of the living body which childhood and early youth imply, in order to the more full illustration of the position. In the first place, the purposes of growth necessarily require a more than ordinary supply of nutriment, and this not only in relation to the augmentation of size and bulk of body, but in order to com-

J. H. Green, Esq.

pensate for the waste produced by the change of materials, and by a highly increased activity of the process of excretion and secretion; but proportionate to these needs, there is a great rapidity of digestion, and for the same reason a great demand for food. Children require not only a large supply of food, but that it should be nutritive, and given frequently; and it is almost unnecessary to say, that in default of these requisites the consequences will be, disorders of the digestive system, diseased mesenteric glands, stunted growth, wasting, weakness, and the liability to, with less powers of resistance to, the inroads of all diseases. Neither, it may be observed, can assimilation be perfectly performed without air and exercise; and whenever the food is abundant, should these be denied, the growing child would droop, and would be the subject of the same consequences, the prevention of which depends upon the appropriation, as well as the supply of nutritive materials. Correspondently to the purposes of growth, there is a quick and energetic circulation of the blood, and this is readiiy increased by external causes of excitement; but the same condition (which we might almost describe as a healthful fever) renders the child peculiarly liable to fevers and inflammatory affections, and requires the avoidance of all those causes which tend to hurry the circulation. I need not say, therefore, if there were no other grounds for condemning the practice, how prejudical dram-drinking must prove. Again, the nervous system is more easily impressible, but at the same time the nervous excitability is sooner exhausted in childhood than after growth has been completed. This readily explains the mobile, restless character of children, who are always acted upon by their immediate sensations; but at the same time it shows the necessity of changing their employments, of alternating occupation with amusement, of allowing them long rest in the horizontal position, and sufficient sleep, the requisite time for which I cannot estimate at less than eight or nine hours, but which must, under a variety of circumstances, be prolonged even to twelve hours. In connexion with the nervous and circulatory system, I may mention that children are extremely susceptible to vicissitudes of temperature, and have not the power, in any considerable degree, of resisting cold; and that hence it is of the utmost importance that they should be warmly clad, and protected likewise in their abodes from cold. Further, the muscular system is capable of, and disposed to, quick and frequent motions; but the muscles have not yet acquired that tone which enables them to perform actions which require strength and persistency of action; and in order to prevent fatigue, it is right that exercise should be varied, not long-continued, nor disproportioned to strength. Lastly, all the parts of the body, especially the bones and joints, are soft and spongy in their texture, and are incapable of resisting mechanical injury. Now if we take all these circumstances into consideration, we shall readily and correctly infer that children are not fitted by nature for laborious or stationary occupation; and that the subjecting of them to business or work which requires strong exertion, or which even being comparatively light, demands uniform, long-continued, and therefore wearisome exercise, must ultimately have an injurious effect on their health. But if, in addition, their food is scanty, supplied only at long intervals, their occupation is not alternated with amusement and exercise in the open air, and their clothing is not warm, disease must be the inevitable consequence of this violent counteraction of all that nature suggests and demands. I may observe, too, that in consequence of the weakness of the muscles, and the yielding texture of the bones and joints, exertion in any unvarying posture will tend to prodnce *deformities*, which are not only unsightly, but are the causative antecedents of disease; and thus even under circumstances in which children are tended with care in other respects, yet where air and exercise are unduly withheld (as in boarding schools), and under other circumstances, we witness as the consequences, frightful curvatures of the spinal column, deformities of the chest, and distortions of the joints of the lower extremities, that diminish or abrogate the healthy natural uses of the body, or lead to destructive disease. Nay, I would add, that if you were to subject the healthiest child to the causes which I have above enumerated, it is impossible that it should not become weakly, emaciated, stunted in its growth, dull, sluggish, and diseased. Take, for instance, a healthy child from an agricultural district in the bloom of health, animated, cheerful, lively, strong, active, and free-limbed, and place it in a factory; let it work for twelve hours a day, in the stifling heat of some of the work-rooms, confined in the impure air breathed by a hundred of others, without any provision for ventilation; let it learn to drink ardent spirits to support its enfeebled frame and depressed feelings under overlabour and harassing tasks, with a proportionate disrelish for wholesome food; let it then be turned out of this heated factory on a damp, foggy November evening, to rest its fevered and debilitated frame in some dank and close cellar of an over-peopled manufacturing town, and shall we be surprised, is it not rather a certain consequence, that the unfortunate child becomes the victim of disease?

J. H. Green, Esq.

7. Should you not think that the employment described, with the circumstances attendant upon it, would have a tendency to produce or increase scrofulous complaints, which it is asserted are distressingly common among children in factories?—There is no disease to which children, both from the constitution of their frame and the various unfavourable circumstances to which it is exposed, are more liable than to *scrofula* in all its multitudinous forms. To the production of this disease, one of the most influential circumstances is, I am persuaded, breathing an impure air; and by purity of air I do not mean anything that can be determined by chemistry; but I refer to the fact, that scrofula chiefly prevails in the children of the inhabitants of densely peopled towns and crowded cities; and it can scarcely be doubted that the operative cause is the vitiation of the atmosphere by the production of some noxious or poisonous agent; and we find that wherever men or animals are herded together without proper ventilation, disease will invariably be generated, as is exemplified in crowded barracks, transports, gaols, kennels, stables, poultry-yards, and the like. So, likewise, the prevalence of the disease is greatly aided by rapid, sudden, and considerable variations of temperature, and by exposure to a cold atmosphere loaded with moisture; and in reference to our present subject, although doubtless the rooms of a factory might be regarded as a preservative, as far as the children are protected from cold and the variations of temperature therein, yet, taken in conjunction with the fact of the children, after working therein, being suddenly exposed, as they often are, to a cold, damp, night air, and then destitute of a warm and comfortable lodging, it will be plain that the causes of disease are not only not removed, but are really heightened. It is indeed most important that protection against cold and damp should be provided in suitable abodes and clothing. The houses of the poor, the improvident, the debauched and vicious, who unfortunately form a large class in all large towns, are crazy, ruinous, filthy, and unventilated; and their unfortunate children, crowded into these receptacles, or thrust forth half naked and in rags, and exposed to the inclemencies of the weather, thus unguarded against its vicissitudes, are subjected to all its deleterious influences, and seldom fail to have the latent seeds of scrofula developed in some of its destructive forms. Some hope of escape might be entertained, if the causes of disease just mentioned were counteracted by abundant and wholesome nourishment; but in those who live from hand to mouth, the supply is uncertain; and although there may be abundance one day, it is counterbalanced by scanty or insufficient fare on the next; and this but too often the consequence of a drunken debauch of the

J. H. Green, Esq.

father of the family; and the same habits are not unfrequently resorted to, or even encouraged, in the children; and instead of nutritious food, the wages of labour or the gains of dishonesty are expended in drams and cordials, or the cravings of hunger stilled by the no less pernicious practice of opium eating. I believe the committee are aware of the fact of the extent to which opium eating has prevailed in the manufacturing towns. Amongst the poorer classes too, from a want of, or from ill directed economy, the children, instead of getting wholesome bread, meat, and beer, are chiefly fed on potatoes, tea, with occasionally a herring, or a taste of bacon. These causes of disease are often, indeed, favoured by original disposition to the disease; but let it not be forgotten, that the child of the most healthy constitution may, by continued exposure to these causes, acquire a disposition to the disease, and become actually the victim of it; whilst those who might have been subjected to it from an original fault of the constitution may, by careful preservation from these causes, remain exempted from the disease. It would be scarcely possible to present in any brief summary the many dire effects of scrofulous disease, but we may mention, first, that the mesenteric glands are often the seat of disease, favoured by the irritation of unwholesome and ill-digested food, shewing itself in weakness, emaciation, protuberant abdomen, and slow fever. Next, the absorbent glands about the neck, the inflammatory swelling, excited, perhaps, by variations of temperature; for the particular seat of the disease, or its development in any particular organ, may be determined by accidental circumstances. Then we find that the disease attacks the skin in the form of scaly eruptions, cracks, spots, ulcerations, and slowly suppurating tubercles. Again, that the eyes become affected in the various forms of scrofulous ophthalmy, that often end in blindness; or the bones, and especially the joints, become diseased, terminating in caries of the spine and white swellings. Then, that the internal viscera are affected with tubercles, as the liver, brain, spleen, &c. And, lastly, that the lungs become the seat of this destructive disease in the form of that incurable complaint of our climate, pulmonary consumption. This is, indeed, a melancholy list of maladies, and one which I am sorry to say might be greatly augmented, as traceable to the neglect and improper management of those whose tender years demand and lay claim to our sympathy and kindest care and attention; and I fear that this country will have much to answer for in permitting the growth of that system of employing children in factories, which tends directly to the creation of all those circumstances which inevitably lead to disease. But if it be melancholy to pass thus in review before us these dire inflictions, how much more melancholy is the reflection, that even the parents are not unfrequently tempted to become accessories to the degradation and misery of their children, and, forgetting their natural affections, themselves inflict the blow that wounds even to death their own offspring. I am not, indeed, prepared to say, that the causes adverted to exist universally in our factories, and I am happy to have an opportunity of mentioning an honourable exception in that of Mr. Ashton, of Hyde, near Manchester; but wherever they do exist, I am quite sure that the results will be, in regard to the health, most destructive, and I think I may venture to add, in regard to morals, most injurious, and that the consequences of this culpable inattention to the physical and moral welfare of the manufacturing class will be a population weak and diseased in body, feeble and degraded in mind, and vicious and dangerous in conduct; and, in reverting to the especial object of this inquiry, I do not hesitate to say, that the objects and purposes of a government desirous of the well-being of the people, and of the class now under consideration, will be very inadequately attained by merely limiting the hours of labour of children employed in factories. Children were not designed for labour, and although in the artificial state of society in which

J. H. Green, Esq.

we live, and considering the imperative demands for sustenance which oblige the poor to employ their children, some labour must be permitted, yet both our conscience and our feelings equally demand that the labour of children should be under such restrictions as will ensure them against their being made the victims of avarice and disease, and as will render it compatible with their physical and moral welfare.

8. Adverting to the present state of society, up to what age do you think the children of the poor ought to be protected from the labour of the factory, and after that age, what are the hours of labour which you would be disposed to sanction as its utmost limits, at least till they had arrived at maturity?—It is difficult, no doubt, to prescribe regulations that shall be universally applicable; but I will not withhold my opinion, that no child should be put to any laborious occupation before he has reached nine years of age; that from nine to twelve the hours of labour should not be more than six in the twenty-four, and that the labour itself should be light and varied; and that from the age of twelve the hours of labour should be gradually increased to twelve hours, including the time for meals, as the utmost average period of labour for the full-grown, strong, and healthy man; and that only thus a sufficient time will be left for rest, healthful recreation, and for intellectual and moral improvement.

9. You have already stated that labour, though it may be denominated "light and easy," yet if continued for any great length of time under the circumstances that have been described, is still injurious to the health of the children and young persons so occupied?—Yes, certainly.

10. Does not the maintaining an erect position of the body itself induce considerable fatigue if long endured?—In order to maintain an erect position of the body, it is necessary that a muscular action should be constantly exerted, therefore it necessarily induces fatigue.

11. Should you not consider that a polluted atmosphere, and one artificially heated, considerably increases the injurious effects of this undue length of labour, so that these circumstances of themselves would furnish a reason for abridging the hours of labour ordinarily undergone in mills and factories?—Certainly I should consider that it would be likely to excite a feverish state of the system.

12. Do you not think that labour pursued in an atmosphere artificially heated, and circumstanced as described, would be more exhausting and pernicious than labour pursued in a natural atmosphere, though at the same temperature?—I do.

13. It follows, therefore, assuming that the children and young persons employed in those mills and factories are not free agents, that they demand legislative protection as much as the slaves in our colonies?—Yes; I am of that opinion.

14. You have alluded to the injurious effects of the system in stunting the growth and weakening the muscular powers of those having to endure it; do you conceive that the deterioration might become hereditary, and even increase from generation to generation, if the causes producing it were to be continued?—Yes, I do; I am of that opinion.

15. It is in evidence that a considerable majority of the young children and young persons so employed are females; do you conceive that the female frame and constitution is as well adapted to long-continued and strenuous exertion as that of the male?—I do not.

16. So that the protection becomes the more necessary, when we refer to the fact of females being the principal operatives in such works?—Yes.

17. Do you not think that the female constitution is particularly susceptible of injury about the age of puberty?—Yes.

18. And that therefore much and permanent injury might ensue from an

J. H. Green, Esq.

undue length of labour endured in a standing attitude at that period of life? —Yes.

19. Has not the labour in question, especially when undergone in a heated temperature, and often under circumstances unfavourable to decency, a direct tendency to anticipate the period of puberty?—Yes, even without indecency; the heat and the constant labour in an upright position would tend to develop puberty prematurely.

20. Do you not think that the premature development of puberty, connected with the consequences of it, is pernicious to the well-being of the parties and the interests of society?—I do.

21. Do you not think that labour is peculiarly pernicious and prejudical when it has to be commenced in the morning; the body not being sufficiently refreshed and recruited by the insufficient sleep of the preceding evening?—Certainly.

22. And that fatigue perpetuated is peculiarly wearisome, of course, to the body, and pernicious to all the functions of life?—Yes, I do.

23. Do you suppose that the mind would be capable of deriving much advantage from any instruction that might be communicated to it after the labour to the extent described to you has been undergone throughout the day? —No, I should think it would be impossible.

24. Should you not conceive that, considered with a view to its physical effects, the confinement of a Sunday-school, after a week's labour like that described, would be additionally cruel and pernicious?—Yes, I should.

25. And less efficacious in conveying instruction?—Yes, as inefficacious as it must be to a mind in a state of stupor and a body in a state of exhaustion.

26. It is alleged by many operatives, that working by artificial lights during the night has a pernicious effect upon their sight; should you be prepared to say that that was a natural consequence of labour so undergone?—It is invariably found that a strong dazzling light affects prejudically the sight, often even to blindness.

27. What effect has the long-continued burning of artificial lights upon the purity of the atmosphere?—It removes the oxygen, and renders it chemically unfit for respiration.

28. Is it your opinion, that early and promiscuous intercourse between the sexes is, ordinarily speaking, attended with prolificness?—I should expect it would tend to produce barrenness.

29. And consequently the absence of illegitimate offspring in the factory districts, were such the fact, would furnish no satisfactory criterion of the degree of morality which prevails in them, as compared with the agricultural and other districts?—No, certainly not.

30. It has been frequently stated that the factory children can be almost universally distinguished from others by the peculiar ghastliness and sallowness of their complexions; should you think that that appearance might result from the system of labour as explained to you?—I should consider it as a necessary result of ill-health, a physiognomical sign, I may say, of a deranged digestion, and the want of energy in the circulation, to which I have already alluded.

31. An official document has been ordered by, and delivered in to, this committee, which shews the great waste of human life, especially in its early stages, in the factory districts, as compared with the mortality in other districts, and even in the great metropolis, commonly supposed to be very unfavourable to life, especially in its earliest stages; should you be prepared to expect, from the description given to you of the factory system, that such would be its results?—Yes, I should.

32. What should you judge to be the effect of this system, as at present

carried on, upon the moral and physical welfare of the community, generally considered?—It would be no less presumptuous than impertinent in me to discuss a topic of so complex a nature, and one which lies beyond the sphere of my professional experience; but it is unfortunately too often forced on my conviction, that vice and civilization, unchecked by moral restraint, are the fruitful causes of disease, for me to remain ignorant that it is impossible to separate the physical welfare of man from his moral well-being; and in considering the effects of the factory system, I cannot but believe that the evils which are now too justly complained of in respect of the health of those employed, admit only of effectual alleviation, or removal, by attention to their moral causes and conditions. I am therefore disposed to think that the question respecting the means of preserving the health of the manufacturing labourer and artisan involves a consideration of topics of far higher import. *Manufactories* and *machinery*, so long as they procure employment for the labouring poor, render the necessaries and comforts of life cheap and easy of acquirement, and are the means of the poor bettering their condition, must be regarded as blessings, and in every way conducive both to the physical and moral welfare of the people. In order to obtain this desirable object, it is, however, necessary that the labourer should participate in the advantages and benefits arising from the employment of machinery, and in diminishing human labour by its use, the only legitimate purpose must be admitted to be that of substituting a machine for the performance of that labour which would reduce man to a mere mechanism, to the end that he may devote the time and leisure acquired thereby to his moral cultivation. It is indispensable, I say, in regulating a manufacturing system, that the labourers employed should never be considered as merely the means to its success, but that their condition, moral and physical, should constitute an essential object of the system, and its success, as the source of wealth and power, be subordinated thereto. But if, instead of this legitimate object, and this wholesome restraint, ruled by the insatiable avarice of gain, the manufacturing system is without check, and has no bound but the possible means of creating wealth, and of making the rich richer; if we find that the population is indefinitely increased, that human beings are called into existence by, and their existence wholly dependent upon, manufactures, the demand for which, in many instances, has only the precarious tenure of caprice and fashion; if, then, in consequence of this multiplication of labourers, wages be lowered till it be simply calculated upon how little life and the motion of a pair of hands can be supported; if we find that these human beings are only regarded as parts of the machinery which they set in motion, and with as little attention to their moral welfare; if we find that these, even to the tenderest age, and without respect to the distinctions of sex, and without regard to decency, are crowded together under all the circumstances that contribute to disease and vice, and all this to add to the wealth of their employers, to minister to the luxuries of the rich, and to make overgrown capitals still more vast and oppressive, whilst the labourers themselves are degraded into the mere negro slaves of Europe; then, I say, that these and all the physical evils incident to such a state, require no medical opinion, but demand unsparing moral correction, or they await the punishment due to depriving man of the birthright of his humanity, of degrading him into the class of means and things to be used, instead of recognising, as the end, his happiness and dignity as a moral and responsible agent.

J. H. Green, Esq.

GEORGE JAMES GUTHRIE, Esq., F.R.S., called in, and examined, 4th August, 1832.

G. J. Guthrie, Esq.

1. What is your profession?—Surgery.

2. Do you hold any office in the Royal College of Surgeons?—I am the vice-president for the present year.

3. Are you surgeon to any of the public hospitals in this metropolis?—I am surgeon to the Westminster Hospital, and to the Westminster Eye Hospital.

4. Your attention has, of course, been directed to every branch of your profession?—Yes, to every branch.

5. Is it not a universally received opinion with the gentlemen of your profession, that moderate exercise or employment, with proper intermissions for refreshment, recreation, and sleep, are ordinarily necessary to the preservation of health?—Certainly.

6. Those advantages you hold to be still more necessary as regards children and young persons?—Certainly.

7. Do you not believe that excessive labour, or labour continued so long as to produce great fatigue in body and mind, and without those intermissions, is prejudicial to human health?—There can be no doubt of it.

8. Would not such a degree of labour, habitually undergone, be still more injurious to children and young persons?—Unquestionably it would.

9. Is not the customary day's labour, whether in agricultural or handicraft employments, established by universal assent in every age and country of the world, namely, twelve hours, inclusive of meal times, or ten hours of actual work, and which term is usually diminished in the winter months, as long, in your opinion, as the human frame, generally speaking, is calculated to sustain with perfect impunity, at any period of its existence?—Certainly.

10. Are not intervals or cessations from labour sufficiently long for taking necessary meals, essential to the health of the labouring classes, generally speaking?—They are.

11. Assuming, then, that the labour undergone in mills and factories almost always greatly exceeds the term already mentioned as that of an ordinary day's work, and extends to thirteen or fourteen hours, and sometimes even to eighteen and nineteen hours a day; can you doubt that, referring now to the less extreme cases, such a system must, in many instances, prove highly prejudicial to the health of those enduring it?—I do not doubt it.

12. When it is undergone without any intermissions whatever, it must be still more injurious?—Certainly.

13. Do you think that when still further extended, so as to be continued to twenty, thirty, or even forty successive hours twice or oftener a week for any length of time, it can by possibility be endured without the most serious consequences?—It cannot.

14. Should you not, without the fullest evidence and assurance of the fact, believe it impossible that such extent and description of labour for months together could be imposed upon or endured by children and young persons?—Certainly.

15. Is not labour pursued during the night instead of the day prejudicial to health?—No doubt of it.

16. Is not any employment, though it may be denominated light and easy, but being one of wearying uniformity, and inducing much fatigue of mind as well as of body, more exhausting and injurious than locomotive exercise, though of a more strenuous nature, undergone for a reasonable length of time, and with due intermissions?—I think so, generally speaking.

17. Is not the exertion necessary to sustain the erect position for a great

length of time, more fatiguing than any other natural posture ordinarily maintained, or moderate exertion equally and alternately exercising the muscles of the body?—Unquestionably.

G. J. Guthrie, Esq.

18. Would not this labour be still more distressing, if it had to be pursued in an atmosphere polluted with dust and flue perpetually evolved from the material manufactured?—It would.

19. Would not labour so circumstanced and continued be still more injurious if undergone in an atmosphere artificially heated to from 70° to 80°, and upwards, in order to facilitate the operations, and from which currents of fresh air must be excluded?—No doubt of it.

20. Is not, then, labour in an atmosphere thus impure and heated artificially more injurious to the constitution and exhausting to the strength than similar labour in the open air, though of the same temperature, but being free, of course, from the impurities in question, and freshened by the natural atmospheric changes and influences?—Yes, certainly.

21. Do not, then, that part of the British community, not being free agents, as compared with the negroes of the crown colonies of this country, and their children, require, in regard of their health, an equal protection as to the duration of their labour?—More so.

22. In addition to other effects it might be likely to produce, does not this labour, when pursued by night, and consequently by gas or other artificial lights, tend, as the operatives allege it does, to injure sight?—Yes.

23. To apply, then, your preceding observations and opinions to the immediate objects of this inquiry, factory children, do you not conceive that the labour in question, continued for the length of time described, without sufficient and often without any intervals even for meals, generally pursued in an erect, or, at least, constrained position, and in a foul and polluted atmosphere, frequently heated to a very high temperature, and many times continued far into or endured all night as well as day, cannot fail to be still more pernicious and destructive to children and young persons than it is to adults in the prime and vigour of life; and does not then the condition of these children, in regard to either justice or humanity, demand legislative protection as imperatively as that of the adult felon or the infant slave?—They require it in a greater degree.

24. Assuming the description of the factory system to be on the whole correct, do you, reasoning from the established maxims and principles of your profession, consider that it is consistent with health?—Certainly not.

25. Various affecting results have been directly attributed to this system of infantile labour by medical men of great experience resident in the manufacturing districts of the country; what, in your opinion, would be the consequences, reasoning from the principles of your profession, and from the analogies that your practice must have largely afforded, which you would expect it in too many instances to produce; and first, what would be the probable effect of long and wearying labour of the nature described, without sufficient intervals for refreshment, or time for sleep, upon the digestive organs of the children and young persons enduring it?—The general result would be, first, a tendency to disease; secondly, diminished power of resisting it; thirdly, greater fatality of disease.

26. What other and particular complaints would that labour and confinement, under such unfavourable circumstances, produce or aggravate?—Struma and pulmonary affections.

27. What effects do you conceive would be produced by this excessive labour, so long continued under the circumstances described, on the osseous system?—In young persons the general results would be deformities of the spine and the lower extremities, especially the joints of the knees and ankles, and also that deformity of the foot which is called flat-footed.

G. J. Guthrie, Esq.

28. Can the peculiarly distressing deformities which have been seen by this committee in several witnesses, and which originated, as they allege, from long standing at their work, and commenced as late as nine or ten years, and in some cases at sixteen or seventeen, the body having been previously perfectly well formed, which cases the committee is assured are lamentably common in the factory districts, be confounded with the disease called rickets? —The children of such persons are predisposed to rickets and all other diseases inducing deformity.

29. It has been stated, as the result of very careful examinations impartially conducted, principally in Sunday-schools, that the stature of children who have worked in factories, compared with those who labour in other employments, is, at corresponding ages, less by upwards of two inches; would a stunted growth be the probable consequence of such a system and degree of labour, as well as a diminution of weight and muscular power?—The consequences of such a system lead to the diminution of stature, of weight, and of appearance of health.

30. An official paper has been ordered by, and delivered to, this committee, which shews the great waste of human life, especially at its early stages, in the manufactory or factory districts, as compared with the mortality at corresponding stages in other towns and places, and even in this great metropolis, commonly supposed to be so unfavourable to life, especially in early youth; should you be prepared to expect such a result from the system as now described?—I should certainly expect a diminution of longevity.

31. It is in evidence, that a considerable majority of the children and young persons employed as described are females; will you state whether the female sex is as well fitted to sustain long exertion, especially in a standing position, as the male, either in respect of the peculiar structure of certain parts of the skeleton or of muscular power?—It is not.

32. Is it not more than ordinarily necessary to give protection from excessive labour, especially to females, when approaching the period of puberty?—Certainly it is.

33. Has not the labour in question, especially when pursued in a heated temperature, and under circumstances unfavourable to decency and morals, a direct tendency to anticipate that period?—It has.

34. Has not that labour, producing, as it generally does, languor and inanition in a very considerable degree, a direct tendency to induce the habit of drinking spirits and tippling of all descriptions?—Certainly.

35. Is not the body in a very unfit state for renewed labour, especially in the case of children, when the fatigue of the preceding day is not removed by sufficient rest, and when therefore the labour has to be resumed in a state of comparative torpor?—It is.

36. Are not the terminating hours of the daily labour of children and young persons, when they are so long as to induce, towards night, great lassitude and weariness, and when it is said to be necessary for those exacting its continuance to stimulate them by constant and severe beating and punishment, more injurious and distressing than all the rest?—Unquestionably.

37. May not the numerous and afflicting accidents, which occur more particularly at the end of the day, and increase towards the termination of the week, be fairly attributed to this over-fatigue, and lassitude, and stupor? —Yes.

38. Do you think a child under nine years of age, of either sex, ought to be doomed to long and continued labour?—No.

39. Are you of opinion that young persons from nine to eighteen, or to about the period when the osseous system is arriving at its full development and strength, ought to labour more than twelve hours a day, including two hours for meals, refreshment and rest?—They ought not.

G. J. Guthrie, Esq.

40. In a word, then, are you of opinion with the late Dr. Baillie, and many other medical men, who gave evidence before preceding committees of both houses on the subject of regulating the labour of children and young persons employed in factories, that ten hours a day actual labour, making, as explained, a day's work to consist of twelve hours, is as long a term as ought, or can, in ordinary cases be endured with impunity by either sex at whatever age, and is in many cases far too much?—I am of that opinion.

41. It is asserted, that children and young persons in other pursuits are sometimes overworked; does that in your mind afford any apology for the system?—No.

42. Are the limitations of the hours of study and application in our public schools and seminaries necessary for the preservation of the health of the scholars and students repairing to them?—Yes, they are.

43. After a week's labour such as described, is not the confinement of the Sunday-school additionally cruel and injurious?—Certainly.

44. The fatigue being in many instances carried into the Sunday, is it probable that children can profit by those instructions as they might otherwise do?—Certainly not.

45. Would not a proper remission of the hours of infantile labour, so as to afford the means of education, be eminently beneficial in that point of view also?—I certainly think so.

46. Have you not been a medical officer in the armies of this country for a considerable length of time?—Yes.

47. In what parts of England, or the world?—In England very little, although I am in the service; I have not been employed since the battle of Waterloo.

48. In what parts of the world have you been principally employed?—In America, and on the continent of Europe in various parts.

49. Alluding to the term of labour or employment in question, would you, with a view to the preservation of the health of the army, sanction for a continuance, soldiers being actually under arms for twelve hours a day for a succession of days?—Such a thing is never done, or thought of; it is not in the contemplation of anybody ever to do it. A soldier is never kept under arms more than four or six hours, unless before the enemy.

50. Have you, on common occasions, ever known soldiers kept under arms for twelve hours in a day?—Never; rarely one-third of the time.

51. What is the usual time that soldiers are kept under arms?—From one to four hours; if it was wished to punish them severely for offences, they used to give them four hours' drill in heavy marching order; but that has, I believe, become antiquated, and they never now, I believe, give them more than two hours of common drill.

52. So that you would consider it inconsistent with the preservation of the health of the army, or the vigour of soldiers, to put them on actual duty, that is, to keep them under arms, for ten or twelve hours a day, and that continued for a considerable number of weeks together?—Unquestionably.

53. In the regulations in the army as to the duty of a soldier when he is put upon guard, do those who have the direction of these things take into consideration the maximum quantity of duty that a soldier ought to perform?—A soldier is never supposed to do more than is consistent with keeping him in a state of perfect health, unless in cases of emergency. If the surgeon of a regiment states to the commanding officer, or if the latter sees it himself, that the duty performed is too severe, it is immediately remitted; but this representation is rarely necessary in the army, because every officer takes as much care of his men as he can possibly do; for instance, if in any garrison, such as Portsmouth, they find the guards come too frequently, and that the men are only one night in bed and one night up, a representation is made,

G. J. Guthrie, Esq.

and a change takes place; if they get two nights in bed and one up, they think they are very well off.

54. Is eight hours out of the twenty-four about the amount of ordinary duty required of a soldier?—Yes; but the ten hours you propose to give to the children in factories is the work you would give to soldiers, even when soldiers are employed upon public works; they would not then be worked more than twelve hours, granting them time for their meals, and for that work they have additional pay.

55. After the statement you have made, that no soldier is required to do more than twelve hours' duty or work under ordinary circumstances, what is considered a fair quantum of duty to be required from a soldier, generally, in a week; how many days' duty will he have to perform where there is the proper number of men?—As for soldier-like duties, they would expect at least one night in bed and one up; and for working duty, they would be twelve hours a day at work, including meals.

56. From your own observation, do you conceive that that extent of military or garrison service is as much as can be undergone, consistently with a due degree of attention to the health of the men?—I would give them rather less than that to keep them in high condition; one night up and one night in bed, for a long continuance, will not keep them in good condition.

57. Speaking of the ordinary duty exacted from a soldier, and averaging the hours per day during the week, do you think that the soldier is required to have arms in his hands half his time?—Certainly not, but they have many other things to do; still, on the whole, a soldier commonly does not work half so hard as a labouring man.

58. Do you conceive, taking all the duties that a soldier has to perform under the circumstances mentioned, not meaning the duty he has to discharge when engaged with the enemy, but under ordinary circumstances, that he is more than half his time employed in the discharge of his duties?—I should think he is not half his time employed on military duty.

59. In that sort of employment connected with his duties?—It is hardly fair to take it in that point of view; when a soldier goes to parade at nine or ten o'clock in the morning, it is in a great measure to keep him in health, and again at four or five o'clock to give him necessary exercise, and to see that he is in a proper state, and not in one of irregularity or idleness.

60. Supposing the children of factories and mills to be employed for fourteen or fifteen hours a day, does that period of labour exceed anything undergone in the army?—It far exceeds anything undergone in the performance of the ordinary duties of a soldier.

61. Would it be thought consistent with the preservation of the health of soldiers if they were so employed?—No, certainly not.

62. Then, as a professional gentleman and medical officer of the army, of great experience, you would say that one way to recover the health of the army, if it was deficient, would be to remit a portion of its duties, if at any time they were excessive?—Feed them well, and give them moderate work, and change their situation, and they soon get fit for any service. My observations hitherto have not applied to service in front of the enemy, in which case necessity has no law, and every man must occasionally work in every way to the uttermost.

THOS. HODGKIN, Esq. M.D., called in, and examined, 30th July, 1832.

T. Hodgkin, Esq., M.D.

1. What is your profession ?—I am a physician.

2. Have you been a medical officer in any public institution?—I was formerly physician to the London Dispensary.

3. Have you been a medical lecturer in any branch of your profession ?—I have given lectures on morbid anatomy at Guy's Hospital.

4. To medical students ?—To medical students.

5. But your attention has been directed to every branch of your profession ?—I believe it has.

6. Then the committee wish to ask you, what your opinion would be, appealing to the general principles of your profession, as to the effects of excessive labour, or labour so long continued as to produce great fatigue of mind and body, and without due intermission for meals or sufficient time for sleep; should you think such a state to be inconsistent, generally speaking, with the preservation of a sound state of health ?—I should think it would impair all the functions, although some more than others.

7. Would not such a degree of labour, habitually undergone, be more injurious, in your opinion, when endured by children and young persons, before they have arrived at a mature period of life ?—Undue exertion of any kind would be more felt by very young children than by adults.

8. Do you conceive that due intermission for repose and sleep are essentially necessary to the health of children ?—Yes, of all individuals, especially of children.

9. Do you, in reference to the general experience of mankind, and the principles of medical science, believe that the customary hours of a day's labour—namely, twelve, including the necessary intervals for refreshment and rest, are, in ordinary cases, as long a term of human labour as is consistent with the preservation of a perfect state of health ?—It seems to be a very rational distribution of labour and rest.

18. Do not you think that the intervals for meals, and the consequent intermissions of labour for short periods during a day's toil, are as necessary to the preservation of health as they seem to be congenial to the wishes and feelings of those enduring human labour ?—Certainly.

11. Then to undergo a great length of labour without such intermissions would, in your opinion, be imminently prejudicial ?—It would produce exhaustion without the means of reparation, and ultimately, if such a state was continued, it would speedily put an end to the machine.

12. Assuming, then, that the labour undergone in mills and factories almost always exceeds, in point of exertion and confinement, a day's labour as ordinarily computed—namely, twelve hours, and often greatly exceeds it, extending sometimes to thirteen, fourteen, fifteen, and even eighteen or nineteen hours a day, or even still more ; can you have any doubt that such a system must, in ordinary cases, prove highly prejudicial to the health and well-being of those that endure it ?—None whatever ; it is difficult to conceive an extraordinary case in which it would not.

13. So that you would think that the preservation of health in any such instance might be regarded rather as the exception which such a system presented than as indicative of a general rule ?—I should.

14. Is it your opinion that labour pursued through the night, instead of in the day, must be more prejudicial to the human constitution ?—I believe that it is so ; I do not doubt it.

15. Reasoning from the atmospheric influences, and the aptitude of the human being, do you think so ?—Yes; I believe light is a very salutary agent, more particularly for the development of young animals of all classes.

T. Hodgkin, Esq., M.D.

16. Should you think that labour or attention, so long continued as has been just described to you, although it might, in ordinary cases, be denominated light and easy, yet still requiring constant attention, and inducing much fatigue, would not produce considerable weariness, and many injurious effects on the human constitution?—I should think that it would—I have no doubt of it.

17. Do not you think that such wearisome and continued uniformity of human exertion, continued for such a length of time, would be more prejudicial, probably, to the constitution, than a more strenuous exertion of the muscles, continued, however, for a moderate length of time, and with due intermissions?—There are two bearings to that question; on the one hand, there is the influence of habit, which would render the occupation, to a certain degree, less injurious; and, on the other hand, there would be the impaired system to endure labour of any kind. The effect of excessive continuance would, I believe, on the whole, counteract any salutary effect which habit would afford.

18. Is there not a considerable muscular exertion of the human body required to maintain the erect position?—There is the exertion of some muscles.

19. And must the exertion be principally laid on a certain set of muscles? —Yes.

20. Not affording, like varied exertion, such alternate exercise of different sets of muscles as would give repose to each alternately?—The weary sensation which every one feels who is accustomed to stand long, must convince us there must be some muscles unduly exerted.

21. Then you conceive the erect position, in which this labour has to be endured, would, generally speaking, give additional severity to that description of labour?—I think that it would; there would be some disadvantage in a constant preservation of the erect posture.

22. Now, alluding to the general term of labour, and to the description of it which has been partially given, would it not be still more distressing and injurious if it had to be pursued in an atmosphere rendered impure by dust and flue, constantly given out from the materials worked up?—I believe that is proved by the result of every occupation which produces much dust.

23. From personal observation, are you acquainted with the effects of the dust that is given out from flax or cotton?—I am not.

24. Arguing from the principles of your profession, and the knowledge of your own practice, you would imagine, if so much dust was given out of the material as almost to load the atmosphere which has to be breathed, that it must be injurious?—It is the notorious effect of a great variety of occupations which produce dust.

25. Would not labour so circumstanced and continued be still more prejudicial if undergone in an atmosphere artificially heated to from 70° to 80°, and upwards, taking the general average of the heat of certain mills now where the operations in question are performed?—A heat above temperate is injurious to, and tends to shorten, life.

26. Would it not, if habitually undergone, and especially as connected with active exercise, tend to weaken the muscular powers of the body? —That must very much be influenced by the purity of the air in conjunction with it; we have instances of quite athletic strength in very hot countries.

27. Then the committee may ask, with confidence, whether labour endured in an atmosphere of equal temperature would not be less prejudicial if that atmosphere was heated from natural causes, rather than by the artificial means that are taken to raise it, as in the instance of many mills

and factories?—Certainly, if the artificial process was adverse to ventilation which it need not be. T. Hodgkin, Esq., M.D.

28. In addition to other circumstances connected with the mill system as now pursued, is not the labour endured during the whole of the night in many instances prejudicial, in your opinion, to the sight?—I think it is extremely probable.

29. So that, upon the whole, in reference more particularly to children and very young persons, whom it is the object of the measure now under the consideration of this committee and of Parliament to relieve, do you not conceive the labour in question (continued for that length of time which has been described, without sufficient, and sometimes without any, intervals for meals, pursued generally in an erect, or at least in a constrained position of the body, and too often in an impure and heated atmosphere, frequently heated to a very high degree of temperature) to be pernicious to the health of children and young persons who have habitually to undergo labour under such circumstances?—There can be no doubt of that.

30. And is it your opinion that, to bring within due limits the hours of labour endured under such circumstances, would, to a considerable extent, and more perhaps than the mere arithmetical proportion of their diminution would seem to indicate, mitigate some of the evils that result from such a system?—There can be very little doubt that the employment of children might be so regulated as to have its evils very much reduced, if not altogether removed.

31. Would not one of the necessary regulations be to abate that excessively long period of human labour?—Certainly; it would be a *sine qua non.*

32. You would conceive, as a physician, that intervals, especially for a remission of labour, like that which has been described to you, are essentially necessary, in ordinary cases, to the preservation of health?—Yes.

33. You have already intimated that, practically, you are not acquainted with the factory system?—I have had no continued observation of the kind, but I have had an opportunity of seeing three young persons who have been employed in a manufactory in the north.

34. The committee would ask you what you conceive would be the results, medically considered, of labour such as that described, and under the circumstances that have been detailed to you in some of the questions that have been propounded?—I apprehend that there would be a gradual injury to the constitution very unfavourable to nutrition, and that under those circumstances the growth would be very much perverted, both by the effect of the assimilated material, and also by the position of the body.

35. Should you conceive that pernicious consequences would result from this system in respect to those cases that are commonly denominated surgical ones?—Some surgical affections might take their origin from these causes.

36. It has been asserted by witnesses before this committee, that diseased and ulcerated legs, especially among the female part of the operatives, have been produced; should you conceive that long standing at the labour in question might produce that effect?—Certainly, by interfering with the circulation.

37. Do you conceive the labour in such an atmosphere as that described would have a tendency to produce pulmonary affections?—The dust and heat would contribute to that effect, and their general impaired constitution would be less able to resist any deleterious cause.

38. Now, alluding to the *osseous* system of the human frame, it has been

T. Hodgkin, Esq., M.D.

asserted that deformities are exceedingly common where the factory system prevails; should you expect, from your knowledge of the human frame, that such would probably be the result of labour continued as described; and under such circumstances as have been already mentioned to you, alluding now to the extreme length of this labour, and to the wearisome and exhausting nature of it?—Various deformities of the legs would, in all probability, be produced, and perhaps of the spine.

39. From the peculiar formation of the human foot, do you think that the standing position would, in many instances deform that also?—I have known instances of the foot being flattened, and the ankle very much depressed and forced inwards, when there has been long standing.

40. Supposing that those deformities have originated as late as from nine to ten, and often as late as at sixteen or seventeen years of age, should not you suppose, if that degree of labour had not been endured, that in such instances the body of the individuals so deformed would have continued to retain its original perfection?—The older the individual, the less likely he or she would be to be distorted; and to be distorted at that age is a proof that the cause must be a powerful one; but I should very much suspect, individuals so altered at that age must be persons of unhealthy constitutions.

41. But should you not think that the original tendency to such a disorder might be counteracted, and the constitution recover, if that severity of labour had not been imposed on the individual?—A person at that age would, in all probability, escape, if that or some other powerful cause did not operate on them.

42. It has been stated by more than one witness before this committee, that a large proportion of those who have been long engaged in mills and factories are deformed in a greater or less degree; one witness, I think, asserted that out of about 2,000 children and young persons that had been so employed, about 200 were found to be deformed; should you not think that to be a large proportion?—It is certainly a very large proportion of deformities to occur among children.

43. Some of them being perfectly straight, as it is alleged on all hands, to twelve or fourteen years of age, and after that being excessively crooked and deformed in all the lower extremities; would you hesitate to attribute such cases as those to the description and degree of labour which they endure?—I think it would be a very obvious inference that their habits had produced it.

44. And not difficult to account for on the principles of your profession?—Not at all; but in that, I apprehend, there would be a variety of circumstances combined, comprising diet as well as occupation.

45. Taking all the circumstances into consideration, should you be prepared to expect the duration of human life to be considerably abridged where that system prevails very extensively?—I think that to be an obvious consequence.

46. Now it being a known fact, that of the children and young persons occupied in these mills and factories a very large majority are of the female sex, will you have the goodness to state whether, in your judgment, the female sex is as well fitted by nature to sustain long exertion, especially in a standing position, as the male, either in respect of the peculiar formation of the female, or the natural capacity of that sex to endure long and severe labour?—Probably before puberty there would be little difference; but one of the evils of this system would be injuriously to hasten that period.

47. Taking into consideration the warmth of the atmosphere where these children labour, and also many of the exciting causes which appear to exist,

and which it is unnecessary to describe, do you imagine the period of puberty may be anticipated in those instances?—I am certain that it must be. T. Hodgkin, Esq., M.D.

48. Now, has not the labour in question, if it is so far pursued as to produce great languor and inanition, a direct tendency to induce those having to undergo it to resort to the habits of tippling and dram-drinking, supposing that they could obtain such stimulants?—I believe it would tend to increase the craving for artificial stimulants to support undue exertion.

49. Is not the body, in your opinion, in a very unfit state to renew its exertion when it has been insufficiently recreated by sleep, and when therefore labour has to be commenced at the beginning of the day with the feelings and signs of weariness still remaining?—Certainly it is; and that is one of the principal reasons which led me to give the answers which I have done to some of the former questions.

50. That accumulated fatigue you conceive to be peculiarly injurious to the constitution?—Yes; without the interposition of intervals sufficient to repair the demand which has been made on the system; this principle applies to a great variety of agents.

51. Is not the labour of children and young persons, when it has to be so long undergone as to render it necessary for those exacting its continuance to stimulate them, especially towards night, by constant and severe beating and punishment, highly injurious?—I submit whether that is not in substance answered by what has gone before; the necessity for such stimulants is one of the proofs of the labour being excessive.

52. Do you think, in a professional point of view, it is safe to excite by other and less objectionable means—namely, by rewards, the children and young persons to an exertion of the kind described; or that such a course can be reconcileable to the health and well-being of those on whom it is made to operate?—That must be very much influenced by the temper of the children so excited; even such means are liable to very great abuse.

53. Do you think that fatigue and drowsiness often occasion those many and distressing accidents which occur in mills and factories?—I can easily conceive that in that state accidents would be more liable to occur.

54. Do you think that a child of either sex, under nine years of age, ought to be labouring in a mill or a factory, for any considerable length of time together, even during the day?—I should be very glad to do anything in my power to discourage their being employed in any labour at so early a period as that.

55. Do you think continued work and labour at that period of life, under any circumstances, consistent with the proper development of the faculties of the mind or body?—I do not think it is.

56. Are you of opinion that a young person from nine to eighteen, or to about the period when the osseous system is arriving at its full development and strength, ought to labour more than twelve hours a day, including two hours for meals and refreshment?—Not habitually.

57. In your judgment, therefore, supposing that a legislative enactment, protecting children and young persons from excessive labour, should be agreed on, is it not unreasonable to extend the protection up to that period?—It appears to be a proper time. It is difficult to say whether, under particular circumstances, it would not be excusable to exceed it, although I see the danger of opening the door.

58. Are you, then, of opinion that, ordinarily speaking, the usual term of labour is quite sufficient for the human body to endure with impunity—namely, twelve hours a day?—Yes; our labouring classes of all descriptions who can procure work are not suffering bodily from having too little.

T. Hodgkin, Esq., M.D.

59. Now, with reference to the proofs of the morality or immorality of this system, as at present conducted, supposing it should appear to you that great immoralities were connected with it, would you esteem it a sufficient disproof of that, if the number of illegitimate children was nevertheless fewer than in other situations?—Not at all.

60. Will you indicate your reasons to this committee for forming that conclusion?—I should say, in proportion to the females giving way to very early and promiscuous intercourse would their fecundity be impaired.

61. Supposing that the labours of the week occupy as many hours as are commonly undergone in these mills and factories—namely, fourteen or fifteen hours a day, exclusive of the time necessary for them to go to and from their work, would you not conceive that the confinement on their only remaining day of rest—namely, on the Sunday, in the Sunday-school, would, physically considered, be an additional cause of suffering to them?—I should be very sorry to say anything which could discourage Sunday-schools, and unless you could prove the children would spend their time in a better way, I hardly know whether we should be doing right in any way to discourage them going to a Sunday-school.

62. Seeing the conscientious convictions you entertain on this subject, and having in view, not the discouragement, but the better support and countenance of those institutions, should you hesitate to state, that if the labour of the week was somewhat contracted, the advantages of those institutions to the lower classes of the community would be more certain and more effectually rendered?—That I do not hesitate to assert. Children who are in tolerable health, and not overworked, would bring sounder intellects and more attention to their Sunday-school lessons.

63. It having been stated to this committee, as well by teachers of Sunday-schools as by numerous witnesses who have themselves been learners at them, that after such a week's labour as that described it is with the greatest difficulty the attention of the scholars can be kept up at all, that they have the greatest aversion to be confined for the remaining day, would you have any hesitation in saying that the mind is not in a proper state to receive the necessary instruction which such institutions would afford, when all the powers, both intellectual and physical, are still remaining under fatigue and exhaustion?—That must be admitted by every one; there can be no doubt of it.

64. Is the limitation of the hours of study and of application in our public schools and seminaries necessary to the preservation of the health of the scholars and of the students repairing to them?—I think it is.

65. Of course, although giving your opinion here as a physician, you feel also as a private individual in relation to the subject of this inquiry; have you any doubt but that such a remission of the hours of labour as, while it would secure perhaps a better chance of health and strength to the parties, would also afford them some better means of moral and mental culture than they can by possibility possess, would not be eminently beneficial to themselves as individuals, and to the interests of the community at large?—I think it would be very beneficial.

66. Your observations are directed to all trades, as well as manufactures, with regard to the hours of labour?—Yes.

67. Do you know anything about the time that children are employed who are apprentices to tailors, shoemakers, or any other trade?—I do not know of any except in the class of milliners' apprentices, and I believe they require as much attention as the children in the manufactories; I have repeatedly seen injurious effects in that class.

68. You think they, in point of regulation, really require as much of the interference of the legislature almost as manufactories?—Very nearly.

69. Then you have no doubt, in innumerable other cases, although they may not have come under your own eye, the same excessive labour and the

same hardships are imposed on youth?—I think it is very likely that they are; but the manufacturing system, I think, rather lends itself to the bringing in of children at so early an age that they can hardly be employed in any other description of work. T. Hodgkin, Esq. M.D.

70. Are not the circumstances often connected with the factory system, still appealing to the principles of your profession, namely, confinement in a heated atmosphere, the impurity of that atmosphere, and other particulars that have been alluded to, calculated to render that description of labour more injurious to the constitution than that sort of *locomotive* exercise pursued in other trades and employments, although it might be continued for as long a term?—Such are additional evils, and of great weight.

71. Are you of opinion than ten hours for labour are more adapted to youth than eleven?—Yes.

72. So you would say that nine are better than ten; you do not mean to say ten is exactly the period fitted for the youthful frame?—I do not consider myself prepared to assign the exact limit which would be the best.

73. Do you think that a bill, the provisions of which were that no child should be allowed to work under ten years of age, but that the limit of labour should be eleven hours, with two hours for refreshment, making thirteen, would be a great benefit to the labouring classes?—I think that would be.

74. Would not a further limitation, in the instance of these young labourers, namely, from thirteen to twelve hours, be a still geater benefit as regards that class of the community?—I should think that it would.

75. So that in preferring thirteen hours to the present unrestricted term of human labour, you are only making a choice of two evils?—The evil is reduced, but is not done away with.

CHARLES ASTON KEY, Esq., called in, and examined, 4th August, 1832.

1. What is your profession?—A surgeon. C. A. Key, Esq.

2. Have you any medical office in any of the great medical institutions of this metropolis?—I am surgeon to Guy's Hospital.

3. Are you a lecturer or medical teacher in this city?—I lecture on surgery at Guy's Hospital, which is not in the city, but in the Borough.

4. Your attention, both in that hospital and in your private practice, has been directed, of course, to the condition of the lower orders of society, in respect to their health and diseases?—Yes, a great deal for the last twelve or fourteen years.

5. Is it not universally received by your profession, that a moderate degree of employment, with proper intermissions for refreshment and rest, is essential to the preservation of health?—I should say, as far as I have learnt from experience, that a moderate degree of exercise, with intervals for refreshment and rest, produces the best condition for the development of the powers of the body.

6. You conceive, that the advantage of moderate exercise, recreation, and rest, is still more necessary to children and young persons during growth?—Yes; more especially for young persons it is necessary.

7. It follows, therefore, as a matter of course, that labour undergone for an undue length of time, so as greatly to fatigue either the mind or body, without having the intermissions alluded to, is, generally speaking, inconsistent with the preservation of health?—I should conceive labour that is performed with so much exhaustion of the mind and body must be very in-

C. A. Key, Esq. jurious indeed; all organs worked beyond their natural powers must be very much injured.

8. You conceive such undue exertion would be still more prejudicial to children during the period of growth?—I cannot but believe, that if persevered in for a great length of time, it must lay the foundation for disease.

9. Do you not think, referring to the established principles of your profession, and more especially to the experience that your practice has afforded you, that twelve hours' labour a day, (to fix a definite period,) with proper intermissions for meals, is, upon the general average of human constitutions, about as much as can be endured with perfect impunity?—I should think that must depend very much upon the nature of the labour; some kinds of labour the body might be subjected to for twelve hours without injury; and, on the other hand, there are other kinds of labour that might be injurious if persevered in for twelve hours a day.

10. Do you not consider that intermissions from labour during the time of taking the meals are essentially necessary to the industrious classes of the community, in regard to their health?—Unquestionably; because, without proper time being allotted for meals, digestion, and the processes preparatory to it, must be imperfectly performed.

11. You consider it to be inconsistent, generally speaking, with the functions of the stomach, that a person should be labouring while taking his meals?—Inconsistent with proper digestion and with the nourishment of the body.

12. Assuming, then, that the labour that is undergone in mills and factories extends to thirteen, fourteen or fifteen hours a day and upwards, should you have any hesitation whatever in concluding that that degree of labour, not alluding particularly to the extreme cases, would be, generally speaking, prejudicial to the constitution?—I should say, it is quite inconsistent with a state of health.

13. Then you would not be surprised at the grievous effects that result from that labour, as deposed to by a series of witnesses that have come before this committee?—No, I should not.

14. Of course, when that labour is occasionally extended to eighteen, twenty, or thirty hours together, or even still longer, the effects cannot fail to prove highly injurious?—There is not one constitution in a hundred that can sustain it with impunity.

15. Do you not think that labour would be more distressing and injurious if it had to be pursued in an atmosphere full of impurities evolved by the materials manufactured, such as the dust from flax, flue from cotton, &c. so prevalent often, in some departments, as to render it difficult for individuals to see each other at some distance?—I should think that impure air of that kind breathed for a great length of time under the circumstances mentioned must be productive of disease, or exceedingly weaken the body, by preventing the natural changes which the blood ought to undergo by respiration.

16. Do you not consider that in those cases where the atmosphere has to be heated to a high degree of temperature, namely, 70° or 80°, the labour in question must be still more prejudicial to the constitution?—Yes; by quickening the circulation of the heart, and thereby tending to shorten life; and also tending to excite inflammation in the finer tissues of the body; for instance, the eyes or the lungs, which we know are influenced very much by a high artificial temperature.

17. You would conceive, perhaps, that another evil would result from working in so high a temperature, when during a considerable part of the year the hands so employed have, on going to their homes, to rush at once

into the cold, when perhaps the thermometer is at the freezing point, or below it?—Children so imperfectly nourished are unable to bear the exceedingly depressing effects of sudden changes from a very high to a low temperature. C. A. Key, Esq.

18. Do you not think that labour pursued in an atmosphere thus artificially heated, connected with the impurities alluded to, is still more prejudicial than labour pursued in an atmosphere of an equal temperature but a natural one, and freshened by the influences of currents of air?—The former must be infinitely more prejudicial than the latter.

19. Do you think that the effects of this labour might be more prejudicial, if it had to be undergone in the night for an equal length of time, and under similar circumstances, than if endured in the day?—It is inverting the natural order of nature, and therefore must be highly injurious both to the body and mind.

20. Would not the artificial lights by which the labour must be pursued have a prejudicial influence upon the health, by destroying to a certain degree the salubrity of the atmosphere?—Undoubtedly they must tend to deteriorate the quality of the air, and render it unfit for the purposes of respiration.

21. Some of the operatives have alleged, that the gas-lights have a tendency to injure the sight when they have to labour by them for so long a period; do you consider that that might be a result produced by labour under those circumstances?—Yes; I conceive that that is a very natural consequence.

22. It is alleged by the witnesses, that the labour in question is very fatiguing and exhausting; it is nevertheless stated by some who apologize for this length of labour, that it is "light and easy;" may I ask you, whether an employment which demands such constant and excessive attention for so long a period of time, and mainly in an erect position of the body, has not a more prejudicial effect than more natural exertions, though somewhat more strenuous, pursued for a moderate length of time and with due intermissions?—I should consider that it matters not of what nature the labour may be, if it is persisted in under an extreme state of exhaustion and fatigue of mind and body; if it be so persisted in, in an erect position, I consider it to be exceedingly injurious to the growth and development of the powers of the body.

23. If pursued in a stooping and constrained position of the body, so as to exercise many of the muscles unduly, it would be still more fatiguing?—Yes, still more so.

24. Have you ever yourself examined mills and factories where this labour is pursued?—No.

25. So that the evidence with which you have favoured this committee is that which has been deduced from the principles of your profession, and from the analogies your practice has afforded?—Yes; by the observation of disease among the labouring classes of the poor of this town.

26. Does not it require considerable muscular exertion to keep the body in an erect position, and does not that position, continued for a considerable length of time together, in itself induce much fatigue and exhaustion?—After a length of time, the body being kept in an erect position, the muscles must become excessively fatigued, and deformity is almost the invariable consequence, more or less. I can speak to that as an effect of constrained position, because I see it in all classes of persons, both in the poor as well as in the higher classes.

27. Bearing in mind the nature of the labour referred to, its average duration, and the circumstances under which it is rendered, above all, that it is generally borne by children and young persons before they have arrived

C. A. Key, Esq. at full maturity, should you not naturally expect that a train of diseases must result from such a state of things?—I should conceive, from what my experience has taught me, that very few human bodies could undergo such a state of labour for any length of time without laying the foundation of disease, or bringing into action diseases of the most serious kind.

28. Considering, then, the nature of the employment described to you, what should you think would be some of the most common results that might be expected to be produced; what do you think are the particular diseases that such a system would produce?—In the first place, digestion must be imperfectly performed; chyle and chyme not being properly prepared in the stomach and bowels, morbid secretions would collect in the intestinal canal, giving rise to the formation of worms, producing irritation in the lining membrane; engendering mesenteric disease, (which I believe to be a very frequent consequence of this kind of labour,) producing scrofulous disease in all the glandular structures of the body, or some of them at least. The effects would also be extended to the functions of all the organs concerned in supporting animal life; the heart, the lungs, and the abdominal viscera must likewise suffer. I have mentioned that the tissue of the lungs and of the eye would also most likely participate in the mischief occasioned.

29. Would not many distressing surgical effects, distinguishing those from the medical cases, be the result of such a system of labour?—It would give rise to the formation of what we call varicose ulcers in the legs; it would also give rise to deformities, what are termed lateral curvatures of the spine, malformation of the chest, yielding of the joints of the legs, the knees and the ankles, and flattening of the arch of the foot.

30. It has been constantly stated by witnesses before this and preceding committees, that factory children can be readily and almost instantly distinguished from children otherwise employed by the peculiar ghastliness and sallowness of their complexion; would such an appearance be reconcileable to those views of the subject you have just detailed?—I think that every occupation, more or less, stamps its own character upon the appearance of the individual, and therefore, in factory children, I can easily believe that the same thing takes place, and that the unhealthy character of the occupation is stamped upon the countenances of the children so occupied.

31. Have you considered that appearance as one of the exterior indications of some of those disorders of the viscera to which you have been alluding?—Certainly; such unhealthiness of aspect is one of the exterior indications of disease such as I have mentioned.

32. Many instances of excessive deformity have been presented to this committee, which have come on at as late a period in youth as fourteen or sixteen years of age; does the disorder usually denominated rickets develop itself, in ordinary cases, at so late a period of life?—Ordinary rickets generally appear soon after birth, at the period when the children are put upon their feet; but I can easily understand that causes that give rise to extreme debility of body would give rise to rickets at a later period of life, or deformity, that is analogous to it.

33. This disorder being very common among the persons so employed, and coming on at so late a period of life, should you attribute such deformity to the nature of their occupation?—Yes, I should attribute it to excessive labour, under circumstances unfavourable to the nourishment of the body.

34. The deformities in question have mainly affected the joints of the individuals suffering under them; is not that a proof that they have been produced by the labour in question?—Certainly; I should take it to be the natural effect of such labour.

35. The deformity has been accompanied with exquisite pain; is not that another proof that undue straining and exertion has been endured by the individuals so affected?—I should, from the description given, consider it as a kind of joint disease, the product of such excessive labour. C. A. Key, Esq

36. The results of several examinations, very carefully and impartially conducted, are these: first, that the height of children employed in mills and factories, compared with those differently occupied at the same ages, is considerably less; and, secondly, their weight much smaller; would not those results be the consequence of a system of labour like that which has now been described to you?—We know, by experience, that a body, if kept in a horizontal position for a great length of time, will grow very fast; and we have reason to believe, that if a contrary course is pursued, and the body is kept for a great length of time in an erect posture, a contrary effect will be produced.

37. Do you think that a child under nine years of age ought to be subjected to the labour of the factory or the mill?—I should say no child ought, at any age, if the labour is of the nature described to me.

38. Speaking of the limitation of the ages and of the hours of labour proposed by the bill before us; do you not think, after children are admitted into a mill at the time specified, namely, at the age of nine, that between that age and eighteen, twelve hours' a day confinement, involving ten hours a day of actual labour, is as much as their constitutions can sustain on an average?—I should say, from what I know of the human body, it is even more than they could sustain with impunity.

39. To what extent should you think that children ranging from nine to thirteen or fourteen years of age ought to be laboured, generally speaking, in mills and factories?—I should say that eight or nine hours is quite as long as a child ought to be confined in a factory during the day.

40. From that age to eighteen you have already stated you think ten hours' labour, involving as it will twelve hours' confinement, is quite as long a day's labour as can be endured with impunity?—Yes, the health must suffer if the labour be prolonged beyond this period.

41. As the business of a mill is to be conducted with the utmost possible degree of regularity, and therefore the time of labour must be most precisely conformed to by every individual in the establishment, that is perhaps an additional reason why the time mentioned should be diminished, instead of being increased, because it has to be conformed to by the most weakly part of the hands, as well as by the most robust?—Certainly.

42. Do you conceive that the osseous system of the human frame is completely matured and developed much before the age of eighteen?—I believe it is never perfectly developed or formed at eighteen, or till the person becomes of age, and even then the bones are not quite set.

43. So that that furnishes an additional reason for a more limited duration of labour?—Certainly, as regards young persons.

44. Perhaps you are of opinion, with many eminent men of your profession, that the term of labour in question, namely, twelve hours, is as much as ought, upon a general average, to be imposed upon human beings at any period of life?—I should say that that is the utmost that ought to be imposed upon any persons of any age or sex, or any condition.

45. It is a well-known fact that a considerable majority of children and young persons employed are females; do you consider that the female sex is, generally speaking, as well calculated to endure labour and fatigue as the male sex?—Much less able to endure labour than the male sex.

46. Is it not peculiarly prejudicial to inflict excessive labour upon females when approaching the age of puberty?—Excessive labour about that

C. A. Key, Esq. period will tend to prevent a proper development of the pelvis, and also of the organs concerned in generation.

47. Would it have a prejudicial effect upon the general health of the party enduring it at that particular period?—Yes, it would influence the general health by interfering with the functions of that part of the system.

48. Will you state your opinion whether labouring in a heated atmosphere would anticipate the period of puberty?—I apprehend, from what we know of the effects of warm climates, that artificial heat would perhaps more particularly tend to anticipate that period.

49. Might not that period be still further anticipated from the circumstance of the indecencies and immorality said to prevail in such establishments?—I should think that such an effect might be expected to be the result of immoral habits, such as have been described; that they would tend to develop the sexual passions previously to the full maturity of the body.

50. Is very early and promiscuous intercourse between the sexes ordinarily attended, as far as the female is concerned, with prolificness?—I should be inclined to think to the contrary.

51. Has not the description of labour which produces much languor and an inanition of body and mind, a great tendency to generate and continue the habit of tippling, said to exist where the system in question prevails?—I should imagine that the labour could hardly be endured without having recourse to some unnatural stimulus of that kind.

52. Is not the body in a very unfit state to renew its labour in the morning, when the fatigue of the preceding day has not been removed by sufficient refreshment and rest?—Such a system of extreme labour, under exhaustion and want of rest, is quite inconsistent with the health of the body.

53. Towards the termination of the day, when it becomes necessary to scourge the children and young persons in order to keep them up to the degree of labour required from them, must not continued exertion then have a most pernicious effect upon the health and feelings of the parties?—It must be totally destructive of the moral feelings, as well as the health of the children.

54. Has not a sense of degradation and hopeless suffering, and a long unvarying prospect of the continuance of such a state, a tendency to injure health?—Extreme depression of mind produced by such exhausting labour, and by so hopeless a condition, must also assist in injuring the child's health very much.

55. Do not you conceive that it would be additionally pernicious to children to send them to a night-school, if such were established, after they have been already confined fourteen hours a day in mills and factories?—It would be adding to the real mass of labour, without the child receiving any proportional improvement of the mind; because, after such fatigue, the child could not be in a fit state to undergo this mental exertion.

56. When, in consequence of the labour of the preceding week, a sense of weariness and exhaustion continues throughout the Sunday, can the institution of a Sunday-school be as advantageous as it otherwise might be, with a view to the mental and moral improvement of the children?—A child so exhausted would feel the Sunday-school as an addition to its daily labour and confinement; and also be incapable of receiving so much instruction as a child in ordinary health would be capable of receiving.

57. Should you not consider, speaking professionally, that the physical

sufferings of children so circumstanced were increased by having to attend a Sunday-school on their only remaining day of leisure?—It appears to me that it would be an additional infliction of hardship to children so circumstanced. C. A. Key, Esq.

58. So that a due remission of the hours of labour on the week-days would tend to restore to such institutions the utility and advantage they are calculated to bestow upon the lower classes of society?—I believe they could only be useful under such circumstances.

59. Is it not found necessary, medically speaking, to limit the hours of study that young persons in schools and seminaries have to endure?—It is known that children who are made to apply too closely to their studies have their health very much impaired, as well as their bodily powers diminished.

60. On the whole, then, you think that a due remission of the hours of labour, as regards children and young persons, would tend much to the security of their health, and be eminently advantageous to them in both a moral and physical point of view?—I believe that, without such remission, health cannot be maintained.

61. The provisions of the bill in question protect children and young persons, beginning with the age of nine and going up to eighteen years of age, from a greater length of confinement than twelve hours, and of actual labour than ten hours; do you sanction with your medical authority the reasonableness of such a limitation?—Most fully.

WILLIAM LUTENER, Esq., called in, and examined, 16th June, 1832.

1. Where do you reside?—At Newtown, in Montgomeryshire. W. Lutener, Esq.

2. What is your profession?—Surgeon and general practitioner in medicine.

3. Is there any manufacturing carried on in Newtown?—A considerable one in Newtown and the neighbourhood, in which the finest and best flannels in the kingdom are made.

4. Are they reckoned the best of that species of manufacture in the kingdom?—Yes.

5. It is a place that has considerably increased of late years, has it not?—Very much.

6. A considerable portion of the population are employed in that manufacture?—Yes, the chief of them; in fact, all, except the usual tradespeople of a country town.

7. Will you state to this committee whether that manufacture is principally carried on in mills and factories, or otherwise?—Chiefly in factories.

8. What are the usual hours of labour in those factories, in the first place directing the inquiry to the labour of the adults?—Twelve hours.

9. With what intervals for rest or refreshment?—An hour and a half, I believe, is the time allowed.

10. What are the hours of the labour of the children?—The same number of hours.

11. Is the labour of the children in busy times of the year ever lengthened beyond twelve hours?—The children work by night in that case.

12. Do the same set of children occasionally work all night after having worked during the whole of the day?—Constantly; at the present time the trade is rather slack, but still it is done now, in one mill in particular.

13. During the last year how many nights do you think that any particular mill that you know of has worked?—I should think every night; it

W. Lutener, Esq.

is my belief so. There might be an exception of a few nights, but the ordinary practice is to work by night when there is a demand for goods. I may explain it by this circumstance, that our power is entirely water power; we have very few mills; the rents are enormously high, and without the tenants of those mills work all night, so as to get twenty-four hours' labour out of the machinery, they cannot pay the rents.

14. So that it is the ordinary, nay, usual practice, to labour the same set of children during the day and all the succeeding night?—It is; just as I left home I made a good many inquiries upon the subject, and I rather think there is less work done now than there has been for the last few years; but they are at work at this time of night at one mill I know.

15. Do the same children work both night and day?—Invariably.

16. Have they no rest?—Yes; they do not work every night.

17. Will you state in what manner a child would work that is employed in the times to which you have referred?—Before I answer that question I should state, that there are two methods of working the children; some of the children work two nights a week, and some three.

18. Will you have the goodness to state the hours of labour of the child that works two nights a week?—It begins on the Monday morning at six o'clock, labours all day till six, with about one hour and a half interval for meals; it then has, I believe, an hour, or half an hour, and returns to its work at seven, and works till six in the morning, having no interval whatever; it is then ready for its daily labour on Tuesday morning at six; its night's labour has ceased at six.

19. It commences its Tuesday labour at six?—Yes.

20. And works during Tuesday till six at night?—Yes.

21. With the same interval for meals during that day?—Yes.

22. On Tuesday night it has a night's rest?—It goes to bed.

23. State its labour on the Wednesday?—It goes to work at six o'clock, and labours till six in the evening, with the usual period for meals.

24. What does it do that night?—It goes to bed; at six o'clock it is released.

25. On the Thursday, state the commencement of its labour, and its termination?—On the Thursday morning it begins to work again at six, and labours till the evening; and either that night, or the next night (it is not a fixed term that they have) I believe it again labours through the night, and the succeeding day.

26. Twice in the week it works two entire nights, thirty-six hours continuously?—Yes; allowing about four hours for meals.

27. Will you have the goodness to state the custom in respect of the labour of those children who have to work three nights a week?—They then begin at six in the morning, the same as in the other case, and work till six the next morning; they then rest all that day, and all the following night; they work three days a week, and three nights together.

28. They work twenty-four successive hours three times a week?—Yes, with about an hour and a half or two hours cessation in the day for their meals, as in the other case.

29. So that each of the children so employed works for twenty-four successive hours, and has afterwards an intermission of an equal number of hours?—Yes, I believe so.

30. That labour and that intermission recurring three times a week?—Yes; but they have the same hours rest.

31. Is this the ordinary and usual practice of those mills, when they are in full activity and in employment?—Certainly.

32. How long have you known that practice to continue without interruption in any one year?—I really cannot speak with any great certainty;

but I know, being up by night, that I have seen them up all night. But in the summer it frequently happens that the river is low—it is the Severn—and they cannot work then beyond a few hours in the day; they dam the water up in floods; they are prevented from working both day and night, and then there is a cessation of working. When they have the power of working they do; they are also occasionally prevented working by the machinery requiring repair. W. Lutener, Esq.

33. How long have you known this to continue, without interruption, at this rate?—I think it is the constant practice.

34. For three fourths of the year?—Decidedly. I am by no means connected with the business, therefore I cannot speak with certainty of it.

35. When they are not interrupted from the causes which you have mentioned, or by any particular stagnation of trade, such is the custom which you have described it to be?—It is the general, constant, custom.

36. Can you state the description of labour?—It is of a very light, easy, kind, and I will describe it as well as I can. I made a point before I came up of visiting a mill, in order to satisfy myself on the point. They are employed in two ways; the night-work is what is called feeding, it is putting the wool into the machine which cards it, and turns it out into what is called rolls, being the first process in spinning, that is their employment by night entirely; but by day there are a great many more children employed than at night, because they are employed in pieceing those rolls, the wool that is turned out in process to the others, they are what are called piecers and feeders; so that they do not want nearly so many children by night as by day. The labour is of a very light kind; it is not so much labour as confinement.

37. It is one that requires perpetual attention and vigilance?—Yes; and if they do not attend to their work by night, but go to sleep, the employer perfectly well knows how much work ought to be done by the morning, on account of the machine.

38. Is the temperature required for that manufacture above the ordinary temperature?—It is not a great heat, but there is an unpleasant smell, on account of the oil that is used.

39. Do you know anything about the rate of the heat?—I do not; I never ascertained it; it is rather high. On account of so many people being brought together, it is necessarily much increased.

40. That lighter part of the work is principally conducted by children and young persons, what is called the feeding?—Yes.

41. Will you state whether those children are under proper superintendence or not during the time of their night labour?—I believe there is nobody with them in general.

42. No adult persons taking the superintendence over them?—No, not generally, I think.

43. Are children of both sexes employed in this work?—Yes.

44. Between what ages do you think?—There are instances of their going in as young as seven, but I fancy from seven to eight they go in generally.

45. To what age do they continue in that employment?—When they get to fifteen or sixteen they turn to weaving, or some other branch of manufacture, by which they get more money.

46. Are the sexes so employed mixed up together in that labour during the night, without the least superintendence over them?—Yes, without the least regard to sex.

47. What are the wages of the children so employed?—From one shilling and three-pence to half-a-crown a week, varying according to their ages. I stated that they have five-pence or six-pence extra for the night-work.

W. Lutener, Esq.

48. It is almost unnecessary to ask you as to your impression of the moral effects of a system of labour like that upon the rising generation ?—I can say very little more than must be obvious to every one, that the morals of society must suffer greatly by children being thus thrown indiscriminately together, alone, and that by night.

49. Have you any establishments to counteract the consequences of such a state of things; any police ?—No; we are without a police, without a constantly-resident magistrate, and without a charity-school of any kind, except Sunday-schools.

50. Then the results are obviously such as they might naturally be expected to be from such a state of things ?—They are as bad as can be imagined.

51. For how many years has that labour to that excessive length of time gone on ?—I have been in Newtown nearly twelve years, and it has been carried on nearly all that time.

52. Have not the manufacturers increased since that time ?—The houses and factories have increased much; the mills have not increased very much.

53. You have stated that there is no charity-school in the place, and if there were one, under that system of labour would it not be impossible for children to attend ?—Those children that worked by night of course could not.

54. Is not the state of education lamentably deficient and low in that place ?—Nothing can be more deplorable than the state of education; we have only some Sunday-schools.

55. Is it obviously deficient; as people grow up do they appear wanting in general information and knowledge, as contrasted with other members of the community not similarly employed ?—Very much.

56. Will you explain to what extent you have made your observations ? —They seem to have no conception of law and control; they spend a good deal of their time in public-houses, of which we have a great number; and I have frequently observed, being up late and out late, coming home at different hours, the doors of public-houses open, and people drinking, revelling, and dancing in them at two and four o'clock.

57. That seems to be the result of the circumstance of the children being kept up in the night by the labour being pursued ?—Yes; the commissioners for settling the boundary under the Reform Bill have stated that the poor's rates are very low in Newtown in consequence of the children being employed in the manufactory.

58. So that the advantage of the diminution of the poor's rates is purchased by the excessive labour and the brutalizing ignorance of those poor children ?—It appears to be so; they of course have had better opportunity of inquiring into the state of the poor's rates in different places than I have.

59. Is the population very disorderly as well as immoral, generally speaking ?—Very bad, as you may suppose, without any one to regulate them. About a year and a half ago some men were committed by a magistrate for an act of vagrancy, calling at his house, and begging; he thought them very unworthy objects, and he refused to relieve them; at least, he committed them, or remanded them for another examination, for begging at his door, and being, I believe, impertinent. There was a meeting in the town of the operatives, who thought this was a very hard case; they went in a body of three hundred or four hundred, seven miles distant; this was to the gentleman; and said, "We come to insist on the release of our fellow-men." The gentleman was so perfectly aware that they would commit some act of violence if he did not give up the men, that he signed a release to bring these men out of gaol, and they went to the gaol, and returned triumphing in their success.

W. Lutener, Esq.

60. So that this state of ignorance has introduced disorders and insubordination into that district?—Completely.

61. What effect do you think such a length of labour, endured by those poor children, must necessarily have upon their constitution physically?—It must certainly be injurious, I think; independently of any professional knowledge, the common sense of any man must convince him that no children can so labour without their constitution being injured.

62. Have you ever seen the children at the end of thirty-six hours' labour?—Yes, I have, within the last fortnight; I have made inquiries of the children when I have met them.—"Where have you been working?" "So and so." "Did you work last night?" "Yes." "Did you work yesterday?" "Yes." "Are you going home now?" And I must say, that I feel astonished that the children do not seem more fatigued and exhausted; they seem to be joyful in going home; I think that may arise, in a great measure, from the joy of the relief.

63. You have never been able to discover the difference of the effect of this labour upon the two children?—The same child does the one and the other.

64. Do you mean that the children employed in those manufactories are not generally unhealthy?—I have never seen any particular disease amongst them; they are not those rosy, healthy, children that our agricultural children are; they are thin and sallow looking, and exceedingly dirty. It is impossible that human nature can support this labour, though there are no obvious bad results at present.

65. Have you made any remarks of the longevity of children so employed, as compared with other children?—No; neither is our manufacture of such an extent that we could very well ascertain it.

66. Do you happen to know the population of Newton?—Newton and Llanllwchairn, which is another parish, and separated from Newton only by the river, contain together, I think, between 6000 and 7000 souls.

67. You state that the greater proportion of the population are engaged in manufactories; do you happen to know the proportion of children so employed?—I believe from 260 to 270, as far as I have been able to make inquiries; when I say children, I should not reckon those older than sixteen.

68. Is this a kind of work that could only be done by children?—It might be done by adults, but it would be more expensive.

69. There is nothing in the machine itself which makes the employment of children necessary?—No, certainly not.

70. Have you had reason to remark, in your professional duties, that accidents occur in this overlabour when the children become overfatigued?—We have had frequent accidents, because the children get sleepy at night, and get their hands in the work. I and my partner have had frequently to amputate the hands and fingers of children.

71. Does this frequently happen during the night?—During the night and day; but the child who has been up all night would be no better the next day.

72. Then you have traced this to the over-fatigue?—A child who was working those twenty-four hours, and got his hands entangled in the machinery, I should say was probably half asleep.

73. Do you think that the constitutions of children can become more habituated to this length of labour than adults would be, supposing that they were suddenly taken to it?—No, I think not.

74. So that your impression is that the constitution gets gradually debilitated by these exertions?—It really must; it is impossible to put it on any other ground; common sense must convince any man.

W. Lutener, Esq.

75. As a medical man, no facts have come under your knowledge to convince you of such a result?—No; I should not like to state what I fancied would result, without I was able to speak from facts.

76. Then you speak rather from general reasoning than from actual experience yourself of what has happened?—Yes.

77. Does not your practice principally lie among the higher class of society?—I see all the bad cases; I have a junior partner who attends to all the poorer classes in the town. There is only one other medical man, and that a naval surgeon; therefore our business is the principal one.

78. You would not naturally know the results of this system of labour from your personal inspection so fully as though you were daily engaged in attending it?—No, I should not.

79. You have not, however, the slightest hesitation in saying that this degree of labour must of necessity be highly prejudicial to the human frame? —Most certainly it must.

80. As a medical man you hesitate not to say that it must debilitate the constitution, and shorten the lives of those who are exposed to it?—Certainly.

81. Does this kind of work produce any species of deformity?—I am not aware that it does; I have not had occasion to observe any instances of the kind.

82. The children are standing on their legs during the whole of the time they are employed?—Yes.

83. And the effect of this continued labour on the children is such, that you would always be able to point out a manufacturing child from a child employed in any other way?—Yes, I should say so, if it had been employed any time.

84. What length of labour do you think would be consistent with the degree of health which we ought to wish to maintain among the manufacturing population under eighteen years of age?—I should say eight or nine hours.

85. Will you have the goodness to state the proportion of children in those manufactories, as compared with the number of adults employed, if it is in your power so to do?—I am afraid it is not.

86. You are not sufficiently acquainted with the internal arrangements of the manufactories to know that?—I have made many inquiries, more particularly with regard to children.

87. Are the parents in full work during this time?—Except the women. The men are generally at full work, except they are drinking at public-houses, where they spend a good deal of time and money.

88. What effect has the labour of the children upon the feelings and conduct of the parents in certain cases, the parents receiving the wages for the labour of their children; does it not induce the parent to live a life of profligacy?—Of course, if the children are enabled to get money, there is more money for the parent to spend at the public-house.

89. Are you aware whether the mortality of Newtown is greater than that of any other part of the country?—We increase in population very much; and I am not aware that the mortality is greater than in any other part of the country. In fact, I should say that our mortality was not very great, for we have fine air, and a river running through the town, and of course that tends very much to the preservation of the health of the inhabitants.

90. Is not the increase of the population attributable, in a great degree, to the accessions that it receives from other parts where manufactures are not pursued?—Yes; we have had a good many strangers latterly coming from Yorkshire and Lancashire.

91. Upon the whole, you think that regulating the time of labour of those

poor children whose condition you describe would be eminently serviceable to the happiness, and the morals, and the good order, of that place?—Indeed, that is my decided opinion. W. Lutener, Esq.

92. Do not you think that there might be an opportunity, which the inhabitants of the town and neighbourhood would avail themselves of, for the establishment of some school for the instruction of the rising generation, if there were some restriction in the hours of labour?—Yes; and indeed I have taken some steps, and have had some correspondence with Lord Clive, who has kindly offered me the use of the room over the Market Hall, and assistance for that purpose; but as long as the children are so employed, and get so much money, I think I should have an up-hill battle to fight with them.

93. You think that this excessive work in the children is prejudicial to the good order of the town?—Yes.

94. You said that you considered that eight or nine hours' labour was as much as persons below eighteen could bear without injury to their health? —That would be my opinion; you may take it perhaps from sixteen years of age.

95. Suppose they were laboured for a longer period than that, do you imagine you should be able to detect that by their general appearance?—After some years' continuance at it I think I should.

96. But you conceive that ten hours' labour of the children, after some years' continuance, would be prejudicial to their health?—I do not go so far as that; but I should say generally, that eight or nine hours is a sufficient duration of labour for the children.

97. You would say that twelve hours would be too long?—Yes.

98. Did you make any application to be examined as a witness before this committee upon the Factories Bill?—No; but, on the contrary, I felt a strong objection to do so, and tried to avoid coming here; but having an order from the committee, I was of course compelled to appear.

99. This long continuance of labour of the children partly proceeds from the high rent demanded at the mills?—Certainly.

100. Is there not plenty of water-power in that neighbourhood, so that additional mills might be very easily erected?—The whole of the mills do not extend a mile and a half along the Severn.

101. So that the liberation of those children from this extravagant length of labour would only involve an expense of additional mills and machinery?—That must clearly be the case, if they cannot work the machinery otherwise; but there can be no reason why they could not work the machines if they had adults to work them. I wish to observe, that some mills were sold a few months ago for 5000 guineas, whieh let at 600*l*. a year.

102. Was the machinery included?—The water-wheels, but not the internal machinery.

103. What did that mill cost in building?—Ten or twelve thousand pounds.

104. But at a much dearer time than at present?—I do not know; it has not been built a great many years.

MALYN, JOHN, Esq., called in, and examined, 23rd July, 1832.

1. What is your profession?—I am a surgeon. J. Malyn, Esq.
2. Where and under whom did you receive your surgical education?—

J. Malyn, Esq. At the Manchester Infirmary, under Mr. Simmons, the late senior surgeon of that institution.

3. Was he not regarded as a man of great eminence in his profession?—I should consider him second to none I ever met with in the world.

4. Did not Mr. Simmons direct his particular attention to the health and condition of children employed in the mills and factories of that town and neighbourhood?—Not to this point exclusively; but I know that the number and importance of the facts which presented themselves in his hospital practice caused him to form a very strong opinion as to the influence of factory employment on children.

5. You are aware of his views upon the subject, namely, that he expressed an opinion that more than ten hours' labour was improper to be imposed upon any children and young persons in mills and factories, and that that, in many cases, was too much?—I have frequently heard him express that opinion; an opinion which, I believe, he as honestly entertained as he fearlessly expressed.

6. And that, perhaps, may import, that he was further of opinion, as a medical man of great knowledge and experience, that it was doubtful whether, beyond that term of labour, an adult constitution could be worked with perfect security?—I am aware of that fact, more from having seen it reported in the evidence which he gave before the House of Lords, in the year 1819, than from any distinct recollection of his personal observations; but, so far as the question relates to youth of either sex, he, while the question was pending, drew the attention of those about him to cases corroborative of what he had expressed to their lordships too frequently, and expatiated on them too much at length, to allow his auditors ever to forget his opinion on that point.

7. Had you any opportunity yourself, besides that which you derived from your connexion with him as his pupil, of becoming acquainted with the diseases of the operative classes of Manchester?—I had the best opportunity that could be enjoyed, having held the important and responsible appointment of physicians' clerk to the Manchester Infirmary from the year 1823 to the year 1826.

8. How many patients annually passed through that infirmary when you held that situation?—As far as my recollection serves me at the moment, I believe about 13,000.

9. Will you state the duties of that situation in the great infirmary of that place?—The physicians' clerk is the locum tenens of the physicians to whom he is attached; and, in that character, it is his duty to prescribe for and undertake the management of the extensive classes of home and out-patients, or such portions of them as those physicians may choose to assign to him.

10. And did, therefore, the prescriptions of the physicians who attended upon any particular cases occasionally undergo your inspection?—They did, to a certain extent; but these had little connexion with the real duties of the office. The town was divided into six districts, and a physician was attached to each; to aid them were the two physicians' clerks. The rule (and that was not imperative) was for the physician to visit patients the first time, and, if he pleased, then to consign them to the care of the physicians' clerk; who, by this arrangement, had three districts to superintend in addition to three corresponding out-districts, the patients in which were from the first entirely under his jurisdiction. Furthermore, it was for him to prescribe, on alternate days, for the out-patients in the absence of the physicians, or, when they attended, to assist them in that duty.

11. So that, in point of fact, you acted under the control of the physician, of course, but having a considerable part of the medical responsibility de-

volving upon you?—I was always subject to their control, had they chosen to have exercised their authority. They might have visited their cases the first time, or they might have attended them throughout, had they been so pleased; but those who hold such appointments usually become too much involved in private practice, and the zeal which characterizes the exertions of newly appointed officers is of too fugitive a nature to admit of their being very anxious about the performance of arduous duties, more especially when the scene of those duties is one of filth, and poverty, and wretchedness. J. Malyn, Esq.

12. On the whole, you have had the best possible opportunities of observing upon the health of the operative classes of society in that great manufacturing town?—I have.

13. Have you also studied the profession elsewhere?—I have, to a considerable extent; not because knowledge could be acquired elsewhere better than in the schools and institutions of Manchester, but a legalized authority to practise being required in every candidate for the appointment to which I have just referred, I devoted two sessions to study in the metropolitan schools that I might obtain the necessary qualification. On resigning that office, I resumed my studies here, where, with short intermissions, I have since continued to pursue my professional researches.

14. Have you seen the medical practice in other schools besides those of England?—Not to any very great extent. I have witnessed the practice of the hospitals, and partaken of the advantages so liberally afforded by the institutions of Paris, in which city I resided for some time during the year 1826.

15. Have you attended hospital practice since you completed your education?—A medical man can be hardly said ever to have completed his education; hence I have constantly availed myself of every opportunity of so doing.

16. Perhaps you are in the habit of teaching the principles and practice of your profession?—I am, regularly, to medical pupils who apply to me for that purpose.

17. Are you also, at present, in the practice of your profession in this city?—I am.

18. You can then say, from observation and experience, if there be not a great difference in the health and constitution of the operative classes of society, which is induced by the difference of the length and kind of labour which they have to undergo?—As a practitioner, and as an observer of facts, I can affirm that there exists a marked difference between the operatives of the metropolis and those whose condition you are now investigating. As a physiologist, and, as such, reasoning on those facts, I have been led to connect that difference with the nature and duration of their employment.

19. Is your recollection precise and accurate regarding the nature and extent of the diseases in the great manufacturing district of Manchester, as those diseases must have necessarily come before you in the capacity of pupil to Mr. Simmons, and physicians' clerk, as you have mentioned?—I have a lively recollection of the events of that period. Indeed, the impression which was made upon my mind by what I witnessed is so indelible, that I deceive myself if either time or circumstances can ever obliterate it.

20. Are you aware of the hours of labour in the mills and factories of Manchester?—I cannot speak decidedly upon the point; but my impression is, that while I resided there they were twelve hours and upwards, exclusive of the time for taking refreshment and rest; or fourteen hours and upwards out of the entire day of twenty-four hours.

21. Is not the labour sometimes denominated light and easy which they have to undergo in those factories?—It is, I am aware, so denominated. It

J. Malyn, Esq. is alleged to be both light and easy; but undeviating uniformity may render the simplest act extremely fatiguing. For instance, the short period which is occupied in the celebration of public worship would be passed in a much less satisfactory manner, in the kneeling or the standing, or the sitting posture, than it is by the judicious admixture of the three. And the reason is obvious; for the muscles which are called into action to maintain for a long time any particular position, or to perform continuously the same act, become fatigued from the want of due alternations of activity and repose.

22. Does not the maintenance of the erect position become, when endured for a considerable length of time, peculiarly fatiguing to the bodily frame? —Unquestionably, exceedingly fatiguing, as, I imagine, every person in his own experience must have ascertained when he has been obliged to stand for even a moderate length of time. By trying to place a dead body on its feet, you will find, that as the joints of the lower extremities are flexed in opposite directions, the weight of the trunk will press them to the ground; and that, as the flexure of the spine is forwards, the head above and the viscera in front will force the body powerfully in that direction. This will give you some idea of the immense muscular power necessary to counteract these different flexures, and to maintain erect that which has a natural tendency to fall.

23. So that the labour, whatever may be its nature, considered abstractedly, becomes severe and fatiguing, from the uniformity and the nature of the position in which it has ordinarily to be pursued?—It does; and that in proportion to the unvarying nature of the act. The same set of muscles is of necessity in constant activity; and by this being urged too far their tonicity is destroyed.

24. Are you aware that this labour has to be endured, as it respects more particularly the cotton and flax factories, in an atmosphere polluted by flax and flues, and other matter rendering respiration difficult and often very oppressive?—That such is the fact I can have but little doubt, from the frequency with which it was asserted by my patients; and the few personal observations which I made corroborated their statement; though, I was not so sensibly struck with the existence of the flue, of which you speak, as by a noisomeness of, and stimulating property in, the atmosphere, which I could not otherwise account for than by the decomposition of the oil with which the machinery was lubricated, and by the attrition of the metal of which the works were formed.

25. So that, upon the whole, the atmosphere was other than pleasant, or, as you presume, healthy?—Decidedly so.

26. Would not the insalubrity of such an atmosphere be still greatly increased if it had to be heightened to a very considerable temperature?—I conceive that its insalubrity mainly depends on this high temperature, whereby the decomposition, to which I have recently alluded, is effected, and the atmosphere thus vitiated. It is, further, my opinion, that artificial heat is of itself far more detrimental than a much higher natural temperature, the purity of the air not being thereby affected, and considerable advantage being derived from its undulations.

27. So that you would conclude that labour is more oppressive and unpleasant when pursued in an artificial temperature than in a natural one, though the thermometer should stand at the very same point in each instance?—I desired so to express myself.

28. Can labour be endured with impunity under those circumstances, and for the length of time that you have expressed in ordinary cases, even by the adults themselves?—It might be endured with impunity for a certain time, because, by withdrawing from the cause before disease is established, the effect will cease. But beyond this short, you may call it experimental,

period, I feel assured that labour, under such circumstances, is incompatible with perfect health, even in an adult. J. Malyn, Esq.

29. Is not that length of labour especially prejudicial to the health of the children and young persons who have not arrived at years of full maturity and vigour, that are so employed ?—That is my opinion ; and indeed, as far as an opinion can approach to certainty, I have no doubt on the subject.

30. In what respects do you conceive that it particularly interferes with the health of the children and young persons ?—During the time that the human body is being developed, and is undergoing those important changes that unceasingly proceed from birth to maturity, the power of withstanding fatigue is diminished in proportion as the subject is removed from that state, that is, the state of maturity; and, in proportion to this inapitude for labour, so is the necessity for repose increased. It follows, therefore, that labour which exceeds the capability of resistance in the frame, must thwart the intention and the efforts of nature, and entirely derange the animal economy.

31. Is that degree of labour that you have been describing still more prejudicial when it is connected with a warm and a polluted atmosphere ?—It is: my last answer had reference to the injury which the unperfected body sustains from fatigue alone; but when it is endured in a hot and polluted atmosphere, the danger is considerably augmented.

32. Will you state the general effects of that labour so pursued on the entire system, connecting it with the circumstances to which reference has been made ; and will you, as far as you can, trace the connexion between the nature of the employment and the effects you have attributed to it, so as to inform the committee regarding your opinions upon that subject ?—I have already alluded to the pernicious effect which is produced upon the animal structure before it arrives at maturity, by the substitution of long periods of labour and short intervals of repose, for moderate exercise and considerable rest. I will explain to you the reason: the osseous system, though hard and unyielding in the adult, is pliable in the young; it acquires its hardness by a gradual deposition of earthy matter in a flexible nidus, which process is not completed before the age of seventeen or eighteen years. In the early stages of this process, the lower extremities are incapable of supporting the superincumbent weight of the body for an immoderate time without giving way; and the obliquity of the thigh bones (produced by the intervention of the pelvis between their upper ends and their approximation to the centre of gravity below, and which obliquity is much greater than it appears when the bones are covered with teguments), this position of the bone increases that tendency. Deformities therefore are the frequent result. The yielding nature of the bones, before the process of ossification is perfected, is exemplified in London in the instance of bakers' boys, one knee of whom is frequently directed inwards by the distorted lateral position they are obliged to assume in carrying their heavy baskets. Independently of other circumstances, which I shall presently notice, a long continuance in the erect position must therefore be full of jeopardy to the young, especially of the female sex, whose pelvis may suffer a, perhaps, not very obvious distortion, but one which might be of perilous import at the period of parturition. The injury which an animal body sustains from the locomotive organs being called too soon and too constantly into requisition, is exemplified in the abridgment which the life and utility of the horse suffer by its early employment; for, among that class of animals, how many are there, that, in the brief period of a dozen years, run their career from the carriage to the hackney coach, and from the hackney coach to the slaughter-house! when, by treating them as Nature intended they should be treated, their utility might be extended far beyond twenty, and instances

J. Malyn, Esq. are not wanting of its exceeding forty years. Adverting to the consequences resulting from high temperature, it is necessary to premise, that the heart, the centre of the circulation, from which the blood is propelled, and to which it returns, has its propulsion and reception of that fluid, in a state of health, duly balanced; if this balance be disturbed, the blood must accumulate in some part of its course, and, unless the equilibrium be speedily restored, disease must be the consequence. The effect of heat is to stimulate the surface of the body, and to induce a greater flow of blood into its vessels, which relieve themselves by rendering the perspirable matter sensible. Cold applied to this excited surface, constringes these vessels, and contracts their diameter; preventing, or, more correctly speaking, impeding the egress of the blood from the heart; while the returning blood passes readily onwards, until it is congested in some part of its course; from this cause evils innumerable arise. Consider then, for an instant, the effect produced on an individual whose cutaneous circulation has been much excited by the high temperature of a mill, and who, in that state, is suddenly plunged into a cold and chilly atmosphere. It is familiar to us, and in the mouths of the most ignorant, that to plunge into water while in a state of perspiration would be to court sickness, if not death; and it remains for others to shew why an exception should be made in favour of cold air. That it is done with apparent impunity for a length of time may be true; but I doubt much if such impunity is real in any one single instance. Rheumatism, both chronic and acute; bronchial and pulmonary affections; effusions into the serous cavities, and diseases of the liver, common and so constant among this class of operatives, sufficiently proclaim the misery connected with such exposure. I would here call the attention of the committee to the fact, a fact which must have appeared in evidence before them, that sallowness of the skin is a distinguishing characteristic of those individuals; so much so that, conjoined with their stunted growth and often emaciated appearance, it is possible to select them from among a multitude. Although this circumstance is, at first view, of trifling importance, it assumes a different aspect when it is known that it indicates functional if not organic derangement of parts which are essential to the preparation of the material which supplies the waste of the system; and on the perfection of which matter the integrity of the body depends. It may therefore not be amiss to stay a moment at this point, and see whether we can find anything connected with their employment to which we can rationally attribute this universal appearance; first inquiring what is the proximate cause of this change of colour. This sallowness arises from the biliary principle not having been properly eliminated from the blood, but circulating with it, and thereby tinging the surface of the body. It is not abstracted from the blood, owing to torpidity or to disease of the liver; and this torpidity of that organ is thus produced:— The body suffers repeated collapses in passing daily from a high to a low temperature. It is a provision of nature against atmospheric vicissitudes, that when the blood is prevented issuing from the heart by the collapse of the cutaneous vessels, the returning blood may, by meeting with a similar state of the vessels of the liver, be prevented passing through it to gorge and oppress, and possibly to annihilate, the functions of the heart. This, constantly repeated, produces that torpidity of the liver which disqualifies it from separating the bile when the collapse is removed, and the blood circulates through its accustomed conduits. You will presently see that the continuance of the principle of bile in the blood is subversive of every law by which Nature regulates her proceedings. I shall have to revert to this collapsed state of the hepatic vessels, as it is connected with the process of digestion, in which this provision has been made for the purpose of assisting the functions of the stomach. I am not aware of the time allowed to operatives for the purpose of taking sustenance, but I suppose it to be short; for I have repeatedly witnessed severe forms of dyspepsia, arising in a great measure from, or at least considerably aggravated by, swallowing their food without mastication, in which state it was never intended it should have been swallowed. People who do not, or will not, make use of the teeth wherewith they are provided, are sure to pay the penalty by establishing a species of rumination analogous to that of some animals; so that the badly triturated mass is returned to the mouth for the purpose of undergoing a proper mastication. The importance and necessity of this process is shewn by the number, and the form, and the gradual development of the teeth; and daily occurrences demonstrate, that whether the negligence may have arisen from vitiated habits or deficiency of time, this trituration cannot be dispensed with. J. Malyn, Esq.

33. You conceive, then, that the operatives having not sufficient time left them for meals, in numerous instances not having any time for more than one meal, and in some no interval for any meal at all during that long labour, must be a circumstance highly injurious to the health, and tending to the derangement of the digestive organs?—Certainly; those people not only labour under the disadvantage of swallowing unmasticated food (itself a fruitful source of disease), but they have another difficulty to contend with; to explain which I must beg to refer, as I said I should, to the collapsed state of the vessels of the liver, which collapse is always coexistent with cutaneous collapse. Most persons are sensible, that after a full meal there is a cool, chilling sensation creeping over the body. This is a healthy action as far as it goes, although it is analogous to the derangement of which I have spoken as arising from the exposure of a person, in a state of excitement, to a low temperature. But in this case, it is a provision made by nature, for the purpose of insuring the digestion of the food. The collapse takes place upon the surface of the body; there is a coincident collapse of the vessels of the liver; the blood endeavours to return to the heart, but, arriving at the liver, it is impeded, and accumulates about the stomach, facilitating the operations of that organ, and perhaps itself contributing to reduce its contents to that condition in which it is called chyme. This having been effected, the digested matter passes into the duodenum; the circulation returns to the surface, the blood passes directly through the liver, the bile flows to the duodenum, and from the chemical process which there ensues, that healthy nutritious matter, called chyle, is formed. Now the high temperature of mills is opposed to this salutary process, for the phenomena I have described cannot then take place.

34. So that, on the principles of your profession, you might reasonably expect that appearance and those diseases of the digestive organs and other parts of the body which you have already described to ensue from that degree of labour, even if you had not experience regarding it?—Yes, that degree of labour, under such circumstances, would of itself be prejudicial. Nature requires, at least would desire, to have a short period of repose after taking a full meal, that the phenomena I have described may not be interfered with. This period of rest ought to be proportioned to the oppression produced by the accumulation of blood about the stomach. I was before proceeding to observe that this is the natural process of digestion; and, being essential to the well-being of the individual, it must be sufficiently apparent how prejudicial it must be for a person who has taken unmasticated, or, we will call it, imperfectly masticated, food into the stomach, to pass immediately into a high temperature, the stimulating effects of which determine the blood to the surface of the body, and consequently allow its transit through the liver. It is evident that then the necessary congestion cannot take place about the stomach. Here, then, the subjects will labour under two disad-

J. Malyn, Esq. vantages; first, they have a mass in the stomach which nature did not intend to be there in that imperfectly masticated state; secondly, they are deprived of the heat which, under judicious circumstances, is provided for the assistance of the digestive organ; and, consequently, it is impossible that the product can be the healthy material which nature intended for the restoration of parts, and to make up for the exhaustion which the body is undergoing at every instant of time. Naturally, from this view of the question, we may conclude that improperly prepared food will produce a diseased state of the body, for it would be strange if there could result a healthy compound from matter of that description. Diseases from this cause declare themselves in various forms, and their connexion with the cause is so obvious that they must be known to almost every person who has taken the trouble to think about the matter, whether he be connected with the profession or not. There is a strong analogy between the appearance of persons exposed to tropical climates, and the appearance of those of whom we are speaking. In both you have the sallowness of skin, the torpidity of the liver, and the numerous forms of derangement in which dyspepsia presents itself. The former, however, have this advantage, that though subject to the vicissitude of temperature, they have not that foul, that impure atmosphere, which is respired by the latter class. The effect of a combination of such causes, the degree to which it enfeebles and undermines the constitution, is just what might have been anticipated; and hence, the appearance of scrofula in its most distressing and aggravated forms is anything but a matter of surprise.

35. Is not that disorder peculiarly prevalent in Manchester?—It is to a great degree; nor have I, in the observations I have made elsewhere, ever seen it so common as it is in that town. Speaking from general recollection, I should say, I have not witnessed, since my departure from that place, one-tenth of what I observed there; although I have ever made it a rule (when otherwise unoccupied) to continue my researches among the poor.

36. Will you resume your examination?—In reference to the effects of this system upon the health, there is one other point to which I would allude; it is the effect of the flue, about which you asked me in a preceding question. I have ever been inclined to attribute to that cause (although I must honestly confess that I have not witnessed the flue myself), in conjunction with other irritating properties of the atmosphere, sundry forms of bronchial diseases which are of common occurrence, and which are, in my opinion, produced by the irritating matter passing down the trachea, and stimulating the bronchia and their glands to inflammatory action. The nature and the extent of the concomitant fever, and the phenomena which they exhibit, cause them often to be mistaken for phthsyis; and hence I have known many, who have witnessed the practice of that institution, remark on the strangeness of so many cases of consumption being cured. The fact is, they were not consumption; and with the increased means of diagnosis which we now possess, the difference would be readily appreciated. Consumption is a very different disease; it is one of the substance of the lungs, in which cavities of matter are formed; and though this is produced frequently in the case of factory people, it is different from the bronchial affection to which I am now alluding, and which I can readily account for, on the supposition of the inhalation of flue, dust, and such like irritating matter.

37. Might not that disease also prove fatal, if there were no intermission given to that description of labour, and no medical relief afforded?—It would prove fatal; in fact, when I say they recover frequently, I by no means wish to have it understood that they at all times get well; for, although it is not phthysis, it is a very dangerous affection. There are great inroads made by

it upon the constitution, and both skill and perseverance are required in its treatment. J. Malyn, Esq.

38. And they would have less chance of recovery if they had not medical treatment, and did not cease the injurious employment alluded to? —True.

39. You have alluded to the deformities that take place in many instances, do you think that those deformities can, in any considerable number of cases, be attributed to the disorder sometimes called rickets?—Decidedly not; a practised eye will at once recognise the distinction; there is not the possibility of mistake, for you find, not merely the inferior extremities giving way, (and it is to be borne in mind, that in what I have said, I have spoken only of the influence on the lower extremities of the superincumbent weight before the completion of ossification, and of their tendency to yield in proportion to the incompleteness of that state,) but in rickets, the back, the chest, the arms, partake of the deformity. Rickets is a disease for the most part peculiar to infancy. In all the bones there is an imperfect deposition of that earthy matter which in the adult gives them strength and durability. Now the committee will see at once, that if the bones continue flexible while the muscles increase and exert their power, not only will the whole body sink, the thighs and legs become curved, the spine distorted, the ribs depressed, and the sternum projected, but, even in the arms, the form of the bones will be altered by the action of the muscles; for, if I take any yielding body (this piece of paper will serve to illustrate it), and attach a cord between two points, the tightening of that cord will cause those attachments to approximate; and, from want of firmness in the material acted on, it will bend in the direction in which the force is applied; thus, by the influence of muscular action, the upper and fore-arms are twisted outwards. These cases, then, are easily distinguishable from those which are the effect of labour inappropriate to the age; and I think that any one present might go into the next room, where specimens of both might be assembled, and, after this explanation, select the one from the other without fear of mistake.

40. Does not the disease called rickets almost always develop itself at a very early period of life, and rarely so late as nine, or ten, or fifteen, or sixteen years of age, which has been the case in the instances of deformity that have come before this committee?—There are instances when it has manifested itself at a later period of life; but, generally speaking (and we have exceptions to every rule), it is a disease confined to infancy.

41. Preceding medical witnesses have stated, that other disorders affecting the lower extremities than deformities of the bones result from these long hours of labour; do you conceive, for instance, that ulcerated legs would be no uncommon effect of that labour?—Yes; I should not be at all surprised at finding such to be the case. By maintaining the erect position of the body for a length of time, the blood does not return so quickly from below as could be wished, whereby the tone of the parts is diminished. To aid its ascent, valves are placed at regulated distances in the veins through which it has to pass to the heart. A column of blood resting on these valves, one will sometimes give way, and the valve next below it will then have a greater column to support, and will consequently be the more readily destroyed. This incapability of resistance will be increased as each successive valve is broken down, and in course of time you may expect to find the legs presenting, from the groin downwards, a mass of tortuous veins, owing to their being bulged out in the spaces between the parts where those valves existed. Ulcerations then frequently take place, owing to the torpidity of

J. Malyn, Esq the circulation in the limbs, and they are rarely curable so long as that posture which first induced them is maintained.

42. You have no hesitation then in stating, that in those numerous cases where the legs of those, and especially of the females, working in factories are affected with ulcers, it is attributable to the length of their labour?—Wherever I met with cases of ulceration under such circumstances, I could not avoid attributing them to the peculiarity of the labour, and I should recommend a change of employment; but I beg to observe, that the last three years of my connexion with the Manchester Infirmary having been passed in the medical department, cases of that description did not come so frequently under my notice as they would have done had I been continuously engaged in the surgical department. I should have very little hesitation however in asserting, from general reasoning, that ulceration of the legs must be frequent among persons so employed.

43. Should you judge from theory, or determine from your own experience, that a stunted growth is also the result of this labour?—I believe it is admitted by all, at least I never heard it disputed, that these persons are a smaller race than corresponding classes of society otherwise occupied; they have ever appeared to me to be contracted in their growth; nor do I see well how it could be otherwise, if they enter upon it at an early period of life. Of course this depends upon their being so employed before the body has acquired a power of resistance; for a person commencing at a later period of life would bear considerably more than he could have done at that age, had it been his original occupation.

44. Do you think, also, that the strength of the human body and its bulk is diminished, to a corresponding degree, by the same labour?—I do. I intended my remarks on the imperfect manner in which their digestion must be carried on, to bear such a construction. I do not see any reason why they should be smaller, if the period of rest or refreshment corresponded with the exigencies of their case. It may be long enough, for aught I know, and they may not avail themselves of it; or, it may be, that they are not enabled to avail themselves of it, by reason of their living at too great a distance. Wherever the blame may be, the effect is certain.

45. Upon the whole, do you conceive that the state of the body produced by that labour predisposes to disease, generally speaking?—I do; there is a feebleness of the vital power, in which the capability of resisting disease is much diminished, and by which their tendency to dissolution, when attacked, is greatly increased.

46. When those individuals, so circumstanced, are victims to disease, do you apprehend then that there is a diminished capability of resistance?—I do; if the opinions I have previously given be correct, we can draw no other conclusion.

47. And, consequently, that the cases of mortality, as it respects such, are numerous?—Just so. I often remarked the few old men and women employed in factories, for there were very few persons that I observed above fifty. I have nothing to refer to in proof of this assertion, and I speak only from my general recollection. Indeed I have an indistinct recollection of its having been stated that they are not capable of doing anything beyond the meridian of life and that they are kept by their masters more as an act of charity than for their services. I am now reassured on this point, for it just occurs to me, that a few days since I heard of a case where the spinners of a gentleman in the country turned out for wages: he expressed himself as being very glad of it, for he thus got rid of a number of old spinners, who, though they could not have gotten employment elsewhere, he had felt himself bound to keep up to that time, they having passed their youth with him.

This early incapacity I should consider as anything but an evidence of the good effects of the system upon their constitutions. J. Malyn, Esq.

48. But have you heard the fact that few do survive to an old age whose youths have been spent in the labours of the factories?—Yes.

49. Is it your opinion that these consequences may be perpetuated and increased in successive generations if the labour continues as already stated, producing the effects which you have attributed to it?—Certainly; I believe that such may be, and is the case. I believe that is one great cause why there is so great an amount of scrofulous diseases.

50. Do not you think that that labour, pursued during the night, is more unhealthy than that which is undergone in the day-time?—I should conceive that such would be the case, reasoning from theory; but I cannot speak as to the actual fact.

51. Do you conceive that the evils that result from the factory system are such as would fall with still greater severity upon the female sex, as compared with the male?—I think they would, from the greater degree of delicacy of the female frame, and from their having less resiliency than man when acted on by disease. I have already, in an early part of my evidence, stated the probable effect of too early employment on the pelvis of that sex. With respect to the last question but one which was put to me regarding night-work, although I cannot speak to the fact, having never witnessed it, I should judge that the evil would be greater than in the day. The condensation of what has been abstracted from the earth during the day cannot improve the atmosphere which they breathe; but, on the contrary, will make it still more prejudicial to those who may have to be exposed to it; not only is night air more noxious than that of the day, but, from the temperature being lower, more calculated to act injuriously on their excited bodies when they issue from the mills. The rays of light are also afflicting to the eye when they proceed from a luminous body of an artificial kind; they are too pungent to be agreeable to the human eye; they require modifying and softening in the same way as the rays of light coming from the sun are attenuated and rendered mild by the blue atmosphere through which they pass. The irritation thus prôduced might render the nervous system more sensitive in general; by means of which the noise might be, in a slight degree, a source of evil. A great evil I should fear would be, that if they do work by night they will not sleep sufficiently during the day, but will be tempted to deprive themselves of rest to partake of the enjoyment which the busy day affords.

52. Several of the operatives complain of the effect of gas light upon the sight, especially when they have to pursue their labour for a considerable number of hours and during the night: should you, as a medical man, conceive that that would be the result from what you have already advanced, of labouring in so brilliant a light so near the organ of vision?—I am not at all surprised at their complaining of it; it is just what I should have expected.

53. It has been stated by a medical witness before this committee, that the labour pursued in these excited rooms has a direct tendency to anticipate the age of puberty; do you conceive that such might be the effect?—I think so, not only from reasoning, but from observation. They are more precocious, as far as I recollect. I am not alluding to their talent, but to the development of their animal propensities; yet I do not think that it would be at all favourable to the propagation of the species; because the vicious ardour of such premature development is calculated to defeat that object.

54. Supposing it were held forth as an argument, that few illegitimate children were the result of premature and illicit connexion, would you hold

J. Malyn, Esq. that the small proportion of illegitimate children led to anything like a proof of the morality of the system?—Decidedly not, for the reasons already assigned.

55. Have you yourself observed much immorality, then, as connected with the factory system, while you were engaged in your professional duties at Manchester?—Yes; their morals have always appeared to me to be exceedingly degraded; indeed, from the accumulation of them together in an excitable atmosphere, we should naturally expect to find them deficient of moral restraint.

56. Do not you think the sense of weariness and fatigue has a direct tendency to induce tippling, in order to give the body an artificial stimulus?—It would have that tendency; and at the same time I should expect that their systems would be less sensible to the influence of that stimulus.

57. That circumstance, therefore, naturally inclines them to the use of ardent spirits?—Such would be my opinion of the tendency of this occupation; I cannot say, however, that I have witnessed more decided cases of drunkenness among those who have worked in factories than among other persons.

58. But you have not directed your attention to instances of that nature?—No; because when it was my duty to visit my patients, those who were well were at work in their factories, and therefore removed from my observation.

59. Do you think, in the case of the persons so exhausted by fatigue and labour, and exposed to such an atmosphere, of course causing perspiration, the effect of largely indulging in ardent spirits, though destructive to the constitution, might not exhibit itself in intoxication?—I certainly conceive that such might be the case. The diminished energy of their nervous system would require a greater quantity to stimulate them to inebriety than would suffice for another person not so circumstanced.

60. Of course, speaking in a medical view, those large stimulants would very much injure the constitution?—Decidedly; they would increase the affection under which I conceive them to labour, that is, torpidity, in most instances, of the liver.

61. You have stated that the number of patients passing through the Manchester Infirmary was about 13,000; your situation, therefore, in that instititution gave you the most ample opportunity of forming the opinions which you have now expressed to this committee?—Certainly.

62. You have been in the habit of continuing your researches since you have completed your education, in addition to what is ordinarily done in merely pursuing a course of general practice?—I have. I have done it not less from inclination than from necessity; because, as I before said, I have been constantly engaged in the education of individuals who are intended for our profession, in fact, of some who are engaged themselves in the profession; and I should, consequently, have been obliged to continue my researches for the purpose of benefiting them, had I been less sensible than I am of the importance of the obligations which, as a surgeon, I am bound to fulfil.

63. In fact, you are now engaged, in addition to your medical duties, as a medical instructor?—I am.

64. Do you frequently attend the hospitals of this city?—Repeatedly.

65. You have already stated how fully Mr. Simmons had made up his mind upon this important subject, namely, the necessity of a legislative restriction of the hours of labour of children and young persons in mills and factories?—I know that he felt very strongly on the subject, that he never cloaked his opinion, and never hesitated to defend it. In fact, it was his custom to draw the attention of those about him to the different diseases which he mainly attributed to causes of this description.

66. Was his opinion in regard to this important subject in the slightest degree changed at the latter period of his life?—Not in the slightest, so long as my connexion with that gentleman gave me the opportunity of knowing his opinions. J. Malyn, Esq.

67. How long is it since he died?—About two years since, but I cannot speak of his opinions since the period that I left the infirmary.

68. All that you should deduce from the principles of your profession, sanctioned by the experience which you have had in it, which has been necessarily very extensive, brings you to this conclusion; that the labour of the factories, as at present endured by children and young persons, is prejudicial to their health, destructive, in many instances, of their welfare, and has a tendency to shorten their existence?—That is the only conclusion I can draw from the facts I have seen.

69. Do you think that it would be proper to take children into a factory at all before the age of nine years?—Decidedly not; or into any other establishment where the child would have to maintain for any length of time an uniform erect position of body.

70. According to your own observations and belief on this subject, do not you think that ten hours a day actual labour, to which is to be added the proper intervals for rest and refreshment during the day, are obviously sufficient to be endured by children of either sex between the ages of nine and eighteen years?—Of that description of employment it is sufficient for the latter age, and too much for the former.

71. You have already said that its effects would be more pernicious to the females, and consequently that they need at least an equal protection, and that, coupled with the circumstances of the majority of those employed in mills and factories being females, strengthens you in your conclusion as to the necessity of a legislative measure on behalf of those individuals?—Yes, for three reasons; first, they are naturally of a weaker constitution than the male; secondly, injuries during growth might be of serious moment at the time of parturition; and thirdly, their propensities are developed earlier than in the male; which might be counteracted, if time or means were afforded for mental cultivation.

72. Should you not in many instances, in those crowded districts, attribute the early and premature marriages to the factory system?—I have not thought much on those points, but it strikes me as natural that such marriages should so occur.

73. Was it ever part of your professional practice to attend women in childbirth?—I have attended them as a student many years since, but it does not belong to the department to which I have devoted my attention.

74. Have cases come under your notice of misfortunes in childbirth, in consequence of what you conceive to be the over-exertion of women?—Not under my own notice.

75. Do you happen to know whether the regulations that are prescribed for the regulation of the cotton factories were evaded or observed in Manchester when you were there?—To inquire into such a question never entered into my mind; for I never, that I am aware of, made it a subject of conversation. I appear here merely to give my opinion of the effects of labour under certain circumstances, and that I do under some disadvantages; for it is only a few days ago that I was brought under notice of Mr. Sadler, by whom I was requested to give my evidence.

76. Do you conceive that the operatives generally at Manchester are less robust and less healthy than persons of the same class whom you meet with in London?—I have no doubt at all upon the subject.

77. Does the mill system, as at present pursued, occasion many accidents among the hands employed?—I know that it does produce those accidents

J. Malyn, Esq. inseparable from machinery. It is possible that, at a late period of the day, a state of languor and lassitude may be brought on, whereby they would be less able to guard against the accidents which would naturally occur in such large establishments, and with such machinery.

78. It is sometimes said, that those accidents occur often at the termination of the day's labour, but generally at the termination of a week's labour, when the person becomes fatigued; does that consist with your experience?—Yes, the latter part of it does; for on the Saturday there was always a number of accidents, but generally trifling, being rarely more than the loss of the end of a finger or fingers; that is, as far as relates to the cleaning of the machinery at the time specified; but I cannot say that these ever appeared to me to be connected with or produced by reason of the languor to which the question refers; although, as I have already observed, I can conceive that listlessness might deprive them of that caution which with their full energies they would exercise.

JOHN MORGAN, Esq., called in, and examined, 30th July, 1832.

J. Morgan, Esq.

1. What is your profession?—A surgeon.

2. Are you a medical officer in any of the great institutions of this metropolis?—I am a surgeon to Guy's Hospital.

3. Are you a public lecturer in any of them?—I lecture on surgery at Guy's Hospital.

4. You have paid attention to every branch of your profession?—I believe I have.

5. The committee beg to direct your attention to the effects of undue labour, especially when endured in confined situations, on the constitution, health, and limbs of those undergoing that labour; and first, whether regarding any class of the community, and at any age, excessive labour, or labour so long continued as to produce great fatigue of body and mind, without leaving sufficient intervals for recreation and refreshment, is not highly prejudicial to the human constitution?—Certainly it is.

6. And would not that degree of labour, habitually undergone, be still more injurious to children and young persons?—Certainly.

7. The committee would ask, whether, according to the general principles of your profession, sanctioned by experience, the ordinary term of human labour, namely, twelve hours, allowing the proper intervals, is not considered, on the average, sufficient for human beings to endure, and whether that term could be much transgressed without, in most instances, prdoucing mischievous effects on the constitution?—I think twelve hours of labour, generally speaking, for a continuance, is as much as any person ought to undergo who values his health; but I think that those hours may be deviated from occasionally, because a great deal must depend on the weather in which the work is done, and on the situation in which it is performed, and also upon the health of the person himself; but, as a general law, I should say it was sufficient.

8. So that what has settled itself down into the almost universal custom of the industrious classes in agricultural and handicraft pursuits, seems, in your opinion, to be sanctioned by the principles of medical science?—Yes, it does.

9. Are not due intervals, or cessations from labour, sufficiently long for taking the necessary meals, essential, generally speaking, to the health of the labouring classes of the community?—Decidedly so.

10. Is there not something in the process of digestion that seems to de-

mand a cessation of labour as connected immediately with the time of taking meals?—Long-continued fasting is one cause of indigestion. J. Morgan, Esq.

11. Assuming, then, that labour undergone in mills and factories almost always exceeds the term alluded to as one of an ordinary day's work, and sometimes extends to thirteen or fourteen, nay, sometimes seventeen or eighteen hours a day, could you have the slightest hesitation in saying, that, in the great majority of instances, such protracted labour must be more or less prejudicial to the human constitution?—I should have no hesitation in saying that it must.

12. Now when that term of labour is extended, as it sometimes is, to twenty or thirty successive hours, or sometimes even more, still with but slight intermission, can you have any doubt that such a degree of human exertion must, in many cases, be speedily attended with the most pernicious effects?—I can have no doubt at all of it.

13. Should you think, as a general principle, that to confine children to their labour during the whole of the night, and to send them to rest in the day-time, is either a natural or a safe course, in reference to the preservation of their health in ordinary cases?—I think the health is preserved better by labour performed during the day than by that performed during the night.

14. Leaving out of consideration the inducement there is for such children and young persons to play during the day, and consequently to defraud themselves of the rest necessary to an exhausted frame, still other circumstances combine, in your mind, to induce you to think that nightly labour cannot be so wholesome as that pursued during the day?—Certainly, that is my opinion.

15. Should not you think that labour continuing for thirteen, fourteen, or fifteen hours a day, although it might in itself be properly denominated light and easy, yet requiring constant and anxious attention, would be necessarily prejudicial to the human constitution?—I think it would.

16. If that labour had to be ordinarily endured in a standing position, would not that necessarily increase the fatigue?—Most probably it would; but if the labour were to be performed in a walking position, it would be more prejudicial than standing.

17. Then as this labour, more especially that pursued in many of the spinning-mills, requires, in some parts of the process, incessant walking, do you consider that that is another circumstance which would heighten the fatigue from labour so undergone?—It would increase fatigue, and consequently increase debility.

18. Would not the labour in question be still more injurious if it had to be endured in an atmosphere rendered impure by the circumstances attending the manufacture, such as dust and flue?—There can be no doubt of it.

19. And would it be still more injurious if that atmosphere were artificially heated to the temperature, for instance, of 70 or 80 degrees, or upwards?—I consider that it would.

20. Does not the circumstance of persons so employed, when leaving their work in this heated atmosphere, having to plunge into the open air, which in our climate is for a considerable part of the year cold and chilly, approaching to the freezing point, most of them being lightly and insufficiently clad, greatly aggravate the injurious effects of labour so pursued?—Yes, it does.

21. Would not night labour, pursued as it of course must be by artificial lights, have a tendency to injure sight?—Such must be the result occasionally, and such I know occasionally to be the result.

22. Certain complaints seem to affect the children and young persons

J. Morgan, Esq.

engaged in mills and factories in a greater proportion than others differently occupied; have the goodness to state to this committee what you believe would in many cases be the consequence of labour so long continued, under the circumstances already indicated to you?—I am not prepared to say exactly what number of diseases might be produced by it; I can only state as to the number I myself have seen produced by it; I have seen diseases of the lungs, scrofulous diseases, diseases of the bones, diseases of the joints generally, diseases of the eyes, and I may have seen many others that I cannot exactly at present recollect.

23. Should you be prepared to expect that diseases of the legs, especially in the female sex, would result from very long standing at their labour?—I should consider it as a necessary consequence.

24. Have you personally examined any cases of deformity which you have been assured have resulted from the factory system?—I have seen three; and I could attribute the deformity entirely to the labour to which they had been subjected, and the life which they had led.

25. Numerous similar cases have come before this committee; it is stated that deformity came on after they had been a considerable length of time at the work described, and it was constantly accompanied by very great pain in the joints and limbs; must not the exhaustion and suffering be great, in your opinion, to induce such striking deformities as those alluded to?—Certainly they must.

26. It has been stated, as the result of several examinations most carefully conducted, that the growth of children employed in factories is stunted as compared with those engaged in other pursuits; should you conceive such a consequence to be naturally produced by the system in question?—Yes, I should.

27. Regarding their weight also, the same comparative observations have been made, as the result of equally careful examination; should you conceive that the muscular powers of those so occupied must have also suffered, compared with those engaged in other and less unwholesome pursuits?—It is what I should naturally expect.

28. An official paper has been delivered in to this committee, by which it appears that the duration of human life is considerably abridged, or, in other words, the proportion of mortality much heightened, where the mill or factory system prevails; should you be prepared for that consequence also, from what has been already detailed to you as the practice of mills and factories?—Yes, I should, certainly.

29. It is a known fact, that a very considerable majority of the young persons employed in these factories are females; will you have the goodness to inform the committee whether you think the female constitution is as capable of sustaining long and continued labour, especially in a standing position, as that of the male?—Not after a certain period; not after the age of from ten to fourteen, as a general rule.

30. Should you think that the female requires protection from an undue degree of labour at the time that the constitution is advancing to a state of puberty?—There can be no question about it.

31. Do not you think that period, namely, the period of puberty, is in many cases anticipated by labour pursued as described, and in a heated atmosphere, and under the excitements that are known to exist while pursuing it?—I am quite certain it is occasionally anticipated by those causes.

32. Is not the anticipation of the period of puberty prejudicial to the future health and longevity of the party who has been subject to such an unnaturally early development of the powers of the body?—With regard to the longevity, I will not speak on that point; I should think it would tend

to injure the general health for a considerable time after; whether it would tend to shorten life, I cannot speak positively to it; it must depend on many other circumstances,—whether that excitement is continued, whether the same cause alluded to produces the same effect throughout the rest of life. The mere circumstance of puberty being anticipated would not, of itself, be sufficient to shorten life. J. Morgan, Esq.

33. Do you not think that a considerable degree of fatigue and weariness, habitually undergone, has a natural tendency to encourage the injurious practice of tippling, said to prevail so greatly in manufacturing districts?—Most certainly it has a tendency to do so.

34. Artificial stimulants seem to be more required by the individual so exhausted than they would under ordinary and more favourable circumstances?—Extreme exertion requires artificial stimulants; and if these children are exerted beyond their power, I should naturally expect they would look for artificial stimulants.

35. Do not you think that the body is in a very unfit state to renew its daily labour when the preceding evening's sleep has been insufficient to remove a sense of weariness and fatigue?—Certainly, in a very unfit state.

36. And the labour which has perpetually to be stimulated by punishment, such as the lash or the scourge, still more injurious?—I think it must be.

37. Will you give the committee your opinion, whether a child under nine years of age should commence working in mills and factories, considering the usual length of time which this bill sanctions, namely, twelve hours, including meal-times?—I do not think a child could, at nine years of age, without injuring its constitution, work constantly all the year round for twelve hours in the day.

38. Including intervals of two hours for rest and refreshment?—Including those intervals.

39. Are you of opinion that a young person under eighteen, and consequently before the bones are, generally speaking, completely developed, and have acquired their full stability, is capable of sustaining more than twelve hours of labour habitually, with impunity to the constitution?—I should think, as a general rule, that twelve hours would be sufficient, but I am not prepared to say that an occasional deviation from that might, or might not, be attended with injurious consequences.

40. Do you not think that deviations, if frequently or habitually occurring, before the body had attained its full maturity, would be unquestionably more prejudicial at than after that period?—Yes, I do.

41. Is it your opinion that a child of either sex, under nine years of age, should habitually and constantly labour at all?—I do not exactly understand what the committee may mean by the term labour.

42. To be engaged in the employment of mills?—They could not safely be engaged in such work as they are now employed in in factories.

43. In reference to the morals of these mills and factories, and assuming them to be otherwise than correct, would you take the circumstance of their being but, comparatively speaking, few illegitimate children, were such the case, as proving that much immorality did not in fact exist?—That would depend, in a great measure, on the immorality of the persons themselves; if they had promiscuous intercourse, it would certainly tend to prevent the rapid production of children; excessive sexual intercourse is one of the causes of barrenness.

44. Supposing that the labour of the week is such as to induce very great fatigue, and to afford no opportunity whatever for mental or moral improvement, either at home or in evening schools, do you suppose that the confine-

J. Morgan, Esq.

ment in the Sunday-school, on their only remaining day of leisure, is quite consistent with the health of the children?—That must depend entirely on the way in which the Sunday would be spent, provided they were not sent to Sunday-schools; it would be better to send them to Sunday-schools than allow them to spend their time in dissipation and drunkenness.

45. Supposing an undue degree of labour is imposed on children and young persons in the week-day, do not you think, medically speaking, that recreation and rest would be good for the children on the only remaining day of leisure?—I think it would certainly be necessary for them.

46. But supposing that the labour of the week were to be so abated as to allow them, without a sense of fatigue, to attend to the duties enjoined on them by the Sunday-schools; can you have a doubt but that they would avail themselves more completely of the advantages of such institutions, in obtaining a degree of moral and mental improvement which they cannot now derive from them?—I have no doubt at all on the subject, that they would.

47. Do not you think that the limitation of the hours of study and application in our schools and seminaries of learning is necessary to the preservation of the health of students and others repairing to those establishments? —I am not very well acquainted with the usual hours.

48. Is not some limitation of the hours, even of study and application in public schools necessary, generally speaking, to the preservation of the health of the children?—Certainly it is.

49. Supposing that the remission proposed in the average duration of the confinement of the labour of children in factories were devoted to teaching them, on evenings, for a short time, the rudiments of a decent education, would not the very change in the employment, and other circumstances connected with it, still operate beneficially on the health of the children in question?—I think it would.

50. Has not the feeling of mind, especially when the person supposes himself to be suffering under a great many privations, and subjected to many indignities and hardships, a considerable effect on the human constitution? —In some constitutions I think it has, certainly.

51. So that a sense of misery and hopelessness would tend to aggravate the mischievous consequence resulting from the factory system?—I think it certainly would.

52. Your observations relative to the protracted labour of children are directed, of course, against all trades, and not confined to manufactories?—In some manufactories they would be more prejudicial than others, on account of the atmosphere which children have to breathe; but, as a general rule, I should say that about twelve hours a day were sufficient; for instance, the employment of a chimney-sweeper, I should consider that more injurious, perhaps, than even these manufactories.

53. Are you not aware that milliners' apprentices, and several other trades, are equally over-worked, and equally require legislative interference?—How far they may require legislative interference it is not for me to say; but I am quite aware that they spend too long a time at their business, and suffer in health.

54. Do they suffer from the same cause?—Not exactly; theirs are sedentary habits; one of the most common effects is a disease in the eye, which leads to blindness.

55. Do you mean that ten hours are more adapted to the strength of children, and you would say that nine would be better than ten?—You can only generalize; I should say that it was, generally speaking.

56. Do you think that a bill, the provisions of which were that no child should be worked under ten years, and that the limits of the hours of labour

should be confined to eleven hours, and two hours for refreshment, making altogether thirteen hours, would be a very great benefit to the labouring classes?—I do not exactly understand what is meant by the term *labour;* I think children under ten years of age might be put to some work, but I should not call it labour. I think that a bill which would limit the hours of labour to eleven hours for children from ten to eighteen, would be a great benefit to the manufacturing classes. J. Morgan, Esq.

57. Do not you think that a bill limiting the confinement and labour to ten hours would be a still greater benefit?—I do.

58. You have been asked in respect of milliners, whether you do not think a legislative measure necessary to regulate the work of milliners; does it follow, in your opinion, as a man of observation, that because it is impossible to make a legislative act to restrain the working hours of milliners employed by people of fashion, there is anything to prevent the legislature passing a bill to regulate the dreadfully long protracted labours in manufactories?—I see no reason why it should *not* be so.

59. You think the undue and pernicious labour of one part of the community is no apology for inflicting the same fate, and all its miserable consequences, on others?—No apology whatever.

PETER MARK ROGET, Esq., M.D., F.R.S., called in, and examined, 1st August, 1832.

1. You are a physician?—I am. P. M. Roget, Esq., M.D.

2. Where have you practised?—In Manchester, first; but principally in London.

3. Did you hold any medical office in Manchester?—I was physician to the Infirmary at Manchester.

4. Have you still any medical appointment, distinct from your private practice in this metropolis?—I am at present consulting physician to the Northern Dispensary, of which I was acting physician for eighteen years; I am also consulting physician to Queen Charlotte's Lying-in-Hospital. I was appointed one of the physicians of the Penitentiary of Milbank, during the prevalence of the epidemic disease in 1823.

5. Your attention has been directed to diseases incident to the human constitution, and in good measure to those suffered by the lower classes of society?—It has.

6. Is it not a generally received opinion in your profession, that some degree of relaxation, and in the open air, and a reasonable quantity of sleep, is necessary for the maintenance of health?—Those circumstances are highly conducive to the maintenance of health.

7. Regarding children and young persons, sleep and recreation are more peculiarly necessary, are they not?—Certainly.

8. Is it not a fact generally acknowledged by your profession, that excessive labour without due intermissions, or a sufficient time for sleep and relaxation, is ordinarily very prejudical to the constitution?—Excessive labour, of course, must be prejudical to the constitution.

9. Can the human constitution, during the period of its rapid growth and development, namely, in childhood and youth, endure with impunity that degree of labour, whether as it regards its duration or strenuousness, which human beings can undergo at a more mature age?—Less labour can be sustained with impunity during the period of growth than at any other period, I should conceive.

10. Will you favour the committee with your opinion, whether the term

P. M. Roget, Esq., M.D.

of labour which seems to have been established in every community, and, generally speaking, in every industrious pursuit of life, namely, twelve hours a day, with due intermissions for meals, is not ordinarily considered about as much as the human constitution is capable of sustaining with impunity?—It appears to me that the period mentioned is quite as much as the human frame is calculated to endure for any length of time, even in the adult state.

11. Are not intervals for the taking of meals, during which the labour of the individual shall cease, generally necessary to the preservation of the health of the industrious classes?—I should conceive them quite necessary.

12. Particularly for the recovery of the tone of their animal spirits?—Certainly.

13. Supposing, then, that the employment in mills and factories extends in many cases to thirteen, or fourteen, and in some to fifteen or seventeen hours a day, or even longer, with but short intervals for refreshment, and in some cases none at all; can you have any doubt as to such a state generally interfering with the health of those who are subjected to it, and producing pernicious consequences on their constitutions?—I should think none but robust constitutions could resist the exhausting effects of such labour as that described.

14. You would conceive that in the instances in which that labour is continued without intermission it would be still more prejudical?—Certainly.

15. Then when it is endured for twenty or thirty successive hours, with but very short intermissions, you would not have the slightest doubt in asserting that the human constitution is not calculated to bear any such exertion and fatigue?—I should think such excessive labour could not be endured for any length of time without risk of serious injury to the constitution.

16. Have you formed an opinion upon the subject, whether labour pursued during the night is not more prejudical than if it were undergone in the day-time?—I have no doubt that labour pursued during the night is more exhausting than during the day.

17. If the pernicious effects of night-labour are connected with the known fact, that those who have to work during the night are often tempted to give so much of the day to recreation and amusement as to deprive them of sufficient rest, you would consider that as rendering the night-work still more prejudical, would you not?—The privation of rest would increase the pernicious consequences, and it would be a question whether the injury from that cause would be balanced by the benefit derived from recreation.

18. Do you think that sleep taken during the day, either casually or as a general practice, restores the exhausted powers of mankind so much or so well as if taken during the night?—I have already stated, that labour during the night is more exhausting; I should find it difficult to answer the question whether sleep is more refreshing during the day or during the night.

19. Is labour that is sometimes called "light and easy," if it is continued for a great length of time, and is of such a nature as to produce great fatigue both of mind and body, less prejudical than more strenuous exertion when pursued for a shorter period with due intermissions?—Amusement would of course lighten all labour; and it would be a question of degree with regard to the amount of injury.

20. Is not the exertion necessary to sustain the body in an erect position for a great length of time considerable?—Certainly; more especially if the person stands while he maintains the erect position.

21. Would not the labour of a factory, if it has to be pursued in a polluted atmosphere, and one mixed up with, and almost saturated by dust and flue, be still more pernicious?—Undoubtedly, an atmosphere of that description would be very insalubrious.

22. Would not labour of the description alluded to, both in point of

duration and the circumstances connected with it, be more injurious if it were undergone in a temperature very highly heated, such as 80°, and the atmosphere also polluted by the rooms being crowded, and by the effluvia arising from the attrition of metallic machinery, and the oil used in it, and by being without proper ventilation?—Certainly.

P. M. Roget, Esq., M.D.

23. Would not labour undergone in an atmosphere artificially heated and polluted as described, be more unwholesome than labour pursued in an equal temperature in a tropical climate?—I should think it more injurious than labour in an equal temperature in a tropical climate.

24. To apply the preceding observations more especially to children and young persons, should not you think that the labour undergone for the length of time described, and under the circumstances detailed to you, would be more prejudical to them even than to adults employed in the same way?—Undoubtedly.

25. The results of the factory system have been described by many medical and other witnesses; assuming that it exists now as has been already detailed to you, what do you think would be the probable consequences, medically considered, of that degree of labour under the circumstances explained?—I should expect it would have a tendency to check growth, to enfeeble the powers of the system, and perhaps to induce scrofula and rickets.

26. It might have an effect on the osseous system, when the constitution, under ordinary circumstances, would escape from deformity?—I think it would have a tendency to induce deformities of different kinds.

27. It has been stated to this committee that as great a proportion as one in ten of those engaged in factories of a certain description are deformed in a greater or a less degree; do not you conceive that proportion is excessive with reference to other classes of the population of this country?—It is much greater in my opinion.

28. Numerous cases of deformity have been mentioned to the committee as having occurred after a period of ten or twelve, or sometimes fifteen or sixteen years of age, the body having been previously perfectly straight and well formed; do not you think that such instances may fairly be attributed to the factory system, and not to the disorder usually called rickets?—I think they may, inasmuch as rickets usually commence at an earlier period of childhood.

29. The deformities in question have been almost always preceded and accompanied by excessive pain in the parts principally affected; would not that indicate an undue degree of labour and stress upon that particular part of the body?—I think it would.

30. You have already stated that you might expect from such a system of labour stunted growth; observations have been already made that verify that fact; it has been observed, also, that a considerable diminution of weight takes place, compared with children otherwise circumstanced; may not that also be attributed to the effects of the same system?—I should expect the diminution of the weight to be referrible to the same cause as the impediment of growth.

31. It is known that a considerable majority of persons employed are females; do you think the female constitution is as well adapted to labour of the description alluded to as that of the male?—I think it is not as well adapted, certainly.

32. Do not you think that a female approaching the age of puberty is peculiarly liable to injuries by over-exertion, especially in an erect position?—Certainly.

33. Has not the labour in question, especially when endured in a heated atmosphere, and under circumstances unfavourable to morals and decency, a direct tendency to anticipate the period of puberty?—Yes.

P. M. Roget, Esq., M.D.

34. Is not the body in a very unfit state for labour, especially that of a child, when the rest of the night has not sufficed to abate the fatigue of the preceding day's labour?—Undoubtedly.

35. Would not that sort of fatigue, perpetually felt at the commencement of every day, be at length peculiarly pernicious to the constitution?—Most assuredly.

36. Do you not think that labour which it is necessary to stimulate generally towards the conclusion of the day by severe and continued chastisement must be very exhausting and pernicious?—I think there can be scarcely a stronger criterion of its being exhausting.

37. Do you think that a child under nine years of age ought to be subjected to the labour of a mill or factory?—I should think not.

38. What degree of labour do you think a child under the age of nine years may be capable of enduring, consistently with a proper attention to health?—I should say four or five, or at the utmost, six hours.

39. Are you of opinion that a young person from nine to eighteen, and especially a female, or in other words, till about the period when the growth is completed, and the bones acquire their utmost degree of consistency and strength, ought to labour more than twelve hours a day, including the time necessary for refreshment?—I think not.

40. Perhaps you are of opinion with Dr. Baillie, as well as many other eminent men, who gave evidence before preceding committees of both Houses upon this subject, that twelve hours a day labour, inclusive of two hours for meals and refreshment, is, in ordinary cases, quite enough for the human constitution to sustain?—I am of that opinion.

41. Alluding to the alleged immorality attendant upon the system of mills and factories, perhaps you would not conceive that a comparative degree of barrenness would be a complete disproof of that accusation?—Certainly not.

42. You would perhaps conceive that the most profligate are very rarely the most prolific of the species?—Generally the least.

43. Considered with a view to its physical consequences, is not the confinement on the Sunday, when added to such a week's labour as that which has been described to you, pernicious to children and young persons?—I think so, certainly.

44. Do you not think that the Sunday-schools of this country might be made much more efficient in communicating necessary instruction and knowledge if the hours of labour during the week were so mitigated as to leave the scholars without the fatigue and exhaustion of which they complain as continuing during the whole of the Sunday?—I can have no doubt of that.

45. Do you not consider that a constant sense of oppression, and the irritated state of feelings resulting from it, continued for a length of years, may have an unfavourable influence upon human health?—Certainly.

46. Would not the proper remission of such hours of labour, so as to bring them within the limits that humanity and custom in other pursuits have already prescribed, and leaving time for improvement on the week-days, by the means of evening or other schools, be beneficial in a moral and political, as well as in a medical, point of view?—That would be a better system in every respect.

WILLIAM SHARP, jun., Esq., called in, and examined, 20th June, 1832.

1. Where do you reside?—At Bradford, in Yorkshire,

2. What is your profession?—A surgeon.

3. Are you acquainted with the public medical institutions at Bradford?—I am surgeon to the Dispensary.

W. Sharp, jun. Esq.

4. Do you know whether there are many spinning factories and mills in that town and neighbourhood?—Yes, there are.

5. In which a very considerable number of children are employed?—Yes.

6. Are they admitted into those mills at an early age?—Yes.

7. Do you happen to know whether a considerable proportion of them are females?—They are.

8. Are you acquainted with the number of hours that they work generally? —I think from twelve hours to some longer period than that; I cannot speak positively to the number, but generally more than twelve.

9. You do not, generally speaking, know the details of the management of many of the factories at Bradford?—No, I do not.

10. Is there any factory in Bradford of which you have an accurate knowledge?—Yes, there is one.

11. State whose mill that is?—It is Mr. John Wood's mill.

12. Is that a worsted manufactory?—Yes.

13. What has given you the opportunity of being acquainted with the internal arrangements of that factory?—I have been requested by Mr. Wood to pay particular attention to the health of the children employed in his mills, and have therefore, for some time past, regularly visited them, generally once a week, to ascertain the state of their health; walked through the rooms in the mill, and inquired of the overlookers, and took all the pains I could to ascertain whether any of them were ill or not.

14. You were professionally employed by him?—Yes.

15. Consequently you paid particular attention to whatever concerned the health of the establishment, generally speaking, and to the recovery of those who were indisposed?—Yes; at least so far as the children were concerned.

16. Has anything struck you as to the general management and arrangement of those mills?—That they are particularly cleanly, and made as comfortable as they can be; that there is every attention paid to the comfort, and cleanliness, and health of those employed in them, as far as possible, under the present system.

17. The system then is made, on the whole, in your judgment, as consistent with the health of those employed as it can be, considering the number of the hours they are engaged?—Yes, I think it is.

18. Are the mills well ventilated?—Yes, I have endeavoured to direct my attention to that as much as I could.

19. Kept as clean as possible?—Yes.

20. Means adopted and opportunity given, for instance, to enable the children to keep themselves clean?—Yes.

21. And to wash themselves at their pleasure?—Yes, I think there is.

22. Do you happen to know whether seats are provided for them, so that they may in some measure abate the fatigue that too long standing and exertion occasion?—There are; Mr. Wood has taken pains also on that point, to provide them all with seats, that they may sit down as they have opportunity, a minute or two at a time sometimes; but it is rarely for any length of time that they can do so.

23. Does it consist with your knowledge, that the hours in that mill also are quite as short as in any similar establishment, and shorter than in many? —I believe shorter than in any at present.

24. Will you state the number of hours they are employed in Mr. Wood's mill?—From six till seven, with half an hour for breakfast, and forty minutes for dinner.

25. And from your knowledge of Mr. Wood, and from your observation of the entire management of that concern in all its parts and details, do you believe that it is Mr. Wood's earnest wish and endeavour to conduct that mill

W. Sharp, jun. Esq.

with an especial view to the health and welfare of those whom he employs?—Yes, I do.

26. But, after all, do you think that those hours of labour, with the intervals that you have mentioned, are consistent with the general health of those employed, more particulaly of the children and young persons so engaged?—I think the present number of hours is too long a period of work.

27. Do you then consider that those hours of confinement and that length of attention to their business is inconsistent with the general health of those employed?—Yes, I do, if continued for too long a period.

28. And you have already stated that you do consider that term of confinement too long?—The present term; yes.

29. Upon what do you ground your opinion as to that term of exertion being too long?—Partly from the general appearance of the children, which is not that of robust health, and partly from the number of cases of sickness which have occurred, and which do occur, amongst them.

30. Have you taken notice for any length of time past of the cases of sickness which occur in that mill, and more especially of those cases which you can pretty clearly trace to the length of confinement that is there required?—Yes, at least since the 1st of January in this year, I have kept a regular account of all the cases which have come under my notice.

31. Would you have any objection to furnish the committee with that account?—No; I have it here, at least a summary of it; the individual cases it is not necessary to specify.

32. Just read over cursorily the results?—Number of children and young persons employed in the mill, five hundred and fifty.

33. Under what age are those five hundred and fifty reckoned to be?—It includes all the children that are employed in the spinning department, and some persons that are rather older, but I do not know the particular ages; it includes what are called the reelers; not, however, all the reelers; there are only a part of them I have had to do with;—Mr. Wood has upwards of one hundred reelers, and I have only put down here fifty, which is as many or more than I have to attend;—but the rest are all, literally speaking, children.

34. Can you say whether there are five hundred of them children?—The only reason why those fifty are entered is because a few, but very few of them, are mixed up in these cases; but there are five hundred children.

35. But those fifty to whom you have alluded might be very properly denominated young persons?—Yes, they might.

36. And that account is founded on five hundred and fifty, including the fifty reelers?—Yes, who are generally fifteen, sixteen, or seventeen years old.

37. That is, you would state it as your full impression that they are under eighteen years of age, the age contemplated by the bill?—Yes.

38. Do you know whether any of the reelers are married?—None.

39. Are the warpers included in that number?—No. Then out of this number there have been sick, from the 1st of January to the 1st of this month (June), that is, the five months which I have attended, exclusively of several that have been slightly indisposed, and that have not come to me, particularly among the children, many of whom have stayed at home a few days, perhaps with the measles, or something of that kind, and whom I have had nothing to do with; but of those that I have attended there are one hundred and sixty-eight cases.

40. Since when?—Since the 1st of January.

41. Will you state to the committee the nature of those cases generally; and first, how many of those cases which you can clearly attribute to the

labour of the mill are surgical?—There have been seventeen of them accidents, generally slight ones, injuries from the machinery, nothing of any very serious consequence; six have had inflamed and ulcerated legs, sore legs. W. Sharp, jun. Esq.

42. Should you suppose that to be a common result of too long standing at that period of life?—I think it is; at least this is a large number to have occurred in so short a time, of serious cases; two or three of these were really very serious cases of ulcers in the legs.

43. What were the other cases?—There are two cases of deformed legs that I have seen during that space of time.

44. Were they deformed when you first saw them?—They were.

45. Do you know whether they had been long in the mill?—I cannot answer to that point.

46. Do you know how the deformity had arisen?—I believe from their employment.

47. Then it was not a fact under your own observation that it had arisen from their employment?—Yes, I think I might say so; they have been employed, I am pretty sure, some time in the mill, and their legs have gradually become deformed.

48. Were they originally straight?—They were.

49. Will you now state to the committee the number of medical cases amongst those that you have attended, that you also attribute to the over-labour of these children?—I cannot say precisely, but there are several bad cases of scrofula, seven of consumption, and several of fever and of general debility, which I should say, without any doubt, had been caused, in a great measure, by their employment.

50. Should you consider that such employment, too long continued, had a natural tendency to generate scrofula?—Yes, inasmuch as it weakens the constitution, and predisposes to scrofulous affection, and aggravates that taint of the constitution where it orginally exists.

51. Regarding pulmonary complaints, should you also suppose that the employment in question, as at present continued, has a great tendency to occasion those maladies?—It appears to me to contribute to them in the same way as it does to scrofula, by weakening the general health.

52. In that number of cases how many deaths have occurred?—Five.

53. To what date is this account continued?—To the 1st of this month (June), being for five months.

54. Do you also attribute that excess of mortality, for such it certainly is at that period of life, partly to this long-continued employment?—Yes.

55. Are you aware that there is a bill before Parliament proposing to limit the labour of children and young persons employed in mills and factories to ten hours a day, exclusive of the times allowed for refreshment and rest?—Yes.

56. Do you think that that is as long a term of labour as could be sustained with impunity by the children and young persons employed in mills and factories, speaking in general terms?—I think it is as long, and perhaps longer, than is consistent with their general health.

57. Do you consider that excessive labour, or labour too long continued, has a direct tendency to produce disease and debility, and to shorten life?—Yes, I do.

58. Do not you think that the worst effects are produced by the terminating hours of a long day's labour?—Yes.

59. That the shortening, therefore, of the present hours of employment, by an hour or two, would very greatly diminish the number of medical and surgical cases that are under your care, if it would not entirely do away with the mischievous effects of too long-continued exertion?—Yes, I think it would.

W. Sharp, jun. Esq.

60. Do you conceive that children and young persons ought to have some opportunity for rest and recreation, in order to secure present health and future strength of constitution ?—Yes, I certainly do.

61. Will you state whether it is not quite necessary, even after the period of infancy, and up to that of more mature age, to guard against too long a continuation of labour in any pursuit whatever, alluding more particularly to young persons at a critical period of life ?—I do.

62. Do you conceive that the human constitution at the time of life to which the committee have been distinctly alluding is more susceptible of injury, from external and other causes, and especially from long-continued exertion, than it is at a more advanced and more mature age ?—Yes.

63. So that, upon the whole, you think it would be even more proper to diminish the length of labour proposed in the bill, than to extend it beyond that period ?—Yes.

64. And that is your opinion as a medical man ?—Yes.

65. Now, living in a manufacturing district, as you do, do you think it is possible to conduct a concern with greater attention to the health and welfare of those employed than is used by Mr. Wood in his establishment in Bradford ?—I do not think it would be practicable to take more care or pains with respect to them than he does.

66. And you are in the habit of visiting his establishment for the purpose of a general inspection of his hands ?—Yes, by his request; I visit them generally once a week.

67. Periodically ?—Periodically; that is, as regularly as my other engagements will allow me.

68. Of course his hands then have the advantage of medical treatment at the commencement of any illness ?—Yes; that is the object of my visit.

69. Do not you consider it a very great advantage to those that are indisposed to have medical assistance at an early stage of their illness ?—Yes, a very great and important one.

70. So that you consider that, in point of medical treatment, the hands employed by Mr. Wood have a still greater advantage than they would have if they were mixed indiscriminately with the labouring and poorer classes of society in the town and neighbourhood ?—Yes, I think they have.

71. All cases that are at all marked, and which it is apprehended may become serious, you are in the habit of paying instant and particular attention to ?—Yes; the overlookers are requested to observe for themselves, and to acquaint me with any indisposition which they may see amongst the children; and the children themselves, and the parents of those children, all know that they can have medical assistance.

72. Now with regard to the education of children, speaking in a medical point of view, would it not increase the inconveniences under which these children at present labour, and would not the consequences be still worse, if they were to be confined two or three hours after their present day's work is over, in a night-school or any establishment for the purpose of giving them education ?—After the present hours ?

73. After the present hours ?—I think the additional confinement would be still more detrimental to their health.

74. Speaking as a medical man of considerable practice in that town and neighbourhood, do you not believe that the long labour of children in factories, confined as they must necessarily be to constant attention and work, and for so great a length of time, with so few and short intermissions, has a direct tendency to injure their health, to cripple their limbs, and to shorten their lives ?—Yes, I do; but if I may be allowed, I will add a remark to that answer. I think amongst the ill effects produced, the injury to the general health is of more consequence than any particular deformities; I would not

lay so muchstress on the deformities as on the breaking up of the constitution; the injury to the general health. W. Sharp, jun. Esq.

75. Do you think that the injuries which may be inflicted on the constitution at an early period of life, may, for a time, be often sustained with perhaps little visible injury, and nevertheless be real, and felt severely at a more advanced age, in consequence of their having exhausted the strength and forestalled the energies of nature?—Yes, by breaking up the constitution, and, of course, injuring the health during the remainder of life.

76. Do you consider that the system of working in factories is in itself unwholesome, even if separated from the abuses which are said now to attend upon it?—No, I should not think so.

77. Do you think that the general health of children so engaged ought to be as good as those engaged in other occupations?—Yes, I think it should.

78. For what number of hours would you consider it consistent with the health of children to be so engaged?—It is difficult to answer that question, because it would merely be an opinion; but I think I can safely say that ten hours is the greatest period for which they ought to labour.

79. Are you disposed to think it should be less?—Yes, rather than more, certainly.

80. Have you an extensive practice amongst the working classes in Bradford?—Yes, I think I may say that, partly from my situation at the Dispensary, where I must, of course, see a great many; partly from my connexion with Mr. Wood's mill; and partly from private practice.

81. What are you in the Dispensary at Bradford?—Surgeon; that is the only medical institution we have in the town; the only public one.

82. From your observation in that situation, are you disposed to think that the children, and the families to which those children belong, who are employed in factories, are in a better or worse condition of health than those of the rest of the population?—I do not know whether I am qualified to give a direct answer to that question.

83. Perhaps you would state your impression?—It requires a great deal of observation to give a correct and decided opinion on a point like that; a great deal.

84. Should you say that there was more sickness in the factories, or less, than in an equal number of individuals of the same class of life in the same town?—I should think there is more.

85. More illness in factories?—Yes, decidedly.

86. Has this been a healthy or an unhealthy year at Bradford?—There has been a good deal of sickness.

87. Has there been any epidemic or prevalent disorder in Bradford this year?—We have had much more of some affections (such as measles) and of bowel complaints than usual. The committee will only find two or three cases there of what are called eruptive fever, that is, small-pox and measles; and in general they are so slight in the country that parents do not think it necessary to have medical attendance for the infants.

88. Out of these 168 cases, there are twenty-two cases of bowel complaints, are there not?—Yes, I believe there are.

89. Should you attribute that in any degree to the employment in a factory, or to the long hours?—Perhaps I should, and in this way; it appears to diminish the appetite, to injure the powers of the stomach, and that account includes stomach complaints as well as bowel complaints; they lose their appetites, and that leads indirectly to disease from their then taking improper food. If children have a good appetite they will eat proper food; if that is not the case they get things that are improper for them.

90. Do you think that the fever, of which there seems to have been thirty-four cases during this time, was any more to be attributed to the hours

W. Sharp, jun. Esq.

of working in the factory than that those hours produced general debility in the constitution ?—Yes, I think some of the fever cases might be fairly attributed to over-exertion in long confinement.

91. Was there much fever this year in Bradford and in other places ?—No, not since January ; there was at the latter end of last year.

92. Throughout Bradford ?—Yes, throughout, in the early part of the winter, but it quite disappeared before that account commences.

93. Then the fever in Bradford, generally, had disappeared, whilst it still continued in the mill ; do the committee understand you to say that ?—No, I do not mean that, but merely to say that there was not more fever than there generally is in the town during that time.

94. Can you state, from your experience, whether cases of fever are more common in mills, or less common, than they are in similar classes of society out of mills ?—I think they are more common among the mills.

95. Have you never heard it stated that there were fewer cases of fever which occurred in mills ?—No, I never heard any such statement.

96. Did you ever read Dr. Percival's paper upon the establishment of the Fever Hospital of Manchester, in which he held the direct effect of factory labour to be the generation of fever complaints ?—No, I have not read that.

97. Now, with respect to the scrofula, is not scrofula very often a hereditary disorder ?—It is generally considered so.

98. Does a warm and sheltered atmosphere of occupation increase scrofula, or, generally speaking, diminish it, as compared with exposure to the air and harder labour ?—I think it generally increases it.

99. Are you aware that parents having scrofulous children ever send them to the mill under a notion that it is better for them ?—No, I am not aware that there is any such notion.

100. Are you aware that any mill-owner, from a contrary notion, refuses to receive such children ?—I believe that some do.

101. Do you know whether there was a medical gentleman in attendance on Mr. Wood's mill previously to the 1st of January ?—Yes, I attended it some time before that, but I have only since then kept a regular account of the cases.

102. You had attended it for some time previously ?—Yes.

103. From your experience, should you say that these cases, during the last five months, were more or less numerous than usual, as far as you can recollect ?—Not more, I think, than they were previously ; and perhaps I may be allowed to observe that that account was not kept with any reference to this question ; it was merely my own private account, which I kept for myself.

104. Are you acquainted with any other mills in Bradford ?—Not directly ; not so as to know anything precise about them.

105. Have you ever attended people employed in those mills in your private practice, or have you relieved them at the Dispensary ?—Yes, very frequently.

106. What should you say were the cases most prevalent which came under your notice ?—Scrofulous affections, ulcers of various kinds, and affections of the chest, leading, I should think, to consumption.

107. From what cause do you think that these cases of ulcers arose ?—When they are on the legs, as they generally are, apparently from too long standing ; evidently so in many cases.

108. Have you had any experience at all regarding a country population in the course of your practice ?—Yes, daily.

109. Is there a very great difference between the health of children in the country and that of those employed in the factories ?—Yes, I think in general they are much more healthy.

W. Sharp, jun. Esq.

110. Now, speaking generally of their condition with reference to the manner in which they are fed and clothed, and so on, is there a superiority in all those respects, do you think, in the country children?—I do not know, I am sure, what to say to that. Mr. Wood is particular with his children, in obliging them to have clean and more decent clothing than, perhaps, many others have; so that I may not be able to form a correct opinion upon that subject.

111. Have you ever been able to take a comparative view of children employed in mills in the country, and of children employed in mills in town? —I have frequently had cases from mills in the country, but I do not see sufficient of them to form any correct comparison between them. A great many of Mr. Wood's children live in the country, and come to his mill; but there is as much sickness among them as there is amongst those that live in the town, and perhaps more.

112. Do you consider that the cases of inflammation of the eye, and other cases of inflammation, are attributable to their peculiar position in a factory? —No; I do not think they are more exposed to them than in most situations in life.

113. Would the same observation apply to rheumatism, abscesses, diseases of the skin, tape worm, and fits, enumerated in the paper?—Yes; I think in general those cases are not peculiar to their situation.

114. Do not you think that the working of children by night, and especially by gas-light, has a tendency to inflame the eye, and consequently to injure the sight?—It may have; I have had some complaints from persons in that way; they have thought that the gas-lights have injured their eyes.

115. Is it your opinion that they have?—I think it is quite possible, but I cannot say that I am sure of it.

116. What are these cases of slight indisposition to the number of nineteen?—They have generally been cases of girls that have been obliged to stop at home three or four days from their work, perhaps with a bad cold, or a little affection of the stomach, which merely kept them from their work for that time; and then they became well again, at least fit to return.

117. The other cases were of a more marked nature?—Yes, of decided illness.

118. Have you any observations to make upon the subject of the heat or temperature of the rooms in which these people work?—I think it is generally below 60°.

119. That is, the temperature is regulated with a regard to the comfort of the hands and their health, their work not requiring, like some branches of the cotton trade, any peculiar degree of heat?—No.

120. Have you ever observed a disposition in the hands to keep the room hotter than you thought proper?—No; the overlookers generally do as I wish them with respect to that.

121. Are the rooms heated by any artificial means?—No.

122. Is there not a steam-pipe?—Yes; I beg pardon, I mistook the question; of course there are the steam-pipes.

123. Are they properly ventilated?—Yes.

124. Has Mr. Wood baths upon his premises?—Yes.

125. In which the hands may bathe as frequently as is thought advisable or pleasant to them?—Hitherto they have not been provided on any very great scale; but Mr. Wood is now making a number of baths for them; there have been a few that they have had an opportunity of using.

126. Alternately?—Yes.

127. Is not scrofula promoted by poor living and sedentary employment? —Yes.

128. In-doors employment?—Yes.

W. Sharp, jun. Esq.

129. Do you think that the large or the small factories are more hurtful to the health of the children, and what are the reasons for your opinion?—I do not know that I can answer that question; they are all large ones with us, I think.

130. Do you consider that scrofula is almost the endemic complaint of the factory?—Yes, I think I may say so; there are very few that are completely free from it; I have only marked down nine cases in my paper, but they were the exceedingly bad ones, in which the whole body was completely affected with scrofula.

131. Do you think that the growth of the children is stinted by this confinement?—I should think it must, inasmuch as the constitution is weakened generally by it.

132. Perhaps you think more with reference to bulk and solidity than to height?—I am not aware that there is any great diminution in the stature—in the height.

133. But you have not made any particular remarks to that effect?—No, I have not.

134. Have you observed that there is a peculiar debility and attenuation of body, generally speaking?—The children, generally, are by no means strong, not as children in the country are—not robust and strong.

135. You mentioned that there were two cases of deformity that have come under your care since the 1st of January; were there any other cripples in the mill previous to that?—No, I do not remember any that have any connexion with the mill; I think not; not with the system.

136. You perhaps had not previously prepared yourself, by direct observations, for an inquiry of this sort?—Not at all.

137. So that your evidence on this occasion is mainly from the result of observations which have naturally presented themselves to you in the course of your practice, and not particularly noted for the purpose of giving evidence on this subject?—Quite so; and the facts noted in the paper were for my own personal information.

138. Have you paid any attention to the state of the houses of the poor?—Yes, of course, while visiting their children I could not fail to see that.

139. Are these children near, or at any considerable distance from the factory of Mr. Wood?—For the most part they live in the town; but Mr. Wood's mill is at the edge of the town on one side, and, of course, a great many of his children live a mile or more out of it.

140. And what is the general character and condition of the houses of the poor themselves?—In general they are not very comfortable; I think they do not pay that attention to their comforts which they have it in their power to do. Latterly, more attention has been turned to that subject by the town, on account of the prevalence of bowel complaints and the apprehension of the cholera; so that, for the last year, they have been kept in a cleaner and more comfortable state than they were before, I should think; and they have been whitewashed.

141. Some attention has been paid to that subject by the authorities or inhabitants of the place?—Yes, there has; I have no doubt that much fever has been prevented this last winter by that; by cleaning the streets, and by making drains where there were none before.

142. Do you know whether Mr. Wood is in the habit of making up what is called lost time, and thereby lengthening the confinement and labour of his hands, at particular times, in his mill?—No, I believe he never does that.

143. Do you not conceive that that would be a source of very great cruelty, as well as injury, to the children, if that custom were adopted by him?—Yes, it would indeed.

144. So that, upon the whole, he has taken every means to preserve the health and provide for the comfort of his children?—I think he has. W. Sharp, jun. Esq.

145. But state whether you do not think that it is desirable and necessary that the hours of labour of the children and young persons should be limited as proposed, in order to give full effect to his intention, and also to aid you, as his medical attendant, in preventing the disorders to which you have alluded?—Yes, I do.

146. Are accidents common in this factory?—Slight accidents of the fingers; the children get their fingers cut in the machinery.

147. The machinery of a worsted-mill is not so constructed as to occasion, generally speaking, those very serious accidents that occur elsewhere?—Not as they are guarded now.

148. Is asthma a disorder common to the work-people in factories?—No.

149. Not even among the carders?—I am not aware that it is.

150. Are pulmonary diseases common?—What we call phthisis is very common indeed.

151. Is not phthisis as serious an affection of the chest as asthma?—It is much more serious; the most serious complaint there is of the lungs is phthisis—tubercles of the lungs.

152. Is that the paper to which reference has been made?—It is.

[*The same is as follows:*]

Mr. Wood's Mill, Bradford, Yorkshire, 1832:

Number of children and young persons employed in the mill, about . 550

Number of the above sick, and attended by Mr. William Sharp, from January 1st to June 1st, 1832 168

Of these, there were,—

Accidents, generally slight wounds of the fingers by the machinery	17	
Fevers: Simple fever 25; Typhus ditto 6; Eruptive ditto 3	34	
Affections of the head	7	
Ditto of the chest	14	
Ditto of the stomach and bowels	22	
Inflammation of the eye	10	
Ditto . of the arm	3	
Ditto . of the legs	6	
Ditto . of the throat	1	
Acute rheumatism	1	
Abscesses	3	
Diseases of the skin	5	
Scrofula (bad cases of)	9	
Phthisis	7	
Chorea	1	
Incontinence of urine	1	
Tape-worm	1	
Fits	2	
Deformed legs	2	
General debility	3	
Slight indisposition	19	
		168

W. Sharp, jun. Esq.

Of this number—

Died	5	
Recovered, about	146	
Been much relieved	3	
Are remaining under treatment	14	
		168

Of the deaths, there were—

Of phthisis	3
Of typhus fever	2
	5

The surgical cases more directly connected with the factory system, are these:

Accidents	17
Inflamed and ulcerated legs	6
Deformed legs	2

The medical cases are those of debility, scrofula, and phthisis, and many of the acute affections.

June 20, 1832. *William Sharp, Jun.*

SAMUEL SMITH, Esq., called in, and examined, 16th July, 1832.

Samuel Smith, Esq.

1. What is your profession?—I am a surgeon in Leeds.

2. How long have you practised in that profession?—I have been a member of the College of Surgeons upwards of twenty years, and I have been in actual practice upwards of nineteen years.

3. Have you had extensive practice, particularly among the poor, during that period?—Yes, because having succeeded, in the first instance, to the practice of the gentleman with whom I served my time, of course I was in full practice from the first day of the commencement; shortly after that period I was appointed the parish surgeon of Leeds, the duties of which situation I filled for some time, and I have now been upwards of thirteen years one of the surgeons of the Leeds Infirmary, the largest hospital in the county of York, an institution through which about 5,000 patients pass aanually, about one-sixth of which number pass under my own immediate care.

4. And in the consultations of a surgical nature of course you assist?—Yes, at them all.

5. Is Leeds a manufacturing town?—It is decidedly so.

6. Do you know whether there are considerable numbers of children employed in the mills and factories of Leeds and the neighbourhood?—There are great numbers; so much so that at the period when they are dismissed from, and return to, the factories, the streets, in certain situations of the town, are actually crowded with them.

7. Is the appearance of factory children upon such occasions easy to be distinguished from that of the children composing the rest of the labouring population differently employed?—I think it is; when we compare the appearance of children working in factories with others who are not accustomed to employments of that nature, it is easy to see that they have not that healthy appearance; they appear languid, weak, and debilitated.

8. What are the hours of labour, as far as your observation and information have extended, which the children employed in the factories of Leeds and the neighbourhood have to endure?—I think in all cases the hours of labour for young children are too long, and, in some cases, most unreasonably so.

9. What is the shortest period of labour, as far as your observation has extended, that the factory children have to undergo in brisk times in the town of Leeds ?—I think they have to work generally about twelve hours as far as I have seen; I have not known of any instances where they have worked less than eleven hours and a half. Samuel Smith, Esq.

10. And in some instances you have perhaps known and heard that that term of labour has been greatly extended ?—I know that in a great number of instances they have to labour fourteen, fifteen, sixteen, and seventeen hours in a day.

11. You have already stated that you think the shortest term of labour you have mentioned is unreasonably long ?—I did not state the shortest period as being unreasonably long; I stated that in all cases I thought they were too long, and in some cases unreasonably so.

12. Is not the labour of children in mills and factories denominated "light and easy;" do you so regard it in reference to its effects upon the constitution ? —It is often described as such, but I do not agree at all with that definition; for admitting, in the first place, that no strenuous exertion were required during the period of their labour, though certainly such is not the case; or admitting that they laboured in a wholesome atmosphere, which also is far from being the fact; or, thirdly, that the materials upon which they work were not deleterious, which in general they are, for there is often much dust and other impurities given out from them ; still the exertion required from them is considerable, constant attention is demanded, and, in all the instances with which I am acquainted, the whole of their labour is performed in a standing position.

13. Do you consider the very uniformity of the exertion would, in all probability, occasion fatigue, and abate the energies of those who have to endure it ?—Even supposing no labour whatever were required under such circumstances, the merely having to sustain the erect position of the body for so long a period is harassing in the extreme, and no one can have an adequate idea of it unless he has himself been subjected to it.

14. You think, for instance, that the standing position is still more fatiguing than the walking one, and that that moderate exertion which the more equal exercise of the muscles produces is preferable to either ?—I do; I once had an opportunity of seeing an illustration of the effect produced by the standing position. About twenty years ago I was an officer in a regiment of local militia, and I noticed, within the first two or three days of training, men (who were mostly unaccustomed to military exercise) were constantly fainting before they had been drilled three hours; what makes me to particularly remember it is, that I was one of them myself who fainted, though at that time extremely strong and healthy.

15. Upon the principles on which you would reason as a professional man, does not the effort to keep the body in an erect position require a constant and complicated action of certain muscles, so as to occasion more fatigue than more strenuous exertion ?—It does, peculiarly so; and moreover there is another circumstance that I would allude to, namely, the increased action of the heart that is required when that position is long sustained. It is necessary that the brain should be supplied by a certain quantity of blood which the heart has more labour in performing in that position than in any other.

16. So that, upon the whole, you would consider as a matter to be taken into serious consideration, the position in which many of those exertions have to be made, namely, in a standing position ?—I think it is of great importance, and it is only surprising to me that those poor creatures can sustain it for so long as they often do, without serious injury.

17. What are the effects of this system, first directing your attention to the

Samuel Smith, Esq.

general health of the parties enduring it?—The effects are generally very plain and obvious. In the first place, those poor children are subjected for too long a period of time to continued labour; they have not a sufficient time allowed them for sleep and for refreshment; and again, from the length of time that they have to labour during the day, and the short intervals allowed for sleep, there really are few moments for play or recreation. All these circumstances combined, produce, in a short time, certain effects. In the first place, we find the digestive organs materially impaired in their powers. This, after a short time, produces extreme debility and lassitude; so that although the body is not reduced to a state of actual disease, and although there may not be any decided organic change in any principal viscera of the body, yet still it is in a very different state from that of health. I do not know that I can give a more proper illustration of the state that the body is then reduced to, than by describing it as a jockey would, that of being "out of condition;" although when the body is reduced to that state, there is no actual disease present, yet there is a continual tendency to disease. There is a diminished power in the body of resisting the attack of disease, and when disease does invade the body, it is always, under such circumstances, attended with greater facility. I make not the least doubt that in a short time our experience on this point will be more considerable than it is at present. I have no hesitation in saying, that if a number of factory children, reduced to the state of body which I have described, should be attacked by the prevailing epidemic (I am alluding to the disease which is now prevailing in Leeds, the cholera) compared with a corresponding number of children of the same age, but whose bodies have not been reduced by this extreme labour, the mortality in the factory children would be much greater, and much more sudden than in the others. I know that many factory children were attacked by this disease previous to my leaving town. In fact, only two hours before I left Leeds, I was called into a factory in making my rounds in the town, to a girl who was suddenly attacked with it there.

18. On the whole, then, have you found it your duty to recommend the withdrawal of children from such an excessive degree of labour?—I have very frequently found it necessary. There is never a year passes, but I see several instances where children are in the act of being worn to death by thus working in factories. After having had such cases under my care for a certain time at the Infirmary, and finding that my services were of very little avail, I have occasionally requested the poor child to come on the next out-patient day, accompanied by its father or mother. I have on such occasions addressed the father or mother in such language as the following:—"Your child will not live long if it continues to work at the factory; I can do no more for it, unless you take it from that situation. If you will remove it, it will immediately begin to thrive; I have now done my duty in making this statement to you, act as you think proper; but I think you will not do your duty unless you immediately remove the child."

19. Then it is your decided opinion, that that employment, pursued to the extent that you have described, increases and generates disease; and on the whole, certainly shortens human existence?—It has undoubtedly a manifest tendency to shorten life and create disease.

20. What are the peculiar diseases which those long hours of labour, and of the kind you have described, most commonly produce?—My own observation and experience has been more directed to the surgical diseases induced by this system than to the medical; because my practice now does not lie much amongst those individuals who send their children to factories. I can, however, have no hesitation whatever in stating that pulmonary diseases and scrofula are a very common consequence of the long labouring of young children in factories. All those diseases also which originate in extreme

Samuel Smith, Esq.

languor and debility, those diseases to which a person who is out of condition will be subject, are frequently induced by this system.

21. You have said that your attention has been mainly directed of late to the surgical cases that have been produced by this system; will you have the goodness to describe the effect of the labour, considering it in regard to that branch of your profession that surgeons usually have under their care?—Long standing has a peculiar effect in producing a varicose state of the veins, particularly in the female; now as that is a disease which is commonly consequent upon pregnancy, I wish to make a distinction, which is this; that although in that state it is so common as to be called the child's evil, I have never, under any circumstances, seen it in a young female who has not borne children, excepting only in those who have been accustomed to work at a factory. If it be necessary, I can explain the circumstance more at length to the committee.

22. Will you have the goodness to do so?—I may then state to the committee, that the blood which supplies the lower extremities is transmitted from the heart by the arteries; it is returned by the veins; these veins, if they were not supplied by valves, would have to sustain the weight of the whole column of blood from the heart, in fact to the feet; but every few inches the veins of the lower extremities are provided with valves, so that the vein has only to sustain the weight of a column equal to a few inches. Now, in pregnancy, in consequence of the enlargement which takes place in the womb, the upper part of the veins are so pressed upon as to retard the return of the blood from the extremities to the heart; hence we find, frequently, under such circumstances, a great enlargement of the veins from this pressure, and after they are enlarged to a certain extent, they become tortuous, giving rise to that disease which is called a varicose state of the veins, and which being so frequently produced by pregnancy, is, as I said, called the child's evil. Now, I have repeatedly seen that disease, as I have before stated, in young girls who have not borne children; upon several occasions, having merely received the parent's word for it, I have suspected the fact, and thought that they must have had children, and in consequence of that suspicion I have examined the body of a patient, to ascertain whether it were really the fact; for, from the state of the veins, I could not believe that she had not borne children, but I have ascertained that to be the fact; I have merely examined the skin of the body, in which, in a woman who has borne a child, there are certain cracks remaining in consequence of the great distention of the skin which has taken place. Having now explained why a varicose state of the veins is produced in the female by pregnancy, I may state that in young women who are subjected to remain twelve, fourteen, or sixteen hours in the day in a standing position, the same effects are produced by those veins having so long to sustain the weight of the column of blood; the valves become weakened and lose their power in returning the blood to the heart.

23. Inducing that disease which you call a varicose state of the veins?—Yes; I wish to state also that I am aware this is a common disease in the male and in the female beyond a certain age, but I have never, under any other circumstances, seen it in individuals under twenty.

24. You would fix the limit at twenty?—I have never seen a varicose state of the veins under that age in young women who have never had children, excepting in factory workers.

25. Should you state as your opinion, that such a varicose state, with the exception of cases of pregnancy, never occurs in young persons under the age of twenty, as far as your experience leads you to give an opinion?—I should think it could not by any possibility occur in any individual under that age, excepting those who have been subjected to the same kind of causes.

26. Will you go on to state the further effects of standing to labour?—It

Samuel Smith, Esq.

has also frequently the effect of producing an ulcerated state of the legs; this I have seen repeatedly; in fact, a few months ago I was quite struck with the circumstance, having three cases under my care of young women under twenty-two years of age, in one ward of the Infirmary, and all from one factory, and that a factory where very long hours were not worked.

27. Was there any hesitation at all in your mind as to attributing that state of the body to the labour to which they had been subjected?—I have no hesitation at all on the subject.

28. Will you state to this committee what are the effects of this description and degree of labour upon the limbs and the bones of the children and young persons who have to undergo it?—Long standing has a singular and very striking effect in producing extreme deformity of the lower extremities and of the spine.

29. Have you turned your attention more particularly to that branch of your profession which you term osteology?—I have particularly so; it is a very important point of anatomy in reference to surgery, and being a subject upon which lectures can be given without the necessity of procuring dead bodies, I have frequently been in the habit of giving lectures upon the structure, formation, fractures, dislocations and diseases of the bones to the pupils of the Leeds Infirmary, as well as to my own.

30. Have you observed the effect of the labour in question upon the bones of the human body at an early period of life?—I have had great opportunities of seeing it; in fact in Easter week I was requested by the operative committee in Leeds to attend at the county meeting, which was held at York; and I delivered my opinions upon that subject there.

31. Are your opinions the same as they were upon that occasion?—They are.

32. Will you have the goodness to state the substance of them to this committee?—Certainly. It will be necessary for me, before I can satisfac-factorily explain the effects produced upon the bones, to say something as to the structure of them; and I do not know that I can do that in a clearer way than by supposing that we shall perform two experiments with a bone. If, then, we take a bone, and throw it into the fire, it will burn, and in the course of a short time it will appear merely a hard white mass, because the animal part of the bone is then entirely destroyed; its veins, arteries, nerves, marrow, and membranes are completely destroyed by the fire; nothing remains but the earthy portion of the bone, which is principally composed of phosphate of lime. Now, in this experiment, we have destroyed the animal part of the bone, and retained the earthy part alone. If we reverse the experiment, take another bone, and, instead of throwing it into the fire, put it into acid, and let it remain there a few days, the acid takes away from the animal part the whole of the earthy constituent, leaving the arteries, veins, nerves, membranes, and marrow untouched. A bone, under those circumstances, will have precisely the same external appearance as it had before it was thrown into the fire; it will have exactly the same form, but it will have entirely lost its strength; it will have become soft and yielding; so much so, that one of the principal bones of the body, under those circumstances, may be bent and twisted in any direction without breaking; it may be even tied in a knot; and to illustrate this part of anatomy, it is usual, when giving lectures upon this subject, to have a preparation of a bone, so tied in a knot, put up in a bottle with spirits. Having now stated that bones are composed of an animal and of an earthy matter, I may add that it is the earthy matter to which the bone principally owes its strength. Now the bones of young children do not contain nearly so much of the earthy constituent of the bone as those of adults. In the very first formation of bone, I mean whilst the child is still in the womb, it contains no earthy particle whatever; even at the

Samuel Smith, Esq.

period of birth, the bones are so soft that they will bend in any direction; even up to twelve or thirteen years of age, under the infliction of an injury, which in the bone of an adult would cause fracture and displacement, they are merely bent. The bones of children at that age certainly will occasionally break; but they will also frequently only bend; but we never see an instance of the bones of an individual upwards of sixteen or seventeen years of age thus bending. From the circumstance of their containing more of the earthy portion, they are rendered more brittle; and consequently they will more readily break. I do not know that I can offer a fairer illustration than by comparing the bones of a child to a green willow, and those of an adult to a dry stick. We all know the effect produced in breaking those two substances; the dry stick will snap immediately asunder; the green willow will bend without its particles being detached from each other.

33. Is there a progress going on in the human bone that continues to increase the strength of the body up to a certain period of life, and beyond the term that you have mentioned?—Yes; the human bones are not completed in their formation till from the twentieth to the twenty-third year, and it may be proved in this manner: the cylindrical bones of the body are formed in three distinct portions; first, the bony matter is deposited in the centre of the shaft of those bones; that of course spreads: the shafts of the bones have a considerable quantity of bone deposited in them before any ossific deposition takes place in the heads; there is of course an upper and a lower head, and the shaft is the intervening portion of bone. Now, betwixt the head of a cylindrical bone and the shaft, there is a layer of cartilage intervening; this is called the epiphysis of the bone; it separates the head of a cylindrical bone from the shaft; and bone is not deposited in this layer of cartilage so as to render it an entire bone, until about the twenty-first, twenty-second, or twenty-third year. Now, if we compare that with the progress of the formation of bone in the horse, we find in the horse that the ephysis disappears at seven or eight years of age; so that the process of ossification is completed in the horse in nearly the third part of the time which is required in the human being.

34. Corresponding with the difference in the average life of the two animals?—Exactly so.

35. Is the human being, previously to the full development and perfecting of the bone, so capable of long-continued and strenuous labour as after that period?—It is probable it may be for a short time previous to that completion.

36. Will you state the effect of too long continued labour upon the bones of the human body, when endured at evidently too early a period of life, or for too long a time together?—Before I can state that, I ought also to mention, in a description of the formation of bone, that the heads of bones are formed of a different material from the shafts; the shafts of the cylindrical bones have the bone deposited in laminæ upon them; the bone there is as compact and as hard as ivory, but the heads of the bones are formed of what we call a cancellated structure, composed of regular cells of ossific matter, instead of layers, therefore they are much softer than the shafts of the bones, so much so, that it is very well known that a ball, in striking the head of a bone, will penetrate it, will make a hole and lodge in the middle of the head of a bone; but it is quite different when it strikes the centre of the shaft of a bone, there, from its being so much more brittle, it will immediately break up into a number of fragments, and be shattered. Now, having so far stated the difference between the formation of the head and the shaft of a bone, I may state that the bones of the foot are formed of the same cancellated structure as the heads of the cylindrical bones. The foot is formed of a very beautiful

Samuel Smith, Esq.

arch of bones of a wedge-like shape, but still of a softer nature than the cylindrical bones; at the centre of this arch of bones, of which the foot is composed, rests the main bone of the leg, and, consequently, these arches have to sustain the whole weight of the body. I am now frequently in the habit of seeing cases in which this arch has given way, from long continued standing in factories; the wedge-like form of the bones is lost, and consequently the arch falls flat to the ground, rendering the foot deformed for life. I ought to state, however, that a person, on timely application, may be relieved from this state by wearing an artificial arch. I am in the habit, at the Infirmary, of placing an artificial arch under the foot, by which means the wedge-like bones are pressed upwards, and after wearing this support for a certain length of time, new bone becomes deposited, and sometimes a partial cure is effected. Long continued standing has also a very injurious effect upon the ankles; it relaxes the ligaments of those joints, and renders them very imperfect. But the principal effects which I have seen produced in this way have been upon the knees. By long continued standing the knees become so weak that they turn inwards, producing that deformity which is called "knock-knees;" and I have sometimes seen it so striking, that the individual has actually lost twelve inches of his height by it, which may be proved in this manner: a well-formed man will, in general, stand the same height as the length of his arms when extended; I have seen individuals of that class whose arms, thus extended, have measured nearly six feet, but who only stood about five feet high.

37. Is the deformity thus produced sometimes of a very striking character, so as to render a person incapable of the usual exertions necessary for a man that has to earn his bread, and so as to strike with an extraordinary degree of sympathy those who have to witness the instances so occurring?—I have seen a great number of cases where individuals have been entirely unfitted for any active employment in consequence of the extreme state of deformity to which they have been reduced.

38. Have not these kinds of deformity been so common as to have called the attention of the medical officers of your institution to their frequency, as sent from those parts of the country more particularly where mill-labour universally prevails?—I remember being very much interested in this inquiry about twelve years ago, for I then saw a great number of cases that came from Bradford; at that period I did not understand those cases as I do now, but upon investigating them, I found that they were all factory cases, and that in the factories of that town at that period they worked much longer hours than they did in Leeds; I therefore attributed it to that cause.

39. Speaking of the extraordinary cases of deformity to which you have been adverting, may not some of those have been occasioned by the disease commonly called rickets?—No; and I have been very much interested in making that distinction; if I had been asked that question a few years ago, I should have answered in the affirmative; but my experience in these cases is now much greater, and I think I am able to explain to this committee the great difference there is in those cases of deformity produced by factory working, and in the cases of deformity produced by the disease which has been mentioned: rickets is a disease which appears in the very earliest infancy, frequently, nay generally, during the period of suckling; the deformities which it occasions, although often closely resembling factory deformities, are generally commenced and completed before the period at which children are sent to work in factories; that disease produces a particular effect upon the whole bony system; it deforms the lower extremities, the spine, the superior extremities, and the skull. Now, in all the cases of factory deformity which I have seen, there has been no effect produced upon either the skull or the superior

extremities, excepting in one single instance where there was a slight enlargement of the wrist. Samuel Smith, Esq.

40. So that you have directed your especial attention to that distinction in reference to this important inquiry?—I have taken a great deal of notice of cases of this kind, and I find that it is a distinction which I can now readily draw between those cases and cases of rickets. I have been interested very much in this inquiry, and I have found cases where I have been satisfied the deformity has not at all commenced till after the body had been perfectly formed for twelve or fourteen years.

41. Is it a commonly, if not uniformly received opinion in your profession, that the first appearance of the disease called rickets is never so late as at that period of life?—I have never known it so; I conceive that that state of the bones cannot take place in the human body so late, unless it has been subjected to the same causes as factory working, or in cases of extreme debility from disease; but in many of the instances which I have seen, I wish it to be remarked, that deformities have taken place in well-formed, strong, muscular boys; in fact, I do not know that I can give a better illustration than the case of James Kirk, who, I think, has been before this committee. The committee perhaps will remember, that he was a very stout boy, with a fine broad chest, one who I have no doubt, if he had not been subjected to such extreme labour as he had been, would have been a fine, stout, manly young fellow.

42. In a book of Dr. Black's, on Medical Statistics, he states, that after the age of three years, the disease called rickets is not known to commence; is that, generally speaking, the impression of the profession at present?—I think I should rather enlarge that period myself; but I would not extend it beyond five years.

43. You would then by no means extend it so far as to include those cases of extraordinary deformity that are known to commence as late as nine, ten, twelve, fifteen, or even seventeen years of age?—I have not the least hesitation in stating, that if the individuals that I have seen deformed by those causes, had had proper hours for sleep and relaxation, with proper food, those deformities would not have taken place.

44. So that whatever the disease might be called, it is occasioned and developed by what is now termed the factory system, as at present pursued; is that your opinion?—It is; for the last year or two I have taken especial care to make inquiries into these circumstances, and I have not seen a single instance contrary to my opinion.

45. Do you consider that this labour at so early a period of life, and so long continued, has the effect of retarding, and ultimately stunting, the growth of the children and young persons exposed to it?—I have not the least hesitation in stating that it has a peculiar effect in retarding the growth of children; as far as my experience goes, I think children who have been laboured five or seven years in a factory, will generally be found to be about a year and a half, or two years in height, health, and strength, behind children of the same age who have had to endure moderate labour, with proper intervals for sleep, rest, and refreshment.

45.* Do you think the muscular strength of the body is impaired by this early and excessive labour in those instances?—There is no doubt whatever of that; because we always find that in factory children the limbs are more slender than they are in others, and that the skin is more relaxed.

46. Do you think that this deterioration of the appearance, the strength, and the health of those exposed to this degree of labour has been progressive, and is likely to be perpetuated if the system be not regulated?—I have no doubt whatever of it, because I have lived myself long enough to have seen with my own eyes that the manufacturers of twenty-five years of age in

Samuel Smith, Esq.

Yorkshire, of the present day, are an inferior race of beings, when we compare their bodily strength to that of manufacturers of the same age five and twenty years ago; I was then seventeen years of age; and of course I am competent to speak to that point.

47. You yourself have observed a very material difference since the mill system has become so universal, and been conducted as it at present is?—I know it from experience; I am as perfectly satisfied of it as I am of any fact; and I have no hesitation in stating to this committee, that if these individuals are subjected to the same causes for a few more generations, the manufacturers of Yorkshire, instead of being what they were fifty years ago, as fine a race of people as were to be found throughout the country, will be a very diminutive and degenerated race.

48. In the observations you have made, you are instituting a comparison between the mill system and the domestic manufacturer, are you not?—I am instituting this comparison, from the circumstance of young children being now allowed to labour twelve, fourteen, sixteen, and eighteen hours in the day.

49. Will you have the goodness to state whether, twenty-five years ago, the period to which you just adverted, there were any factories in existence, or whether they were so widely extended as they are at this moment?—There were then very few factories; I think almost all the long chimneys in Leeds have risen up since that time.

50. And in those factories the time of labour did not extend to that unmerciful period which is now exacted from the children? — Certainly not; it has been gradually encroaching for the last nine or ten years.

51. Does not another effect of the factory system demand the serious attention of the legislature of the country—namely, the numerous and distressing accidents that take place within these establishments?—There are many and very serious accidents take place in the factories; numerous instances of individuals losing limbs. About seven or ten years ago, Sir Charles Bell paid me a short visit; and he was very much struck with the number of accidents of the kind alluded to which we had in the infirmary; so much so, that he collected from me evidence regarding that circumstance; and he visited other hospitals with that view, thinking that it would be an excellent provision if Parliament were to interfere in order that some kind of pension might be allowed to individuals so losing their limbs.

52. Are some of those accidents of a very fearful character, so as totally to disable them in after life?—I have frequently seen accidents of the most dreadful kind that it is possible to conceive; I have seen cases in which the arm had been torn off near the shoulder joint; I have seen the upper extremity chopped into small fragments, from the tip of the finger to above the elbow; I have seen every extremity in the body broken; I have very frequently seen the most shocking cases of lacerations that it is possible to conceive.

53. Have the goodness to state to the committee in what manner you think that a legislative enactment could be made beneficial for the prevention of accidents from machinery?—I have no doubt that a great number of accidents might be prevented by some act to compel the owners of mills to have such horizontal and upright shafts as revolve with great rapidity, in situations, where children are placed near them, sheathed and covered with square boxes of wood, which may be done at a very trifling expense, and which I understand is often neglected.

54. Do you happen to know whether there are any such precautions in any of the factories?—There are; and I have no doubt also that a very great proportion of the accidents which take place will be avoided by shortening

the duration of the labour of children, because I am thoroughly convinced that many of those accidents take place during the time at which children are exhausted, sleepy, and tired, from the long period for which they have continued their labour; they are in that state of lassitude and fatigue that they cannot keep their eyes open; and I believe frequently the fingers become involved in the machinery whilst they are in that helpless state. Samuel Smith, Esq.

55. Do you happen to know whether a considerable proportion of the children and young persons employed in mills and factories are females?—I believe a great many of them are females, because I see many of them coming out of the mills.

56. Are not the females still less capable of sustaining this long labour than males would be of a similar age?—No doubt whatever of it; because in the female neither the bony nor the muscular system is so strongly developed as it is in the male; in fact, the whole body is more delicately formed.

47. Is the peculiar structure of the female form so well adapted to long-continued labour, and especially labour which is endured standing, as is that of a male?—No, it is not; because in the female the pelvis is considerably wider than it is in the male; in consequence of that, the heads of the thigh-bones are much more separated from each other than in the male; this, of course, causes a greater obliquity in the thigh of the female, causing what we all know in them to be a proper formation, a slight tendency of the knees inwards; from this tendency of the knees inwards, or rather in consequence of the heads of the thigh-bones being further separated from each other, they have greater difficulty in maintaining the centre of gravity; the knees turn more inwards, and of course they are weaker than they are in the male, and therefore not so well calculated for sustaining the body for a length of time in that position.

58. The fact, then, is, that the bones of the lower extremities of the female are not placed so perpendicularly under the superincumbent weight of the body as those of the male?—They are not, from the causes which I have described. With regard, however, to that question respecting the female, there is one circumstance which I think of paramount importance to any other; I allude to the change which takes place in the form of the female pelvis at a certain period of life. In early youth, up to about twelve or thirteen years of age, the female pelvis is formed in the same manner and nearly of the same shape as the male; but at that period of life when certain changes take place in the female constitution, the upper part of the pelvis, instead of being what it was, a round ring of bone, becomes expanded, and when properly developed, of an oval form. This is a wise provision of nature to allow of the expulsion of the head of a child during the period of child-bearing. Now this alteration in the form takes place at the particular period of life to which I have alluded; therefore, at this time the ring of bone which I have described has to bear the whole weight of the body at the back part of it; it has to sustain the lower part of the spine; this, of course, will press considerably upon the bone at that part. At the sides, this ring of bone forms the sockets for the reception of the heads of the thigh-bones, and therefore, under these circumstances, in the case of a female in delicate health, and having for an extremely long period to sustain the upright posture, the pelvis is prevented from being properly developed; and, in many of those instances, instead of forming an oval aperture, it forms a triangular one, the part supporting the spine being pressed downwards, and the parts receiving the heads of the thigh-bones being pressed inwards. This occasionally produces the most lamentable effects in females under these circumstances, when they are expecting to become mothers; in consequence of the deficiency of this development of

Samuel Smith, Esq.

the bones of the pelvis, there is not actually space sufficient for the exit of the child which is within the womb. Under these circumstances, it is often the painful duty of the accoucheur to destroy the life of the child in order that he may preserve the more valuable one of the mother. I have seen many instances of this kind, all of which, with one single exception, have been those of females who have worked long hours at factories, at the period when the pelvis becomes developed, and for some years afterwards.

59. Then it is your opinion, considered with a view to the principles of anatomy, that such would be the effect of too long standing, and that opinion is confirmed by your own actual practice in relation to those cases?—It is.

60. It has been occasionally suggested that the protection of children should extend to the age of twelve, so as to exempt them from labour; would that, in your opinion, be a sufficient protection for the children, and more especially to the females so engaged?—In reference to the females, I should conceive that they would require protection more at that particular period of life than at any other, for it is very well known that after the age of twelve years certain changes begin to take place in the female constitution, to which it is unnecessary for me further to allude than merely to state that for some considerable time they remain in an extremely delicate state of health.

61. One in which it is likely that labour too long continued, and especially in the standing position as you have described, would be injurious, and in many cases destructive?—Very injurious.

62. Do you also think that protection from excessive labour should be extended to pretty nearly the period of full maturity, both in regard of males and females?—I think myself that it is very injurious to overwork an individual during the period of his growth; therefore, I should say that protection was required till the time, or nearly so, at which the body becomes fully developed.

63. Do you think that the protection, as extended in both sexes to the age of eighteen, is an unreasonable or an unnecessary protection?—I think not, because many individuals will arrive at maturity much later than others; some, for instance, will have arrived at the full growth of the body at fourteen or fifteen, while other individuals do not arrive at that maturity till twenty-one or twenty-two years of age.

64. But in those cases where the body arrives at its full stature at the age of fourteen or fifteen years, now alluding to the male sex, do you suppose that the strength of the parties has also matured at so early a period as that?—I think not, because we in general find that those individuals who arrive the earliest at maturity are the first to fail in strength.

65. You think, then, that the idea is founded upon just physiological principles, and consistent with experience, that a person who grows rapidly, and who early attains to his full stature, indicates weakness rather than strength of constitution?—There is no doubt whatever of that.

66. So that that should afford no reasonable or proper ground for subjecting them to excessive labour at an anticipated period of life?—Certainly not; it is just the same with a human body as it is with a plant, those that are the longest in arriving at maturity flourish the most, and survive to the latest period.

67. It has been occasionally argued, that a remission of only one-sixth part of the usual day's labour could very little relieve the operative classes, the children and young persons more particularly, or at least not relieve them beyond the apparent difference of the hours of labour; is that your opinion?—There is no doubt whatever that if we are to suppose the hours of labour

at present to be twelve, by deducting two hours, although you only take away one-sixth part of the duration of labour, still you will remove considerably more than that amount of actual fatigue. I should say, that although you take away only one-sixth of actual labour, you would, perhaps, diminish the actual amount of fatigue three-fourths; and I will argue in this manner: we all know that exercise has a tendency to increase the strength, when it is carried short of producing actual and considerable fatigue; but when it is pushed beyond that point, it has a directly contrary tendency. Now we may make here an illustration, by supposing a horse to be laboured a certain number of hours every day, he will go through those hours for a certain length of time without suffering, or without being put out of condition; but increase the amount of his labour considerably, and you will find that he will very shortly lose condition; but I am confident that many of these poor factory children have really to work much longer hours than it would be possible for a horse to sustain. I believe if horses in this country were put to the same period of labour that factory children are, in a very few years the animal would be almost extinct among us. Every gentleman who is in the habit of using horses well knows the effect produced upon them by too long continued labour; you may give them what keep and corn you please, but nothing will counteract the effects of too long continued labour.

68. As you have appealed to comparative anatomy in previous parts of your evidence, I may ask you whether you think the analogy would hold good respecting the effect upon the future strength and the longevity of the human being, as compared with the horse, in regard to early as well as to long continued labour; that is, in the human being, do you think it would, as strikingly as in the horse, destroy the shape, and diminish the strength, and curtail the life of the animal?—It is a very well ascertained fact, and known to every gentleman who understands horses, that when they are put to work at too early an age, their joints become relaxed and weak, their health suffers, and they never can be made perfect animals afterwards.

69. The damage done in both instances is irreparable, is it not?—I think the injury inflicted in both instances is irreparable.

70. Have you observed the moral effects of the system as at present pursued, namely, that of too early and too long continued labour?—I have not particularly noticed that.

71. Will you state, then, to the committee, what you conceive would probably be the effect of labour in a heated atmosphere more especially, and endured at too early a period?—I think labour in a heated atmosphere occasionally has a tendency to produce premature puberty, although, at the same time, I must mention, that in other instances it produces exactly the contrary effect; in those instances where the body is so extremely reduced and weakened, it will retard puberty, in others it frequently occasions premature puberty.

72. Supposing it was stated, that a greater number of illegitimate children were born in agricultural districts than in those districts where the manufacturing system very generally and extensively prevails, would you suppose that that would be a sufficient reason for concluding that the agricultural district was the less moral and the manufacturing the more moral of the two, or otherwise?—I should certainly think that there was more morality in the agricultural districts than in the manufacturing.

73. But supposing it were stated to you that there were, nevertheless, a greater number of illegitimate children in the agricultural districts than in the manufacturing districts, what conclusion would you draw from that fact?—I certainly should not, because there were more illegitimate children in the agricultural districts than in the manufacturing, draw the conclusion

Samuel Smith, Esq.

that therefore there was more immorality of the kind alluded to in the former instance; although, at the same time, I may be allowed to state, that I do not believe the fact of there being more illegitimate children in the agricultural districts than in the manufacturing; but even supposing that to be a fact, still I should not draw that inference from the circumstance, because every one possessed of common sense and of any knowledge whatever upon such subjects as these, knows very well that promiscuous intercourse has a direct tendency to produce sterility; therefore I would say, that in the agricultural districts, the circumstance of intercourse not being followed by conception was the exception; but in the manufacturing districts, where intercourse was followed by conception, I would say that that was the exception.

74. Supposing that the factory system has a tendency to produce unnaturally early intercourse, do you not think that that would have an effect to interfere with human prolificness?—I should think it would, both in the male and the female.

75. Have you yourself examined any mills or Sunday-schools with particular reference to this question, or have your observations been the result of your general practice, so as not to be taken from individual instances?—I may state to this committee that the information which I have obtained upon this subject has been fairly obtained in my professional practice. I have never visited a single mill, or a single factory, or a school, to obtain information upon this subject; I have never sought out a single case with a view to assist this question. When I say that I have never visited a single mill or factory, I mean at the request of those who were interested in bringing forward this bill; I have only seen one mill and one school, and that was done at the request of individuals who were opposed to the provisions of this bill. I have frequently been requested by those who favour this bill to go over schools and factories, but I have invariably refused.

76. What was the result of your observations in that school or factory which you say you were induced to visit?—The school that I visited was a Sunday-school, which is kept by a gentleman in a village near Leeds, and he particularly requested that I would go to see the children there, in order that my opinions might be changed as to the effects of factory working, for he stated that I could not see a finer, healthier set of children anywhere. At his request I went; he stated that he had not a single deformed child in the school. The two first boys that I saw were boys of about sixteen, who were evidently deformed, both in their spine and lower extremities, by factory working. I thought the greater part of the children also to be unhealthy looking.

77. So that your observation there corroborated your previous observations and opinions upon the subject, as now expressed?—Certainly; when I went with the gentleman into his house he asked me my opinion; I told him that I was thoroughly convinced, from what I had seen in his school, that my opinions were perfectly correct, and I put it to him in this way: I said, "Sir, if you go into a market to buy a horse, you pay from twenty to eighty guineas for that animal, and the difference in the price is occasioned by his greater strength, size, beauty, and activity; now the comparison that I will make with these children is this, that if we were to compare them to horses, they are all the lowest priced animals that it is possible to conceive." With respect to the one factory which I visited, I had made a statement that line spreading had a tendency to produce lateral curvature of the spine. I was requested by a gentleman to visit his mill to convince me to the contrary. When I saw the girls at their employment, I found they had continually to stand with the spine twisted to one side; consequently I was convinced that my opinion was correct.

Samuel Smith, Esq.

78. Did you make any personal examination at that mill?—I did not.

79. Had your conclusion respecting the deformity of the spine, as connected with the factory system, been derived from cases that had come before you in your professional capacity?—From great numbers of patients that have come to the infirmary with such diseases.

80. Supposing you were to go into a mill at the request of anybody to ascertain, from personal inspection, the state of the health and of the longevity attending that employment in that particular establishment, do you conceive that you could, by possibility, obtain from such data any satisfactory results whatever?—No, I have never obtained any information upon the subject in that way; my information has been obtained in a very different manner. If I wanted to ascertain the results, I should visit the hospitals, the dispensaries, the poor-houses, and the sick beds. I conceive that no satisfactory result could be obtained from any statement of mills, unless that statement were to include every individual that had ever worked in that mill for twenty years, and the subsequent course of life followed by those individuals after leaving the mill. Most of the cases of disease and deformity which I have seen produced by this system have been in individuals who, at the period when they have been under my care, were totally unfitted and disqualified for going into a mill.

81. So that that class of persons would be off the books of the mill almost entirely?—No doubt.

82. And the accounts of that mill regarding health and longevity would be entirely free from the most striking and numerous cases of sickness, death, and deformity, which that work has unquestionably occasioned?—No doubt whatever.

83. You have expressed yourself regarding the effects of this system, do you give to this committee your observations with the utmost degree of confidence, as resulting from the thorough conviction of your mind, and the long and extensive practice that you have had among that extensive population where your professional practice lies?—In stating the evidence which I have upon this question, I know that in some respects I shall suffer by it; but in the course which I have taken regarding it, I have been entirely guided by a wish to do good and to serve the cause of humanity. I am a person of very retired habits, never having on any occasion in my life placed myself before the public, excepting upon this question.

84. You have thought it your duty to give the result of your experience regarding this inquiry?—I did; I could not withhold it; it was a very painful thing for me to attend the meeting that was held at Leeds, and the one that was held at York, but I conceived it my duty so to do; and I am not aware that ever, in the course of my life, I made a speech before the public, excepting on those occasions.

85. Have you seen hospital practice in other places as well as Leeds?—I have.

86. Where?—I was house-pupil with Sir Charles Bell, in London, two years, and attended the hospitals of London during that period; I was also one winter in Edinburgh, and occasionally attended the hospitals there.

87. Do you remark that the difference in the cases in the hospitals of the metropolis and of Edinburgh, as constrasted with those in the manufacturing district of which you are now a medical officer, confirms all the observations which you have made to this committee?—The cases are very different in the London hospitals; bruises and severe contusions bear a very great proportion; in our infirmary, lacerations and cuts inflicted by machinery are more common.

88. On the whole, do you think that it is perfectly necessary to the health and welfare of the population that the hours of labour of the children

Samuel Smith, Esq.

and young persons should be limited?—I do think so; and I am also satisfied that it is quite necessary to their happiness; for there is now such a state of excitement and agitation produced upon this question, particularly in the county in which I reside, that I am sure they will never be satisfied until they have obtained the ten hours' bill.

89. You think that the number of the cases you have seen furnishes you with a fair criterion of the general effect upon the factory system?—I have seen a great deal of it; more than it is possible to conceive.

90. Do you know whether all the medical profession of Leeds entertain the same views of the general effect of the system as yourself?—I have asked every particular friend in the profession since I attended the York meeting, these questions: "Have you read what I said at York; is there a single observation that I made that you can for a moment dispute?" and I have never met with one individual yet but approved of every word that I stated.

91. Did you ever know instances of deformity or disease which you consider the result of excessive labour in the case of persons that have never worked over-hours; that is, who have never worked above eleven hours?—In most of the cases which I have seen, the hours of labour have exceeded eleven.

92. What, generally speaking, have those hours been?—Twelve and thirteen hours; I have seen deformities frequently produced by twelve and thirteen hours' actual labour; but I must also state to the committee, that in many of the cases which I have seen the hours have been longer than that.

93. Will you state some of the longest hours: what are those longer periods of time to which you allude?—In the last case that I saw, which was about a month ago, a young girl of about fifteen or sixteen years of age went to consult a physician in my neighbourhood, and, finding that it was a surgical case, he sent her over to me; she was a patient who went for gratuitous advice; she was very much deformed in her knees and her ankles, and also very much reduced in her health and strength. Upon investigating her case, I found that she had worked from five o'clock in the morning to nine or ten at night; but in order that I may not exceed, I will say nine, though I believe it was ten at night; for six months in succession she had worked for those hours, and during the whole of that period she had not been allowed a single minute for food, for rest, or for recreation. She was obliged to take her breakfast as she followed her work; she was obliged to take her dinner as she followed her work, and the same with her drinking. Now when she stated this case to me, I thought it most monstrous, and I would not believe it on her statement, but she was accompanied by her mother, who was a respectable widow, and the mother substantiated to me every word that the daughter had said. The very next day, or within one or two days after I had seen this case, the reform dinner was held at Leeds, and afterwards a meeting, at which certain candidates for the borough of Leeds appeared. One of those candidates, on the question being put to him, stated to the meeting that he would support the ten hours' bill, provided it could be proved that children were over-worked. The same evening I visited the Commercial Buildings, and got to a table where there were three or four friends of this candidate, and I stated to them how monstrous it was to express such a doubt in the town of Leeds. I then stated my case to prove to them that it was true that children were overworked, and I was well laughed at, and offered bets to a considerable amount of three to one, that it was impossible that case could be true. In consequence of this I was determined that I would investigate the matter further, and being myself extremely busy at the time, I went to the short-

time committee, stated the case, the name and address, and requested that they would investigate the case for me. The following day I called, and had the matter fairly explained, and was assured that what I had stated was perfectly true. I forget now the occupation of the girl.

94. What was her name?—Her name was Knott; she lived at Holbeck-moor-side. This girl did not work at the same occupation as all the other girls in the mill.

95. Was it a flax-mill?—No, it was a woollen-mill; when I mentioned the case to the short-time committee they understood it directly, they said, "she is so and so," mentioning the name of the occupation, which I forget.

96. Was she a scribbler?—I believe scribbler was the term; then of course all the scribblers in that mill worked the same hours. Now that case I must say gave me a new light; I was much shocked and distressed to think that it was possible that a human being could be placed to labour from five o'clock in the morning to nine at night,—though I believe it was ten, but I will say nine,—without ever having an opportunity of going out of the mill to get her meals, or to go to play, or to go to rest.

97. Had she been working those long hours till within a short time of her being ill?—Till within a few months.

98. You stated that they worked longer hours generally at Bradford than at Leeds, but you have never yourself seen any of the mills to which you have alluded at Leeds?—Not a great many; I stated they worked longer hours at that period; whether they do now or not, I do not know; I rather think they do not, for I think the hours at Bradford have been diminished, and at Leeds they have been increased since that time.

99. You stated that in your opinion the labour in factories was injurious from the uniformity of the exertion?—Yes.

100. Does not every muscular exertion become more easy to the party performing it, in proportion as he has become accustomed to it?—It certainly does in some respects; no doubt of it.

101. Then you would, in some degree, abate the fatigue arising from the exertion itself?—In some degree, no doubt, it will; but we find there is nothing has a greater tendency to preserve one from fatigue than variety in muscular exertion.

102. Cannot a greater proportion of work be done, and that work done more easily, by one person applying himself to one department of it continually, than by a person shifting from one department to another?—Yes, it is possible he may do more work, but he will not do it so pleasantly to himself.

103. With respect to those persons whom you mentioned now, who were drilled for three hours, and you said fainted away in consequence, did not you find in a few weeks they could be drilled for six or seven hours without suffering at all?—No doubt of it.

104. And when they became accustomed to those positions, they did it with no trouble at all?—No, I would not say so; the human body can never be brought to maintain the erect position without fatigue, because there is a certain class of muscles that are put into action, and those muscles will, after a certain time, become fatigued; so if you place a person to stand, you will, after an hour or two, find that he will want constant supports and helps, first on the one side and then on the other; it is the most fatiguing thing a person can do to stand for a long time.

105. In all mills and factories there is a considerable degree of working as well as standing, is there not?—Yes.

106. Have they not to bend and raise the body in order to feed and attend the machines?—I know very little indeed, I am very glad to say,

Samuel Smith, Esq.

about factory employment; but this I do know, that in all those cases which I have taken the trouble to inquire into, they have no chairs or seats.

107. Are you aware that in worsted mills they have chairs allowed them?—No, I am not.

108. You stated, also, that persons who work in mills are probably still more injured by the labour, because they are not in a condition to do the work; from your knowledge of their habits and diet, do you think that the money is expended in purchasing food of a nature well fitted to sustain them in that employment?—I know very little of the private habits of such individuals; but I do know this, that if they had three times the amount of wages that they have, and provided that that money were all expended in obtaining the very best and most nutritious food, the great length of time for which many of them would have to labour would still keep them in a state far from healthy.

109. But you cannot state to the committee whether their earnings are at present expended on food in a judicious manner or otherwise?—I do not know that, indeed; but if it were, in many instances, particularly of the children, I know this, that as food is sold at the present day, the amount of money which they obtain for their labour is quite inadequate to supply them with a sufficient quantity of proper food.

110. Do you mean to state that where a family is employed, the wages earned by that family are at present inadequate for its support?—I do, for the children.

111. Is that supposing the case where the parents only are employed, or where the parents and children also are employed?—I think in many instances the amount of the earnings of the child is not sufficient to obtain food for that child.

112. Are you at all familiar with the periods at which children usually begin to be employed in other pursuits, not in factories; do children, for instance, who engage themselves in other mechanical occupations between the age of ten and fifteen, earn sufficient wages for their own subsistence?—I think, in general, they earn more than they do in the factories, and, in general, they are placed to labour of that kind at a later period of life than the children are sent into factories.

113. Do you believe that they earn more when they begin to labour, and that they begin to labour later than the children who go to factories?—They begin to labour later, and they generally receive more remuneration for that labour than the factory children of the same age.

114. Can you state the relative proportion of wages in a factory and any other trade?—I understand those subjects very little indeed, though from what little I have known and seen upon the subject, I think the statement I have made is correct; but I have paid very little attention to that part of the subject.

115. You have not considered the working of this question with regard to the effect of a diminution of the wages upon the factory population?—I beg to state to the committee that I have never entered into the question in any other way than in my professional capacity; I have never interested myself about it in other respects.

116. When you stated to parents that they ought to remove their children if they wished to preserve their health, what answer did the parents make on those occasions?—In general they have been very much affected by the plain manner in which I have stated it to them, and they have taken them away; in some instances it has been necessary for me to make this statement, in order to induce parents to allow children to become in-patients of the infirmary; under such circumstances it is very common for us to take

poor factory children into the house, and to put them to bed; allow them only to get up in the afternoons; place them under very little medical treatment, but give them full diet, beer, and wine.

Samuel Smith, Esq.

117. If the committee understand you rightly, you are of opinion that labour in factories, owing to the standing position in which the work-people are obliged to work, and owing to the atmosphere which they must there breathe, and the dust by which their breath will be affected, must always be more or less unwholesome?—It will, under any circumstances, be more or less so.

118. Under the most favourable circumstances, then, you would not expect a population so engaged to enjoy the same degree of health as an agricultural population?—Certainly not; if the hours of labour were very much shortened, and the remuneration increased, which would enable them, of course, to get more nutritious food, and they lived in the same situation with regard to wholesome air, then I think it is possible that their health might be very fully equal to that of the agricultural labourers.

119. What period of labour would you assign as that in which those children, for instance, who are affected by the provisions of the bill now before the committee, could labour with safety to their health?—I should think myself that ten hours was the extreme for children under a certain age; I should prefer myself if that time even were shortened; and I think it very probable that many will suffer even under the ten hours' labour.

120. Do you think that at that age standing for ten hours in a mill would produce those effects upon the bones which you have mentioned to the committee?—I think it will; if not in so striking a degree, I think it will in some degree, particularly in the youngest children.

121. In considering the limitation of hours to be enacted by Parliament, ought not the committee to consider the diminution of wages, and consequently the diminution of the comforts arising from them, as well as the injury done to health by the greater length of the hours?—Certainly, I think that consideration should be taken into view; but, although I understand very little of such subjects, still I have a strong impression that Mr. Sadler's bill will have a tendency to increase wages rather than to diminish them.

122. Can you state whether that is the general impression among the working classes in Yorkshire?—It is; they have that impression in some degree.

123. Do you think that they have formed any of their opinions regarding this bill on the impression or anticipation that it will serve them in wages?—No, I think not: for the general impression is that it will, *in the first instance*, reduce their wages for a certain time, and afterwards have a contrary effect.

124. You stated that females were not as competent to sustain the labour of the factories as males of the same age; is it not considered that females attain to maturity and full strength much earlier than males?—They do.

125. And would they not be as able to do the labour proportioned to their strength as the males of the same age?—No, I think not; the female is altogether a more delicate being than the male.

126. Do not you think the constitution of the female naturally inclines to more sedentary occupations than that of the males?—Yes.

127. Are you not aware that a number of females have been introduced into factories owing to combinations which sometimes prevail among the workmen?—I think I was informed some years ago that Mr. Gott had employed a great number of females in that department which is called "drawing;" that is the only knowledge I have upon the subject; that is an occupation which is generally followed by men.

Samuel Smith, Esq.

128. Will you have the goodness to state from any documents in your possession, either from your private practice or from your practice in public institutions, what proportion of cases of instrumental labour or embryotomy, whether attended with the death of the child, or with the safe deliverance of the child, have come under your observations ?—Merely making a guess, I would state that that kind of instrumental labour in which the life of the child is preserved, takes place about once in 150 or 200 cases, but that species of labour in which it is necessary to destroy the life of the child probably does not occur above once in 1,000 cases.

129. Are those the proportions which would exist in ordinary cases, or the proportions which exist in a town where labourers in factories are mixed up with the rest of the population ?—I have no decided document to go by. If I had been aware that I should have been asked that question, I could have been provided to answer it in a satisfactory manner.

130. Can you furnish to the committee such documents ?—I can, by obtaining the results of the practice in the different Lying-in Hospitals of the kingdom. I believe it will be impossible to obtain such a document from the town of Leeds, for this reason, the Lying-in Hospital of Leeds has only been established seven or eight years, and it has very few patients annually. As far as my own experience goes, I may state that I have attended nearly 4,000 cases of midwifery; that in those cases I have performed the operation alluded to about seven times. The reason why I stated the average as one in a thousand was, because in two individuals I have performed the operation three times each.

131. Then the operation to which you allude is embryotomy, namely, destroying the infant in the womb to save the life of the mother ?—Yes.

132. Is it not an universally admitted axiom in your profession, that, under ordinary circumstances, parturition is much more easy and much less fatal in the labouring classes of the community than in the more opulent and luxurious classes of society ?—Certainly it is; but at the same time I ought also to state, that I believe healthy women in the country have often more tedious labours than manufacturers in towns, although their recoveries are much safer.

133. So that if any documents were produced to this committee which should make it out that the number of cases requiring the distressing operation to which you have alluded was not relatively greater in the lower ranks of society, but nearly equal to that which occurs in the higher ranks, you would still conceive that factory labour had been the means of increasing the difficulty and danger of parturition in the lower classes ?—I have to observe that in general the labours of those females who have been much worked at factories, and in whom the pelvis is perfectly formed, are much more easy and much quicker than those of females under other circumstances; their bodies are in a weak, relaxed state, and their labours are consequently quicker than those of the healthy women in the country, though perhaps not always so safe.

134. Do you apprehend that a weak state of body is always conducive to an easy delivery ?—No, but it very frequently is so; we find no women suffer so little during labour as females who are actually dying of consumption.

135. You stated that no correct information could be obtained regarding the health of persons employed in factories from any returns furnished by those factories themselves; but you stated, on the other hand, that such information could be derived from returns taken from infirmaries and hospitals, &c. Is it in your power to furnish this committee with any document shewing the number of cases brought into the Leeds Infirmary; and what proportion of those cases were persons employed in factories ?—I do

not know that I could; I think that I could bring forward a document of that kind, as far as regarded the accidents; but I think not as far as regarded the common cases; and I wish it to be understood that I have but little knowledge of the mere medical cases; my observations have been entirely confined to the surgical cases. Samuel Smith, Esq.

136. But if such a document could be procured, should you consider it as a fair criterion of the relative sickness of the two parts of the population?—I cannot say, unless another document were at the same obtained, stating accurately the number of each portion of the population.

137. It is your opinion, which you have already expressed, that the fairest way would be to trace the individuals engaged in mills and factories for a considerable length of time, and see how their health was affected, and what ultimately became of them?—Yes; but my experience goes so far as this, that when I see a particular case of disease, I can very frequently point it out and say, that is a factory case; and in general it is very easy to distinguish such individuals as they pass in the street; their complexion is quite different from that of individuals who are in other employments; in a word, their condition, in every point of view, is one that demands the serious attention of the Legislature.

CHARLES TURNER THACKRAH, Esq., called in, and examined, 18th July, 1832.

1. What is your profession?—A general practitioner in medicine and surgery, at Leeds. C. T. Thackrah, Esq.

2. Is it the custom in Leeds very frequently to unite the two branches of the profession, namely, that of medicine and surgery?—It is a general custom.

3. It has been the custom amongst the most eminent of the profession that have resided there?—It has.

4. How long have you paid attention to the condition of individuals employed in mills and factories, and other laborious pursuits?—At intervals, since the year 1823.

5. From that period you have been in the habit of renewing the subject in your own mind, and continuing your observations upon it?—I have.

6. Have you written upon the subject?—I have.

7. A work expressly confined to the consideration of the effect of arts, trades, and professions, on health and longevity?—I have.

8. A work that has attracted a considerable degree of notice in your profession?—I believe it has.

9. You have seen no cause to alter any of the opinions you have expressed in that work?—None, materially; seeking truth only, I have had to correct errors in the details of particular departments, but I do not remember that I have had to alter any general principle or deduction.

10. What is your opinion of the effects produced upon the health, the welfare, and longevity of those employed under the factory system as at present pursued?—My opinion is, that the factory system reduces the nervous power, in other words, the vigour of the constitution, that it renders persons more feeble, more subject to suffer from attacks of disease; and finally, that persons constantly so employed are shorter-lived than others. There may be other points, but these strike me at this moment as the principal.

11. Have you made calculations substantiating the latter fact, namely,

C. T. Thackrah, Esq.

the shorter duration of human life in manufacturing than in agricultural districts?—I have, from the census of 1821.

12. You found that the facts published in that document entirely confirmed your opinions regarding the effect on human longevity of those particular pursuits?—They did, decidedly. My attention was first called to the subject of employments as affecting health, from remarking the sickness and disease that prevailed in large towns, and especially in certain occupations. This led me to inquire into the cause, and my inquiries convinced me that a greater amount of disease exists in manufacturing than in agricultural districts; and it was at a subsequent period that I referred to the population returns; these confirmed my views, by exhibiting a greater mortality in the manufacturing than in the agricultural.

13. So that the opinions which you, as a medical man, had previously formed, were fully substantiated by the facts which you found in the census of 1821?—Yes.

14. You have stated, in your publication, those facts in reference to this particular inquiry?—I have; I compared the three Ridings of Yorkshire; but if the first, the East, were thrown out, the contrast would be more distinct, since the East Riding contains some evils peculiar to itself, and unconnected with my inquiry. In comparing the West Riding, the manufacturing district, with the North, the agricultural, it appears that in the former the number of persons between forty and fifty years of age in 1,000 is less than in the North; and when we go to other ages further on, from fifty to sixty, sixty to seventy, and so on, the proportion of persons living in the West Riding very greatly diminishes; in other words, that the people in the West Riding have decidedly shorter lives than the people in the North Riding.

15. Do you think that children suffer more than adults from the factory system?—I would say, on the whole, that they suffer considerably more; but there are some employments in which, at first sight, an exception might be made; children appear to bear dusty occupations with much less annoyance than adults; difficulty of breathing is rare among them; but this exception from disease I conceive to be more apparent than real; the children I believe to be considerably injured; and although they do not shew disease in the lungs, or any great change in their general health at an early period, yet such individuals rarely become strong adults; and at a subsequent period they are more liable to consumption, and other serious affections of the lungs, than persons of like station who are not employed in such dusty occupations.

16. So that you think that the constitution may be undergoing very serious and even permanent injury without the magnitude of that injury being decidedly apparent in the youthful period of existence?—Certainly, you express my meaning fully; in fact, from childhood to puberty diseases are not frequent. This period of life is comparatively healthy.

17. The diseases of that period you do not consider to be so constant and so fatal as at other periods?—I do not. This observation applies to society at large, and accounts for the comparative freedom from disease of children placed in unhealthy situations.

18. Do you conceive that at that particular period the human constitution is more tenacious of life?—It is very tenacious of life.

19. You would not therefore conceive it would be any just answer to those who, like yourself, declare that the factory system, as it is ordinarily pursued, is injurious to health and tends to shorten life, if it were asserted that at the period of existence alluded to no great excess in mortality should become apparent?—Certainly not; that would by no means satisfy me; I should consider it no answer whatever to the general statement on the subject; I am well aware that the actual extent of mortality in factories

C. T. Thackrah, Esq.

is small among children; I have been often told, in reference to particular mills, that there is but a small per-centage of deaths in the course of the year; but this statement has not answered my objection, being aware that mills in general do not produce immediate and direct mortality; their chief effect on the operatives, in my opinion, is the undermining the health, the destroying the constitution, and the rendering people liable to attacks of disease to which they would not have been subject, or under which they would not have succumbed, if they had been in other situations. With few exceptions, the diseases developed in mills are chronic rather than acute.

20. In allusion to the period of life in which it would be expected, according to the opinions which you have delivered, that the effects of the mill system would become very apparent and distressing; have you made any personal examinations so as to ascertain, as far as you could, whether those opinions are substantiated by facts?—Yes; particularly in reference to the more dusty occupations; I have found that the lungs are sooner or later seriously altered in their capacity; that the power of respiration is diminished; that after middle age inflammatory affections, or changes of structure the effects of inflammatory affections, are found in the lungs and air tube; and a number of maladies of other parts or systems are connected with, or result from, these diseases of the pulmonary organs. I have had a great number of mill people come to my house at different times for examination; a party, perhaps, of ten at once, who were in different departments of the mills; I found those men who had attained the age of from forty to fifty almost universally diseased; I am now speaking of the dusty occupations; I do not apply this to mills where there is no dust. I would wish to be distinctly understood in that respect.

21. Will you please to state to the committee upon what principle you conducted the examination to which you have referred?—The first thing was to examine the capacity of the lungs, to see, in fact, whether the lungs can take in as much air as they ought to do, and as much as they are wont to do in persons of similar size, age, and sex. For this purpose I have used a glass jar, which I call a pulmoneter, filled and inverted over water. The person subjected to the experiment merely blows out as much as he can at one expiration or effort through a tube, the lower end of which is placed under the glass jar. The bubbles of air rising up displace a certain quantity of water. At the top of the jar, the water subsiding, indicates, by the cubic inches marked on the glass, the quantity of air that the person throws out at one expiration. This examination by the pulmoneter does not shew the *nature* of the disease, it only shews the extent of the disease. The next mode is to ascertain the nature of the disease, and that is by the use of the stethescope, an instrument now generally known to the profession. I may state that I do not recollect that I ever applied the stethescope to any person who had been twenty years in a dusty mill, in whom I did not find decided marks of diseases in the lungs or air tube.

22. Do you think that arises only from the dusty employment, or from the day's work being continued beyond a reasonable quantity of hours?—This effect I ascribe only to the dust, but of course the longer the persons are exposed to that dust the greater the effect.

23. Will you state to this committee what you conceive to be the general effects of labour too long continued in the atmosphere of mills and factories, generally considered, leaving out of the question any particular dusty manufacture?—I should say, a reduction of vital power proportionate to the length of that confinement, and with this reduction of vital power a series of evils to the constitution; chronic maladies, and an inability to resist acute ones, and a shortening of life.

C. T. Thackrah, Esq.

24. And in the case of attacks of acute disease, you do not think the constitution, under such circumstances, is as capable of resistance ?—Decidedly not.

25. You have already expressed the opinion, that the general result which you have detected, both from personal observation and examination, and from adverting to the censuses of the two Ridings, is, that the life is inevitably shortened by that species of labour ?—Yes; as a general observation. Besides this, there is a general and marked effect in the production of chronic maladies; we hardly ever see a clean tongue in a crowded factory; the power of digestion is certainly diminished. Of this, I may mention a little illustration. A man, who, with his seven children, had been employed in the home manufacture of cloth, removed into another neighbourhood, and sent his children to a cotton mill; here he soon found a considerable change in their appetites. Before, they were cheerful and contented with plain food, and at the usual times; now they want food much more frequently, cannot eat plain meat, but must have dainties.

26. You would consider that as indicating a diminution in the powers of digestion ?—Certainly I should; first, because they could not go so long without food; and secondly, because they required something of a nice or piquant kind.

27. Should you attribute part of the pernicious effects upon the constitutions of those employed to their being deprived of fresh air ?—Certainly; the long-continued labour and the want of fresh air are the two principal causes of the general effects to which I have alluded.

28. So that you comprehend under the system as at present pursued the circumstance of their not having fresh air and proper recreation ?—Yes.

29. Do you not attribute great weight, also, to the circumstance of their not having sufficient sleep and rest ?—I think that is a very material circumstance, in many instances.

30. Is it not your opinion, that children require much longer time for sleep than adults ?—Certainly.

31. Do you think that this labour has a considerable effect upon the constitution in this way, namely, by unfitting the person who may arrive at mature age from pursuing with equal activity and success any labour that he will then have to undergo ?—Yes; because on account of the want of fresh air at the period when fresh air is particularly required, and the excessive or long protracted labour at the period wheu such labour ought to be avoided, it is rare that the muscular system, or indeed any system, is fully developed.

32. Do you think that this practice of overlabour of children produces present suffering, as well as threatens future consequences to their health and longevity, and that, independently of the professional views that you have given, it is a system of considerable privation and suffering ?—I do.

33. You have witnessed, perhaps, that the children in going to mills and coming from them appear as if they were experiencing a considerable degree of suffering ?—I feel convinced that that is the case; I would add, moreover, that if we ask the children in mills, "Are you pretty well?" they say, "Yes;" they have not any particular ailment; but if we examine them, they have not that degree of health, that muscular power, and that buoyancy of spirits, which we find in children not confined and congregated in mills.

34. In your publication you have alluded to the undue waking of children in the morning, and their retiring too late at night to their rest, and the consequent insufficiency of the period of sleep as amongst the

C. T. Thackrah, Esq.

cruelties of that system?—Assuredly I think it is a very great cruelty of the system.

35. And you think that that circumstance must tend to injure their health, combined with the others to which you have alluded?—Assuredly.

36. How many hours a day do you think children might work without any injury to their health?—That is a question which I should feel great difficulty in answering. My opinion is, that young children ought not to work at all; that the period of growth ought not to be the period of labour; and hence, if I am asked as to time, I should say that the least time is the best time; because I consider that all labour is injurious at an early period of life.

37. Up to what age?—The human body is not fully developed till the adult period; till puberty, at any rate, no strong or long-continued labour should be enforced. Before this period, such labour, or even longc-ontinued attention without it, robs the constitution of that nervous energy which is necessary for the proper growth and full development of the body at large.

38. You would think it, therefore, not sufficient to protect children from what might properly be denominated strenuous exertion, but also from continuing too long at one uniform wearying employment?—Decidedly, and on the principle I allude to, that whatever exhausts the nervous energy draws off the supply of the body at large, and thus produces debility.

39. Should you conceive eleven hours a day too long for children to labour?—I should.

40. If, on account of the unfortunate state of society and of opinion, you should find it impossible at present to obtain a ten hours' bill, do you not think that some benefit would be obtained, and a large benefit, by an eleven hours' bill strictly enforced? — I should think this better than the present state of things, but I should be very ill content with such a measure.

41. You have already said that you have given great attention to this subject as one that concerns your profession as well as your feelings as a man; and your objection to the different limitations of the hours of labour of children and young persons, as hitherto proposed, has been to the restrictions not going far enough, rather than to their being too stringent in that respect?—Decidedly.

42. You could hardly sanction with entire cordiality the proposition of limiting the hours of labour of children and young persons to ten?—I would much rather say six. I speak as a medical man and a friend of humanity.

43. You think, viewing the question in those lights, that the hours of labour in respect of those children and young persons ought to be reduced, as far as is at all reconcileable with the state of our trade and commerce, external and internal?—Decidedly.

44. Do you not think that the mischiefs of such a system will increase through successive generations if persevered in?—I think that the evils would increase; for I find, from circumstances in trade, and greater demands for goods, that the intervals for food and refreshment have been progressively shortened, and though occasionally these intervals have been again lengthened, the improvement has been very temporary. There is, in fact, a greater disposition to maintain these short intervals than to return again to long ones.

45. The question had principally reference to your professional opinion, namely, whether the evils of this system, if persevered in, will not, physically considered, be perpetuated and increased?—Yes, I conceive that they must be so.

46. It is your opinion that at present we have only seen the commence-

C. T. Thackrah, Esq.

ment of such a system, especially in certain districts, and that we can hardly contemplate the mischief that we may apprehend from it, should it remain unregulated, as it hitherto has done?—Decidedly, I should say so.

47. Will you be so good as to give to the committee your opinion as to the age at which you think that children might, under proper management, and with due recreation, be set to work?—That is a question I feel great difficulty in answering, because my opinion as a medical man is, that young children should not work at all; but if they must work, the later the year of commencement the better.

48. Do you conceive that the bill now under the consideration of Parliament, forbidding the working of children and young persons in mills and factories beyond the period of ten hours a day, is necessary?—I think it is highly necessary; but, in my opinion, it does not go far enough.

49. You have observed a considerable difference in the nature of the employment as to the effects that it produces upon the health of those employed?—Yes, a very great difference.

50. You have made distinctions in favour of certain branches in your work; do you still maintain those distinctions?—I do; I have not changed those opinions. The confinement of a number of persons for a number of hours, with such labour and attention, in a confined atmosphere, which one would find is a circumstance attending mills in general, produces the disorder of the digestive organs, reduces the nervous power, and causes the general effects to which I have alluded; but not the diseases of the lungs, to which I referred at the early part of this examination; for such diseases of the lungs result from the dust in the employment.

51. That is a superadded evil, and is confined to dusty employments?—Yes.

52. In what mills do you think there is such a degree of dust as to be prejudicial?—I have examined cotton mills at Manchester, at Leeds, and at a village near Leeds, and in those I found comparatively little dust; I should say, scarcely dust enough to be pernicious. But I speak only from my own personal observation, and I may not have seen the worst cotton mills. The flax appear to me the most injurious of factories. With respect to woollen and worsted there is scarcely any dust in any department of those mills; the woollen manufacture, in general, I consider a healthy occupation.

53. Are you acquainted with the silk mills?—Silk I have also seen at Manchester.

54. Is there any dust in those mills?—There appears to be little or none.

55. But still your general objections are to the system, as involving too long a period of labour, and in places where there is a considerable number of persons; and taking into consideration the whole circumstances of mill labour, your general observation would apply to silk as well as to other mills?—Yes; in consequence of the long confinement. The same term of labour in mills I conceive to be more injurious than it would be in private houses or the home manufacture.

56. Occasionally remarks have been made in favour of the mill system, founded upon the ground of the seclusion of the operatives from the inclemencies and variations of the seasons; and consequently it has been argued that such occupations must be more wholesome than those that are pursued in the open air; what is your opinion upon that?—I think that view quite erroneous. I consider that persons pent up for twelve or fourteen hours a day in a close apartment, will be infinitely more liable to suffer from transitions than persons that are out in all weathers and under all circumstances.

57. Supposing that the committee were to assume that the wages of the children employed in mills and factories, and consequently the income of

C. T. Thackrah, Esq.

their parents, would be reduced in proportion to the contemplated diminution of the hours of labour, do you conceive that they would suffer as much in health from the reduction of some of their comforts and necessaries, as they suffer from those long hours of labour?—The first point in the question is rather one belonging to political economy, a subject with which I am unacquainted; but if I might venture an opinion, I would say that an universal reduction of the hours of labour, by lessening the production, would enhance the price of the commodity; and if the price were enhanced, I presume that wages would rise, and a person receive as much for ten hours as he before received for eleven or twelve hours' labour. With respect to the second part of the question, I infer, from what I know of the management of food among the poor, that if their wages were judiciously applied, a reduction of them in many employments would not, in reality, prevent a due supply of food. It appears to me that the poor, as well as the rich, have numerous artificial wants, and that they not unfrequently lay out that money in superfluous dress which might be better applied in providing food and comforts. I suspect, too, that where wages are high, or where children earn money by excessive labour, this money is not well applied. In many instances the parents turn it to the support of debauchery. I do not think, in fact, that the children are generally well fed in proportion to their excessive work; I do not think that they have a better quality, or greater quantity, of food, in consequence of their excessive labour; many instances I have known of the reverse.

58. Do you think, then, that the excess of the wages that may be obtained by this excessive labour is an inducement to expend the necessary wages in other ways than in procuring the comforts and the necessaries of life?—Yes.

59. Do you think that the working classes could sustain any diminution of the present wages?—Of course there is a very great variety in wages; I am sure that many classes of artisans could well bear a diminution, particularly if their work were of shorter duration; high wages, moreover, very often, if not generally, lead men to intemperance. There are, however, some operatives who could not bear reduction of wages; weavers, for instance, who earn but ten shillings or twelve shillings a week.

60. Have you observed considerable fluctuation in the demand for labour in your manufacturing district?—I have.

61. That at one period there has been an excessive demand for it in comparison with the hands to be employed and the number of hours in which they ought to be laboured; and at another period there have been numbers of them thrown out of employment?—Yes.

62. Supposing that their wages correspond with such fluctuations, do you not, in the first place, think that the excess would be disadvantageous to the morals and health of those obtaining them?—Decidedly; because it does not induce them to lay money by for future need; but, on the contrary, leads them to intemperance; and their families are left in more destitute situations than the families of those who have uniform work.

63. It then, of necessity, initiates them in improvident and profligate habits?—Decidedly.

64. Does it not therefore plunge them into additional distress, when they find themselves deprived of the ordinary time of labour?—Yes; the constitution, accustomed to a stimulant diet and a great quantity of liquor, suffers much more from deficiency of food than in persons accustomed to a moderate diet.

65. So that, in both cases, you consider that the effect upon health is decidedly pernicious?—Certainly; both the extremes and the fluctuations between them are pernicious.

C. T. Thackrah, Esq.

66. You are aware that there is considerable connexion in mills and factories between the labour of children and that of adults; supposing that the reduction in the hours of the labour of children should incidentally have the effect of diminishing the labour of the adults in mills and factories, which you are quite aware is often carried to an immoderate length, especially in brisk times, should you regard that as an objection to the passing of the bill which is proposed?—Decidedly not; I should consider it a great collateral advantage; for excessive labour is the common fault of this country.

67. Will you give your opinion as to the effect of carrying the labour to such an excessive degree in reference to the entire population of the working classes considered indiscriminately?—Excessive labour assuredly diminishes life, and therefore, in proportion to that excess of labour, will not only the present comfort be diminished, but the duration of life shortened; persons, in fact, are prematurely "worn out" by labour.

68. Do you know it as a fact, which you deduce from your studies as well as your observations, that the labouring classes of the community live a much shorter length of time than those that are absolved from the necessity of such excessive toil?—I have understood that to be the case, but I do not know it from my personal observation; I have learnt it from my medical reading.

69. Do you conceive that, without a legislative enactment, the mere feelings of humanity, and the knowledge of the effects produced by this system, as detailed in your work, and as pretty generally now comprehended by society, at least to some extent, will be sufficient to regulate the system? —Certainly not.

70. Will you state the reason why you think that such would not be the result?—In the first place, I think that mill-owners, even humane and enlightened mill-owners, are never fully sensible of the evil of long-continued confinement and the other injurious agents in their employ. We none of us properly appreciate the evils that we see daily; and in point of fact it is proved that some legislative enactment is necessary, by the experience in reference to the shortening the intervals of labour to which I have alluded before; for when a manufacturer has been pressed by his engagements, he has very often shortened the intervals of labour. Another thing is, that if one manufacturer increases labour, another manufacturer must do the same, to compete in the market with the first. This encroachment on the time which the work-people require for rest and refreshment has occurred again and again; the evil is progressive, and can only, I conceive, be prevented by legislative enactment.

71. So that a legislative regulation in that view of the question, would be equally a protection to the humane master and to the over-wrought operative?—Decidedly.

72. Have you paid particular attention to the medical department of your profession?—I have.

73. Have you not given lectures frequently, since you have completed your education, yourself, and are you not a lecturer at the present moment in the north of England?—I am; I lecture on anatomy and physiology, and on surgery.

74. Have you been in the habit of giving your pupils lectures on medicine?—Yes, on medicine particularly.

75. Do not the factories themselves illustrate the principles you teach, and do not your observations there well agree with the doctrines of medical science?—Yes.

76. So that, in point of fact, you can, in your case and experience, confirm the necessity of a legislative protection for children and young persons engaged in the mills and factories of this country?—Decidedly; with

reference to the children and their parents, allow me to make a remark; I find that even when the masters are not wishful for their regular hands to do the night-work, or to perform excessive labour, the cupidity of the work-people themselves leads them to undergo this excessive or protracted labour; and parents, I believe, sometimes urge their children to the work; consequently the children require protection from themselves and protection from their parents as well as from their masters.

C. T. Thackrah, Esq.

77. Do you conceive that it would be either proper or even possible to establish night-schools, and connect them with the labour now endured in mills and factories in this country, presuming that the hours of labour are, including the necessary meals and refreshment, thirteen or fourteen?—No. And if the present term of labour were reduced, I would, as a medical man, rather that the two or three hours a day taken from labour were devoted to sports and pastimes, which are necessary for the preservation of their health. Though intellectual and moral education should be regarded, and I should be glad if there were time for both, I think, in the present state of things, that physical education, or the improvement of health, is most urgently required. Children want that fresh air and recreation which could not be enjoyed in school.

78. And the enjoyment of which is now perfectly inconsistent with the hours of labour imposed upon them?—Yes.

79. Consequently, to add education to the present period of their labour would increase their physical sufferings?—It would certainly increase them.

80. Do you not regard the confining children in Sunday-schools as having that tendency?—Sunday-schools, however valuable in other respects, are objectionable in this. Surely one day in the week children so much confined should have for walking out and enjoying the fresh air and green face of nature; but I should think that Sunday-schools would produce unalloyed good if children had two or three hours more for recreation during each week-day.

81. If, therefore, there were a regulation which, upon the whole, would give an opportunity for a little recreation and for education, it would, in your judgment, be one of the greatest benefits conferred upon that class of the community?—I am decidedly and fully of that opinion.

BENJAMIN TRAVERS, Esq., F.R.S., called in, and examined, 7th August, 1832.

B. Travers, Esq.

1. What is your profession?—A surgeon.

2. Are you a surgeon to any of the hospitals in the metropolis?—I am senior surgeon to St. Thomas's Hospital, in Southwark.

3. Have you given instructions in the principles and practice of your profession to students in medicine or surgery?—I have formerly lectured on surgery.

4. You have directed you attention to every branch of your profession, have you not?—I have.

5. Is it not universally admitted, that a moderate degree of exercise, with due intermissions for refreshment and rest, are usually necessary for the preservation of health?—Certainly.

6. That undue labour, that is, labour of so strenuous a character, or so long continued as to induce great and permanent fatigue, without due intervals for rest or refreshment, is pernicious to the human constitution?—Undoubtedly so.

Benj. Travers, Esq.

7. Have you any doubt as to that state of endurance being still more prejudicial to children and young persons than it is to adults in the prime and vigour of their life?—Undoubtedly, I should think it so.

8. Would you, as a general observation, be disposed to think that the usual duration of a day's labour, namely, twelve hours, which is the term allowed in agricultural and manufacturing pursuits in this and other countries, is, in ordinary cases, about as much as the human frame is calculated, for a continuance, to endure with perfect safety?—Fully as much.

9. In reference to the working class of society more especially, do you not think that intermissions sufficient for taking food are necessary for the purposes of health?—I should consider them indispensable.

10. You are not, perhaps, personally cognizant of the labour undergone in mills and factories?—No; I have, in visiting the country, occasionally gone into them as a matter of curiosity, but I am not further acquainted with them.

11. As a professional gentleman, your practice has not led you to the factories?—It has not.

12. Assuming that the labour of the factories continues to a much greater length than the term alluded to, and extends in almost all cases to thirteen or fourteen hours a day, and sometimes to sixteen or eighteen, and even upwards, that it is pursued generally in an impure atmosphere, and sometimes in one heated to a high temperature, say from 70° to 80°, have you any hesitation in stating that that kind and degree of labour, not now alluding to the extreme cases, must be ordinarily prejudicial to health?—None whatever; I should consider it permanently injurious.

13. When it had to be endured in early life, under all the circumstances, or any of them, alluded to, it would in many cases be, in your judgment, prejudicial to the development of the bodily functions, and be injurious to health, and tend to shorten human life?—Irreparably injurious.

14. Should you consider that though the labour in question is what some have occasionally called "light and easy," yet if it is such as to demand perpetual vigilance, in many instances constant motion, and is performed in an erect and sometimes in very constrained positions of the body, such labour would fail to be as prejudicial as perhaps stronger exertions, if endured for a shorter length of time, and with due intermissions?—I should think such circumstances would all be great aggravations of the intensity of labour.

15. Bearing in mind the duration of labour, the position in which it is ordinarily pursued, the weariness that it is stated to occasion, and the atmosphere in which it is undergone, what would be some of the results which, in your judgment, would follow labour of that kind long continued, and especially as regards the children and young persons so occupied?—I should think all such results as might be referred to a deprivation of the nutrient faculty; that the circumstances stated must, sooner or later, in many cases, engender scrofula, which, when once engendered, may be considered as the parent of those deformities and vices of growth and those deteriorations of health to which young persons especially are liable.

16. If this labour has often to be undergone in an atmosphere almost saturated with the dust, and what is sometimes called the flues, evolved from the material manufactured, would that have a tendency, in your opinion, to induce pulmonary affections and other diseases of a fatal character?—I should think it would be liable to do so.

17. Several witnesses have appeared before this committee who are labouring under excessive deformity, especially of the lower extremities of the body, which they allege came on at from ten to fifteen or sixteen years of age, with excessive pain, and accompanied sometimes with great enlargement of the joints; should you have any hesitation in attributing those

symptoms to the system of labour alluded to?—I should say none; because I consider such a system as quite competent and very likely to produce them. B. Travers, Esq.

18. You would not confound such appearances as present themselves at that period of life in persons previously well formed, with the disease commonly called the rickets?—Certainly not; it has a different appearance and character.

19. It is stated as the result of many careful examinations, that the growth of children employed in factories, as compared with that of others differently occupied, is considerably less, and that they are, in fact, stunted in stature; would you expect that to be the result of such a system of labour?—Certainly, the direct consequence of it.

20. Similar examinations have taken place with respect to the weight of the two descriptions of children, and it has been found that the factory children were much lighter; do you think the muscular power would be much diminished by the description of labour?—Yes, upon the whole, I should think so; some muscles would be put to extraordinary exertion at the expense of other parts of the muscular system; but upon the whole, I should expect there would be a positive diminution in the muscular power and in the weight.

21. It is alleged that a great majority of the young persons employed are of the female sex; do you think females as competent to sustain labour as males?—I should think not, in general.

22. Do you think females about the age of puberty can safely undergo protracted labour?—Certainly not; it is the most exceptionable of all periods for severe or continued labour.

23. Is it not a strong indication that labour is pernicious when it has to be resumed in the morning with a great sense of remaining weariness and fatigue, which has not been dissipated by the rest of the preceding night?—Certainly.

24. That, long continued, will be, in your opinion, pernicious to the constitution?—Certainly, especially so.

25. If labour towards the termination of the day has to be stimulated by continued chastisement, is not that labour injurious to the body?—Certainly; it is injurious to the body through the mind.

26. May not that consciousness of being oppressed, together with a hopelessness of mind, very materially affect the health?—Much would depend on the natural temperament of the child, but I should think in many instances decidedly so.

27. Do you think that a child under nine years of age ought to be allowed to labour in a mill or factory in the mode in which that employment has been conducted?—I should say not.

28. Do you think that from that age to the period when the human frame is arriving at its perfection, or about eighteen years of age, more than ten hours of actual labour, involving twelve hours of confinement, ought to be imposed upon those children?—Certainly not; I think it only wonderful so much can be endured.

29. You are perhaps then of opinion with many very able medical men, who have been before previous committees, to whom has been submitted the consideration of the question of the labour of young people in factories, that not more than twelve hours of labour, upon the average of human constitutions, can be borne with impunity at any period of life, or by either sex?—I am entirely of that opinion.

30. Your opinion is, that the people of this country labour in general longer than is consistent with their health?—Yes.

31. And that with relation to children and young persons, this becomes a most flagrant violation of the principles of humanity?—Certainly; because

B. Travers, Esq.

that age involves the most important period of existence considered in its relation to after life. And I consider that, next to wholesome food, pure air and wholesome exercise, by which I mean regulated variety of motion, are the principal agents in the establishment of corporeal and mental health.

32. You think that over-labour without due intervals is inconsistent with the proper development of the faculties of both mind and body?—Certainly.

33. Have you any hesitation in saying this deterioration would be perpetuated and increased if the causes already adverted to should be continued? —None at all; I am sure predisposition is hereditary, if not disease.

34. It is your opinion, that after a long day's labour producing great fatigue, the mind is incapable of obtaining benefit from a system of evening instruction?—Certainly.

35. And that the advantage which might result from the institution of Sunday-schools is greatly abridged if the person sent to them is still labouring under a sense of fatigue, induced by the over-exertion of the previous week? —Certainly.

36. Have you any hesitation in thinking that an abridgment of the labour of children and young persons from the extravagant length described, so as to reduce it within the moderate limits proposed by this bill, would be eventually advantageous to society at large, as well as to those in particular whom its provisions are designed to protect?—I have no hesitation in acceding to that opinion.

SIR GEORGE LEMAN TUTHILL, F.R.S., called in, and examined, 3rd August, 1832.

Sir G. L. Tuthill.

1. What is your profession?—I am a physician.

2. Have you been an officer in any of the medical institutions of this city? —Yes; I have been physician to the Westminster Hospital for the last twenty years, and to Bethlem Hospital for the last fifteen years.

3. Have you given medical lectures in any of those institutions?—I have given medical lectures for many years, but not at an hospital.

4. Is it not among your profession a universally received opinion, that a moderate degree of exercise, with due intermissions for refreshment and rest, are essential to the preservation of health?—It is.

5. Are not rest and sleep, duly alternating with exercise, even more necessary to children and young persons than to adults?—I think so.

6. It follows then, perhaps, as an acknowledged principle in your profession, that excessive labour so long continued, or of such a nature as to induce great and continued fatigue, whether of body or mind, and without the intermissions in question, would be prejudicial to the human constitution? —Certainly it would.

7. Would not excessive labour, without due intermissions, be peculiarly prejudicial to persons during their growth?—I think it would.

8. May the committee ask whether you are not of opinion, that an ordinary day's labour, meaning that which, by the universal assent of mankind, is the term usually undergone in the agricultural districts of this and other countries, as far as we are acquainted with them, and also the term to which handicraft workmen and mechanics conform, namely, twelve hours a day, including intermissions for meals, is, ordinarily speaking, a sufficient term of human labour to be endured with perfect impunity?—I think it is.

9. It appears in evidence before this committee, and has likewise been proved before preceding ones appointed to examine into the same subject,

that the labour endured in mills and factories very greatly exceeds that term, and extends, in many instances, to fourteen, sixteen, or eighteen hours a day or upwards; would there be any difficulty in pronouncing that degree of labour to be prejudicial; I mean as respects the effects on the average number of those enduring it?—I think no difficulty at all.

10. Then when that labour is extended, as it is sometimes known to be, for a considerable length of time together, to thirty or forty hours more than once in the same week, consequently leaving but few intermissions indeed for rest, can there be any doubt that it must, in a great plurality of instances, be most imminently prejudicial?—I think it must.

11. Alluding now to the custom of many mills and factories where children are employed all the night, their rest being assigned to them in the day-time, have you, upon consideration of the subject, or from personal experience, made up your mind as to the effects of labour pursued in the night, compared with that which is undergone in the day-time?—I think the effect of severe or long-continued labour would be worse during the night than during the day.

12. You think night-labour less congenial to the constitution than day-labour, and the atmosphere during the night less calculated to sustain the ordinary occupations of manual labour than the atmosphere during the day?—I think so.

13. The labour in question has been sometimes denominated "light and easy;" but, continued, as it is usually is, for a very great and excessive length of time, compared with that which has to be undergone in other pursuits already referred to, should you not consider that long-protracted attention, occasioning much fatigue to both mind and body, would be more prejudicial to the constitution than a shorter term of labour, though more strenuous, if undergone with due intermissions?—I think it might; but that would depend upon the degree in which the other labour was strenuous, and on its duration.

14. It would be a matter of comparison?—Yes.

15. But it would, upon the whole, be injurious, in your opinion, to the constitution for any work demanding constant attention to be pursued for a great length of time without intervals?—Certainly; but very severe labour for a short time might also have a bad effect upon the constitution.

16. Referring only to that labour that is commonly undergone in ordinary pursuits, would not that more strenuous exertion, endured for a moderate length of time, be less prejudicial than the labour pursued for that extravagant length of time in mills and factories?—I think it would.

17. Is the muscular effort to sustain for a great length of time together the erect position of the body very fatiguing?—Certainly.

18. So that such labour, so pursued, would be still more exhausting?—I think it would.

19. It would, in your judgment, perhaps, be still more distressing and injurious if it had to be performed in an impure atmosphere?—Certainly.

20. And still more so, if that atmosphere were heated to a high degree of temperature, say from 70° to 80°?—Yes, it would.

21. Would not the circumstance of children and young persons compelled to labour in such an atmosphere, and consequently being much heated and emaciated, having, on leaving it, to plunge nightly into the cold air during many months of the year, necessarily produce many disorders of the system?—I think such sudden transitions would be very hurtful.

22. Making a comparison between the labour so pursued in a heated atmosphere, and often an impure one like that in mills and factories, as high in temperature perhaps as 80°, would not labour pursued under such circumstances be still more deleterious to the constitution than if it had to be under-

Sir G. L. Tuthill.

gone in the open air, though at an equal temperature, but without its impurities, and subject to the freshening influences of the atmosphere?—Certainly; and I suppose that, at the temperature mentioned, the free access of the external air is not admitted, otherwise the temperature could not be preserved.

23. Should you not conceive, if the same degree of labour, meaning that it should require the same muscular exertion or mental attention, had to be pursued in an artificial atmosphere like that described, that it would be more prejudicial than labour pursued in the natural atmosphere, in a tropical climate?—I should think it would, to persons of the same constitution; there being, in the former case, no free admission of the atmospheric air, and, consequently, the air of the factories being charged with a variety of impurities.

24. You have already given your opinion as to the labour pursued during the night being probably more prejudicial than that undergone in the day-time; do you not conceive that the having to work by artificial lights during the night may also tend to render labour more insalubrious, and also prejudicial to the eyes, as the operatives now begin to allege?—Working during the night involves the necessity of using artificial light; and artificial light contributes to render unfit for respiration the air of the factory in which the light is used. Where artificial light is used to any great extent, there must be a considerable quantity of carbonic-acid gas mixing with the air of the apartment, which is prejudicial to health when it exceeds a certain limit.

25. Then with reference to the effects of working by gas-light, and for a considerable number of hours, do you conceive that a light placed near the eyes might have the effect of injuring the sight?—It might do so; but that will depend upon its intensity; I do not know how the light is placed; but whether it were gas-light or any other light, it would have the same effect upon the air, the only great products of combustion in common artificial light being water and carbonic-acid gas, the latter of which is fatal to life if breathed in a state of purity, and prejudicial to health if mixed in considerable quantity with common air.

26. It consumes the vital part of the air?—Yes; the oxygen of the air is converted into carbonic-acid gas; and oxygen is the respirable principle, which is thus changed in its nature.

27. It is said that children working in mills and factories are very easily distinguished from children differently employed, not only by their weak and emaciated appearance, but more particularly by the unnatural ghastliness and sallowness of their complexion; do you conceive that that might be attributed to the factory system, or any causes that it generates?—Certainly; I conceive that long-continued labour in a vitiated atmosphere, and privation of rest, would contribute to disorder all the functions of the system; the functions of the stomach in particular, and of the different organs that are connected with it, would be thrown into disturbance; and no function of the frame whatever would be performed as in health.

28. It is in evidence, that to maintain, or rather to enforce, the regularity that is demanded in certain establishments, the times for making water are few and limited, and are arbitrarily allowed, not exceeding three times a day; would not that, with children, occasionally lead to pernicious consequences?—Yes, it would.

29. Would not limiting the number of times in which children and young persons are allowed to ease nature be often necessarily pernicious?—Certainly; and the more so the longer the interval prescribed.

30. To make the observations that have been already elicited from you to bear more directly and entirely upon the objects of this inquiry, namely, factory children; do you not conceive that the labour in question continued

Sir G. L. Tuthill.

for the length of time described, often without any interval for meals, generally pursued in an erect or constrained position, in a foul and polluted atmosphere, often heated to a high temperature, and continued throughout the night, must be in its effects more pernicious and destructive to children and young persons than it is to adults in the prime and vigour of life; and does not the condition of those children, therefore, if it be as described to you, demand, in your judgment, a legislative protection in favour of their health and welfare?—I have no doubt of it being an immense evil, as it is described to me; but whether it can be corrected in any other way than by a legislative enactment I do not know.

31. A succession of witnesses have, from their own experience, attributed many distressing results as commonly produced by it, results that have been confirmed by medical authorities, whose practice lies in the manufacturing districts; should you, assuming that the description of factory labour is correct, hesitate to trace to such a system many of the disorders and complaints destructive of health, and abridging the duration of human life?—I think it must have those effects.

32. What would be some of the most striking effects that you think labour so continued, under the circumstances described, must be supposed to produce, reasoning upon the principles of your profession?—I think it would produce universal weakness, if pursued in very early life; and that the different parts of the body could not be properly developed under such circumstances.

33. That weakness would render the person more susceptible of other disorders?—Yes, if exposed to the exciting causes of them.

34. Less capable of resisting disease when attacked by it?—Certainly.

35. Do you not conceive that labour of the description and extent alluded to might have a very pernicious effect upon the osseous system?—In early life it might.

36. Many cases of distressing deformity have appeared before this committee, and the committee are also assured that an extraordinarily large proportion of such cases occur where the factory system is pursued, which deformities came on, in many instances, after ten years of age, and sometimes at sixteen or seventeen; should you hesitate in those instances in attributing, generally speaking, such deformities to the length of labour described to you? —Certainly not; especially if that labour involved a particular position.

37. You think that a constrained attitude of the body, or, in other words, that labour that demands an uniform position, is, perhaps, more fatiguing and more injurious to the human frame than the alternate exercise of the different muscles of the body involved in varying attitudes and positions?—Certainly; and I think that continued labour in early life in a fixed position would tend to destroy the symmetry and just proportions of the frame, and be likely to produce a tendency to deformity.

38. Have you ever been in a factory?—No, I have not.

39. So that your observations are deduced from the principles of your profession?—Certainly.

40. Various examinations have taken place in Sunday-schools, by which it would appear that the growth of children is impeded, and their stature stunted very materially by this long-continued labour; might that result be reasonably expected from such a state of things?—I should think it might.

41. That involves another fact that has been observed regarding factory children as compared with others, namely, that they are considerably less in weight than children of a similar age otherwise occupied; do you think that might be a consequence of such employment?—Yes, I think it might; I have stated that I should not expect, under the system described, that the parts of the body would be perfectly or fully developed.

Sir G. L. Tuthill.

42. An official paper has been presented to the committee, by which it appears, that where this system prevails human life is much abridged, and that there is a very considerable increase of mortality occurring at the initial periods of life; should you be disposed to consider that also as the result of the system described?—Yes, I should.

43. Supposing the existing generation, or, at least, those of them that are exposed to this long and pernicious labour, to be deteriorated as described, might not that deterioration become hereditary, and be continued and increased, supposing the same causes were still in operation?—I should suppose that, under such circumstances, speaking of the females principally, there would be a feeble evolution of the fœtus, and that a robust race would not be propagated by such feeble parents.

44. It is in evidence that a very considerable majority of children and young persons employed in mills and factories are females; do you conceive that the constitution of the female is as well calculated to sustain long and fatiguing labour as that of the male?—I think not.

45. Do you think that persons, generally speaking, ought to be subjected, during the period of their growth, to long and exhausting labour?—I think they ought not.

46. Is it your opinion that females more especially, when arriving at the age of puberty, should be protected from long and exhausting labour?—I think they should.

47. Might not labouring in a standing position, continued for a great length of time, have a very pernicious effect upon certain parts of the body, so as to render gestation more painful, and parturition more dangerous?—If any deformity of the pelvis were produced, that would be true; I do not know how that fact is in the manufacturing districts.

48. What would be the effect, generally speaking, upon children and young persons under eighteen, of a series of years employed as described, and of their working from thirteen to eighteen hours a day during the week, and being obliged to resume their toil without sufficient nocturnal rest to recruit the powers of the body?—I should think all the powers of the body would necessarily be enfeebled, and general weakness would be apparent in the individuals thus labouring, so that they would have less muscular power, and less energy of every kind, than if employed under more favourable circumstances.

49. Do you think that a child under nine years of age ought, ordinarily considered, to be a labourer in one of those mills and factories?—I think not.

50. Do you not conceive that, from the age of nine to eighteen, twelve hours of daily labour, including the intervals necessary for taking meals, leaving ten hours as the term of actual labour, is as much as children and young persons of either sex can endure with impunity during that period of life?—I think it is.

51. Do you think a child from ten to twelve years of age is capable of undergoing ten hours' labour?—I should doubt it very much, even although the labour be light; but different children differ very much in the powers of their frame at that period; it is difficult to apply a general rule to the cases of children.

52. When you get up to sixteen or seventeen, would you not allow a boy to work in a factory for more than ten hours, provided that labour was not of a very exhausting character, and consisting merely of the motion of his arms and legs, and being confined altogether for twelve hours?—I should not; because I think that would be as much as should be imposed upon him.

53. Is the human constitution so well fitted for labour during the period of its growth as when the system is matured and perfected, especially in relation to the bones?—I should say certainly not.

Sir G. L. Tuthill.

54. So that it is your opinion, that a person of sixteen or seventeen is not capable of performing such long and strenuous labour as a person of twenty or twenty-five ?—I think not.

55. When does the growth generally stop ?—It varies with different individuals ; I should think generally about eighteen or nineteen.

56. Any severe labour of the kind described as existing in the factories would be prejudicial to persons during the period of their growth?—Yes, more prejudicial.

57. Admitting that the labour in question takes up the whole time of the children and young persons employed on week-days, and leaves a great sense of the languor and fatigue upon them during Sunday, do not you think that confining children in a heated atmosphere, and obliging them to pay the necessary attention to instruction and learning on the Sunday, does, under such circumstances, inflict upon them additional physical suffering ?—I should think that it must inflict upon them additional suffering, and that the majority of them must fall asleep.

58. It is in evidence, that it is with the greatest difficulty they can be dragged to those schools, or, when there, kept awake ; is the mind in a fitting state to receive instruction under those circumstances ?—Certainly not.

59. Do you not think that such institutions might be rendered far more serviceable to the poor children, as well as to the public, if the hours of labour were so limited as to leave some opportunities of receiving instruction on the week-days, so as to leave the mind more at leisure, and the body some rest, on the Sunday ?—Certainly.

60. Do you think that this constant and unremitting labour, which induces the children to fall asleep as soon as it is over, and renders it difficult to arouse them to resume their task, has any effect in destroying the capabilities of the mind to receive instruction ?—Yes, I think it has.

61. Is not the degree of labour that can only be extorted from the persons in question by severe punishment and chastisement peculiarly injurious to those having to endure it ?—Certainly.

62. When the energies of the body are wasted, and when sleep and stupor can hardly be resisted, and punishment is required at the latter part of the day almost incessantly, would not that, in your judgment, permanently and severely injure the constitutions of the individuals so treated ?—Certainly.

63. In every point of view, whether in respect of humanity, public advantage, or the health of the individuals to whom the attention of thi~ committee is more specially directed, do you think a proper modification of the hours of labour would be essentially beneficial ?—Yes, I do.

64. How many hours would you allow ?—Not more than ten hours.

65. After nine years of age ?—From nine to eighteen.

66. Would you not allow an increase after fourteen ; is not a boy of sixteen more capable of bearing labour than a boy of ten ?—A boy of sixteen is certainly more capable of bearing labour than a boy of ten.

67. Does not it follow from thence that the labour of a boy of ten ought to be different to a boy of sixteen ?—Yes ; but I doubt whether you could have a rule for every year.

68. Does not it follow, therefore, that the degree of detriment arising from excessive labour must vary according to the ages of the persons subjected to it ?—Yes, supposing the constitutions to be the same.

69. Therefore it would be more desirable to make a distinction in the labour of children to be determined by their ages ?—Yes, by their ages and constitutions, if practicable.

70. As the very essence of the manufacturing system requires an entire uniformity in the time of going to labour and leaving it, and as it could not possibly be conducted in any other way, or, however desirable it might be,

Sir G. L. Tuthill.

subjected to a variable duration of labour, whether calculated upon the ages of the children and young persons employed, or their different capacities to endure labour, you have stated that up to eighteen years of age you do not think beyond ten hours of actual labour ought to be imposed?—I think not.

71. Perhaps you are of opinion with the late Dr. Baillie, that, ordinarily considered, the usual term of labour, namely, ten hours, exclusive of the time for meals and refreshment, making up twelve hours, is enough for persons of either sex, of any age, and under any circumstances whatever?—I think it is.

72. Do you not think a boy of sixteen is better able to bear twelve hours' labour, with due intervals, than a boy of ten is to bear ten hours' labour?—It is very likely he may; and I have no doubt that, if it were possible, a graduated scale, according to the age and constitution of the persons to be employed, would be a right scale; but from the questions put to me, I conceived it was necessary to fix some one period of labour that might be least objectionable when applied to all ages between nine and eighteen.

73. Why do you take the limit of eighteen?—The question put to me expressed that limit.

74. Supposing we went beyond eighteen?—I suppose, for a certain period of life, this labour may be considered compulsory, the persons employed being children, and not free agents; the moment they are of age to determine for themselves, if they choose voluntarily to work for fourteen or sixteen hours, it would be a different thing.

75. You think that, ordinarily considered, the human constitution is approaching to a period of strength and perfection up to about eighteen years of age?—Yes, I do; and probably somewhat longer.

THOMAS YOUNG, Esq., M.D., called in, and examined, 18th July, 1832.

Thos. Young, Esq. M.D.

1. What is your profession?—A physician.

2. Where do you reside?—At Bolton, in Lancashire.

3. Is that a place where there is a considerable number of mills and factories?—It is a very considerable place; the population is nearly 50,000, but I am not aware of the proportion engaged in manufactures.

4. But you know that there is a considerable number so employed?—I do.

5. Do you happen to know the usual hours of labour in those establishments?—Never less than twelve hours, exclusive of meals, as far as I have been able to learn; but sometimes more.

6. As a physician, do you believe that even the shortest hours of labour which you have mentioned are too long to be consistent with the health and welfare of the individual so employed?—I do.

7. Do you consider that that length of labour has a more pernicious effect when endured by children and young persons?—Certainly.

8. Does not the effect become still more injurious when the labour has to be pursued in a polluted, and often in an excessively heated atmosphere?—Certainly it does.

9. Will you please to state to this committee what your opinions are as to the medical effects produced by these hours of labour, deduced as well from the principles of your profession as from your actual observations?—The first effects appear to be upon the digestion; the appetite suffers, the digestion is impaired, and consequent emaciation and debility are induced. Scrofulous diseases are common; I am not aware that this disease would be produced in a sound child, born of healthy parents, but if a predisposition to scrofula

existed in the constitution, the disease (which might otherwise perhaps have remained dormant in the system) is likely to be called into action.

Thos. Young, Esq., M.D.

10. Have you observed whether pulmonary complaints are the frequent result of such labour?—They are; for example, consumption and asthma; the latter, however, I have more frequently observed in adults than in children.

11. But so as clearly to be traceable, in your judgment, to the particular nature of the employment to which reference is made?—Certainly; to the transition from excessive heat to cold, and the inhalation of dust and cotton-flue.

12. Have you observed any other effects, as resulting from this undue labour?—I have observed cases of convulsions in children; they drop down, apparently exhausted, while engaged at their work, and are affected with fits resembling epilepsy.

13. But still such as, in your judgment, are attributable to the nature and excess of their employment?—Certainly; pains in the head are often complained of by the children, attributable to the excessive heat and confinement, and cases of typhus fever are common.

14. It has been stated frequently, that febrile complaints are more general under such circumstances than they would otherwise be; do you think them more severe also?—I cannot say that I have observed that.

15. Your branch of the profession does not exclude, but rather demand, the study of anatomy; does it not?—Certainly.

16. You have to be acquainted, then, with many cases that are technically denominated surgical ones?—Certainly; though we do not prosecute anatomy with a view to surgical operations, anatomy and physiology enter into the education of a physician.

17. Will you have the goodness to state what you conceive to be the effects of this long-continued labour upon the structure of the body?—I will: ossification not being complete at that early age, the bones yield under the weight of the body, and distortion is thus produced.

18. Are the ligaments much affected by long standing?—They are.

19. Do you think that deformity is the frequent result of factory labour, as now undergone?—I do.

20. What part of the body have you particularly observed to be affected by long standing under those circumstances?—The lower extremities are chiefly affected from the cause, I presume, that I have stated, namely, the bones giving way to the superincumbent weight of the body before ossification has been completed.

21. Have you observed deformities in the foot to result from this system?—I have.

22. Do you believe that those swellings in the ankles, and the enlargement of the bones of the ankles, that sometimes occur, and also of the knees, and which have been described as incidental to the factory system, are to be attributed, in many instances, to the labour which it imposes?—I do. Sometimes, without evident disease of the bones at all, the joints are distorted from relaxation of the ligaments; this most frequently happens in the knee, and the patient becomes bow-legged, as it is called, or in-knee'd, according as the external or internal lateral ligaments of the joints are affected. I have also frequently observed that pains in the bones are complained of without distortions of any kind, of which, however, I have no doubt that those pains are often the precursors; at least all the cases of actual distortion that I inquired into were preceded by such pains.

23. That distortion is occasionally excessive, is it not?—It is.

24. How do you, in attributing that disease in many instances to the factory system, distinguish it from the disorder called "rickets?"—Rickets is a

Thos. Young, Esq., M.D. disease of early infancy, and in the cases referred to the patients were perfectly straight at the time they entered the factory, in every instance which I inquired into. Besides, I should expect the distortion in rickets to be more general, and to be found particularly in the chest and spine; and whereas it was confined to the lower extremities in the cases I refer to, I can only attribute such cases to long standing, the bones not being completely ossified, and yielding to the superincumbent weight of the body.

25. This work in factories is now and then attempted to be justified by being denominated "light and easy;" will you state whether it would obviate those effects even if the work were proved to be, as it is denominated, "light and easy?"—I think not. The employment cannot be considered a laborious one in itself, or for a short period; but it is one which requires constant attention, it is irksome and fatiguing from its uniformity, the length of time it is followed, and the postures of body required; it may be rather denominated fatiguing than laborious; it is not hard labour. To illustrate it, let us suppose a female doomed to thread needles as fast as possible, in constant succession and incessantly for twelve hours a day; to thread a needle is by no means a laborious operation, but the continued and unvaried employment would be irksome and fatiguing in the extreme.

26. Would it not in some measure exhaust the nervous energies, or at least fatigue the mind as well as the body, and occasion consequently as pernicious an effect upon the health, and sometimes more so, than if the labour were more strenuous, and at the same time more varied?—It certainly would.

27. Do you think that this anticipated labour, in reference to the age of the children and young persons who endure it, combined with the length of time that it has to be pursued, has the almost necessary effect of abridging the active period of life when it does arrive, and in which labour ought principally to be undergone?—I conceive that it has that effect.

28. Do you observe, that persons who have advanced to about the meridian of life, and have been engaged in such avocations very soon, become unfit for them?—I have not made that observation in any particular instance.

29. Is it the general observation in your neighbourhood, that those employed in cotton-mills and factories are, in many instances, inadequate to their labour after they have become forty or fifty years of age?—I have frequently heard that remark.

30. And it consists with your observation that such may be the case?—Yes.

31. Have you any doubt that life is shortened by that excessive degree of labour thus early imposed upon human beings?—I have had no means of estimating the mortality of spinners as compared with that of the general population; but I cannot doubt that factory-working tends to shorten life, inasmuch as it tends to produce disease.

32. Have you yourself examined any factories?—Three in Bolton and three in Manchester; but my examination was very cursory.

33. Was it sufficient to enable you to make certain observations of the appearance and health of the children so occupied?—The object of my visit to the factories was chiefly to take the temperature of the rooms; as to any information to be acquired respecting the state of health of the children, I trusted rather to the Sunday-schools for that information.

34. Speaking with reference to your main object, did you find the temperature of the rooms such as in your opinion was inconsistent with the health of the children?—Yes; and, with the permission of the committee, I shall state the temperatures as I found them.

35. Under what circumstances did you happen to visit those factories?—I went of my own accord, with a view to this inquiry.

36. Was there any difficulty made to your examination?—None what-

ever; I ascertained that the masters of those factories were favourable to the measure now proposed, otherwise I should not have applied to them. Thos. Young, Esq., M.D.

37. Will you state the temperatures of the mills which you examined?—In the first mill that I examined, the temperature of the card-room was 68°; of the weaving-room 68°; of the lowest spinning-room 73°; of the higher 75°; and of the highest of all 75°; of the dressing-room, as it is called (which is above that), 82°. I ought to mention, however, that no children are employed in the room last-mentioned. In the second factory I examined I took the extremes; the temperature of the card-room was 76°, and that of the highest room 80°; I was informed by the manager of this mill that the temperature of the highest room is generally about 90°; I was informed at this mill also, by a spinner, that the temperature of the factory was lower on that day (Monday), the steam having been off from four o'clock on Saturday till that morning. In the third mill that I examined in Bolton, the temperature was low, as they spin coarse numbers; a very great quantity of dust and flue were flying about, but the temperature was low, no steam being employed except in winter; I was informed that for spinning high numbers a high temperature is required, but the labour is comparatively light; in spinning low numbers only a low temperature is required, but the work is more laborious; in the mill I now speak of the numbers were excessively low; the work was very coarse, and no steam, in fact, was employed at all, so that I did not take the temperature. The fourth mill I examined was in Manchester, on the 28th of June: the temperature of the card-room, 70°; of the first spinning-room 78°, the second 82°, the third 80°, the fourth not employed, the fifth 82°, and the sixth 86°. This factory seems to be well regulated, great attention is paid to cleanliness, and the moral and religious instruction of the people. The heat, to my feelings, was oppressive, and caused profuse perspiration. The fifth factory that I visited was also in Manchester; the temperature of the card-room 65°, the first spinning-room 75°, the second 77°, the third 80°, the fourth 79°. In the sixth factory (the last I examined in Manchester) the temperature of the card-room was 73°, first spinning-room 75°, second 78°, third 77°, fourth 78°. I was informed that the steam had been turned off that morning, but in the staircase the heat was extreme, owing to a large circular chimney passing through the middle of it; and the atmosphere of the rooms was impregnated with offensive effluvia, arising from the conveniences within the factory.

38. Do not you think that the heat which you have described as necessary in those mills and factories, is an additional and a very momentous reason for shortening the hours of labour which are undergone in them?—Certainly it is.

39. Do you think that it is possible for the constitution, generally speaking, to maintain unimpaired those hours of labour in an atmosphere heated to the temperature which you have described to the committee?—In general it is not.

40. You state also, that there are other circumstances that are pernicious and injurious in certain of those mills which ought to be obviated, as rendering confinement in them still more injurious and oppressive?—Independently of the temperature of the atmosphere being heated, it is also very impure from the breathing of a number of people in close and unventilated apartments; and also from another circumstance which I before alluded to, namely, the conveniences within the factory, which give rise to very offensive effluvia.

41. Was that universally the case?—I believe generally; not in all the mills that I visited; but I believe it is generally the case.

42. Would not the bill now proposed, therefore, in that point of view, as they would have occasion to be less resorted to than they are now, if oppor-

Thos. Young, Esq., M.D.

tunities were given to the children to leave the mill more freqnently, be of considerable advantage?—No doubt it would. The object, I believe, is to save time; to prevent the people from going out.

43. Did you yourself examine any of the children, and had you an opportunity of questioning them particularly?—I observed upon their general appearance; but my examination was unavoidably very superficial, for the children were busily engaged at their work, moving from one place to another; and the noise of the machinery was also an obstacle to minute examination; in addition to which, the heat was so oppressive to myself, that I wished to make my stay as short as possible; particularly as I intended visiting the Sunday-school, with a view to ascertain more particularly the health of the children.

44. All that you understood and saw in those mills confirmed you in the opinion which you have already expressed, that the hours of the labour of the children were too long, and that it was necessary, in regard to their health and improvement, to abridge them?—Yes.

45. Did you observe any cases of deformity in the mills in passing thus cursorily through them?—I did; I desired cases of that description to be pointed out; but I found that the information to be derived in this way was not to be relied on, having discovered cases which had not been mentioned to me by my attendant. I therefore had to trust to my own discrimination, and I have no doubt that many cases escaped my notice.

46 So that, after having asked for those places to be placed under your inspection, and the attendant professedly doing so, you still detected deformed children that were not submitted to your notice?—I did.

47. Which, had it not been for your own casual observation, you would not have known to be so?—I should not.

48. Do you therefore think that a cursory visit of those mills and factories by any means gives a full idea of the condition in which the children are, regarding their general health and welfare?—Certainly not; I could only observe upon the general appearance of the children present, which was unhealthy in the extreme. I was informed that they are frequently changed, but I could get no information on this point from the masters, who were even ignorant of the number employed in the factories, as they are engaged and dismissed, not by the master, but by the spinners themselves. For these reasons the examination of factories must be considered a very fallacious test.

49. You state that you examined certain Sunday-schools with the same view, that is, with a wish to ascertain the effect of the factory labour upon children so engaged?—I examined three in Bolton.

50. Will you give to this committee the result of your observation in those schools; could you distinguish those who worked in factories?—I believe I could have distinguished them, but I did not attempt to do so. I desired the teachers to separate them from the other children, and with the permission of the committee, I will read the notes I took at the time I made my visit. The largest school I visited was in Ridgeway Gates, belonging to the Methodists. The number of children present under sixteen was 446 boys and 458 girls; of those, 165 boys and 171 girls were employed in factories. I began my examination with the girls: twenty-seven out of the 171 had met with accidents from machinery, including one fracture of the finger, one permanent contraction of the finger, two fractures of the wrist; the remaining accidents were slight: five out of 171 girls were deformed in the knees; nineteen complained of cough; thirty-one of loss of appetite; and nine were scrofulous. Of the boys, forty-six had been caught by the machinery; three had lost fingers; one had had a finger dislocated; the remaining accidents were slight: three were deformed in the knee, seventeen complained of cough, generally attended with expectoration of flue, six of loss of

Thos. Young, Esq, M.D.

appetite, and ten were scrofulous. The great majority of those children complained of pains in the limbs and back from long standing. I also visited the school in Maudsley-street; it is on a smaller scale, and belongs to the people called Independents; number of children stated at 400; I did not count them, that was the number given me by the master; fifty-five out of the number present were employed in factories, namely, twenty-four boys and thirty-one girls. Sixteen boys stated that they had met with accidents from machinery, but on examination they were found to be slight; four out of seven boys working in the card-room were affected with cough; three out of the remainder were similarly affected. Eight had recently been confined with fevers, attributed to confinement in the factory. Of thirty-one girls, six complained of pains in the limbs, attributed to long standing; and five of loss of appetite. On the same day I visited the school in Fletcher-street, also belonging to the Methodists; present 280. Of those, forty-one were employed in factories: of twenty boys, one showed the cicatrix of an injury of the finger which had been caught in the machinery, and the other had sustained the loss of a finger from the same cause; two suffered from cough with expectoration of flue: of twenty-one girls, three complained of pains in the limbs and joints, increased towards evening, and attributed to long standing; four of pain in the head, attributed to the heat and confinement; one of loss of appetite, and two of cough, with expectoration of flue. I have made a general note here. The general appearance of the children in those schools is extremely unfavourable as contrasted with those engaged in other employments; they have a sallow and unhealthy aspect; many of them have a peculiar flatness of the foot, and are much stinted in their growth. With regard to accidents from machinery, I have met with much more serious cases than those referred to in the factories and Sunday-schools, in my private practice; two of those cases I have recorded.

51. Do you think that the labour in factories has a tendency to interfere with and diminish the growth of the children employed in them?—I do.

52. Do you think also it would have a tendency to diminish the muscular power of persons so occupied?—Certainly.

53. Will you state whether the female can bear labour as well as the male?—I think females cannot endure labour so well as the males.

54. Then any regulation in regard to the labour of children in factories ought, of course, to have reference to the capability of the female to endure it without injury?—Undoubtedly.

55. Of course the effects that you have stated as the general effects of the system were as apparent in the females as in the males?—Certainly.

56. Will you state whether there were any other effects to which you have observed the female liable from which perhaps the male would be in some measure free?—I am not aware of any, except such as arise from the peculiar delicacy of the female frame.

57. Do you think such effects would be more marked in the female than in the male?—Yes. I would take the liberty of making one observation here: when it was understood that I was coming here as an evidence, a statement was sent me from the parish-church school in Bolton, and which I feel myself bound in candour to take notice of: this statement represents the health of the factory children in that particular school as highly favourable. I consider the examination of Sunday-schools a fallacious test, for this reason, that comparatively few children, I am sorry to say, attend them, and those few we may suppose are favourable specimens; they are the children of religious parents, of parents who take an interest in the religious and moral instruction of their children, and I found their appearance in the Sunday-schools very much superior to what I had anticipated from seeing them in the factories.

Thos. Young, Esq , M.D.

58. And those who are weaker and indisposed would not attend?—No, of course not.

59. You would then suppose that an examination of a mill or factory, or even a Sunday-school, merely for the purpose of observing those who might be in tolerable health at the time being, would not be a sufficient proof that the employment was healthful, or that it had not produced such mischievous effects upon the constitution?—Certainly not; in fact, no satisfactory calculation can be made unless we know the proportions of children who have been disabled or sick, or who have been absent from the school within a given time, which I had no means of ascertaining.

60. Those children who have been totally disabled from pursuing their labour by sickness and by accident, and some that die from severity of exertion, would disappear from the books of the mill, would they not?—Undoubtedly.

61. The most striking effects of the system would be those which would not be a matter of record, even if there were any statement made regarding the health of the hands employed in the establishments in question?—Yes; and as to Sunday-schools, a very low degree of health and vigour is perfectly consistent with attendance upon schools, and there may not be any formal disease, but at the same time a general deterioration of the health which would not preclude the attendance upon Sunday-schools.

62. Do you think that, considering the peculiar constitution of the females, especially under certain circumstances, the labour of that sex carried to such a degree is peculiarly prejudicial?—I do consider it so.

63. Have you known any instances in which you have traced consequences resulting from that labour which you think would not have ensued had it not been so continued?—Abortion I have certainly witnessed. One case of that is within my recollection, which I shall state. I remember the case of a girl who was seized with the pangs of labour while engaged at her work; she was carried to her home by her companions, but before she reached it she had given birth to a child, apparently about the sixth month.

64. What do you think are the moral effects of this system as pursued at present?—The morals of the factory children at Bolton are very bad, I am sorry to say.

65. Do you observe any breaches of decency, both in language and in conduct, that lead you to make that observation?—I do; the animal propensities are early developed in the mills, and very frequently before the development of those moral feelings which would restrain their indulgence; there is little modesty among the females. These remarks must, of course, be received with some limitation; there are, of course, some exceptions, and I hope many. I have often observed them coming out of the factory; their conduct was indecorous, and their language gross and obscene. I have been informed that illegitimate children are rare, but I beg to suggest that the very circumstance of the frequency and promiscuousness of intercourse which has been reported to me, must operate against conception.

66. So that, considering their state of manners and morals, the circumstance of there being a proportionately small number of illegitimate children would operate in your mind to induce you to believe that morals were in a very depraved state, rather than the reverse?—I know that morality is at a very low ebb.

67. What do you think is the direct cause of early breaches of morality, especially in relation to the improper connexions of the sexes, in those particular pursuits?—I attribute this state of things to the cause that I have mentioned, namely, the early development of the animal propensities from the high temperature of the factories, which is not sufficiently checked by the moral and religious education of the children.

Thos. Young, Esq., M.D.

68. Do you think premature puberty is induced by the over-employment pursued?—It is, though I have seen instances of the very reverse, namely, where that period has been very much delayed by it.

69. Is it consistent with your general observation that the females sooner arrive at the period at which they are likely to become mothers than they otherwise would?—They do; in fact, the factory has the same effect in that respect as a warm climate would produce; it is well known that females in a warm climate sooner become mothers than in a cold one.

70. Then the best result from such a state of things would be early and premature marriage?—Certainly, that is the best effect that can be anticipated.

71. Do you think that the protection which this bill affords in point of age, namely, up to the age of eighteen, is a necessary protection, or might that protection be somewhat abated, and the term be brought lower down?—I think not.

72. Are there not periods of the constitution during youth that require more peculiarly than others this necessary protection from extreme labour?—Certainly.

73. Do you conceive that ten hours a day labour is as much as the constitution usually can bear, especially in reference to children and young persons under eighteen?—I think that it is the extreme limit that ought to be allowed; that it is quite enough for the healthy and robust, and too much for the feeble and delicate.

74. Are the regulations of Mr. Hobhouse's bill observed in the mills at Bolton?—I am not aware of that; nor do I know the provisions of that bill with respect to mills.

75. Are you aware of the provisions of the old bill for factories, Sir Robert Peel's bill?—I understand that that bill restricted the hours to twelve.

76. Do you know whether those provisions are adhered to?—I believe not. Indeed I know they are not, from instances that have come to my knowledge.

77. Upon the whole, then, you consider a limitation, such as that proposed in the bill, would be beneficial to the health and happiness of the children, and would subserve the interests of society at large?—I do.

78. Is that, generally speaking, the impression of the people in the town where you practise?—It is.

79. Is such a limitation sought after by those mainly interested in the question, namely, the operatives?—It is, to my certain knowledge.

80. And are not several of the mill-owners themselves willing, nay, professedly anxious, for a limitation of the hours of labour?—They are, on this condition, that the regulations contemplated shall be made binding on all masters, so that no one shall be able to take advantage of another.

THE ENTIRE DEBATE,

CONSISTING OF THE

SPEECHES

OF

MICHAEL THOMAS SADLER, Esq.,

THE CHANCELLOR OF THE EXCHEQUER, AND OTHERS,

In the House of Commons, Friday, March 16, 1832,

ON MOVING THE SECOND READING OF

THE FACTORIES REGULATION BILL.

Speech of M. T. Sadler, Esq.

SIR,—In rising to move the second reading of the bill which I have had the honour of introducing into Parliament, for regulating the labour of children and young persons, not being free agents, employed in the mills and factories of the country, of whatever description, I shall, as far as is consistent with the high importance of the subject, and the great and general interest which it has excited throughout the entire community, compress the arguments and facts upon which I found the necessity of this measure; and I shall not misapply many moments of the time which I must still occupy by allusions personal to myself, however much I may have been provoked to such a course. I will merely say, that in bringing forward this measure, I make no pretensions to a degree of humanity beyond that which I share with the people at large; still less am I influenced by any views adverse to the prosperity and extension of our manufactures and commerce: least of all can I be governed by feelings otherwise than cordial to those embarked in these great concerns. On the contrary, in pursuing this course, I am acting under the impression, at least, that the measure which I propose will advance the true and permanent interests of the manufacturers, the cordial encouragment and support of many of the most humane and best-informed of whom I regard as the strongest proof of the necessity of the measure, and the surest presage of its success. As to the imputation cast upon me by others of a different description, who wish to defeat this attempt, as they have hitherto done preceding ones of a like nature, not only by thwarting the designs, but by maligning the motives, of those who make them; and who, therefore, accuse me of being instigated by a mean desire of popularity, in now undertaking a cause which, nevertheless, some of them know well enough I advocated as strongly long before I was in Parliament, as I can do on the present occasion: I say, as to this imputation, I should have passed it over in silence, only that it affords me an opportunity, which I will not neglect, of proving, even from the mouths of its opponents, that the measure is popular—popular in the fullest and best sense of the term,—and the House, I think, has seen, from the petitions which have already loaded its table, signed by magistrates, clergy, and professional men, as well as by immense numbers of the operative classes, that its popularity is founded upon the principles, the intellect, and the feelings of the British community, and that those who resist it must reckon on contempt and indignation.

The bill which I now implore the House to sanction with its authority, has for its object the liberation of children and other young persons employed in the mills and factories of the United Kingdom, from that over-exertion and long confinement which common sense, as well as experience, has shewn to be utterly inconsistent with the improvement of their minds, the preservation of their morals, and the maintenance of their health;—in a word, to rescue them from a state of suffering and degradation, which it is conceived the children of the industrious classes in hardly any other country have ever endured.

I am aware that some gentlemen profess, upon principle, a great reluctance to legislate upon these matters, holding such interference to be an evil. So, I reply, is all legislation, upon whatever subject, and an evil only to be tolerated for the purpose of preventing some greater one. I shall, therefore, content myself with meeting this objection, common as it is, by simply challenging those who urge it to shew us a case which has stronger claims for the interposition of the law, whether we regard the nature of the evil to be abated, as affecting the individuals, society at large, and posterity; or the utter helplessness of those on whose behalf we are called on to interfere; or, lastly, the fact, which experience has left no longer in doubt, that, if the law does not, there is no other power that can or will adequately protect them.

But, I apprehend, the strongest objections that will be offered on this occasion will be grounded upon the pretence that the very principle of the bill is an improper interference between the employer and the employed, and an attempt to regulate by law the market of labour. Were that market supplied by free agents, properly so denominated, I should fully participate in these objections. Theoretically, indeed, such is the case; but, practically, I fear, the fact is far otherwise, even regarding those who are of mature age; and the boasted freedom of our labourers in many pursuits will, on a just view of their condition, be found little more than a name. Those who argue the question upon mere abstract principles seem, in my apprehension, too much to forget the condition of society; the unequal division of property, or rather its total monopoly by the few, leaving the many nothing but what they can obtain by their daily labour; which very labour cannot become available for the purposes of daily subsistence without the consent of those who own the property of the community,—all the materials, elements, call them what you please, on which labour can be bestowed, being in their possession. Hence it is clear that, excepting in a state of things where the demand for labour fully equals the supply (which it would be absurdly false to say exists in this country), the employer and the employed do not meet on equal terms in the market of labour; on the contrary, the latter, whatever be his age, and call him as free as you please, is often almost entirely at the mercy of the former;—he would be wholly so, were it not for the operation of the poor-laws, which are a palpable interference with the market of labour, and condemned as such by their opponents. Hence is it that labour is so imperfectly distributed, and so inadequately remunerated; that one part of the population is overworked, while another is wholly without employment; evils which operate reciprocally upon each other, till a community which might afford a sufficiency of moderate employment for all, exhibits at one and the same time part of its members reduced to the condition of slaves by over-exertion, and another part to that of paupers by involuntary idleness. In a word, wealth, still more than know-

Speech of M. T. Sadler, Esq.

ledge, is power; and power, liable to abuse wherever vested, is least of all free from tyrannical exercise when it owes its existence to a sordid source. Hence have all laws, human or divine, attempted to protect the labourer from the injustice and cruelty which are too often practised upon him. Our statute-book contains many proofs of this, and especially in its provision for the poor. The anti-truck bill of last year is an instance of this benevolent kind of interposition; and that sacred institution which has been adopted and legally enforced, as far as the limits of civilization extend, and which justifies its claim to divine origin by its humanity and mercy—the institution of the Sabbath—is a constantly-recurring example of interference between the employer and the employed, solely and avowedly in favour of the latter; and I cannot help regretting, that almost every other red-letter day has been long ago blotted out from the dark calendar of labouring poverty, whose holidays are now too "few and far between" to cheer the spirits or recruit the health of our industrious population. It was promised, indeed, and might have been expected, that the great inventions of recent times would have restored a few of these,—would have somewhat abridged human labour in its duration, and abated its intensity: and it is only by effecting this that machinery can justify its very definition, as consisting of inventions to shorten human labour. I look forward to the period when machinery will fully vindicate its pretensions, and surpass, in its beneficial effects, all that its most sanguine advocates have anticipated: when those inventions, whether so complicate and minute as almost to supplant the human hand, or so stupendous as to tame the very elements, and yoke them to the triumphal car of human industry, shall outstrip our boldest expectations, not so much, indeed, by still further augmenting the superfluities of the rich as by increasing the comforts and diminishing the labour of the poor; thereby restoring to the mass of our fellow beings those physical enjoyments, that degree of leisure, those means of moral and mental improvement, which alone can advance them to that state of happiness and dignity to which, I trust, it is their destiny to attain. Hitherto, however, I repeat, the effect has been far different. The condition of the operative manufacturers has been rendered more and more dependent and precarious; their labour, when employed, is in many cases so increased as to be utterly irreconcilable with the preservation of health or even life; infancy itself is forced into the market of labour, where it becomes the unresisting victim of cruelty and oppression; while, as might be expected from such an unnatural state of things, the remuneration for this increasing and excessive toil is regularly diminishing, till at length multitudes among us are reduced, in their physical condition at least, below the level of the slave or the brute. In proof that this is no singular or overcharged view of the present effect, or at all events of the ultimate consequences, of this dreadful system, I shall appeal to the language of a benevolent and enlightened individual, formerly a member of this House, and an ornament to it and the country—I mean the late Sir Robert Peel. His deliberate judgment upon this important subject is thus recorded in a document which he delivered to the committee on the bill he introduced in 1816:—"Such indiscriminate and unlimited employment of the poor, consisting of a great proportion of the inhabitants of the trading districts, will be attended with effects to the rising generation so serious and alarming, that I cannot contemplate them without dismay; and thus that great effort of British ingenuity, whereby the machinery of our manufacturers has been brought to such perfection, instead of being a blessing, will be converted into its bitterest curse."

Neither in quoting this passage, nor in making the observations which introduced it, would I be understood to recommend any interference with the efforts of human ingenuity, or with the market of labour, as supplied by

Speech of M. T. Sadler, Esq.

free agents. But in shewing how far even adults are from being free agents, in the proper meaning of the term, and, on the contrary, how dependent for their employment, and consequently their daily bread, upon the will of others, I have prepared the way for the conclusion, that children, at all events, are not to be regarded as free labourers; and that it is the duty of this House to protect them from that system of cruelty and oppression to which I shall presently advert. The common-place objection, that the parents are free agents, and that the children therefore ought to be regarded as such, I apprehend has but little force. It is, however, so often and so confidently urged, that I shall be excused for giving it some attention.

The parents who surrender their children to this infantile slavery may be separated into two classes. The first, and I trust by far the most numerous one, consists of those who are obliged, by extreme indigence, so to act, but who do it with great reluctance and bitter regret: themselves perhaps out of employment, or working at very low wages, and their families in a state of great destitution; what can they do? The overseer, as is in evidence, refuses relief if they have children capable of working in factories, whom they refuse to send thither. They choose, therefore, what they probably deem the lesser evil, and reluctantly resign their offspring to the captivity and pollution of the mill. They rouse them in the winter morning, which, as a poor father says before the Lords' Committee, they "feel very sorry" to do; they receive them fatigued and exhausted, many a weary hour after the day has closed; they see them droop and sicken, and, in many cases, become cripples and die, before they reach their prime; and they do all this, because they must otherwise suffer unrelieved, and starve, like Ugolino, amidst their starving children. It is mockery to contend that these parents have a choice; that they can dictate to, or even parley with, the employer, as to the number of hours their child shall be worked, or the treatment it shall be subject to in his mill; and it is an insult to the parental heart to say, that they resign it voluntarily;—no, "Their poverty, and not their will, consents."—Consents, indeed! but often with tears, as Dr. Ashton, a physician familiar with the whole system, informed the committee, a noble member of which, indeed, observed, to one of the poor parents then examined, who was speaking of the successive fate of several of his children, whom he had been obliged to send to the factory—"You can hardly speak of them without crying?" The answer was, "No!" And few, I should suppose, refrained from sympathizing with him, who heard his simple but melancholy story. Free agents! To suppose that parents are free agents while dooming their own flesh and blood to this fate, is to believe them monsters!

But, Sir, there are such monsters: unknown, indeed, in the brute creation, they belong to our own kind, and are found in our own country; and they are generated by the very system which I am attacking. They have been long known, and often described, as constituting the remaining class of parents to which I have adverted. Dead to the instincts of nature, and reversing the order of society, instead of providing for their offspring, they make their offspring provide for them; not only for their necessities, but for their intemperance and profligacy. They purchase idleness by the sweat of their infants, the price of whose happiness, health, and existence, they spend in the haunts of dissipation and vice. Thus, at the very same hour of night that the father is at his guilty orgies, the child is panting in the factory. Such wretches count upon their children as upon their cattle;—nay, to so disgusting a state of degradation does the system lead, that they make the certainty of having offspring the indispensable condition of marriage, that they may breed a generation of slaves. These, then, are some of the free agents, without the *storgè* of the beast, or the feelings of the man, to whom

Speech of M. T. Sadler, Esq.

the advocates of the present system assure us we ought to entrust the labouring of little children. One of these "free agents," a witness against Sir Robert Peel's bill, confessed that he had pushed his own child down and broken her arm, because she did not do as he thought proper, while in the mill. The Lords' Committee refused to hear him another word. And shall we listen to those who urge us to commit little children to such guardianship? We have heard, in a late memorable case, a *dictum*, uncontradicted I believe in any quarter, stating that, by the constitution of England, the first law officer of the crown, representing the sovereign, is the guardian of all children, of whatever rank, improperly treated by their parents; but that that court is limited in its interference by the circumstance of there being property under its control. Will it be contended, then, that in these extreme cases of cruelty ond oppression (for such I shall call them), where protection is far more imperatively demanded, that poverty should be a bar against the course of British justice?—If so, let us boast no longer of the impartiality of our laws! Why, if in a solitary instance a parent were to confine his child, or a master his apprentice, in a heated room, and knowingly keep him at his labour more hours than nature could sustain, and at length the victim were to die under the tyrannous oppression, and a coroner's inquest were to return a true and just verdict upon the occasion, what would be the result? . . . . And are the multiplication of such gradual murders, and the effrontry with which they are perpetrated, to become their expiation?—If not, it is high time that the legislature should interfere and rescue from the conspiracy of such fathers and such masters, instigated by kindred feelings, these innocent victims of cruelty and oppression.

There are other descriptions of children, also, whom I should be glad to know how the objectors to whom I am alluding make out to be free agents. I mean, first, poor orphan children—a class which the system is a very efficient instrument in multiplying, very few adult spinners, as it is often alleged, and as I shall prove, surviving forty, in many instances, therefore, leaving their children fatherless at a very early period of life; indeed, so numerous are these, that a physician, examined on the occasion to which I have so often alluded, was painfully struck with the proportion. Are these orphans free agents? Again, there is in all manufacturing towns a great number of illegitimate children, and these also are very much increased by the system in question. I am aware that a celebrated authority has said, these are, "comparatively speaking, of no value to society; others would supply their place,"—yet still I cannot but regard these as objects of the deepest compassion. To this list of free agents I might also add the little children who are still apprenticed out in considerable numbers, often, I fear, by the too ready sanction of the magistrates—whose hard, and sometimes fatal, treatment has been the subject of many recent communications which I have received from individuals of the highest credit and respectability. But, as the objectors to legislative protection for the factory children can make it out to be unnecessary, because their parents are "free agents" for them, when they have any surviving, so also it is quite as clear, probably, in their apprehension, that the parish officer is as good a free agent for the poor orphan, the illegitimate, or the friendless little apprentice, who may be under his special protection.

But I will proceed no further with these objections. The idea of treating children, and especially the children of the poor,—and, above all, the children of the poor imprisoned in factories,—as free agents, is too absurd to justify the attention I have already paid to it. The protection of poor children and young persons from those hardships and cruelties to which their age and condition have always rendered them peculiarly liable, has ever

Speech of M. T. Sadler, Esq.

been held one of the first and most important duties of every Christian legislature. Our own has not been unmindful in this respect; and it is mainly owing to the change of circumstances that many of its humane provisions have been rendered inoperative, and that the present measure has become the more necessary. I had meant to take a short review of these various efforts, down to the time of the benevolent Hanway; but, interesting as the subject is, and applicable to the present discussion, I must forbear, in respect to the time it would occupy, to do so. It was the introduction of Sir Richard Arkwright's invention that revolutionized the entire system of our national industry. Previously to that period, the incipient manufactures of the country were carried on in the villages and around the domestic hearth: that invention transferred them principally to the great towns, and almost confined them to what are now called factories. Thus, children became the principal operatives; and they no longer performed their tasks, as before, under the parental eye, and had them affectionately and considerately apportioned, according to their health and capacities, but one universal rule of labour was prescribed to all ages, to both sexes, and to every state and constitution. Such a regulation, therefore, it might have been expected, would have been adapted to the different degrees of physical strength in the young, the delicate, and especially the female sex. But no!—I speak it with shame, with horror—it was stretched, in many cases—I had almost said in nearly all—beyond what the most athletic and robust of our own sex, in the prime and vigour of life, can with impunity sustain,—to the ultimate destruction, in a vast majority of instances, of the health, the happiness, and the very life, of the miserable victims. Our ancestors could not have supposed it possible—posterity will not believe it true—it will be placed among the historic doubts of some future antiquary—that a generation of Englishmen could exist, or had existed, that would labour lisping infancy, of a few summers old, regardless alike of its smiles or tears, and unmoved by its unresisting weakness, eleven, twelve, thirteen, fourteen, sixteen hours a day, and through the weary night also, till, in the dewy morn of existence, the bud of youth faded, and fell ere it was unfolded. "Oh, cursed lust of gold!" Oh, the guilt which England was contracting in the kindling eye of Heaven, when nothing but exultations were heard about the perfection of her machinery, the march of her manufactures, and the rapid increase of her wealth and prosperity!

Early, however, in this century, the late Sir Robert Peel, knowing well the enormities of the factory system, and finding, from his own experience, that nothing but a legislative enactment could remove them, obtained the first act for the protection of poor children employed in cotton factories. About fifteen or sixteen years afterwards, he carried another measure, of a similar, but more comprehensive nature. Lastly, the right honourable member for Westminster obtained another act, last session, having the same benevolent object in view. But, on all these occasions, the attempt, by whomsoever made, or whatever was its character, was met with the same strenuous, or, as I might well call it, vehement opposition. Whether it was proposed to limit the labour of infants and young persons, besides the time necessary for their meals and refreshment, to ten, eleven, or even twelve hours a day, it was all one; the proposal was scouted and resisted. The motives and conduct of those engaged in attempting to obtain this protection were maligned. The universal humanity of all those, in every pursuit, whose power over these children was unrestrained, was boldly asserted; the superior health, happiness, and even longevity, of those employed were always maintained. Whatever was the nature or duration of the employment which these young persons, whether daily or nightly, pursued, it was contended that no injury, but abundance of good, was done to them. On every occasion this opposition has virtually succeeded, so as to defeat the original intentions of those who have successively proposed these measures. It has succeeded in lengthening the term of infantine labour, in limiting every act to one particular branch of business, in introducing provisions which have rendered them liable to constant evasions, and it is well known that the whole of them are evaded, and rendered little better than a dead letter.

The very same opposition that has so long and so often triumphed over justice and humanity, is again organized, and actively at work, and will proceed as before. Every branch of manufacture proposed to be regulated claims in turn to be excepted; a committee of inquiry is again demanded, and, I fear, in order to postpone, if not finally to defeat, the present measure. The nature of the evidence that will be brought forward is perfectly familiar to those acquainted at all with the subject. Certificates and declarations will be obtained in abundance, from divines and doctors, as to the morality and health which the present system promotes and secures. I cannot refrain from giving a sample of what may be expected in this line, and I think it will prepare us for, and arm us against, whatever may be advanced in favour of so unnatural and opppressive a system. I mean not to impeach the intentional veracity or the learning of the witnesses who appeared in its favour, and whose evidence cuts a very conspicuous figure in these ponderous Reports; it furnishes, however, another proof of the strange things that may be, perhaps conscientiously, believed and asserted when the mind or conduct is under a particular bias. They have said that the children who were worked without any regulation, and consequently according to their employers' sole will and pleasure, were not only equally, but more healthy, and better instructed, than those not so occupied; that night-labour was in no way prejudicial, but actually preferred; that the artificial heat of the rooms was really advantageous, and quite pleasant; and that nothing could equal the reluctance of the children to have it abated. That so far from being fatigued with, for example, twelve hours' labour, the children performed even the last hour's work with greater interest and spirit than any of the rest. What a pity the term was not lengthened! in a few more hours they would have been worked into a perfect ecstacy of delight. We had been indeed informed that the women and children often cried with fatigue, but their tears were doubtless tears of rapture. A doctor is produced, who will not pronounce, without examination, to what extent this luxury of excessive labour might be carried without being prejudicial. I must quote a few of his answers to certain queries. "Should you not think (he is asked) that, generally speaking, a child eight years old standing twelve hours in the day would be injurious?" The doctor reverses, perhaps by mistake, the figures, but his answer concludes—"I believe it is not." "Supposing (it was again demanded) I were to ask you whether you thought it injurious to a child to be kept standing three-and-twenty hours out of the four and twenty, should you not think it must be necessarily injurious to the health; without any fact to rest upon, as a simple proposition put to a gentleman of the medical profession?" "Before I answer that question (the doctor replies) I should wish to have an examination, to see how the case stood; and if there were such an extravagant thing to take place, and it should appear that the person was not injured by having stood three-and-twenty hours, I should then say it was not inconsistent with the health of the person so employed." "As you doubted (said a noble Lord) whether a child could work for twenty-three hours, without suffering, would you extend your doubts to twenty-four hours?"—"That was put to me as an extreme case (says the doctor); my answer only went to this effect, that it was not in my power to

Speech of M. T. Sadler, Esq.

assign any limits." This same authority will not take upon himself to say whether it would be injurious to a child to be kept working during the time it gets its meals. Another medical gentleman is "totally unable to give an answer" whether "children, from six to twelve years of age, being employed from thirteen to fifteen hours in a cotton factory, in an erect position, and in a temperature of about eighty degrees, is consistent with safety to their constitution." Another boldly asserts that he does not see it necessary that young persons should have any recreation or amusement; nor that the constant inspiration of particles of cotton is at all injurious to the lungs. Reports of the state of particular mills are also given on medical authority, but the reporters seem to have totally forgotten that they had examined a body of persons constantly recruited, from which the severely sick, and those who had "retired to die," were necessarily absent; and not to have suspected that many of these mills were also previously and carefully prepared for such inspection. Still, I observe, it is allowed "that many of them (the children) were pale, and apparently of a delicate complexion;" but "without any decided symptoms of disease." What did that paleness and delicacy, in the rosy morning of life, indicate? Why, that disease, though not decided as to its symptoms, was fastening, with mortal grasp, upon its victims; that already early labour and confinement had, "like a worm i'the bud, fed on their damask cheek;" that the murderous system was then about its secret, but certain and deadly, work. In corroboration, however, of all that these learned persons have advanced, and in full proof of the excellency of the entire system, bills of mortality of certain places and works were adduced, in some of which it was made to appear that, in a mean number of 888 persons employed, the annual mortality had, during eight years, averaged $3\frac{875}{1000}$, or one in 229 only! This sort of evidence suggests many ludicrous ideas, which, however, I shall suppress as unsuitable to the subject; it will, doubtless, be again adduced in great abundance before another select committee. Physicians, divines, and others will be still found to testify to the same effect. But I will take the liberty of shewing, before I sit down, the true value of such certificates. The Parliament, indeed, did not much regard these champions of the factory system on a former occasion; and, after what I shall advance, I hope the house will not trouble them again.

I shall now proceed to shew the necessity of a general measure for regulating the labour of children and young persons employed in mills and factories, of whatever description, the protective acts already obtained having been confined in their operation to one branch of manufacture only, and in that almost entirely defeated as to their original intention and design.

I need not inform the house that the great invention of Sir Richard Arkwright, originally used for the spinning of cotton, has at length been applied, with the necessary adaptations, to a similar process in almost all our manufactures. Now, the fact that Parliament has several times, notwithstanding the severest opposition, seen it necessary to regulate the labour of children in the former pursuit, proves the same necessity to exist regarding those other factories, now so numerous, which have been hitherto entirely exempted from all such control. It would be the grossest injustice, as well as insult, to argue that those engaged in the cotton-trade were one whit less humane and considerate, and consequently required legislative interference one whit more than those engaged in spinning any other material; and if it be contended that the labour of the latter is, in many cases, either less unhealthy or less immoderate than that of the former, I meet the assertion with a direct negative. Nor, in contending for the necessity of this measure, do I implicate the conduct of the mill-owners generally, many of whom, I am well convinced, are among the most humane and considerate of employers; on the contrary, the interests of these, as well as the welfare of the children, equally demand legislative protection.

And, first, in reference to one description of spinners, from some of whom I am now meeting with opposition of every kind,—I mean the spinners of flax,—I would seriously ask any gentleman who has himself gone through a modern flax mill, whether he can entertain the slightest doubt that the occupation, as now pursued, must, in too many cases, be injurious to health and destructive of life. In many departments of these mills, the dust is great, and known to be highly injurious. In those in which fine spinning has been introduced, the air has to be heated, as in some of the cotton-mills; the flax has also, in one of the processes, to be passed through water heated to a high temperature, into which the children have constantly to plunge their arms, while the steam and the spray from the bobbins wet their clothes, especially about their middle, till the water might be wrung from them, in which condition they have, during the winter months, to pass nightly into the inclement air, and to shiver and freeze on their return home. In the heckling-rooms, in which children are now principally employed, the dust is excessive. The rooms are generally low, lighted by gas, and sometimes heated by steam; altogether exhibiting a state of human suffering the effects of which I will not trust myself to describe, but appeal to higher authority.

I hold in my hand a treatise by a medical gentleman of great intelligence, Mr. Thackrah, of Leeds, who, in his work "On the Effects of Arts and Trades on Health and Longevity," thus speaks of this pursuit:—"A large proportion of men in this department die young. We find, indeed, comparatively speaking, few old persons in any of the departments of the flax-mills." —"On inquiry, at one of the largest establishments in this neighbourhood, we found, that of 1079 persons employed, there are only nine who had attained the age of fifty; and besides these only twenty-two who have reached forty."

It may, perhaps, be here remarked, that this factory census does not indicate the rate of mortality, but merely shews that few adults are required in these establishments. If so, then another enormous abuse comes into view; namely, that this unregulated system overlabours the child, and deserts the adult; thus reversing the natural period of toil, and leaving numbers without employment, or the knowledge how to pursue it if they could obtain any, just at the period when the active exertions of life ought to commence. Why this is to realize, in regard of these victims of premature labour, the fate of the poor little chimney-sweeper, whose lot, once commiserated so deeply, is now, I think, too much forgotten, and whose principal hardship is, not that he is of a degraded class, but that when he has learnt his business he has outgrown it, and is turned upon society too late to learn any other occupation, and has therefore to seek an employment for which he is unqualified. So far, then, this unrestricted factory system perpetrates the deepest injury, not only upon individuals, but also upon society at large.

But to return to Mr. Thackrah. He says that a visitor cannot remain many minutes in certain rooms without being sensibly affected in his respiration. Also, that "a suffocating sensation is often produced by the tubes which convey steam for heating the rooms." He examined, by the stethoscope, several individuals so employed, and found, in all of them, "the lungs or air-tube considerably diseased." He adds, that the coughs of the persons waiting to be examined were so troublesome as continually to interrupt and confuse the exploration by that instrument. He says, "that though the wages for this labour are by no means great, still the time of

Speech of M. T. Sadler, Esq.

labour in the flax-mills is excessive. The people are now (November 1830) working from half-past six in the morning till eight at night, and are allowed only an interval of forty minutes in all that time. Thus human beings are kept in an atmosphere of flax dust nearly thirteen hours in the day, and this not one but six days in the week." "No man of humanity," he observes, "can reflect, without distress, on the state of thousands of children,—roused from their beds at an early hour, hurried to the mills, and kept there, with an interval of only forty minutes, till a late hour at night—kept, moreover, in an atmosphere loaded with noxious dust." "Health," he exclaims, "cleanliness, mental improvement—how are they regarded? Recreation is out of the question. There is scarcely time for meals. The very period of sleep, so necessary to the young, is too often abridged. Nay, children are sometimes worked even in the night! Human beings thus decay before they arrive at the term of maturity." He observes elsewhere, "that this system has grown up by a series of encroachments upon the poor children; that the benevolent masters are not able to rectify these abuses. A legislative enactment is the alone remedy for this as well as the other great opprobrium of our manufactures—the improper employment of children." Such are the opinions of this medical gentleman upon this subject, written long before the present bill was before the House; and founded upon daily observation and experience.

I might add the opinion of another very excellent practitioner of the same place, Mr. Smith, respecting the cruelty of the present system, and the misery and decrepitude which it inflicts upon its victims; but his opinions, given with great force and ability, have, I think, been already widely disseminated by means of the press. The other surgeons of the Leeds Infirmary—all men of great professional eminence—entertain, I believe, precisely similar opinions. One of them, Mr. Hey, a name that at once commands the highest respect in every medical society of this country, or indeed of Europe, presided, as mayor of Leeds, at an immensely numerous meeting of the inhabitants of that borough, when a petition from that place, in favour of the bill, was unanimously agreed to, and afterwards received the signatures of between 18,000 and 20,000 persons.

In silk and worsted mills, and especially in the former, the nature of the employment may be less prejudicial in itself; but then its duration is often more protracted, and it falls in a larger proportion upon females and young children. In many spun-silk mills, in which a different operation from that of silk-throwing—and one conducted upon Arkwright's principle—is carried on, the practice of working children at a very tender age, and often all night, prevails. In some of these, I am informed, they commence at one o'clock on the Monday morning, and leave off at eleven on Saturday night; thus delicately avoiding the Sabbath, indeed, but rendering its profitable observance, either for improvement, instruction, or worship, an utter impossibility.

In the worsted mills, the greatest irregularities, as to the hours of working, have existed, and therefore occasional oppression, in these departments, has long prevailed. Let the following extract suffice, from a document drawn up by a gentleman in this branch of business, Mr. Wood,—to mention whose name is to kindle at once the most enthusiastic feelings in the bosoms of the honest operatives of the north, and to whom is due the honour of originating and supporting this attempt to regulate the labour of children; and who, while he has conducted his own manufacture with the greatest humanity and kindness, has still earnestly sought to ameliorate the general condition of the labouring poor. This gentleman gives the ages of 475 persons, principally females, employed at a worsted-mill, which, it appears, average about the age of thirteen; and adds—"Children of these years are obliged to be

Speech of M. T. Sadler, Esq.

at the factories, winter and summer, by six in the morning, and to remain there till seven in the evening, with but one brief interval of thirty minutes, every day except Saturday, ceasing work on that day, in some factories, at half-past five, in others at six or seven P.M. Not unfrequently this labour is extended till eight or nine at night—fifteen hours—having but the same interval for meals, rest, or recreation: nay, such is the steady growth of this overworking system, that children have been confined in the factory from six in the morning till eight at night—fourteen hours continuously, without any time being allowed for meals, rest, or recreation; the meals to be taken while attending the machines, and this the practice of years."

This picture, sufficiently appalling, has also to be darkened by the addition of frequent night-labour. Such is the practice at Bradford and the neighbourhood. But to shew that the evils are not confined to any particular neighbourhood, and that they prevail wherever unprotected children are the principal labourers of the community, I shall next advert to their treatment in the flannel manufactories in the Principality of Wales. I quote the following account, which I have received from the most respectable quarter:—

"With certain fluctuations in the degree of labour, resulting from the difference in the demand of manufactured goods, the children here work twenty-four hours every other day, out of which they are allowed three hours only for meals, &c. When trade is particularly brisk, the elder children work from six in the morning till seven in the evening, two hours being allowed for meals, &c., and every other night they work all night, which is still a more severe case: for this additional night-labour they receive five-pence. There is another lamentable circumstance attending the employment of these poor children, which is that they are left the whole of the night alone; the sexes indiscriminately mixed together; consequently you may imagine that the depravity of our work-people is indeed very great. The adults are employed in feeding the engines. Independent of moral considerations, the accidents that occur to these poor little creatures are really dreadful; the numbers of persons to be seen with mutilated and amputated limbs are quite distressing, and this will ever be the case till some better regulation is carried into effect. There is not a single place of charitable education for a population of about 8,000 souls, beyond a Sunday school."

As to woollen mills, they are not, generally speaking, injurious to health; though such is the case in certain departments of them, especially since the introduction of the rotatory machines. Here I might argue that the lightness of the labour, which is the reason usually urged against an interference with excessive hours, no longer applies, as in woollen mills the labour is, in general, much more strenuous than that in most of the before-mentioned factories. But I disdain to avail myself of an argument, however plausible, which I believe to be fallacious, and I will here observe, once for all, that it is not so much the degree of labour which is injurious to these work-children (how revolting the compound sounds!—it is not yet admitted, I think, into our language; I trust it will never be familiarized to our feelings);—I say, it is not so much the degree, as the duration of their labour, that is so cruel and destructive to these poor work-children. It is the wearisome uniformity of the employment,—the constrained positions in which it is pursued,—and, above all, the constant and close confinement, which are more fatiguing, to the body as well as mind, than more varied and voluntary, though far stronger, exertion. I dwell upon this point, because it is the sole possible plea for the long and imprisoning hours of the present laborious system; though, when properly considered, it is one of the most powerful arguments against it. Light labour! Is the labour of holding this pen, and of writing with it, strenuous? And yet, ask a clerk in any of the public offices, or in any private counting-house, when he has been at his employment some half-dozen hours in the day less than one of these children, whether he does not think that he has had enough of this light labour—to say nothing of the holidays, of which he has many, and the child none. Ask the recruit, recent from the plough, whether an hour of his light exertion is not more fatiguing than any three he ever endured in the fields. Ask his experienced officer how long he can subject even the veteran to this sort of slight but constrained exertion, though in the open air, with impunity. I might appeal to the chair, whether the lingering hours which have to be endured here, though unaccompanied with any bodily exertion whatever, are not "weariness to the flesh." But what would be the feelings of the youngest and most active individual amongst us, if, for example, he were compelled to pace that table, engaged in some constant and anxious employment, stunned with the noise of revolving wheels, suffocated with the heat and stench of a low, crowded, and gas-lighted apartment, bathed in sweat, and stimulated by the scourge of an inexorable task-master? I say, what would be his ideas of the light labour of twelve or fourteen hours in such a pursuit, and when, once or twice in every week, the night also was added to such a day? And how would he feel, if long years of such light labour lay before him? If he be a parent, let him imagine the child of his bosom in that situation, and then judge of the children of thousands who are as dear to the Universal Parent as are his own to him! Let him think of his own childhood, and he will then remember that this light labour is the fatigue of youth, and that strenuous exertion, when the buoyant spirit exercises the entire frame, is its sport. I might quote authorities on this subject, but it is unnecessary. Common sense and common feeling at once decide the point, and confute this disgusting plea of tyranny for the captivity of youth. Hence the late Sir Robert Peel, in bringing forward his last measure, emphatically observed, that "it was not so much the hardship as the duration of labour, which had caused the mischievous effects on the rising generation." But if, after all, honourable members choose to argue the question on different grounds, and wish to establish a variation in the duration of the labour of children in mills and factories, in reference to the nature of the employment,—be it so. Confident in my own mind that the bill proposes the utmost limit which the youthful constitution can safely bear, in any pursuit, or under any circumstances, I can have no objection to that period being abridged in the more pernicious and strenuous employments of the country.

Speech of M. T. Sadler, Esq.

I shall not attempt at present to give any precise account of the length of labour generally borne in different mills and factories; it varies according to the humanity of the employer, and the demand for his goods at particular seasons. But let me here remark, that these variations constitute one of the main reasons for a legislative protection; otherwise the humane masters will be driven out of the trade: for these, it is quite clear, cannot control others less feelingly disposed. They are, indeed, in the present state of things, as little free agents as the children whom they employ; and, moreover, the want of a due regulation throws the effects of those fluctuations to which trade and manufactures are subject, in an undue and distressing degree upon those who are the least able to sustain their effects. Thus, if the demand and profit of the employer increase, the labour of the operatives, most of whom are children, augments, till many of them are literally worked to death: if that demand diminish, the children are thrown partially or wholly out of work, and left to beggary and the parish. So that their labour, averaged throughout the year, as some mill-owners I perceive have calculated its duration, does not appear so excessive. For, at the very moment that a strenuous opposition is being made against the curtailment of infantile labour, the masters themselves, in certain flax-mills in the North, have curtailed it to some purpose—having, if I am not misinformed, diminished the employment in some mills, and shut up others entirely. And I have no doubt but that, at this particular moment, abundance of evidence might be adduced before a select committee to shew that the hours mentioned in the bill are observed, and indeed a much stricter limitation enforced. But then if it be right that the owners should be allowed to throw out of employment all these children at a few days' notice, is it proper that they should be permitted to work them for an unlimited number of hours, the moment it suits their purpose? If the effect of this bill were, in some measure to equalize the labour of these poor children, and thereby prevent those fluctuations which are so distressing to them in both its extremes, it would so far accomplish a most beneficent object. It might, I think, transfer a little of the fluctuation from the factory to the stock-room, with great advantage to the operatives, and consequently to the public at large.

It is impossible to furnish any uniform account of the hours of labour endured by children in these factories, and I am unwilling to represent extreme cases as general ones, although it is the bounden duty of Parliament to provide against such, as it does, for example, with respect to atrocious crimes, which are extreme cases in civilized society. I shall, therefore, only give one or two instances of the extent of oppression to which the system is occasionally carried. The following were the hours of labour imposed upon the children employed in a factory at Leeds last summer:—On Monday morning, work commenced at six o'clock: at nine, half an hour for breakfast; from half-past nine till twelve, work. Dinner, one hour; from one till half-past four, work. Afternoon meal, half an hour; from five till eight, work: rest for half an hour. From half-past eight till twelve (midnight), work: an hour's rest. From one in the morning till five, work: half an hour's rest. From half-past five till nine, work: breakfast. From half-past nine till twelve, work: dinner; from one till half-past four, work. Rest half an hour; and work again from five till nine o'clock on Tuesday evening, when the labour terminated, and the gang of adult and infant slaves was dismissed for the night, after having toiled thirty-nine hours, with brief intervals (amounting to only six hours in the whole) for refreshment, but none for sleep. On Wednesday and Thursday, day work only. From Friday morning till Saturday night, the same prolonged labour repeated, with intermissions, as on Monday, Monday night, and Tuesday; except that the labour of the last day closed at five.—The ensuing day, Sunday, must, under such circumstances, be a day of stupor, to rouse the children from which would only be to continue their physical sufferings, without the possibility of compensating them with any moral good. Clergymen, Sunday-school masters, and other benevolent persons, are beginning to feel this to be the case; physicians, I find, have long observed it; and parents, wishful as they are that their offspring should have some little instruction, are yet more anxious that they should have rest. Sunday schools have long been rendered appendages to the manufacturing system, which has necessarily emptied the day-schools of the poor wherever that system prevails: not content with monopolizing the whole week with protracted labour, the Sabbath itself is thus rendered a day of languor and exhaustion, in which it is impossible that due instruction can be received, or the solemn duties which religion enjoins duly performed; in fact, it is a mere fallow for the worn-out frame, in order that it may be able to produce another series of exhausting crops of human labour. If some limits, therefore, are not prescribed to these constant and cruel encroachments, our labouring population will become, ere long, imbruted with ignorance, as well as enslaved by excessive toil.

I now proceed to shew the physical and moral consequences of this dreadful system; and on this important part of the subject, as I am aware that I shall be at issue with its supporters and apologists, I shall appeal to authorities which none will be disposed lightly to dispute; and to facts—decisive, as I think, of the whole question—facts which I challenge them to controvert or evade. The authorities to which I allude are such as the late Doctor Baillie, Sir Astley Cooper, Sir Gilbert Blaine, Doctor Pemberton, Sir Anthony Carlisle, Sir George Tuthill, and many other physicians and surgeons of the highest eminence and celebrity throughout the profession, especially for their physiological science. The deliberate opinions of these distinguished persons on the subject under our consideration are contained in these volumes (minutes of evidence before former parliamentary committees). Time would fail were I now to read them, or even to make selections. They ought, however, to have been carefully consulted by certain honourable members before they had cried out for another select committee, ignorant, I would fain hope, of the affecting evidence published by preceding ones. Let it suffice that these high authorities are strong and unanimous against the system of early and protracted labour for children and young persons. I appeal to the whole of them in favour of this bill, or rather indeed of far more binding limitations than it proposes; for when I advert to their deliberate declarations, I feel a growing dissatisfaction at its provisions. Other medical authorities of great eminence have since then appeared, whose views I know, in many instances, to be, if possible, still more marked and decided upon this important point; and can it, I would ask the house, be a question whether we should tolerate a degree of infantile labour, which our highest medical authorities assure us the human frame is utterly incapable of sustaining with impunity?

If it be objected that these individuals, however great and distinguished, had no practical knowledge of the factory-system and its effects, I will turn to another description of evidence,—namely, to that of professional gentlemen practically acquainted with it, and long residing in its very seat and centre—Manchester. The first is a name equally dear to philosophy and philanthropy,—long at the head of the profession in that part of the empire—Dr. Perceval. He saw the rise, progress, and effects of the system, and closely connected as he was with many who were making rapid fortunes by it, still he expressed himself upon the subject, as a professional man and a patriot, in terms of the strongest indignation. He says, even of the large factories, which some suppose need little regulation, that they "are generally injurious to the constitution of those employed in them, even when no particular diseases prevail, from the close confinement which is enjoined, from the debilitating effects of hot or impure air, and from the want of the active exercises which nature points out as essential to childhood and youth, to invigorate the system and to fit our species for the employment and the duties of manhood. The untimely labour of the night, and the protracted labour of the day, with respect to children, not only tend to diminish future expectation as to the general sum of life and industry, by impairing the strength and destroying the vital stamina of the rising generation, but it too often gives encouragement to idleness, extravagance, and profligacy, in the parents, who, contrary to the order of nature, subsist by the oppression of their offspring." He goes on to deplore the impediments which the system throws in the way of education, and asserts the necessity of establishing "*a general system of laws for the wise, humane, and equal government of all such works.*" I regret that time will not permit me to quote him more

Speech of M. T. Sadler, Esq.

at large; but who, that is acquainted with general literature or philosophy, can be ignorant of the writings of Perceval of Manchester?

I will refer the House to another authority, belonging to a different branch of the same profession, and scarcely a less celebrated one,—I mean the late Mr. Simmons. After nearly thirty years' experience in the General Infirmary at Manchester, and being also at the head of other charitable institutions in that town connected with his profession, few men, I should conceive, were more competent to speak as to the effects produced by the factory system than himself. I must again deeply regret that I cannot quote his opinions at length. His description of the consequences of this species of overexertion is most appalling; and he adds these emphatic words:—"I am convinced that the hours of employment are too long to endure at *any age*." Speaking of the evils of the system, he says, "I shudder at contemplating them!"

I might multiply these authentic and affecting testimonies to almost any extent. I will, however, present, in as few words as possible, the effects, as described by medical men, of these long hours of confinement, without sufficient intervals for meals, recreation, and rest, and continued often through the night, in rooms artificially heated, and lit by gas; the atmosphere being otherwise so polluted and offensive as to render respiration painful, even for a few minutes. They describe the consequences to be in many cases languor and debility, sickness, loss of appetite, pulmonary complaints, such as difficulty of breathing, coughs, asthmas, and consumption; struma, the endemia of the factory, and other chronic diseases; while, if these more distressing effects are not produced, the muscular power is enfeebled, the growth impeded, and life greatly abridged. Deformity is also a common and distressing result of this overstrained and too early labour. The bones, in which the animal, in contra-distinction to the earthy, matter is known to prevail in early life, are then pliable, and often cannot sustain the superincumbent weight of the body for so many hours without injury. Hence, those of the leg become bent; the arch of the foot, which is composed of several bones of a wedge-like form, is pressed downwards, and its elasticity destroyed; from which arises that disease in the foot only lately described, but common in factory districts. The spine is often greatly affected, and its processes irregularly protruded, by which great deformity is occasioned. The ligaments, also, fail by over pressure and tension. Hence the hinge-joints, of which they are the main support, such as those of the knee and the ankle, are overstrained, producing the deformity called knock-knees and lame ankles, so exceedingly common in mills. Thus are numbers of children distorted and crippled in early life, and frequently rendered incapable of any active exertion during the rest of their days. To this catalogue of sufferings must be added, mutilation of limbs, or loss of life, by frequent accidents. The overworking of these children occasions a weariness and lethargy which it is impossible always to resist; hence, drowsy and exhausted, the poor creatures fall too often among the machinery, which is not, in many instances, sufficiently sheathed, when their muscles are lacerated, their bones broken, or their limbs torn off, in which cases they are constantly sent to the infirmaries to be cured, and, if crippled for life, they are turned out and maintained at the public cost; or they are sometimes killed upon the spot. I have myself known, in more instances than one, the arm torn off,—in one horrible case both; and a poor girl now exists upon a charitable subscription, who met with that dreadful accident at one of the flax-mills in my neighbourhood. In another factory, and that recently, the mangled limbs of a boy were sent home to his mother, unprepared for the appalling spectacle. I will not describe the result. It is true that a great majority of these accidents are of a less serious nature, but the admission-books of the infirmaries in any manufacturing district will shew the number; and their accounts of the expense of buying irons to support the bending legs of the young children who become crippled by long standing in the mills, will also prove the tendency of over-confinement and early labour to produce deformity. Dr. Ashton and Surgeon Graham, who examined six mills in Stockport, in which 824 persons were employed, principally children, have reported the result individually, and the list seems rather that of an hospital than a workshop. The particulars are deeply affecting, but I must only give the totals. Of 824 persons, 183 only were pronounced healthy; 240 were stated to be delicate; 258 unhealthy; 43 very much stunted; 100 with enlarged ankles and knees; and among the whole there were 37 cases of distortion. The accidents by machinery are not, I think, noticed; but I find that Dr. Winstanley, one of the physicians of the Manchester Infirmary, on examining 106 children in a Sunday-school, discovered that no less than 47 of them had suffered accidents from this one cause. I have this morning received, from one of the most eminent surgeons of this metropolis, a letter, in which he informs me, that on making a tour through the manufacturing districts some years ago, he was painfully struck with the numerous cases of mutilation which he observed, and which he attributed to this long and wearying system of labour in mills and factories. Of the mortality which this system occasions, I shall speak hereafter.

Can anything, then, darken the picture which I have hastily drawn, or, rather, which others, infinitely more competent to the task, have strikingly portrayed? Yes, Sir, and that remains to be added which renders it the most disgusting as well as distressing system which ever put human feelings to to utmost test of endurance. It has the universally-recognised brand and test of barbarism as well as cruelty upon it. It is the feebler sex principally on which this enormous wrong is perpetrated. Female children must be laboured to the utmost extent of their physical powers, and indeed frequently far beyond them. Need I state the peculiar hardships, the disgusting cruelty, which this involves? I speak not, poor things, of the loss of their beauty,—of the greater physical sufferings to which their sex exposes them. But, again taking with me the highest medical authorities, I refer to the consequences of early and immoderate labour; especially at the period when the system rapidly attains its full development, and is peculiarly susceptible of permanent injury. Still more are the effects felt when they become mothers, for which, I fear, their previous pursuits have little qualified them. It is in evidence, that long standing has a known tendency—how shall I express it?—*contrahere et minuere pelvem*,—and thereby to increase greatly the danger and difficulty of parturition, rendering embryotomy—one of the most distressing operations which a surgeon ever has to perform—occasionally necessary. I have communications upon this subject from persons of grêat professional experience; but still I prefer to appeal to evidence before the public; and one reference shall suffice. Dr. Jones, who had practised in the neighbourhood of certain mills, in favour of which much evidence was adduced, which indeed it is rarely difficult to procure, states, that in the "eight or ten years during which he was an accoucher, he met with more cases requiring the aid of instruments (that circumstance shewing them to be bad ones) than a gentleman of great practice in Birmingham, to whom he was previously a pupil, had met with in the whole course of his life." Abundance of evidence to the same effect is before me. But I forbear. I confess, therefore, that I feel my indignation roused when I see papers put forth in which it is stated, as a recommendation forsooth of the present system, and as a reason why it should by no means be regulated, that in certain mills girls are principally employed. This is a matter of ex-

Speech of M. T. Sadler, Esq.

ultation! I would ask those who so regard it, in the language of the poet, "Art thou of woman born, and feel'st no shame?"

Nor are the mental, any more than the physical, sufferings of these poor young creatures to be overlooked. In the very morning of life, when their little hearts yearn within them for some relaxation and amusement, to be thus taken captive, and debarred the sports of youth, is almost as great—nay, a greater cruelty than to inflict upon them, thus early, the toil of advanced life. Their fate, alas! reverses the patriarch's pathetic exclamation, and their infant days are labour and sorrow. I perceive that I excite the risibility of an honourable gentleman opposite. What there is to smile at in these just representations of infantile sufferings, I am really at a loss to imagine. I will venture, however, to give him and the House a few more of these amusing facts before I have done with the subject.

It may be thought almost impossible that children should be assembled so early, and dismissed so late, and still kept through the whole period in a state of active exertion. I will attempt to explain this. First, then, their early and punctual attendance is enforced by fines, as are many other regulations of a very severe character; so that a child may lose a considerable part of its wages by being a few minutes too late in the morning: that they should not leave too soon is very sufficiently provided against. Now, this extreme punctuality is no slight aggravation of the sufferings of the child. It is not in one case out of ten, perhaps, that the parent has a clock; and as nature is not very wakeful in a short night's rest, after a long day's labour, the child, to ensure punctuality, must be often roused much too early. Whoever has lived in a manufacturing town, must have heard, if he happened to be awake many hours before light on a winter's morning, the patter of little pattens on the pavement, continuing perhaps for half an hour together, though the time appointed for assembling was the same. Even then the child is not always safe, however punctual; for, in some mills, two descriptions of clocks are kept, and it is easy to guess how they are occasionally managed. So much for the system of fines, by which, I am told, some mill-owners have boasted that they have made large sums annually.

Then, in order to keep the children awake, and to stimulate their exertions, means are made use of, to which I shall now advert, as a last instance of the degradation to which this system has reduced the manufacturing operatives of this country. Sir, children are beaten with thongs prepared for the purpose. Yes, the females of this country, no matter whether children or grown up,—I hardly know which is the more disgusting outrage,—are beaten upon the face, arms, and bosom, beaten in your free market of labour, as you term it, like slaves. These are the instruments.—[*Here the honourable member exhibited some black, heavy, leathern thongs,—one of them fixed in a sort of handle, the smack of which, when struck upon the table, resounded through the house.*]—They are quite equal to breaking an arm, but that the bones of the young are, as I have before said, pliant. The marks, however, of the thong are long visible; and the poor wretch is flogged before its companions; flogged, I say, like a dog, by the tyrant overlooker. We speak with execration of the cart-whip of the West Indies—but let us see this night an equal feeling rise against the factory-thong of England. Is it necessary that we should inquire, by means of a select committee, whether this practice is to be put down? and whether females in England shall be still flogged to their labour? Sir, I should wish to propose an additional clause in this bill, enacting that the overseer who dares to lay the lash on the almost naked body of the child, shall be sentenced to the treadwheel for a month; and it would be but right if the master who knowingly tolerates the infliction of this cruelty on abused infancy, this insult upon parental feeling, this disgrace upon the national character, should bear him company, though he roll to the house of correction in his chariot.

Speech of M. T. Sadler, Esq.

But the entire system, as now conducted, has not merely its defenders, but its eulogists. Hence, in a celebrated Review in which its rise and progress are discussed, we find this opinion delivered with the utmost confidence:—"We scruple not to say that the health, morals, and intelligence of the population have all gained by the establishment of the present system." That this improvement ought to have been the result I have already said: that it will be so I confidently hope; but it can only be the case by adopting regulations of the nature now proposed. Health, morals, and mental improvement, can never consist with constant confinement and excessive labour imposed upon any class of the community, especially upon its youth. That the general intelligence, however, has increased, I fully believe; so it would have done under any system; so it has in almost every country, whether manufacturing or otherwise: and I claim this increasing intelligence in favour of the regulation which I propose, for it has declared in behalf of this measure with a force and unanimity rarely known on any other subject. But as to public morals, alas! what has been gained by this excessive slavery of the juvenile part of the manufacturing population? It is during the present century, as the article in question shews, that the system has so greatly increased. The number of the criminal committals has been annually furnished since the year 1805; let us, then, advert to those important returns for proof upon this subject. In the metropolitan county, including London, in which, if anywhere, the number is likely to be excessive, and the increase great, the average of the first three years was 1192, that of the three last 3491, nearly three-fold! But in Lancashire (including its rural hundreds) the average of the same period had increased from 369 to 2088, nearly six-fold! The mean proportion of the committals of England, exclusive of these two counties, being now about one in 1255, that of Lancashire as one in about 550 annually. Where is this to end?

As to that species of immorality not cognizable by law, there are no means of obtaining equally precise information, otherwise I fear there would be found in this respect as little ground for exultation. Not to mention minor offences, the practice of tippling and drunkenness has astonishingly increased, and has been accompanied by a revolting indecency not formerly known among us, women and children now publicly indulging in this vice: such are the degrading effects of the system. The great increase of debauchery of another kind it would be absurd to deny; I never did hear it denied, that many of the mills, at least those in which night-working is pursued, are, in this respect, little better than brothels. The science of human physiology has been, I may say, disgustingly advanced, having been able to demonstrate how extremely near the confines of actual childhood the human female may become an unhappy mother, from the disgraceful scenes which have occurred in some of these mills and factories. Indeed, it is in evidence, on the authority of medical men conversant with that state of society, that the period of puberty is unnaturally anticipated, But not to dwell upon the effects of this precocity, I will proceed to consider the alleged improvement in health. I shall determine this important point, by referring to the place selected by the strenuous advocates of the system, and its principal seat, Manchester, the surprising longevity of that town having been over and over again asserted in proof of the incalculable advantages of the factory-system as now conducted.

Speaking of the general health of that town, I of course refer to that of the operatives, and not to that of the higher and more opulent ranks: and, in determining this, I shall still refer to medical authority and statistical facts, and not to loose and unfounded opinions, by which the public has been too long imposed upon and misled. I take, then, a tract just published, entitled "Remarks on the Health of English Manufacturers, and on the need which exists for establishing for them convalescent retreats." It is written by Mr. Roberton, a gentleman well known in the medical world, and author of previous works of deserved celebrity. He rebuts Mr. Senior's assumption, founded on a series of gross mistakes, as to the great improvement which has taken place in our manufacturing population, and says of these so-much-improved operatives, that "the nature of their present employment renders existence itself, in thousands of instances, in every great town, one long disease." He states, regarding Manchester, that "during the last year, 1830, the patients admitted at the four great dispensaries amounted to 22,626," independently of those assisted by other charitable institutions, such as the Infirmary, &c, amounting in all to at least 10,000 more. To this he adds other calculations, which bring him to the conclusion, that "not fewer, perhaps, than three-fourths of the inhabitants of Manchester annually are, or fancy they are, under the necessity of submitting to medical treatment." He describes at some length, and with great force and feeling, the evils of the factory-system, and attributes to it, to use his own words, "the astounding inebriety" of the population, many of whom have recourse, after long and exhausting toil, to that means of kindling a temporary sense of vigour and comfort. He states the lamentable effects, in other points of view, which are thus produced, and the want of moral and mental improvement with which they are necessarily accompanied. He says, that the present manufacturing system "has not produced a healthy population, neither one well instructed and provident, but one, on the contrary, where there exists always considerable, and sometimes general poverty, and an extraordinary amount of petty crime,—that, in several respects, they are in a less healthy and a worse condition than at any period within the last two centuries."

I will give an appalling proof of this general misery and degradation. It appears that, during the last year, there were delivered by the lying-in charity of Manchester no less than 4562 poor married women; far more than half, therefore of the mothers of Manchester are assisted by public charity; in a word, nearly three-fifths of the children of that town are branded with the stigma of pauperism at their very birth.

If it be argued that this institution is too indiscriminate in its charity, and consequently that its operations afford no just indication of the extent of the distress actually endured, I will again quote, in reply, the authority of Mr. Roberton:—"An overwhelming majority of the persons so relieved are in a state of incredible destitution!" I proceed to prove his assertions by still stronger facts.

The main, and, as it has been hitherto held, triumphant defence of the present system of excessive infantile labour has been placed upon the assumed longevity of Manchester; and, had what has been asserted in this respect been true, or at all approaching to the truth, the argument would have been doubtless settled in its favour. Thus we find it stated, over and over again, that the mortality which had kept diminishing for half a century, had, in 1811, fallen as low as one in seventy-four; and that the proportion in 1821 was still smaller. It is asserted, I see, in a petition from the mill-owners of Keighley against this bill, that this proportion is one in fifty-eight, while that of Middlesex is one in twenty-six. I am glad the opponents of the bill have given this sample of their intelligence. But to return. It has been long remarked by statistical writers, that every community to which large numbers of immigrants, principally in the active period of life, are constantly added, will exhibit a corresponding diminution in the proportion of deaths, without that circumstance at all proving any real increase in the general health and longevity of the place. But the above proportions were, nevertheless, so extravagant, and the argument founded upon them so important, that I determined to give the subject the most careful and impartial examination. I have done so; and these are the results. Taking the whole parish of Manchester,—and so far, therefore, doing great injustice to my argument, as that parish contains, I think, nearly thirty townships and chapelries, some of which are principally agricultural,—but, taking the whole parish, I find that, in the collegiate church there, and in six other churches,—in the two churches of Salford, in those of Charltonrow, now part of the town, and in the eleven chapelries, including the Roman Catholic and other dissenting burial-grounds, there were interred between the years 1821 and 1830 inclusive, 59,377 individuals. The mean population of the whole parish (i. e. the geometric mean, in order to be as exact as possible) was, during the same period, 228,951. Now, the number of burials is defective, one church (that of Peter's) being omitted, and I think other burial-grounds also; but does the number actually returned give a proportion of 1 in 74? or even as the Keighley petitioners reckon, 1 in 58? No, Sir, it gives a proportion of 1 in $37\frac{9}{10}$ths, as the annual mortality of the extended district included in the entire parish of Manchester! In Salford the number of deaths during the same term was 996, the mean population having been 32,421, or 1 death in every $32\frac{1}{2}$: and this in a population, let me again repeat, increasing immensely by immigration.

Speech of M. T. Sadler, Esq.

But a further calculation has to be made before the subject under consideration can be properly understood. A vast excess of this mortality, we may be assured, rests upon the poor; for nobody disputes that the longevity of the wealthier classes has, in the mean time, greatly improved. Thus, in Paris, a large and unhealthy city, where the mortality, however, is, I think, less than 1 in 42, Dr. Villermé found, that in the first arrondissement, where the wealthier inhabitants principally reside, only 1 in 50 died annually; while, in the twelfth, principally inhabited by the poor, the proportion was as great as 1 in 24. Apply this to the mortality of Manchester, and then let us hear what can be said respecting the longevity of the poor manufacturers of that place. It proves all that has been advanced concerning the effects of infantile and long-protracted labour, which not only enfeebles, but sweeps to their untimely fate, so vast a proportion of the population.

But, on a point of such paramount importance, and so entirely decisive of the argument, no species of evidence ought to be wanting. I have therefore examined the last census of Manchester with great care,—I mean that part of it in which the registered burials are given, together with the ages of the interred; and I have compared the proportion of those buried under the age of forty, and those buried above that age, with the corresponding interments of the immensely larger cities of London and Paris, taking the last ten years in the former instance, and one intermediate year in the two latter, without any selection whatsoever. These are the results:—to every 100,000 interments in each of these places under forty, there would be above that age in London, 63,666; in Paris, 65,109; in Manchester, 47,291 only; in other words, 16,375 fewer would have survived that period in Manchester than in London, and 17,818 fewer than in Paris. Can anything, then, be more true than the complaint of the operative spinners, that few of them survive forty; and where is the man that dares to oppose the effectual regulation of so murderous a system?

In the census of 1821, the population of England was generally given according to the different ages of the people. Manchester, however, furnished no such information; otherwise, I am persuaded, another argument might have been adduced demonstrative of the same melancholy fact. But I have examined the census of the hundred of Salford and of Macclesfield, which includes the towns of Macclesfield and Stockport, containing a great number of mills, and I have compared them with other places; and, taking the

M. T. Sadler, Esq.

number of children under five, and the numbers which arrived at the different divisions, and especially at the more advanced periods of life, I find that wherever the present system prevails the most melancholy waste of human existence is clearly demonstrable. I may also observe, that calculating the mean duration of life from mortuary registers, it is, in the metropolis of England, about 32 years; in Paris, 31; in Manchester, $24\frac{7}{10}$ years only! In other towns, where the same system prevails, it is still less: thus in Stockport it is 22 years only, that town not having increased by immigration quite so rapidly as Manchester. A comparison with Liverpool in this respect would be manifestly inconclusive, so large a number of the adult inhabitants of that great sea-port being constantly absent on maritime pursuits.

Other calculations I hold in my hand, but I will not weary the House with giving them in detail. I cannot, however, refrain from presenting the results of one of them, as it disposes of the confident assertion of the improvement of the manufacturing community. In 1780, the celebrated Dr. Heysham enumerated the population of Carlisle with great care, and separated the individuals into classes according to their ages. In 1821, a similar enumeration took place. In the mean time, the great discovery of vaccination had been made,—of such immense importance in these calculations; but it will be seen that even that, and all the acknowledged improvements in medical science, fail to compensate, in the amount of human lives, for the baneful effects of our manufacturing system as at present pursued? Calculating on the first division of the population, namely, the children under five years of age, and assuming them to be 1000, there were, in 1780, from five years old to twenty, 2229; in 1821, 2107; between twenty and forty, in the former period, 2143; in the latter, 1904; above forty, in 1780, 2084; but, in 1821, 1455 only!

Again, between the years 1779 and 1787 inclusive, Dr. Heysham gives the number of interments in Carlisle; they amounted to 1840, of which 1164 were of persons under forty years of age, and the remaining 676 above that period of life. On examining the census, I find that between the years 1821 and 1830 inclusive, there have been buried in the same place 3025 under forty, and 1273 above that age. In the former period, therefore, there would be to every 10,000 deaths under the age of forty, 5808 above that age; whereas, in the latter, the proportion has been 4208 only; showing how much smaller a number survives that age than formerly. The only way to evade this conclusion, is to suppose that the population has advanced with far greater rapidity since the former period; but this supposition would imply gross ignorance of the facts. The population of Carlisle was enumerated in 1764, and it is clear that it had been increasing even more rapidly before 1780 than it has done since. But in Carlisle, as everywhere else, the greatest proportion of mortality falls upon the poor, the expectation of life, as far as regards the upper classes of society, having evidently increased. What then becomes of the statements of the great improvement among our manufacturing poor with facts like these before our eyes?

I could multiply these proofs of the effects of the system to a great extent; but time will not permit, nor can it be necessary. I have taken up the challenge regarding the effects of the system as now pursued, given by its eulogists, and have contested the cause of humanity—as I trust I may call it —in the very arena, and with the weapons, which themselves have chosen,— the prosperity, health, and longevity, of the operatives of Manchester. I have shewn that the same results accompany the same system wherever pursued, namely, slavery, profligacy, crime, disease, and death. Transfer the system to the whole country, and then contemplate its effects; those effects are seen rapidly developing themselves as it advances. Infantile labour leads to premature marriage, which crowds the generations upon each other, and this

Speech of M. T Sadler, Esq.

circumstance, together with the great discoveries in medicine, may have increased the numbers of the people; but as far as the system has prevailed it has diminished the relative number of the athletic and active, and given us in their stead a weak, stunted, and degenerate race; and thus lessened the proportion of those who have to bear the burdens, and fight the battles, of the country. On this latter point I am furnished with some striking facts, and among the rest, the comparative difficulty of passing recruits wherever this system prevails, but such evidence is unnecessary; I would rather point to its more obvious and general effects. Look, then, I say, at the miserable condition of the feeble beings which are at once its instruments and its victims. Supposing that it has augmented our numbers, "it has," in the emphatic language of the Sacred Volume, "multiplied the people, and not increased the joy." I invoke, therefore, the justice, the humanity, and the patriotism of this House: the feelings of the country are already roused. I call upon Parliament to assist, by this measure, to lighten the load of an oppressed people, which bows them and their very children to the dust;—I call upon it to snatch the scourge from the taskmasters of the country, and to break the bonds of infant slavery.

The principal features of this bill for regulating the labour of children and other young persons in mills and factories are these:—First, to prohibit the labour of infants therein under the age of nine years; to limit the actual work, from nine to eighteen years of age, to ten hours daily, exclusive of the time allowed for meals and refreshment, with an abatement of two hours on the Saturday, as a necessary preparation for the Sabbath; and to forbid all night-work under the age of twenty-one.

In this bill I have omitted many important provisions which I had intended to insert, in order to obviate, if possible, multiplied objections, and to secure the attainment of its main object. Thus, I had drawn up a clause subjecting the mill-owner or occupier to a heavy fine when any serious accident occurred, in consequence of any negligence in not properly sheathing or defending the machinery. I had intended to propose a remission of an hour from each day's labour for children under fourteen, or otherwise of six hours on one day in every week, for the purpose of affording those who are thus early and unnaturally forced into the market of labour some opportunity of receiving the rudiments of instruction and education, the expense of which, upon the modern system, would have been comparatively nothing, especially if shared between the mill-owner and the public. Above all, I had contemplated a clause putting down night-labour altogether. None of these propositions, I think, are half so extravagant, if duly considered, as the demands now made upon infantile labour, involving, as they too often do, the sacrifice of happiness, health, improvement—nay, life itself. But not to endanger the principal object which I have in view, and regarding the present attempt as the commencement only of a series of measures in behalf of the industrious classes, all I propose to the House on the present occasion is, the remission of labour to the extent already explained.

And, first, as to the period of life at which this bill permits children to be worked in factories—namely, nine years old. I will only observe that our ancient statutes—not always peculiarly favouring the condition of poverty—have not been neglectful of this matter. The 23d of Edward III., if I mistake not, assumes that a male child under fourteen is *non potens in corpore;* and the same of a female child under twelve. In the fifth of Elizabeth, however, I think that period was fixed at twelve years for both sexes, previously to which they were deemed *non potentes in corpore;* and I may further observe, that another humane provision of certain of these statutes was the hiring of these young persons by the year, so that they might not be turned adrift on every fluctuation in the demand for their labour, nor even discharged in sickness, at the pleasure or convenience of their employers. The late Sir

Speech of M. T. Sadler, Esq.

Robert Peel's bill originally fixed upon the age of ten, which was ultimately reduced to that of nine, where the present measure leaves it. Where is the individual, if disinterested and humane, that will contend that this is not early enough for these poor creatures to commence their career of labour and sorrow?—Then, as to the daily duration of their labour before they arrive at years of discretion, namely, these ten long hours, which, with the necessary intervals, stretch the term to at least twelve, is not this sufficiently severe? And at a period when the professions of humanity are so much louder than formerly,—when the penal code of crime has been revised so much, and mitigated so often—that there should be no amelioration proposed in the penal code (for such I am sure I may term it) of infantile labour seems indeed strange. The term proposed is that fixed upon by that truly benevolent and enlightened individual just alluded to. His opinion, on introducing his last measure in 1815, I quote from the *Parliamentary Register* of that period:—"What he (Sir Robert Peel) was disposed to recommend was, that no child should be so employed under the age of ten years, and the duration of their labour to be limited to twelve hours and a half *per diem*, including the time for recreation and meals, which would leave TEN HOURS for laborious employment."

I think it would be almost an insult to the House to appeal to authorities in proof that this is labour enough for any age to endure; abundantly sufficient for infancy and youth. A few medical authorities, however, I will quote, who delivered their opinion before Sir Robert Peel's committee. Dr. Jones, a physician of much experience, asserts that "*eight* or *nine* hours are the longest period which he could sanction." Dr. Winstanley, physician of the Manchester Infirmary, affirms that "*eleven* hours *could not be endured without injury.*" Mr. Bautflower, an eminent surgeon of the same place, says that "*ten* hours are amply sufficient." Mr. Ogle, another experienced individual of the same profession, says that "*eight* or *nine* hours are sufficient." Mr. Simmons, the senior surgeon of the Manchester Infirmary, declares that the hours of working of a person under sixteen, exclusive of the time allowed for meals, ought not to exceed "*nine* hours in winter, and *ten* in summer;" and he adds that to which I particularly call the attention of the House: "It may become a question," says this eminent and experienced person, "whether with impunity the *strength of adults is capable of much longer exertion than this.*" Sir Gilbert Blaine says, that under ten he "should have no objection to sanction *five* or *six* hours, and at a more mature age *ten* hours, if not sedentary." I might multiply these authorities at pleasure, but I will close them with the opinion of a physician, after whom I conceive few could be quoted with advantage—I mean the late Dr. Baillie. He says, eyond *ten* hours a day there ought to be *no increase* of labour. "I think," says he, "*ten* hours of confinement to labour, as far as I can judge, is as much as is compatible with the perfect well-being of *any constitution.*"

These appeals are to me most affecting. Is there not, Sir, something inexpressibly cruel, most disgustingly selfish, in thus attempting to ascertain the utmost limits to which infant labour and fatigue may be carried without their certainly occasioning misery and destruction;—the full extent of profitable torture that may be safely inflicted, and in appealing to learned and experienced doctors to fix the precise point beyond which it would be murder to proceed? Are we to treat innocent infants, then, like the criminal soldier, who receives his punishment under the eye of the regimental surgeon lest he should expire beneath the lash? But, horrible to relate, these eminent men *have* stood over fainting infancy, and long since *have* forbade the infliction; yet it has been continued till thousands have expired under it. They discharged their duty; they said ten hours were fully sufficient. That term has been exceeded; and I have shewn the fatal consequences.

But if it be still necessary to sanction the term of labour they have fixed,

Speech of M. T. Sadler, Esq.

I will appeal to general experience, which, after all, is, on a point like this, better than all authority. In every other description of labour, twelve hours a day, deducting at least the intervals which this bill also prescribes, is the utmost duration of labour that the master thinks of demanding, or the workmen of enduring. Such is the case in agriculture. Look at the county reports, and you will find that these are the utmost limits in summer, except during a few weeks in harvest, and that they are diminished in winter, so as to correspond with the days as they become shorter; averaging, perhaps, throughout the year, not more than eight or nine hours of actual work. The same rule obtains among all those workmen and artisans whose pursuits are most essential to human existence, such as masons, bricklayers, carpenters, and others. Nor is it either humane or patriotic, as an eminent medical writer has remarked, to tempt them to those protracted exertions, which are often endured at the expense of health, and bring on premature decay.

This natural regulation prevails everywhere, and has been observed in all ages. Thus, if we open the most ancient of volumes, we shall find that twelve hours, including the necessary intervals, constitute the longest day of human labour, which is still curtailed as the natural day shortens. And as this limitation is dictated by the law of nature, so also is it affirmed by the law of God. Hear, then, the divine institution in this matter:—"Thou shalt not oppress an hired servant that is poor and needy, whether he be of thy brethren or of thy stranger that is in thy land within thy gates. At his day thou shalt give him his hire; neither shall *the sun go down* upon it; for he is poor, and setteth his heart upon it; lest he cry against thee to the Lord, and it be sin unto thee." "Are there not twelve hours in the day?" is reiterated by still higher authority. "The night cometh when no man can work." Yet, though man cannot—will not,—yet, says this system, children SHALL. I ask, whether it be right to work weak and helpless children in England longer than adult labourers and artisans in the prime of their days have consented to toil in any age or country.

I make another appeal on behalf of these children. I appeal to the utmost labour imposed upon criminals and felons sentenced to expiate their offences,—often of an atrocious character, and almost always perpetrated by the daring and powerful, in the vigour of life,—in the jails, houses of correction, bridewells, or other places of confinement and punishment in this kingdom. The law, even regarding these, "in wrath remembereth mercy," and after the fullest investigation, doubtless, lest justice should degenerate into cruelty, limits the power of its own ministers thus: "Every prisoner sentenced to hard labour shall, unless prevented by sickness, be employed so many hours a day, NOT EXCEEDING TEN, exclusive of the time allowed for meals, as shall be directed by the rules and regulations to be made under this Act, excepting on Sundays and other holidays." I have examined the whole of these regulations, as established in every prison in England, and I find that the average labour imposed falls far short of the limits prescribed by this bill. Even the convicts at the hulks, I am informed, are only worked in winter from eight o'clock in the morning as long as daylight lasts; and in summer from seven in the morning till six in the evening, from which time is deducted, in both seasons, about one hour and a half for meals; making, therefore, the duration of their actual labour, in summer, nine hours and a half, and in winter, perhaps, about two hours less. I ask, then, whether it is right, or even politic, thus to give a premium to crime,—to protect guilt and prosecute innocence,—to work unoffending children longer than the law permits in the case of adult criminals and felons, whose labour constitutes their punishment.

Lastly, I appeal, in behalf of these children, to the protection afforded to the slaves of our West Indian Colonies. By the Orders in Council, bearing date the 2nd of November last, the labour of the slaves, in all the crown colonies of England, is regulated as follows:—By section 90 of those orders, no slave, of whatever age, is to be worked in any agricultural or manufacturing labour in the night, but only between six o'clock in the morning and six o'clock in the evening. By section 91, all such slaves are "entitled to an entire intermission and cessation of every description of work and labour from the hour of eight till the hour of nine in the morning, and from the hour of twelve till the hour of two in the afternoon, of each and every day throughout the year." Hence no slave can be worked more than nine hours in any one day. So much for the adult slaves. But by the succeeding section (92) it is ordered, that "no slave under the age of fourteen, or above the age of sixty, shall be compelled or required to engage in, or perform, any agricultural work or labour in any of the said colonies, during more than six hours in the whole, in any one day." Passing over many other beneficent regulations, such as allowing forty holidays annually, exclusive of Sundays, and the prohibition of the labouring of pregnant females—although I see witnesses for the factory system assert that to work white females up to the period of their confinement is not at all improper or injurious—I say, passing over every minor consideration, I can hardly restrain my indignation within due bounds, while I appeal to the regulations regarding these slaves, and see that those proposed in favour of the British children are so vehemently opposed. I compare not the English child with the African child; but I ask this House, and his Majesty's Government, whether it would not be right and becoming to consider the English child as favourably as the African adult. You have limited the labour of the robust negro to nine hours; but when I propose that the labour of the young white slave shall not exceed ten, the proposition is deemed extravagant!

I might further appeal to our treatment of the brute creation. Acts of Parliament have protected these from cruelty infinitely less than that which this system tolerates. And yet the selfishness of man acts as the guardian of his cattle, and renders such laws almost unnecessary. The gentleman will not ride his hunter before he is full grown, nor does the farmer yoke his yearling foal to the plough, and scourge it forward as many hours, and even more, than the full-grown colt would bear. No! it is the factory-child alone that is thus treated. By what term shall we designate that state of the law which permits the labour of the helpless infant hours after it would have interposed in behalf of the panting brute,—hours after the driver has released his youthful slave?

But I contend that, even as regards the poor factory-child, a being esteemed so utterly worthless, it is detrimental to the mere interests of the employer to pursue so cruel a course. Beyond certain limits human industry and attention cannot be profitably stimulated. When those limits are passed, what remains but that imperfect service, equally distressing to the employed and unprofitable to the employer; and the oppressor may be well addressed in the words of the poet:—

> Thou canst not take what nature will not yield,
> Nor reap the harvest though thou spoil the field.

This fact is already in evidence in the sober language of prose: manufacturers have confessed that this excessive labour has been rarely profitable, though they have been urged to such a course by the rivalry and competition which the system both creates and continues. But were it ever so profitable, no gains, however ample, no prosperity, however permanent, could justify a practice so cruel and destructive.

I have not time to enter upon the arguments which the advocates of the present system advance in its favour, nor to refute the objections which they perpetually urge against its proper regulation. They are precisely the same as those put forth by the planters in the crown colonies against the amelioration, to which I have already referred, of the condition of the blacks and their children. The capital engaged,—the competition to be feared,—the irritation which will be produced,—the superior condition of those to be protected,—are in both cases strenuously urged. The capability of children above a certain age to endure long labour as well as adults is equally asserted; and, above all, the coincidence of the request from both quarters for a select committee is also remarkable. But how different have Government met this proposition in each case? As regards the slave, they have declared that inquiries have been pursued long enough, and that they need no further information to enable them to legislate on "the eternal obligations which religion founds npon the law of God." Such are the words of the noble secretary for the colonies. Would that this reasoning and this feeling were transferred to these poor, oppressed children, slaves in all but name, who now demand our protection! I conceive that then the eternal obligation which religion founds upon the law of God would suffice to teach us our duty, without confiding our consciences to the keeping of a select committee. We can hardly be sincere in making this solemn appeal, if, after having determined that six hours' labour are enough for a negro child, we doubt whether ten are fully sufficient for a British one.

Another objection of some of the opposing mill-owners I will briefly notice. They cannot consent, forsooth, to an abridgment of the long and slavish hours of infant labour because of the corn-laws. Why, these individuals (some of them not originally perhaps of the most opulent class of the community) have, during the operation of these laws, rapidly amassed enormous fortunes; yet, during the whole period, they could seldom afford either to increase the wages or diminish the toil of these little labourers, to whom, however forgetful they may be of the fact, many of them owe every farthing they possess: they have generally done the reverse. And they talk of corn-laws as their apology! "This is too bad." Can any man be fool enough to suppose that, were the corn-laws abolished to-morrow, and every grain we consume grown and ground in foreign parts, that such individuals would cease to "grind the faces of the poor?"

But their opposition to this measure, it seems, is grounded on philanthropic considerations alone. The loss, say they, would fall upon the poor, whose wages would be diminished just in proportion as their extravagant labour was moderated. I rather doubt this conclusion: indeed, I think we are warranted, not only by the dictates of common sense, but also by the principles of political economy, in denying it altogether. Nothing can be clearer, as a general axiom, than that the wages of labour are necessarily affected by the quantity of its products in the market. Thus a great authority in that science says, "When the demand is given, prices and values," consequently wages, "vary inversely as the supply." In full conformity with this doctrine, we have constantly heard the great and frequently-recurring distresses of the operatives in certain branches of our manufactures attributed to *over-production*. This was asserted again and again by the late Mr. Huskisson, and repeated, I think, on a very recent occasion, by the present vice-president of the board of trade. How, then, can it be, that if over-production has the effect of lowering wages, the moderating of this over-production could have any other effect than that of raising wages; and if so, how, I would ask, can that over-production be so well moderated as by regulating the excessive labour of infant hands? If it be said in reply, that the employment of many of our mills and factories depends on a demand that is governed by foreign competition, I fearlessly reply that this is not the fact. First, as to every free market in the world, I maintain that the most formidable competition is not that between British and foreign spinners, but between rival British ones only, which, when traced to its ultimate consequences, will be too often found a competition in cruelty and oppression, of which these innocent little labourers are the victims. Then, as to other and restricted markets, and particularly those for our cotton yarns, the competition is not, as is too often stated, between British labour and foreign labour, but between British labour and foreign imposts; which, as our goods have progressively cheapened, have, upon the protective principle, been proportionally advanced, and would be still further raised were we, under the sanction of the learned doubts already quoted, to work our unprotected children for three and twenty hours out of the four and twenty. Thus our cotton-twist is subjected to enormous duties in Russia, Austria, the United States, and elsewhere; while in France and Spain it is prohibited altogether. Now I would seriously ask those who raise this outcry regarding foreign competition, whether these heavy duties and prohibitions are not of themselves a sufficient answer to their entire argument. As to the idea of regulating the labour of our infants and children according to the fluctuations of foreign tariffs, which we may rest assured will be still fixed on the protective principle, it is as ridiculously absurd as it is infamously cruel. A very little attention, however, to this part of the argument of our opponents will shew its utter fallacy. The cry concerning competition was as loudly raised against the bill of the late Sir Robert Peel as it can be against the present measure; and the ruin of the spinning trade was as confidently pronounced if it were carried. It passed; and what has been the result? Why, the export of our yarns has nearly quintupled since he brought in that bill, while the trade of our then greatest competitors in that branch of manufacture, the French, has hardly kept pace even with the slow increase of their population. But the idea of French competition in cotton-spinning is a farce; their machinery, exceedingly inferior to ours, is also dearer to at least an equal degree; which circumstances, with many other obvious disadvantages, far more than counterbalance their supposed advantages in some other respects. Still more ludicrous is the novel assumption as to the rivalry of the United States in this respect. The country beyond all others where labour is dear and land is cheap, and where the mass of the population is so thinly scattered, turning cotton spinner to any extent is too absurd for a moment's consideration. Let our import duty on the raw material, and theirs on the spun yarn, amounting together to perhaps 40 per cent., dispose at once of this plea of tyranny and oppression.

But the whole argument, or rather objection, as applied to many other branches of spinning, equally fails. The export of worsted yarns, for instance, is small. In silk spinning, (not meaning by that term silk throwing,) I believe the competition with foreigners is not felt; and yet, in many of the mills of this description, the shameful practice of working young and helpless children by night, and within an hour of the Sabbath morning, is unblushingly pursued. But I shall discuss the point no further. The same argument, I have already said, was urged against Sir Robert Peel's bill; to which he replied, that "no foreigners were then known to work the same number of hours" even as he proposed. Nor do I believe that they then consented, if they do now, to labour their children in the heated atmosphere in which so many of ours sicken and perish. If they have commenced the practice of long hours and excessive labour, it is our competition, and not theirs, that stands chargeable with these evils; nor will I be too confident that our cruelty has not been contagious. Let us, then, enter into another competition with them—a competition of humanity and justice—and I firmly believe that they will still be our rivals. At all events, let us, as was strongly urged at the introduction of the free-trade system, so called, set them the example.

In corroboration of the views I have taken, that excessive labour has a tendency the reverse of increasing its remuneration, I might appeal to the great

Speech of M. T. Sadler, Esq.

and constant declension in the declared or real, as compared with the official or assumed and stationary value of our exported goods, of almost every description, especially of those more particularly alluded to. Thus, since the peace (1815), though our exports of cotton yarns have increased in quantity nearly five-fold, their real value has advanced little more than one-third; a fact which, together with many others of a similar nature, indicates but too clearly the increasing labour and diminishing remuneration of our industrious and over-laboured population. But this subject would carry me into too wide a field; and one, moreover, on which it is unnecessary for me, on this occasion, to expatiate. There are few operatives of this country, I fear, who are not made aware of this fact by the most infallible of all teachers—bitter experience. Those who contend that the amount of wages follows the degree of labour, however great or protracted, seem, in my apprehension, totally to lose sight of one of the most simple and beautiful provisions in the economy of nature—that whereby human wants and necessities are nicely and beneficently balanced by the means of supplying them; but in this striking adjustment neither infant nor excessive labour is contemplated, and where it is called forth, it will sooner or later derange the whole frame of society. Finally, I would ask those who are still disposed to hold a contrary notion, and who therefore calculate, with such arithmetical precision, the exact loss that the operatives will sustain if their labour or that of their children were to be properly regulated, whether, upon their principles, the Sabbath is not a public nuisance and private injury, especially to the labouring poor. According to their views, it must certainly have the direct effect of diminishing the wages of the industrious classes just one-sixth, and of reducing their comforts therefore in the same proportion. Dr. Paley has argued to the contrary, and shewn, that in the present state of society, the labourers of the community receive as much for their six days' toil as they would for the seven. I think he might have said more, for reasons obvious enough, but upon which I shall not now enter. He has thus demonstrated that sacred institution to be one in all respects favourable to humanity. The entire question is, I think, too plain to require this attempt at elucidation, and its final solution appears to be this:—that degree of labour will be ultimately most profitable to mankind which is dictated by their necessities, proportioned to their strength, and consistent with their welfare and happiness. In concluding these remarks, I cannot refrain from observing how uncandidly the very same set of reasoners deal on these occasions with the labouring poor. Over-production is to account for their low wages and consequent distress; diminished production, on the other hand, is to produce the very same effect, and still further lessen their means of subsistence. These arguments, or rather excuses, remind one of the poor lamb in the ancient fable, which found it impossible to satisfy the wolf: whether it drank above or below at the same stream, it was equally in fault, and suffered accordingly.

I might here notice, that an entire series of advantages which would incidentally result from this measure for shortening undue labour, is overlooked by its opponents; for instance, the giving of employment to idle hands, and the affording of additional activity to many industrious pursuits; but I shall pass these entirely by; nor shall I even insist upon the validity of the argument I have been just urging; on the contrary, admitting all that has been said as to the fall in the wages of these children and young persons, and assuming the accuracy of the computations so confidently put forth on the subject, still, I say, the question stands upon other and more sacred grounds. It is one which the public will no longer permit the interested parties to decide by mere pecuniary calculations. And were the work-people threatened (as indeed they now are) that their wages would be diminished, and their means of subsistence still more abridged;—that, severe as are their present distresses,

Speech of M. T. Sadler, Esq.

the furnace of their affliction would be "heated one seven times more than it is wont to be heated,"—they have taken their firm resolve: they will not bow down to the golden image,—they will not sacrifice their children to Moloch;—and where is the father, the patriot, the Christian, that does not glory in that resolve.

I must now apologize to this house for having so long occupied its time and attention. I owe, however, a deeper apology to those whose cause I have attempted to advocate, for having, after all, left untouched many important claims which they have earnestly pressed upon my notice. But if honourable members will consult their own bosoms, they will find them there. We are about to deal with the strongest instinct and the holiest feelings of the human heart. The happiness and tranquillity of the present generation, and the hopes of futurity, depend, in no slight degree, on our resolves. The industrious classes are looking with intense interest to the proceedings of this night, and are demanding protection for themselves and their children. Thousands of maternal bosoms are beating with the deepest anxiety for the future fate of their long-oppressed and degraded offspring. Nay, the children themselves are made aware of the importance of your present decision, and look towards this house for succour. I wish I could bring a group of these little ones to that bar,—I am sure their silent appearance would plead more forcibly in their behalf than the loudest eloquence. I shall not soon forget their affecting presence on a recent occasion, when many thousands of the people of the north were assembled in their cause,—when in the intervals of those loud and general acclamations which rent the air, while their great and unrivalled champion, Richard Oastler (whose name is now lisped by thousands of these infants, and will be transmitted to posterity with undiminished gratitude and affection)—when this friend of the factory children was pleading their cause as he alone can plead it, the repeated cheers of a number of shrill voices were heard, which sounded like echoes to our own; and on looking around, we saw several groups of little children amidst the crowd, who raised their voices in the fervour of hope and exultation, while they heard their sufferings commiserated, and, as they believed, about to be redressed. Sir, I still hope, as I did then, that their righteous cause will prevail. But I have seen enough to mingle apprehension with my hopes. I perceive the rich and the powerful once more leaguing against them, and wielding that wealth which these children, or such as they, have created against their cause. I have long seen the mighty efforts that are made to keep them in bondage, and have been deeply affected at their continued success; so that I can hardly refrain from exclaiming, with one of old, "I returned, and considered all the oppressions that are done under the sun, and beheld the tears of such as were oppressed, and on the side of the oppressors there was power, and they had no comforter!"

I trust, however, that this house, whose peculiar duty it is to defend the weak and redress the injured, will interpose and extend that protection to these defenceless children, which is equally demanded by the principles of justice, mercy, and policy. Many have been the struggles made in their behalf, but hitherto they have been defeated; the laws passed for their protection have been avowedly and shamefully evaded, and have therefore had little practical effect but to legalize cruelty and suffering. Hence, at this late hour, while I am thus feebly, but earnestly, pleading the cause of these oppressed children, what numbers of them are still tethered to their toil, confined in heated rooms, bathed in perspiration, stunned with the roar of revolving wheels, poisoned with the noxious effluvia of grease and gas, till, at last, weary and exhausted, they turn out, almost naked, into the inclement air, and creep, shivering, to beds from which a relay of their young work-fellows have just risen. Such, at the best, is the fate of many of them, while, in numerous instances, they are diseased, stunted, crippled, depraved, and destroyed. Sir, let that pestilence, which no longer walketh in darkness among us, but destroyeth at noon-day, once seize upon our manufacturing population, and dreadful will be the consequences. A national fast has been appointed on this solemn occasion; and it is well:—let it be one which the Deity himself has chosen,—let us undo the heavy burdens, and let the oppressed go free.

Speech of M. T. Sadler, Esq.

Sir, I have shewn the suffering, the crime, the mortality, attendant upon this system; consequences which, I trust, parliament will at length arrest. Earnestly do I wish that I could have prevailed upon this house and his Majesty's government to adopt the proposed measure without the delay which will attend a further and, as I shall ever maintain, an unnecessary inquiry. Would that we might have come to a resolution as to the hours that innocent and helpless children are henceforth to be worked in these pursuits, so as to render the preservation of their health and life probable, and the due improvement of their minds and morals possible! Would that we had at once decided, as we could wish others to decide regarding our own children, under like circumstances, or as we shall wish that we had done, when the Universal Parent shall call us to a strict account for our conduct to one of the least of these little ones! As the case, however, is otherwise,—as we are, it seems, still to inquire and delay, I will now move the second reading of the bill; and afterwards propose such a committee as, I hope, will assist in carrying into effect the principle of a measure so important to the prosperity, character, and happiness of the British people.

The CHANCELLOR of the EXCHEQUER.—I do not rise for the purpose of opposing the second reading of this bill, because I am quite ready to admit that the honourable member has established a case for sending the bill to a committee. I must, however, declare that, in my opinion, many of the observations and facts which the honourable member has submitted to the House do not apply strictly to the measure before us. I have not the least doubt that the honourable member has endeavoured, to the utmost of his ability, to satisfy himself of the truth of the circumstances he has stated, but I cannot avoid saying, that some of them appear absolutely incredible. The honourable member says, that children of nine years of age begin their work on Tuesday morning at five o'clock, and continue their labour till twelve o'clock on Tuesday night; that then, after the interval of one hour, they work again till five o'clock on Wednesday morning, and that, after the interval of another hour, they work again till seven o'clock on Wednesday evening. I have not the least doubt that the honourable member believed what he was stating to be matter of fact; but it appears to me incredible that any person should insist upon such labour being performed, or, if they should, that any children could be found to execute it. Facts of this kind would go far to prove that some regulation is necessary in the factories in which children are employed; but it is requisite that the facts should be proved. Inquiry will shew the rate of mortality which must ensue from such prodigious labour. The honourable member has stated only one side of the question, and I admit that if the facts which he has adduced be true, they would be quite sufficient to justify us in legislating on the subject; but we must look at the other side of the question, and inquire what will be the effect upon the labouring classes and the children themselves, of preventing the employment of children in manufactories. I am of opinion that the effect of a measure, such as that proposed by the honourable member, must, necessarily, be a fall in the rate of wages, or, what is more probable, that children would cease to be employed at all in manufactories. Now I appeal

Chancellor of the Exchequer.

Chancellor of the Exchequer. to the honourable member whether a measure which would prevent children from obtaining any employment in factories would not be more injurious than beneficial to the labouring classes?—I admit that some regulation is required to prevent cases of extreme hardship and oppression, but it is necessary that the House should proceed in this matter with the utmost caution. The greatest criticism which I shall make on the honourable member's bill is, that it really imposes restrictions on adults—on persons arrived at an age at which they are as capable of exertion as at any period of their life. The honourable member dwelt with much force upon the mortality which results from the unwholesome nature of the employment in cotton factories. As long as we have a manufacturing population in the kingdom it will be impossible to render their occupation as wholesome as that of agricultural labourers, or persons engaged in out-door labour. This is an evil which does not admit of remedy. It is too late now to argue about the unwholesome nature of manufacturing employment. We have got a manufacturing population, and it must be employed. Any measure which shall have the effect of diminishing the means of employment to labourers engaged in manufactures will produce extensive misery. These are the views which I entertain on this subject. I hope and trust that the honourable member will consent to some alterations being made in his bill, for I think that, as it stands at present, I cannot possibly support it.

I am not prepared to pledge myself in any way with respect to the measure. Even if the committee should report favourably of the bill, I will not pledge myself before the report be made to support a measure of such vast importance. I do not feel myself justified in saying that, whatever the opinion of the committee may be, I will abide by their decision. The honourable member has, I think, done perfectly right in referring this measure to a select committee. Although he seems to think that a more public investigation would be desirable, I am inclined to believe that the examination of a select committee is more likely to elicit the truth than an examination at the bar of this House. I hope that by the exertions of the committee the whole question will be fairly brought under the consideration of the House.

J. T. Hope, Esq. MR. JOHN T. HOPE.—I must trespass upon the attention of the House for a short time whilst I state my reasons for doubting the expediency of the measure now under consideration. I fully appreciate the active benevolence and general philanthropy of the honourable member for Aldborough, who has undertaken to advocate the cause of the children employed in spinning-manufactories; but he must allow me to say, that whilst I do not yield to him in the earnest desire which I entertain to promote the happiness and comfort of this class of the community, I doubt whether he can attain the end which he has in view by the means which he has advised the House to adopt. I will not now enter into any discussion as to the propriety or impropriety of interference with free labour. I believe it is admitted on both sides of the house that such interference generally is unwarrantable; nay, I am willing to admit that the labour of children must, in some degree, be considered of a compulsory nature; but I contend that those very circumstances which give such labour the character of compulsory, carry with them a remedy for the evil of which the honourable member complains, in the protection of their parents. I cannot comprehend how we can, by legislative enactments, supply the place of parental affections in behalf of the child. If these natural ties be unavailing, legislative enactments will prove equally useless and unavailing. I am, however, prepared to argue the question on more general grounds than these. I doubt, in the first place, whether a case of necessity for Parliamentary interference be fairly made out. I admit that many cases of great individual hardship occur in manufactories. The honourable member quoted many such cases, and, aided by his eloquence, they doubtless would produce a great effect upon the House. The honourable member contends, generally, that children are worked greatly beyond their strength, and that the hours of labour are extended through the night; and that these circumstances combined produce a bad effect upon their morals and education, are injurious to their health, and in many instances occasion premature death. With reference to this proposition, I beg to call the attention of the House to some facts which are contained in the reports of a committee of this House which sat in 1816, and of committees of the Lords, which sat in 1818 and 1819.

J. T. Hope, Esq.

The first point to which I will allude is the age of the children. The honourable member is entirely mistaken in supposing that a large portion of the children employed in factories are of a tender age. I find by a return, which is verified on oath, annexed to the lords' report of 1819, that out of a body of work-people amounting to 12,461, employed in cotton factories in England, there were only 196 under nine years of age. From other returns in my possession, it appears that of the children employed in the manufactory at Kirkland, in Fifeshire, there are only one in seventeen under twelve years of age, and one in four under fourteen. I am therefore entitled to say, that the argument which the honourable member has founded upon the assumption that a great portion of the children employed in cotton factories are of very tender age, is not entitled to much weight. Now, with respect to the hours of employment, I find from the returns annexed to the lords' report of 1818, that the average number of working hours in the week, taken from fifty-seven instances, is seventy-two; a number not little exceeding that which was established by the bill which was introduced by the right honourable baronet the secretary-at-war, last session. The average number of hours per day, on the total amount of hours and persons, deducted from the returns of 325 cotton factory establishments, from a table put in and proved upon oath, and inserted in the appendix to the evidence of 1819, is twelve and a fraction.

In considering this subject, it is material that we should look at the number of hours in which children are employed who are engaged in other trades and manufactures, to which protection is not extended. I find that children employed in the earthenware and porcelain manufactures work from twelve to fifteen hours per day; file-cutters, nail-makers, forgers, and colliers, work for twelve hours per day; those employed in the manufacture of hosiery, and in lace-manufactories, work for twelve, thirteen, fourteen, and fifteen hours per day; those engaged in calico printing work for twelve, fourteen, fifteen, and sixteen hours per day; needle makers, manufacturers of arms, and pin-makers work for thirteen, and fourteen hours per day. I contend, therefore, that the children employed in cotton and other spinning factories are not subjected to greater labour than those who are employed in other manufactures, to whom no protection is extended by this bill.

I shall now proceed to examine what is the effect of employment in cotton-factories on the education, morals, and health of the children. I have in my possession a certificate from several clergymen residing in the neighbourhood of some mills in the county of Fife, in Scotland; and these gentlemen certify that the children employed in those mills, and the grown up persons also, are as well conducted, in a moral point of view, as the agricultural population. The Reverend Mr. Jones, of the parish of Holywell, who gave evidence before the lords' committee in 1819, states that such of his parishioners as are employed in the cotton-factories are more orderly and moral characters than any others. With respect to education, it is in evidence, that all the persons engaged in the cotton-mills at Cupar are able to read, and the greater part able to write. Out of eight-hundred and ninety-one persons employed

J. T. Hope, Esq

in the mills in the parish of Catrene, in Ayrshire, there were very few who were not able to read. At Holywell it is stated that the children employed in cotton factories are better instructed than those engaged in other kinds of labour, on account of the opportunities afforded them of attending Sunday-schools. I will next proceed to consider the very serious statement which the honourable member has made respecting the health of the children employed in cotton-factories. It is but natural, as the noble Chancellor of the Exchequer has observed, that persons engaged in manufactures should not enjoy the same degree of health as falls to the lot of those employed in agriculture. However, I shall endeavour to shew that the honourable member has given too exaggerated an account of the ill health of the former. It appears from the evidence given before the lords' committee in 1819, that out 610 children employed in the Holywell mills, the average number of sick was six; whilst, in a regiment of infantry of 600 men, twenty to twenty-five is not considered a large proportion of sickness.

It the factory of Mr. Bisley, where there are 549 children employed in the mills, the sick are one in seventy; in Pendleton mills, where there are 531 employed, the average number of sick is eight. It appears by the evidence of Mr. Lee, in the commons' report of 1816, page 341, that there is in his factory at Manchester a sick fund, supported by one two-hundredth of the wages of children, and the books of this institution shew that the sick are, on the average, only as one in a hundred. But the evidence from the records of the Infirmary in that town is of a still more convincing nature. It appears from the statements produced by Dr. Holme, before the lords' committee, 1818, that whilst the population engaged in spinning mills forms one-sixth of the total population of Manchester, the proportion of cotton spinners relieved was in the following ratio to persons engaged in other occupations:—in the home patients, as one in twenty-three, or one in twenty-four; amongst the in-patients, as one in fourteen; in the Fever Hospital, as one in twelve; and in the out-patients of this last establishment, as one to twenty-three. I will now state some facts connected with parish relief. The honourable member for Aldborough has mentioned, as a proof of the poverty which prevails generally in Manchester, the large number of females who had been admitted to the Lying-in Hospital, but he did not tell us what were the proportions of the cotton-spinners' wives, and of those of persons engaged in other occupations, the only true criterion of their comparative health and state. This remarkable fact appears in the evidence of the lords' committee in 1818, that out of 1797 persons who applied for parish relief in Manchester, only twenty-eight were cotton-spinners. The other applicants were 576 weavers and 221 persons engaged in other trades. From the same evidence it appears that out of 582 who received parish relief at Preston there was not one cotton-spinner. I now come to the question of mortality, and I beg, in the onset, to refer to a passage in the speech of my honourable friend, the member for Aldborough, whom I regret not now to see in this place. The honourable member threw great discredit on a statement given in evidence before the lords' committee, that out of 866 spinners the deaths in ten years did not exceed fifty-seven. From the exact coincidence in the figures with the calculation which I have extracted from that evidence, I am disposed to conclude that the witness to whom he alludes is Mr. Buchanan, before the lords, 1818, page 63. Now, Sir, I can state not only that this gentleman was a person every way worthy of credit in himself, but that his statements are confirmed by the clergyman of the parish of Lorn, in Ayrshire, in which the factory is situated, and that the letter of the reverend gentleman is printed in the appendix to the report. By the evidence of Mr. Houldsworth before the commons' committee, 1816, it appears, that out of 10,000 persons employed in factories, there died in the same space of time thirty-seven, whilst of 3224 persons employed in other trades there died fifteen persons, being a proportion of thirty-seven to forty-eight in favour of factories. In the parish of Holywell I find that the proportion of deaths in factories is stated to be one in two hundred and seventeen, in the whole parish one in fifty-eight on an average of three years. J. T. Hope, Esq.

I now proceed to a more general statement, which bears out these particular facts. By a table compiled from the sworn evidence before the lords' committee, it appears that if 100 families be selected who are employed, both parents and children, in factories, that in a given time there are 527 children born, and 38 deceased out of each 100 born; whilst in 100 families, of which neither parents nor children are employed in factories, 517 children are born, and 43 children dead out of each 100 born. The proportion in Scotland is nearly the same, giving a result of a greater number of births, and a smaller number of deaths, in favour of persons employed in factories. I shall conclude this part of the subject with some general statements, drawn from the population returns. Before I do so it is, however, right to state, that the returns referred to, in drawing up these tables, were those of 1811. But this circumstance, as the calculations rest on comparison, will not effect their correctness as applied to a later date.

In the population of the four agricultural counties of Bedford, Bucks, Berks, and Cambridge, the proportion of deaths to 100 births was 64.3; the number of persons in proportion to whom one was buried, 51.1; whereas, in the population of the manufacturing county of Lancaster, the proportion of deaths to 100 births was 59, and the number of persons in proportion to whom one was buried, 51. Taking, again, the population of all England, as compared with the parishes of Manchester, Salford, and Wigan, the respective results are, all England, deaths to 100 births, 64.8; numbers in proportion to one buried, 52.3. Manchester, &c. deaths in 100, 56.71; numbers in whom one buried, 49.5. From which it appears, that whether we compare the county of Lancaster with agricultural counties, or the spinning districts with those counties, or with all England, there is a less degree of mortality amongst persons engaged in spinning-factories. I will now state, shortly, a return of the height of the militia, from a return given in to the adjutant-general's office with respect to the growth of persons engaged in these factories. I have selected the militia of the agricultural counties before referred to, and compared them with regiments raised in counties where factories are generally established, and I find the average heights to be—

Of Bedford . .	65.5	Of Lancashire . .	69.02
Of Bucks . . .	66.4	Of West York . .	67.37
Of Cambridge . .	66.7	Of Cheshire . . .	67.01
Of Berwick . .	66.8	Of Lanark . . .	68.03

shewing, both in England and in Scotland, a result in favour of the spinning population.

This fact is not, perhaps, very material, but as the honourable member referred to the stunted appearance of persons engaged in the cotton manufacture as a proof of the necessity of subjecting the factories to some regulations, I may appeal to it to shew how necessary it is that we should have accurate information before we pass this bill.

I will now proceed to call the attention of the house for a few moments to another part of the question, namely, the inexpediency, in a national and commercial point of view, with regard to the interests of the master manufacturers, and even of the children themselves. In the first place, it is an objection to the honourable member's bill that it presses unequally on different kinds of mills. In steam-mills, the works can be carried on at any period, but in water-mills the times of working are uncertain and fluctuating.

J. T. Hope, Esq.

During the summer, the working of water mills is impeded by droughts, and occasionally by floods. By the provisions of this bill the proprietors of water-mills will have no opportunity of regaining the time which they lose by these casualties. If, therefore, this bill should pass, the consequence will be, that all persons who have embarked their capital in the machinery of water-mills will be obliged to remove to great towns. This will be attended with unfortunate results, for water-mills are generally established in agricultural districts, and are productive of the greatest benefit, by affording employment to a considerable portion of the population.

It is obvious, that if you limit the hours of labour, you will, to nearly the same extent, reduce the profits of the capital on which the labour is employed. Under these circumstances, the manufacturers must either raise the price of the manufactured article or diminish the wages of their workmen. If they raise the price of the article, the foreigner enters into competition with them. I need not remind the house of the immense capital which is embarked in the cotton manufacture, amounting to about £15,000,000, or of the great quantity of manufacturing produce which is exported to foreign countries. I am informed that the foreign cotton-manufacturers, and particularly the Americans, tread closely upon the heels of our manufacturers. If the latter should be obliged to raise the price of their articles, the foreign markets would, in a great measure, be closed against them, and the increased price would also decrease the demand in the home market. To avoid these ruinous consequences the manufacturers will, in all cases where it is possible to dispense with their labour, cease to employ children at all, and thus employ a greater number of adults than before. The honourable member for Aldborough seems to consider this an advantageous course; but I cannot concur with that opinion, because I think that the labour of children is a great resource to their parents and of great benefit to themselves. But I understand that, in some branches of the cotton-manufacture, it is impossible to separate the labour of children from that of adults. If, therefore, the manufacturers comply with the provisions of this bill, they must, in some instances, reduce the labour, and consequently the wages, of adults, in the same proportion as those of children.

I am of opinion, however, that the manufacturers would endeavour to evade the provisions of the bill by employing two sets of children, one to work in the morning and the other in the evening. They will, of course, pay them only half wages, and will, therefore, be put to no additional expense. Moreover, we may not unreasonably expect that the workmen, who commonly hire their own piecers, will exchange the children from one mill to another, and so work them twice a day; and whenever that shall happen, the latter will, of course, derive no benefit from the honourable member's bill; and this is the more to be apprehended,—if, indeed, the doctrine of the honourable member be true,—that whilst the upper and middling classes are considered fit and sufficient guardians of their own children, the parental affections of the lower orders are not sufficiently powerful to protect their offspring. It is, therefore, on these grounds—because, in the first place, I doubt whether parliament can protect children as effectually as their parents; secondly, because I am of opinion that a case for parliamentary interference has not been made out; and thirdly, because I believe that the bill will be productive of great inconvenience, not only to persons who have embarked large capital in the cotton manufactures, but even to workmen and children themselves—that I feel it my duty to oppose this measure.

Lord Viscount Morpeth.

LORD VISCOUNT MORPETH.—I rejoice that the bill has met with no opposition in its present stage, but, at the same time, I must express my satisfaction that the subject is to be referred to the investigation of a committee. The bill is of peculiar interest to the inhabitants of the county which I have the honour to represent, and is of the utmost importance to the manufacturers of the country generally. I think that the honourable member for Aldborough has no option in this case, but is bound to go to the committee, where anything that may appear objectionable in the bill may be removed.

Mr. James.

MR. JAMES.—I have no doubt that the honourable member for Aldborough is actuated by the best intentions and motives in relation to this bill, but I think that the course which he pursues will fail in attaining the object which he has in view. Undoubtedly the system which is pursued in these manufactories relating to the working of young children is a great evil; but it appears to me that the remedy which the honourable gentleman proposes to apply is worse than the disease. There appears to me to be only a choice of evils—the children must either work or starve. I say, therefore, that it is the least evil of the two to go on with the present system. If you decrease the hours of labour in the manufactories the workmen who now receive six shillings will only receive five shillings a-week. In point of fact, the adoption of this measure would operate to the injury of our manufactures, and as a premium to the foreign produce. The manufacturer is to be prevented working his mill for more than a certain number of hours together, and he will often be unable to execute the orders which he may receive, and consequently, the purchaser must go to foreign countries for a supply. The result will be, that you will drive the English capitalist to foreign countries, where there is no restrictions upon the employment of labour and capital, but where there is open competition, and the master and labourer are permitted to make their contracts without the interference of the legislature. It appears that the mania for legislation is particularly strong on subjects of this nature, and in cases where interference is calculated to do the greatest injury. This is only another clumsy attempt to bolster up the present system of taxation. If the honourable member desires to benefit the children engaged in these manufactories, why does he not propose that the taxes which press heavily upon the productive classes should be repealed, and, above all, the tax on corn. This would be a much better course than interfering between the employer and the employed. I contend that the root of the evil is heavy taxation and the corn laws. Whilst these exist, it will be impossible for us to compete with other countries without having recourse to extraordinary exertions. I deprecate legislative interference between the masters and their workmen, because I consider it an attempt to bolster up a bad system. We never can successfully compete with the foreign manufacturer until the people are enabled to procure a cheap food, which can only be done by a repeal of the corn laws. The only reason that I can conceive for keeping up the present system is, that it is beneficial to honourable and right honourable gentlemen whom I see around me.

Mr. Schonswar.

MR. SCHONSWAR.—I will not trespass on the time of the house for more than a very few minutes; but I cannot support the second reading of this bill without stating some of the reasons which induce me to do so. I, in common with the great body of the community, must feel anxious to get rid of a system under which so much suffering is experienced. I think, however, it is desirable that, previously to our legislating on the subject, an inquiry should be made into the system, and that proper steps should be

Mr. Schonswar.

taken to prevent any injury being done to our manufactures. I have been requested by a large body of my constituents to support inquiry, so that the truth of the statements made on this subject may be ascertained. If on investigation it should appear that one-tenth part of the circumstances stated by the honourable member for Aldborough be correct, such a system should not be allowed to continue. If it be true that human beings are thus destroyed, and that thousands of children are subjected to this species of slavery, I think that, for the character and honour of the country, we are bound to get rid of the system, whatever may be the consequence. If any of the circumstances stated by the honourable member are founded in fact, we ought, as men and as Christians, to interfere in behalf of a helpless class of persons, who ought peculiarly to have the protection of the legislature. I would call upon honourable gentlemen, as fathers, to consider what would be their feelings if their children, at a tender age, were to be subjected to this treatment, and were to be confined in a hot and fetid atmosphere, and kept standing on their legs for several hours together.

What would be the consequences if the children of those in the higher classes of society, who have "all the appliances and means to boot" of warm clothing and plenty of food, were thus treated? I must say, that the man who can regard the sufferings of these poor children with indifference has a degree of hard-heartedness the possession of which I do not envy. To reject everything like exaggeration from the discussion, it must be admitted that the system is vicious, and requires an immediate remedy. We are now called upon to examine, in a committee, the evidence upon which the honourable member supports his motion, to separate the good part of the system from the bad, to preserve the good, and cut away that which is bad. I shall only add, that I trust that the proceedings before the committee will be conducted with such a degree of speed as to allow time for some measure on the subject to pass into a law before the expiration of the present session.

Mr. Mackenzie.

MR. STEWART MACKENZIE.—I shall follow the example that has been set by other honourable members, and shall trouble the house with but very few observations. I feel called upon to make one or two remarks, in consequence of having been entrusted with the presentation of a petition to this house from the town of Glasgow, in which the petitioners deny many of the allegations previously made with respect to the treatment of the children in the manufactories of Scotland. It is the desire of the manufacturers that the subject should undergo the fullest investigation, and they assure me that they will satisfactorily rebut many of the charges urged against them. I am told enough to say that, notwithstanding all I have heard, and all the investigations that have taken place, I am not satisfied of the correctness of the statements that have been made on this subject; and, above all, I do not believe that it will be possible to remedy the evils the honourable gentleman complains of without adopting a prejudicial interference with our manufactures. I think that, by attempting too much, the honourable member will fail in attaining any greater protection for the children employed in these manufactories than they have at present. I will not enter into an examination of the details of the bill, but I do not think that a case has been made out to justify it. The evil which is complained of is not carried on to any great extent, and I am sure that the honourable member would not be able to prove that children are employed in the manufactories in Scotland under nine of years of age.

From 1802 to the present time many attempts were made to regulate cotton-factories by Parliamentary enactments, but they all failed; and I do not anticipate that the present bill will be more successful. The right honourable secretary at war introduced a bill on the subject, and between the measure and Sir Robert Peel's bill no less than six enactments were passed on the subject. I am convinced that the present measure will never remedy the evils complained of, as the penalties are so severe that it will become inoperative, or that it will interfere so much with our manufacturers as to prevent the law being carried into effect. The former measures only applied to cotton-factories, but the bill of the honourable member is to include in its operations all other manufactories. From 1802 to 1831, a period of twenty-nine years, no complaints were ever made of the over-working of children in the northern silk and flax factories. Under the bill of the honourable gentleman, a person twenty years of age is just as much under control as a child of ten years old. I cannot help feeling that this and the other evils complained of arise from the superabundant population in our manufacturing districts. By interfering in the manner proposed in this bill, you will lessen the profits of the manufacturer, and the result will be that a large portion of the capitalists of the country will be induced to remove to other countries. In France, Russia, and Spain, there are no prohibition laws against the introduction of our cottons. In many parts of the continent there are manufactories of cottons, which produce goods at nearly as cheap a rate as they can be produced in this country, and a bill of this nature is only calculated to increase the difficulties of the English manufacturer and prevent him competing with the foreigner. I cannot help feeling that if the bill of the honourable gentleman is passed in its present form, there will be a great risk of its inflicting considerable evil. I am glad that the subject is to be referred to a committee, when I trust that many of the objectionable clauses will be struck out. Mr. Mackenzie.

Lord WILLIAM LENNOX.—I have great pleasure in supporting the measure of the honourable member for Aldborough, for I feel it to be the bounden duty of every well-wisher to society to exert his utmost influence to put an end to the present injurious practices; not only with a view of promoting the great cause of humanity, but of removing from this country the stigma that must be necessarily attached to it by an encouragement of so nefarious a system, a system which, as now carried on, is a disgrace to a civilized community. Lord William Lennox.

An Honourable Member.—No, no, no!

Lord William Lennox.—The honourable member may say no, but I can prove my words. The system, instead of laying the basis for health, cheerfulness, industry, longevity, independence of parochial aid, is calculated to bring human beings either to an untimely grave, or doom them to a protracted life of disease both of body and mind, bringing pauperism and misery upon themselves, and making them a burthen to society. In order to bring to maturity the faculties of the body and mind, sufficient time must be allotted for meals, inhaling pure air, rest, and recreation. And how could that be done when, in some manufactories, children work seventy-nine hours per week, have scarcely an hour for meals, and perhaps only six or seven hours for rest. I want no evidence to tell me that the present system is not as pernicious to the children's health as it is destructive to their morals. In addition to the instances brought forward by the honourable member for Aldborough of the absurd evidence given in the committee in 1816, I will merely state two cases. One medical gentleman was asked whether a child could keep standing twenty-three hours out of twenty-four. He replied—"That was a question of very great doubt." Another medical gentleman said that the inhaling of cotton fumes was not injurious, because the effect was removed by constant expectoration. When asked whether constant expectoration was not pernicious, "Oh," answered the practitioner, "that

Lord William Lennox.

depends upon a variety of facts." I cannot think the master manufacturers devoid of all feeling, or that a chain of self-interest can predominate over a feeling of humanity; I therefore trust that they will hail this bill with pleasure, a bill that goes to extirpate English slavery,—I call it by no other name; —for can the annals of West India slavery produce a case of more touching interest, more pathetically described, than the one of the little girl, who, fearful of the punishment of her cruel-hearted master if she was late, was afraid to trust her eyelids to repose; deprived of her natural rest, badly fed, badly clad, she dragged on a miserable existence, and was brought to an untimely grave—a victim of cruel slavery. I hope whilst we all wish to ameliorate the state of the foreign slave, we shall not be wholly unmindful of the condition of our own.

Mr. Wilbraham.

Mr. WILBRAHAM.—As the representative of one of the counties in which it is stated that the system is carried on, I feel called upon to make a few observations. I certainly think that some exaggeration has taken place with respect to the treatment of children in the manufactories, but, at the same time, if the children undergo anything like the suffering described by the honourable member, some measure ought to be passed for their protection. In my opinion, the manufacturers would not oppose a measure which, at the same time that it offered protection to the children did not interfere unnecessarily in the conduct of the mills. I am sure that the manufacturers are too humane to object to any improvement of this nature if it did not injuriously interfere between the employer and the employed. The present state of things has been produced by the competition which has existed, either of capital or labour, at home and abroad. We know that in times of excitement on any particular subject, the most exaggerated statements are put forth, and are believed without hesitation. I cannot help feeling that this has been the case with regard to this subject, but at any rate some alteration should be made. This is a subject, however, which requires the greatest caution in legislating on. I consider that the principles of the bill is already conceded, but, with all deference to the opinion of the honourable member for Aldborough, I could not consent to hardly any clause in the bill, as it stands at present, without going into committee. We are not about to send the principle of the bill to the consideration of the committee, for the principle is admitted; but it is desirable that the greatest caution should be used in framing a measure which is of such vital importance to the manufacturers of this country. At the same time we should not forget, that the great end of all government is the protection of the weak against the strong, and we are now called upon to throw the mantle of British law over an unprotected class of the community, who will be the future producers of our wealth and the guardians of our shores.

Mr. Lennard.

Mr. LENNARD.—As I presented a petition on this subject from the borough of Maldon, I feel called upon to say a few words. I trust that the appointment of the proposed committee will not create such delay as to prevent the passing of the bill during the present session. The public feel a deep interest on the subject, and justly so; and I consider that what is stated in the preamble, and what therefore must have been proved up stairs, is sufficient to induce the house to give the bill a cordial support. The statement that in many cases children are employed during the whole night in these manufactories is quite enough to make it imperative on the house to interfere on their behalf. I cannot think that any one can fairly complain that the bill goes too far, when it allows children to be worked as many hours as a felon condemned to hard labour is allowed to be worked in our gaols. But it is not only on the ground of humanity that I support the bill of the honourable gentleman opposite. I think that the children of all classes, and more especially in our crowded towns, should have some leisure for mental improvement. This cannot be where they are liable to be worked fourteen, or sixteen, or eighteen hours out of the twenty-four. It has been said that former acts on this subject have been evaded. I hope this will be guarded against by my honourable friend, and if this object cannot be otherwise secured, I hope that the bill, when it again comes before the house, will enact that the mills shall cease to work after a certain hour. This will be a sure means of accomplishing the object which the honourable member has in view. Mr. Lennard.

MR. HEYWOOD.—This is a question in which the interests of a large portion of my constituents are so deeply involved that, notwithstanding the lateness of the hour, I hope I may be permitted to offer a few observations. I cordially concur in the importance of some legislative interference to restrict the hours of labour for children in factories. I gave my sincere support to the act of the last session, which had that object in view: and as the bill now before the house has the same object, I shall vote for its second reading. The provisions of this bill, however, appear to me to be very much at variance with its object, and I have no hesitation in saying—in reference to cotton-factories, at least—that if it pass in its present shape, one of two results will attend it: either it will be wholly inoperative, and leave the children entirely unprotected, or, if enforced, it will be productive of the most injurious effects to the trade, and of severe suffering to the working classes. Mr. Heywood.

I believe that it will be wholly inoperative, and I found this opinion upon a careful observance of the operation of the act of last session: that act is, in many instances, openly evaded; and, though it is more generally observed, its observance depends upon the good sense and good feeling of the masters, who have been so much calumniated in this house, and not upon any difficulty they would have in evading it. I speak the more confidently on this subject, from my knowledge of the difficulties which an association formed in Lancashire for the very purpose of enforcing the observance of this act have met with. They see the transgression of the law, but have no power of checking it. The present bill affords no better security for its observance than the act of the last session, and it affords an increased and irresistible temptation, both to the masters and work-people, to take advantage of its defects, by its diminution of the profits of the one and the wages of the other. On the other hand, if the provisions of this bill could be enforced, and if all persons under eighteen years of age were prevented from working more than ten hours per day, great distress amongst the working classes would be the inevitable consequences. They are paid by the quantity of work done, and their wages would be diminished in proportion to the lessened time allowed them. I know the case of a single family—a father and eight children, all upwards of fourteen years of age—whose earnings were diminished by 13*s.* in the first week after the act of the last session came into operation. Now they had before worked thirteen hours and a half in the day, and were then reduced to twelve. Under the present bill, reducing the hours to ten, this same family would suffer a further reduction of 15*s.* a week.

The subject is one surrounded with difficulties, and I am rejoiced to think that it is about to undergo a thorough investigation. It is one upon which much misapprehension prevails; and my fear is, that, from mistaken notions of humanity, we may inflict upon the working classes a deeper wound than that we propose to cure. We must remember that food and clothing are as essential to health as air and exercise; and take care that while we give the latter we do not take away the former.

Mr. Howard.

Mr. PHILIP HOWARD.—Notwithstanding the lateness of the hour, I cannot refrain from offering my tribute to the sincerity and perseverance with which the honourable member for Aldborough advocates the cause of the labouring classes; I think he has made out a very strong case for the restriction of the hours of labour of children under the years of twelve or fourteen who work in mills or factories; but I must say that, in most of the other provisions of the bill, his zeal as a philanthropist has exceeded the discretion of the legislator; and those two offices, with an enlightened view to the welfare and interests of the community, can never be severed.

With reference to what has fallen from my honourable colleague, the member for Carlisle, I must observe that no reduction of the price of bread, consequent upon a repeal of the Corn Laws, could take place without a decrease of wages; and though some savings in the public expenditure might still be effected, yet, consistently with the maintenance of faith to the public creditor, upon the security of which the poor man's savings and the rich man's capital alike depend, I do not think any great or material reduction could be effected—sufficient, I mean, to prove of any essential benefit or relief to the labouring classes; so that we must not allow those subjects to divert our attention from the calm and earnest consideration of the measure now before the house. I must be allowed to vindicate the manufacturers from that imputation of hard-heartedness which has been cast upon them. At a meeting held at Carlisle, I beg to state that a resolution was adopted expressive of their feeling that the whole subject should be maturely considered in a select committee of this house, with a view to promote the happiness and comfort of the operative and his children, as well as the interests of the manufacturer. Mr. Furguson, a gentleman of Carlisle, with whom I conferred to-day, has confirmed me in my opinion that such is the general wish of the manufacturers of that large and populous city; so that with the sanction of their judgment I support the principle of this bill, though I am aware it goes too far, and stands in need of many alterations before it can prove a real boon to those whom it is meant to serve.

Mr. Strickland.

Mr. STRICKLAND.—I cannot suffer the question to be put without saying one or two words. I have been informed that in several of the manufactories in the neighbourhood of Manchester it is not an uncommon thing to employ the children thirteen or fourteen hours a day. If this be the case, I appeal to the feelings of any man of common humanity, whether this is not over-working children of from nine to twelve years of age. I think that the honourable member has made out a case of great hardship, and that we ought not to hesitate to apply a remedy. I will merely state one circumstance to the house which occurred before the West India Committee. A witness, in answer to a question put to him, stated, that the owners of slaves in Demerara never would think of such a thing as employing their slaves for thirteen or fourteen hours a day, as they knew that it would destroy them in a very short time. It was said, that the slaves, under no circumstances, are called upon to work more than nine or ten hours a day; and the reason assigned was, that if they worked longer their constitutions would be destroyed in the course of a few years. I think it is desirable that the penalties for an infraction of the act should be greatly reduced; for if you have such heavy penalties they are likely to become inoperative.

Mr. Hunt.

Mr. HUNT.—If this were a mere question of profit and loss there might be a doubt on the subject; but it is a question of cupidity against humanity. The question is, whether the children of the manufacturing poor should work for more hours than human nature can sustain. I can confirm many

Mr. Hunt.

of the statements of the honourable member for Aldborough from my own observations; and I am sure, if honourable members would take the trouble to inspect the workshops, no evidence of medical men would induce them to believe that the employment of children in such places is not prejudicial to their health.

It is very well for honourable gentlemen to remain quietly at home and legislate upon the question, but if they were to see hundreds of the poor, unfortunate wretches employed in the cotton-mills in Lancashire, they would feel the absolute necessity of adopting an active interference. The honourable member very truly stated, that such a system as exists in civilized England could not be carried on in any part of India. I am convinced that not even the American savages—that not even the cannibals—would suffer their children to be worked in this way. Indeed, I will go further and say, that there is not a brute beast who would suffer his children to be treated with so much inhumanity. [A laugh from several honourable members.] Honourable gentlemen may laugh, but I say that the brute creation have too great a regard for their young to suffer them to meet with such treatment. The honourable member very truly stated, that necessity alone could induce the labouring classes of England to let their children experience such treatment as they meet with in our manufactures. A father or mother who would willingly consent to suffer a daughter to be treated with the heartless brutality which children have to undergo in our manufactories, must be utterly devoid of the feelings of humanity. The honourable member has described an instrument used in the punishment of the female children; now I would ask any one whether a blow from that instrument is not almost as bad as the infliction of the dreadful punishment of the knout, which is used on the criminals in Russia? A father who would patiently suffer his daughter to be treated in this way, would prostitute her, and live upon the wages of her shame.

If any political economy notions should induce the house to consent to the continuance of the practice, I say that a disgrace will be inflicted on the country. With regard to going into committee, I confess I can see no necessity for doing so, as we have already ample evidence to shew the revolting cruelty of the present system. If, however, a committee be appointed, I hope that it will not be like that named on the silk trade, which is a perfect mockery, and, indeed, an insult to the country, as it is well known that many of the members of it are pledged to the doctrines of free trade. An honourable member has expressed a hope that the penalties to be inflicted by the bill will not be heavy. Now I hope they will be as severe as possible, and that one half will be given to the informer.

I trust that steps will be taken to render the working of this bill as secure as possible. I have been given to understand, that in consequence of the manner in which it was drawn up, the bill of the right honourable the secretary at war has already become almost inoperative. It has been said, that this bill will be injurious, as it will prevent the manufacturer keeping his mills at work as long as he does at present, and therefore, that he will be obliged to refuse many foreign orders. I say, let the manufacturer keep double the number of workmen, but do not let him destroy the health of the rising generation. At present, the children earn only two shillings and sixpence, or three shillings a week, and of course there could not be a material diminution in this amount of wages. I much fear that the appointment of this committee will lead to much delay, and will prevent the bill passing into a law this year. I know that many of the manufacturers of the north of England are opposed to this bill, but I trust that the committee will proceed with all practicable speed in the investigation of the subject.

Mr. Hunt. I am sorry to detain the house at this late hour, but as above ten thousand persons, in the town which I represent are engaged in the cotton manufactories, I feel called upon to state my opinion on the subject. The whole of my constituents are in favour of this bill, even if it should lead to a reduction in the wages of the children employed in the mills. My constituents have instructed me to support this bill; but even if they had desired me to oppose it, I should have refused, as I would rather have resigned my seat in this house than have done so.

Mr Kearsley. MR. KEARSLEY.—I think that I prove myself to be one of the best friends of the manufacturing poor by opposing this bill of the honourable member for Aldborough. The bill of the right honourable secretary at war, which was passed last session, met with the concurrence of both parties in this house, but it has proved totally inoperative. As, however, the right honourable secretary is absent, I will say nothing about him or his bill. I think that that bill went quite far enough, but this measure would lead to a most improper and injudicious interference between the master and servant. I deplore that any system similar to that described by the honourable member—who, I cannot, however, help saying, appeared to me to have made rather an exaggerated statement—should exist in this country; but I fear that it cannot be completely put a stop to in the present state of the country. I trust that this measure will not be submitted to the investigation of a packed committee. As for the observations of the noble lord, the member for Lynn, he must excuse me for saying, that he knows nothing whatever about cotton-mills, nor, indeed, about anything else. He has dared me to come forward, and I thus meet his challenge. I live among a manufacturing class of people, and I will say, that they are a much better and more honest set of people than any other portion of the community. I say that the integrity and honesty of all these persons, high or low, is much greater than is to be met with in the circles in which I have to mix. I will not trespass longer on the time of the house, but I hope that God will turn the hearts of many whom I see around me, and will lead the misguided into the right path. I cannot sit down without protesting against the right of any one to take me to task for expressing my opinion in any way that I may think proper.

Lord William Lennox. LORD WILLIAM LENNOX.—I rise in explanation to say, I complained of the honourable member for Wigan's interruption. It would be far more parliamentary to expose my mistakes in one of his speeches, so full of eloquence, and infinitely preferable to the unknown tongue of groans which he so often indulges in.

Lord Nugent. LORD NUGENT.—I think it inexpedient, after we have agreed to the principle of the bill, that we should proceed with the present discussion. If we were to yield to our feelings, I am sure there is not a man in the house who would not be induced to support this measure after the extremely able and eloquent speech of my honourable friend. At the same time, I must say that there is great difficulty in legislating on this subject. By unnecessarily interfering with our manufacturers we may injure them, because we may prevent them competing in the foreign market with the foreign manufacturers, and thus lead to a reduction in the rate of wages of labour. The question is one of great importance and difficulty, and I trust that the gentlemen who are appointed on the committee will exert themselves in the investigation of the subject. The honourable member has performed a great service by bringing in this bill, and I trust that he will be successful in carrying such a measure into effect, which, at the same time that it effectually destroys the present system, will not in any degree injure our manfacturers. I think that it is desirable to discuss the principle of the bill no further, but send it at once to the committee. Lord Nugent.

MR. BLAMIRE.—I trust that I may be allowed to make one observation. I formerly thought, with many other persons, that the state of mortality in the manufacturing towns is much lower than it is in the country districts, but on referring to the returns which have been laid on the table of the house, I find, on comparing the number of illegitimate children born in the large towns and the agricultural districts, that the proportion, to the disgrace of the peasantry be it spoken, is much larger in the latter. Mr. Blamire.

I rejoice that some measure similar to this is likely to be passed into a law; for I am sure, that if anything is calculated to demoralize the people, it is the continuance of a system like the present, I am happy that the honourable member has consented to send the bill to a committee up stairs, where, I have no doubt, such alterations will be made as to render the measure palatable to the manufacturers, and at the same time not destroy its efficiency.

MR. CURTEIS.—I cannot help feeling that there is some degree of inconsistency in the conduct of the noble lord on the present occasion. I recollect that formerly the noble lord was the ardent supporter of a measure having for its object an interference between the employer and the employed in the agricultural districts; but he now comes forward and objects to the application of the same principle to the manufacturers. With respect to the bill of the honourable member, I agree in the principle of the measure, but I object to most of the details of it. Mr. Curteis.

[The bill was then read a second time; and, on the motion of Mr. Sadler, was ordered to be committed to the following members: —Mr. Sadler, Lord Viscount Morpeth, Mr. Strickland, Mr. Heywood, Mr. Wilbraham, Mr. George Vernon, Mr. Benett, Sir Henry Bunbury, Mr. C. Poulett Thomson, Mr. Dixon, Sir John Hobhouse, Mr. Horatio Ross, Mr. Robinson, Mr. Meynell, Mr. Perceval, Mr. Boldero, Lord Nugent, Mr. Shiel, Sir George Rose, Mr. Attwood, Mr. Ridley Colborne, Mr. Kenyon, Mr. Fowell Buxton, Mr. Estcourt, (Oxford,) Mr. John Smith, Mr. John Weyland, Lord Viscount Lowther, Mr. Hope, Mr. Moreton, and Mr. Lennard.]

MR. LABOUCHERE.—I have no wish to interfere with the appointment of the committee, but I merely wish to observe that there is no member connected with Somersetshire in the list just read, and the bill will greatly affect many manufacturers in that county. I am not opposed to the principle of the bill—namely, to limit the number of hours in which children are to work in the manufactories, but I fear that some of the clauses in this bill will either render it inoperative in the west of England or destroy some important branches of trade. I recollect that last year, in consequence of the effect the right honourable the secretary at war's bill would have had on the silk manufacturers of the west of England, he was obliged to confine the operation of his measure to the cotton-factories. In the west of England the engines in the silk-mills are propelled by water; whereas they are propelled by steam in the north. I am sure that if the bill was to pass in its present form it would close all the former mills; and I know that it has excited great alarm in the neighbourhood of the place which I have the honour to represent. I believe that my constituents are not opposed to the principle of the bill, but they would be obliged to close their factories if some of the clauses were to be carried into effect. It is Mr. Labouchere.

Mr. Labouchere. the anxious wish of the manufacturers of the west of England that some honourable gentleman connected with that part of the country should be placed on the committee. My honourable friend, the member for Somersetshire, informs me, that he is already too much engaged in committees to be able to attend. This is also the case with myself, but my honourable friend and colleague has promised me that he will attend if the honourable member will consent to his being placed on the committee. I shall conclude with proposing that Mr. Bainbridge be added to the committee.

Mr. Sanford. MR. SANFORD.—I beg to second the proposition of my honourable friend the member for Taunton. It fell last year to my lot to shew how the bill of the right honourable the secretary at war would affect the silk manufacturers of the west of England; and I am sure that if this bill was carried into effect in its present form, it would do incalculable mischief in those districts. I am not opposed to the principle of the bill, but I am desirous that it should be fully discussed; and I therefore rejoice that it is to be referred to a committee. There are some clauses most objectionable in their present form, which I trust will be altered in the committee.

Mr. Sadler. MR. SADLER.—I can only say, that I wish to make the committee as efficient as possible, and shall not, therefore, oppose the addition of the honourable member's name. I think, however, since the house has agreed to the principle of the measure, that it would be inexpedient to increase the number of members on the committee.

The motion was then agreed to; five to be a quorum.

FIRST REPORT

OF THE

Central Board of His Majesty's Commissioners,

APPOINTED TO

Collect Information in the Manufacturing Districts,

AS TO THE

EMPLOYMENT OF CHILDREN IN FACTORIES,

AND AS TO THE

PROPRIETY AND MEANS

OF CURTAILING

THE HOURS OF THEIR LABOUR.

HIS Majesty having been pleased, in compliance with an humble address of the knights, citizens, and burgesses, and commissioners of burghs, in Parliament assembled, to appoint a commission under the Great Seal to collect information in the manufacturing districts, so as to enable Parliament to legislate during the present session—

"As to the employment of children in factories;"

"As to the effects of such employment both with regard to their morals and their bodily health;"

"As to the propriety and means of curtailing the hours of their labour;" —and also,

"In what respect the laws made for the protection of such children have been found insufficient for such purpose;"—and what may be

"The further provisions necessary for their protection:"

And his Majesty having in this commission enjoined the several commissioners to obey all directions touching the premises which might from time to time be given to them or any one or more of them by one of his Majesty's principal secretaries of state; we, whose hands and seals are hereunto affixed, having, by a letter of the 20th instant, been directed by his Majesty's principal secretary of state for the Home Department to submit to him the whole of the evidence in such a form that it might be laid before the House of Commons in the early part of the ensuing week; and also to state to him any opinions or recommendations which we might be able to form on consideration of the whole of the evidence; do, in obedience to that direction, herewith transmit the evidence, and beg to submit the opinions and recommendations we have founded thereon.

Before we set forth the conclusions which appear to us deducible from the evidence collected up to this period, under the several heads specified in the terms of the commission, we beg to state the nature of the course taken to collect it.

Four districts were traced out, comprehending the seats of each of the principal branches of manufacture in which any large proportion of infant labour is employed. Two civil commissioners and one medical commissioner were appointed to each district. We, the undersigned, were appointed to act as a central board, to frame instructions for the district commissioners, and to prepare the general report. At the time when this duty was confided to us, the only official information we received as to its nature, independently of the terms of the commission, was a verbal intimation that the inquiry was expected to be into "the whole truth respecting the employment of children in factories." We thought it our duty in the first instance to invite communications and suggestions as to the course of inquiry from the parties favourable to the bill for the regulation of factories, recently introduced into Parliament, as well as from those opposed to it; and we did so, as a first step, with the view of rendering the inquiry as extensively satisfactory as possible, and of adopting such suggestions, offered by either party, as might appear to us conducive to that object. We were not favoured with any assistance from any of the supporters of that bill. Several of the principal manufacturers, and of the representatives of manufacturing districts directly affected by the inquiry, waited upon us, and gave us all the information which we requested of them.

Our next step was to frame a set of queries, which were printed, and transmitted in some instances directly to the owners of manufacturing establishments, and in others to the magistrates of the towns in the manufacturing districts, who were requested to forward copies to all the mill-owners and manufacturers within their jurisdiction, with directions to be filled up, and delivered into the hands of the district commissioners on their arrival in the neighbourhood. The intention of these queries was to aid the inquiries of the commissioners, by procuring for them, on their entrance into any district, some previous accounts of the manufactories which they might have to visit, and to obtain from those which they might not have opportunities of examining replies to queries which it appeared desirable to have extensively answered. We also transmitted, with these queries, blank tabular lists, to be filled up with the names and other particulars relating to the workmen employed at the several places in question. The object proposed by these lists was partly to aid, as above, the inquiries of the commissioners, by stating the names of workmen maimed or affected with sickness who might be examined as witnesses; partly to obtain the particulars included in the lists, as tests of the accuracy of any statements relative to the workmen employed; and partly to obtain particulars as to their ages, the numbers who are married, the numbers who are educated, the loss of time from sickness, and other particulars which could not be obtained with equal fulness in the form of general answers.

It was required of the proprietor of the works, that he should cause to be placed in one of the rooms frequented by the workpeople a copy of the queries, open to view during one whole day.

These queries and lists were necessarily general, in order to include every description of manufacture. Other queries and tabular forms for statistical returns, adapted to the particular manufactures of each district, have been drawn up and circulated by the several district commissioners.

With the aid and advice of Dr. J. Mitchell and Mr. Finlaison, actuaries, we also framed queries to be put by medical men, and directed inquiries to be made, with the view of determining the amount of sickness or mortality prevalent amongst the manufacturing population of the classes in question as compared with other classes.

Having taken these steps, we proceeded to frame for the guidance of the district commissioners instructions as to the objects of the inquiry and procedure which they should adopt to obtain information. We here only quote such portions of those instructions as relate to the selection and examination of witnesses. On this subject we stated to the commissioners—

"On arriving in any town or district, you will as a general rule obtain an interview with the chief magistrate, the town clerk, or other public officer of the district to whom the distribution of the queries and the lists has been confided, unless you have previous information of any witnesses of the working class who have not hitherto been examined, and who are desirous of giving evidence.

"You will inquire of him, and of any other magistrates you may meet, whether any and what complaints have been made before them, arising out of the extent of the hours of labour, or the treatment of the persons employed in the manufactories within their districts. You will inspect the minutes of evidence taken on those occasions; and if any of them appear to be of importance, you will procure copies, as they may be of service to you in directing your examination of the parties referred to.

"You will also inquire from such officers as to any facts within their knowledge, illustrative of the comparative moral characters of the youth employed in the places in question in the several branches of manufacturing industry, and of those who are not employed in any manufacture. In this, as in other instances where any general allegations are made with respect to the character or conduct of parties, you will take care to ascertain how far those general allegations are justified by the number and the frequency of the instances; you will guard against any precipitate conclusions from extraordinary or anomalous cases; you will ascertain and shew how far any irregularities of conduct are within the proper control of the employers or the parents of children, or whether any other and what control is available for their repression. After the perusal of such returns as you may obtain, you will proceed to those manufactories which the replies or other information transmitted to you may indicate as the places where inquiry may be the most usefully directed.

"Except when any circumstances of peculiar difficulty arise which may induce you to consult your colleagues, you will take your examinations separately. Each civil commissioner will endeavour, as far as may be practicable, to examine the condition of the children employed in different classes of manufactures.

"The medical commissioner will, in the first instance, proceed to inspect the children employed at the factories, making inquiries with relation to such peculiar cases as may be described in the list, or as may be observed on inspection. He will endeavour to accompany one of the civil commissioners in the first visit to any large manufactory or work of an important class. After the inspection of the physical condition of the persons employed, and the circumstances of each place as to ventilation and other points adverted to in the portion of the instructions especially addressed to the medical commissioners, he will proceed to inspect the residences of the workpeople, whether children or adults, leaving the civil commissioner to make the detailed inquiries respecting the hours of labour, the factory regulations, the treatment of the children, and other points not strictly medical. Each commissioner will, for the sake of despatch, however, avail himself of any information connected with either department of the inquiry which may come before him, or examine a witness of any class, when time or opportunities would be lost by transferring the witness to another commissioner.

"When witnesses come before you to give evidence, you will, in the first instance, examine the labouring men apart, or in classes; and, if the case should require it, each individual singly. If any petition with relation to the Factory Bill have been presented from any workmen in any town or district which you may visit, you will endeavour to learn who have been the chief movers in the petition, and will examine those who may best express the opinions of the petitioners.

"It may be desirable that you should examine such witnesses, when known, before visiting the magistrates or other local authorities; but when you meet with any witnesses who have given evidence before the Committees of Parliament, the more desirable course will be to defer their examinations until the examinations of the other witnesses have been concluded. One objection to the evidence given before the committees being that it has not been taken on oath, you will on no account omit to call those witnesses before you, and re-examine them on oath in relation to the points as to which they have previously been examined, and on any other points.

"You will next examine singly and separately such of the children as you may take indifferently, or as for any special reason you may select.

"In examining children and workmen, you will not omit to question them as to whether since the date at which discussion of the bill before Parliament commenced, and since that of the appointment, any alteration has been made in the regulations of their employers, affecting their welfare. If those alterations are substantial improvements, you will endeavour to ascertain whether any measures may be usefully taken to secure their permanence.

"You will next endeavour to examine separately the parents of the children employed. You will inquire into the general character of the parents of children, and the influence of that character in the control of their offspring, and as to the extent to which any legislative securities would probably be seconded by the exertions of the parents; whether they have complained of the treatment of their children, and if not, how the absence of complaint is to be accounted for.

"It is especially desired that you should endeavour to obtain information from as many witnesses unconnected with the employment of infant labour as you may have opportunity of examining; witnesses unconnected with any parties to former proceedings relative to the subject, as clergymen, medical men, and others. In the cases of such witnesses, you may occasionally find it expedient to relieve them from the fear of displeasure on the parts of persons implicated in their evidence, by receiving their information without subjecting them to any regular form of examination or any publication of their names.

"You will in general take the evidence of the employers of labourers the last; you will, however, should any material contradiction be made to the evidence of the preceding witnesses, re-examine them, or give them an opportunity of answering any new allegations made in evidence.

"In examining the witnesses, it is desirable for despatch, where it may be done, that you should examine one in the presence of the others of the same class; as, workmen with their fellow-workmen; and require their concurrence or dissent to all or any part of the deposition of the examinant. You will examine the parties on oath, unless in particular cases, in which, from the peculiarity of the circumstances, it may appear to you expedient to depart from that practice: you will also in general require them to sign their depositions. If testimony which you deem important be positively refused, or if any obstructions be wilfully placed in the way of an important course of inquiry, you will immediately transmit to the central Board an account of the circumstances of the case, in order that steps may be taken to meet the exigency.

"With relation to workmen or the parents of children whom you may examine, you will endeavour to obtain for them all protection for giving evidence freely, by preventing, where practicable or expedient, their names being made public, or by requiring from their masters some public assurance or pledge that they (the witnesses) shall in no way be prejudiced by any evidence which they may give.

"It is desirable to prevent the presence of reporters at the examinations.

"You will explain to them, that the object of the commission in pursuing this course is, first, to prevent the inquiry receiving prejudice from the partial publication of evidence, which publication must necessarily be piece-meal and imperfect; and next, to protect the free communications of the witnesses.

"The object of the commission being to obtain the most authentic, accurate, and complete information within the shortest time, the central board entrust to your discretion the adoption of any additional other means for the better attainment of that object, or even the liberty of departing from the general course of proceeding thus hastily and imperfectly traced out for your direction, in such cases as experience will not fail to suggest as you advance.

"It is desired, however, that you should give them the earliest information of any occasional departure from the line of your instructions.

"We beg urgently to request that you will closely examine the proprietors of mills or factories, the overlookers, the children, or the other persons employed, as to whether any and what recent alterations have been made in the mode of conducting the business of the factories, and endeavour to obtain an account of persons who have recently been removed from them, and to determine whether such persons have been removed for the purpose of preventing complaint or inspection. You will also let the examinations on oath as to this point, and the result of such examinations, appear conspicuously on your minutes.

"It has been suggested, as a method for facilitating the access of witnesses of the working class, who might be reluctant to press through a crowd to offer their testimony, that you should publish by advertisement, on the eve of your arrival in a town, an invitation to such persons to send in written statements of any particulars which they may have to communicate, or to forward the names of witnesses whom it may be desired to examine, or who may be desirous of being examined, along with statements of the facts with regard to which they be competent or desirous to afford their testimony. We are favourable to some such course of proceeding as that which is above proposed, as a means of ensuring a full hearing to those classes of persons who will not have so good an opportunity as the master-manufacturers of making themselves heard; the time and mode of adoption must be left to your discretion. Probably the best mode would be that of notification by advertisement in the local papers, 'that all communications addressed to his Majesty's commissioners, and left at the post-office before will meet with attention.'

"Where statements are made respecting the health of children and persons employed in factories, and such statements appear to be corroborated by the tabular returns, the medical commissioners will nevertheless compare those statements with the statements which they will receive from collateral and independent sources, as at sick clubs, hospitals, dispensaries, &c., where they will ascertain what proportion of the patients registered were employed in factories, &c.

"The civil commissioners will also follow the same course with relation to their branches of the inquiry."

We are informed that the mode of taking evidence under this commission has been objected to on the ground of the non-employment of short-hand writers.

The question of the expediency of employing short-hand writers was taken into consideration at the commencement of our proceedings, when it was finally decided in the negative. The grounds of this decision were, that the employment of a short-hand writer for each of the twelve district commissioners would occasion great additional expense and delay, besides producing an undigested and useless mass of notes and minutes. The method in general followed by each commissioner, after writing down the statement of the witness, has been to read it over to him for his correction. In whatever degree this mode of taking evidence may be considered to be liable to objection as untrustworthy, that which has hitherto ever been in use with the judges of the land must be considered to be equally so, insamuch as it is in effect the same, excepting that the notes of a judge are not usually read over to the witnesses for correction or addition. A large portion of the evidence has been taken in a more laborious manner, by the commissioners first writing out the questions at length and reading them to the witness, whose deliberate answer has then been taken down, and afterwards read to him. This mode of taking evidence has appeared to us not only to have all the advantage of oral examination, such as that of putting questions suggested by previous answers, but also to secure a degree of care and exactness in the questions, and of deliberate expression in the answers, unattainable by means of a short-hand writer. With respect to the degree of publicity which has been admitted in the examinations taken under this commission, the same course has been followed which has been usual under other commissions, as well as in committees of both Houses of Parliament. We have met with no objections to that course which have not appeared to us merely captious or declamatory, especially when made by parties whose presence was requested instead of being repelled by the commissioners. We have, on the other hand, been made acquainted with circumstances which would have justified a more complete suspension of publicity than has at any time been adopted in the course of this inquiry. While secrecy in some cases, as it had been anticipated in the instructions which were issued by the central board to the district commissioners, was deemed requisite for the protection of operative witnesses from the probable displeasure of their employers, the extreme excitement and violence which, in such places as Leeds and Bradford, was fomented by the influence of delegates and committees, would have rendered the attempt to hold an open court impracticable, had any such unprecedented course been in contemplation. The inaccurate accounts which have been sent abroad of such proceedings of the commission as were necessarily open, the exclusive selection of hostile statements for daily publication, the wilful and persevering misrepresentations of the scope and end proposed to the inquiry, lead us to believe that no other course than that which has been adopted, in accordance with the practice of all former commissions of this nature, could have preserved our whole proceedings from the most mischievous falsification.

The present investigation has led to a full exposition of the condition of the factories in the different districts of the kingdom. We shall endeavour to present a view of that condition, derived from the evidence in relation to the regular hours of labour, the time allowed for meals, the extra hours of labour, the age at which children begin to work, the nature of their employment, the state of the buildings in which that employment is carried on, the treatment to which the children are subjected, and the ultimate effects of their employment on their physical and moral condition.

We shall endeavour to give an abstract of the evidence received from the different districts, in relation to each of these subjects.

Stuart, 3.
Mackintosh, 36.
Mack. 6. 7. 10. 11. 12. 15. 16. 21. 22. 25. 27. 29. 31. 35. &c. &c.
Ibid. 1. 9. 14. 17. 20. 34. 40. 41. &c.

In relation to the regular hours of labour, it appears from the evidence that in Scotland there are two or three factories in which the regular hours of labour do not exceed from ten to eleven daily, but that in general they are from twelve to twelve hours and a half, while in several districts they are not less than thirteen.

Ibid. 1. 17. 20. 31.

It is customary to leave off work on the Saturdays in some places one, and in others two hours earlier than on the other days; but the time thus lost on Saturday is sometimes made up by working a quarter of an hour later on the other days. Mackintosh, 1.

In England, in the north-eastern district, in a few factories, the regular hours of labour do not exceed eleven. In general, both at Leicester and Nottingham, they are not less than twelve. "Eleven hours is called a day at Leeds;" but it is seldom that in this district the hours are really less than twelve, while occasionally they are thirteen. In Manchester the regular hours of work are twelve There are many places in the western district, as at Coventry and Birmingham, in which the regular hours of labour do not exceed ten; while it appears that some of the workpeople labour upon an average not more than nine hours daily. In these towns, indeed, there is no factory labour properly so called, for the operatives, with few exceptions, work at their own houses. But in some of the factories in the great clothing district the hours of labour are the same, seldom if ever exceeding ten. In general, however, they are somewhat longer; both in the carpet and in the clothing factories they are seldom less than eleven and scarcely ever more than twelve; this is the average; for there is considerable irregularity in both; in the carpet factory, partly on account "of the dissipated habits of many of the weavers, who remain idle for two or three days, and make up their lost time by working extra hours to finish their piece on Saturday," and partly because "the weaver has often to wait for material from the master manufacturer where particular shades of colour may have to be dyed for the carpet he is weaving; while the clothing factories, being for the most part worked by water power, cannot of course be carried on with regularity. One of the witnesses, a proprietor, states that owing to the want of a due supply of water the workpeople sometimes cannot work more than three hours a day in summer; and that on an average they do not, in the summer season, work more than six hours a day. Another witness, an operative, deposes that his children in the factory in general go away after nine hours' work, and that they play so much that he does not think they really work above four or six hours. And a third witness, a proprietor, (chairman of the woollen-manufacturers of Gloucestershire,) deposes that in his own factory, in those parts in which children are employed, the regular hours are from nine in the morning until four in the evening, deducting an hour for dinner; and that for the last three years the children have worked only seven hours daily. In all the districts these hours are exclusive of the time allowed for meals, and of time lost from the machinery going wrong, and from holidays.

Pow.1. Drink.20. Drinkw. 5. 16. 20. 38. 40. Pow. 2. 16. Horner & Woolriche, 7. 5. Ibid. 9. Hor.&Woolriche. Ibid. 43. 54. 59. Ibid. 14. 16. 47. Hor. & Wool. 11. Ibid. 54. Ibid. 64. Ibid. 53.

In some factories, in the several districts, there is no intermission of the work, day or night. In such cases two sets of workpeople are employed, each set commonly working twelve hours. Occasionally there are three sets, and then each set works eight hours.

It appears that the time allowed for meals differs considerably in different districts, and in the same district in different factories. In one or two factories in Scotland, the meal hours are one hour for breakfast and one hour and a half for dinner. In a few others three quarters of an hour is allowed for each of these meals; but in the great majority of cases the time allowed is half an hour for breakfast and half an hour for dinner, with no stoppage for tea, or drinking, as it is termed. In the north-eastern district, the practice in some factories, as at Leicester and Nottingham, is to stop half an hour for breakfast, one hour for dinner, and half an hour for tea; but in others only a quarter of an hour is allowed for breakfast and half an hour for dinner; sometimes there is no stoppage either for breakfast or tea, but only for dinner. in some factories for an hour, in others, and this is the more general rule, for half an hour. At Leeds they sometimes stop half an hour for breakfast, one hour for dinner, and half an hour for drinking;

Mack. 36. Mack. 11. 40. 41. 45. 23. Ibid. 5. 7. 10. 13. 14. 16. 17. 19. 20. 21. 22. 23. &c.&c. Drinkwater, 22. 24. 26. 33. Ibid. 48. Ibid. 26. 23. 48. 46. 66. Ibid.

[Drinkwater, 65.] but this is very unusual. It is seldom that they stop more than forty minutes for dinner, and often not at all either for breakfast or drinking. There is, however, much difference in this respect in different factories, and in some it is pretty evident that practices have been resorted to, to cheat the workpeople of a portion of their meal hours, which cannot be too strongly reprobated. [Drinkwater, 79.] "We could see the clock in the yard. I have heard it said that the minute-hand used to tumble when it got to the top at dinner-time; it very seldom used to tumble at any other time. I've seen it drop myself, happen five minutes; so that when it was really twelve o'clock it would drop to five minutes after twelve. This was in the dinner-hour. I can't tell what it was for. We always considered among ourselves it was to shorten our meal-times. We had got wind of it, and one day a dozen of us looked at a window just at the time, and it was so. That was the only time I saw that. I have heard others talk that they had seen it on other days. The overlooker told the time-keeper that all the hands in the mill were grumbling about it, and they would not [Drinkwater, 86.] stand it." "They are all apt enough to think they have time gained on them when there is really no such thing: it may have happened in some singular places. I have heard tell of many doing so, but I never saw it rightly proved. There are plenty of clocks about to check them if anything of that kind was practised. I have worked with a speed-clock; it went with the water-wheel; if they lost any time with the speed by reason of the work not spinning well, so as to make them run a little slower, we worked by the speed-clock; but if the work went well and could be run quicker, then we had to work while it was seven by the real clock. The difference might be sometimes ten minutes or a quarter of an hour. We would have been content with either if they would use the same both ways; but it seemed hard on us to be all loss and no gain." It is stated, however, by many witnesses, that of late some improvement has taken place in Leeds. "Within the last eighteen months several [Power, 45. Drinkwater, 86.] mills have considerably shortened their time of daily labour, and in many instances fifteen minutes are allowed for breakfast, and ten minutes for afternoon drinkings, where formerly no time whatever was allowed for these meals. In many instances they have also given ten minutes, and in some twenty minutes more for dinner than was formerly allowed." "Quite lately, since the agitation of this question, some have had rest for breakfast and drinking, as well as dinner. Mr. Marshall and some others gave it about the time that Hobhouse's bill came out; Walker and Hives, and Atkinson and Stains gave it then. Most of them have not altered to this day. I believe that with those that gave the time, the work went better for the hands before any was given than with those who have never given any time. I mean it was at the best mills that this time has been given. It has not been given where it was most wanted."

[Cowell. Tufnell.] In some factories in Manchester the only stoppage during the day is an hour for dinner. No time is allowed for breakfast or tea: but this is not common. The usual practice is to stop half an hour for breakfast and an hour for dinner; it is seldom that there is any time allowed for tea.

[Hor. & Wool. 13. 16. 18. 39. 46. 49. 58. 51.] In the western district, in all the branches of manufacture, it is customary to allow an hour for breakfast, an hour for dinner, and half an hour for tea; though in some factories only half an hour is allowed for breakfast. [Ibid. 15. 23. 51.] In the fulling department also the work goes on continuously; but it is stated by some of the witnesses that this work requires no superintendence during the meal hours, and by others that two or three of the workpeople are sometimes detained in charge of it; children are never thus employed.

[Mack. 3. 9.] In many factories it is not an unusual practice for the workpeople to stop [Drinkw. 43. 28. Mack. 20.] during a part of the dinner-hour to clean the machinery, this sometimes occupies them half their dinner-hour, at other times not more than ten [Ibid. & Drinkw.] minutes. The children commonly stop to clean their own work. In some factories care is taken on the part of the proprietors to secure to the workpeople the whole of the time allotted to meals; while in others this time is encroached upon without scruple. It will be seen in the evidence that there is the greatest difference in this respect in different factories.

Occasionally, but not often, the work continues without intermission during the whole of the meal hours; the engine never stopping excepting about ten minutes to be oiled, and the workpeople "eating how they can." "Did not stop for meals; used to eat how we could. Never stop to take our meals except at dinner; has gone on so this six years and more. Did not stop for breakfast or tea. Sometimes the breakfast would stand an hour and a half; sometimes we'd never touch it. Many a time I've brought mine out and never touched it, because I hadn't time. All in my room would rather stop, because the breakfast got covered with lint so. No time for breakfast, but got a bite now and then as we could. No time for breakfast or tea; took it as they could; a bite and a run; sometimes not able to eat it from its being so covered with dust." By some of the witnesses it is stated, that this practice of continuing the work during the meal hours is sometimes adopted at the request of the workpeople, who by this means are enabled to leave the factory an hour earlier in the evening, which they prefer to stopping an hour for their dinner. Drinkw.&Power, Leicester and Nottingham. Mack. 34. Drink.

Sometimes it appears to depend in some measure on the nature of the work. Where the practice of working during the whole of the meal hours prevails, the workpeople never leave the factory from the time they enter it in the morning until they have finished their work in the evening. What food they take is either prepared for them in the factory or brought to them already prepared by their friends. In some factories conveniences are gratuitously provided by the proprietors for cooking the food, and detached rooms are fitted up for the workpeople, in which they wash, dress, and eat. Sometimes the school-room in which the children are taught in the evening is appropriated as a dressing and eating room for the adults during the day. But in many cases there is no washing, dressing, or cooking room. There are no conveniences for cooking, except the steam-engine fires; and there is a deplorable want of comfort and cleanliness. In many districts it is the general practice for the workpeople to leave the factory at dinner; sometimes, but not often, at breakfast; and seldom or never at tea. In the best regulated factories the workpeople change their dress on entering and leaving the rooms in which they work; putting off their best things on entering the mill, and resuming them on going out. Whatever be the practice relative to the hours of labour, and the cessation or continuance of the work during meal hours for the adult labourers, to these the youngest children must of necessity submit. Drinkw. 24. 48. Ibid. 51. 58. 62. 61. Hor. & Wool. 14. 16. 20. Drinkwater. Ib. 61. 48. 51. 58. Barry, Mackintosh, Stuart.

In order to regain the time lost by stoppages, whether from the breakage of machinery, from the want of a due supply of water, or from holidays, it is the custom for the people to work sometimes half an hour, at other times an hour, and occasionally even as much as two hours daily, until the whole of the lost time be made up. When the children do not clean the machinery out of the hours allotted for their meals, they clean it at extra hours. In Scotland this appears to be the general practice. Then it is stated by the workpeople that the children clean the machinery at extra hours. "In all the mills in this town (Kirkaldy) the children stop fifteen minutes after the mill sets (after seven o'clock) to clean the machinery." Mackintosh, 5. Ibid.

The account given by the operatives themselves of the extra hours they work to make up lost time is as follows:—"Make up lost time half an hour a day." "Never works more than twenty minutes extra when making up." "Works sometimes a quarter, sometimes half an hour, to make up time." "Lost time made up at the rate of half or three quarters of an hour per day." Ibid. 2. Ibid. Ibid. Ibid. 14.

Makintosh, 22. "For nine months together they have worked an hour extra, making up time."
Ibid. "When water has been frozen, she has been obliged to work until ten or eleven at night to night to make up lost time."

Ibid. 39. Ibid. 41. Ibid. Ibid. The overlookers give the same account. "The time made up at the rate of half an hour a day; and a great deal of the time is not made up." "Seldom make up time, but when they do, it is at the rate of half an hour per day." "Time made up at the rate of from a quarter to half an hour." "Stoppages made up at the rate of half an hour a day."

Idid. 22. Ibid. 27. Managers state, "Option always given to the people whether they will make up holyday time, or have it deducted from their wages, and they generally choose the former." "Lost time is made up at the rate of half an hour per day for four days in the week; make up all holydays but Christmas-day."

Drinkwater, 46. The same practice prevails in the north-eastern district. Thus a proprietor at Nottingham states, "We have got up time when an accident has happened to machinery, as much as three hours a work." And similar statements will be found in the depositions made by the workpeople in every part of the evidence from this district.

Horner & Woolriche, 11. In the western district the practice of working extra hours has arisen partly from the irregular supply of water, and partly from the irregular habits of the operatives, occasioned in some measure by the irregularity with which they are furnished with material for their work. In both cases this irregularity tells very severely on the children, and more especially in the carpet factories on the drawers (always children), who must attend the weaver at whatever time he is at work, and who are often called up at three or four in the morning, and kept on for sixteen and eighteen hours. The drawers are entirely under the control of the weavers, both as to their time of work and payment, the master neither engaging them nor exercising any further control than requiring them to be dismissed by the weaver in cases of misconduct.

For additional labour to make up lost time from stoppages arising from any of those sources, with scarcely a single exception, no additional wages are paid, and the workpeople, young and old, perform this labour with reluctance.

On the other hand, when from any cause there is a press of work, requiring extra hours of labour, for which extra wages are paid, there seems to be no limit to the period for which the people will continue at their employment; sometimes indeed reluctantly, but more often not only without grudging, but with thankfulness, looking upon the permission to do so as a privilege and boon. To the young persons, and especially to the young female workers, this extra labour is often extremely irksome and harassing, and the younger the age the more injurious the consequences, as will appear from facts hereafter to be stated.

SCOTLAND. Mackintosh 13. Ibid. 42. Ibid. 5. This practice of working extra hours is peculiar to no district, and to no description of factory; but is more or less common to all. "Have sometimes and do now occasionally work sixteen hours." "Before that (two years ago) very commonly worked fourteen or fifteen hours through the whole winter, and got extra wages. It was our pleasure; we did it many a time just for the sake of the wages." "Worked all last night [I found her working at a quarter before six]; worked from a quarter before six yesterday morn; will work till six this evening; thirty-four hours, exclusive of two hours for meals; did this because the hands were short, and she should gain an additional shilling; has at this moment, although she has been standing already twenty-four hours, no pain in her knees or ankles; is not tired, or else would not do it." "It's all our pleasure; they do not force us to do it; would prefer the present hours and pay to a reduction of both."

NORTH EASTERN DISTRICT. Power, 14. "Am twelve years old. Have been in the mill twelve months. Begin at six o'clock, and stop at half past seven. Generally have about twelve hours

and a half of it. Have worked over-hours for two or three weeks together. Worked breakfast-time and tea-time, and did not go away till eight. Do you work over-hours or not, just as you like?—No; them as works must work. I would rather stay and do it than that anybody else should come in my place. I should lose the money. If over-hours are put on next week, shall you be glad or sorry?—It wont signify. I shall be neither glad nor sorry. I hear you are to have an hour a day overwork next week, do you mind about it?—No. Don't you play sometimes after work is over?—Yes, sometimes. Well, are you not sorry to lose that?—No; I don't mind about it. I had rather work as I do than lose any of my wages. I go to a Sunday-school sometimes; I went first about a month ago. I have been every Sunday since. I can only read in the alphabet yet. I mean to go regular. There is no reason why I should not. I wants to be a scholar." "Have worked here (Milne's) two Ibid. 19.
years; am now fourteen; work sixteen hours and a half a day. I was badly, and asked to stop at eight one night lately, and I was told if I went I must not come back." "I have worked till twelve at night last summer. We Drinkwater, 48.
began at six in the morning. I told book-keeper I did not like to work so late; he said I mote. We only get a penny an hour for over-time." "We Ibid. 26.
used to come at half past eight at night, and work all night, till the rest of the girls came in the morning. They would come at seven. Sometimes we worked on till half past eight the next night, after we had been working all the night before. We worked in meal-hours, except at dinner. I have done that sometimes three nights a week, and sometimes four nights. It was not regular; it was just as the overlooker chose. Sometimes the slubbers would work on all night too, not always. The pieceners would have to stay all night then too. They used to go sleep, poor things! when they had over-hours in the night." "In 1829 they worked night and day. The day set used to Drinkwater, 18. 19.
work from six till eight and nine, and sometimes till eleven or twelve. The children who worked as pieceners for the slubbers used to fall asleep, and we had much trouble with them."

Overlooker at Milne's says, "We have forty-five children. Our regular day Power, 21. 22.
is from six to seven. It should be an hour for dinner, but it is only half an hour. No time allowed for tea or breakfast: there used to be a quarter of an hour for each; it's altered now. We call it twelve hours a day. Over-time is paid for extra. When we're busy we work over-hours. Our present time is 'till half past nine (beginning at six). It has been so all the winter, and since to this time. We have some very young ones; as young as eight. We occasionally find it necessary to make a difference as to the time of keeping some of the children. Master has said, Pick out the youngest, and let them go, and get some of the young women to take their places. At the factory of Messrs. Mills and Elliott they go on working all the night as well as day. I believe they have done so for the last year and a half; they have left it off about a week."

Commissioner states, "We have distinct information that in Milne's factory Ibid. 16.
(Nottingham), the clearing children are kept to their work constantly during a period of sixteen hours."

"Am eleven years old; used to go to the factory a little before six, some- Drinkwater, 32.
times at five, and work on till nine at night; never came away before nine, without they were on the spree of it, getting a drop; had half an hour at breakfast, an hour at dinner, and half an hour at tea; did not always have full meal-times. They sometimes rang the bell before the time, and we were forced to go. I worked all night one night with Castles. That was for our own good. We chused it ourselves. We wanted to have something to spend. We had been working from six in the morning the day before; and Castles asked us whether we were all agreeable to work all night on, because it was he Christmas week; he wanted something himself. There were two of us

together; we were both agreeable. We were tired, but he let us have a lie down in the course of the night a little while. We went on working till nine o'clock the next night. We were very tired then. I fell asleep the second day once or twice. Castles did not strap me for it that day; he was very good-humoured then, because we had been working all night. I have known the other hands work all night three nights in a week."

Drinkwater, 51. 52. FOREMAN.—"We are now working half an hour over-time. We reckon it as nearly as we can to their wages for over-hours. Those who have three shillings get a halfpenny per hour. They never grumble about over-hours; sometimes ask for them; sometimes ask to give over. Most of the last year we have been working night-work; now we mean to keep to fifteen hours and a half, and twelve and a half on Saturday. The hands were not satisfied when we worked only twelve. They wanted more; there was a general cry amongst them. May have had some that left us from over-work, but not many."

Power, 61. OPERATIVE.—"Like the sixteen hours as well as the twelve; it is very acceptable when the pay comes. We were asked when it was altered to sixteen hours, and the hands took their choice. Manager stood at the door
Drinkwater, 57. and asked every one as they came out."—"We are working now as long hours as we ever worked, barring half an hour; don't know that the time ever did
Ibid. 61. us any harm, and am sure the money did us a deal of good."—"There's one bad thing here (*laughing*), we have no over-hours. I've heard of your coming down, and what it was about. I think it a very good thing. I hope you wont make us work shorter hours though.

From the causes already assigned, namely, the irregularity with which the operative is supplied with material for his work, the irregularity of the power by which the machinery is driven, and the dissipated habits of the workers, favoured, if not induced, by the occasional idleness growing out of the two first causes, it appears that in the carpet-factories it is the constant practice,
Horner & Woolriche, 23. and in the clothing district the frequent practice, to work extra hours:—"It is very much the case with some sort of men to go idle part of the week and to work extra hours the rest. In such cases I have known men to work from three o'clock in the morning till ten o'clock at night; the drawers must work the same hours; they must always go together; they can't do without one
Ibid. 24. another." "It is the practice for the weavers to be idle and dissipated part of the week and to work extra hours the rest. We abound with that evil; we witness it every week round; even the regular workmen must often be idle part of the week, from the irregularity of the work coming in. It is very op-
Ibid. 18. pressive indeed to the children." "I have known instances, in the depth of winter, of drawers being called up to work by four o'clock in the morning,
Ibid. 19. and earlier. I believe it is the common practice for the idle weavers to place their draw-boys in the loom, and to employ younger boys or girls as drawers, to make up for their own laziness or dissipation. The weavers are in general idle the early part of the week, and they afterwards work from eighteen to twenty hours to make up their lost time, during which the draw-boy or draw-girl must attend them. I have known frequent instances of their commencing work at two or three o'clock in the morning."

Ibid 46. 50. In the clothing district both workmen and masters agree in stating that if extra work for extra pay were refused when a press of business comes, the
Ibid. 46. 53. workmen so refusing would lose their situations; both also concur in the statement, that it is the constant practice for parents, and even for children
Ibid. 59. themselves, to apply to the masters for extra work for additional wages, and cases have been detailed in which children have worked upwards of fourteen hours.

It appears that parents encourage their children to make the extraordinary efforts of which we have given some examples, by leading them to consider the wages which they thus earn as peculiarly their own, although a cheat is often practised upon them even with regard to these extra wages. While all the witnesses agree in the statement, that whatever the child earns by its regular hours of labour is uniformly appropriated by the parent, it appears that a large portion of the additional wages earned by extra hours is also taken by the latter. Boy, twelve years old, states, " We are paid for over-hours at the rate of two-pence for three hours; I have always that for myself. What do you do with it?—I save it for clothes sometimes. I put it into a money-club for clothes. I have worked nine hours over in one week. I got for that five-pence halfpenny. I gave it my mother, and she made it up to sixpence, and put it into the money-club. She always puts by sixpence a week from my wages for that. Then your mother gets what you earn by the over-hours, don't she?—No; I gets it for myself. Does your mother like you to work over-hours?—No; she don't like it. She never asked for me to be excused. She knows it would'nt be no use. Sometimes mother gives me a halfpenny to spend. What do you do with it?—I saves it to buy shoes. Have never saved above a shilling for that; mother put more to it and bought me a pair. I have sometimes bought some good stuff with it." Power, 14.

" Was very nigh nine years of age when I first went to piecen. Got two shillings and sixpence a week at first. Think I was a good hand at it. When I had been there half a year got three shillings. Am now eleven. Get four shillings. Always pay my wages to my mother. Never spent a penny in my life. My father takes my wages from the pay table, and gives to me to take home to my mother. My mother used to give me a halfpenny or a penny again; she said to buy me apples, or what I'd please to tell her, for me to eat on Sunday. She knew of it before-hand [his working all night on, after having worked from six in the morning the day before.] I told her at my meal time. She let me keep what I got so."—" All the younger children give their earnings to their parents."—" Children constantly apply to the masters for extra work, when such work is to be got, because it is usual for them to get the extra wages as pocket money." Drinkwater, 31. Ibid. 10. Horner & Wooliche, 50.

It appears in evidence, that in some rare instances children begin to work in factories at five years old; it is not uncommon to find them there at six; many are under seven; still more under eight; but the greater number are nine; while some, but comparatively few, branches of manufacture do not admit of the employment of children under ten years of age. Drinkwater. Power. Horner & Woolriche.

The present inquiry has led to a very complete exposition of the nature of the labour in which children are employed in the different factories of the kingdom; for an account of which, as relates to Scotland, we beg to refer to the Reports of Sir David Barry, pp. 2. 3. 4. 5. 35. et seq., to those of Mr. Mackintosh, pp. 5. 12. 4. 5. 8. 15. 21. 11. 13. 17. 23. 26. 28. 29. 36. 37. 39. 18. 14. 9. 2. 15. 18. 42. et seq., and to those of Mr. Stuart; relating to the north-eastern district of England, to the Reports of Mr. Drinkwater, pp. 2. 3. 5. 13. 17. 32. 33. 38. 43. 48. 51. 52. 61., and to those of Mr. Power, pp. 3. 12. 13. 15. 16., in whose Reports will also be found an account of the nature and state of what may be termed the domestic manufactures of Leicester and Nottingham, as compared with the factories in the same places, 4. 5. 10. 17. et seq.; as relates to Lancashire, to the Reports of Messrs. Cowell and Tufnell; and as relates to the western district, to the Reports of Messrs. Horner and Woolriche, Birmingham, pp. 1. 3. 4. 5. 7. 8. 9. 10., Kidderminster, 11. 12. 14. 15. 27; clothing district, Stroud, Gloucester, Frome, Tiverton, &c., 30. 37. 50. 51. 59. 27. 60. 68.; and relating to the potteries, to the Report of Mr. Spencer.

The present inquiry has likewise brought together a large body of evidence relative to those various circumstances connected with the state of factories which concur with the nature of the employment in exerting an important influence on the health of the workpeople, whether children or adults, but which more especially affect the health of the former. Such concurrent circumstances are, the situation of the factory, the state of the drainage about the building, the size and height of the work-rooms, the perfect or imperfect ventilation, the degree of temperature, the nature and quantity of the effluvia evolved, whether necessarily or not necessarily, in the different processes of manufacture, the conveniences afforded to the workpeople for washing and changing their clothes, on leaving the factory, and the habitual state both of the factory and of the operatives as to cleanliness. Details, which place in a striking point of view, on the one hand, the conservative influence of careful and judicious attention to such concurrent causes in the general arrangements of the establishment; and, on the other, the pernicious consequences that result from inattention to them, will be found in the account given of the state of individual factories in most of the Reports of Sir David Barry, in the Reports from Scotland in general, and in many parts of the Reports from Leicester, Nottingham, and the western district. In relation to all those circumstances, the Reports of the Commissioners agree in shewing that the large factories, and those recently built, have a prodigious advantage over the old and small mills. The working-rooms in the large and modern buildings are, without exception, more spacious and lofty; the buildings are better drained; more effectual expedients are adopted to secure free ventilation, and to maintain a more equable and moderate temperature.

Stuart, 38. Gilchrist's mill at Bervie. Ibid. 47. Mackintosh, 23. Stuart, 21.

It is of the old and small mills that the Report pretty uniformly is—" dirty, low-roofed, ill-ventilated, ill-drained, no conveniences for washing or dressing, no contrivance for carrying off dust and other effluvia, machinery not boxed in, passages so narrow that they can hardly be defined, some of the flats so low that it is scarcely possible to stand upright in the centre of the rooms;" while the account of the recent structures and the large establishments in general is—" infinitely better managed in respect to ventilation, height of roofs, and freedom from danger to the workers near the machinery, by the greater width of the passages in the working-rooms, and by the more effectual boxing in of the machinery, than those on a small scale." There are not wanting establishments in which every advantage of this kind is combined in an almost perfect degree, of which the following may be cited as examples:—

Stuart, 64.

Deanston Cotton-mill Factory, near Doune, in Perthshire.—" This is one of those beautifully-situated and admirably-regulated great manufacturing establishments which it is a pleasure to see, on account of the general arrangements of every department of this extensive work, as well as the happiness which a numerous population, engaged in the pursuits of industry, apparently enjoy. The apartments in the mill first erected, are not equal in height nor in other respects to those of the works lately erected; but the whole are clean, well ventilated, and have the machinery well fenced. The preparing-rooms in the lately erected part of the work are, owing to the superior construction of the fanners, which blow the whole of the dust to the open air, more thoroughly freed from the impurities generally prevailing in the preparation-rooms than those in any factory where we have hitherto been. Indeed, I ought more properly to have said, which was literally the case, that there was no appearance of dust nor of impure air in those preparing-rooms. Even in the web-dressing-room a fanner is most usefully employed in dissipating the noxious heat and moisture. It seems strange that those fanners have not yet found their way into the flax-spinning establishments which we have seen, and where they are so very requisite on account of the quantity of dust and refuse of the material floating in the room, to such an extent as almost to obscure the nearest objects. The windows, instead of being constructed in the usual way, in many of the mills which we have seen, so that only a single pane of glass in each window can be opened, are so hung that the whole of the upper part of each window may be let down from the top, and a free current of air admitted. The general heat of the apartments is from 65° to 70°. A greater degree of warmth is never required, excepting in the web-dressing-room, where the thermometer to-day stood at 80°. The temperature of the atmosphere yesterday, in the shade, at the period of our inspection, varied from 65° to 68°. There are here apartments for the females to dress and undress in, and a pipe of water in each story, and every arrangement is adopted throughout the work that tends to the convenience and accommodation of the persons employed. The workers live at the distance of about a mile from the works, with the exception of about a hundred of them, for whom the company have built houses, let to them. I can hardly say whether the construction of those houses, or the ingenious contrivances with a view to the convenience of the people which Mr. Smith has put in execution, or the cleanness and neatness with which the interiors of those nice cottages are kept by the workers, are most to be admired. There are bits of garden ground attached to each of the houses, and a drain has been constructed for carrying off every sort of filth. The whole arrangements about this extensive factory, at which cotton-spinning, power-weaving, iron-founding, and machine-making are carried on, are obviously made with a view, as far as possible, to the substantial comfort of the people; and a more cheerful, happy-looking set of industrious men and women, and of young people, is seldom, if I am not mistaken, to be found. There is abundance of room throughout the whole work; no appearance of human beings crowded on each other in any part of it. There are forty spinners in an apartment eighty-two feet long by fifty-two in breadth."

Medical report of the cotton spinning & weaving factory of Mesrs. James Finlay and Co. Barry, 38.

" The rooms are ventilated in the old mill by means of windows, and in the new mill by means of openings between the windows into chimneys, in addition to windows opening up and down. The drainage is perfect. The water-closets have water-traps fitting into moveable receptacles, which are removed every morning. Rooms, with water-cocks, for washing and dressing, are being prepared in the new mill. The general atmosphere of the rooms is clear and well ventilated. There are no offensive smells. *Dust fans* are employed, revolving in large tubes, which draw up all the dust with considerable force, and keep the atmosphere of the rooms light, fresh, and agreeable. These machines are highly worthy of general adoption in all manufactories."

Stuart, 49.

COTTON-MILL of MESSRS. BANNERMAN.—" A splendid work, erected only a few years ago, the size over walls two hundred and thirty feet by fifty-three. There is here an ascending and descending room, moved by steam; but what pleased me most, on going through the extensive apartments of this establishment, was to observe the sufficient space which each worker enjoys, so that, even in an atmosphere generally heated to about seventy degrees, there is at least apparently absent the effluvia created by any crowded number of human beings. There are about seventy-four workers in each of the spinning apartments, two hundred and twenty-five feet long by forty-eight and a half wide; not more, relatively, than are often found in the drawing-room of persons assembled for a private dinner party. Dressing and undressing rooms are provided for the female workers on each floor, in which their working clothes are kept. There is also a pipe of spring water for drinking in each apartment, a large pipe of water for extinguishing fires on each story, large fanners in the preparing room to free it from dust. The machinery is well fenced. The workers smiled when I asked if they had any fault to find or complaint to make."

Stuart, 92.

COTTON-MILLS OF NEW LANARK.—" Still under the same excellent management, with a view to the health, education, and general comfort of the workers, which prevailed during the proprietorship of the late philanthropic

Mr. David Dale, of Glasgow, who founded the establishment, and of his son-in-law and successor in the mills, the well known Mr. Robert Owen. At the mills, three in number, employment is given to nine hundred and thirty workers. The number of working apartments is twenty-four. I found that above two hours were spent in walking through them as quickly as I could, occasionally talking to the workers, to enable me to select a few of them as witnesses. A fourth mill is at present in course of being erected. The whole of the apartments, the walls, the floors, and the machinery, were thoroughly clean, and no unpleasant smell could be detected anywhere. The windows all open from the top; they are furnished with linen blinds where necessary; and all the working-rooms and passages of the buildings seemed to be perfectly well ventilated, and the machinery sufficiently fenced; the floors, walls, and the machinery are as clean and as bright and as free from dust as one should expect to see in a first-rate drawing-room."

See also Barry, 1. 6. Mackintosh, 31. 39. 45. 52.

While not a few of such signal examples are recorded of a beneficent care exercised over the workpeople, yet it must be admitted that there are too many instances in which an utter disregard is shewn, not only to their convenience and comfort, but even to circumstances which must influence, in no inconsiderable degree, their moral feelings and habits. Had the fact not been established by indubitable evidence, every one must must have been slow to credit, that in this age and country the proprietors of extensive factories could have been indifferent to the well-being of their workpeople (for the matter is not one *merely* of convenience and comfort) to such a degree as is implied in the following statements:—" But one water-closet for both sexes, which children and men and women use indiscriminately." " Privies situated in view; common to males and females: this in his (witness's) opinion has a tendency to destroy shame, and conduces to immorality." " Workers complain of smells from the water-closets." " Picking-rooms pretty well freed from dust by effective fanners; but there is considerable annoyance to the workers from the effects of the water-closets; the effluvia must be unpleasant in warm weather; it made the walking through the apartments to-day very disagreeable in several places;" and yet this is the account given by one of the commissioners of a factory stated to be " remarkable as that at which the finest cotton is spun in Scotland," and as having " the greatest number of spindles in Glasgow, about 43,000." It would appear that there is one district in England in which the privies are in a condition no less disgusting and disgraceful.

Stuart, 77. Mackintosh, 9. Stuart, 78. Spencer's report on the Potteries.

It will appear from the evidence annexed to this report that the commissioners have everywhere investigated with the utmost care the treatment to which children are subjected while engaged in the labour of the factory. These inquiries have obtained from the children themselves, from their parents, from operatives, overlookers, proprietors, medical practitioners, and magistrates, such statements, amongst others, as the following:—" When she was a child too little to put on her ain claithes the overlooker used to beat her till she screamed again."—" Gets many a good beating and swearing. They are all very ill used. The overseer carries a strap." " Has been licked four or five times." " The boys are often severely strapped; the girls sometimes get a clout. The mothers often complain of this. Has seen the boys have black and blue marks after strapping." " Three weeks ago the overseer struck him in the eye with his clenched fist so as to force him to be absent two days; another overseer used to beat him with his fist, striking him so that his arm was black and blue." " Has often seen the workers beat cruelly. Has seen the girls strapped; but the boys were beat so that they fell to the floor in the course of the beating, with a rope with four tails, called a cat. Has seen the boys black and blue, crying for mercy."

SCOTLAND. Stuart, 18. Mackintosh. Ibid. 3. Stuart. Ibid.

"The other night a little girl came home cruelly beaten; wished to go before a magistrate, but was advised not. That man is always strapping the children." "The boys are badly used. They are whipped with a strap till they cry out and shed tears; has seen the managers kick and strike them. Has suffered much from the slubber's ill-treatment. It is the practice of the slubbers to go out and amuse themselves for an hour or so, and then make up their work in the same time, which is a great fatigue to the pieceners, keeping them 'on the run' for an hour and a half together, besides kicking and beating them for doing it badly, when they were so much tired." "The slubbers are all brutes to the children; they get intoxicated, and then kick them about; they are all alike." "Never complained to the master; did once to his mother, and she gave him a halfpenny not to mind it, to go back to work like a good boy. Sometimes he used to be surly, and would not go, and then she always had that tale about the halfpenny; sometimes he got the halfpenny, and sometimes not. He has seen the other children beaten. The little girls standing at the drawing-head. They would run home to fetch their mothers sometimes."

North-Eastern District. Power, 18. Drinkwater, 12. Ibid. 14. Ibid. 27.

"Hears the spinners swear very bad at their piecers, and sees 'em lick 'em sometimes; some licks 'em with a strap, some licks 'em with hand; some straps is as long as your arm; some is very thick, and some thin; don't know where they get the straps; there is an overlooker in the room; he very seldom comes in; they wont allow 'em if they knows of it; (child volunteered this last observation; asked how she knew that the overlookers would not allow the spinners to lick the little hands, answers, "Because I've heard 'em say so.") Girls cry when struck with strap; only one girl struck yesterday; they very seldom strike 'em."

Lancashire District. Cowell.

"There is an overlooker in the room, who is a man; the doffer always scolds her when she is idle, not the overlooker; the doffer is a girl; sometimes sees her hit the little hands; always hits them with her hand; sometimes the overlooker hits the little hands; always with her hand when she does; her mother is a throstle spinner in her room; the overseer scolds the little hands; says he'll bag 'em; sometimes swears at 'em; sometimes overseer beats a 'little hand;' when he does it is always with his open hand; it is not so very hard; sometimes on the face, sometimes on the back; he never beats her; some on 'em cries when they are beat, some doesn't; he beats very seldom; didn't beat any yesterday, nor last week, nor week before; doesn't know how long it is ago since she has seen him strike a girl. If our little helper gets careless we may have occasion to correct her a bit; some uses 'em very bad; beats 'em; but only with the hand, and pulls their ears; some cry, but not often; ours is a good overlooker, but has heard overlookers curse very bad; the women weavers themselves curse; has never cursed herself; can say so honestly from her heart."

"Drawers are entirely under the control of the weavers; they must obey their employer; if they do not they are sometimes beat and sometimes discharged." "I chastise them occasionally with a light whip; do not allow it by my workmen; sometimes they are punished with a foolscap; sometimes with a cane, but not severely."

Western District. Horner & Wooliche, 11. 12. 5. 7.

It appears in evidence that in Scotland, and in the eastern district of England, where the harshest treatment of children has taken place, the greatest number of bad cases occur in the small obscure mills belonging to the smallest proprietors, and that the bad treatment is inflicted by violent and dissipated workmen, often the very men who raise the loudest outcry about the cruelties to which children are subject in factories. A striking picture of a mill of this class, one of the very mills indeed in which various witnesses depose that the treatment is oftentimes harsh and brutal, is given by Mr. Stuart:—"It seemed more to resemble a receptacle of demons than the

Report of Mr. John Gilchrist's mill at Bervie, in Kincardineshire, 38. 47.

workhouse of industrious human beings. We saw the workers, it must be admitted, at a moment not propitious for them, when they were only regaining their senses after the bacchanalian orgies of the former evening and night, which had too obviously been spent in the riotous debauchery following a market or fair held at Bervie on the preceding day. The appearance and language of the workers, both men and women, proved the state of demoralization which exists here. The house of Gilchrist, the mill-owner, presented a picture of filth and want of comfort of every kind, such as I have rarely seen anywhere else. Those engaged in vicious courses, to whom the regularity and excellent regulation in many respects of the Montrose establishments would of course be irksome, doubtless find a fit asylum here; but it was painful to find in the bothy, the eating and sleeping room of such a nest of profligates, two or three young females without a parent or relation there or in the neighbourhood to look after their conduct, or to make any attempt to rescue them."

Ibid. 24.

A mill of the same class, likewise inspected by Mr. Stuart, is one specified by Alexander Dean, in the printed Report of the Committtee of the House of Commons of 1832, page 368 and sequel. "The name of the persons to whom the mill belongs," says Mr. Stuart, "is Braid, and they both appear to be persons of violent, ungovernable temper, and of habits very unlike those of the mill-owners whom we have seen here: in short, the management of this mill is an exception almost to anything which we, or at least I, have seen elsewhere." "The apartments in the spinning-mills are, as it appeared to me, the dirtiest and most low-roofed we have yet seen; several of them very damp, windows so constructed that they cannot be opened, and the smell of whale-oil and tar very disagreeable. I complained to Mr. Wilson, junior, of the bad air and smell, as being to me almost intolerable, but he treated my opinion very lightly, telling me that the smell of oil was peculiarly healthful, and that he preferred it to that of the carnation."

Stuart's report of Messrs. Wilson and Co's woollen mills at Bannockburn, 67.

In considering the statements of the severe and cruel treatment of children it would be injustice not to bear in mind that it is established by the most abundant evidence that in Scotland at least the small mill is the only factory in which such treatment ever takes place in the present day, and that there are many even of the smallest mills in that country honourably distinguished for a kinder treatment of their workers; but the great mass of the people employed in factories, and especially the young, are in establishments of which such descriptions as the following are given: "We reached Catrine, the great manufacturing establishment of Messrs. James Finlay and Company, yesterday morning; I had great pleasure in walking through the eighteen apartments of the spinning-mill and power-loom weaving establishment, and witnessing the admirable order of the works, and the apparent happiness of the people employed, which is quite as remarkable and as obvious as at any of the other great factories situated in country districts. The windows open from the top; the rooms are thoroughly ventilated: there is a clock and a thermometer in every room; no unpleasant smell in any part of the work; the utmost cleanness and neatness prevail throughout; the works are well provided with fanners in the preparing rooms; there are between 800 and 900 workers, all occupying houses originally built by the company, of a very different and superior description from those generally occupied by persons of the same situation in life in this country. They have a chapel and every establishment necessary for their accommodation. The population of the village amounts to 4,253 persons, one half of which is a population engaged in manufactures; yet in the last twenty years the landed proprietors of the parish have only been called on to pay for the poor 212*l.* 14*s.* 1*d.*, not much more than 10*l.* per annum."

Stuart, 93.

Stuart, 50.

The general appearance of the workers at Stanley Mills, consisting of about two thousand persons, who were on our account dismissed for dinner at an earlier hour than usual, that we might have a good opportunity of seeing them, in passing us in small numbers at the gate, was very gratifying in point of good looks, health, apparel, &c. The porter at the gate, who was a worker at the mills from the period when they were set a-going, is eighty-four years of age, and in the enjoyment of good health. So is his wife, though now eighty-eight. One of the female workers, who has been at this work for many years, emphatically replied to my question, how she liked it? "real weel."

With scarcely a single exception, in all the other factories which may be considered large, though not on such a magnificent scale, the care taken of the workpeople by the proprietors is equally paternal, and the treatment of the children equally considerate, gentle, and beneficent. "We cannot," says Mr. Stuart, "send off our report respecting Mr. Craig's factory, without adding our testimony to that of Sir David Barry as to the admirable management of this factory, and the uniformly kind and benevolent treatment which the persons employed declare they receive from Mr. Craig and his family. The workers evidently took pleasure in making it known to us. The workmen some time ago held a meeting with a view to petition Parliament that their hours should be curtailed, but they separated without taking any step, when some of the people present reminded them that they were safe in Mr. Craig's hands, that he would do nothing contrary to their interest, and had himself expressed a wish that the hours of labour should be shortened." Stuart's report on Mr. Craig's dry flax-spinning mill at Prestonholme near Edinburgh, 1.

"HUGH GRANT, twenty-nine years old, has been at this mill for about twelve years. Is now an overseer. In extricating some tow from the machinery, the rollers caught his hand, and drew it in. He was off work after losing his hand for fourteen weeks. Mr. Craig provided medical assistance for him. He and Mrs. Craig are most attentive to the health of all, and 'would not permit any thing to be done to hurt the comfort of the people.' They came and saw him after the accident, and offered him everything in their house that could be of use to him. 'He could not say but he was well done to.' Ibid. 3.

"The arrangements in the great spinning establishment of Messrs. Richards and Company at Montrose, and in their spinning-mill at Logie on the River North Esk, a few miles from Montrose, are highly creditable to the active and excellent superintendence of Mr. Jameson, the intelligent manager of the work. The machinery is boxed in so completely that those mills may serve as models in that respect to all similar establishments in this country. The passages in the spinning-rooms are large, seats are in every case provided for the spinners, the apartments are thoroughly cleaned and ventilated, and a corresponding air of comfort and cheerfulness pervades the workers." Ibid. 38.

Reports of factories in which corporal punishment is strictly forbidden, and, as is proved by the testimony of all classes of witnesses, is never inflicted, will be found also in Mr. Mackintosh's report at pages 3. 14. 15. 17. 18. 19. 20. 21. 27. 28. 32. 33. 42. 36. 37. 40. 41. 42. 43. 45. By all classes of witnesses it is stated, that "strapping was more customary in former times than it is now;" (Mackintosh, 19.) "that, as far as he sees, the system is very much changed;" (Ibid. 27.) "that formerly there was a great deal of strapping, but there is very little now;" (Ibid. 28. Ibid. 31.) "that the usage of the children is very different; they are not now beat;" (Ibid. 35.) "that he has seen boys severely beat when he was a young man, but not for a number of years;" "that he does not use a strap now, though he did formerly." (Ibid. 41.) Mr. Stuart reports, "I am glad to be able to state, that in the course of my survey here not one recent case of cruelty, or of positive bad treatment of any of the workers, young or old, has been laid before us. And this is the more remarkable, after all that has been said; in the first place, because the operatives publicly and by advertisement invited all persons aggrieved to come forward; and secondly, because, although several cases of tyrannical conduct on the part of the mill-owners or overseers were communicated to the committee of the (Stuart's report of Dundee, 22.)

House of Commons by the witnesses before mentioned from Dundee, all of whom, except Smart, now deceased, appeared before us, and confirmed on oath their evidence, as contained in the committee's report, the mill-owners, while they made great exertions with that view, through their solicitors, Messrs. Shiell and Small, have not had much if any success in shaking the credibility of those witnesses in important particulars. I need not add, that all the witnesses on the subject of ill-treatment who offered themselves to us, or were suggested, have been examined. I have never failed, in every mill which I have inspected, to ask several of the operatives whether they have any ground of complaint against their employer, and have almost always received answers in the negative, excepting as to the long hours of labour."

Stuart's report from Perth, 51.

" If anything like abuse in respect to punishment now takes place, it is in the smaller mills, where a strap, as it appears from the evidence, is not unfrequently in the possession of the overseer; but I doubt very much whether any such abuse exists, or has of late years existed, in this country, in any degree worthy of notice. My impression, founded on previous knowledge, and on what has transpired in this investigation, being decided, that country schoolmasters in Scotland are far more apt than mill-owners or overseers to exert their authority, by applying the laws with undue severity. The accusation of cruelty or severity, which was in the course of the parliamentary investigation of last year brought against the owners and overseers of flax-spinning mills in this country, whatever may have happened formerly, seems to me to be utterly and entirely unfounded at the present day."

Ibid.

" The workers here seemed very healthy and happy-looking. A few of the male workers have formed themselves into a band of instrumental music, who were, on account of their proficiency, I presume, allowed a holiday while we inspected the work, and serenaded us while there, and from the work to the pier, where we re-embarked. There is here an excellent understanding between the manager and the workers; which, wherever it takes place, is attended with good effects to both parties."

Report from Greenock, 119.

Ibid. 121.

" It is established by evidence, altogether incontrovertible, that the charges made against proprietors of factories, as having authorized the infliction of severe punishments on the young workers, or the exercise of oppressive measures of any kind towards them, have no foundation whatever. At Dundee, where instances of improper treatment had, in 1832, been specified by witnesses examined before the committee of the House of Commons, the operatives, before and while the commissioners were there, by public advertisement, invited all those who had been ill-used to come forward and state their complaints to us, but not one case of that description, of recent date, was brought forward; and all the respectable witnesses throughout Scotland agree in declaring, that whatever may have happened in the beginning of the factory system, at a period when coercion was far more resorted to even in public schools than now, they are ignorant of any recent instance of punishment attended with severity, or with anything like unpleasant consequences. Mr. Steele, who appeared at the head of the deputation of operatives in Glasgow, and who was frequently in communication with us there, expressly declared on oath, in his evidence taken by Mr. Mackintosh, that he knew of no case where the proprietor of a factory had shewn any disposition to treat any of the workers cruelly or oppressively. On the other hand, very many of the proprietors of factories out of towns, such as the proprietors of the Stanley, Deanston, Lanark, and Catrine Works in the country, and such as Messrs. James and William Brown of Dundee, Messrs. Richards of Montrose, Messrs. Bannerman of Aberdeen, Messrs. James Oswald and Co. of Glasgow, and many others in the towns, have, for long periods of years, voluntarily incurred a great expenditure, entirely with a view to the health, the education, the religious instruction, the morals, or the general comforts of the great population employed by them. Is it then fit, even supposing the chief enactments of the bill to be necessary, that the phraseology of Lord Ashley's bill, and its severe enactments by penalties, and their payment to common informers, should be applied to individuals who have acted, and are acting, in this most liberal, disinterested, and benevolent manner; or generally against the other proprietors of factories in the great manufacturing districts of Scotland?"

Statements to the same effect are made by the other commissioners for Scotland, Mr. Mackintosh and Sir David Barry. "Whatever may be thought of the nature or duration of the work to be performed by children, the general tenour of the whole body of evidence will be allowed sufficiently to disprove the existence of any system of corporal punishment as a stimulus to exact it, some very conclusive admissions (Mr. Steele, p. 67, Mr. Henderson, p. 81), negative satisfactorily the existence of anything which can be called habitual cruelty practised upon children. These admissions are made by men who have had extensive experience of the whole interior management of factories, and their avowed predilections would not lead them to conceal any deformities in the system. We had, I believe, during our progress, no one intimation, even anonymously, to direct our inquiries to any quarter where any habitual ill-usage of children was insinuated to exist at present. The facts which come nearest to such treatment will be found, with one exception, to be of some years standing. Of the whole number of children I have seen only one, a little girl, which had a mark of a blow visible, and that, upon inquiry, was inflicted by her own father. Mackintosh, 95.

In like manner, from the statements and depositions obtained under the present inquiry in the several districts in England, and from all classes of witnesses, it appears that in the great majority of cases, corporal punishment is prohibited by the proprietors, while it is proved on oath by several witnesses, that operatives and overlookers have been suspended and even dismissed from their employment for disobeying this command. It is impossible to read the evidence from Leeds, Manchester, and the western district, without being satisfied that a great improvement has taken place within the last few years in the treatment of children. What ill-treatment still exists is found chiefly in the small and obscure factories, while both in the large and small factories in England it is inflicted by workmen over children whom they themselves hire and pay, and who are completely under their control. In Scotland, personal chastisement when inflicted is inflicted by the overlooker; in England, by the workpeople. Among the indications of a desire on the part of the proprietors to promote the comfort and health of the workers in general, and of the young people in particular, we cannot help referring to the medical examination by Dr. Hawkins of the factory at Belper and Milford, belonging to the Messrs. Strutt, in which two thousand workpeople are employed; it is stated that a man who is a good swimmer is employed by the proprietors to conduct the lads twice a day in summer to a pond appropriated to their use for the purpose of bathing; and of a factory belonging to Mr. William Newton at Cresbrook mills, Tideswell, remarkable for the large number of apprentices which it contains, there is the following statement; Mr. Newton receives nothing with the children, and gives them nothing more than board, lodging, washing, and raiment, and a monthly allowance of pocket-money to each apprentice, which varies from *6d.* to 1*s.* 6*d.* Mr. Newton provides two individuals who act as Sunday schoolmasters. They do not go to church, because the nearest church is three miles off, but prayers are read to them twice on the Sunday. "I must state," adds Dr. Hawkins, "to the honour of Mr. Newton, that after a very minute and unexpected examination of his establishment, and of the apprentices in private, I could ascertain no point in their treatment that savoured of niggardliness nor of Power, 41.

harshness. The remoteness of the situation, the distance from public opinion, the absence of parents and relations, all afford an opportunity for abuse; but that opportunity is not seized. I particularly questioned the children separately, and obtained from them the following particulars of their diet. Their breakfast is of milk porridge and bread, as much as they please; their supper is the same. They have meat six days in the week for dinner, and as much as they choose, with potatoes and broth. There are separate eating as well as sleeping rooms for the boys and girls. The girls also have a separate piece of ground to play in. They have clean sheets once a fortnight, and clean shirts and shifts once a week. The beds are clean and neat, and not too many in a room; three little ones sleep in one bed, and two of the older in one bed. The greater part remain and marry in this establishment. Mr. N. affirms, that during the last twenty-four years only one of his apprentice girls has been pregnant before marriage, and that during the same period only four such cases have occurred in the whole of his mills."

Having thus considered the general treatment of children in factories, and the collateral circumstances under which their employment is carried on, and which influence in no inconsiderable degree the effects of that employment, we come now to consider what those effects really are, as far as they are ascertained by the evidence collected under the present investigation.

The effects of factory labour on children are immediate and remote; the immediate effects are, fatigue, sleepiness, and pain; the remote effects, such at least as are usually conceived to result from it, are, deterioration of the physical constitution, deformity, disease, and deficient mental instruction and moral culture.

1. The degree of fatigue produced on children by ordinary factory labour may be gathered from their own account of their feelings, and from the statements of parents, adult operatives, overlookers, and proprietors.

The statements of the children, and more especially of the younger children, as to their own feeling of fatigue, may be said to be uniform. The intensity of the feeling is influenced, without doubt, by the age of the child, and the constitutional robustness or feebleness of the individual; but the feeling itself is always the same, and differs only in degree. The expressions of fatigue are the strongest and the most constant on the part of the young children employed in the factories in Scotland, because there the ordinary hours of work are in general longer by an hour or an hour and a quarter than in the factories of England. We have been struck with the perfect uniformity of the answers returned to the Commissioners by the young workers in this country, in the largest and best regulated factories as well as in the smaller and less advantageously conducted. In fact, whether the factory be in the pure air of the country, or in the large town; under the best or the worst management; and whatever be the nature of the work, whether light or laborious; or the kind of treatment, whether considerate and gentle, or strict and harsh; the account of the child, when questioned as to its feeling of fatigue, is the same. The answer always being "Sick-tired, especially in the winter-nights." "So tired when she leaves the mill that she can do nothing."
Stuart. "Feels so tired, she throws herself down when she gangs hame, no caring what she does." "Often much tired, and feels sore, standing so long on her
Mackintosh. legs." Often so tired she could not eat her supper." "Night and morning very tired; has two sisters in the mill; has heard them complain to her mother,
Ibid. and she says they must work." "When the tow is coarse, we are so tired
Ibid. we are not able to set one foot by the other." "Whiles I do not know what to do with myself; as tired every morning as I can be."

Young persons of more advanced age, speaking of their own feelings when younger, give to the Commissioners such representations as the following:— "Many a time has been so fatigued that she could hardly take off her clothes at night, or put them on in the morning; her mother would be raging at her, because when she sat down she could not get up again through the house." "Looks on the long hours as a great bondage." "Thinks they are no much Mackintosh. better than the Israelites in Egypt, and their life is no pleasure to them." "When a child, was so tired that she could seldom eat her supper, and never Ibid. awoke of herself." "Are the hours to be shortened?" earnestly demanded one of these girls of the Commissioner who was examining her, "for they Stuart. are too long."

The truth of the account given by the children of the fatigue they experience by the ordinary labour of the factory is confirmed by the testimony of their parents. In general the representation made by parents is like the following:—"Her children come home so tired and worn out they can hardly eat their supper." "Has often seen his daughter come home in the evening so fatigued that she would go to bed supperless." "Has seen the young Ibid. workers absolutely oppressed, and unable to sit down or rise up; this has happened to his own children."

These statements are confirmed by the evidence of the adult operatives. The depositions of the witnesses of this class are to the effect that "the Stuart. younger workers are greatly fatigued;" that "children are often very swere (unwilling) in the mornings;" that "children are quite tired out;" that "the long hours exhaust the workers, especially the young ones, to such a degree that they can hardly walk home;" that "young workers are absolutely Mackintosh. oppressed, and so tired as to be unable to sit down or rise up;" that "younger workers are so tired they often cannot raise their hands to their head;" that "all the children are very keen for shorter hours, thinking them now such bondage that they might as well be in a prison;" that "the children, when engaged in their regular work, are often exhausted beyond what can be expressed;" that "the sufferings of the children absolutely require that the hours should be shortened."

The depositions of the overlookers are to the same effect; namely, that though the children may not complain, yet they seem tired and sleepy, and happy to get out of doors to play themselves. That "the work overtires Ibid. workers in general." "Often sees the children very tired and very stiff-like." "Is entirely of opinion, after real experience, that the hours of labour are far too long for the children, for their health and education; has from twenty-two to twenty-four boys under his charge, from nine to about fourteen years old; and they are generally much tired at night, always anxious, asking if it be near the mill stopping." "Never knew a single worker among the children that did not complain of the long hours which prevent them from getting education, and from getting health in the open air." Ibid.

The managers in like manner state that "the labour exhausts the children;" Ibid. that "workers are tired in the evening;" that "children inquire anxiously for the hour of stopping;" and admissions to the same effect, on the part of managers and proprietors, will be found in every part of the Scotch depositions.

In the north-eastern district the evidence is equally complete that the fatigue NORTH-EASTERN DISTRICT. of the young workers is great. "I have known the children," says one wit- Drinkwater, 19. ness, "hide themselves in the stove among the wool, so that they should not go home when the work was over, when we have worked till ten or eleven. I have seen six or eight fetched out oft he stove, and beat home; beat out of the mill, however. I do not know why they should hide themselves, unless it was that they were too tired to go home."

"Many a one I have had to rouse in the last hour when the work is very LANCASHIRE DISTRICT. slack, from fatigue." "The children were very much jaded, especially when Tufnell. we worked late at night." "The children bore the long hours very ill Ibid. indeed." "Exhausted in body and depressed in mind by the length of the Ibid. Ibid.

Tufnell. hours and the height of the temperature." "I found, when I was an overlooker, that after the children from eight to twelve years had worked eight or nine or ten hours, they were nearly ready to faint; some were asleep; some were only kept to work by being spoken to, or by a little chastisement, to make them jump up. I was sometimes obliged to chastise them when they were almost fainting, and it hurts my feelings; then they would spring up and work pretty well for another hour; but the last two or three hours were my hardest work, for they then got so exhausted." Ibid. "I have never seen fathers carrying their children backwards nor forwards to the factories, but I have seen children apparently under nine, and from nine to twelve years of age, going to the factories at five in the morning, almost asleep in the streets."

Western District. Horner & Woolriche, 8. 22. 24. "Some children do appear fatigued and some do not." "I have noticed the drawers exhausted beyond what I could express." "Many times the drawers are worked beyond their strength." There is, however, a striking contrast in the statements of all the witnesses relative to the fatigue of the children in the factories of the western district, in which the hours of labour for children are so much shorter than in the other factories of the kingdom.

2. Children complain as much of sleepiness as of fatigue. "Often feels so sleepy that he cannot keep his eyes open." "Longs for the mill's stopping, is so sleepy." "Often falls asleep while sitting, sometimes while standing." "Her little sister falls asleep, and they wake her by a cry." "Has two younger sisters in the mill; they fall asleep directly they get home." "Was up before four this morning, which made her fall asleep when the mill was inspected at one to-day by the Factory Commissioners; often so tired at night that she falls asleep before leaving the mill."

North-Eastern District. Drinkwater, 19. "I always found it more difficult to keep my piecers awake the last hours of a winter's evening. I have told the master, and I have been told by him that I did not half hide them. This was when they worked from six to eight." "I have seen them fall asleep, and they have been performing their work with their hands while they were asleep, after the billey had stopped, when their work was over. I have stopped and looked at them for two minutes, going through the motions of piecening fast asleep, when there was really no work to do, and they were really doing nothing. Ibid. I believe, when we have been working long hours, that they have never been washed but on a Saturday night for weeks together." "Children at night are so fatigued that they are asleep often as soon as they sit down, so that it is impossible to waken them to sense enough to wash themselves, or scarcely to eat a bit of supper, being so stupid in sleep. Manchester. Tufnell. I experience it by my own child, and I did by myself when a child, for once I fell asleep, even on my knees to pray on my bed-side, and slept a length of time till the family came to bed." Overlookers and managers in innumerable instances depose to the same effect.

3. Pains in the limbs, back, loins, and side are frequent, but not as frequent as fatigue and drowsiness. The frequency and severity of the pain uniformly bears a strict relation to the tender age of the child and the severity of the labour. Pain is seldom complained of when the labour did not commence until the age of nine, and was not immoderate. Girls suffer from pain more commonly than boys, and up to a more advanced age; though occasionally men, and not unfrequently young women, and women beyond the meridian of life, complain of pain, yet there is evidence that the youngest children are so distressed by pain of their feet, in consequence of the long standing, that they sometimes throw off their shoes, and so take cold. Scotland. Stuart. "Feet feel so sair that they make him greet." "Was quite well when she went to the mill, but the confinement brought on a complaint in her head, and her left side is now pained." Mackintosh. "Many nights I do not get a wink of sleep for the pain." "At first suffered so much from the pain that he could hardly sleep, but it went

off." "Knee failed from excessive labour; severe pains and aches would come on, particularly in the morning; it was better in the evening; felt no pains in any other parts. There were two or three complaining at the same time of their knees aching." "I have seen children under eighteen years of age before six at night, their legs has hurt them to that degree that they have many a time been crying." Power. Tufnell.

4. Swelling of the feet is a still more frequent source of suffering. "Obliged to bathe her feet to subdue the swelling." "The long standing gives her swelled feet and ankles, and fatigues her so much that sometimes she does nae ken how to get to her bed." "Night and morning her legs swell, and are often very painful." That this affection is common is confirmed by the concurrent statements of parents, operatives, overlookers, and managers. SCOTLAND. Stuart.

5. That this excessive fatigue, privation of sleep, pain in various parts of the body, and swelling of the feet experienced by the young workers, coupled with the constant standing, the peculiar attitudes of the body, and the peculiar motions of the limbs required in the labour of the factory, together with the elevated temperature and the impure atmosphere in which that labour is often carried on, do sometimes ultimately terminate in the production of serious, permanent, and incurable disease, appears to us to be established. From cases detailed in the evidence, and the accuracy of which has been strictly investigated, we do not conceive it to be possible to arrive at any other conclusion. The evidence, especially from Dundee and Glasgow, from Leicester, Nottingham, Leeds, and Bradford, from Manchester and Stockport, in a word, from all the great manufacturing towns, with the exception, perhaps, of those in the western district, in which there is little indication of disease produced by early and excessive labour, shews that grievous and incurable maladies do result in young persons from labour commenced in the factory at the age at which it is at present not uncommon to begin it, and continued for the number of hours during which it is not unusual to protract it.

6. From the same evidence it appears, that the physical evil inflicted on children by factory labour, when commenced as early and continued as long as it now is, is not the only evil sustained by them. From the statements and depositions of witnesses of all classes it appears, that even when the employment of children at so early an age, and for so many hours as is customary at present, produces no manifest bodily disease, yet, in the great majority of cases, it incapacitates them from receiving instruction. On this head the statements of the children themselves must be admitted to be of some importance; and it will be found that the young children very generally declare that they are too much fatigued to attend school, even when a school is provided for them. This is more uniformly the declaration of the children in the factories of Scotland than in those of England. The evidence of other witnesses, both as to the capacity of the children for receiving instruction, and as to their actual state in regard to education, is conflicting. Few will be prepared to expect the statements that will be found on this head in regard to Scotland, where the education of the children is neglected to a far greater extent than is commonly believed; where only a very small number can write; where, though perhaps the majority can read, many cannot; and where, with some honourable exceptions, it seems certain that the care once bestowed on the instruction of the young has ceased to be exemplary. The reports of the commissioners for Scotland, who will be found to have kept this subject continually before their view, are decisive on this head. "Many of the persons sworn could not write nor sign their depositions. The reports mark the signatures in every case where the parties could write. I suspect the want of education so general on the part of these people, which has surprised me, is to be attributed to their being for so long a period of the day SCOTLAND. Stuart's report, Dunfermline, 3.

Stuart's report, (Perth) 51. confined to the factories." "The overseers of the small mills, when the proprietors are absent, almost uniformly, as the central board will notice, declare their aversion to the present long hours of working, as injurious to the health of the workers, and as rendering their education impossible."

Ibid. 122. "Still the employment of workers in factories cannot, where proper regulations are attended to, be in most cases with propriety termed an unhealthy one; and it would therefore seem that the long confinement of labour is more injurious to them, in preventing them from being sufficiently educated, and of course sufficiently instructed in their moral duties, than in other respects. Here, too, although there is abundance of evidence from clergymen, as well as from teachers, of a conflicting description, I think it upon the whole impossible to doubt, that the young workers must be so much fatigued with the very long hours of labour, that they cannot be so fit to receive instruction as other young people, and that they have too little time for being at school, even to enable them to learn to read, write, and to understand accounts tolerably. Want of education cannot fail to have an unfavourable influence on their morals."

Mackintosh, 9. "The number that can write is very small; many here, as elsewhere, asserting, while under examination, they could, and afterwards being unable to write their own signatures."

Ibid. 25. "Thinks as a class they are very deficient in learning. Knows many twenty years old that are incapable of reading a chapter of the New Testament. Many cannot write at all. In some cases evening schools have been established, but from the long hours the girls are too exhausted to derive much advantage. On some occasions when he has requested parents to send their children, they have answered, that they were too tired, and were more fit for their bed than for the Bible. Has often heard the working people say that they are prepared to forego some of their wages for the sake of obtaining

Ibid. 40. the time of educating their children." "Not changed in the opinion given by him, (No. 214 of the Report of Evidence,) that the great length of labour during the day in a great measure precludes all adequate instruction in the evening, and that there are a great many children who would willingly go to school if they were earlier dismissed. Believes that there is not a single school carried on by a manufacturer in Aberdeen; that the only opportunities afforded to the children were set up within these seven years by two clergymen; that in these two schools the girls who attend regularly are generally superior in character to those who do not so attend; that therefore the circumstance of a number of those being able to read and write ought not to be taken as a specimen of the advancement of the whole; at the same time a considerable portion of those who do attend are unable to read, and a still

Ibid. 66. larger are unable to write." "Has been for the last sixteen years well acquainted with the working population of Glasgow, and has perceived a great change effected for the worse in the moral habits and education of the children. About twelve or fourteen years ago it was a very rare occurrence to meet with a child that could not read; whilst very much the contrary is the case now. About three years ago a number of girls from Messrs. Muir, Brown, and Co.'s, of the ages of sixteen or seventeen, attended for the first time his school, when, to his surprise, he found that they could scarcely read at all. He found that each of these had formerly learnt to read, but had forgotten their knowledge. After a short attendance, shame at not being equally instructed with much younger children drove them away again. The present body of master manufacturers, with few exceptions, seem to entertain a very different idea of the necessity of having schools attached to their establishments than that which influenced them a few years back. Knows of only one evening-school attached to a factory in Glasgow; and the result of that disregard of education is, that the children, with some exceptions, employed in public works, drink, smoke, curse and swear, and are generally very profligate."

Dr. Stevenson Macgill, professor of divinity in the university of Glasgow, states "that the time for education is in general too short; that were it longer, fatigue renders the scholars capable of learning but little; that these observations apply principally to the manufactories in populous cities and their neighbourhood. That in several large manufactories and cotton mills carried on in healthy situations, where the people form small villages, have their houses and small gardens, are all known and superintended, and have their chaplains and teachers, a very different order of things takes place; at the same time he must express his opinion even respecting these, that the children in general are obliged to work at too early an age; that the health of body and mind requires a longer period to be allotted for a good education, and for those relaxations which are necessary to the well-being of early youth." Mackintosh, 73.

One of the great evils to which people employed in factories are exposed is, the danger of receiving serious and even fatal injury from the machinery. It does not seem possible, by any precautions that are practicable, to remove this danger altogether. There are factories in which everything is done that it seems practicable to do to reduce this danger to the least possible amount, and with such success that no serious accident happens for years together. By the returns which we have received, however, it appears that there are other factories, and that these are by no means few in number, not confined to the smaller mills, in which serious accidents are continually occurring, and in which, notwithstanding, dangerous parts of the machinery are allowed to remain unfenced. The greater the carelessness of the proprietors in neglecting sufficiently to fence the machinery; and the greater the number of accidents, the less their sympathy with the sufferers. In factories in which precaution is taken to prevent accidents, care is taken of the workpeople when they do occur, and a desire is shewn to make what compensation may be possible. But it appears in evidence that cases frequently occur in which the workpeople are abandoned from the moment that an accident occurs; their wages are stopped, no medical attendance is provided, and whatever the extent of the injury, no compensation is afforded.

From the whole of the evidence laid before us, of which we have thus endeavoured to exhibit the material points, we find:—

1st. That the children employed in all the principal branches of manufacture throughout the kingdom work during the same number of hours as the adults.

2nd. That the effects of labour during such hours are, in a great number of cases,

Permanent deterioration of the physical constitution:

The production of disease often wholly irremediable: and

The partial or entire exclusion (by reason of excessive fatigue) from the means of obtaining adequate education and acquiring useful habits, or of profiting by those means when afforded.

3rd. That at the age when children suffer these injuries from the labour they undergo, they are not free agents, but are let out to hire, the wages they earn being received and appropriated by their parents and guardians.

We are therefore of opinion that a case is made out for the interference of the legislature in behalf of the children employed in factories.

4th. In regard to morals, we find that though the statements and depositions of the different witnesses that have been examined are to a considerable degree conflicting, yet there is no evidence to shew that vice and immorality are more prevalent amongst these people, considered as a class, than amongst any other portion of the community in the same station, and with the same limited means of information. Distinguished from other classes by being collected together (both sexes, young and old) in large numbers, the language and behaviour common to uneducated people, under such circumstances, is found to be checked in no inconsiderable degree by the presence of fathers, mothers, and brothers; and for any evil of this kind which may nevertheless exist, the proper remedy seems to be a more general and careful education of the young people.

5th. In regard to the inquiry "in what respect the laws made for the protection of such children have been found insufficient for such purpose;" we find that in country situations the existing law is seldom or never attempted to be enforced, that in several principal manufacturing towns it is openly disregarded, that in others its operation is extremely partial and incomplete, and that even in Manchester, where the leading manufacturers felt an interest in carrying the act into execution as against the evasions practised by the small mill-owners, the attempt to enforce its provisions through the agency of a committee of masters has for some time back been given up. On the whole, we find that the present law has been almost entirely inoperative with respect to the legitimate objects contemplated by it, and has only had the semblance of efficiency under circumstances in which it conformed to the state of things already in existence, or in which that part of its provisions which are adopted in some places would have equally been adopted without legislative interference, as there is reason for presuming, if we advert to the fact, that such provisions have actually been adopted in the progress of improvement in other branches of manufacture unrestricted by law. On the other hand, the large classes of workpeople who come within the provisions of the recent as of all former acts, have been familiarized with contempt of the law, and with the practice of fraud, evasion, and perjury.

Stuart, 49, 127.

Cowell, p. 1 et seq. 58, 81. Tufnell, 11. 12. 18. 65. 78. 90. 113. 134.

It will be seen that the evidence here quoted relates chiefly to the condition of children. A large body of evidence has, however, been collected, tending to shew the actual condition of the manufacturing, compared with that of the non-manufacturing, population of the country, including children and adults, and both sexes. The evidence received relates especially to the state of health, to the amount of sickness, and to the rate of mortality of the former as compared with the latter class. This evidence has been placed in the hands of the actuaries whom we have consulted, who, from the examination which they have made of it, have given it as their opinion, that it affords materials for determining the actual as well as the comparative condition of the working classes of the country, far superior to any which had heretofore been collected. It has not been possible, however, as yet, to put the returns already received (and more are expected) in such an order and form as to obtain the results with exactness, but we propose to include the returns themselves, together with the results they afford, in a supplementary report.

Having defined the evil arising from the present employment of children in factories, we have proceeded to investigate what measures admit of being taken for their future protection. In this stage of the inquiry we have found the attention of the public, as well as of the parties concerned, so powerfully directed to the plan for the regulation of labour in factories known under the title of the Ten Hour Bill, that a large body of evidence has necessarily come before us as to the probable operation of such provisions of the proposed plan as have become matter of public speculation. We proceed to submit some portions of that evidence, having first stated as our conclusions, formed on a view of its whole tenour:—

1. That this bill does not accomplish the object at which it purports to aim. Its professed object is the protection of children; but it does not protect children. For the same evidence which shows that the legislative protection of children is necessary shows that the restriction of the labour of children to ten hours a day is not an adequate protection.

Mr. Osburn, junior, deposes:—"I have been one of the principal persons Drinkwater, 92.
in Leeds concerned in preparing the case for the advocates of the ten hours' bill. I was chairman of the short-time committee from March last year till about the time Parliament was prorogued. I examined most of the witnesses before they went to London. Both Mr Sadler and myself considered that the evidence for ten hours rested chiefly on the medical men." Now the medical witnesses of the greatest professional authority, selected by Mr. Osburn to give their evidence before the committee of 1832, declare as follows:—
(Mr. Samuel Smith.) "I think it very probable that many will suffer, even Report 1832.
under the ten hours labour. Disease of the bones will be produced in some 10,448.
degree."—— (Mr. Brodie.) "I think ten hours a day too much for children of
ten or twelve years of age."——(Sir G. L. Tuthill.) "I doubt if a child from 11.124.
ten to twelve is capable of enduring ten hours of even light labour."—— 11,341.
(Mr. Key.) "Eight or nine hours is quite as long as a child aged from nine to
fourteen ought to be confined in a factory."—— Mr. Samuel Smith has been 11,435.
re-examined by one of the commissioners, and deposes.—"I was examined Drinkwater, 44.
before the committee in 1832. I have since once or twice read over the evidence I then gave, and I am now willing it should stand as confirmed by me upon oath. I do not think that the limits of ten hours labour is a sufficient protection to children between the ages of nine and twelve. I think about Ibid. 46.
eight hours actual labour, with proper intervals, would insure protection to that period of childhood from such effects as I have been speaking of, I mean to say, under ordinary circumstances."

2. This bill, making no provision for the occupation of any part of the time of children for their own benefit, either before or after their hours of labour, and taking no charge of their education, elementary or moral, leaves the removal of a most important portion of the evil under which children suffer unattempted.

3. While this bill does, and attempts to do, so little for children, its operation, if it could be carried into effect, would be to restrict the labour of adults, as well as that of children, to ten hours.

Independently of the objection which there appears to be in principle to any compulsory interference with the hours or terms of adult labour, we find reason to anticipate very serious practical evils from imposing any such arbitrary restriction on the operations of so large a proportion of the manufacturing industry of the country.

The most direct and undisputed consequence of the passing of the ten hours' bill would be the general limitation of the labour of adults within the same hours as those assigned to children and adolescents. We are spared the labour of weighing conflicting testimony on this point, as it is generally admitted or assumed on both sides of the question. On the part of the manufacturers it is generally taken for granted that such will be the first effect of the measure under discussion, and that assumption is made the basis of reasoning as to its ultimate issue. With the operatives the same assumption is prominently put forward in the arguments of most of the leading advocates of the measure, and is generally dwelt upon as forming a principal item amongst the benefits which they expect to derive from the passing of the measure. It may be sufficient to adduce instances of explicit statement of this expectation, in which the conflicting parties are agreed, before proceeding to those ulterior consequences on which their views are naturally opposite.

Power, 34. Benjamin Bradshaw, Leeds, with reference to one of his former answers before the committee of 1832, states—"It is almost sure in some cases that the ten hours' bill would stop the work in the whole factory. To what extent it will do that I am not prepared to say; it is my hope it will do so altogether. It is the general belief of the operatives themselves that it will have that effect, and it is their general wish that it should be so."

Drinkwater, 71. William Swithenback, Leeds, states—"I believe the opinion of the work-people generally is to work only ten hours; they think it is long enough. I believe it is the opinion of most that it will be the means of stopping their labour too."

Ibid. 93. Mr. Wm. Osburn, Leeds, says—"I have conversed with the operatives on the probable effect that the proposed bill may have on adult labour; they have generally the opinion that it will shorten their hours also. They look forward universally to work only ten hours themselves. I feel no doubt whatever on that point. I have heard some say that it will throw the labour more into the hands of unrestricted persons. I believe they expect at first that the wages will fall, and then that they will return to their old level. That agrees with my own opinion."

Tufnell, 29. Mr. James Bradley, Stockport, in answer to the following question, "Would turning away all hands under eighteen, after ten hours work, compel you to turn away the older hands at the same time?" says, "Certainly, as one half of our hands are under eighteen. It would be impossible to keep the spinning on without the piecers, and who are always young, being in general from nine years old to sixteen; and it would of course be impossible to keep the power-looms going without the mules."

Ibid. 3. Thomas Hodges, of Manchester, gives the following answers on this head:—"Would turning away those under eighteen years of age, after ten hours work, compel the master to turn away the grown-up workmen also, and stop the machinery?—I think it would compel them to stop their mill in a general way."

"Are there any factories where they would be likely to go on without those who are under eighteen?—I should think not. I should think in the mill that I was engaged in one third of them would be under eighteen."

Ibid. 9. Another Manchester operative makes the following statement:—"I must admit that the generality of operatives do anticipate that they shall be relieved from their toilsome labour by not being able to continue it when the children are stopped."

Ibid. 15. Josiah Hunt, overlooker in Mr. Black's mill at Manchester, in answer to the following question, "Would turning away those under eighteen years of age, after ten hours work, compel the master to turn away the grown-up workmen also, and stop the machinery?" says, "Well, in most places it would; it would with us; with power-looms in general it would."

Ibid. 19. Mr. J. Bell Clarke, a master spinner at Manchester, states as follows:—"If Lord Ashley's Bill were passed, it would be almost equal to a moving-power bill."

The point at which the opinions of the masters and the operatives begin to diverge is in the estimate of the ulterior effects to be anticipated from the passing of the proposed measure, and from the consequent reduction of the hours of work for children and adults, attended, as the majority of witnesses agree must be the case, with a corresponding diminution of production.

It appears to be the general opinion of the operatives, that though wages may in the first instance fall, from reduction of the hours of labour, the artificial scarcity of commodities thus occasioned will effect a rise of prices, and a consequent rise of wages, as well as an increase of work for hands which are now partially out of employ, by occasioning the erection of new establishments to supply the deficiency of production caused by the diminution of labour.

James Brierly states, "There was but one opinion on one part of the subject, that there would be a reduction of wages for a time. The opinion of the men generally was, that it would be only for a time; that working long hours has a tendency generally to bring down wages: the masters thought that wages would keep down."

Edward Sansome states, "I have signed the petition for the present bill. I have had a sheet lying for signature at my house. Our operatives are not quite all of the same mind about it, but few have refused to sign it; the man who made most opposition to it signed it himself. Seale was his name. His impression, and that of those who thought with him, was, that whereas now they work twelve hours in a regular way, taking two hours from that time, and limiting it to ten, would reduce the weekly amount of wages, and that the families would be worse off in consequence. That is all the ground of opposition. Drinkwater, 16.

"Now, for your opinion, and that of those who support the bill; do you think that would not be the case in the first instance?—I believe it would in the first instance. But allowing they do only work twelve hours now—which I believe is not so, but those hours are greatly exceeded in many cases—my opinion is, that the reduction of hours will operate as if you took two men out of every twelve, and the effect would be to make labour scarcer, and that in three months of the passing of the bill they would get the same amount of wages for the labour of ten hours as they now get for twelve; with this further advantage to the community at large, that if the same demand for goods produced continued as exists now, it would call the unemployed into employment, and lessen the amount of poor's rates. That is my motive, and my sole reason for supporting the thing. I don't think I have anything more to add upon this. I do not contemplate herein a rise of wages, but the same wages for a less degree of toil." Power, 8.

John Hannam, Leeds, says, "I think it advisable to restrict machinery so as not to throw people out of employment. That is my real opinion: many differ with me: they think machinery should be done away with altogether. Some say only such machinery as takes away manual labour should be done away with; but I am an advocate for machinery myself; but I want the benefit of it to be that the people who work at it should get part of it. That is one great reason why I want the hours shorter, to give a chance of work for those who are willing and able to work. I think it would be better than the present bill, if the hours were made shorter for every body, men as well as children. I think in many places this bill will have that effect; where the children work with men, where the one are stopped the other will stop too." Drinkwater, 83.

This witness follows his principles out to their full extent. With reference to Question 6872 of the committee of 1832: "I mean, if it should be found, when the hours are brought down to ten for all, that many are still out of work, that they should be brought down to nine or eight, so as to give employment for all. I think it is better that all should work for eight hours than that half should work for sixteen. I mean that twice the quantity of machinery should be used. I have said that bringing the hours down to ten will rather raise wages than lower them: it would continue in the same way if it was brought down to eight hours. I think that would hold all through if you brought it down to one hour. The shorter time a man works, the more he will get for his labour for that in porportion. Certainly it would be a greater expense to the master to lay out for more machinery than now, if they were working such short hours. He would have that expense to bear, and the expense of greater wages too. As soon as a master found that did not answer, he would give over buying machinery, and let the labour be done by hand. I conceive in that way machinery would find its level. Our bill would not touch those that work by hand, and so he might get liberty to work more hours than those that work by machinery. I don't think this bill would stop the progress of machinery much; it must be a shorter bill than this that would do that. When machinery had found its level in this way, the prices of goods must rise: there is plenty of room for that. I think that would be the end of it, that prices would rise; because it has just been the reverse, that prices have lowered while we have been working long hours. I don't think that would take so many customers out of the market as it would bring in amongst those who can't afford to buy. There are various opinions, but I think this is the general one among us that I have told you."

The process by which not only the operative supporters of the ten hours' bill, but some otherwise well-informed persons, appear to have been led to the conclusion that a restriction of the working hours for adults, and a consequent increase of the cost of production in this country, could have the effect of raising and permanently maintaining a range of comparatively high prices, appears to have been the general application to all products of labour of a proposition true with respect to particular articles only, which are kept in a state of artificial scarcity by monopoly.

There might, indeed, if the restriction of hours affecting the productive power of machinery in this country to the extent that would be effected by the proposed ten hour bill were immediately enforced, be a temporary scarcity of manufactured goods, so as in the first instance to occasion a rise of prices, which might, as long as it lasted, allow of the maintenance of wages, notwithstanding the reduced hours. But this rise could not in the nature of things be of long duration. This temporary rise of price, combined with the permanent advantage to the foreign manufacturer of the increased cost of production in this country, would inevitably operate in producing an extension of the existing works and the erection of new ones abroad. The increased production from these and from the extended works in this country, stimulated by the advance of price, would, after no long interval, restore the former proportion of the supply to the demand, and the consequences would then be most disastrous to the English manufacturer and workman. The smaller and less favourably situated manufactories would be swept away. Even the more opulent and best situated establishments would have great difficulty in maintaining their ground, workmen would be thrown out of employ, and wages must, under such circumstances, inevitably fall to the lowest point consistent with the most bare subsistence of the working class.

These effects would vary in degree in the different branches of the manufacture to which the artificially restricted production in this country would apply; but the ultimate result in all would be a general reduction of profit and wages.

Of the probable effects of foreign competition in interfering with and eventually supplanting the manufactures of this country, on the supposition of an increased cost of the production by the ten hours' bill, some of the witnesses express themselves to entertain no apprehension.

Mr. William Osburn, Leeds, says, "I expect that the masters will still be able to give the same amount of wages that they do now, when the whole labour of the factory population is reduced to ten hours. I do not expect that the same quantity of goods will be produced, but that labour will rise in value, in consequence of there being less in the market. I expect the difference will not be made good out of profits, but out of the price; that prices will rise. I consider prices now factitiously low. I must modify that opinion. I expect that some part will be made good out of profits. I have no knowledge (other than general) of the proportion of the home and foreign trade of Leeds. My impression is, that the home trade is by far the most important. I think also, as they are fencing us out with high duties in almost all countries, that they also could give us an advance. I have a relation in Charlestown, in South Carolina, and when I last saw him here two years ago, he remarked how much cheaper and better the English cloth was than the American. I understood him to mean that, notwithstanding the duty, the English cloth could compete very well in that market. He is not in trade there."

To the same effect are the statements of John Adams and Patrick M'Gowan, spinners, Glasgow, who declare, as the result of their observation during a stay of fifteen months in the United States, that they are persuaded that the Americans will never, in their time, compete with our cotton goods in the market. The grounds, however, on which such statements are made appear to us to be vague and unauthenticated. On the other hand, all the most eminent manufacturers who have been examined consider that a reduction of hours would give a sufficient advantage to foreigners to induce them to extend their manufacturing establishments. Some of the latter class of witnesses give the precise data for calculating the relative cost of production, and all of them concur in the general fact of the great and increasing competition which the cotton, woollen, and linen manufactures of this country are experiencing abroad, and the silk manufacture in the home market.

It appears, from calculations made by Mr. William Rathbone Greg (see Tufnell, p. 14,) from sources in the authenticity of which he attests on oath his belief, that the only advantages in manufacturing industry possessed by this country over some of the continental nations, are the lower rate of interest of fixed capital, and the less amount of capital sunk in machinery, owing to more work being done in the same time by the same machinery. Every other item in the cost of production is heavier in this country than in Germany, France, or Switzerland; and the passing of a measure such as that now before Parliament, by restricting in effect the moving power within limitations of time unparalleled in the law or practice of any other industrious people, would involve the loss of the sole countervailing advantage at present derived from the greater speed of superior machinery.

The following statements with regard to the results to be apprehended from an increased cost of production in this country proceed from gentlemen connected with each of the principal branches of manufacture, whose evidence has been taken by the Central Board.

Mr. James Hadden, flax-spinner, Aberdeen, states, in answer to the question, "How do you consider the increased cost of the manufactured article by the increased charge from the interest of the plant before spoken of would affect your power of competing in the foreign market?—It would affect it in the same proportion as the increase of the cost. I have no hesitation in saying, that we are now in a position in which very small advances would produce very great effects, the extent of which it is scarcely possible to see. It must be recollected, that it is only by the superiority of the machinery of this country that we are able at all to compete in the same markets abroad with the cheap hand-labour of those countries from whence we obtain the raw material, and it must be kept in view that from those countries it is that we obtain the raw material, flax."

Mr. Henry Ashworth, cotton-spinner, Bolton, states, in answer to the question,—"What would be the effect of the limitation of the hours of adult labour in mills to ten hours per day?—The result would be a considerably diminished rate of profit to our manufacturers, an addition of five to seven per cent. on the cost of the article produced; and if an increase in its selling price to that amount were obtainable, which might be the case for a time, it would add so much to the profits of foreign manufacturers, but nothing to those of our own; hence it would lead to the extension of rival manufactories abroad, and entirely prevent that necessary increase of our own which at no distant period will be required for the employment of our rapidly-increasing population. It would also tend to encourage the baneful practice of night-working in mills, already so much deprecated, unless specially provided against by law; and would greatly stimulate the introduction of machinery to supersede human labour; and the unavoidable consequence would be, diminished wages and want of employment. I may further state, as my opinion, that such a limitation of the hours of labour could not be enforced except by way of experiment; for when the law was found to be at variance with the mutual interests of masters and workmen, they would no longer forbear to break it, and the public would tolerate the infraction.

"Is it then your opinion that twelve hours' labour per day is requisite for the manufacturer to make a fair profit, and in order that we may maintain the superiority which at present we possess over our rivals, the foreigners?—It is my opinion that twelve hours' labour is indispensable to remunerate the manufacturer for the outlay of his capital in mills and machinery, a large portion of which is of a perishable nature, and, generally speaking, however judiciously applied, is, when offered for sale, of a very uncertain and depreciated value. With respect to our rivalry with foreign manufacturers, if it is really the desire of the people of this country that the cotton manufacture, which contributes more largely than any other to the exigences of the state, should continue and flourish here, it must not be fettered with a ten hours' bill, nor any other injurious restriction; we have already to contend with rivalry abroad of serious extent, and still rapidly extending, entirely unrestricted in their time and mode of operation, and greatly fostered and encouraged by legislation wherever it exists. If, therefore, we should injudiciously restrict our own industry and enterprise, we shall promote the success of these rivals, not only by improving their markets, by reason of our increased cost of production, but by the consequent depression of our trade, affording them a very material assistance in promoting the prosperity and extension of their own manufactures out of the result of our distress, which would drive away the most enterprising and valuable of our artizans, as well as capitalists, and lead them to seek that protection, profitable employment, and better estimation, in other states, which had been denied then in their own country; and however much this mistaken policy might hereafter be regretted, the recovery of our manufacturing superiority would be totally impossible."

John William Partridge, Esq., woollen-manufacturer, Stroud, states, in answer to the question, "Do you manufacture much for exportation?—"The larger proportion of the coarse cloths is manufactured for exportation; a proportion of the fine cloths is also exported."

"What would be the effect of the ten hours' bill upon foreign competition?—For the last eight or ten years the American manufacturers have been competing with us, and pressing us very hardly in their own markets, and many of our best workmen have emigrated there. The French, the Flemish, and the Prussians, are also competing with us very strongly, and altogether they have gained the Russian trade from us. We have now greater difficulty to keep the lead as exporters than we ever had, and profits are lower. The effect therefore of any restriction would be still further to increase the difficulties. If we were to do the same amount of work under the ten hours' bill, it must be with an increased outlay of capital in machinery and mills, which is the worst outlay we can make, and would tend greatly to increase the cost of production. Wages remaining the same, there would be an increase of the cost of production from the increased amount of the interest on capital, and the wear and tear of machinery. In our trade, the interest, and wear and tear of machinery, can hardly be calculated at less than twenty-five per cent."

Joseph Grout, Esq., silk-manufacturer, London, states, in answer to the following question, "What is the present state of the foreign competition which you have to meet?—We have to meet the competition of countries of Europe where the price of labour is exceedingly low, and the raw material raised on their own soil. We are also exposed to the competition of Bengal and China, where the price of labour is still lower. In Bengal the price of labour for such children as we employ is not more perhaps than a penny per day. In China the labour is equally cheap. Labour in France is certainly not more than two thirds the price of labour in this country. In Switzerland the price of labour is still lower. There are no goods of our manufacture exported, and there are very large quantities imported. By the custom-house lists it appears that the importation is increasing.

"Is not your machinery superior?—I do not think we have any advantage in machinery; in fact, we are now copying the French machinery; and I learn that in France very great improvements have recently been made.

"What are the hours of work abroad?—I believe that they are unlimited. The people in France, Switzerland, and Italy, work from daylight to dark in summer-time. This, however, I do not state from my own knowledge."

The opinions entertained by manufacturers vary in some degree with regard to the immediate effects which would be produced on wages by the expected operation of the ten hours' bill on production. Mr. Robert Hyde Greg, of the firm of Samuel Greg and Company, in answer to the question, "What do you think would be the effect of the ten hours' bill on wages?" replies, "No distinct answer can be given without knowing what would be the effect on *quantity*, (which would practically depend on the provisions of the bill proving efficient, and of which I doubt the possibility,) and what on *price*, both which would depend on such a variety of causes as to baffle any anticipation of the actual result. If the price rose in proportion to the reduced quantity (an impossible supposition), masters and operatives might get the same profits and the same wages, though they would pay more for their clothes; if the masters got the same interest and profits, without which he would not invest capital in mills, and if, owing to foreign competition and other causes, the price did not rise at all, the reduction of wages would amount to from twenty to twenty-five per cent. In coarse spinning the fixed charges and contingencies amount to as much as the wages; and if the former remained the same, and price stationary, the reduction on the latter must be double. The actual effects would, doubtless, be divided amongst the different parties. The premium given to the increase of foreign establishments would, of course, be proportionate to the advance of price caused by our diminished production, so that even were wages kept at their level for a time, it would only be to postpone the evil, and to make it come at last with double effect and more fatal consequences. I do not touch upon the mischief produced, most serious and most extensive, by the diminished consumption of our mills, upon all trades connected in any way with the factories. If we work one sixth less time we consume one sixth less cotton, coal, oil, iron, lead, wood, &c. &c., and one sixth less shipping, carriage, &c. &c., would be required, so far as regards our consumable materials. This view of the question is independent of the effect the reduction may cause upon wages and prices, and must ensue if the quantity be lessened.

"Is it not a well ascertained fact, that any rise of price always causes more than a proportionate decrease of consumption in any article, and *vice versa?*—Without being able to state the particulars in our own trade, we know as a matter of fact, that an increased price is attended by a diminished consumption in a greater ratio than the rise of price; and, on the contrary, that a fall in the price is accompanied by a more than a proportionate increase of consumption. To the truth of the latter part of this proposition the whole history of the cotton trade bears testimony."

With regard to the ultimate effects on wages of the proposed artificial check to production, there is no variation of opinion amongst those of the witnesses whose minds are not pre-occupied by a notion which we have endeavoured to shew is equally untenable in reasoning as it is destitute of foundation in experience—the notion, namely, of a permanent rise of price as necessarily consequent on a diminished produce in this country resulting from reduced hours of labour.

Mr. Henry Ashworth, of the firm of Henry and Edmund Ashworth, in replying to the question, "What would be the increased cost," &c., says, "The increased cost of production, in our business, for a diminution of two hours per day, would be from five to seven per cent. for use of capital and standing charges, supposing that wages were ten-twelfths of what they had been for the longer time.

"Is it then your opinion, that in the event of a restriction of the hours of work to ten per day, that the wages would fall in proportion of twelve to ten?—Yes, they would fall in that proportion, or more."

Mr. Joseph Grout, whose evidence has been above cited, states:—"We should certainly reduce wages in proportion to the reduction of hours. We ought, in justice to ourselves, to reduce wages in a still greater proportion, to cover the loss of interest on capital which would be occasioned by the restriction of the use of the machinery and plant; but we could not deduct the whole of this loss of interest from wages, because they could not bear such a reduction. The cost of production would therefore be increased, and our competition with foreigners would be rendered more difficult."

Some individuals, even amongst the operatives, extend their view farther than immediate effects, though the opinions of such individuals do not appear to be the leading ones. Aaron Jackson, spinner, Manchester, on being interrogated, "Do the operatives advocate the bill?" replies "They do; but they do it in this view; they expect the same wages, and they are in a grand mistake about it; for the masters will not be able to give the same proportionate wages for ten hours as twelve, and rent and taxes will continue the same, and one sixth less to pay it with, and wear and tear, and everything else will be nearly the same. And a spinner employing four piecers, and working ten hours under the bill, his wages would be reduced one half.

"Then you think that, in general, operatives will have to work only ten hours, but lose more than one sixth of their wages?—Yes; they will certainly. If the mill stops after ten hours they will lose more than one sixth, as the master has less work done, so his profits must certainly be less to pay his rent with.

"Is not your opinion different from that of the majority of operatives?—Yes; by woful experience of having a family that has taught it me in the two last bills; when they reduced my labour four hours per day, it reduced my wages 10*s.* a week; these operatives have no experience like me, but they think that they shall benefit in wages by it; but when I point it out to them in the following manner, they agree with me, and are as thick on it as I am. I am a spinner employing four piecers; the two hours a day being taken off, proves it is not in my power to give them their usual wages, and if I cannot pay four, how can the masters pay four hundred. This clearly proves that it will be a loss to every one, and it makes them very afraid of its passing when they understand it in its true light, and wonder what must become of them, as they can hardly do on their present wages. Every one will feel it, because, so much less being paid a week, any one that keeps a shop will have less demands; it will take two thousand pounds a week less to pay the hands with in this town, and their labour will be lost."

One principal objection which has been brought against the proposed measure is the arbitrary mode in which the periods of labour are defined, without reference to any circumstances which may make occasional deviations from any fixed practice desirable or even necessary. In districts, such as the clothing district of Gloucestershire, where the power employed is principally water-power, it appears from the inquiries of the commissioners, as well as from the statements of the manufacturers, that great evil would result from any legislative prohibition against adapting the hours of work to the irregular supply of water. It sometimes happens that the mills situated upon the lower parts of the streams are not supplied with water till two or three o'clock in the afternoon, and in this case, according to clause eight, they could not work more than five or six hours a day. It is further to be observed, that the eighth clause of the proposed bill, either through inadvertence, or with intentions which are not apparent, makes no provision for factories worked partly by steam and partly by water. It is obvious that the inconvenience resulting from the interruptions specified in the above clause is not confined to factories where the propelling power is water only, but extends in a greater or less degree to all factories in which water is the principal moving power.

Another defect generally complained of in the proposed measure is, that it makes no provision for existing contracts. It is not unusual for master spinners to make contracts for long periods to furnish a given quantity of yarn within a given time. The effect of the bill may be to prevent the fulfilment of these contracts on the part of the spinners, as they might not be able to furnish the given quantity within the proposed restricted time of working. Tufnel, p. 68.

Leases of factories commonly contain a clause that the lessee shall, at the expiration of the term of the lease, pay to the lessor a sum equal to any depreciation which may have taken place in the value of the machinery, which is a fixture on the premises. In such cases the proposed act of the legislature would occasion an artificial depreciation in property of this description, and inflict a corresponding loss on the occupier, which, at the time of taking the lease, he had no reason to anticipate.

In Manchester and the neighbourhood it is customary to let parts of factories, with a portion of the power of the engine sufficient to turn a limited quantity of machinery at a limited speed for a certain number of hours per week, usually for sixty-nine hours per week. An enactment of which the effect should be to limit the working hours to a smaller number than is provided for by the terms of the lease would injure the occupier to the extent, at least, of that part of the rent paid for the fuel and use of the engine during the hours exceeding those prescribed by the legislature. The imperfection stamped on the measure referred to by the absence of all provision for contracts of this nature may be determined by the statement of a single case. "Your tenant has let out two rooms in your factory to an under-tenant; this under-lease contains a clause binding the lessor to turn the mill-gearing in these two rooms sixty-nine hours per week. Should the ten hour bill pass, he will be prevented working his own factory more than fifty-eight hours per week: will he not then be compelled to keep the steam-engine going for the simple purpose of turning the gearing in these two rooms, if the tenant should require it?—Yes, certainly. This lease also contains a clause, that if Mr. Pooley's engine should be stopped by accident, the time so lost to his under-tenant should be either made up by working it extra hours, or the

Ibid, p. 69.

Ibid, p. 70.

rent, during the time of such suspension, shall be calculated at the rate of 361*l.* per annum, instead of 843*l.* 10*s.* If the engine be stopped by the wilful neglect of Pooley, an abatement of the rent after the rate of 3*s.* 3*d.* per spindle per annum is to be made, and compensation in damages to undertenant.

"What will be the legal effect of the proposed ten hour bill as respects this clause?—As the bill now stands, it would be a casus omissus."

Having stated our opinion with regard to such important and leading provisions of the proposed bill as have come before the public, we do not feel ourselves called upon to enter into a minute consideration of the whole of its subordinate details. It may be sufficient to observe that some of its clauses afford instances of unnecessary interference with the convenience of the workpeople, as well as with that of their employers. And it may generally be stated, that the penal clauses of this bill are of a nature so vexatious and so arbitrary as, if sanctioned by the legislature, would create a serious objection to the investment of capital in manufacturing industry in this country. On the whole, we are disposed to give credit to the statement made by several manufacturers, that they would rather support the bill as it now stands than any less injurious modification of it, because they believe that if it were attempted to be enforced, it would put a stop to so many works, and produce such an extent of suffering, as to bring about its very speedy repeal, after affording a lesson likely to prevent the repetition of interference with the voluntary engagements between employers and workpeople.

The most active if not the best instructed supporters of this measure have manifested a spirit of hostility to the progress of the present inquiry, to which we believe that few parallel instances are upon record on a subject of grave national importance. We refer the principal part of the agitation on this subject to that class of men who entitle themselves, unfortunately with some truth, the delegates of the workpeople, whom the repeal of the combination laws released from all restrictions in the disposal of their own property (labour), and who now seek to impose restrictions equally vexatious on the disposal of the property of others. We refer it to that class of men who, while stating the present inquiry to be merely whether children ought to work more than ten hours a day, are exerting their whole efforts for the restriction of adult labour, and for the arbitrary stoppage of the moving power.

It is deserving of attention, that while the protection due to children is sought to be extended to persons of eighteen and of twenty-one, the topics which have been constantly urged to engage the public sympathies in favour of the above-mentioned measure have borne exclusive reference to the claims of children, properly so called. There is not one of the motives of persuasion which have been commonly urged by the friends of the ten hour bill which affords a colourable plea for extending the protection of the legislature to the labour of adults or adolescents. It might be sufficient to establish the correctness of this observation, if we simply referred to the general impression which is felt by the public with regard to the proper objects of compulsory interference as described by the supporters of the ten hour bill. The language held in every place where popular excitement has been directed against the sober investigation of this subject shews clearly, that, whatever may be the real views of those parties who have made themselves most prominent in agitating the question, they knew how to select the ground where the real strength of their case lay when it came to be pleaded before the public tribunal. Accordingly, peculiar stress was laid upon such instances as those, which appear not wholly unknown in the west riding of Yorkshire, of parents carrying their children to mills in the morning on their backs, and carrying them back at night. The deficiency of time for education and of time for play, and the cruelty of immuring children of tender age in factories, have been principally dwelt upon as the hardships of the present system. Processions of children, and studied compositions presented by children, have been in all cases the machinery employed to divert public attention from the true state of the question.

We should not feel ourselves warranted in suppressing our conviction that the interests of the children, which alone supply materials for popular excitement on the subject of the proposed measure, are, of all other considerations, that which appears to enter least into the councils of the operative agitators for that measure.

It ought to be remarked, in the first place, that such acts of severity and cruelty towards children employed in factories as are still found of occasional occurrence, are for the most part chargeable neither on the masters nor on the overlookers, but on the spinners or the slubbers themselves. It is the practice of these latter parties to engage the children who work under them, and corporal punishment, when it is inflicted at all in factories, is administered to a child by the hands of a parent, or at least on the child of a working-man by a working-man, in most cases himself the parent of children in like circumstances. It farther appears in evidence, that sometimes the sole consideration by which parents are influenced in making choice of a person under whom to place their children is the amount of wages, not the mode of treatment to be secured to them.

Cowell, p. 51. 53. 76. 79.

Mr. Rowland Detrosier, a witness who is extensively acquainted with the cotton manufacture, replied to the question, "Is not correction sometimes substituted for fines?—Yes; but that takes place principally amongst a distinct class of the children. It is necessary to premise, that the children employed in cotton-factory labour are not all under the control of, or employed by, the proprietor. A very considerable number is employed and paid by the spinners and stretchers, where there are stretchers. These are what are called piecers and scavengers, the youngest children being employed in the latter capacity, and as they grow up, for a time, in the double capacity of scavengers and piecers. In coarse mills, that is, mills in which low numbers of yarn are spun, the wages of the scavengers is commonly from 1*s.* 6*d.* to 3*s.* 6*d.*, according to size and ability. The men do not practise the system of fining, generally speaking, and especially towards these children. The sum which they earn is so small it would be considered by many a shame to make it less. They do not, however, scruple to give them a good bobbying, as it is called, that is, beating them with a rope thickened at one end, or perhaps with a strap, or, in some few brutal instances, with the combined weapons of fist and foot.

"But this severity, you say, is practised towards the children who are employed by the men, and not employed by the masters?—Yes.

"And the men inflict the punishment?—Yes.

"Not the overlookers?—Not in these instances.

"But how do you reconcile your statement with the fact that the men have been the principal complainers of the cruelties practised towards the children, and also the parties who are most active in endeavouring to obtain for the children legislative protection?—My statement is also fact. I do not profess to reconcile the apparent inconsistency. The men are in some measure forced by circumstances into the practice of that severity of which I have spoken."

Mr. Detrosier further replied to the following question, "You state that exaggerations have been prevalent amongst the supporters of the recent legislative measure; have you thought it your duty to remonstrate with them on the use of those exaggerations?—I have not publicly remonstrated, except on one or two occasions at lectures; but I have stated to parties actively engaged, that they were making use of isolated cases as examples of a general practice. I told the parties that such conduct would ultimately injure their cause, which I considered stood in need of no such aid.

"Were those parties, parties who have protested against the commission? —I cannot positively state that to have been the case.

"What is your belief on the subject?—I believe that in one or two instances parties who have protested against the commission were present.

"Were the parties with reference to whom you express your belief, persons who have addressed public meetings on the subject?—I believe two of the parties have.

"Were they parties who have declared that further inquiry was unnecessary?—I believe that one of the parties did make such a declaration.

"On your oath, have you no reason to believe that the opposition was made under an apprehension that the exaggerations of which you have spoken might be developed and displayed by further inquiry?—I have no knowledge on which to ground such a belief. The parties to whom I declared that they were exaggerating, by making use of isolated cases as evidence of general practice, declared that they could prove those isolated cases, which they conceived made out the case for the ten hours bill.

"Must not considerable bodies of working men be fully cognizant of such exaggerations?—Decidedly not considerable bodies of working men; the isolated cases are proved to them, and seeing those facts, they are prepared to believe the general assertion founded upon them; men working in factories must of necessity know that cases of extreme cruelty of conduct are not general.

"Are you aware of any honest ground of opposition to further inquiry?— On the contrary, my conviction has always been on the subject of the commission, that though further evidence was not necessary to prove that twelve hours were too long for children to work, such an inquiry would be productive of good to the working classes, as the facts of the case could not by any possibility be overturned, even by adverse inquiry, but must be more fully and clearly established, and with such conviction I could not but regret the opposition to inquiry.

"By children, what aged persons do you understand?—Those from the age of nine to fourteen."

An apology is made for the conduct described in the foregoing statement, by supposing that the workmen are constrained to this severity by the pressure of the system upon themselves. This, however, is rebutted by the fact given in evidence, that in numerous well-regulated establishments the infliction of corporal punishment is effectively prohibited, and that in these the amount of production is apparently not less than in others where coercion is allowed on the part of the operatives.

In support of these indications, that the interest of the children is really not at the root of the agitation of this question, excepting amongst benevolent individuals in a higher sphere, we might cite innumerable cases extracted
Tufnell, 71. 73. from the evidence. It appears that although the case of the children is in-
76. 94. 99. 109. variably put forward as the plea for restriction in all appeals to the public, it
120. 124. 134. 142. is hardly so much as mentioned in the meetings or discussions of the operative
body themselves, or if mentioned, it is only in connexion with the anticipated curtailment of the working-time for adults.

The men who have placed themselves at the head of the agitation of this question are the same men who, in every instance of rash and headlong strikes, have assumed the command of the discontented members of the operative body, and who have used the grossest means of intimidation to subju-
Cowell, 68. gate the quiet and contented part of the workpeople. It is established by a
Stuart, 126. mass of concurrent testimony, such as rarely has been brought to bear on any

point of inquiry, that the former proceedings of these men have in hardly any case been successful, so far as regards the object ostensibly aimed at by agitation. This uniform result affords a presumption, which is confirmed by as much of direct evidence as the nature of the case admits, that their leaders have other objects in view besides those to which their efforts are professedly directed. It appears that agitation is the trade by which they live, and that success in the attainment of the objects at which they profess to aim would involve the loss of their actual occupation, which consists in keeping up discontent at such a height as to secure distinction and profit for themselves at the expense of their fellows.

George Lees, manager in a factory at Stockport, stated, in answer to the question, "Do you know how the delegates are supported?—Sometimes I have asked questions of our operatives, and I have found that they have paid each 1*d.* or 2*d.* a week, and in some mills as much as 6*d.*; but lately I think they have given it over, as the well-informed have found the delegates out to be deceivers. **Tufnell, p. 113.**

"How have the delegates deceived them?—I believe the delegates tell them, if they work shorter time they will get better wages; and I am also of opinion, from my own practical observations, that in the present framing of the bill, both to the masters and operatives, it is expected there will be perpetual disputes, owing to the hands working different hours, and these disputes will always find employment for the delegates; and this is anticipated by them."

It is further given in evidence, that the delegates live on the operatives, doing no other work than advocating the ten hours bill. **Ibid. p. 122.**

The following further statements with regard to the general character of the operative leaders were made by Mr. Rowland Detrosier, a portion of whose evidence before the central board has been quoted:—

"Have you paid attention to the circumstances which affect wages?—I have; and have had frequent opportunities of discussing this subject with some of the more intelligent of the working men, during the periods of their turns-out.

"Do you mean by the more intelligent of the men those who most influenced them, or their leaders?—No; I mean those who, in my estimation, possessed a more than usual degree of knowledge. But I have also had discussions and conversations with their leaders.

"Are they always or usually of the most intelligent?—Not always; they are sometimes more characterized by their extreme opinions than by their positive knowledge.

"What proportion of their measures have been attended with success?—Comparatively few indeed.

"What proportion of their measures have been attended with cost of privation to the operatives?—I am compelled to say, that in almost all cases the results have been unfavourable to the working men themselves.

"And was this so in plans for matters unconnected with direct disputes with masters, as well as contests for wages?—Yes.

"Specify some instances.—A newspaper was established in Manchester by the operatives. It was called the Voice of the People. It was mismanaged, and failed. A sum of about 2000*l.*, I believe, was lost in this speculation. In addition to this, a very laudable scheme, as I consider, was carried into effect by the dyers of Manchester; they commenced a dying establishment, the profits of which were to be devoted to the purposes of their trades union.

"What were those purposes?—I think that union combined the two objects of a benefit society and a protection of their wages; but of this I am not quite positive.

"What were the results?—It failed; more, however, from a want of union amongst themselves, than from any lack of ability to carry it on.

"Can you give any other instances?—I have heard of other instances, but I am not able to specify them particularly.

"Have you heard of any instance of success?—I know of no instances of success connected with trades unions.

"Neither in trading speculation nor in strikes?—I know of none.

"Were not the funds for these experiments made up of pittances sometimes hardly spared?—They were certainly hardly earned, and in many instances they could be ill spared.

"Do not these instances tend to shew that concerns requiring the labour of numbers cannot be beneficially conducted if the wills of many are suffered to prevail in the management; the difference of temper as well as of opinion being incompatible with the unity of purpose and efficiency of control which is essential to success?—Undoubtedly. I attribute the failures of the various schemes of the working classes to benefit themselves to a want of knowledge, and consequent prevalence of mere prejudice and individual wills.

"The great evil of the working class you consider to be the want of education, or their ignorance?—Yes.

"And is not passion and ungovernable will too frequently found, on such occasions, the substitute of knowledge?—Yes; that is what I mean by my former answer.

"Is not the confidence of some of the leaders, on such occasions, frequently in direct proportion to their ignorance?—I have known some leaders of that character. But I have also known some leaders who, though very imperfectly informed on the questions in which they were more especially engaged, as they really meant well, were willing to be informed, and who frequently admitted that there was too much of passion mixed up with the proceedings of both masters and men. There is not a more ignorant set than some of the masters, nor men more governed by prejudice.

"Is it not the most confident of the leaders, those who promise most, who obtain the most influence?—Not in all cases. The repeated defeats of the men in the great question of wages have made them somewhat wary in that respect.

"Are those men whom you mention as willing to be informed persons of influence?—Yes, among the better informed of the working men.

"What has been the governing opinion amongst the working men on the subject of wages?—There is no question on which the prevalence of prejudice has had a more injurious effect than on that of wages. The men seem hitherto to have acted on the supposition, that they could control the rate of wages independently of the supply of labour. They seem scarcely ever to have considered the subject of supply and demand as applicable to the labouring population; nor am I aware that any of those who are called their leaders have endeavoured to enlighten them on this subject. Indeed, it appears to me that prejudice rather than knowledge has been appealed to on these occasions. Often suffering reductions of wages, which appeared to them to have their origin only in the wills of their employers, it is not to be wondered at, that, in their imperfect state of knowledge, an appeal to their feelings should almost always be the most successful course.

"Can you state which of their leaders has, in writing or otherwise, done any thing which you can adduce as evidence of a capability to enlighten them on this subject—anything on which you think you could state, 'this shews that he clearly understands this important subject?'—Of the leaders with whom I have been personally acquainted, I do not know of anything upon which I should pronounce so decided an opinion.

"What is the usual remuneration of a delegate?—I think it is the sum they would earn if working at their trades, with the addition of their travelling expenses.

"What have most of the delegates been ?—Operatives; but of what particular trades I cannot say.

"Are the persons who are now leaders of the trades unions the same persons who have always been so ?—I think not in all cases, though there are some who have remained in that situation, and have retained the confidence of the people.

"Are not the trades unions the most powerful aud influential of the associations of the operatives connected with the cotton trade ?—Yes, undoubtedly.

"Is it not the firm belief of these leaders, that under a ten hours bill they will be able to obtain the twelve hours wages ?—I am not prepared to answer that question, not having conversed with them.

"Have you taken a great interest in the proceedings of the operatives, and attended to their discussions on this subject ?—Not at all with reference to the short-time bill.

"Can you state that foreign competition has, on any particular occasion, been discussed as an element that might affect prices, and wages, and work ?—I know that it has been discussed by the intelligent portion, but on what particular occasion I cannot state.

"Do the majority of working men take it into account ?—The majority certainly do not."

The effects of the proceedings of bodies under such guidance have been represented by several manufacturers in much the same terms as those used by Mr. Ashworth in his evidence to the central board. He states, in reply to the question, "Have you any objection to state your reason why you are not inclined to extend your works, as you appear to have done so heretofore ?—We are not disposed again to extend our works; but the reason is not an agreeable one to assign. Since entering into business we have several times expended the last shilling we possessed in the extension of our mills and machinery, and have had a considerable degree of enjoyment in the profit afforded to ourselves, and a pleasure in observing the increase and extension of comfort thus afforded to our numerous classes of workpeople; but during the past three years the rate of profit has been greatly diminished, and the contentment and good order of the workpeople has been seriously disturbed, chiefly by the interference of mischievous agitators; and during most of this period the promoters of time-bills have threatened still further inroads upon our profits, by proposing limitations of the time of working, and many vexatious penalties and restrictions; and they have endeavoured, with singular ingenuity and audacity, to fasten upon us, as a body, in the eyes of the public, the most unjust imputations of avarice and cruelty. We have therefore become indisposed to make any further extension to our works, although we are young men, having young families, and strong desires to make ample provision for them, and as much disposed as ever we have been to extend the prosperity and comforts thus diffused to our neighbours and our workpeople."

The pernicious notion of the propriety aud necessity of legislative interference to restrict the hours of adult labour is mischievously sanctioned by some persons engaged in manufactures, and by gentlemen connected with them, who may be served by popularity, and whose judgments may thereby be biassed in favour of the doctrine. On examination, however, we invariably find that the exact time to which adult labour might be restricted, as each of these manufacturers conceive, is the exact time which he works his own mill. Thus the person whose machinery usually works eleven hours, or eleven hours and a half, considers that eleven hours and a half ought to be imposed by law as the maximum of adult labour; the person who works his mill twelve hours is the advocate for a limitation of the adult labour to that period. The working over-hours by water-mills to make up lost time by failure of water, or the working over-hours to make up a sudden order for exportation, when adults and even children usually perform the extra work with alacrity for the sake of extra pay, and connive at evasions, would by these advocates for restriction be equally prohibited. Each of these supporters of restrictions is ready to concur in a limitation of labour which will not interfere with his own convenience, and is willing that the activity of all others of the same trade, and of all other trades, should be restricted to his own pace.

In recommending legislative restriction of the labour of children, as not being free agents, and not being able to protect themselves, we have been careful not to lose sight of the practical limits within which alone any general rule admits of application. We have not found these limits in the greater or lesser intensity, or in the greater or lesser unwholesomeness of infant labour in factories. It appears in evidence, that of all employments to which children are subjected, those carried on in factories are amongst the least laborious, and of all departments of in-door labour, amongst the least unwholesome. It is in evidence, that boys employed in collieries are subjected, at a very early age, to very severe labour, that cases of deformity are more common, and accidents more frequent amongst them than amongst children employed in factories. Hand-loom weavers, frame-work knitters, lace-runners, and workpeople engaged in other lines of domestic manufacture, are in most cases worked at earlier ages, for longer hours and for less wages, than the body of children employed in factories. Proofs of these statements might be multiplied from that part of the evidence before us which bears on these collateral points of
Power, 43. inquiry. One witness is even anxious to have his opinion recorded on the necessity of extending legislative protection to many other descriptions of labourers besides those in factories, and refers to the case of apprentices of milliners and shopkeepers, and even to that of school-girls in finishing schools.

With reference to the foregoing opinion in favour of the extension of legislative interference beyond the limits of factory-labour, we are induced briefly to state the grounds which appear to justify that interference with factories, as distinguished from collieries, or from establishments of a domestic nature.

Children employed in factories, as a distinct class, form a very considerable proportion of the infant population. We have found that the numbers so employed are rapidly increasing, not only in proportion to the increase of the population employed in manufacturing industry, but, in consequence of the tendency of improvements in machinery to throw more and more of the work upon children, to the displacement of adult labour. The children so employed are assembled together in large numbers, and in buildings of peculiar construction, which cannot be mistaken for private dwellings. Their daily entrance into, and dismissal from, the factories, take place with the regularity of military discipline.

These assemblages, therefore, so situated, may easily be subjected to regulations which could not, even if the motives for interference were otherwise equally strong or even stronger, be applied to children in other employments, without such an extent and expense of police, and such a vexatious scrutiny of private dwellings and occupations, as could not be borne.

Having stated the grounds on which labour in factories is distinguishable from other modes of employment, we proceed to state the grounds on which we consider that legislative protection is required on behalf of children, properly so called, employed in factories. We find that in all the principal branches of manufacture throughout the kingdom they are forced to work during the same hours of labour as the adults. We find that the effects of the labour during such hours upon children in numerous cases are—1. A permanent deterioration of the physical constitution, and the production of various diseases, often wholly irremediable. 2. Exclusion from the means of obtaining education, elementary and moral, or of profiting by those means by reason of excessive fatigue.

The grounds for legislative interference in behalf of children under such circumstances appear to be—1. That at the age at which the children in question are put to labour they are not free agents, inasmuch as they are let out to hire, and do not receive the wages they earn, but those wages are appropriated by their parents or guardians. 2. That the labour they perform is not proportioned, and is not pretended to be proportioned, to their strength, but is regulated solely by the duration of the labour of adults.

Cowell, 42.

The restrictions we venture to propose with regard to children are, that children under nine years of age shall not be employed in mills or factories, subject, however, to the considerations hereinafter stated. That until the commencement of the fourteenth year the hours of labour during any one day shall not in any case exceed eight. That until the commencement of the fourteenth year children shall not in any case be allowed to work at night; that is to say, between the hours of ten at night and five in the morning.

The grounds on which we recommend the above restriction on hours of labour to be limited to the commencement of the fourteenth year, are—1. That at that age the period of childhood, properly so called, ceases, and that of puberty is established, when the body becomes more capable of enduring protracted labour. It appears in evidence, from the statements and depositions of all classes and witnesses, including the young persons themselves, that the same labour which was fatiguing and exhausting at an earlier period is in general comparatively easy after the age in question. 2. That from the comparative infrequency with which serious and permanent disease appears to have been produced when labour did not commence before the ninth year, and was not immoderate, there is reason to conclude that the restriction now suggested will afford an adequate protection. 3. That, in general, at or about the fourteenth year young persons are no longer treated as children; they are not usually chastised by corporal punishment, and at the same time an important change takes place in what may be termed their domestic condition. For the most part they cease to be under the complete control of their parents and guardians. They begin to retain a part of their wages. They frequently pay for their own lodging, board, and clothing. They usually make their own contracts, and are, in the proper sense of the words, free agents. For all these reasons we conceive that this is the natural period when young persons may be placed on the same footing as adults as far as regards the disposal of their labour.

In proposing the foregoing limitation of the labour of children, we admit that while a certain proportion of those who have passed the prescribed age might advantageously share in the benefits of restriction, a certain proportion of those who are included in that restriction may, without injury, work longer hours. Cases of this kind, however, being cases of exception, do not appear to come within any general measure of legislation. The grounds on which we abstain from recommending more than one limit are, the facilities which every additional restriction would afford to fraud and evasion, and the impossibility, according to the evidence, of employing three sets of hands with different limitations. The nearer the approximation of the hours of one set to the hours of another, the greater would be the facilities and temptations of the lower set to run into the working hours of the higher set. If the usual working hours of the adults were eleven and a half, or twelve hours daily,

Central Board, 26.

the inconveniences of another set ceasing to work at the end of the tenth hour would be so considerable that they would go on working to the twelfth. If we suppose three sets of hands employed in a factory, one of which should be restricted to eight, a second to ten, and a third to twelve hours work, the eight hours set would be apt to run to ten hours, and the ten to twelve. Moreover, whatever may be the objections on the part of the manufacturers to double sets, would apply with increased force to three sets.

The great evil of the manufacturing system, as at present conducted, has appeared to us to be, that it entails the necessity of continuing the labour of children to the utmost length of that of the adults. The only remedy for this evil short of a limitation of the labour of adults, which would in our opinion create an evil greater than that which is sought to be remedied, appears to be the plan of working double sets of children. To this plan there have been intimated to us great objections on the part both of masters and of workmen: on the part of the masters, because it will be attended with inconvenience and somewhat higher wages: on the part of the workmen for various reasons; 1st. Because when working by the piece increased expense in hiring, or increased trouble in teaching children, will necessarily diminish their net earnings:—2d. Because by a more general limitation to ten hours they expect to get twelve hours wages for less work:—3d. Because the reduction to half wages, or little more, of the children reduced to six or eight hours work, must necessarily, in so far, tend to reduce the earnings, and consequently the comforts of the family.

There can be no doubt, from the whole tenour of the evidence before us, that the plan of double sets will be productive of more or less inconvenience and expense to the manufacturer. It has appeared to us, however, that the same objections must attach more or less to any change of the present modes of working; but we consider the object aimed at by the working of double sets, namely, that of counteracting the tendency to an undue employment of infant labour, to be such as more than compensates for the sacrifice to be made in attaining it. And no other mode of effectually accomplishing that most desirable object has occurred to us likely to be attended with so little evil or suffering as that which we have ventured to recommend.

The suggestion of the plan of double sets by some of the most considerable manufacturers, and the acquiescence in it, reluctant though it be, by others, appears to us to afford a guarantee of its practicability. In one instance, viz. that of Mr. Marshall at Leeds, it has been acted upon as an experiment, limited, however, to an age below that which we have suggested. We extract the following evidence on this head, taken before the central board, from several leading manufacturers.

Mr. James Hadden examined:

Central Board, 1. "Assuming that it is necessary to have some restriction of the working hours of children, what do you consider would be the effect of restricting the working hours of all children from the ages of nine to the end of the thirteenth year not exceeding eight hours, as compared with the restriction of the working hours of all under eighteen years of age to ten hours?—Such a reduction to that of eight hours of children under the age of fourteen would be preferable, inasmuch as it would not render necessary a proportionate reduction of the working hours of adults. The restriction of the working hours of all under eighteen would affect so large a proportion of our working people, that the labour and wages of the adults must necessarily be diminished by it. Although I do not approve of legislative interference, as it will be attended with injury to the working classes themselves as well as to the manufacturer, I think the restriction of the working hours of children under fourteen years to eight hours the preferable course, as it would the least diminish the productive industry of the country and the rate of wages to the children themselves."

Mr. Henry Ashworth examined:

"You have proposed that no child under eleven or twelve years of age should be allowed to work more than six hours in a day; would less difficulty be found in obtaining double sets of hands to work at six hours, than two sets to work eight, and four hours each?—These arrangements would vary in their accommodations with the different families to whom the regulations applied; it would be convenient and advantageous to some families who had two children at work under twelve years of age that they should work six hours each, so that one of them might attend school whilst the other was at work; and there are some cases where it might be more agreeable that a boy of twelve years of age should work with the eight hours set, and a younger one with the four hours set, instead of a fixed period of six hours each; but I think there is danger of this accommodation being counterbalanced by its interference with the regularity of the schools.

"If this limitation of children's labour should have the effect of compelling the mothers of families to resort to the mills to make up the loss of wages thus occasioned, would it not then be desirable that the labour of children should not be confined to six hours, that the mothers might work with the four hours set, and not be taken too much from their domestic occupations?—If legislation should be the means of forcing the mothers of families back to the mills, which every one must exceedingly regret, it is certainly desirable that they should be allowed to work with the four hours set."

Mr. John William Partridge examined:

"Assuming that a restriction of the hours of infant labour to below ten hours is found absolutely necessary, what would be the valid objections to a restriction compelling the adoption of two sets of hands, as thus: that children from nine to the end of thirteen years of age should only be allowed to work eight hours; the labour of adolescents above that age, and of adults, being left entirely free, and the trade being freed from all restriction as to meal-times and working-hours; this restriction also being progressive, those under eleven years of age only, for example, being included in the restriction during the first year, those within twelve being included at the end of the next year, and so on?—The first objection would, I conceive, be, that we could not get a sufficient stock of hands for double sets.

"What is the age at which children are the most valuable to you?—The proportion of those below twelve years of age who are valuable is very small.

"Consequently the inconvenience of the restriction in the first instance would be very small, would it not?—If it must be, it would be the least so.

"To meet the next restriction, would you not have the means of obtaining a full stock of hands?—It would be very difficult to find enough of hands.

"From what facts within your own experience do you infer this difficulty?—I know that at present eight out of ten of those children who are capable of working are now employed.

"Are those children now employed children from the immediate neighbourhood?—Yes.

"Almost entirely?—Yes.

"Then you have not been accustomed to search for children beyond the immediate neighbourhood?—No.

"Consequently cannot speak as to what supplies of children distant neighbourhoods might afford?—No.

"When you spoke of the difficulties of obtaining children, you spoke entirely from your experience of the present state of the population?—Yes.

"And did not take into account the probable progress of population during the next three years?—No; I spoke of the present state of the population.

"How many children under fourteen do you commonly employ?—About eighty.

"Of that number, how many are indispensable, on account of their youth, for particular processes of your manufacture?—I do not think that any of them are indispensable under that age merely on account of their age.

"The work might then be done as well by young persons of fourteen and upwards?—Yes; but at a greater expense."

Mr. Joseph Grout, examined:

"Assuming that a restriction of the working hours of children, from nine to the end of the thirteenth year, is necessary, to eight hours each day, so as to compel the employment of double sets, what do you consider would be the inconveniences or obstacles to the adoption of such a measure?—The main objection would be, the difficulty of obtaining a sufficient stock of hands. The extent of our works at Norwich, Yarmouth, Millinghall, and Ditchingham, was limited by the numbers of such hands whom we could find. After having erected mills at one place, with the view of obtaining hands, we have been obliged to go to another, having found it difficult to obtain a sufficient number of young females for our purpose. The persons whom we employ are nearly all females.

"Could not adult labour be substituted extensively?—The wages we are at present obliged to pay are so very low that no adult labour could be got for the money. I think, however, that if the labour of all those above thirteen were left free, that there would be no objection to legislation with respect to those under that age: when they are thirteen, the restrictions should, however, be imposed very gradually."

Mr. Anthony R. Strutt states, in answer to the question—

"Assuming that a restriction of the hours of the labour of children is necessary, what is your opinion of the practicability and expediency of limiting the work of all children under fourteen to six or eight hours, so as to enforce the employment of them in double sets?—Our's being a country situation, we should experience great difficulty in getting the number of hands requisite to replace those which by such a regulation would be reduced to half work; and it would be attended with considerable expense and inconvenience."

Mr. John Marshall, m.p., examined:

"From your experience do you believe that in the employment of double sets of children fraud in working them at another mill during the same day may be completely prevented?—Certainly; a child could not come to us in the morning, and work with another in the afternoon, without our soon finding it out. Our direct interest would ensure our taking efficient measures to do so.

"Could you ensure the attendance of children at a school during the hours when they were unemployed?—We actually do it; and we insist upon it at present.

"Do you combine the working of the double sets with any regulations as to education?—Yes.

"Would you have the goodness to state the circumstances under which you commenced the practice, and the mode in which you now carry it on?—The circumstances which occasioned our employment of two sets in the day, of young hands between the age of nine and eleven, are these: our attention was drawn to the subject of legislating on the ages of children in factories in 1830. On the consideration that we then gave to the subject, we thought that nine years old was too young to admit children into the mill to work the full time. In January, 1831, we ordered that no children under ten years old should be admitted into the mill; and on the 1st January 1832 we again raised the limit to eleven years old. During the year 1832, our head over-lookers frequently mentioned to us that the parents were much dissatisfied with this restriction on the labour of their children; many of them, whose entire families were working with us then, and had been doing so for many years, complained, and stated that they would not keep them unemployed, but would take their young children (under eleven years old) to work elsewhere, though they should do so reluctantly. We at once acknowledged the correctness of their complaint, and saw that we were causing considerable inconvenience to them by our restrictions. We saw that, unless the children were engaged in their education they were no better for this state of non-employment; and it was also evident to us that they ought to have acquired all the education suitable for their station in life before they were nine years old: we therefore determined to give employment to the children of nine and ten years old, whom we had excluded from the mill by our former regulations; but we restricted the regulation to such as were the children of our old workpeople, and who lived in the neighbourhood of the mill; we arranged that a certain number of them should work one half of the day, and the remainder the other half of the day. The experiment was begun on a small scale in January 1833, and is now only practised by a comparatively small class in our mills at Leeds; about forty, I think. We at the same time made it compulsory upon these children that they should attend our school in Holbech during that portion of their time that they were not working in the mill. We gave them 1*s.* 6*d.* per week, out of which they had to pay 1*d.*, 2*d.*, or 3*d.* per week. The half fee for learning to read, 1*d.*; read and write, 2*d.*; or read, write, and account, 3*d.*, respectively. It was the duty of the man who paid the wages in the manufactory to see that these children were actually in regular attendance at the school. The plan has worked well for the children, and has been much approved of by the parents.

He further states, in answer to the question—

"Do you not consider that in proportion as the restricted hours of one set of workmen approached to the hours of another would be the facilities and temptation to fraud?—Thus, if you had three sets of hands, the adults working twelve hours, the adolescents restricted to ten hours, and the very young to eight hours, would there not be a temptation and facility to the eight hours set to get into the ten hours set, and the ten hours set to get into the twelve hours work?—Yes, there would certainly be that temptation; but I cannot contemplate the possibility of working a mill with three sets.

In point of principle, if interference be justifiable at all on the grounds we have stated, it cannot stop short of the limit we have proposed, viz., the completion of the thirteenth year. At the same time, such has been unfortunately the tendency of improvements in machinery progressively to increase the proportion of the labour of children, that, according to all the witnesses examined, there would be a great difficulty, if not an impossibility, of immediately getting the number of hands requisite, if all under fourteen were immediately to be reduced to six or eight hours. This difficulty would apply probably to all mills, but in the greatest degree to those which are not situated in populous towns or districts; and even if the difficulty on this head were less, the loss of earnings by the operative class, from a sudden reduction of the working hours of their children up to their fourteenth year, would often entail severe suffering. The measure therefore which we venture to recommend, however salutary in its tendency, might fail altogether on the score of the alleged difficulty, if not impracticability, as regarded the manufacturer on the one hand, and of the distress it might occasion to the working class on the other, if it were to be immediately enforced, instead of giving (as should be done in all great changes such as this would be) ample time for preparation. We would therefore, to meet the difficulty so prominently put forward by several of the master manufacturers of finding a sufficiency of hands with which to replace those whose working hours would be reduced by the proposed regulation, suggest that the limitation to eight hours work should, in the first instance, be applied only to children up to the commencement of the twelfth year of their age; that this limitation should take effect in about six months from the passing of the bill, and should be extended, by stages or intervals of six or twelve months each, as might hereafter be determined in framing the enactment, to children under the thirteenth and the fourteenth years of their age respectively.

By the application of this limitation in the first instance to children under twelve, the most urgent part of the evil proposed to be remedied is met; none of the hands of twelve and thirteen will be displaced or reduced so as to occasion immediate distress to their families. Time will be given for judging of the experiment as far as it goes, for in so great a change it is desirable that the measure should in the first instance be considered as merely tentative.

But there can be no pretence for setting up the plea of impracticability to double sets under twelve. Among the memorials presented to us from the manufacturers, some propose to exclude altogether from their works children under ten, and others would exclude children under eleven; while there are individual witnesses who would not object to a total prohibition under twelve. Now, surely, giving them the option of employing such children for half time must be a facility compared with total prohibition.

In a set of resolutions recently communicated to us by the committee of master cotton-spinners of Lancashire, assembled in London, the practicability of double sets for all under twelve years of age is distinctly recognised as one of two alternatives to which they declare that they will not present any
Central Board, 27. opposition. (See Resolutions, &c.)

The proprietors of the Burley Mill, near Leeds, write to us in the following terms:—" Observing in the newspapers that the factory commissioners have suggested, and that Lord Althorp approved of, the plan of limiting the period of labour for children under fourteen years of age to eight hours, and apprehensive that my replies to some of the printed queries might lead you to suppose that such a plan would be injurious to my mill, I beg to state, that although I have one hundred and fifty children under that age, and could not in any degree accomplish a double set, yet, provided that all above that age might work twelve hours, I think I could make such an arrangement as would prevent the plan in contemplation from materially, if at all, affecting my productive power; and there cannot, I think, be a doubt but that the plan would be of great benefit to the younger children."

In order to give increased facility for getting a sufficient number to constitute double sets under twelve, there would not in our opinion be any great objection to reduce below nine years the age of absolute prohibition; because, the hours of work being limited to six or eight, time being thus afforded for school and recreation, the occupation moreover not being laborious, it is not clear that children at the age of eight years might not harmlessly at least to themselves, and to the convenience and relief of their parents, be admitted into the factories.

If the difficulty is likely to be so great, as it is by some manufacturers represented to be, of getting a number of hands even under the age of twelve equal to the number now engaged with a view to make double sets, each set being of the same number as that now employed, or of getting half the number above the restricted age, so as to answer the same purpose, it proves the more strongly the magnitude of the evil, and the urgency for a remedy, inasmuch as it shows how very large a proportion of infant labour is now tasked to the same hours as that of adults.

It is stated in the resolutions of the master cotton-spinners, (See appendix to evidence before Central Board) "That any system of relays, unaccompanied by a restriction for persons above fourteen years of age, will indubitably cause a great extension of the hours of labour to those who are above fourteen."

This allegation of the indubitable increase of the working hours of adolescents and of adults has been introduced in the arguments addressed to us by manufacturers, who have been heard though not examined by us. This allegation is founded on a proposition which has been usually addressed to us in these terms:—

The interest upon the capital sunk in the machinery constitutes a smaller charge in the cost of production in proportion to the greater quantity produced in a given time, and it is consequently to the advantage of the manufacturer, and he will be stimulated by competition, to work the longest hours.

Now the inference founded on this proposition is entirely rebutted by the fact, that no such results as those predicated do ensue, even in the business of these parties. If this reasoning were valid, competition would by this time have led to universal night-work in their own and other manufactories where large capital is invested in machinery; there would have been no such thing as a mill worked for fewer hours than those worked by law; yet we find cotton-mills, in which considerable capital is invested, such as that of Mr. Ashton, of Hyde, which are worked for eleven hours and a half, or even below that time. On the other hand, we find, that in other large branches of manufactures, such as those of flax, woollen-yarn, and worsted, where there are considerable investments in machinery, and active competition, freed from all legislative restrictions, the hours of labour throughout the country may be stated as about twelve daily. Mr. Power, for instance, states, with regard to the manufactures comprehended in the north-eastern district, "In the flax-mills we found the ordinary hours about eleven and a half; in the woollen-yarn mills and cloth-dressing mills, about twelve; in the worsted-mills alone we found the ordinary hours amounting to thirteen. Of the latter branch it is observable, as distinguished from the others, that the earlier processes are not carried on (except in some slight degree) by the assistance of power, but by hand-labour of the combers, sorters, &c., the mill-wages being from that circumstance little more than one-fourth of the whole wages; hence, the interest of fixed capital will be proportionally less in this branch, where the hours of work are the greatest."

The instances of working long hours are exceptions to the general practice, and are only resorted to on such occasions as those to which we have adverted; namely, to execute sudden orders, or, in the instances of water-mills, to make up lost time. But it appears in evidence that the practice of working more than the ordinary hours of the trade in each branch cannot be turned to account to any such extent as these manufacturers have assumed. When the work is protracted beyond those hours, the workmen become less efficient, the quality of the work is injured, the amount of waste is augmented, and, moreover, an additional expense for light is incurred.

There appears to be a general tendency in all manufactures to settle down the extent of the labour in their operatives to about twelve hours daily. All

See the evidence of Mr. Greenough, Tufnell, p. 134.

the witnesses agree, that during the last twenty years great improvements have been made in the operations of the larger branches of our manufactures, by which the severities in the mode of performing the labour, as well as the duration in time, have been considerably reduced, not by means of any legislation, but mostly by the voluntary care of more intelligent manufacturers, influenced, it may be, in some degree, by the action of public opinion. As the size of a manufactory is increased, so is the necessity for order and facility in the operations, to which filth and disorder, injurious to the health of the workpeople, are prejudicial. As the larger manufactories have the advantage in competition, the smaller ones must, by the natural course of commercial operations, be expected to give way.

To hold out, in the face of the course of operations to which we have adverted, that there is a powerful tendency throughout all branches of manufacture to increase the severity of adult labour, or that any regulation for the employment of double sets "would indubitably cause a great extension of the hours of labour to those who are above the age of fourteen," is, to propagate a manifest delusion, which it appears to us can only have been caused by commercial jealousy having blinded the parties to the countervailing circumstances constantly under their own observations.

The following extracts from the examinations of witnesses by Mr. Tufnell will show with what degree of consideration for the circumstances of other manufacturers some manufacturers have advocated measures tending to fix arbitrarily the hours of work in all places.

The witnesses were advocates for the ten hours bill, but the principles of the restriction for which they contended is the same as that to which many of the opponents of the bill are favourable.

"Mr. John Cheetham, of the firm of George Cheetham and Sons, cotton-spinners, examined:

"Would it be possible to make an efficient bill?—Yes, by restricting the moving power, or prohibiting any machinery to be worked after a certain hour in an evening, or before a certain hour in the morning, and no lost time to be worked up; and in our opinion any legislative measure that does not restrict the moving power, and include all ages, will be found inoperative. I think a mill-inspector, elected in every township, whose business it should be to see that the law was observed, would be the best means of preventing its evasion.

"Would you include water-mills in such an enactment?—Yes.

"Would that not be an injustice?—I think not; it is the duty as well as interest of a water-proprietor to make a provision by means of reservoirs for a constant supply, and then he is in the situation of a steam-proprietor.

"Suppose he cannot make these reservoirs without incurring such an expense as would absorb all his profits?—In such cases the proprietor of the water ought to make the tenant compensation.

"Suppose the tenant is proprietor, who is to make compensation—to whom?—Then I should say he has enjoyed a previous advantage over his competitors in trade which ought to compensate him for his present increased outlay.

"Why do you think he has till now enjoyed an advantage over his competitors in trade?—Because it is a well-ascertained fact that water-power is cheaper than steam.

"Then if a mill-owner wishes to set up a manufactory, he can always do it cheaper by purchasing a water-fall than a steam-engine?—Yes; if he does not pay too high for his water.

"Suppose he does not pay too high for his steam-engine, would he be in the same condition?—No; because the price of fuel is a greater object than the price of a steam-engine.

"Why is it cheaper to purchase a waterfall than a steam-engine?—On this ground—the constant supply of water is much cheaper to turn an engine with than the supply of coal.

"Then will not any one who has to sell a waterfall ask a higher price for it, and a purchaser be willing to give a higher price for it, inasmuch as you say it is cheaper to work by water than steam?—Yes; but then an individual will not resort to it unless he gives a lower price for it.

"Then you think that the proprietor of a waterfall is willing to sell a certain amount of power at a less rate than the same power would cost if obtained by a steam-engine?—Yes."

"Mr. Bell Clarke, of the firm of Clarke and Sons, examined.

"You are of course aware of the effect this bill will have in preventing any person, of any age, from working more than ten hours daily, between six in the morning and seven in the evening?—Yes, I am. I consider all mill labour family labour; the present plan being for children and wives to go to the factory before their husbands, if they happen to be mechanics or labourers, and to leave later.

"Are you aware that in some factories moved by water-power it sometimes happens that it is impossible to get the necessary supply of water before a late hour of the day?—Yes.

"Then might not such establishments be sometimes prevented from working more than five or six hours daily?—I can't say. I think if such is the case they should not have built their mills in such a situation.

"Suppose they built their factories many years ago, on the understanding that a ten hours' bill would never have been passed?—A man ought to have sufficient foresight to have known that it was unnatural to work persons unreasonable hours, and therefore he ought never to have built a mill there. In all cases where such a mill is short of water he should put down a steam-engine to assist.

"Suppose he pays a large yearly rent for the use of the water, would justice require that he should continue to pay the rent, when he gets no benefit from his lease?—He hired the premises with a knowledge of the quantity of water, and therefore he has no claim for a reduction of rent.

"Do you think there is any difficulty in the case, when a single room in a factory is let out, and the lessor has contracted to turn the mill-gearing in that room sixty-nine hours a week?—If an efficient act of parliament is put in force, all mills will work together. If the lessor produces less he will get a better price for the less quantity produced, which will make up for the loss he may sustain on account of rent."

Whilst on the one hand we have received from the employers of the largest numbers of children such testimony as that which we have quoted as to the feasible nature of the measure we recommended, we have been met, on the parts of other manufacturers, by allegations of the "impracticability" and the "impossibility" of obtaining, or working mills by means of, double sets of children. We do not for a moment lose sight of the inconvenience to which this and every new regulation which deviates from the ordinary practice of business must subject those on whom that regulation is imposed, and we have therefore adopted as a principle, that no regulation should be imposed, unless a strong case of necessity for its establishment were first made out. We find, however, in the course of our inquiry, that the words "impracticable" and "impossible" are too commonly attached by many of the manufacturers to any regulation which may subject them to expense or to temporary inconvenience. A number of the lists of the persons employed at the mills, which were requested to be made out by the mill-owners, have not been made out, as it was declared utterly "impracticable" and "impossible" to comply with the request, while in a greater proportion of instances of manufacturers similarly situated, those same lists were speedily and satisfactorily filled up with all the requisite information. Some of those who have made representations to us have declared the employment of double sets impossible; meaning, as it appeared on inquiry, that the second set could not be found on the morrow, or within a few days, after the legislative enactment coming into immediate operation, which statement we believe, and, to provide against the difficulty, have recommended the progressive operation of the measure at different stages; others declare the employment of double sets to be impossible, and it appears, on examination, that they found their allegation on the fact that all the children of their own neighbourhood who are of the requisite age are already employed, but it also appears that they have never extended their search for hands beyond the immediate neighbourhood, and have made no applications for children from the crowded workhouses, or from the agricultural parishes, where the complaints are so loud of the pressure of population beyond the means of employment.

The employment of children from distant neighbourhoods is objected to on the grounds that the wages to be given to them would be too small to induce the parents to remove; and, in the absence of the natural guardians, it would be necessary for the employers to become guardians, and take the children as apprentices, which in the first place would burden with settlements the parish where the manufactory is situated, and next would impose on the manufacturer "the trouble of taking care of so many children." This last objection (which is a very pregnant one) may be dismissed with the observation, that although it may occasionally be a hardship to encumber the manufacturer with the duties of a guardian, yet the treatment of apprentices in those instances we have met with, where numbers are employed by large capitalists, has been such as to favour a measure tending to ensure their employment under circumstances where their treatment is more exposed to observation, and which impose a more regular and systematic management than can be expected from smaller capitalists or tradesmen. The second objection, we trust, will be met by the adoption of the measures which we understand will probably be recommended to the legislature by his Majesty's commissioners for inquiring into the administration of the poor laws. In the course of this inquiry we meet with striking exemplifications of the chief evil of the present poor law system; namely, the obstruction which it offers to the circulation of labour. It appears that, under the operation of that system, the population is penned up in petty districts, in some of which they increase beyond the demand for their labour; and the children of the paupers are kept in indolence, which unfits them for steady industry as adults, whilst in other districts the demand for the labour of children appears to be such as to occasion them to be worked beyond their powers, and to impair their capacity and well-being as adults. The fact that the general wages of children and youths in the manufacturing towns are double the wages of children and youths in the agricultural districts, whilst in the latter the workhouses are full of unemployed persons, affords an indication of the working of the system. Yet the certainty of an adequate unrepulsive parochial allowance, happen what may, impairs the habits of economy of those of the manufacturing classes in the towns who receive such high wages as would enable them to provide against temporary cessations of employment, and the casualties of sickness and superannuation, for which they now claim provision from their parishes.

In a report of his Majesty's commissioners under that inquiry it is stated, that the progress of the arts and manufactures tends greatly to the division of labour, and (for the purpose of that division and united action) to the aggregation of large bodies of labourers in particular employments in small localities. In the same progress, and especially in the progress of the manufactures dependent on fashion, the demands for labour are liable to extreme fluctuations. Particular manufactures leave one part of the country, and are established in another; some branches of manufacture expire whilst new ones spring up. The present administration of the poor laws, and in some degree the state of the law itself, frequently operate most mischievously, by indisposing workmen to follow the demands for employment into new districts, and also by weakening the motives to seek new employments when old ones have altogether ceased. The subdivision of the country into small districts or parishes has occasioned very serious burthens to be thrown upon the parish where the manufactures which are subject to the fluctuations have been located.

We admit the force of the objections to giving employment to any of the unemployed stock of hands in distant parishes, but we trust it may be presumed that the present system of the poor laws will not be allowed by parliament to remain a barrier to the wholesome circulation of labour, and continue to operate to the prejudice of the two great classes of the community; with regard to the one, by imposing on children a ruinous excess of labour, and the other, by creating habits of idleness equally mischievous to the whole community.

By many of the objectors to the employment of double sets no account is taken of the contribution which may be made to them by the employment of adults for those processes for which infant labour is at present exclusively requisite. By many others of the objectors the contributions from adult labour is admitted, but it is stated as an objection that the restriction of the hours of infant labour would compel the mothers of families to work in the mills; a consequence which is deprecated as extremely mischievous; but they would usually be employed to make up half sets; and we cannot admit that an arrangement which would compel the mother of a family to work six hours in a mill is as mischievous as one which compels her child to work twelve hours, to the prohibition of its education, and its own misery, and the injury of the community, in after life.

It appears that in a family of the average number only two children would come within the period of restriction. In the chief manufacturing districts, where there is a considerable demand for infant labour, a large family is frequently considered in the light of an important source of income to a working man. Although cases of hardship and suffering must be expected from the reduction of the earnings of the younger members of families, yet that reduction would leave an average family in full employment in the manufacturing districts, in the receipt of better wages, and in a better condition, all circumstances considered, than the families of many other classes of operatives in towns, and of the mass of the agricultural population.

Objections have been brought against the employment of double sets of children founded on the anticipated effects of an extra number of hands being ultimately thrown by it on the market of adult labour.

We would in the first place observe, that no extraordinary encouragement would be held out by the plan which we recommend to the employment of large numbers of children, inasmuch as no great increase of the present amount of wages would be thereby provided for distribution amongst them. The tendency of the measure would, as we are led to believe, be rather to restore to adults a part of the employment which the progress of machinery is fast transferring to children. It does not appear probable that the temptation offered by such wages as the manufacturer can afford to give for double sets of children will be such as to cause a surplus population to be brought up expressly for the purpose of supplying hands to the factories. It is more likely that young hands, at present employed, will be collected from country parishes, at present overstocked, to share in the moderate labour and wages provided by this arrangement. It appears to be perfectly well understood by the labouring class that factory employment is not to be considered as a trade which can be permanently relied upon, excepting in the superior departments, and there seems to be no reason to suppose that the difficulty of finding sufficient employment for adults will be aggravated by the circumstance that an additional number of children will acquire the habits of industry from temporary employment. Under the proposed arrangement, as at present, the employment of children would, we conceive, continue to be deemed advantageous until they should have attained an age which would fit them for other occupations. For the reasons above stated, we do not consider that the burthen of an increased stock of adult labour would be created in anything like the apprehended extent. Neither do we consider that the anticipation of such an increase ought to be set against the enormous evil at present existing, of excessive labour imposed during the period of childhood, and permanent injury thereby entailed on so large a class of the population.

The short time which has been allowed us by the constant accessions of evidence to consider and define the leading principles which we have recommended as a basis for legislation on this subject has not permitted us to enter, with any degree of completeness which would satisfy ourselves, into the detail of the measures which may be found requisite for carrying those principles into practical operation.

The measures for consideration, with reference to this object, are, first those to be framed for the provision of securities with regard to the age of the parties taken into employment in factories, and next, those which relate to the duration of their daily labour.

Any measures by which the enforcement of the law shall be made chiefly dependent on those who have an interest in breaking it may be expected to prove as inefficient as the provisions of the existing law. On the part of the parent, who, under the existing law, is called upon to give the certificate of the age of the child, (which certificate forms at present the main security against evasion on this point,) we find a strong interest in the commission of fraud, amounting sometimes almost to a necessity; on the part of the immediate agents or overlookers, probably the friends of the parent, a willingness to connive at it; and on the part of the masters no especial motive to exert vigilance. This state of interests in opposition to the law is only met by the occasional attempts of informers, or of persons engaged specially for that purpose without the mills, to obtain information and evidence of what is passing within.

Cowell, 4. Thomas Worsley, a witness who had considerable experience as agent for the masters of Manchester, and for the operatives at Stockport, states, amongst the difficulties of obtaining a prosecution, that under the existing law "you have got to produce the parents to prove the actual age of the child" (which proof, we would observe, is to disqualify him for employment); "then you have to produce a person who worked near the child in the same room, to prove that on the day named in the information the child worked more than the legal hours; all this has to be proved in the first instance: then, with reference to Sir John C. Hobhouse's bill, if you lay the information against the master, the overlooker or manager can still depose that he received orders from the master not to work children more than is allowed by the act, and his oath to this effect is still sufficient to cause the information to be quashed, notwithstanding you may have succeeded in proving the former part of your case."

Tufnell, 65. Mr. W. Foulkes, a solicitor at Manchester, who was appointed by the committee of masters in that town for enforcing the provisions of Sir John Hobhouse's Act, enumerates, amongst other difficulties encountered—"First the difficulties of strangers introducing themselves into factories, either to see the parties at work, or to calculate their time of work; next, the unwillingness of operatives who are in full employment to give evidence, and frequently both the parents and the children, when summoned, giving accounts that have been gross fabrications, and which I have known to be so proved. I have known cases where the fathers have directed the children to swear, as the children have stated, that they were upwards of twenty-one years of age, and before the week was over it has been discovered, and they have admitted, that they have not turned sixteen. One instance in particular, where the mother stated the father would have been fined 5*l.* if she had not said that their child was of full age.

"Fined by whom?—My impression is, that she alluded to the clause which makes the parents liable if they give false certificates. I have known cases which, from the account that the witnesses had previously given, would have led to a conviction of the parties summoned; but when examined upon oath, their evidence has been at complete variance with their former statement, and so improbable, that the magistrates, three in number, one and all, declared their disbelief of it, and regretted they could not punish them for their conduct in court.

"Are you aware that, by the proposed bill, mill-owners are not to be liable to punishment for employing children too young, or more than the legal time, if a certificate, unobjected to, of their being old enough to be without the provisions of the act, be delivered to the mill-owner by their parents or guardians, and the child is not manifestly under age?—Yes. But there would be great difficulty in objecting; no one would take the trouble; and a decision has been come to here, that the certificate was not an effectual answer to an information for working persons under the legal age, when it appeared manifest that if due care had been exercised it would have been evidently detected to have been an imposition, and at the same time when, upon the appearance of the person so certified, all parties must have known, if they had chosen to use fair discretion, that the certificate was false, and that the child could not be of the age represented.

"Then if that decision be law, this provision of the proposed act will not be more stringent than the corresponding one of Sir John Hobhouse?—Certainly not.

"Then you think that the act would be defeated with the connivance of the parents? I think it very often has been the case, not only with the connivance of parents with the overlooker or manager, or persons employing them, but also by an imposition of the parents on the factory owners or their managers. I know of one instance where the parent applied for work for her child at a factory in this town, where she stated the age to be under fifteen; of course she was refused for night work: within a few months that child was certified at another factory to be turned twenty-one, was taken into employment, and upon that the manager was subsequently convicted, because it appeared to the magistrate so evident that due caution had not been used on the part of the manager, and the child, though the mother swore her to be twenty-one, was proved to be fifteen."

One witness in the course of his examination by Mr. Tufnell is asked, "Then your children have never worked thirteen hours and a half?—No; only my eldest son Thomas before he was sixteen worked at a mill where they worked night and day; and night-piecers being short at that mill, I was necessitated to signing twenty-one, and my son was then employed as such, or else he must have lost his bread, and me too."

Mr. Cowell observed to the witness Worsley, "The evidence which you have latterly afforded produces an impression on my mind that Sir John Cam

Hobhouse's bill has had the effect of holding out temptations of more or less strength towards perjury, but of sufficient strength to overcome the average degree of honesty to be found amoung overlookers, operatives, parents, and perhaps children and employers; but this impression is produced by the details of a very few cases. Have you any further facts within your own knowledge which you can produce, and which in your opinion are calculated to weaken or to strengthen such impression?—I am sorry to say that I cannot mention, from my own experience, one single fact or incident (unless the case of Fisher can be considered as such) calculated to weaken any such impression, if it has been produced on your mind, neither do I know any such facts by hearsay or otherwise. I can mention other facts occurring within my own experience of a nature to strengthen such an impression; and I will state for myself that it is now fourteen months since I felt so fully convinced that, by acting under Sir John Cam Hobhouse's bill, I was doing an injury, by placing people between the choice of either starvation or perjury, that I determined never to bring any information under the law before the magistrates which required any evidence out of factories to support it."

It appears that he came to a similar determination with respect to proceedings founded on the evidence to be obtained from within the factories. "What was the general impression produced among the operatives, as far as you are able to speak to it, as to its being prudent or safe for operatives to give evidence of a nature to convict manufacturers infringing the law by the publicity of the consequences of giving such evidence to those who had been thus far brought forward as witnesses?—It was very difficult to get any witnesses to come forward at all, and even to get any information of what was going on in a mill. I have applied for information to persons working in mills who were well known to me, and who made me promise before they would give it me that I would not bring informations against their masters, so that they (my informants) would be compelled to be witnesses in the case.

"Do you mean to say that the operatives, as a body and as individuals, felt themselves unwilling to lend assistance in enforcing the law?—I mean to say that they were unwilling to render assistance in enforcing the law from the fear of being dismissed from their work, and of finding it difficult to procure work elsewhere afterwards, and I made up my mind at that time never to lay another information till there was an alteration in the law."

On the consideration of the difficulties displayed in the testimony of these witnesses, we are prepared to recommend, that it should be declared unlawful to employ any child of the prescribed age without a certificate from a surgeon or medical man resident in the township where the mill or factory is situated, who shall certify, on inspection of the child, that he believes it to be of the full growth and usual condition of a child of the age prescribed by the legislature, and fitted for employment in a manufactory. This certificate should be given in the presence of a magistrate, by whom it should be countersigned, provided that he also were satisfied that the child was of the average condition of a child of the prescribed age. The age would be fixed by the legislature, as one of the means of determining the physical condition, which alone is the proper qualification for employment. Unless a discretion of this nature were given to the parties certifying they might feel themselves bound to certify to the age of the child on the production of copies of baptismal registers, which are easily forged, or on the evidence of parents, who would be under temptations to perjure themselves, such as have already been described.

The duties performed in this instance by the parties certifying are similar in their nature to those performed by two magistrates in binding parish apprentices. If the medical certificate alone were required, it is to be apprehended that in many neighbourhoods practitioners would be found whose practice is dependent on the labouring classes, and who would sometimes find a difficulty in refusing certificates to children below the proper standard as to age or condition.

The most important period for the exercise of vigilance is, however, that of admission to the period of full work. We propose that it should be guarded by a similar but a more special examination and certificate or indenture, to be given by an inspector, who should also certify, on examination, that the child examined has received an elementary education of the nature which may be hereafter prescribed. This last provision may be made to serve as a check against evasions of such regulations as may be adopted with regard to attendance at schools, or misapplications of the previous time allowed by the reduced labour.

We shall not at present treat of the collateral securities and penalties with which it might be requisite to accompany the foregoing regulations.

As one security against children being worked beyond the time prescribed, we propose that the proprietor of a mill shall be liable to a penalty on proof of a child having been within the mill more than eight hours; one half the penalty should be paid to the informer, the other half we propose should be applied to the use of the school where the child went, unless it were a school attached to the mill. This provision we consider would obviate many technical difficulties, and the necessity of obtaining the evidence of workmen in the mill, or of parties interested, and we conceive would facilitate conviction. As we propose that the only facts which it shall be requisite to prove to support the information shall be, that the child was seen to enter and was seen to leave the mill at a distance of time from the hour of entrance exceeding the time allowed by law, the onus should be thrown on the mill-owner of proving that any of the children or young persons in his employment are duly certificated.

Several eminent manufacturers have represented to us, that the only certain method of ensuring obedience to any legislative measures on this subject would be by the appointment of officers charged with the powers and duties requisite to enforce their execution. The necessity of some appointments of this nature has indeed been urged from all parts of the country.

In general it is conceived that the officer ought to be resident, and should be charged with exclusive jurisdiction of complaints relating to the infraction of legislative regulations of manufactories. The prominent objection to such an establishment of resident officers is chiefly the expense; for the manufactories being spread all over the country, such officers must necessarily be very numerous and expensive, if they are adequately paid for their services. We consider that by giving to the magistrates a concurrent jurisdiction on complaints made before them, a comparatively small agency would suffice.

The necessity of the appointment of inspectors has been most urgently stated by those manufacturers who have had chiefly in view the restriction of the hours of labour in other factories to the level of their own. The greater necessity of the appointment of some special agency for the enforcement of the measures we have recommended must be admitted, when it is recollected that they relate solely to the children, and are not directly conducive to the immediate interests either of the master manufacturers, or of the operatives, or of any powerful class, and are not therefore likely to receive continuous voluntary support. On the whole, we recommend the appointment by the Government of three inspectors to go circuits of the chief manufacturing districts, at intervals as short as may be practicable, and exercise the functions with which they may be invested for carrying the law into force. For this purpose each inspector should have the right of entering all manufactories where children are employed, and of ordering machinery to be fenced off, and directing arrangements of a sanitary nature, compatible with the execution of the manufacturing processes; and he should also have cognizance of the arrangements for the education of the children employed. He should have power to hear and determine all complaints of infraction of the provisions of the law, to give directions with relation to them to peace officers, and fine for neglect. It should be the duty of the inspectors to meet as a board, to report periodically to the government for the use of the legislature as to their proceedings and as to any amendments of the law which they might find requisite, or which might be called for. For this purpose they should be invested with the power of examining witnesses on oath, and of compelling their attendance.

In several of the most important manufacturing districts, the resident magistrates are manufacturers; and the appointment of officers of the character and the concurrent jurisdiction we have recommended would enable a complainant to reserve his complaint, if he thought proper, until the period of the visit of the inspectors. Some mills are so remotely situated in solitary places apart from towns that it would be impracticable to visit them with the same frequency. But in these places the difficulty of finding a magistrate who was not a manufacturer, before whom a complaint might be made, probably would not exist.

We consider that the performance of the function of reporting periodically to the government, by persons whose duty should be to examine the evidence on which allegations of abuse were founded, and to whom all complaints might be referred for examination, would be attended with considerable advantages, in the security it would give against the occurrence of practices inconsistent with humanity, aud in the protection which on the other hand it would extend to the master-manufacturers against groundless complaints.

It appears in evidence that in other countries care is taken to afford education to the people, and to prevent those who are engaged in labour at an early age from being deprived of this advantage.

Central Board, 2. Mr. Peter Kennedy (sworn), proprietor of a cotton-mill at Feldkirch, in the province of Tyrol, in the Austrian dominions :—" Why do you not take any under that age?—Because it is an inconvenience to lose their labour in the winter. By the law of the country all children are obliged to go to school until they can read and write to the satisfaction of the priest; they usually go to school in the winter, when other employments fail, and it generally takes three winters for their education. The priest takes care of the education. If he found that a child was at the factory, or not sent to the school, he would remonstrate with the parent, and, on failure of sending the child to school, he would cite the parents before a judge, who would compel obedience by imprisonment, or very effectual means. It is to avoid these interruptions of the labour by the education that children above eleven or twelve years of age are taken.

" Then all you employ are usually educated?—Yes; I do not know that we have any workman who cannot read or write."

Mr. James Kempson of Philadelphia, cotton-manufacturer :—" What proportion of the persons employed are natives of the United States?—Throughout the New England, which are considered the manufacturing states, above eight tenths of the persons employed are natives of the United States.

" Are many of the remaining two-tenths English workmen?—The greater proportion of them; but, as a general rule, they do not like to take English workmen in the New England factories.

" Why do they not like the English workmen?—Because they are so dissipated and so discontented.

" Is this their general character in the United States?—Yes. After they have been some time in the country, they are noted as the greatest drunkards we have. The wholesale price of whiskey is with us nine-pence a gallon, and they appear not to be able to overcome the temptation. Our own workmen are better educated, and more intelligent and more moral, and refrain more from sensual indulgence.

"How does the discontent of the English workmen of which you have spoken usually manifest itself?—In the workmen becoming masters; in strikes and demands for wages, almost always ill considered, with which the master cannot comply, and which grievously interfere with his commercial operations. Their ignorant expectations generate ill-will and hostility towards the masters.

"Are no jealousies entertained by the American workmen towards their masters?—In America we never hear the word master; they usually speak of the manufacturer by name, or as their employer, and view him rather as a tradesman to whom they dispose of their labour, than as a person having a hostile interest. There are no jealousies between American master and workman of the nature of those which appear to prevail between the English workmen and their employers.

"Are there no combinations to keep up wages in America?—None amongst the American cotton-manufacturers.

"Are there no combination laws?—None.

"To what do you attribute this state of things amongst the American workmen?—To their superior education, to their moral instruction, and to their temperate habits.

"Have you any national system of education?—We have public schools, supported partly by state funds and partly by bequests. All children have the privilege of attending.

"Do they, in point of fact, very generally attend in the manufacturing states?—They universally attend; and I think that information is more generally diffused through the villages and the whole community of the New England states than amongst any other community of which I have any knowledge.

"What is the general view taken of these schools by the manufacturers and persons of wealth in America?—From their experience they deem them of the greatest importance to the welfare of the state. They are encouraged by the state government, and all the leading persons of the state.

"How do the children whom you employ obtain education?—The manufacturers are always anxious that the children should absent themselves from the manufactory during two or three months of the year to attend the schools. The manufacturers very frequently suggest to the parents the necessity of the children being taken to school. The sending the child to school is generally an inconvenience to the manufacturer.

"Is the inconvenience of the children going to the school such as to increase the cost of production?—I do not think it does increase the cost of production. The only inconvenience is in the trouble of getting other hands. We think the advantage of their being educated more than counterbalances that trouble.

"What is the nature of this national education?—It consists of reading, writing, arithmetic, grammar, and geography."

The liberal provision already made by some of our own manufacturers for the education of their workpeople affords grounds for expectation that this object would be promoted rather than thwarted by the enlightened owners of factories in this country.

Mr. Anthony R. Strutt examined by the Central Board, p. 24.

"Do you apprehend that, in the event of employing two sets of children, there would be any danger of the morning set being sent by their parents to work in the afternoon in any other factory ?—In our situation they could not go to any other factory; but I believe that they would be employed in lace work, and in seaming stockings, &c.

"Would not a regulation, on your part, requiring the children employed in your factory to attend school, the one set after and the other before their working hours, obviate in a great degree, if not entirely, such a practice ?—I should think that a compulsory regulation for such attendance at school ought to accompany the restriction proposed of the hours of labour, as being essential to the proper working of the plan suggested; and at all events it would be adopted by us. Indeed we require at present all children, before they are admitted to our mill, to have attended an elementary school of our own, attached to the works.

"Do they attend school after they have been admitted into the factory ?—All the boys that leave work on Saturday at four o'clock we compel to attend school for about a couple of hours, not only for their improvement, but to prevent the mischief which we found, before the adoption of this regulation, they were apt to get into from mere wantonness, and of which the neighbourhood began to complain. The girls are allowed to go home to assist their mothers in sewing and household work. All the young persons under twenty are expected to attended Sunday-schools.

"Do you consider that an educated adult, supposing equal native capacity, is likely to be more useful to you, as well as a more orderly and better member of society, than one who, from being incessantly employed in labour from infancy, has had no opportunity of obtaining any school instruction ?—Most certainly."

Since the whole of our recommendations have for their object the care and benefit of the children, we have been desirous of devising means for securing the occupation of a portion of the time abridged from their hours of labour to their own advantage. We think the best mode of accomplishing this object will be the occupation, suppose of three (or four) hours of every day in education; we are the more disposed to recommend this, since it will secure two ulterior objects of considerable importance: first, it will be the best means of preventing the employment of the same child in two different factories on the same day, or in any other kind of labour likely to be injurious to its health; and secondly, it will better qualify the persons so educated to adapt themselves to other employments, if in after-life the vicissitudes of trade, or other causes, should render it desirable that they should find other means of support.

As a means of securing that the prescribed portion of every day should be devoted to the purpose of education, we recommend that every child on entering a factory be required to produce a ticket certifying that such portion of time has been spent in school; the afternoon set certifying that they have been at school during the prescribed number of hours on the forenoon of the same day; and the morning set that they have been at school during the same number of hours in the afternoon of the preceding day. And we further recommend, that the inspector be required and empowered to direct the execution of such regulations and securities, adapted to local circumstances, as he may deem requisite for the accomplishment of this object.

From the evidence collected it appears that in many of the mills numerous accidents of a grievous nature do occur to the workpeople. It appears also that these accidents may be prevented, since in some mills, where more care of the workpeople is in general displayed, they are prevented. It appears further that whilst some manufacturers liberally contribute to the relief of the sufferers, many other manufacturers leave them to obtain relief from public bounty, or as they may.

The refusal to contribute to the expense of the cure of those who have been maimed, is usually founded on the assertion that the accident was occasioned by culpable heedlessness or temerity. In the cases of the children of tender years we do not consider this a valid defence against the claim for contribution from the employer. We cannot suppose an obligation to perpetual caution and discretion imposed on children at an age when those qualities do not usually exist. The indiscretion of children must, we consider, be presumed, and guarded against as a thing that must necessarily, and to a greater or less extent, be manifested by all of them.

But the accidents which occur to the adults are of themselves evidence (unless they were wilfully incurred in a state of delirium) that the individual used all the caution of which he is capable; and it may be presumed that the loss of life or limb, or the infliction of severe pain, would rarely be wantonly incurred.

Some of the manufacturers have proposed that the inspectors, who they think ought to be appointed to ensure compliance with any legislative regulation, should have power to inspect the factories, and direct what parts of the machinery should be fenced off, and that after such directions have been complied with, the manufacturer should be relieved from further responsibility.

We concur in the proposition for giving such power to inspectors, but we do not concur in the proposal to relieve the manufacturer from responsibility.

We apprehend that no inspector would probably be so fully conversant with all the uses of every variety of machinery as to be acquainted with all the dangers which may be provided against; and also, that whilst there is much machinery which does not from its nature admit of being boxed off, there is much that could not be made entirely safe without the reconstruction of whole manufactories.

Excluding from consideration the cases of culpable temerity on the part of the adults, and assuming that the aid to be given, when accidents do occur, shall afford no bounty on carelessness, the cases which remain for provision are those of adults, which may be considered purely accidental. Taking a case of this class where mischief has occurred in the performance of the joint business of the labourer and his employer; the question is, by which of these parties the pecuniary consequences of such mischief shall be sustained.

We conceive that it may be stated as a principle of jurisprudence applicable to the cases of evils arising from causes which ordinary prudence cannot avert, that responsibility should be concentrated, or as closely as possible apportioned on those who have the best means of preventing the mischief. Unless we are to impose on the workman the obligation of perpetual care and apprehension of danger, the nature of the injuries inflicted are of themselves evidence that all the care which can be taken by individuals attending to their work is taken by them; it is only the proprietor of the machinery who has the most effectual means of guarding against the dangers attendant upon its use.

If such an extent of pecuniary responsibility for the accidents which are incidental to the use of the machines is imposed upon him, those consequences will be more likely to be taken into account, and to be guarded against at the time of the erection of the machinery. The workmen are not prone to regard immediate dangers, still less dangers which are remote and contingent, and many of the accidents are of a nature apparently too uncertain to form data for insurance. It could hardly be expected that a workman in entering a manufactory should object that any portion of the machinery is dangerous, and that it ought to be boxed off. But the proprietor of the machine is necessarily the person who can best foresee all the consequences incidental to its use, and can best guard against them. By throwing upon him a portion of the pecuniary responsibility for those mischiefs, we combine interest with duty, and add to the efficiency of both.

If the pecuniary consequences from unavoidable accidents were considerable, the imposition of the proposed responsibility may be met by the master, or by a deduction from the wages. Considering the defective nature of most existing modes of provision against sickness and casualties by benefit or friendly societies, and also, unhappily, the large proportion of those who, from improvidence, do not take advantage of these or other means (of which some portion of the working class avail themselves in so exemplary and admirable a manner), if we were to devise a form of insurance against the casualties in question, available to all classes, we should recommend that measures should be taken to secure from the master the regular deductions of the amount of the contribution of the persons employed.

We propose that, in the case of all accidents whatsoever from machinery occurring to children under fourteen years of age, the proprietor of the machinery shall pay for the medical attendance on the child, and all the expenses of the cure, until medical attendance is no longer required; and also, during the same period, shall continue to pay wages at the rate of half the wages enjoyed by the individual in question at the time of the occurrence of the accident.

We are of opinion that persons above that age, in all cases where the injury was received from accidents in the ordinary course of business, where there was no culpable temerity, should receive similar treatment at the expense of the employer, and should also be allowed half wages until the period of cure, as we believe that an allowance of full wages would occasion considerable fraud in the protraction of that period, especially in the cases of accidents of a less serious nature.

We think that the remedy should be given on complaint before a magistrate or the inspector.

With regard to fatal injuries occasioned by wilful negligence, we have at present no new remedies to suggest as substitutes to those at present afforded by the common law.

We trust that, in consideration of the extremely short period to which we have been limited in the performance of the task assigned to us, allowance will be made for the unavoidable imperfection of the report which we now respectfully submit.

(L.S.) THOMAS TOOKE.
(L.S.) EDWIN CHADWICK.
(L.S.) THOMAS SOUTHWOOD SMITH.

25th June 1833.

SECOND REPORT

OF THE

Central Board of His Majesty's Commissioners,

APPOINTED TO

Collect Information in the Manufacturing Districts,

AS TO THE

EMPLOYMENT OF CHILDREN IN FACTORIES

AND AS TO THE

PROPRIETY AND MEANS

OF CURTAILING

THE HOURS OF THEIR LABOUR.

THE printing of the reports and evidence of the medical commissioners has been unavoidably delayed for a few days.* Some portion of this evidence, consisting chiefly of statistical details, remains in the hands of the actuaries whom we have consulted. We have deemed it desirable to postpone the publication of all documents of this kind until the general results which they afford have been obtained.

We were unable, within the limits of our first report, to direct attention to a large mass of evidence which had come before us illustrative of the physical condition of the manufacturing population. Among the striking and decisive

* We avail ourselves of the present occasion to supply some passages of our first report, which, in the extreme haste with which it was necessary to carry it through the press, were overlooked. The first of the passages which was thus accidentally omitted is the following statement of Sir David Barry, relating to the treatment of children in the factories of Scotland, which should have been inserted in the first report at page 24, at the end of the second paragraph:—"I must state to the central board, and I trust that it will reach his Majesty's government, that no case of cruelty, gross oppression, or of punishment attended with corporal injury, inflicted by owners of mills upon their workers, has come to my knowledge during my investigation as a factory commissioner, in Scotland; whilst, on the contrary, many traits of almost parental kindness on the part of the master, and of corresponding gratitude on the part of the servant, have been brought before me in the course of my inquiries." Sir David Barry's General Report, 73.

The second passage, likewise accidentally omitted, relates to the evidence that the fatigue, drowsiness, and pain produced in young children by ordinary factory-labour begin to diminish about the eleventh or twelfth year, and wholly or nearly cease at the age of adolescence or puberty.† The following evidence to this effect should have been inserted at page 52, line 15 from the bottom:—Boy twelve years of age, deposes, "Used to be very tired when younger, not now." Girl, same age, deposes, "Never tired now; never falls asleep" Girl, same age—"Never hurt any way; it's no very sair work." Girl, same age—"Has been eighteen months at work in mill, is never tired." Girl, fourteen— Stuart. "Does not find her work hurt her" Girl, same age—"Has been in mill one year; likes it fine." Girl, same age—"Never tired; likes it brawley." Two girls, fifteen —"Never tired; like over-hours." Girl, same age—"Have had my health very well; should not like shorter hours; should like over-hours better." Mackintosh, 26, & 25 to 55. Girl, turned fifteen—"Been here since I was quite a child, health very good; the girls generally have very good health. You see, sir, we have plenty of air. There's one bad thing Drinkwater.

† According to physiologists, both English and French, the age of adolescence or puberty commences earlier in the female than in the male, beginning in the former sex at eleven or twelve, and in the latter at fourteen.—T. S. S.

statements, both of the civil and medical commissioners, we now beg leave to refer to the following:—

SCOTLAND
Stewart's report, Dumfermline, 3.

"The men and boys are generally pale, far more so than the females, several of whom at most of the mills have not lost a ruddy complexion. In the country part of this parish the people have quite as healthy an appearance as the agricultural population. The women are not only robust, but retain their rosy complexions."

Ibid. Kirkaldy, 10.

"The people engaged in the factories in this town appear to me, in general, as healthy looking as persons engaged in other work within doors. No stranger could see the population there, employed at the mill and bleach field, without noticing them as a fine race of people. The robust appearance of two or three females who have been employed at this work for very long periods, forty years, or nearly that time, is striking."

Ibid. Dundee, 21. 22.

"The young people belonging to the factories could not, I am persuaded, have been distinguished in appearance from children employed at any other work within doors; their complexion, however, certainly is not to be compared, in point of freshness and ruddiness, with children employed in the open air, or in agricultural labour."

Ibid. Glasgow, 83.

"Notwithstanding the degree of heat required, many of the females, although they have worked for considerable periods, retain the bloom of health, and all the appearance of being robust. The male workers, which is universally the case, were the most pale and cadaverous looking. My own inspection of factories comprehends ninety-one, and I have examined and taken the depositions of above 500 persons on oath. It certainly does not appear to me, from anything I have seen, or from any proof adduced, though on this subject there is much contradictory medical evidence, that the employment of children above nine years old in those occupations in a factory for which alone they are fit is necessarily attended with risk, or effects unfavourable to health. The heated and impure atmosphere which generally prevails to some extent in every factory, unquestionably, soon deprives most of the workers (for all, it is thoroughly proved, are not affected in the same way) of ruddiness of complexion and of robust appearance. But it has not been shewn that even the young workers,

here (*laughing*), we have no over-hours. I've heard of your coming, and what it was about. I think it is a very good thing. I hope you wont make us work shorter hours though." Girl, seventeen—"Been standing twenty-four hours; has no pain in knees or ankles, else would not do it." Girl, nineteen—"Canna say I'm very tired; we've a good master."

Stuart.

Girl, same age—"Took ill with it at first, but I am well enough now." Twenty, and upwards—"Tired and sleepy at first; but does not now feel the work too severe. "Should not like to be tied down to get less wages." "Should not like to work shorter hours if I get less paid." Aged sixty-three—"Has worked forty-one years in the mill; tired sometimes, but always took well with her work, and has had no accident." Occasionally an individual is found to whom the labour of the factory is irksome, but in the immense majority of cases, especially in the large establishments, the young people, when they approach the age at which the labour is no longer exhausting, have made to the commis-

Stuart.

sioners such statements as the following:—Boy, eleven years old—"Tires me standing a little now and then, but not half so much, no, nor a quarter as much, as I used to be getting in 'tatoes with my back. Used to get leathered a deal when I was getting in 'tatoes. I

Mackintosh.

like this a good deal best." "Before coming to the mill used to play about in the streets; like being in the mill as well as playing in the streets." Girl, twelve—"Well enough pleased with the hours I have." "Likes the mill quite well." "I am very well contented." "We have hantel of freedom; doesn't feel tired at night; we are accustomed to it." "Didn't like it at first, but likes it very well now." "Thinks she should like working

Ibid.

as she does better than in the fields." "Prefers mills to service; sair work service." "Likes being in the mill better than being in service, because she is sorer confined in service than in the mill. At the mill we have Sunday, and we have the time after the mill sets; we hadn't that at service." "These are a hantle of very contented and happy people, to tell the truth." "The children are quite cheery." "They all seem perfectly well and happy."

Stuart.

In like manner all classes of witnesses concur in stating that the older children in general experience no pains in the limbs or other parts of the body, and have no swellings of the legs. "The children's legs never swell." "Never saw any children with swelled legs." "Her legs stand it brawly." "Never find her legs swell, though she has sometimes stood twenty-four hours at a time." "Never saw the legs of the workpeople swelled."

whose hours of labour are at present regulated by Sir John Hobhouse's act of 1831, that is, who work for twelve hours for five days, and for nine on Saturday, are generally unhealthy. Boys, in all cases, become pale in factories sooner than girls; and Sir David Barry has, I know, reported favourably on the subject of the health of the latter,—on this, among other grounds, that the girls brought up in factories, and married from thence, have almost always large families of children. The due ventilation of factories will, I am persuaded, tend much to improve the health of the workers. Still the employment of workers in factories cannot, where proper regulations are attended to, be in most cases, with propriety, termed an unhealthy one; and it would therefore seem that the long continuance of labour is more injurious to them, in preventing them from being sufficiently educated, and of course sufficiently instructed in their moral duties, than in other respects."

Sir David Barry, whose medical evidence on this subject is peculiarly valuable, because it is, in every instance, the result of personal inspection and strict examinations, reports as follows:—

"The health of the operatives in general appears excellent. Some few look rather delicate, but seem to work cheerfully. The appearance of by far the greater number was healthful, robust, fully grown for age. Did not see even one case of distortion or narrow pelvis. Many of the girls were beautifully formed, who had been from ten years to maturity in the mill. Dr. Stephenson, who has been the chief accoucheur to the factory women for the last fifteen years, has not had a single forceps case amongst them, although in that time he has attended upwards of thirty first labours. Abortion is not more frequent than amongst other women. Dr. Stephenson knows of three or four slightly curved spines amongst the girls of the mill, which they are anxious to conceal; but in the whole course of his midwifery practice amongst them has never seen a case of narrow pelvis produced by factory labour. He states that parents send their least promising offspring to the mill, and their most robust to trades or agriculture. With regard to the physical appearance of the young persons, I went round the village whilst they were at dinner, and saw no squalid, emaciated, nor stunted individuals. I noticed five sisters, from thirteen upwards, all employed in the mill from their childhood, every one of whom might be termed a fine-grown girl; some of them remarkable for symmetry and strength." "This day examined carefully and individually one hundred and eleven girls of the classes stated, with a view to find, if possible, a case in which the plantar-arch had been broken down by continued standing, as is stated in the evidence lately printed to occur sometimes in factory workers. Found many beautifully formed feet in those who had worked the longest. In one case, a woman, aged forty-three, who had worked from the age of seven, the foot was remarkably small and high in the instep. In no case did the plantar-arch seem to have been in the slightest degree disturbed. The pelves of the adult girls were remarkably well formed, with strikingly well-developed glutæi. Nothing but the evidence of my own senses could have induced me to believe that girls, indeed any human beings, worked as stated from nine upwards, could yet possess in maturity the apparent extreme of high health and vigour, with finely-proportioned forms. The masters are unanimous in asserting that girls, and they alone are trained to flax-spinning, never become expert artists if they begin to learn after eleven. I observed two girls, for some time in Mr. Malcolm's mill, about thirteen each, in the same pass or space between two frames; one attended to sixty wet spindles, or the spinning of sixty threads of yarn, of five ounces to the hank, the other to fifty spindles. The first had 11*d.*, the other 10*d.* per day. The range which each girl had to move over along her spindles, or the length of the pass, was about twenty-two feet. It is quite impossible to give an adequate notion of the quickness and dexterity with

Sir David Barry's Medical Report, Dunfermline, 1. 2.

Ibid. Dunfermline, 3.

Ibid. 5.

which these girls joined their broken ends of threads; shifted the pirns; screwed and unscrewed the flies, &c. To supply the place of such artists by new hands would be utterly impracticable, and difficult in the extreme to find a relay of hands equally expert, under present circumstances. There is no sameness of attitude—no standing still; every muscle is in action, and that in quick succession."

Dunfermline, 5. "Not having been able to detect any deformities by individual personal
Aberdeen, 32. examination, and suspecting that they might have been kept out of my sight, I have latterly adopted the plan of swearing one or two of the oldest overseers, as to whether they have known any mill-worker to have become deformed during their superintendence, and also whether they know of any person actually in the mill at the time suffering from ill-health, injury, or deformity of any kind, without reference to the cause of these misfortunes. This mode of proceeding has produced two or three spinal curvatures, as many cases of swelled ancles and feet, and one shortened thigh from disease of hip-joint, with some cases of flax-dust dyspnœa; but the deformities, upon being investigated, were all found to have occurred previously to mill-service. I cannot pass over the Montrose mills without bearing testimony to the excellent management of those seen by me. Even Mr. Braid expressed his admiration of some of them, and exclaimed that he might return to Dundee as soon as he liked, for that there did not seem to be the slightest inclination amongst the operatives to assist in his views towards effecting an alteration in the present mode of working."

Glasgow, 39. "The most remarkable persons employed in this splendid establishment (Messrs. H. Monteith and Co.) are twenty adult females, termed stove-girls. They hang up the prepared webs to dry in the stoves, and afterwards take them down. They earn 7*s*. 6*d*. per week. I have been in the stove, and seen them at work around me, whilst the thermometer in my hand marked 140° of Fahrenheit. I was informed by the overlooker that it often stands higher. As the wet cloth is drying the temperature sinks a few degrees. These girls are constantly passing through the open air from one stove to another, but remain only a few minutes in each. Mr. Rodger, the benevolent manager of the works, informs me, that candidates for this department are never wanting: tall girls, however, and rather thin, are preferred. Each is provided with fine flannel chemises by the proprietors, which are constantly worn. Some are very fine-looking girls, and all appear to be in perfect health. They work barefooted, and often have leisure to sit. Mr. R. states that they are as healthy as any girls in the establishment, and that when any of them happens to catch cold they are very soon cured by going into the stove again."

It appears from the whole of the evidence collected in the northern district, that, whatever may be said about the delicacy of the female organization, and the inability of the female operative to endure fatigue, the female, as a child, an adolescent, and an adult, bears factory labour better than the male, and, in regard to her own peculiar constitution and health, sustains no appreciable injury from it.

Sir D. Barry, Glasgow, 53. "At the Blantyre mills the spinners are all males. I visited the dwellings of nine of that class, without making any selection. Found that every one of them was married, and that the wife had been in every instance a mill girl, some of these women having begun factory work so early as at six and a half years of age. The number of children born to all these couples was fifty-one; the number now living, forty-six. As many of these children as are able to work, and can find vacancies, are employed in the mill. They all live in rooms rented from the owners, and are well lodged. I saw them at breakfast-time, and the meal was composed of the following:—viz., porridge and milk for the children; coffee, eggs, bread, oaten cake, and butter for the father. I have the notes taken on the spot before me, but think it needless to transcribe them at full length. All the adult male spinners are pale and thin; they are subject to capricious appetite and dyspepsia; but being prohibited from smoking, they generally chew tobacco largely, and often commence that pernicious practice at an early age. As all the spinners have been piecers, and bred up in the mills from their childhood, it is fair to conclude, that their mode of life is not favourable to the development of the manly form, seeing that few, or none of them, are tall, athletic men. Females are much less deteriorated in their appearance by mill-work than males. Amongst some thousand young women whom I have now carefully observed both in and out of their factories, and after having examined upon oath those who had known them longest as to the existence of deformities amongst them, I have not met with one distorted or narrow pelvis. If there be any difference between factory and other adult girls relative to that portion of the female form, I would say, that in the former, in this country, it is more fully developed. Of all the married women who had been mill-girls from their childhood, whom I visited at their own dwellings, and inquired about from their husbands, there are but two unfruitful. The husbands of all were spinners. The children were numerous for the time the couples had been married, and as healthy looking as those of any class of the community. Spinners almost always marry young, and select girls from seventeen to twenty-two, who immediately quit the mill upon being married; sometimes a little before that event. It is a general rule in factories, that married women living with their husbands are neither employed nor retained. Both adult males and females, whose work obliges them to stand constantly, are more subject to varicose veins of the lower extremities, and to a larger and more dangerous extent than ever I have witnessed even in foot soldiers. The females are more subject than the males to evening swellings of the feet and ankles, but I know of nothing which unfits them from becoming prolific and healthy mothers, if married at a suitable age. There is one thing I feel convinced of from observation, that young persons, especially females, who have begun mill-work at from ten till twelve, independently of their becoming much more expert artists, preserve their health better, and possess sounder feet and legs at twenty-five, than those who have commenced from thirteen to sixteen, and upwards."

General Repo Greenock, 71.

The account of the physical condition of the manufacturing population in the large towns in the north-eastern district of England is less favourable. It is of this district that the commissioners state, "We have found undoubted instances of children five years old sent to work thirteen hours a day; and frequently of children nine, ten, and eleven, consigned to labour for fourteen and fifteen hours." The effects ascertained by the commissioners in many cases are "deformity," and in still more "stunted growth, relaxed muscles, and slender conformation;" "twisting of the ends of the long bones, relaxation of the ligaments of the knees, ankles, and the like." "The representation that these effects are so common and universal as to enable some persons invariab., to distinguish factory children from other children is, I have no hesitation in saying, an exaggerated and unfaithful picture of their general condition; at the same time it must be said, that the individual instances in which some one or other of those effects of severe labour are discernible, are rather frequent than rare. This was more particularly observable of the children in the "radford mills than elsewhere." "It may have happened from a certain degree of obstruction which was opposed to the prosecution of our inquiries at Leeds, or it may have happened from other causes, that no instance came under my observation of deformity traceable to a daily period of twelve hours work endured after the age of twelve; but I may say, that no instance even approaching to such a state of facts has come before me,

North-Eastern District

Power, 74.

Ibid.

Dr. Loudon, 25

Power, 74.

Ibid. 72.

though I have used every diligence in investigating that point. That such cases, however, may and do occur, is not impossible; but I doubt their frequent occurrence, unless accompanied by the circumstance of very weak constitution or insufficient sustenance." "Upon the whole, there remains no doubt upon my mind, that under the system pursued in many of the factories, the children of the labouring classes stand in need of, and ought to have, legislative protection against the conspiracy, insensibly formed between their masters and parents, to tax them to a degree of toil beyond their strength."

Power, 74.

Dr Loudon, 23.

"In conclusion, I think it has been clearly proved that children have been worked a most unreasonable and cruel length of time daily, and that even adults have been expected to do a certain quantity of labour which scarcely any human being is able to endure. I am of opinion that no child under fourteen years of age should work in a factory of any description for more than eight hours a day." From fourteen upwards I would recommend that no individual should, under any circumstances, work more than twelve hours a day; although if practicable, as a physician, I would prefer the limitation of ten hours for all persons who earn their bread by their industry."

With regard to the foregoing opinions of Dr. Loudon it should be observed, that they are stated without any exclusive reference to labour in factories, and without any limitation of the age at which restriction should terminate. Dr. Loudon, speaking merely "as a physician," would, "if practicable," recommend, "from fourteen years upwards," "the limitation of ten hours for all persons who earn their bread by their industry." This conclusion, if regarded in a purely medical point of view, in which alone it is stated, may be subscribed to without weakening the force of the moral and social considerations which discountenance all legislative interference with the disposal of labour extending beyond the age of childhood, properly so called. Indeed, the unlimited range of the foregoing recommendation affords a more conclusive proof than any which have yet been adduced of the necessity, adverted to in our first report, for seeking some other principle of legislative interference than one which, as we have there stated, does not appear more applicable to labour employed in factories than to the mass of other occupations which are followed as a means of subsistence.

First Report, p. 51.

First Report, Power. p. 76.

We beg to refer, as directly bearing on this part of the subject, to the following passage, extracted from the report on Yorkshire by Mr. Power:—

"It may be desirable to say a few words upon a proposition very widely entertained, and very unhesitatingly advanced, that ten hours a day is a quite sufficient period of labour for any person of whatever age. That may be very true of all classes of labourers; professional men, tradesmen, artisans, agricultural labourers, domestic manufacturers, and lastly, of operatives *whose labour is assisted by steam or water power;* and the same observation may apply to a term of daily labour much less than ten hours; but under what pretext, or with what face of pretension, the latter class can lay claim to a peculiar privilege of exemption from more than ten hours' daily labour is hard indeed to conjecture. They compare their condition with that of the small class, comparatively speaking, of labouring artisans, such as carpenters, stonemasons, bricklayers, &c., who, they say, work only from six to six; a class, however, in this respect distinguished from the operatives, that their work is done entirely by hand labour, and after service of apprenticeship, accompanied with some outlay; but what do they think of the numerous classes of domestic operatives, the frame-work knitters, the hand-loom weavers, the wool-combers, the lace-manufacturers, and a variety of others, who work, and work hard, from twelve to fifteen hours a day to earn a bare subsistence; and this frequently from a very early age, and in a state of confinement which may be truly called injurious to the health? The consideration also is probably overlooked by these interested reasoners, that by reducing the hours of labour, and thereby the amount of subsistence derivable from the less objectionable occupations, a corresponding increase may ultimately be produced of competition for employment in the more objectionable ones, and thus the whole labouring community become injured by the wanton abandonment of its fair advantages. Yet is this conviction that 'a ten hours' bill will make a ten hours' mill' at the bottom of much excitement that prevails among the operatives for this subject.

"It appears to me, then, that a greater degree of protection may be afforded to those properly called children, without involving the stoppage of the moving power, by reducing the limit of hours from ten to eight, and by lowering the period of life to be protected by that limit to a point much below the age of eighteen."

Dr. Hawkins' Report.

The statements relative to the condition of the factory population in the Lancashire district are, to a considerable extent, conflicting. Dr. Hawkins says:—

"I believe that most travellers are struck by the lowness of stature, the leanness, and the paleness which present themselves so commonly to the eye at Manchester, and above all among the factory classes. I have never been in any town in Great Britain, nor in Europe, in which degeneracy of form and colour from the national standard has been so obvious.

"It is impossible not to notice the total absence of public gardens, parks, and walks, at Manchester; it is scarcely in the power of the factory workmen to taste the breath of nature, or to look upon its verdure, and this defect is a strong impediment to convalescence from disease, which is usually tedious and difficult at Manchester. The medical police is also capable of much improvement.

"If I am to confide in my own observation, and in the accounts furnished to me by workpeople of every age in private conversations frequently repeated, I must arrive at the conclusion, that the proprietors are generally anxious to promote the convenience and comfort of their dependents as far as the system admits, and they usually endeavour to prevent acts of harshness and of immorality; that if such cases arise, it is mainly owing to their absence, or to their neglect of personal superintendence; and that there are not a few among them who really act a paternal part, and receive the recompence of respect and gratitude. Their situation is a difficult one; but the more closely they assume the character of the observant master of a great family, and the more narrowly they investigate, appreciate, and purify the composition of their family, the more likely is every factory to become respectable and happy."*

We think it right to introduce in this place some portions of the evidence on

* "In some factories I could not trace the slightest desire for change; neither the name nor the necessity of the ten hours' bills appeared to have penetrated so far. In particular, I must cite the mills of the Messrs. Strutts, at Belper, as uttering one voice of satisfaction, and as presenting no dark shades which I had the power of discerning. These gentlemen allow to all who choose to avail themselves of the offer, a pint of good tea or coffee, with sugar and milk, for one halfpenny; and those who accede to this plan obtain medical assistance gratis. They have provided a dancing-room in their mill, where festoons of flowers are suspended, and a band of music is heard on holidays, as a substitute for the public-house to their female youth." "The deformities which I witnessed below the age of eighteen are not numerous; they occurred to about one in a hundred of all whom I examined. Deformities, indeed, appear to be far less frequent at present than formerly. The greater part of the instances in which I have seen them were of the adult or middle age; there is at present more space in many of the mills, the machinery itself is improved, and it is also better fenced or protected than at former periods. On the other hand, the young people whom I examined complained very frequently of sore feet, and of weakness in the knees, which they imputed to long standing. The long standing is, indeed, the most common theme of complaint: the mere labour is not otherwise great."

this part of the subject, whether of a favourable or unfavourable character, extracted from the minutes of the civil commissioners, as well as of the medical commissioner at Manchester. Dr. Hawkins' examinations appear to have been taken in a great measure conjointly with his colleagues. The main part of his separate investigations, with the exception of the short minutes annexed to his report, have been recently laid before us in the form of statistical tables, which are now in the hands of the actuaries consulted by us, and will form a portion of our supplementary report.

See First Report.

Samuel Holt, Joseph Gaskell, and John Rowbotham, overlookers at Birley's mill, examined:—

Cowell.

"Mr. Rowbotham has three children employed in Mr. Birley's mills, and one in Messrs. Hatfield and Frost's at Wallington. Those in Mr. Birley's are employed from half past five till seven, and have been regularly employed, without intervals, excepting at Whitsuntide and Good Friday, and other casual times when the mill may stop. The two eldest from the age of ten, and the youngest, who is a girl, from the age of thirteen; the eldest was employed in the card-room till he was fifteen years old, and has since been employed in the warehouse.

"Q. Have you observed any bad effects on the health of your children produced by their employment in factories, and particularly in the card-room?

"(*Mr. Rowbotham.*)—None whatever; I think they are as healthy as they would have been in any employment that I know of.

"Q. Is it the case that children who, from scrofula or weakness, or ill-health of some kind or other, are unfitted for many of the usual occupations at which children are placed, are fit for some descriptions of factory work?

"(*All.*)—This is the case.

"(*Mr. Gaskell.*)—Children who are naturally deformed, others who are weak in their limbs, others who have weakly health, are in Mr. Birley's employ, and I do not see what they could do, or what they are fit for, except some work in factories.

"(*Mr. Rowbotham.*)—I have seen delicate healths that have improved after they came into mill. Those that require warmth and dryness have improved.

"(*Mr. Gaskell.*)—I have seen such cases in my own observation; there is generally something that a child can do in a cotton-mill.

"(*Mr. Rowbotham.*)—If they can't do one thing, they can do another.

"(*Mr. Holt.*)—I concur in these opinions, but I recollect no particular instances. We have a few at our works who are unfit for anything else but factory work,—grown-up people who have lost their limbs.

"Q. Have you any idea how many children in every one hundred engaged in your factory would be unfit for other work, from original weakness, deformity, or ill-health?

"After some consideration and conversation the witnesses agreed, that if they were to examine the mill over they should be enabled to find instances of this kind, but how many they could not say, as the point never had occurred to any of them before. There are one thousand five hundred hands in the mill, and Mr Gaskell thought he could enumerate twenty adults , and children at once, Mr. Rowbotham between twenty and thirty, and Mr. Holt about the same number as Mr. Gaskell.

"Q. Do you think that the factory life has a prejudicial effect upon the health, bearing in mind that the question supposes that the hours of work have hitherto been nearly fourteen hours a day, and that the temperature, and ventilation, and cleanliness, have not been so well attended to in mills as they might have been?

"(*Mr. Gaskell.*)—Comparing a factory life, under the circumstances that you have mentioned, with other occupations generally, I do not think it less healthy than the generality. That is my deliberate opinion.

"(*Mr. Rowbotham.*)—My opinion is the same. While you have been putting the question, and Mr. Gaskell has been giving his answer, I have been running over in my mind many branches of occupation in Manchester, and I cannot think of one which occurs to me as more healthy than a factory life.

"(*Mr. Holt.*)—I have never been in any factory but Messrs. Birley's, and am now forty-one. I cannot speak of factories in general; but I consider my own health very good, and I do not know of anything to make me think that a factory life is more unhealthy than any other."

JOHN BETHEL, GEORGE LEE, THOMAS EWART, spinners, JAMES BELSHAW, managers in Mr. POTTER's mill, Manchester, examined. **Tufnell.**

[*Two of the witnesses delivered in the following papers*]:—

"*John Bethel*, forty two years of age, is a mule-spinner. Commenced as piecer at ten years of age. Has been always amongst mules; in all, thirty-two years. Has a wife and six children. Four of the children employed in cotton-mills. Neither himself nor any of his family have ever had a day's sickness, except from the small pox or measles. Considers the accidents which occur in cotton-mills to result, either from carelessness, or else from the machinery not being properly boxed off. Has generally worked twelve hours per day, but has worked fifteen hours per day for more than a year and a half. Would object to a bill limiting the working hours to ten per day if he thought it would occasion a corresponding reduction in wages. His wife worked in cotton-mills from an early age, until the family became so large as to require her presence at home.

"*James Belshaw*, thirty-seven years of age, manager of a cotton-mill, commenced as a mechanic. Has been acquainted with cotton-mills and the regulations in them twenty-five years. Has a wife and five children, but none of them employed in cotton-mills. Has enjoyed good health since he has been acquainted with the cotton business. Believes, from information he has received, that some mills in the country work as long as fourteen or fifteen hours per day. Considers these hours too long even for adults. Considers that twelve working hours per day are long enough, and would prefer an effective bill for those hours to limiting the time further. When children are ill-treated in cotton-mills, it is almost universally by the spinners who employ them, and is contrary to the regulations of any mill that he has been acquainted with, and remembers several operative spinners discharged for it. The scavengers and creel-fillers are not employed continuously, but have numerous opportunities of resting themselves during the day. Many of the younger hands attend Sunday-schools, and those of sixteen and eighteen years are frequently teachers in the same. Considers, from many years' observations, that the chastity and morality of individuals employed in cotton-mills is quite as good as those of others employed in other branches of trade. Believes the greatest number of accidents have arisen through carelessness and inattention. Knows one instance of a man having his arm pulled off with attempting to unlap a strap from a round shaft while it was running, and he had been cautioned to let it alone. In the weaving department we employ none under fourteen years of age, and those of that age are employed as helpers. Their work is to shuttle cops. Has heard some of the operatives say that if the hours of labour are shortened, it will reduce the production in the market, and consequently will cause an advance upon the price of the commodity, so as to enable the employers to give them the same wages. Believes this is the general impression amongst the working classes, and is of opinion that the disappointment they will meet with will produce discontent, and make them more dissatisfied. [*The papers having been read over to them.*] Do these papers contain true statements ?—Yes."

RICHARD JOHNSON and GEORGE WOLSTENHOLME examined:

[*The two witnesses, previous to examination, delivered in the following papers. Each gave his assent to the answers of the other.*]

"*Richard Johnson*, M.R.C.S. He has been ten years in private practice in Bolton, seven of which he has been one of the honorary and consulting surgeons to the Bolton Dispensary. He served an apprenticeship of six years, in one of the most extensive coal-mining districts in this country, to two gentlemen of great practice. The children in Bolton working in cotton-mills suffer more from want of personal cleanliness, and from sudden change in the temperature of the atmosphere, without proper attention to clothing, than from the length of time at work. In some cotton-mills want of ventilation may be considered injurious. The children in Bolton are as moral and in as good physical condition as anywhere else where men, women, and children work under the same roof in any other employment, and are better grown than domestic drudges, miners' children, weavers, winders, dress-makers, rag-dressers, tailors, and girls that are educated at second-rate boarding-schools. Richard Johnson does not consider cases of midwifery more difficult in mill-workers than others. Scrofula is not more frequent in spinners than others. Severe accidents rare considering the number employed.

"*George Wolstenholme*, surgeon, Bolton. Was four years house-pupil in the Liverpool Infirmary, five years house-surgeon and apothecary to Bolton Dispensary, and subsequently six years honorary surgeon to the same institution. Has been nine years in private practice. The class of persons applying for relief at Liverpool quite as sickly in their general appearance as those applying at Bolton dispensary. Has had considerable practice amongst the working-classes in Bolton and the neighbourhood. Does not consider the persons employed in cotton-mills generally an unhealthy class. Attends many cotton-spinners' families, but has seldom seen any of them sick, except the wife and young children, who do not work in mills. The health of factory people is much better than their pallid appearance would indicate to any person not intimately acquainted with them. Their health fully equal to tailors, shoe-makers, letter-press printers, plumbers, watchmakers, file-cutters, dress-makers, straw-bonnet-makers, and some domestic servants. Parturition amongst women employed in factories not more difficult than in any other class. In about one thousand cases in my own practice I have only had one fatal midwifery case, and that a person in the country who had never worked in any factory. Scrofula not very frequent; it generally appears in children of the poorest parents, and is more probably developed by deficient food and clothing than by employment in mills. Imperfect ventilation more injurious than high temperature, and probably a very principal cause of the sickly appearance of factory people. In Bolton, some mills attended by a surgeon for one halfpenny per head per week. Severe accidents from machinery not frequent."

A. B. examined:

Tufnell.

"Are there many children in your factory deformed from overwork?—Not deformed in our mill. We have one in our mill that was deformed at another mill. Our system of spinning saves the children from deformity, by putting up the wheels ourselves without the children assisting us in the pushing up, which is practised in some mills, and tends much to the deforming of children."

Joseph Shepley, machine-maker, formerly piecer, examined : Tufnell

" Were any deformed ?—Not one in the mill as I know of, but our sort of machinery was not of a nature to make us deformed. We was always in a perpendicular position.

" Were the scavengers in that position ?—No ; they creeped under the carriages ; but there was nothing to hurt them.

" Why was your machinery less liable to produce deformity than other machinery ?—I can tell you what kind of machinery makes cripples, that is, the old kind of water-frame, which was built very low in the spindle-boxes to accommodate young children.

" And are they higher now ?—Yes, the machine is improved.

" Then your factory had not this kind of low water-frames ?—No.

" Have they been introduced since then ?—No. There were many in different mills then, but not in our place.

" Are these low spinning-frames used now ?—Many of them have destroyed them, and put in new machinery, what they call the throstle machinery ; some have done away with them entirely, and put in mules.

" Then do you mean to say the present machinery is less liable to cause deformity than forty years ago ?—Yes.

" Is that the general opinion ?—Yes, that is the general opinion.

" Then why do we now hear of so much talk about deformities caused by working at the machinery ?—They cannot prove it now. You may rip up a deal of these old hands what were deformed at these old machines, but I look upon it that the cotton trade now makes fewer cripples than any other trade now going.

" But people are all saying the children are deformed from overwork ?—Well, it is as I tell you. Those as is took care of are as healthy as at any other trade. I know that the cotton trade has produced as few cripples, according to the number of hands employed, as any other trade in England.

" Then you don't think that the present machinery has any tendency whatever to deform those who work at it ?—Well, I cannot say that it has ; it all depends on good clothes, and good victuals and beds, and then I cannot see that they take any harm."

Mary Y., aged seventeen, examined, said that she had been at factory work all day as a piecer, and was not tired : Cowell.

[We came suddenly into the house, where all the family but one was sitting round the table ; the father was reading in the British Biography the life of Bishop Bonner ; two of the others were reading small books, which had been given for good conduct at Bennett-street school ; a third girl was netting a cap so assiduously that she hardly looked up to answer my questions.]

William Sidebotham, spinner :

" Did they never complain that they could not learn their tasks in consequence of being over-fatigued ?—No ; they sit hours at night getting their Sunday-school tasks off."

Joseph Shepley, machine-maker, formerly piecer, examined : Tufnell.

" Did you suffer much from fatigue ?—No ; we had very great opportunities ; we could go out out every hour and a half to the river side for five or ten minutes.

" Then you had not quite twelve hours labour a day ?—These opportunities took up about thirty-five minutes ; you could have gone out four times a day."

Tufnell. JOHN YOUNG, warehouseman to Joseph Plant and Co., spinners of fine yarn, examined :

"Do the factory children ever complain that they cannot attend to their lessons in consequence of being over-fatigued?—No, not at all. I speak directly to that."

Ibid. MARTHA CHARLESWORTH, roller-coverer, examined:—

"Do the children ever work in the evening after leaving the mill?—Yes; every decent woman sets her child a sewing a chemise or a pinafore, or washing, or something, and some on them sends their children to a night-school.

"Do you ever see the factory children playing in the streets after their work is done?—Yes, every day of my life, at shuttlecock or some game."

Some of the following extracts present the effects of factory labour under a less favourable point of view.

Ibid. ABRAHAM DOCKER, baker, Stockport, formerly slubber, examined:—

"Do you think that one reason for the wish to get this bill passed is, that the parents do not think so much of their children as themselves, knowing that the ten hours bill will prevent grown-up persons working more than ten hours daily, and imagining that a reduction of time will not have much effect on their wages?—No; that is not my opinion. My opinion is, that a child who has to rise before five o'clock in the morning, who has the distance of a mile or three quarters of a mile to go before half-past five, as in many cases the piecers have, to oil the machinery before it starts, and when the child is kept there till half past seven o'clock or eight at night, and has to return the same distance home, the child is so fatigued that their parents in general wishes for a ten hours bill, that they may have their children home a little sooner at night. I made it my business to see last night an overlooker, whom I can rely on his word. He informed me that he beat two children yesterday on account of their not being able to follow their employment. I asked him what number of girls he had; he said he had twelve girls, and three were sick at that time. I asked him their ages particular, and he said he had eight of them under eighteen years of age. I asked him whether they had any time for breakfast or afternoon refreshment; he told me it was not likely, when he had received eight certificates that they were eighteen years and upwards. He informed me he was paid by piece-work, and not by the day. He told me it was his duty to turn off as much as he could, as his wages were bettered the more he turned off. I asked him if he wished to give evidence on such a case as that. He said, 'If I was to give evidence, I must suffer like many in the town of Stockport.' He asked me to think on John Frost, who gave evidence before the lords' committee in 1819, and who was kept out of employment upwards of three years, and sent to his own parish."

Ibid. JOSIAH HUNT, overlooker at power-loom weaving at Mr. J. B. Clark's mill:—

"When they are sent to night-schools, are they drowsy?—Yes, even before they leave the mill. Many a one I have had to rouse in the last hour, when the work is very slack, from sleep."

JOHN PILKINGTON, tailor, aged twenty-nine, formerly piecer, examined:—

"You worked in a factory as a child?—Yes, at six years old.

"Where?—Bellhouse's cotton factory, Oxford Road, was the first factory. I worked there twelve months. Then I went to Appleton's cotton factory, Shepley Street, Bank Top. I worked there four years as a scavenger and piecer. From there I went to Lewis's cotton factory, Bank Top, as a scavenger, and got my arm catched there, and very near taken off, and that was through fatigue of working so long. I stayed there twelve months.

"State what effect the factory labour had on you?—It had the effect to cause my limbs to grow crooked; the heat caused me to grow weak; and the crawling under the wheels caused my limbs to grow in that way, and with the weakness it caused me not to be able to walk a quarter of the way I ought at that age. [Dr. Hawkins was called in. Witness stripped his legs. They were bent inwards, so that they formed an angle of about sixty with the ground; the knees nearly touched, while the feet were about a foot and a half apart. He had this deformity much worse when he was fourteen, at which age he left the mill, and then could scarce walk one hundred yards. A scar was on his arm, the mark of a wound caused by the end pinion.]

"Were you perfectly straight before you entered the mill?—Yes.

"Are you quite certain of that?—Yes, I believe I was. I always heard my mother say so.

"Has your health improved since you have left the factory?—Yes, very much, both in strength and appetite; even my limbs have grown a deal straighter.

"Does the labour which children have to undergo in factories now deform them as much as it did you?—No, I don't think it does.

"Can you say how much less liable they are to deformity?—Because of their age, they are a deal stronger; instead of going to a factory at five or six, they are now eight or nine."

JANE A. B., aged twenty-six, reeler, examined:— Tufnell.

"In answer to various questions:—'I have never felt over-worked, or been unable to sleep at night, or lost my appetite. This is because the work I am at I can please myself with, and work according to my strength; but I am not quite well.'"

TITUS ROWBOTHAM, machine-maker:— Ibid.

"Now I mean to say, that if wages should even be depressed as much as one sixth by a diminution in the hours of work, then the time that the operatives will gain will bring advantages that will do more than compensate such a diminution of wages. Now I will take the case of a young female who gains 9*s.* a week: the hours which she works at present give her no chance at all of doing her own washing, getting up her own linen, mending her own clothes, cooking, or even purchasing her own victuals at the best hand; she is exhausted in body and depressed in mind by the length of the hours and height of temperature, and she has neither time nor inclination to attend to any of the above-named things, or to gain any knowledge of domestic economy. Now I mean to say that she would be a gainer, even in a pecuniary light, if she had time to do these things, even though her wages were in consequence lowered to 7*s.* a week from 9*s.* And since you have said that they will endeavour, after the passing of the ten hours bill, to make up their earnings by unusual activity to what they are at present; will they not urge the children to greater exertions than they undergo at present?—I think, certainly, that the children will have to perform a greater quantity of labour in any given time than they perform at present in the same time, but I do not think that this excess of labour will be considerable or very material; and, in my opinion, it will subtract very little from the benefit which will be conferred upon them by shortening the time of their daily labour, because their employment is all walking or standing, and I conceive that the fatigue of their work consists not so much in the motion as their being compelled to be on their feet for so great a length of time."

Or. Hawkins. BENJAMIN SCHOFIELD, workman:—

"How many hours did you work at Car Hill Mill?—Twelve and a half about.

"How many years did you work at Car Hill Mill?—About two.

"How was your health during those two years?—Not so good.

"In what way did you suffer?—I was fatigued in the evening; often could not eat my supper; and when my parents were trying to learn me to read I used to feel sleepy, and could not mind my book.

"What class of workpeople suffer most in their health; the children, the young men and women, or the elderly folks?—All suffer in their turn; the children get most fatigued; that generally comes on about six o'clock."

Tufnell. THOMAS RALPHS, JAMES HALL, DANIEL CONNELLY, spinners, examined:

"Do many of the children get deformed from their work?—There are instances, but it is not a general thing.

"What particular sort of work is it that deforms them?—In some places pieceing deforms them, where they have to push up the wheel.

"Don't the spinners in general do that alone?—Yes.

"Then the children are sometimes deformed by the spinners making the piecers do what they ought to do themselves?—No; the masters are the cause of it.

"Do you mean to say that the master insists on the piecers doing his work?—It is either the master or the overlooker. The piecers do this mostly in Mr. Birley's mills.

"Did you ever spin with a box-organ wheel?—Yes; they are easier to spin at; we can use more power with less labour.

"Does the carriage require to be pushed just in the same way with a box-organ wheel as without?—Yes.

"Then the labour of a piecer is just the same?—Yes.

"Have you ever seen the piecers push up the carriages in your mill?—No; our manager would stop them very soon if he saw them push up.

"How did you find out that the piecers push up the carriages in Mr. Birley's mill?—(*Hall.*) I have seen them through the windows, and spinners that have worked there have told me that."

MARY WOODHOUSE, midwife:—

Dr. Hawkins. "You have frequent opportunities of examining the persons of factory women; do you often find any sort of deformity, distortion, or crookedness?—Sometimes only. I cannot speak positively.

"Is there any part of the factory work that has appeared to you injurious to women?—There is one part which I have no doubt fatigues and hurts them when they are advanced in pregnancy. I may be wrong; but if it could be avoided I think it would be better. This work is what is called stretching; it is done in the cotton-mills. They stretch backwards and forwards, and are forced to make a great stride. The machine comes forward against them, and then they push it back again. But I cannot describe these things well. I know very little of this from my own observation, but I have heard women speak of it."

Tufnell. REV. JAMES BROOK, Unitarian minister, Hyde, examined:—

"Do you perceive that factory work has a tendency to deform children?—Not unless children of a feeble constitution are put to laborious employments too soon, before they have strength equal to the work. I do not think that factory employments have a greater tendency to do so than other trades where the employment is equally constant.

"Do you then think that the factory employments have no tendency to injure either the health or personal appearance of those engaged in them?—Not when the factory is constructed with proper attention to health and proper conveniences, and the hours of labour are not too long. There is, however, a great difference in cotton-mills in both these respects; some are more favourable to health than others; and the hours of labour very much dedending entirely on the pleasure of the mill-owner. In some mills the hours are so long as to injure, I should think, the health both of children and adults."

RALPH STANLEY, overseer of the poor of Gorton, and governor of the workhouse, examined:— Tufnell.

"Did you ever remark any factory children deformed?—Not there; I think their employment is as healthy as other trades. I was brought up in a calico-printing factory, and I consider their work to be more healthy than that, as in printing they are engaged with minerals and acids, which hurt the health. I think it is not possible for a printer to work the same hours as a spinner, as it is more laborious, and the rooms for some styles of work are hotter than spinners' rooms.

"You think you have seen some children in Manchester deformed by factory work?—Yes; I have seen one or two in Manchester."

JOSIAH HUNT, overlooker at power-loom weaving at MR. J. B. CLARK'S mill, examined:— Ibid.

"Are there many deformed from overwork?—A great many in Manchester; in the back, lock-knee'd, bow-legged, in all shapes."

GEORGE and ELIZABETH SHAWCROSS, superintendents of the children apprenticed to Messrs. GREG'S mills, Quarry Bank, Wilmslow, examined:— Ibid.

"Are any of the children deformed?—No; I don't know that we have ever had one.

"Have you ever heard that the throstle-spinning injures the children?—No.

"Does it injure their legs?—I don't believe it does; but I have heard those that have hurt themselves say that it is their own fault. If they use first one leg and then the other to stop the throstle, there is no harm at all. They have no need to stop always with one leg. It is just as easy to stop with one leg as the other.

"Is there any other work in the mill that has any tendency to injure those employed at it?—No, I never heard of any."

THOMAS LEAMING, superintendent of card-rooms at Salford cotton-mills, examined:— Ibid.

"Did you ever remark that throstle-spinners are not so straight as other people?—Generally they are not; in some mills the throstles are very low.

"Does not stopping the throstles injure the legs?—I am not aware of that."

ROBERT ROBERTS, overlooker at Salford cotton-mills, examined:— Ibid.

"Do you think that factory labour tends to deform children?—No, I believe not at all.

"Do you think that the children employed in your factory are as healthy as at other employments out of factories in the town?—Yes, I believe they are.

"Does not the throstle-work produce deformity?—About thirty years ago I remember it did, when they worked sixteen hours a day; but at present the hours are so short there is no fear of deformity.

"Are you certain that you remember more instances of deformity thirty years ago than now?—Yes; in Manchester there was much more then, when it was the custom to use apprentices. There are some scores of men now who were deformed at that time. I cannot remember more than one or two instances of persons deformed by working in our mill for this last twenty years."

WESTERN DISTRICT.

The medical commissioner for the Western District, who will be found, from a reference to his reports, to have kept the physical condition of the manufacturing population in that part of the country continually in his view, reports as follows:—

Woolriche, Report from Stroud.

"The impression made upon me by a careful inspection of the manufacturing class in this district is generally favourable as to their condition. The inhabitants of these beautiful valleys are nearly all manufacturers of fine cloth, or spinners; these, with their assistants, who are usually children or young persons, make up nearly the whole population, except the necessary artisans, such as stone-masons, bricklayers, carpenters, and blacksmiths, with the less laborious trades of tailors, shoemakers, and the usual shopkeepers and other people who are found in country towns. It is admitted that the adult population is rather under-sized, but they do not appear to me deficient in strength; and, although often pallid, they almost universally state their health to be good, with the occasional exception of some stomach complaint. As regards size, too, there are many examples of well-grown men and women who have lived nearly all their lives, that is, from seven or eight years of age, in factories. I have learned, too, that this has always been a favourite recruiting position, particularly with the marines, who are not likely to have persevered in collecting sickly and unserviceable subjects for that corps. The children in most of the factories I have visited exhibit an appearance which has surprised as much as it has pleased me. Considering the period they labour and their usual food, I was not prepared to see groups of rather unusually healthy-looking children turned out of these much-abused factories. I know not what may be the case in the cotton factories or in the cloth district of Yorkshire, but here in Gloucestershire, the great manufacturers, without any exception that I have discovered, appear most anxious to promote the interest and comfort of their young operatives; and although the lesser factories are less distinguished in this particular, I know of no case of aggravated neglect.

Woolriche and Horner.

"With the exception of the town of Chard in Somersetshire, we have had everywhere reason to be satisfied with the general appearance of the children; they were substantially clothed, and had every sign of being well fed and of enjoying good health. We have seen neither distortions nor deformities, except in a few instances; and these, when inquired into, had taken place before the persons had worked at a factory. To our question, whether at the end of the day the children appear weary and exhausted, the almost uniform answer has been, that in summer they are seen, full of spirit, playing in the fields. Some of the children had undoubtedly a less healthy appearance than the generality; but that would be the case in every assemblage of people, young or old, in any situation. The exception as to Chard, where the children certainly appeared less healthy, and were not so well clothed as in other places, was perhaps made more striking by our having just before visited the factory of Mr. Heathcote at Tiverton, where everything is in high order, where the people look clean, well fed, healthy, and respectable, and are working in large lofty rooms, with plenty of free space for each individual. Upon taking all we have seen and heard into account, and with every anxiety to guard against over-statement, we feel ourselves bound to say, that throughout the whole of the district we have visited we have found nothing in the state of the health, morals, or general condition of the children employed in factories which calls for the proposed legislative interference in their behalf. It has been very agreeable to us to find a kind friendly feeling so generally subsisting between the master and his workpeople, and we are disposed to consider that circumstance as a strong proof that the system upon which the manufacturers conduct their works cannot be at variance with the general welfare of the people whom they employ. We have not been able to discover that parents have made complaints, either to their masters or to others, that their children were over-worked in the factories to the injury of their health, or had asked for a diminution of their hours of labour. The manufacturers, with few exceptions, have said that they would infinitely prefer not taking children before they are nine years of age, and that they are only induced to do so in any case by the urgent solicitation of parents. All the operatives we have examined on this point agree that such a limitation would be desirable, provided the children were kept at school from the time they are capable of running about until they are nine years old, and provided the parents were in circumstances to maintain them well.

Woolriche, General Report, 109.

"There is only one instance (at Bruton) of any factory or manufacturing establishment in the west where there is a medical man appointed to attend; it has been necessary, therefore, to look to other sources for information; these have been the physicians and surgeons of the neighbourhood, the clergymen, magistrates, the adult workmen themselves, as well as the masters or proprietors, who in many instances have given proofs of their anxious attention to the comfort and well-being of those they employ. From these sources the information has been collected, which, after much consideration, has led me to conclude, that the general condition of the manufacturing classes in the districts I have visited is greatly superior to that of the surrounding agriculturists of the lower class. This opinion is confirmed by many depositions; and all the evidence collected goes to prove, that there is nothing in the clothing, silk, or lace manufactures of Gloucestershire, Somersetshire, Devonshire, or Wiltshire, *as usually practised in those counties*, calculated to produce, or that actually does produce, any bad effect on the health of the children or young persons employed in them. In many of the factories the children have an appearance strikingly healthy; some are unusually ruddy, active, and lively; and I state deliberately, that *I saw no case* of disease or distortion which could in *fairness* be solely attributable to the *employment* of the individual. That persons working in factories are not exempt from their full proportion of disease must be admitted; but after the inspection of many thousand operatives, old and young, I feel convinced, that, taking into account the advantages and disadvantages of the manufacturing classes, as regards the confinement on the one hand, and their abundant supply of nourishing food on the other, the balance will be found considerably in their favour, as compared with the badly paid agricultural labourers of the present day. In such an inquiry it is necessary to consider, whether the diseases resulting from a scanty and insufficient subsistence are not pregnant with as much suffering and as much danger as those produced from labour in factories, if indeed any are actually so produced, where the labour is generally so moderate and well regulated. The smallness of the number of cases of distortion of *any kind* which I saw in the clothing districts was surprising, not more than eleven cases in the whole circuit, and several of these very slight. In answer to inquiries made of themselves as to the origin of their complaints, it was stated that none of them had originated in the factories; several of the individuals were of advanced age, and admitted that they had suffered much privation and hardship during infancy, owing to the poverty of their parents, and to which they attributed their complaints. In several places the medical men stated, 'that occasionally autumnal fevers prevailed; the average mortality *very* low.' At Stroud their fever is designated 'mild typhus;' and the average mortality not exceeding one and a half per cent. At Frome one of the surgeons calls the fever, which he states occasionally prevails, 'fever of a mild typhoid character, which rarely terminates fatally;' and another calls it, "simple continued fever, sometimes running into mild typhus;' and adds, 'we have seen it more malignant in the agricultural villages surrounding Frome.' At Bruton the surgeon states, 'fever seldom prevails here; once within the last five years only; out of upwards of two hundred who had the disease, only four died.' At Trowbridge the surgeon states, 'autumnal remittents prevail in a small degree; fevers of a typhoid type are seldom seen in this town; the average mortality is the *smallest possible.*' The reports of medical men on the subject of *consumption* and *scrofula*, with *one* exception, is, that these diseases do not prevail *more* among the *manufacturing* than the other classes; and the general prevailing belief, in nearly all the places we have visited, is, that the manufacturers are as healthy as any other class, rich or poor. From Tiverton, where there is a physician, I have yet obtained no medical report; but from the clergyman, who examined the register of the parish, I procured an account of deaths in that town for the last three years, and I did so from finding that they could furnish a similar account from the great factory of Mr. Heathcote. The result is somewhat remarkable: In the town population of 9,766, the average annual deaths are *one* in *fifty-four;* in the factory, where *eight hundred and fourteen* are now employed, and where there is little variation in numbers, the average annual deaths give only *one* in *one hundred and one;* but it must be recollected, that the factory is exempt from the mortality of infancy."

But evidence on this subject, more satisfactory than any general statements, will be found in the returns obtained (to which we continue to receive fresh accessions) to the inquiries instituted relative to the amount of sickness and the rate of mortality among the manufacturing population, whether children or adults, as compared with the amount of sickness and the rate of mortality among the working classes not manufacturing. We have already adverted to the opinion expressed by the actuaries in whose hands these documents have been placed, that the inquiries instituted under the present commission will have accumulated a mass of evidence on these subjects, of great and permanent value.

We submit additional evidence from Macclesfield and Manchester, which was not placed in the hands of the board till after the first report had been printed. We farther submit an examination of Mr. Henry Ashworth, taken before the central board, which will be found highly to merit attention.

(L.S.) THOMAS TOOKE.
(L.S.) EDWIN CHADWICK.
(L.S.) THOMAS SOUTHWOOD SMITH.

July 13, 1833.

THE DEBATE

WHICH TOOK PLACE

ON MONDAY, MAY THE 9TH, 1836,

UPON

MR. POULETT THOMSON'S MOTION

FOR THE

SECOND READING FOR REPEALING A PART

OF

THE FACTORY ACT.

Lord FRANCIS EGERTON, Mr. HARDY, Sir GEORGE STRICKLAND, Mr. WAKLEY, and other honourable Members, presented petitions for the alteration of this Act, from inhabitants of Leeds—Bradford—Huddersfield—Great Horton—Ashton-under-Lyne—Manchester—Stockport—chairman of a meeting of inhabitants of Halifax—Otley—Bingley—teachers of St Peter's Church Sunday-school, Ashton-under-Lyne—Taunton Sunday-school, Ashton-under-Lyne—St. Michael's Church Sunday-school, Ashton-under-line—Time Bill Committee, Ashton-under-Lyne—cotton yarn dressers of Manchester—operative mechanics, Manchester—conductors and teachers of Mill-lane Sunday-school, Ashton-under-Lyne—workers in public factories in Glasgow—Sunday-school teachers in Chorley—persons employed in the mills of Lewis, Williams, and Co., Manchester—in cotton-mills and factories in Dukinfield—power-loom overlookers of Manchester, Salford, and other places—power-loom weavers of Stockport—operative cotton-spinners and piecers of Manchester—medical practitioners of Bury—operative cotton-spinners in the employ of Mr. Barton, of Manchester—and proprietors of cotton factories in Renfrew.—To lie on the table.

MR. POULETT THOMSON.—I would have merely moved the second reading of the factories act amendment bill, and then waited to hear and answer objections to it, if I had not wished to remove some misconception which has gone abroad respecting its nature. I am the more surprised at this misconception, because the measure seems to speak plainly for itself. It consists of but a single clause; and anybody who knows the provisions of the act at present in force, or who has attended to the discussions in Parliament upon it, must be aware that the only object I have in view is the repeal of one section of that act, and to place the law regarding factories in the same situation as that in which it stood on the 1st of March last. A bill was formerly introduced, in which an attempt was made to restrict the hours of labour of those who might fairly be considered capable of deciding for themselves. It was rejected by the House; and a measure was brought forward by Government, having for its object the protection of children under a certain age. After the 1st of March, 1836, children between the ages of twelve and thirteen years, as well as under, are, by that bill, restricted from working more than eight hours per day. This is the restriction I wish to remove by the bill upon the table; and my object is, to leave the law as it stood prior to the 1st of

March, by providing that children between twelve and thirteen years old, like their seniors, may decide for themselves; and, if they think proper, may work for twelve hours per day. The grounds for repealing the clause in question are simply these. The inspectors of factories have made an unanimous declaration that they have found it almost impossible to enforce the law as it has stood since the 1st of March last, and the inspectors, manufacturers, and all the opponents of the existing act, have stated that if the clause referred to were allowed to continue law, the inevitable consequence would be, that all children between the ages of twelve and thirteen years would be thrown out of employment. The act of 1833 went upon the principle of relays of children, each relay working for eight hours; but experience has shewn that that system has entirely failed in Glasgow, Manchester, and in all large manufacturing towns, although it has been found to work well in certain districts. In large manufacturing towns it has been found utterly impossible to procure children enough for the execution of the relay system. Hence, therefore, the necessity for the bill I am now advocating.

Then comes the question whether it is advisable for the ends of humanity—if humanity alone is to be considered—with a view to the state of employment in the country, at once to put an end to the occupation of 35,000 children, according to the opponents of the law as it stands,—or of 25,000 or 27,000 children, according to the best estimate ministers have been able to make. Upon this point I am ready to meet the noble lord opposite; for I am persuaded, that having duly protected children under twelve years of age, and restricted them to eight hours' work, well-understood humanity requires that those between twelve and thirteen years of age should not be thrown out of employment and cast back upon their parents. I believe that sixty-nine hours labour per week will not prove injurious to children above the age of twelve years. Upon this point I have taken the opinion of forty-eight medical men, forty-three of whom agree, that provided the children were properly clothed and fed it would not be injurious to them, while only five state that they are of a contrary opinion. There is another party who find fault with the measure on entirely different grounds; but I do not consider that this is the stage of the proceeding on which they ought to put forth the strength of their opposition. I allude to those who are for extending the protection to adults as well as to children, and are for limiting the employment of all to, at most, ten hours per day. I appeal to the noble lord whether he would not be defeating his own end by adding his strength to that of such opponents:—for the plan of a ten hour bill carries with it the principle that children are also to work for ten hours, and to have no greater protection than grown persons. Both now, and at all times, I must protest against a course of that kind, since I believe it would inflict the most grievous tyranny upon those who, having only their labour to sell, have a right to make the most of it. Great injury would thus be done to manufactures, but double injury to those employed in them. I have not, hitherto, looked at the subject with reference to the general interests of trade, but upon that point I might fairly rest my opposition to a ten hours' factory bill.

The right honourable baronet, the member for Tamworth, in speaking of the cotton-tax upon one occasion, dwelt upon the impolicy of putting a tax of even 5-16ths of 1*d.* per lb. upon that staple article of our manufactures. What was that restriction in comparison with that of taking off 1-6th part of the labour of those engaged in the mills of this country, which would be equivalent to a tax of 2*d.* per lb. upon cotton? Is that protection? It is not protection: it is tyranny of the grossest kind as regards the operatives—it is as regards our manufacturers perfect suicide; for the end would inevitably be, that capital and industry must find their way to other countries, and that we who depend on foreign markets for the sale of two-thirds or three-fourths of our staple manufacture would be undersold in those markets. In Saxony, Switzerland, and the United States, manufactories are rapidly springing up in every direction, and already with those countries the competition is great and the contest almost doubtful; and it would be too late to remedy the evil when the operatives in all parts of the kingdom were calling for work, and expressing their willingness, if they could obtain it, to labour even beyond the hours at present required of them. What would be the result of adopting the views of the class of opponents to whom I am alluding? The working classes would be crying for employment—they would be willing to work a much longer time than the law allows; but they would receive for answer, when they applied to the masters, that it was useless to repeal the law, and that the advantages which they had once enjoyed had already gone over to other countries. I believe that the operatives themselves are not anxious for the adoption of a ten hours' bill; but they were led away by persons who were anxious to be appointed their delegates, in order that they might come to London, and be "hand-and-glove" with members of parliament. These persons delude the operatives into the belief that they would get twelve hours' pay for ten hours' labour. That, of course, is a proposition which can be treated only with ridicule by every person acquainted with the relation which exists between capital and profits. I will propose the second reading of the bill, and leave the House to deal with it as they may think proper. On grounds of humanity I entreat the House to pass it; for if they do not do so, all children between twelve and thirteen years of age, at present engaged in factories, will be thrown out of employment. If the House should determine that the existing law ought to be enforced, and take upon themselves the responsibility of throwing 35,000 children out of employment, the government must, of course, enforce that determination; but I fear that all parties would have cause to regret the circumstance. If any honourable member should propose a ten hours' bill, I shall be prepared to deal with it; but at the present moment, I am unwilling to re-open the whole factory question. It is most disagreeable to me to be compelled to open a part of this factory question, on moving the second reading of this Bill; and nothing but an imperative sense of duty could have induced me to address to the House the few observations which I have made. I move that the bill be now read a second time. Mr. P. Thomson.

LORD ASHLEY.—I trust that when the House recollects the active part which I took in 1833, with reference to the question that has now been brought under its consideration by the motion and speech of the right honourable president of the Board of Trade, it will grant me its indulgence while I make some remarks upon the subject. I have the less scruple in presenting myself to the notice of the House upon this occasion, because I seldom occupy much of its attention; and, with respect to the question at present before it, I have cautiously abstained from offering any observations during the last two sessions. After I was defeated in 1833, in the attempt to carry the bill which I introduced, I avoided putting questions to the members of the government or moving for returns, because I was anxious that the bill introduced by the government, and adopted by parliament, should have a fair trial. Not such, however, has been the conduct of the government; for within a few days of the time when a clause in my own act was to come into operation, namely, in March, 1836, the right honourable the president of the Board of Trade gave notice, that he would introduce the present bill for the purpose of repealing that clause. This was, in fact, a condemnation of my measure by the right honourable gentleman. The main ground upon which the right honourable gentleman rests the present bill are the reports of the inspectors; but if any honourable member will take the trouble to refer to those documents, he will find that the statements contained in them are totally unsup- Lord Ashley.

Lord Ashley. ported by evidence. The House is now called upon to affirm, by a solemn decision, in defiance of all the evidence obtained from 1802 to the present day, that twelve hours are not too long a period for children twelve years of age to labour. Like the right honourable gentleman, I will not enter into any discussion respecting the ten hours' bill, but will confine myself to the single point which is brought under the consideration of the House by the present bill.

The right honourable gentleman says, that, in bringing forward this measure, he is actuated solely by a desire to benefit the children. I give the right honourable gentleman credit for having that intention, but I doubt whether the provisions of the bill will allow it to be carried into effect. Every argument which was directed against my bill in 1833, the right honourable gentleman has now urged against his own act. The alleged danger from foreign competition was as valid an argument in 1833 as it is at the present period; but in 1833 the right honourable gentleman successfully refuted that and all other arguments founded on the danger of legislative interference between masters and workmen. The right honourable gentleman having then refuted his opponents, now comes down to this House and refutes his refutation. The right honourable gentleman said, that if the House should refuse to pass the bill, 35,000 children would be thrown out of employment; but I have taken some pains to obtain information upon that point, and I have been told (and amongst others by the honourable member for Oldham, who is a high authority on the subject) that it would be utterly impossible for the mill-owners to carry on their business if they were to dismiss the children under the age of thirteen in their employment. If my information be correct, the argument upon which the right honourable gentleman founds his anticipated dismissal of the children must fall to the ground. I will here read some extracts from the reports of the inspectors, to shew that it is at least extremely improbable that the services of children under the age of thirteen could be dispensed with.

Mr. Rickards says, "An influx of fresh hands from the agricultural districts would be no relief to them (the larger mill-owners), for children entering mills for the first time at thirteen or fourteen years of age can never become expert workers; they must begin at an earlier age."

Mr. Horner, in his first report, p. 51, says, "The tendency of improvement in machinery is more and more to substitute infant for adult labour. We have found that the number of children employed are rapidly increasing, in consequence of the tendency of improvements in machinery to throw more and more of work upon children, to the displacement of adult labour."

Mr. Rickards, in his report of February, says, "New mills are now erecting in various parts of the country, and many old ones being at the same time enlarged or improved, more and more hands will consequently be wanted; the demand for children will proportionably increase."

In page 26, speaking of the determination of masters to part with their younger hands, he says, "I cannot, however, bring myself to believe that this determination will be carried to the extent threatened; because it appears to me that, as a general measure, masters will be unable to furnish themselves with the required substitutes; but that it will be generally attempted and partially executed I cannot doubt, and that much inconvenience and injury will be the result."

I appeal to the honourable members who sit behind the right honourable gentleman, and who have experience upon this subject, and I challenge them to declare that they believe that factories could be carried on at all without the assistance of children under the age of thirteen years. If they answer, as they must do, in the negative, there is an end of that part of the right honourable gentleman's case.

I will now briefly advert to the original necessity which existed for legislation upon this subject. The House will recollect that, in 1833, I brought in a bill which was founded upon the evidence taken before the committee, of which Mr. Sadler was the chairman. Many members contended that the case of the operatives only had been considered by that committee, to the exclusion of the case of the mill-owners. A commission was subsequently issued to collect evidence in the country, and the commissioners made a long report, from which I propose now to read a few extracts. I beg the House to bear in mind that I will read only from the report of the commissioners, which may be said to contain the mill-owner's case, and I will not quote one word from the evidence given before the parliamentary committee. Lord Ashley.

"In Scotland (where the hours are somewhat longer than in England) complaints of children uniform—"sick, tired, especially in the winter nights; —feels so tired, she throws herself down when she goes home, not caring what she does. She looks on the long hours as a great bondage; thinks they are not much better than the Israelites in Egypt, and their life is no pleasure to them." "Are the hours to be shortened? (earnestly demanded one of these girls of the commissioner who was examining her) for they are too long." These statements are confirmed by the evidence of the adult operatives. "The young workers are absolutely oppressed, and so tired as to be unable to sit down or rise up; so tired, that they often cannot raise their hands to their heads." "The children, when engaged in their regular work, are often exhausted beyond what can be expressed." "The sufferings of the children absolutely require that the hours should be shortened." An overlooker states, "Hours of labour too long; has twenty-four boys under his charge, from nine to fourteen years old, generally much tired; always anxious, asking if it be near the mill stopping."

This was not the evidence of young children, but of children whose ages range between sixteen and eighteen. If such are the sufferings endured at the age of sixteen or eighteen, must not the sufferings of children of tender years be tenfold greater? In Yorkshire it is stated—

"The children bore the long hours very ill indeed; exhausted in body and mind by the length of the hours and height of the temperature. I found, when I was an overlooker (says one), that after the children, from eight to twelve years, had worked eight, nine, or ten hours, they were nearly ready to faint; some were asleep; some were only kept to work by being spoken to, or by a little chastisement, to make them jump up. I was sometimes obliged to chastise them when they were almost fainting, and it hurt my feelings; then they would spring up and work pretty well for another hour; but the last two or three hours was my hardest work, for they then got so exhausted. I always found it more difficult to keep my piecers awake the last hours of a winter's evening. I have told the master, and I have been told by him that I did not half hide them. This was when they worked from six to eight. I have seen them fall asleep, and they have been performing their work with their hands while they were asleep, after the billy had stopped, when their work was over. I have stopped and looked on them for two minutes, going through the motions of piecening, fast asleep, when there was really no work to do."

Here the general report proceeds: "Pains in the limbs, back, loins, and side, are frequent, but not so frequent as fatigue and drowsiness. Girls suffer from pain more commonly than boys, and up to a more advanced age; occasionally men, and not unfrequently young women, and women beyond the meridian of life, complain of pain; yet there is evidence that the youngest children are so distressed by pains in their feet, in consequence of their long standing, that they sometimes throw off their shoes, and so take cold."

A Leicestershire witness says, "I have seen children under eighteen years of

Lord Ashley. age before six at night; their legs have hurt them to that degree that they have many a time been crying. The long standing gives them swelled feet and ankles, and fatigues them so much, that sometimes they do not know how to get to their bed. Night and morning their legs swell."

The general report says, "That this affection is common is confirmed by the concurrent testimony of parents, operatives, overlookers, and managers."

The extracts which I have read are only a fraction of the evidence which might properly be brought under the consideration of the House. The House may recollect that when I introduced my bill in 1832, I grounded it principally on the evidence of the medical witnesses, and upon the same ground I oppose the present bill. Four medical commissioners were appointed in 1832 to collect evidence, and I will read some passages from their reports. The four medical commissioners were Dr. Southwood Smith, Sir David Barry, Dr. Bissett Hawkins, and Dr. Loudon. Dr. Southwood Smith remained in London on the central committee.

Dr. Loudon reports, "In conclusion, I think it has been clearly proved that children have been worked a most unreasonable and cruel length of time, daily; and that even adults have been expected to do a certain quantity of labour, which scarcely any human being is able to endure. I am of opinion no child under fourteen years of age should work in a factory of any description more than eight hours a day. From fourteen upwards, I would recommend that no individual should, under any circumstances, work more than twelve hours a day; although, if practicable, as a physician, I would prefer the limitation of ten hours for all persons who earn their bread by their industry."

Sir David Barry reports, "Although all the sources of immediate and prospective suffering may be so far remedied or mitigated, as to render twelve hours of factory work compatible with average health and longevity, yet I am of opinion that less labour ought to be required from the infant workers, and that more time should be allowed them for sleep, recreation, and the improvement of their minds, than they at present enjoy."

Dr. Hawkins reports, "I am compelled to declare my deliberate opinion, that no child should be employed in factory labour below the age of ten; that no individual, under the age of eighteen, should be engaged in it longer than ten hours daily; and that it is highly desirable to procure a still further diminution of the hours of labour for children below thirteen years of age.

"Again, as to the reduction of hours for all below eighteen, I feel the less distrust in my own opinion, because it is sanctioned by a large majority of eminent medical men practising in this district (Lancashire).

"This bill does not accomplish the object at which it purports to aim. Its professed object is the protection of children, but it does not protect children. In the same evidence, which shews that the legislative protection of children is necessary, it is also shewn that the restriction of the labour of children to ten hours a day is not an adequate protection."—p. 32.

Again, p. 34, "While this bill does, and attempts to do, so little for children."

III. General report proposes, p. 52, "That until the commencement of the fourteenth year, the hours of labour during any one day shall not in any case exceed eight. The grounds on which we recommend the above restriction on hours of labour, are—1st, 'That at that age the period of childhood, properly so called, ceases, and puberty is established.' 2d. 'That, in general, at or about the fourteenth year young persons are no longer treated as children; for the most part they cease to be under the complete control of their parents and guardians—they begin to retain a part of their wages—they usually make their own contracts, and are, in the proper sense of the word, free agents.' "

Lord Ashley.

From general report: "That this successive fatigue, privation of sleep, pain in various parts of the body, and swelling of the feet, experienced by the young workers, coupled with the constant standing, the peculiar attitudes of the body, and the peculiar motions of the limbs required in the labour of the factory, together with the elevated temperature and impure atmosphere in which that labour is often carried on, do sometimes ultimately terminate in the production of serious, permanent, and incurable disease, appears to us to be established."

Now, referring to the appendix of the report, I find that of thirty-one medical men who were examined, sixteen gave it as their most decided opinion that ten hours is the utmost quantity of labour which can be endured by the children, with the slightest chance of preserving their health; and it is remarkable that the other gentlemen are very cautious and guarded in their testimony, so much so, that there is only one of them who gives a decided opinion; six of them give no opinion at all. Mr. Shaw, of Manchester, says eleven hours; Mr. Hill, of Derby, from ten to eleven hours; Mr. Clarke says, perhaps twelve hours; Mr. Bartley, of Manchester, eleven hours; Mr. Alexander, for all ages between thirteen and eighteen, eleven hours; Mr. Robertson, of Manchester, who has written a treatise on the subject, says ten hours and a half; two others at Manchester say eleven hours; and the only one I find who asserts that children may be worked twelve hours without injuring their constitution, is Dr. Phillips; but I say that, up to thirteen years of age, this is an amount of labour which they cannot endure,—the medical testimony says they cannot up to eighteen; the testimony I have adduced is drawn from the report of the commissioners, which report has been acted upon by the government in the framing of the bill they now require us to repeal. I was myself charged with proposing something monstrous, when I thought of ten hours as the restriction, but in this I am borne out by the report; I was told that my bill would not accomplish its object—the protection of the children—and why? Because the restriction is not an adequate protection. In summing up the evidence, the commissioners say,—

"That the successive fatigue, privation of sleep, pain in various parts of the body, and swelling of the feet, experienced by the young workers, coupled with the constant standing, the peculiar attitudes of the body, and the peculiar motions of the limbs required in the labour of the factory, together with the elevated temperature and impure atmosphere in which that labour is often carried on, do sometimes ultimately terminate in the production of serious, permanent, and incurable disease, appears to us to be established."

They then make a decided proposition to the House, that up to the commencement of fourteen years the children shall not be worked more than eight hours a day. Now it appears that this part of the report is to be treated with contempt, and all protection between eight and thirteen years is to be removed. It is on the behalf of these children I appeal to the House; and the strong ground on which I take my stand in so doing is, that I look upon the bill which has been passed by this House as a compromise between the masters and the children; that the government having brought forward a measure which has been acceded to by the parties interested, they have no right to come to us now and propose to repeal the most essential clauses of it without their consent. If the necessity of this repeal were felt, there would have been petitions presented to this House praying for it; but the only petitions which have been presented are in favour of the continuance of the present, or the substitution of a ten hour bill in its place. We have had one from Manchester to this effect, signed by no less than 33,000 of its inhabitants. The language of his Majesty's ministers, when they introduced this bill to the House, was not at all calculated to lead us to suppose it was only intended to be an experiment, that it was to be in operation for a short time, and then

Lord Ashley. to be repealed; on the contrary, Lord Althorp on that occasion said, "the commissioners having reported that they did not think the protection the bill afforded to the children would be sufficient, he should move a resolution that it be an instruction to the committee to provide that no children under the age of fourteen should be permitted to work more than eight hours in the day." The measure, indeed, could not be intended to be of a temporary nature, inasmuch as the evil itself is permanent.

His lordship further said—"The great object urged by the noble lord (Ashley), and one which every man would willingly give his aid to, is, to afford children the benefit of education, which it is impossible they can enjoy while they are compelled to continual labour throughout every day; and I therefore think that care should be taken that the children should have this advantage during the intervals which occur in their labour. The great difficulty in all these cases is, to carry the law into effect; and although a law has existed for many years, restricting the hours of labour in cotton mills, it is well known that in many instances it has remained completely a dead letter. For this reason, I do not think it would be wise in the legislature to pass any measure of restriction, without securing the means of carrying these restrictions, in every case, into full effect; because the consequence of not doing so would be to give an unfair advantage to those to whom we should be far from wishing to give an advantage, namely, to those who evade the provisions of the law, and do that which the legislature has decided to be both cruel and oppressive."

Now, does the right honourable gentleman think that the country will be satisfied that this pledge has been fulfilled, when they see that scarcely two years and a half have elapsed since the protection for the children was enacted, and that he now comes forward with a proposition for its repeal. I do not mean to say that no new proposition is to be received for the amendment of a measure after it has passed this House—I know that it is impossible so to control the power of parliament, and I consider that it would be unwise if it were possible to do so; but I will say that the House is pledged to the principle, if not to the details, of the bill, and that from it we cannot—we must not depart. There is a clause in the negro emancipation act, which provides that those people shall not work more than forty-five hours in the week—three hours less than those assigned, even by the present bill, to the factory children under the age of fourteen. Now, what would the House—and what would the country say, if his Majesty's ministers were to tell us they repent of having granted the boon of emancipation to the negro, and that they propose to cancel it, and compel them henceforward to work for twelve hours every day. Some may certainly contend that, in the one case, we are dealing with free labour and in the other we are not; but I should like to know what difference there is, in any other respect, between this and the proposition now before the House. A circumstance which has very strongly excited my suspicion of this measure is, that it is only proposed to make one amendment in the bill, notwithstanding the report uniformly says that several amendments are necessary, in consequence of the vast number of difficulties which attend the execution of the powers of the factory bill. The repeal of this clause will be a direct violation of that act, because it will render it impossible to carry into effect one of the main provisions of it. It was intended that two hours every day should be allowed the children for the purpose of education; for this reason two sets were to be employed in the factories. But how will this be done if the proposition of the right honourable gentleman is agreed to by the House?—The school clauses, it is well known, are abominated by the masters, and it must be seen that if they are abandoned with regard to the children under fourteen, the House will be perpetually called upon for repeal after repeal, until every fragment of that salutary provision is repealed; in fact, I cannot but consider that this measure is merely put forth as a "feeler," and that, if it is yielded to, it will become the basis of future encroachments, which will end in the total destruction of the bill. I am confirmed in this view of the subject by the evidence taken before a committee, moved for and appointed by the right honourable gentleman opposite, on the subject of manufactures and commerce. Before that committee, Mr. William Rathbone Gregg was incidentally asked a question relating to factory labour, and his reply was, "he had little doubt that after all these attempts at legislative interference in the matter had been found utterly unavailing, the mill-owners would be quite at liberty, and would not work much more than twelve hours a day." Lord Ashley.

After such evidence as this, I can neither resist the conviction, that this measure is such a "feeler" as I have described it, nor can I, in my conscience, suffer the House to divide upon it without first stating that, although, if it should appear to be expedient, I might be disposed to consent to the obliteration of some particular clause, I cannot consent to the destruction of the bill. If the right honourable gentleman will pledge himself to provide such a substitute for this clause as shall preserve those points which I consider essential in it, I will desist from any further opposition to the second reading; but without some such hope of mitigation I have no alternative. I must move that the bill be read a second time this day six months; as, having to choose between the two evils, I shall certainly prefer that the masters should be inconvenienced rather than that any cruelty be exercised towards the children. I have no such intimation from the right honourable gentleman, and therefore I move, sir, "That the bill be read a second time this day six months."

Mr. POULTER.—I rise to second the amendment; and in doing so, I cannot help saying that I very much regretted to hear the observations of the right honourable gentleman, the president of the Board of Trade, with respect to the necessity of abrogating his own law, because, in so doing, I considered that he was exposing himself to a reproach which is certainly undeserved by the general tenour and liberality of his character.

Sir, I take it for granted, that if any fears of the fatal effects of foreign manufacturing competition—such as the right honourable gentleman assures us he entertains in reference to the possible consequences of our perseverance in that act which he desires to extinguish—had been really entertained by our own mill-owners, the table of this House would have groaned beneath the weight of petitions from Lancashire and the manufacturing districts, expressive of those fears. But as to foreign competition, in my opinion, there is no fear of that. The fear is, that the mill-owners are running a race of competition against each other, in which the interests of humanity are to be sacrificed; and to that sacrifice I, for one, will never consent. I prefer the interests of these poor children, who are without natural protectors, or whose natural protectors have most cruelly deserted them; and having been so deserted, it becomes the duty of the members of this House to stand *in loco parentum*. If the president of the Board of Trade feels himself in the painful situation of sacrificing the great principle of humanity by his proposed alteration of the existing law, I will tell him of a case in which that law has been strictly observed by a mill-owner, who, on the other hand, has sacrificed a little of his worldly interest to the principle of humanity. At a factory in Bradford, 1763 persons are employed, 616 of whom are under eighteen years of age; and of those 567 are females. All the children under twelve years of age have never, since the first of March last, worked more than eight hours a day, to each of them two hours a day being allowed for education and recreation. Those of them between twelve and fifteen years of age work only ten hours a day each, and to each is secured one hour for instruction,—an admirable example, and one which ought to be

Mr. Poulter generally followed. How is this effected? Human power is kept in action eleven hours a day, and about one-tenth of young persons more than is absolutely necessary are employed to conduct the machinery; and in this way that most magnificent establishment is carried on, with every regard to the general principle of humanity. Instead of asking the House to make the proposed reduction in the existing law, the right honourable gentleman ought to call for new powers to enforce it. I can tell him of no less than nine factories in one single town in which his own law was most grossly violated so lately as last Good Friday. By the ninth clause of his bill, that day is reserved as a whole holiday; and by the same clause it is proposed that eight half-holidays shall be secured to the children in the course of the year. Now, if such a day as good Friday be thus taken from them, what is the chance of their having the eight half-holidays? None at all; and the whole clause becomes a dead letter. I can give the names of the children who have been thus unjustly treated.

Again, it is a very prevalent practice to obtain false certificates as to the ages of the children. These children are dressed up so as to appear much older than they really are; and I can mention the names of children who, at eleven years of age, have been certificated as being thirteen years old. I can tell the right honourable gentleman, also, of cases where the children, during the meal hours, have been employed in cleaning the machinery. That is another evasion of the law. I am not personally acquainted with the district in which these things have taken place, but I have seen those who are. These are things the House ought to look to, and the law ought to be enforced, not relaxed. If there be not inspectors enough, let their number be increased; let their visits be more frequent; and let those visits be at uncertain times, when they are not expected. Those are the things that are wanted, and not a reduction of the present law. When I find that out of 616 young persons out of eighteen years of age, no less than 567 are females, this fact suggests a very important and serious consideration in reference to the future welfare of the poorer classes of the people. These are, many of them at least, to be the future mothers of families. If they are to be worked to the extent of twelve hours a day, the constitutional strength and health of the children must be materially endangered. Constitutional affections must result from this species of over-work, destructive of the health of their future offspring, and producing distortion and debility. Baron Humboldt declared that out of 1,000,000 savages whose tribes he had visited, he never met with a single instance of deformity. It is said, in answer to this, that Baron Humboldt does not state how many of those people die in infancy. But I think that deformity is oftener an acquired than a natural defect. Savages have the free use of the elements; they live in the open air, and take much exercise: that is the reason of there being so few instances of deformity among them. If you proceed with your present system, you will run the risk of entailing deformity and distortion, not only upon the children themselves, but upon their future families. If the law of nature is thus permanently violated, she will resent it, sooner or later, by gradually deteriorating large classes of the people. Then, look at the want of education among these children! How very necessary is it to make provision for them in that respect. You cannot bring together young persons of both sexes at a very early age without multiplying greatly their demoralizing tendencies. Nothing can overcome those tendencies but moral and religious instruction, and yet, at the very age most important for the purposes of education, and when you may expect the greatest consequences to flow from good moral and religious instruction, these children are left without the chance of improvement. During the week they have no time, and on Sundays they are too much exhausted to attend any school. When I support political reforms, I do so upon this principle —that such political reforms are always to be accompanied by the moral education of all classes of the people. If not, I do not deny that those reforms may be misused and turned to evil. It is upon this ground alone that I have supported the political reforms which have taken place during the last few years. I have done so in the hope that the privileges which those reforms confer upon the people, they will be taught to use beneficially to themselves. Mr. Poulter.

I trust the House will never consent to the motion of the right honourable gentleman, but will support, without any regard to politics—for I am no politician on this occasion—the amendment of the noble lord. I am very much attached, politically, to his Majesty's Government, but I am a little more attached to these poor factory children. I trust the House will never abandon those great principles of humanity which it has once advocated, and will never allow a system to exist which will turn into evil the blessings of human invention,—which is repugnant to the natural affections of our minds,—and which must tend to bring into a state of moral and physical deterioration the most interesting portion of the human race.

Mr. GISBORNE.—Having been led, from circumstances, to pay great attention to the working of the factory system, and having frequently been called upon to act as a magistrate under the existing factory bill, I am anxious to state to the House the views which I entertain on this subject. It is a great gratification to me to speak in this part of the debate, after the very temperate and candid manner in which the noble lord has brought forward the question; and I hope I shall be influenced by the same spirit, and give no utterance to any party feeling or sarcasm. Whenever the noble lord's bill shall come before the House again, I am sure the house will give it the most serious consideration. But I may perhaps be allowed to doubt whether the legislature, at this protracted period of the session, will be disposed to make so great a change as that of enacting a ten hour bill as the restriction to be laid on the moving power in our factories. But it is at this moment more important to consider what is the present state of the law, and what has been its operation; because, unless the right honourable gentleman who has brought in this bill can establish that the practical effect of the present law will, by inevitable deduction, be to turn out of work all children between twelve and thirteen years of age, this House ought not to consent to the right honourable gentleman's motion. Mr. Gisborne.

I remember that a great many of our manufacturers told us, at the time, that the effect of the clauses imposing restrictions on the labour of children under thirteen years of age would be, that they should turn all such children out of their factories. We did not believe them; or, if we did, we legislated in disregard of their warning. I appeal to the honourable member for Ashton-under-Lyne whether that prediction has not been verified?—whether the system of relays has not failed?—and whether more than five per cent. of such children have been since so employed? I ask him whether the system of relays and education is in operation in any part of the manufacturing district with which he is acquainted?—I ask him whether, in the large establishment with which he is connected, he has not found it impossible to work upon that system; and whether all the children, up to the age of twelve years, have not been practically excluded from those works? I know that the system of relays was tried by persons in that neighbourhood—that they tried it honestly, and with the most perfect intention of conforming themselves with the law, and making the system work well. Mr. Thomas Ashton gave the system a fair trial; but he failed in it—how?—Because all the children left him. They would not work under these restrictions. It is only in country places where the manufacturers have the opportunity of procuring a larger number of children, and where employment is more

Mr. Gisborne.

difficult to be obtained, that the relay system and education clauses have had any effect whatever. This is how the bill has operated with respect to children of the ages of ten, eleven, and twelve years; and the simple question before the House is this—is it for the interest of the children themselves—is it for the interest of humanity—that the bill should so operate with respect to children between the age of twelve and thirteen years? Will you persevere in a course by which every child between twelve and thirteen years of age shall be turned out of work in the factories, and be thrown into a state of idleness, or be compelled to seek employment elsewhere?—If I were to talk on the subject for hours, I could bring it to no other issue than that.

It is well known, that from Buckinghamshire and other agricultural districts, a great many families have lately been removed into the manufacturing districts. Let us consider what it is that operates that removal. It is not that the heads of families can get work. When a man removes with his family from an agricultural district to a manufacturing district, perhaps he may get employment as a labourer, at wages from 12*s*. to 15*s*. a week; but he is not at all fit for labour in the factories. Considering the increased expense of housekeeping in town districts, that rate of wages would never induce him to leave his own situation; but the inducement for him to go is, that his children immediately get employment and good wages. Children between twelve and thirteen years of age, in the manufacturing districts, I understand, will get from 4*s*. to 7*s*. a week each. How great an inducement must that be to persons in the agricultural districts, who are in want of employment for their children, to carry them to a manufacturing district? It is not, indeed, the actual wording of your enactment, but the practical effect of it is, that none of these children shall take employment up to thirteen years of age. If that be the effect of the law, the natural consequence must be to prevent these removals of families, which have hitherto been so beneficial to all parties. I will not enter into a consideration of the general effect of your system on the trade and commerce of the country; but I will only suggest one matter for your consideration. Suppose, by your interference with the system of free-trade, you should, in fact, enable other countries to compete with us, and thereby throw a large body of our own operatives out of employment; ought you not to compare the misery you would thereby create with that which may possibly be created by your allowing children between twelve and thirteen years of age to be employed for twelve hours a day? Why, sir, it is not only the misery that would be created in the manufacturing districts that we have to consider, but we ought also to reflect on the amount of misery which would be created in the agricultural districts, and the distress that would be brought on the landed interest when those operatives, from whom they are now relieved by the manufacturing system, should fall back again on the land for support. It would not require much indiscreet legislation to produce that effect; but the amount of misery it would occasion would be so great that the House ought to pause before it ran any danger of bringing about such consequences. It is upon that ground, chiefly, that I shall give my cordial support to the second reading of this bill.

Mr Benett.

Mr. BENETT.—Living on the borders of a large manufacturing district, I am anxious to explain what I conceive to be the question which is now at issue. It does not appear to me to be a question of profit or loss to the nation—that is a matter of trifling importance in comparison with the effect of the factory system upon the physical and moral condition of the children employed under it. But it is a question of very great importance, indeed, whether those children shall, by the effect of labouring a certain number of hours a day, be destroyed in their bodily, and, consequently, in the mental faculties. I have seen a great deal of the operation of our factory system in towns; and, comparing the great mass of the people in the manufacturing districts with the mass of the people in the agricultural districts, I must say that the contrast is distressing in the extreme. The deformity of body, and moral disposition, and mental constitution, in the one, in comparison with the other, is certainly a contrast of the most painful kind. With respect to the number of hours' labour which children of the age of ten, eleven, and twelve years are capable of enduring, I think the evidence of the medical men who were examined before the commission which was issued a short time ago is conclusive against the system that has so long prevailed. I have made some references to that evidence, and they have convinced me that it is impossible that children can be continuously employed for the length of time which they were in the habit of being before the factory bill passed without great detriment both to their health and morals. We are told that if we restrict the labour of children between twelve and thirteen years of age to a certain number of hours, they will be turned out of work altogether. But how does that accord with the present state of manufactures in this country? I find that new factories are springing up in every quarter; and there is a sort of rivalship going on between the manufacturers in giving high wages, the effect of which is to induce the lowest rate of profit. That such a state of things should exist conjointly with an excess of labour in the market, is so entirely contrary to all the principles of political economy, that it convinces me there is no ground for apprehension that the children will be unemployed, by reason of our continuing the law in its present shape. Mr. Benett.

The price of labour, like everything else, depends upon demand and supply. My belief is, that the demand for labour is becoming so great, that the children will earn more in six or eight hours than they have been accustomed, hitherto, to earn in ten, eleven, or twelve hours. The assertion which the manufacturers make with respect to the anticipated injurious operation on their profits by foreign competition, is so completely overturned by the fact of the great increase of our factories, that I give no sort of credit to it whatsoever. With respect to the evasion of the law, I believe it to be the fact that many impositions have been practised. I have it from the manufacturers themselves that they have not obeyed the law. It is exceedingly difficult, I admit, to ascertain whether a child be twelve or thirteen years of age; but it is for the House to consider whether the age shall continue as fixed by the present law, at thirteen years, or whether it shall, by the proposed measure, stop at twelve? I contend that we ought to let the present law have full operation, or else change it into a ten hours' bill, which would do away with the necessity of any inquiry as to the age of the child. Nothing, in my opinion, can be more fallacious, unjustifiable, and wicked, than to consider wealth, even in a great commercial nation like England, as that which, under any circumstances, ought to be put in comparison with the health and morals of the people. If, indeed, any danger were likely to be incurred, as to the power of our manufactures to compete with foreign productions, then, perhaps, we should have a right to inquire how far an increase in the number of hours' labour by children might ward off that danger; but even in that case, I should prefer the physical health and moral condition of the people to any imaginable increase of the national wealth. Upon these grounds, I shall oppose the second reading of this bill.

Dr. BOWRING.—Sir, the honourable member for Wiltshire, who has declared that he cares nothing about the profit or loss of the manufacturer, but only about the comfort of the labourer, seems wholly to forget that nothing but the profits of the master could give food, or raiment, or enjoyments to those whom he employed. It might, no doubt, be a most satisfactory Dr. Bowring.

Dr. Bowring. result if the same wages could be obtained for a few as for many hours' labour; but such a result is not among the conditions of human existence. I mistrust that interference on behalf of the poor which the poor are themselves to pay for; and will never lend myself to those delusions by which the meritorious classes of society are taught to believe that their wages, like everything else, are not dependent on supply and demand. Let the question be presented honestly and fairly. Let the parents of factory children know—but they know it well—that the diminishing the hours of daily toil must diminish the amount of weekly pay. To protect them, as it is called, is merely to protect them against the comforts which the two hours additional labour would purchase. Certainly, there are cases of hardship and oppression, but I dislike all cases of legislative interference between master and man,—between parent and child. And, morever, all such interference would be unsuccessful. What has every speaker confessed? Why, that the laws are trampled on—that the regulations are not obeyed—that the act of parliament cannot be enforced! Why continue in this course of helpless legislation? Why struggle, perpetually, to maintain a state of things which the common interest overthrows? What is the use of laws to which you cannot give effect? Happily, no law can shake to their foundation the great social interests of mankind at large. There would be no bounds to the absurd freaks of legislators, did not the common interest—the general instinct—check their progress. You prohibit foreign trade—what then? The smuggler comes—a public benefactor, though a breaker of the laws—and tumbles down the barrier you have raised against friendly communication. So your laws to regulate wages, and hours of labour, and conditions of contract for work—they are merely cobwebs broken through at will—because it is the interest of master and servant that they should be broken.

The honourable member has contrasted the superiority of the agricultural children in the country with the manufacturing children of towns. Now, I hold an opinion wholly opposed to his. I am not without acquaintance among both classes; and I venture to assert, that the children in the manufacturing districts are far better instructed, more intelligent, more moral; and so, in fact, is the manufacturing population as a whole. In what part of Europe, for example, is the *minimum* of bastardy to be found? In the manufacturing districts of Switzerland. In our agricultural counties the number of illegitimate births is as 1 to 20; in the parts of Switzerland to which I have referred, as 1 to 44.

It has been said that we have nothing to fear from foreign rivalry—this is a strange error. But if foreign rivalry be not more perilous, it is because many other nations, following our foolish example, have taxed food and raw produce, and built up a system of protection and illiberality of which they are paying the inevitable cost. But look to the countries which are emancipated—look to Switzerland—see her—remote from all the means of supply, but without a custom-house—without a tax on food or labour—without any legislative interference—without factory bills or boards of trade, or protection of any sort—see how her manufactures have found their way to every market of the world—see how her people have grown and prospered in the unbounded liberty of exchange. Are the House aware that in that country half the manufacturing population have become the proprietors of the land on which they live, and the houses in which they dwell? They want no protection but the protection of freedom; and they are formidable rivals, and must be so,—formidable in proportion to their emancipation from an interfering policy. I hope the time is not distant in which the true cause of the distressed condition of our labouring population will be boldly investigated. But it is not by laws which cannot be enforced, nor by prohibitions, nor by restrictions,—it is by a total change of system. Abrogate your corn-laws—

liberalize your commercial system—cultivate commerce with all the nations of the world; thus will you raise wages—thus prevent the necessity for exhausting labour. All other projects are but palliatives,—if, indeed, they palliate,—the source of the evil is untouched by them. Dr. Bowring.

Mr. Brotherton.

Mr. BROTHERTON.—As this is a subject in which I have long felt a deep interest, and with respect to which I have had some experience, I hope I shall be excused if I occupy the attention of the House for a very few minutes. The subject is one of paramount importance to the manufacturing population of this country, and involves many considerations; inasmuch as upwards of 360,000 persons are actually employed in the cotton, woollen, and silk mills of this country, and a far greater number dependent upon them for support. The attention of parliament has, on several occasions, been engaged in considering in what manner protection could be given to the children employed in factories, without running the risk of injuring our foreign trade. Various acts have been passed, and I admit that the condition of children working in factories has been much ameliorated; but much remains to be done, in order to produce that state of society which it is desirable should be attained. I consider it as a fundamental principle, that it is the duty of the legislature to regard, in all their proceedings, the physical, intellectual, and moral improvement of the great mass of the people, and that in legislating on this subject the House ought to keep in view those general principles. I can never believe that the prosperity of this, or of any other country, can depend on the continued violation of the principles of justice and humanity. I do not think that the labour in factories is injurious to health, if it be not excessive. But, under the present system, every man who knows anything of the matter knows that the labour and confinement are excessive. I have no hesitation in declaring, unequivocally, that the labour is too much for children to bear. When we consider that between 400,000 and 500,000 persons are immured in factories from half-past five or six o'clock in the morning till eight or nine o'clock at night, and this not upon a few particular occasions only, but day after day, week after week, and year after year, during their whole lives, or as long as they are capable of following their employment, I would ask, whether any person professing the Christian religion can sanction such a system as that, or desire that such a state of things should continue? Of what use can education be to those who are thus employed? Surely human beings are not to be considered as mere machines, or instruments of labour, out of which the utmost possible degree of exertion should be extorted. But those employed in factories have no time for intellectual improvement, or for acquiring moral imstruction.

I hold in my hand a copy of a bill which was brought into parliament in the year 1815 by the late Sir Robert Peel, whose memory will long be held in grateful remembrance by the labouring classes in the manufacturing districts, for his humane exertions to shorten the hours of labour in factories. The object of that bill was, to limit the hours of labour to ten hours and a half a day. He had to struggle with the same difficulties as those who are advocates of his views have to contend against now. I was a master-spinner at that time, and I believe I stood almost, or quite, alone at Manchester in rendering assistance to those who, with Sir Robert Peel, were endeavouring to accomplish these benevolent intentions. With the exception of myself, I believe all the master-spinners of Manchester were opposed to the measure. What was the consequence? Sir Robert Peel was four years endeavouring to accomplish what he designed; and at last, in the year 1819, he succeeded only in carrying a bill which limited the period of labour to twelve hours a day, or seventy-two a week, for all children under the age of sixteen years. It was soon found that this measure, being limited in favour of children under

Mr. Brotherton.

sixteen years of age, was liable to be evaded; and subsequent experience shewed that it was evaded. The consequence of which was, that in the year 1825 another bill was introduced and carried, limiting the hours of labour to sixty nine hours per week. Again, in 1831, another act was passed, called Sir John Hobhouse's act, and which is the very best that has been passed, because it was the most simple, and could not easily be evaded. That act applied to persons under eighteen years of age working in the day, and prohibited any person under the age of twenty-one from working in the night. Its effects were very beneficial; but it applied only to persons employed in cotton factories. Great excesses and many abuses were heard of in the woollen trade, and also in the silk manufactories; and the late Mr. Sadler afterwards introduced a bill which applied to all these alike.

It was not, however, till the year 1833 that the existing act was brought forward; and although I resisted some of the clauses of that measure, I must still claim the merit of consistency in opposing the bill which is now brought forward for altering one of its most important provisions. I told Lord Althorp, when the bill of 1833 was under consideration, that the relay system would never answer; and as the bill passed through committee, I pointed out the provisions which I was satisfied (and subsequent experience has shewn that I was correct) would be evaded. The provisions of the present act are, that no child under thirteen years of age shall be employed more than eight hours a day, or forty-eight hours per week. That no child above thirteen years of age, and under eighteen years, shall be employed more than twelve hours a day, or sixty-nine hours per week; but a person eighteen years of age may be worked any number of hours. It cannot be denied that the bill has totally failed in many of the great objects for which it was intended. The inconvenience arising from the necessity for relays of children, and the different periods of time which they are required to work—some eight hours, some twelve hours, and others thirteen or fourteen hours—the surgeons' certificates and the schoolmasters' certificates—render the act so complex and inconvenient, that numerous frauds are committed, and the provisions of the act are continually violated. It is a common trick, according to the inspectors' reports, for parents to send children of a healthy appearance and sufficient age to a surgeon to procure a certificate, under a false name, for a younger brother or sister. Anne, for example, is sent to a surgeon for a certificate, and gives her name Sarah; and Sarah, a child of ten, or perhaps nine years of age, is admitted to work in a mill for twelve hours per day in direct violation of the laws; and thus little or no protection is afforded to the children. If the House wish for evidence to shew how the act has been observed, it is only necessary to refer to the return which has been laid on the table of the house of the number of masters who have been guilty of a violation of the law. By that return it appears, that during the last year no fewer than 250 masters of mills have been convicted under the act, and paid penalties to the amount of upwards of 1000*l.* When it is considered that these masters of mills are men of respectability, and presumed to be superior in moral feeling and education to those whom they employ, is it not disgraceful to find them violating an act of parliament so grossly and so constantly as they appear to have done? I repeat, that I consider such conduct most disgraceful. In the majority of instances in which the penalties of the act have been enforced, it was for the offence of compelling the children to work more than twelve hours a day. If the masters are guilty of such infractions of the law, and display such a contempt for the laws of nature and the common rights of humanity, how can we expect the poor and comparatively uneducated artisan to respect the laws? How can the legislature expect the laws to be obeyed by the poor, when they are daily and hourly violated by the rich? Having thus pointed out the defects of the existing system, I beg for a moment to be

Mr. Brotherton.

allowed to consider the remedy. I never like to object to a measure unless I think it is in my power to suggest a remedy. This, then, is the remedy I would propose:—that, for all persons employed in factories under the age of twenty-one, there should be one uniform period of labour, with such security, in the way of inspection, as would prevent the time being exceeded. The system is such that persons of all ages, and all constitutions, old and young, weak and strong, must all necessarily work together at one and the same time. Let the legislature, therefore, fix upon some period of labour within the range and compass of human strength, and make it applicable to all alike, or, at all events, to all persons under the age of twenty-one years. Under the existing law, frauds as to age are constantly committed, and will continue to be committed, so long as the present law remains; because they are connived at, not only by the masters, but the parents of the children, and by the children themselves.

The simple question which has now to be considered is this, whether, after having affirmed the principle that no child under thirteen years of age shall be employed in factories more than eight hours a day, this house is prepared to reverse that decision, and enact that children of twelve years of age shall work twelve hours a day? I cannot agree to reverse the decision of parliament; I should prefer the time for all under thirteen years of age being increased from eight to ten hours,—as is the case in silk mills,—rather than that children just turned twelve should be liable to work twelve hours, and often longer. This plan would remove much of the inconvenience arising from double sets of children. I am, however, of opinion, that ten hours' labour in a factory is sufficient for either man, woman, or child. But when any proposition of this kind is made—whenever the voice of humanity interposes between the master and the labourer—the house is assailed with the cry of "Beware of foreign competition; if you reduce the period of labour, you put a fatal restriction upon the British manufacturer, and render him incapable of competing with the manufacturers of the Continent."

We have it in evidence, that previous to the passing of Sir Robert Peel's act, the usual number of hours for which persons were employed in factories was seventy-seven in the course of the week; and from returns on the table of the house, it appears that it was not unusual for children of seven and eight years old to be kept at work as many as ninety-three hours in the week. Sir Robert Peel's act reduced the number of hours to seventy-two in the week; and when this was done, the legislature was told by those who professed to understand everything connected with the subject, that the possibility of our manufacturers continuing to compete with the manufacturers of foreign countries was completely taken away. But how was this assertion borne out by the fact? At the time of the passing of the late Sir Robert Peel's act in 1819, the exportation of cotton twist from this country amounted annually to 18,000,000 lbs.; and in six years afterwards the quantity annually exported was 45,000,000 lbs. The period of labour was again reduced in the year 1825, and the same argument was used, and nothing but positive and immediate ruin could fall on the heads of the devoted manufacturers of this country. What was the fact? In the year 1834, the exportation of cotton twist amounted to 76,000,000 lbs. Facts prove, then, that this country, with its superior machinery, has nothing to fear from foreign competition. I admit that if labour be restricted to a great extent, the cost of the manufactured article will be increased, or the wages must be diminished. It is the cheapness of food in other countries which excites apprehension in the minds of our manufacturers. But if the people cannot be supported with reasonable labour, food must be rendered cheaper,—the corn-laws must be repealed. The tax on the importation of raw cotton ought also to be taken off. This tax produces about 400,000*l.* per annum, which is

Mr. Brotherton.

from 6d. to 1s. per week taken out of the pockets of every man, woman, and child employed in cotton factories.

I differ from the honourable member for Kilmarnock (Dr. Bowring), and others, who say, "only repeal the corn-laws, and all laws bearing upon free-trade in labour, and then the people will be much better off, and will never be so unmercifully worked." All the inspectors agree that, if there were no legislative interference, the children employed in factories would be over-worked. And when free-labour is talked about, let it be remembered that the inspectors also report, that human labour in union with the machinery of a cotton-mill is not free. Such is the system that it is impossible for a man, or a woman, or a child, to say that they will work only such a length of time. I have been grieved, many times, to see females stand at their work when it would have been much better for them that they should go home; but they dared not leave their employment a moment before the usual time, lest they should lose it altogether. No discretion is left to the operative as to the number of hours he should be employed; he must work the usual time, or he will not be allowed to work at all. The effect of this must be prejudical to the health of the children.

The right honourable gentleman, the president of the Board of Trade, in the course of the observations he made in moving the second reading of this bill, threw out an insinuation which I think was not quite fair. The right honourable gentleman said, "that the reason why the operatives wish for this restriction is, that they may have twelve hours' wages for ten hours' work." It might, with just as much justice, be said that the masters, when they violate the law, and keep the children beyond the time prescribed by the legislature, are anxious to get twelve or fourteen hours' work for ten hours' wages. The fact is, the house is not aware of the great temptation that there is for over-working these persons. To some master manufacturers an additional hour a day obtained from each of the persons he employs would make a difference of from *50l.* to *100l.* per week. Thus an enormous advantage is gained by those who violate the law, and the honest and humane manufacturer is left without a chance of competing with them.

If there were no legislative interference, I have no doubt that many mills would be found working sixteen and seventeen hours a day. Many persons engaged in the manufactories are desirous of making rapid fortunes, and I should as soon expect to extract oil from granite as to obtain anything from the humanity of these worshippers of mammon. I beg to bear my humble but honest testimony against the existing system. Many humane men are no doubt compelled to continue it contrary to their wishes. I have considered the subject, carefully, for a considerable length of time; and I am convinced, and have no hesitation in declaring, as a practical man, that a bill for an uniform time for all ages under twenty-one years, with such security as would prevent the time being exceeded, would not be injurious to the master; whilst the operative would be satisfied, the children protected, the commercial prosperity of the country extended, and the people employed, contented, healthy, and happy.

Mr. G. F. Young.

Mr. G. F. YOUNG.—The cheering appeal which has just been made to the house is rendered even more cheering when we are able to perceive that the philanthropic views of the honourable gentleman are not inconsistent with the policy of free trade enforced by the honourable member for Kilmarnock. I think the house and country have some reason to complain that this question should now be forced upon the attention of Parliament, after the comparatively recent decision which the legislature came to in the year 1833. It cannot have escaped the recollection of honourable members who took an interest in the question at that time, that the result at which the house then

Mr. G. F. Young.

came was in consequence of a wise and judicious compromise of conflicting opinions on a very difficult question. When the noble lord, the member for Dorchester, brought in his ten hours' bill, it was only defeated by a majority of one; and in consequence of that strongly expressed opinion, the house appointed commissioners to make inquiry into this question, alleging that the evidence taken before the committee which sat in the preceding session was not such as could be legislated upon. When he brought in his bill, one part of which we are now asked to repeal, the right honourable gentleman, the president of the Board of Trade, was one of those who took part in the discussion. A right honourable colleague of his told us that he rejoiced that the subject had been taken up; particularly as regarded the number of hours for infant employment, and that after reading the evidence already before the house, two opinions could not be entertained upon the subject; and he now trusted that the question would be once and for ever set at rest. That was the opinion of the president of the Board of Control. A bill was introduced, and was referred to a committee. The right honourable gentleman then told us that the committee was appointed, not to receive evidence, but to see whether the protection proposed by Lord Althorp might not be extended even further. Thus it was clearly the object, on introducing this bill, to carry the principle of the restriction of labour further than the noble lord had ventured to propose to the house. I well recollect the compliment paid to him for having adopted so liberal a principle; and after that, I think it rather hard that we should be dragged into this discussion, before the bill has yet come into actual operation at all. It is alleged that it has been introduced in consequence of the recommendation of the inspectors of factories. The opinion of those inspectors runs almost concurrent in favour of the object of the bill. But that recommendation is but one of a series of recommendations made just as forcibly by these very inspectors. Having, however, given this one recommendation, which is not favourable to the interests of humanity,—that is acted upon; whilst the many other recommendations which they have made on the other side of the question are entirely neglected. I can perfectly understand the principle, that all interference with labour is, in itself, injudicious, prejudical, and contrary to the maxims of sound policy; but I cannot understand upon what principle they are legislating who bring in a bill of this description, by which they leave a considerable portion of infant labour fettered with restrictions, and propose to remove those restrictions with reference alone to the labour of children between twelve and thirteen years of age. If you take this course, you ought to be prepared to shew, by evidence, that the age of twelve years is just the point to which the protection of the law ought to be extended, and that if it be carried one year further it will be pernicious. The *onus* of proof lies with you. Have we had any such proof? None.

I utterly disbelieve that by the rejection of this bill those results would follow which the right honourable gentleman has endeavoured to make the house believe would be the consequence of such a course. If the manufacturers dismiss the children, will the mills continue to be worked? I believe they would; but they could only be continued by the subsititution of adult labour for the labour of the children so dismissed. Now, they could not get that adult labour unless there was a superabundance of that description of labour in the market. If that be so, then they have a choice of leaving unemployed children under thirteen years of age, or leaving unemployed persons above that age. If they are placed in this dilemma to choose between the two, I would rath er that they should leave the children under the age of thirteen to follow those pursuits more congenial to their age, and more beneficial to their physical and moral powers. It is said that the employment is not of that injurious character to the children which it has been described to be. I will only call to the consideration of the house that overwhelming

Mr. G. F. Young.

mass of evidence now before us, all concurring in the same conclusion, that it is impossible for the human frame to endure continued labour more than ten hours in a factory, at an average temperature of eighty degrees, without producing distortion, deformity, and disease, and without ultimately shortening life. I have had my attention drawn to one part of the statement put forward by the master manufacturers, which I must confess does go very far to prove that the employment in those particular factories is not unhealthy; but my honourable friend overlooks one part of that statement. In a note in page 5, there is a return of the number of persons on the sick-fund connected with the factories at Bolton. This statement certainly shews that those who come on the fund from other factories exceed in number those who come from the cotton-mills. But this return being for the three years 1833, 1834, and 1835, shews that the number of sick persons on the fund was progressively and rapidly diminishing. Now, that took place concurrently with the operation of the law, by which young children, in those years, were progressively relieved from a stated number of hours labour; thus proving that their sickness was in a great degree the result of the labour to which they had been exposed. A letter has appeared in a journal of this evening, in which the names of the honourable members for Manchester and Wigan are mentioned, and which contains many important statements. The writer tells his name, and confidently appeals to those honourable members for a corroboration of the statements made by him. With the permission of the house I will read the letter. It is as follows:—

"TO THE EDITOR OF 'THE STANDARD.'

"2, Northumberland-court, Strand,
May 7, 1836.

"SIR,—There is no subject whatever connected with the factory system which persons who have not a practical acquaintance with the interior of mills have more diffculty in forming a correct judgment of, than the amount of actual labour and fatigue endured by the children while performing a single day's work; few, probably, of the masters themselves have sufficiently turned their attention to the point. At a conference which took place last December, at the Albion Hotel, Manchester, between several members of parliament representing Lancashire boroughs, some of whom were mill-owners, and a number of delegates from the operatives, this particular subject of youthful labour was agitated and very fully discussed. The surprise elicited by the statements then made regarding it was very great, but those statements were incontrovertible, however startling. It was there shewn that, in many branches of cotton-spinning, the children had not less than five-and-twenty miles to walk in the mill in a day, independently of their other exertions. In consequence of the impression this assertion, and the facts and calculations with which it was supported, made at the meeting in question, Mr. Mark Philips, M.P. for Manchester, and Mr. Potter, M.P. for Wigan, visited the Chorlton new mills, at which I then worked,—that establishment being one of the best conducted in Lancashire, and its then owner one of the kindest of masters. They came into the wheel-house in which I was working. Mr. Potter took out of his pocket a thermometer, and hung it upon the beam, to ascertain the heat of the apartment; in a few minutes the quicksilver rose to about 82 degrees. I had five children under me at the time, their ages varying from about ten to fifteen.

"After keeping the machinery in motion for awhile, that those gentlemen might be eye-witnesses of the manner in which the children were ordinarily occupied, and the speed being ascertained by a reference to their watches, which proved that the children followed the machine 1560 times a-day, I stopped the machinery, and put it up to give an opportunity for measuring the space over which the children had been travelling. This I proceeded to do with a rule, and, after allowing for the various distances which the children had to walk in all the various directions that the work required, the conclusion was established, perfectly, I believe, to the satisfaction of those gentlemen, that my children, in an ordinary day of twelve hours' labour, independently of fatigue arising from other causes, had to undergo that of travelling fifteen miles three hundred and twenty-five yards in the mill. It is here material to observe, that this was in fine spinning, in which the speed of the machinery is much slower than where coarser numbers are spun; but this advantage is, perhaps, fully counterbalanced by the high temperature which it needs. In the coarser branches of spinning, the speed of the machinery is much greater, amounting in some mills to even four times as much; but, as the threads do not break so often, that increased speed alone will not lead to a correct estimate of the distance the children travel over; but, in many instances, the distance is enormous, far exceeding what I have just stated.

"Close to the particular mill in which the investigation I have described had been made, stands one in which the machinery was then working full four times as fast. I drew Mr. Philips and Mr. Potter's attention to the fact. From the window at which we were standing they looked out, and were able to make the observation for themselves. They were eye-witnesses of the consequent toil of the children. I subsequently made a calculation founded upon what they had thus seen in that adjoining mill, and took it to Mr. Potter's house; neither he nor his lady, each of whom took the trouble of going over the calculation, could alter the result, or discover a flaw in it. That result was so astounding, that I do not think it prudent to hazard the effect its publication would produce; for by those who have not the actual data and all the circumstances before them, I freely acknowledge it would be received as incredible. I think it sufficient to say that the distance it established considerably exceeded five-and-twenty miles; and I thus challenge, openly challenge, contradiction to my statement. I am confident Mr. Potter will confirm it.

"One other fact I beg leave to mention, and which I am ready to prove, as I am all I have already stated, before any committee, or even at the bar of the house. It is, that, since 1825, when Sir John Hobhouse brought in his bill (on which occasion I was in London as a deputy), the speed in cotton machinery generally has been increased, to speak within compass, one-fourth; or, in other words, equivalent to the additional labour of three hours a day.

"Be it remembered, that it is now full twenty years since the late Sir Robert Peel, who had had immense experience of the then system, brought in a bill for limiting the daily labour of young persons in cotton mills to ten hours and a half. If, during all that time, the labour, as is notoriously the fact, has been, step by step, increasing, and if, since 1825, its progress has been such as I, fearless of contradiction, assert it to have been, can parliament now hesitate to pass a ten hour bill for all ages? Is it not a most just and reasonable request, one that, if it errs at all, only errs in this, that it asks too little in the way of protection for the youthful workers, a large majority of whom, in most branches of spinning, are females? Shall the manly feelings of the British nation be in vain appealed to on this occasion?

"It will doubtless be said, by some, that my statements are too extraordinary to be believed. The same was said of the results established by the comparative tables of mortality calculated in Mr. Rickman's office, and from public documents. It was alleged that they were deceptive, because they included places rapidly increased by migration. Mr. Sadler has demonstrably shewn, that any argument fairly deducible from that circumstance would rather strengthen, and certainly in no degree weaken, the conclusion to which those tables inevitably lead. It was also alleged, by Mr. Commissioner

Mr. G. F. Young.

Drinkwater, that the calculations in question should not have been formed upon the mortuary registers (which give the ages of the dead), but upon the population return (which furnish the ages of the living). The late Mr. Sadler, in his last work, entitled 'Factory Statistics,' accepted the challenge, and after making the necessary calculations, established the same result as Mr. Rickman, namely, that 'about as many have died before their twentieh year, where the factory system extensively prevails, as before their fortieth elsewhere.'

"If this is not enough to arouse the attention of the legislature to the wrongs of the factory population, nothing but some violent convulsion, which no one would deprecate and deplore more deeply than myself, can ever have that effect. "DAVID M'WILLIAMS."

In conclusion, my decided opinion is, that on this question the dictates of humanity, and the principles of sound policy, are perfectly compatible; and that the true interests of the manufacturer are identical with the well-being of those whose labour he employs.

Mr. Mark Philips.

MR. MARK PHILIPS.—The honourable gentleman has read to the house a statement which has appeared in the columns of an evening print, to which I have had no access, and therefore cannot answer for or guarantee the correctness of it. I believe I know perfectly well the name of the writer—David M'Williams. It is actually true that the honourable member for Wigan and myself did attend the examination in the mill as stated; but my firm conviction, at the present moment, is, that there is a misstatement as to the thermometer being at so high a degree as is mentioned in the letter. Having been connected for many years with mills in that district, I can take upon myself to say, that there is no one room in any of the largest mills in Manchester, in which the atmosphere was ever known to be so high as is alleged by the writer of that letter. I saw five children employed, and was most anxious to ascertain the number of miles they walked in a given time; but there was such great irregularity in the movements, that I believe it impossible, by any rule, to come accurately at the result of the number of miles a party walked in the course of the day. I believe the distance was proved to be very considerable; and I do not say that the factory system is not open to many serious evils. But although I do not look at it as a perfect system, yet I will not condemn it, when I know that, by throwing that condemnation upon it, I shall be taking bread from the mouths of thousands. We must deal with things as we find them. Amelioration, I admit, ought to be carried to the utmost practicable extent; but, for the sake of nominal amelioration, I am not disposed to adhere to a law which will commit the fatal error of doing injury to those whom it is especially intended to protect. I have seen a letter, written by the honourable member for Ashton-under-Lyne, in which it is stated, that if the thirteen-year old clause continued, 20,000 or 30,000 children, between the ages of twelve and thirteen, would be thrown out of employment. The remark I made to the writer of the letter, which the honourable member opposite has read, was, that if I was not much mistaken in my judgment, 27,000 children, under the age of thirteen years, would be proscribed by law —that law to which he ascribed so much importance.

I do not object to the proposition of the honourable member for Shaftesbury, as to the increased number of inspectors; in principle I agree with him. I have no desire that the effect of their appointment should be nullified. Reference having been made to my visits to the factories, I beg to state to the house, that on one of those occasions I was told, that, by an increased motion, the machinery might be so improved that it would perform as much work in ten hours as it then did in fourteen; and as this is a point very much insisted upon by the friends of a ten hours' bill, and frequently insinuated by them into the minds of the mechanics, I wish to say that that is an entire mistake, and that the machinery cannot be so altered as to produce this effect. It is sometimes argued, that by shortening the hours of labour the increased vigour of the operatives will be such as to counterbalance the deficiency in the number of hours. I have no doubt that, under such circumstances, they would be more active, but that the products of their increased activity would be so extensive I must take upon myself to deny; and I say now, as I said before, that it would be unjust to require that twelve hours' pay should be given for only ten hours' work. I ask the honourable member for Ashton-under-Lyne, —does he not remember a turn-out which took place in the button trade a few years ago, near his own residence; the ruin which was brought upon the children, the misery entailed upon the operatives, and the bad feeling engendered between the masters and the men, which led even to murder itself,—all consequent upon that turn-out; and can it be supposed that, with my eyes open to this fact, I can consent to any mode of legislation which may lead to a repetition of such disastrous scenes? No; I wish for the settlement of the question upon practical grounds, such as those upon which I consider the measure of my right honourable colleague to be founded. We are told that we have not brought forward any petitions from the master manufacturers, praying for this alteration; if we have not, I can state to the house, that before the commencement of the session, statements were made by them to my right honourable colleague, of the evil consequences which would be felt by the operatives, unless such a bill as this now before us were brought into the house; and I think I can even prove the case out of the mouths of those who are opposed to the measure. I was much surprised to hear the honourable member for Ashton-under-Lyne cheer the noble lord, the member for Dorsetshire, when he stated his disbelief of the difficulties the manufacturers are alleged to be labouring under in consequence of the bill of 1833, because I have a written statement of his own with respect to the operation of that act, in which he declares, that if the masters are prevented from employing children under thirteen years of age, from 30,000 to 40,000 of the latter will be dismissed, in consequence of which it will be almost impossible to carry on the manufactories.——

Mr.
G. F. Young.

MR. HINDLEY.—Read the whole of the letter.

Mr. Hindley.

MR. MARK PHILIPS.—I will, then, read the letter, which contains statements to the following effect:—

Mr.
Mark Philips.

"On Friday I put a question to the government, whether it was their intention to enforce the provisions of the factory bill, to which I received only an evasive answer. It appears that the masters have endeavoured to induce Mr. P. Thomson to bring in a bill for the purpose of suspending the operation of his own act; if he succeed in carrying it, my hopes are that a good measure will be substituted for it. Many of the parents will be dissatisfied if their children are not employed in the mills; the spinners will have to pay more to the piecers. I fear that some parties are unwilling to have the clause altered; but it is only by the enforcement of this point the masters will be induced to concede, for if they are prevented from employing the hands which are under thirteen years of age, from 30,000 to 40,000 of them will be thrown out of employment on the 29th of February, and it will then be impossible to carry on the factories; under the pressure of difficulty, the employers may be disposed to make a concession in point of time, or to agree to some restriction in the moving power, if it were to the interest of the operatives to obtain such a result; but I hope that there are none, however inconvenient it may be, who will hesitate to require that the thirteen-hour bill should be repealed, and a new one obtained."

Mr. Mark Philips.

Now, I do not think it will appear that the context of the letter at all weakens the force of the statement I selected from it. I am assured by manufacturers of the borough I represent, that in one instance the number discharged will be 150, in another 70, and in another 50, in that place alone, unless some provision be made to counteract the effects of the clause restricting the employment of children under thirteen. I give my assent to the bill of my right honourable colleague, because I foresee the greatest difficulty arising out of the restriction. I do not advocate it merely on the behalf of the master-manufacturers;—they can run the risk of the experiment for a year; they can stop some of their machinery, and they will thus become acquainted with the extent of their loss; but it is impossible to estimate the loss of those who will be turned out on the world; and I must tell the honourable member for Tynemouth that he does not understand the cotton-spinning business; indeed, this is not to be wondered at—for I do not, myself, although I have lived during my whole life in a district in which it is carried on; if he did, he would not talk, as he has done, of the facility with which the young persons who are discharged may find employment in other branches of the business. The thing may be very well as a matter of theory, but it is impracticable; there is no branch of the business which will absorb the hands thus set adrift.

The noble lord, the member for Dorsetshire, has very prudently kept the question of the ten hours' bill out of sight; and, therefore, I shall not open that wide field of discussion, although, had it been introduced, it was my intention to have argued against it. I shall, however, do so, should any such attempt be made. Now, with respect to the hard labour which is stated to be endured by the operatives, I confess that I like to take examples as near home as I can; and I believe the honourable member for Ashton-under-Lyne will not deny, that, not long since, he caused his own people to be worked half an hour longer than any of his neighbours—those very over-worked people, who, in an extract he read from the report, are compared with the over-tasked Israelites. There is a notice of a motion on the book, to take into consideration the condition of the hand-loom weavers. What would be their condition if the hours of their labour were restricted? It would be tenfold worse than it is. If the cost of twist were to be enhanced, a direct *bonus* would be given to the foreigner; the trade would be driven from the country, and would take root elsewhere. I am so convinced of the correctness of my views, that I wish to state, that if a measure were to be passed which would have this effect—I, who am indirectly connected with mill property, would advertise the whole of my property for sale. I would lend my capital to those who have now the management of the mill. I would say, "take it, and make the best of it you can in countries where there is an open competition;" and I am sure, that so employing it, either in France or in Switzerland, the advantages it would produce would be threefold those it does now. The operatives must feel the continuance of the present bill to be mischievous to their interests, although it has been stated by the noble lord, the member for Dorsetshire, to be a compromise between the masters and themselves. The factory inspectors have shewn that it cannot be carried into effect without greatly injuring them; and I am not disposed to sanction a restriction which, it appears to me, must be followed by the most serious consequences.

Sir Robert Inglis.

SIR ROBERT INGLIS.—It was stated, in a letter read by the honourable member for Tynemouth, that the labour of the children in the factories exceeds the average daily marching of an English soldier; that every child has to walk fifteen miles a day. Does the honourable member for Manchester mean to say that this is the conclusion to be drawn from the evidence?

Mr. Mark Philips.

MR. MARK PHILIPS.—It is an extremely difficult question.

Sir Robert Inglis.

Sir ROBERT INGLIS.—Supposing it, then, to be only equal to ten miles a day; is it fit that children under thirteen, who are subjected to it, should be left unprotected by the law? If ever there was a case in which two and two do not make four, this is one; for it has been clearly proved by the statistics of honourable members, that where the number of hours of labour has been reduced, there has been an accompanying increase in the produce of the labour. The honourable member for Kilmarnock, who, as an advocate for pure abstract political government, appears to regard the children merely as machines for the production of cotton;—he tells us that smugglers are beneficial to a country, inasmuch as they shew us the faults of our legislation. The connexion of this argument with the bill before us I was not fortunate enough to perceive. I am quite willing to admit that we have no right to expect an equal amount of wages for a lower amount of labour; but, at the same time, I believe that if human labour is forced beyond its physical power, the very element out of which the power proceeds is destroyed. The statements which have been made by the honourable member for Salford are uncontradicted by the honourable member for Manchester, and they satisfy me that an economy of human labour is not likely to impair, but will increase, the productiveness of it. Are not the children in our manufactories as much entitled to protection as persons of a different colour on the other side of the Atlantic? What are the statements of the physicians who have given their testimony? They say that no child under eighteen years of age should be worked more than ten hours a day. I trust the house will adhere to the principle they established in 1833, when they passed a measure which was laid before them, as a compromise of the ten hour bill, proposed by my noble friend the member for Dorchester, whose present amendment I shall feel it to be my duty to support.

Mr. Ainsworth.

Mr. AINSWORTH.—I cannot avoid supporting the bill of the right honourable president of the Board of Trade, approving as I do of the arguments which have been advanced in its favour by the honourable member for Manchester; and as the representative of a large and populous manufacturing district, I consider it my duty not to give a silent vote. I have listened to the arguments both of the supporters and of the opposers of the measure, and I cannot but notice how few there are who have attempted to shew that the system on which factory labour is now conducted tends to prejudice either the health or morals of those who are employed in it. The noble lord (Ashley) has not quoted any one instance of the sort, and if the fact were so, it appears singular to me that there are none who have addressed themselves to the point. It has been said, indeed, that no petitions have been sent here by the masters; but a variety of declarations have been made by most respectable individuals deeply interested in this important question, to the effect that the most serious injury will accrue, both to themselves and their property, if the amendment of my right honourable friend, the president of the Board of Trade, be not passed into a law; and when I find that these statements of theirs are fully borne out by the report of Mr. Rickards, who was deputed by his Majesty's government to inquire into the working of the factory system, I think there cannot be the least doubt that they have their foundation in fact.

I live, sir, in a populous district, and have visited a great number of manufactories; and I therefore hold myself competent to say, that these interferences with labour, besides being injurious and vexatious, are calculated to produce anything but the effect intended. If, as appears by the testimony of many honourable gentlemen who have spoken upon the subject, frauds are committed under the present restrictions, how much more numerous will they not become if you narrow in a still greater degree the limits of labour? they will certainly increase in a tenfold degree. Will the house legislate upon

Mr. Ainsworth.

the subject, or shall it be left in its present position? It will be far better to accept the proposition of my right honourable friend than submit to the present restrictions. Some honourable members seemed to scout the idea of danger arising out of foreign competition; but it is easier to deal in general declamation than to grapple with facts. In those countries where labour is not restrictive, the produce will be more abundant than where labour is limited, and the article can consequently be sold at a cheaper rate. Those who can purchase as good an article at a cheaper rate will certainly not come to our dearer market. In France, Germany, and Switzerland, there are no restrictions. I do not say that we should have no restrictions; but I caution the house not to impose such restrictions on labour as would overpower our means of competing with other countries. On the continent they labour for fifteen hours a-day; but with our superior machinery, with our superior skill, talent, and ingenuity, I consider that twelve hours would sufficiently enable us to compete with them. If, however, the restriction were greater, I apprehend that our mechanical improvements and superior skill would not be sufficient protection against continental competition; and therefore I would strenuously urge upon the house the extreme caution with which it should adopt any measure having a tendency to restrict free labour.

Dr. Lushington.

Dr. LUSHINGTON.—I have listened to all the arguments which have been advanced in favour of the proposition of my right honourable friend, but I must confess they have not only failed to convince me, but have not in the slightest degree removed my repugnance to this measure. Though I have no personal or practical experience on this subject, yet am I not therefore precluded from offering an opinion, for it is one of the first principles of legislation that the prejudices and feeling of persons who are intimately interested in any question, are best corrected by the opinions of those who come into the discussion with free minds, and judgments unbiassed by any personal consideration. To those, then, who would impugn my want of information, I will make the reply which I remember Mr. Canning to have made in answer to the argument that danger must follow from persons legislating for the West Indies who were unacquainted with West Indian society; and I will say, "Must I be precluded from legislating as respects the mines of Cornwall, because I have never travelled through the town of Truro?" Legislation will always be improved by experience, but its foundation must be based upon good sense and sound principle, and every body acquainted with the most valuable portions of our enactments will admit that the great bulk of our best laws were the production of persons unconnected with the interests for which they legislated.

Before proceeding briefly to answer the arguments which have been adduced in favour of my right honourable friend's proposition, I shall first simply state the real question, which is this:—By the law, as it at present stands, children under thirteen years of age, are restricted from working more than eight hours a-day; by the proposed measure of my right honourable friend, it is intended to place children of that tender age under a regulation which would permit them to work twelve hours a-day, or, in other words, add twenty-four hours to their weekly labour, and this proposition is said to be borne out by the principles of free trade. Sir, no one has been a more strenuous or uniform supporter of the principles of free trade than I have been; but then, free trade, as I apprehend the meaning of the term, can only exist when both parties between whom it is carried on are equally capable of judging for themselves what is their own interest, and protecting that interest when they have formed their judgment. Under such circumstances as these, it is competent to them to call upon the legislature not to interfere with their proceedings—to suffer them to act upon their own discretion and judgment;—and the result of that non-interference will be, that whilst each individual will pursue the course which tends to his own interest, he will also be acting in a manner that will conduce to the public benefit. These Sir, are the real and wholesome principles of free trade. In adults, the fullest liberty should be allowed to them to bargain, as they best may, for the sale of their own labour. The glass-blowing trade is known to be an injurious one, and greatly destructive of health; but the legislature has no right to interfere with an individual if he think proper to undertake this labour, and thereby procure wages to the amount of 16*s*. per day. Coal-heaving is also a very laborious and destructive employment; but if the adult, who makes 7*s*. 6*d*. per day, chooses to engage in it, the legislature has no right to forbid him. But this is not now the question. The question before the house is, whether or not it will afford protection to persons incapable of protecting themselves? Not only argument but precedent is in favour of the affirmative. Has not the house already sanctioned the principle, by affording the protection sought for in this instance to another class? Have we not passed an act for the protection of infant chimney-sweepers? If these require protection, do not the children in factories equally demand it? The daily experience of natural causes and effects teaches us that infant children cannot labour for twelve successive hours without ruinous consequences. Nature must be consulted in all sound legislature. Indeed, this is a principle of English law—*nunquam aliud natura, aliud sapientia docet.* In legislation, as in everything else, nature and wisdom must go hand-in-hand. The law of England extends its protection to infants, to madmen, and all others incapable of protecting themselves. In these and similar cases, the rules of political economy are inapplicable. Dr. Lushington.

We are told that, unless this restriction be removed, our trade will be ruined by foreign competition. When argument fails, prophecy is easy; and in no instance more than the present has prophecy been so freely indulged. But I too can prophesy; and I do say that, as surely as the sun will rise or set, any greater protraction of the hours of labour for children will be certain destruction to their health—and I will set this prophecy and its consequences against the predictions of those who vaticinate the destruction of our trade. I will further set against it the experience of twenty years, and the testimony of master manufacturers who are opposed to the measure of my right honourable friend. On these grounds, sir, I feel myself called upon to oppose the proposed alteration; and I will ask my right honourable friend, whether he is not, on the present occasion, acting rather as member for Manchester than minister of the crown? I do not impute to my right honourable friend that he is actuated by any improper motive. I impute to him merely that which influences most of the honourable members of this house—the being liable to be too much swayed by the opinions of influential persons amongst his constituents; and I say that, as regards this question, he cannot express his opinions with the same independence that I can. But to return to the question.—It has been stated that the condition of the manufacturer is better than that of the agriculturist. Such was the statement of the honourable member for Kilmarnock—but I very much doubt its correctness. The manufacturer was described to be superior, as well intellectually as physically. The latter assertion refutes itself; and as to the former, although the dwellers in towns may have more minute information on particular points, the information of the agriculturist is more general and varied. I agree with the honourable member for the University of Oxford in considering the bill of 1833 as the result of a compromise between the masters and the men; and I therefore look upon the present proposition as a violation of that compromise—which, if called upon to vote, I certainly will not sanction.

MR. BAINES.—It is an assumption, not supported by fact, to suppose that persons who labour in manufactories are necessarily unhealthy. I have Mr. Baines.

Mr. Baines. passed my whole life amongst a labouring population, and the result of my experience is, that there is no class of children better fed, better clad, better lodged, or more healthy, than those of manufacturers. I make this assertion not upon light grounds ; my opinion is the result of many years' experience. Is it because I happen to be member for Leeds that I should be precluded from giving utterance to that which is my firm conviction upon this subject, or is the opinion of one who has daily opportunities of collecting facts upon which to found that opinion less entitled to regard than the opinions of those who have not similar opportunities of acquiring information? Previously to my taking my seat in this house, and knowing that the subject would be brought before parliament, I visited almost every factory in Leeds, and endeavoured to ascertain the state of health of those employed in them, directing more particular attention to the state of the children, and I never witnessed a greater appearance of health or spirits than they presented. Not satisfied with this, I also called upon the masters, and questioned them as to their opinions with regard to the laws respecting factory labour. The two questions I put were, first—whether it would be desirable to abridge the hours of labour ; or, whether it would be well to permit children under twelve years of age to labour in the factories? The answers given to both questions were in the negative. I went further; I called upon the surgeons who attended those places, and after ascertaining their opinions, I desired them to collect the labourers, that we might get their opinions also. They were called together, the same questions were put to them ; they were also asked,—whether it was prudent or desirable to abridge the hours of labour, or to make any alteration in the age of the children? And their answer was,—we think it would be well to abridge the time of labour for one hour; but not to alter the regulation as regards the ages of children, from twelve to ten years. Such was the information which I collected from the masters, the surgeons, and the operatives themselves, and surely it is not inferior to any testimony which has yet been adduced to the house upon this subject.

Acting then upon this testimony, and undeterred by any dread of popular displeasure, I do think that it is for the interest of children above the age of twelve years not to be too closely restricted in their hours of labour. What will be the consequence of too close a restriction? Why, finding that they are not permitted to earn their accustomed wages, they will betake themselves to other more lucrative but less healthy pursuits. To get their twelve full hours, they will apply for work in the coal-pits, and thus adopt a mode of life far more unwholesome. With regard to the danger which threatens our manufactures if further restrictions be imposed, although it is treated as chimerical, I know it to be real. We need fear no competition with foreign markets whilst we are on anything like an equality with them; but if they can produce an article of the same quality as ours at one halfpenny or a penny less than ours, as soon as that fact becomes known our present customers will abandon us. Let the house consider the situation in which persons, now prosperous, may be placed by restrictions which will have the effect of immediately injuring, and ultimately ruining, this branch of trade. Let not honourable members hope, or fill their minds with an undue assurance, that though a branch of trade may be taken from us, we can again recal it. All experience is against the encouragement of such a hope. Has any branch of trade which we have seen removed ever returned? I will ask honourable gentlemen who are so powerfully swayed by humanity upon this subject, to give me a sufficient reason why children of thirteen years should not be suffered to work twelve hours a day in factories, when we have a law on our statute books allowing parish orphans to be apprenticed at nine years, and placing no limit to the hours of labour to which they may be subjected? In this case, too, the persons thus apprenticed are orphans; they have no parents or friends to look after them; whilst the children employed in factories have their parents on the spot to look after and protect them. I hope the noble lord opposite (Ashley) will direct his humane attention to the condition of these children. In saying this, I do not mean anything that can be possibly supposed offensive. I do believe that gentlemen who take opposite views of this question are influenced by the best and purest motives—that they are merely actuated by principles of humanity; but I must tell them that the course which they are now pursuing, however well intended, will only have the effect of arraying the men against the masters, and the masters against the men. Feeling thus, I shall brave all the odium which may attach to the vote which I shall feel myself bound to give, when I have the consciousness that in doing so I act, as far as my judgment leads me to think, in accordance with humanity, with justice, and with love of my country. Mr. Baines.

MR. POTTER.—The factory commissioners having recommended the alteration in the bill now under consideration, and knowing, as I do, from actual observation, that the bill is habitually violated, I shall certainly vote for the second reading. When the bill goes into committee, other alterations can be proposed, such as additional inspectorships. Foreign competition has been considered of no importance, and that we have nothing to fear from it; but I say I entertain a very different opinion on that subject, and I earnestly beg the attention of honourable gentlemen to the following statement, which is taken from official documents. The importation of cotton-wool into Great Britain in 1815 was 371,000 bales, while the importation and consumption of France and other States of Europe was inconsiderable, and the cotton manufacture of America quite insignificant. In 1822 the consumption of cotton-wool in Great Britain was 865,000 bales; in the same year that of America was 173,000 bales, and of France 272,463 bales, and of other States of Europe 200,000; making in all 646,263 bales, being nearly double what was imported into Great Britain only seventeen years before, and two-third parts of what was imported into Great Britain in the same year of 1832. In 1835, the consumption of the States of America had increased to 216,888 bales. If we suppose France and the other States of Europe to have increased their manufacture only in the same proportion, then the whole consumption of rival foreign States would be 784,000 bales, while that of Great Britain for 1835 is estimated at 944,673 bales. Thus it would appear that in twenty years the foreign rival cotton-manufacture had increased so much in amount as to consume in the last year four-fifth parts of the quantity of raw cotton-wool that was used in Great Britain during the same period. Mr. Potter.

SIR JOHN ELLEY.—If the house divides upon the question, I really feel some difficulty as to making up my mind how I should vote; but with regard to the question of health between the manufacturing and agricultural portion of the population, I find no difficulty whatsoever in at once coming to a conclusion. Were the question to lie between the popular health and the manufacturing wealth, I should find still less difficulty. In Leeds we are told, that amongst the manufacturing classes there is even a redundancy of health. I have not been lately to Leeds; but when last I visited that town, so far from witnessing the plethoric health which I have heard described to-night, I saw nothing of it; on the contrary, I saw nothing but want and misery; and the wretched and squalid race whom I beheld were evidently unfit for any other earthly pursuit but that in which they were engaged. There are times when the manufactories cease to give the persons engaged in them full employment, and the workmen become disposable for other labour. Under these circumstances many of them take to the army; but I must confess, for my part, if I wanted to fill up my companies, I should not from choice go to the manufacturing districts to look for recruits. I have been on many Sir John Elley.

Sir John Elley. recruiting parties; and when on emergencies we sometimes went to the manufacturing towns and populous districts, at periods when we expected to find many unemployed, although many came and expressed their willingness to enter, we found them for the greater part not possessed of sufficient physical power, and, I am sorry to add, deficient also in all mental endowments. Compare them with the agriculturists and those whose avocations lead them to live in the open air, and there is the greatest difference imaginable, both corporeal and intellectual. The difference is strikingly illustrated by this fact, namely, that when the proper officer examines the recruits, he almost invariably rejects five out of ten of the manufacturing classes, whilst of the agricultural he scarcely rejects one out of ten. In making this statement, and voting as I shall do upon this occasion, let it not be supposed that I am actuated by any morbid sensibility, or desirous of adapting myself to the popular feeling on this topic, I do so because it is my fixed opinion, that though this house ought not to interfere with trade to its injury, it is still its duty to see carried into effect, when necessary, some measure calculated to preserve the health of so large a portion of human beings.

Mr. Chas. Villiers. MR. CHARLES VILLIERS.—Though I may hold different opinions, I do not feel myself warranted, neither am I inclined, to impugn the motives of those who propose to restrict the hours of labour by certain regulations. I will admit—indeed I fully do admit—the benevolence of their motives; but it must be remembered, that there is no necessary connexion between wisdom and benevolence. On the contrary, benevolence not unfrequently interferes, through humane motives, to the great injury of the very party whom it intends to serve. Appeals have been made to-night to every faculty but the reasoning one. We are told that we should be to the factory children *in loco parentis*. Every feeling has been touched, every sympathy excited; but then comes in the principle of civil law to aid, in illustration,—and we are told of the incompetent and the lunatic having the shield of legal protection. This argument is used to induce us to believe that it is necessary to interfere between the parent and child. But then see the assumption upon which it is founded. It assumes that the poor care less for their children than the rich; indeed, that they have less natural affection than the lower animals. Are we warranted in making that assumption? On the contrary, does not all experience shew us that their attachments are stronger? If we once admit this as a principle upon which to legislate and interfere so unnecessarily, we shall strike at the root of all domestic well-being. If you thus interfere with the children of the poor, it will, in effect, be saying, "You have so little regard for your children that we must interfere to prevent you making them a sacrifice to your cupidity." It is asserted that children are compelled to work too much, and for too long a time. That may be the case; but I ask, is it from choice they so work, or is it from choice that their parents permit them? No; it is from necessity—from dire and absolute necessity. How then is this difficulty to be met? Is it by passing a law forbidding them to work? Can you provide means adequate to fill up the wants which you thus occasion? Can you say to a family, "You must not earn so much in wages as you have been accustomed to do?" Will you undertake to diminish their income? and if so, how and by whom will you adjust the difference?

Another injurious result of this interference with labour arises out of your system of relays, the effect of which will be to reduce the wages of adults. Let me not be misunderstood as being an advocate of forcing children of tender years to work too hard; quite otherwise. I think the demand for, and supply of labour, will always make their own equable adjustment; and we should take special care lest our unnecessary interference should tend to the injury of those very manufactures upon which the interests of those for whose benefit we seek to legislate depend. If it be really the intention to benefit the productive classes, instead of restricting the hours of labour, let us endeavour to take off the taxes which press upon the necessaries of life, and which at the same time bring nothing to the public revenue. Let us repeal the corn-laws. Let those whose humanity is so prompt as respects the factory children come forward, and instead of supporting a proposition which may gain them popularity for one session, bring forward one which will be of less doubtful and more lasting benefit, and tend to the welfare of the whole population.

Mr. Chas. Villiers.

COLONEL THOMPSON.—If the manufacturing interest had for twenty years together put a tax on home-grown corn, and the result had been that the agricultural labourers had been reduced by competition among themselves to great distress, I wonder whether the agriculturists would have invented no better remedy than asking for an act to regulate the hours of agricultural labour, to prevent the plough-men, and plough-women, and plough-children, from being overworked. I suspect they would have hit upon something much more substantial. I, therefore, rejoice that my honourable friend (Mr. Villiers) has broken the ice upon a topic which I had begun to be afraid had, by common consent, been banished from the debate. I hope the manufacturing interest, both operatives and capitalists, will in time see the advantage of turning the tables on their opponents, and they will not fail to make some use of the parallel he has now applied to them. Having been applied to by the working classes as one they deemed their friend, I feel some difficulty in deciding how to vote, because, whichever way I do it, there are some that will say they expected better things from me. On the one hand, it will be painful to vote against the expressed opinions of the working classes as to what they think best for themselves; and, on the other, I cannot think of doing anything that shall be construed into giving into the miserable delusion which induced the operatives to accept the factory act at the hands of those I must call their adversaries, in lieu of demanding the removal of the prohibition on foreign trade, which was at the bottom of their sufferings. I should certainly have voted against the original factory act; but since that folly has been committed, I do not know why the operatives should not have it in the way they like as well as in the way they do not. If, therefore, the operatives will give me full credit for opposition to the impolicy of accepting the factory act as a composition for their wrongs,—as I should have voted for letting the law alone before, so I will vote for letting it alone now, unless I hear something to change my mind. In fact, the whole matter now in dispute is so small in proportion to the mighty interests that are ruthlessly thrown away—it is such a mere difference between "tweedle-dum and tweedle-dee," in the comparison—that I do not think it worth while to go against the wishes of the working classes by voting for alteration in the law. Much has been said of foreign competition. If the manufacturing capitalists are afraid of the effects which may arise from the difference between working children under thirteen years of age, according to the present law, and the proposed alteration, why do not they set about looking after the effects arising from the prohibition of foreign trade enacted by the corn-laws? They allowed themselves to be told they could compete with foreigners, because they were not at a stand-still altogether. Just so; a man with one leg tied up in a race, might hop at a certain speed; he might find some so maimed or lazy that he could go beyond them; with these it might be proved he could compete; —but is that the sort of competition the manufacturers should be content with? It is plain to every man with his eyes open, that, setting aside this fallacy, foreigners have outrun us, are outrunning us, and would outrun us

Colonel Thompson.

Colonel Thompson.

hereafter. There may be no recovering what has been already thrown away, but that is no reason why something should not be tried for preservation in the future.

Mr. Hindley.

Mr. HINDLEY.—Sir, having presented many petitions on this question, I trust the House will allow me, even at this late hour, to ask its indulgence whilst I give my reasons for opposing the bill of the right honourable gentleman opposite. And in so doing—though I am aware that the question of the propriety or impropriety of legislation is not necessarily involved in the proposition, yet, as many honourable members seem to have doubts upon the subject, I should wish for a moment to advert to it. The opinion of gentlemen engaged in trade I think has undergone a considerable change, since the last agitation of the factory question. At that time, the general cry was *Laissez nous faire;* now, no such exclamation is heard, and it appears to me that legislation is generally admitted by the masters to have been beneficial. The reports of the inspectors fully bear out this view of the subject. Mr. Horner says, as to the regulation of the hours of labour, by putting all parties upon equality, "this part of the act is hailed as a great benefit by a large majority of mill-owners." Mr. Saunders says, "considerable advantage to all concerned must inevitably follow a more regular and uniform system than that heretofore acted on." Nay, so sensible does Mr. Rickards appear to be of the necessity of legislation, that he says, were the act to be suspended "for any time, long or short, I believe it would throw these districts into great confusion. Overworking would be generally, if not universally practised. Children would be devoid of that legislative protection which has hitherto done much good and is calculated to do more. The excesses of former times will be renewed; and at some future period we may have to recommence our labours under increased difficulties and disadvantages."

Again, "Objections have been started to the proposed rule for restraining the working of mills within certain hours, as being an undue interference with the natural freedom of industry. But the objection is, I think, in this instance, unfounded. It is not proposed, for example, to restrain the freedom of adult labour. A man who has the entire mastery of his own limbs and person, may work himself to death, if he pleases, under the proposed law; but not so with children and minors, we legislate freely for their rights, interests, and properties. The law is most anxious to protect these from injury, and various provisions are made to accomplish it. Then why not legislate for their personal labour also, where their bodily health and structure are likely to be seriously affected in the absence of due regulation and control? This is peculiarly the case where human labour works in unison with the steam-engine or the water-wheel. That tyrant power may at any time, and without any effort, cripple or destroy thousands of human beings, if not duly restrained. Human labour in these mills or factories is not on a par with other branches of industry. It is governed by a power that needs neither food nor rest; whenever the avarice of masters, and the stimulus of higher wages working on their men, may, and assuredly will, lead to excessive exertion, of which disease, deformity, or death, are but too certain to be the unhappy results."

Mr. Inspector Howell states, "In visiting the mills in my district, I have been pleased to find instances in which the act has been acknowledged to work benefit, by limiting the hours of the children, and more particularly by putting a stop to their working in the night."

The arguments derived from theory, against legislation altogether, appear to me completely refuted by these results of practical experience; and I ask the House, if they do not fully bear me out in repeating, that legislation has been productive of improvement in the factory system? But gentlemen may ask, if this be so, why have we so many new factory bills? Since 1815, there have been ten or twelve. "Why do you keep troubling us to repeal and amend our previous measures?" My answer is, that certain gentlemen, averse to any legislation, finding they could not prevent a law, have studiously contrived to make it ineffective, in the hope of inducing the House eventually to reject legislation on the subject altogether. Thus, when the right honourable baronet, the president of the Board of Control, brought in a bill likely to attain its end, the mill-owners contrived to reject a valuable clause suggested by my honourable friend, the member for Salford, and, in so doing, made the bill ineffective; yet when the present act was in progress, the honourable member for Bolton, on the part of the masters, declared, "all the legislation required was, the making Sir John Hobhouse's bill effective!" Whatever difference of opinion, however, may exist as to the propriety of a law, there will, I trust, be no difference as to the proposition of the right honourable the president of the Board of Trade. What is the bill? The right honourable gentleman says it speaks for itself; and what does it say? Stripped of all its legal mystification, it repeals the protection now given to children between twelve and thirteen years not to be worked more than eight hours a day, and enacts that they—i. e., children of twelve years of age—shall work twelve hours, which, with two hours for meals, one for going to and returning from the mill, and eight for sleep, made twenty-three hours, leaving only one single hour to the child for improvement, recreation, and the enjoyment of domestic society. Let it be distinctly understood, that this is the proposition of the right honourable gentleman; and can there be two opinions upon it? Is there any member of this House bold enough to affirm, that a child of twelve years of age, travelling, it may be, eighteen miles a day during the course of its work of twelve hours, should have only one single hour for, I would say, the very barest possible consciousness of life itself? And this after the legislature, in 1833, had guaranteed, both to it and the adult negro, four hours more! What would be said, asks the noble lord (Ashley), if this proposition had been made regarding the adult slave? Why, the whole country would have been in arms; and I hesitate not to say that it would have been the downfal of any ministers who persevered in such an attempt. And why is an equally inhuman proposal to be made respecting factory children? For pecuniary considerations? Surely the right honourable gentleman's own colleague (Sir John Hobhouse) has answered these questions in his own speech on the second reading of his bill in 1825.

Mr. Hindley.

He said, "Ought we to allow a portion of our fellow-subjects to be rendered miserable for such a consideration? No. It would be better to give up the cotton trade altogether than to draw such a sum out of the blood and bones and sinews of these unfortunate children. The legislature was bound to protect them."

The House ought to recollect the circumstances under which the present protection was afforded. It was no hasty measure, passed without examination or discussion. After much evidence, the House has affirmed the principle on the second reading of the bill of the noble lord, that no child under eighteen should work more than ten hours. Further inquiry, however, was demanded. Commissioners were sent into the country to examine the condition of the children, who were weighed, and measured, and subjected to the most minute inspection. The result of all this was stated to the House in the words of Lord Althorp:—"The commissioners are of opinion that, with regard to children under fourteen years of age, the bill of the noble lord does not go far enough. They are of opinion that children under fourteen should not work more than eight hours a day."

On this recommendation, though not to its full extent, the present act was passed, children of thirteen only, and not of fourteen, as originally proposed,

Mr. Hindley. being included in its protection, and the noble lord (Ashley) was actually taunted with cruelty, for supposing that children of so tender an age could, by any possibility, be suffered to work ten hours a day. For the convenience of the trade, two years and a half were given before the bill came into full operation. And how have they been employed? In equalizing the hours of work? To some extent they have—the beneficial consequences of which are acknowledged in the reports. But the children have not received that protection to which they are entitled, through the evasions of the system. The return, laid upon the table of the House, of the number of children of each age who had been certified, completely proves this. According to the population returns, there appears to be a mere fractional difference between the number of children of thirteen and twelve years of age; whereas there appear from the factory returns to be from thirty to forty per cent. more children of twelve years of age than of thirteen.

In Lancashire they employ of 12 years of age . .	13,300
Whereas of 13 only	10,200
In Derbyshire of 12	518
of 13 only	391
And it was very singular that in Glossop (a well known place for overworking) the number of children of 12, 13, and 14 years of age employed in factories is	756
Whereas the number of those of 15, 16, and 17, is only .	394
In Yorkshire the number of 12 years of age is nearly . .	8,000
Whereas of 13 only	5,600
In Cheshire, the same result appears.	
of 12	2,600
of 13	1,880
In Scotland, of 12	6,036
of 13	4,494

From which fact it is impossible to avoid the conclusion, that a great portion of those certified to be twelve, are really below that age. In corroboration of this surmise, I shall state a fact to the House which occurred during a recent visit to a mill in the neighbourhood of Glasgow. Mr. Horner, the inspector, was examining the children at the time; and on inquiring for the certificate of a boy apparently not twelve years old, he received one stating him to be twelve and a half. His younger brother then came forward, and his certificate attested him to be twelve! The parliamentary returns present a similar result with regard to young persons of eighteen years of age, who are allowed to work more than twelve hours per day.

In Lancashire, the certificates of 17 years of age are .	6,503
of 18	8,463
of 19	6,772
In Derbyshire, of 17	226
of 18	437
of 19	268
Of which there are, in Glossop, of 17 . .	165
of 18 . . .	354
of 19 . . .	198

Now, it appears exceedingly singular that there should be so many young persons of the age of eighteen in Glossop; a case, however, which was brought before the magistrates against one of the large manufacturers there, for working children, elicits a simple explanation. On examining the certificate-book of that gentleman (a constituent of the honourable member oppo-

Mr. Hindley.

site, Mr. Gisborne, but whose name I will not mention, being anxious to avoid all personalities—a gentleman, however, whose character for perseverance and industry is well known in the district in which he lives) it was found that it contained no fewer than sixty-six cases in which the children of the ages of twelve, thirteen, fourteen, and fifteen years, and so certified in 1834, had in 1835 been all certified as being of eighteen years of age. I will not detain the House with the returns from other parts of the country, which all exhibit the same remarkable majority in favour of eighteen years of age; but I would advert to the fact that the returns regarding children employed in silk-mills, to which the restriction of twelve years of age does not extend, do not exhibit the same wonderful difference with respect to the numbers of the different ages employed. In Cheshire the returns state, that there are employed in silk-mills, of

Children of	12 years of age	790
	13	791
In Leek	12	117
	13	118
In Scotland	12	51
	13	52

With these facts before us, it is impossible to avoid the conclusion that the act is ineffectual for the protection of the class for whose benefit it was stated to have been passed. Indeed, Mr. Horner himself states, in his report, that full one-third of the children employed in his district, with certificates of their having the ordinary strength and appearance of twelve years, have obtained them by deceptions practised upon the surgeons. With all this proof of the ineffectiveness of the present bill, it might have been expected that the right honourable gentleman would have come down to the House to enforce, and not to repeal, the protection given the children. But when I put the question to him on the introduction of the measure, he said it was not his intention to do so, but merely to place the act as it stood upon the 1st of March last. And upon what grounds, and on whose part, does he take this course? Have there been any petitions presented in favour of this measure? No! But this day alone numerous petitions have been presented against it; one signed by 35,000 from Manchester, by 15,000 from Glasgow, by 9000 from Leeds, by 8000 from Stockport, by 4000 from Ashton, by 3000 from Warrington. Surely the noble lord at the head of the Home Department, who called upon the House not to vote for the motion of the honourable member for Southwark (Mr. Harvey) on the pension list, because there were only two small petitions in its favour, will give us the weight of his argument, and say to his right honourable colleague—"The House cannot recognise your proposition, unsupported as it is by any petition, and opposed by the loudly expressed desires of the great body of the people." Oh, but, says the honourable member for Bolton, though there have not been petitions there have been memorials and deputations. True! The parties, whose interests the right honourable gentleman advocates, have had recourse to another mode of proceeding. Instead of publicly making known their wishes to this House, they have privately memorialized the Board of Trade; and, in acceding to their wishes, I cannot help thinking that the right honourable gentleman stands in a position which he ought not to occupy—as the advocate of the masters, instead of the arbitrator between the two parties. But perhaps the right honourable gentleman will say, that it is not upon these memorials, but upon the reports of the inspectors, that he has been induced to bring forward his bill. If so, I ask him, why he does not adopt other alterations they propose, equally important? Why does he not take the advice of Mr.

Mr. Hindley. Rickards, and do away with the distinction between two classes of children, which that gentleman says cannot be observed?

"I believe, too, that the limitation of one class of children to a certain number of hours, and another class to another, in the same mill, can never be put in force by legal or official means. Evasion is so easy in the interior of mills, and detection so difficult, that when private interests combine, the vigilance of public officers, if not always on the spot, may, and will be, continually defeated."

Why does not the right honourable gentleman do away with the education clauses, which are, on all hands, allowed to be impracticable?

Mr. Horner says, "I have found a universal feeling that the education clause of the act, as it now stands, is utterly impracticable; and I am persuaded, that if it be attempted to put it in force, the manufacturers, however reluctantly, will instantly dismiss every child liable to the restrictions."

And Mr. Saunders, in regard to this clause, says, "The machinery (if I may adopt the term by which the schooling is attempted to be enforced) is universally considered to be the great error of the act; and its effect will render it impracticable, at any rate, for general purposes, to employ such children, whilst the occasional absence or truancy of a child involves the master in a heavy penalty, either by losing the child's labour, or for employing it illegally."

And, above all, why does the right honourable gentleman not limit the most unconstitutional powers which the inspector at present possesses, and which Mr. Rickards himself declares to be contrary to principle?

He says, "Offences, in nine cases out of ten, perhaps ninety-nine out of one hundred, are only likely to be brought to light by the visits, or by the vigilance, of the inspector himself, or his own superintendents. In such cases he might be placed in the awkward situation of being informant, witness, and judge, in the same cause; or judge, when the informant and only witness may be his own servant, acting under his own directions. The inspector is also authorized to make rules, orders, and regulations, which, on due notification and publicity, have the force of law: he is, therefore, a legislator for the mills within his division, and consequently not the properest judge that could be selected to decide on alleged breaches of his own rules."

Why, if the right honourable gentleman professes to proceed upon the reports of the inspectors, and not upon the representations of the masters, does he not propose these and other alterations suggested by the official gentlemen? The reason is evident. He is anxious to relieve the manufacturers from an immediate and pressing difficulty, and he fears, lest, by introducing other alterations, public attention may be directed to the subject. I implore him to alter his course. Let me advise him to open up the whole question of factory labour, and to introduce such a bill as will be practicable—not injurious to the masters, and just and beneficial to the operatives. In doing this, I shall be happy to lend him my humble but hearty assistance, and, by that means, contribute to settle this great and important question, once and for ever. But to his present proposal, I feel it my duty to give my strenuous opposition; for I am quite at a loss to tell what new arguments have been or can be advanced, which were not known before the passing of the present act. Are we now to be told that parents and masters would concur in endeavouring to obtain, the one as much money, and the other as much work, as possible out of the children? This had been long ago proved from the commissioners' report.

Mr. Drinkwater says, "The smaller children of course pay over to their mother their week's wages, but the parents seem to have devised a scheme by which the children delude themselves with the idea that the produce of their hours of extra labour are more peculiarly their own."

Mr Hindley.

And Mr. Power, another of the commissioners, says,—"Upon the whole, there remains no doubt on my mind that, under the system pursued in many of the factories, the children of the labouring classes stand in need of, and ought to have, legislative protection against the conspiracy insensibly formed between their masters and parents to tax them with a degree of toil beyond their strength."

And again:—"Up to a certain period of life, the children are absolutely dependent on their parents for support; and before that period it is that a tyranny is often imposed upon them beyond their physical powers of endurance."

And the central board state in their report, "That it appears that parents encourage their children to make the extraordinary efforts, of which we have given some examples, by leading them to consider the wages they thus earn as peculiarly their own, although a cheat is often practised upon them, even with regard to these extra wages. While all the witnesses agree in the statement, that whatever the child earns by its regular hours of labour is uniformly appropriated by the parent, it appears that a large portion of the additional wages earned by extra hours is also taken by the latter."

Nor is this desire on the part of the parents to derive a profit from their children at all times dictated by necessity, as the honourable member for Wolverhampton would have us to believe. I have myself known instances in which men, who were earning 30*s*. or 40*s*. per week, were extremely anxious to have their children introduced into the mill before the age permitted by law. "And why not?" ask some honourable gentlemen. "It does the children no harm; they are as happy and as healthy as children can be."

The noble lord, the member for Dorsetshire, has read to the House extracts from the medical evidence, which might lead to an opposite conclusion. I shall, in addition to those extracts, only trouble the House with the result of the careful examination made by Dr. Hawkins, of two schools in Manchester.

The Doctor says,—"In order to ascertain the state of health of the youthful factory classes, compared with youth in other conditions, I made a careful examination of the Bennett Street Sunday-school, at Manchester, in which abundance of all trades exists. I accordingly took an account of 350 of both sexes not engaged in factories, and of 350 of both sexes engaged in factories. Of the former several remain at home and do nothing; some are in service; some are dress-makers; some engaged in warehouses and shops. Their age varied from nine years to twenty for the most part. Of the 350 not in factories, 21 had bad health, 88 had middling health, and 241 had good health; but of 350 in factories, 73 had bad health, 134 had middling health, and 143 had good health.

"Again, at the St. Augustine's Sunday-school, at Manchester, I compared 50 boys engaged in factories with 50 boys not in factories; some of whom lived at home doing nothing, while others were engaged in shops and in various trades. Of the 50 not in factories, one had bad health, 18 had middling health, and 31 had good health; but of the 50 in factories, 13 had bad health, 19 had middling health, and 18 had good health.

"It will be seen that the advantage of health is at least double at these institutions on the side of those young people who are not engaged in factory work. The information afforded towards this comparison by the registers of sick clubs or benefit societies is not conclusive, since these sick clubs usually contain all classes indiscriminately. The average quantity of illness for every member of the Bennett School sick club during the year 1832, was one week, one day, and six hours; but about one half of this club is composed of youth not engaged in factories. It appears to me, also, that factory children are

Mr. Hindley. usually very slow in coming on the sick list of these clubs; they usually go on working to the last possible moment, so eager are the parents to secure their wages!"

The weight of this evidence, however, is now attempted to be shaken by the masters, in what is called an authentic document, which has been sent round to honourable members—being the

Statement of a Sick Fund connected with a Sunday-School at Bolton.

	Subscribers.		Relief.								Deaths.	
	Cotton Mill Operatives.	Other Trades.	Cotton Mill Operatives.	Amount.			Other Trades.	Amount.			Cotton Mill Operatives.	Other Trades.
1833	284	268	58	45	14	6	89	75	17	3	3	3
1834	277	272	42	32	10	3	95	75	17	1	1	5
1835	283	244	40	32	3	0	75	51	19	0	0	1!!

Why, truly, there is no new thing under the sun. Similar statements were made twenty years ago, and in answer to one of them, the right honourable member for Tamworth indulged in a strain of irony so good humoured and so pointed, that I will venture to read it to the House:—

"The instances produced from the evidence were certainly strong enough to support the most unqualified of the assertions which had been made as to the healthiness of cotton-mills. One of the instances was that of a mill at Glasgow, in which, he believed, an honourable gentleman opposite (Mr. Finlay) was concerned. It was given in evidence, that in this mill, 873 children were employed in 1811, 871 in 1812, and 891 in 1813. Among the 873 there were only three deaths; among the 871 two deaths; among the 891 two deaths; being in the proportion of one death in 445 persons. So very extraordinarily a small proportion had naturally excited the astonishment of the committee, and, therefore, as was to be expected, they questioned medical gentlemen as to the proportion of deaths in different parts of the kingdom. When this statement was shewn to Sir Gilbert Blane, he expressed his surprise, and observed, that if the fact was not asserted by respectable persons, he should not believe it; and being asked why he distrusted it, he said, that the average number of deaths in England and Wales, was one in 50 (in 1801 there had been one in 44). There were favoured spots certainly, Cardigan, in which the deaths were as one in 74; Monmouth, in which there was one in 68, Cornwall, one in 62; and Gloucester, one in 61; yet, in cotton-factories, they are stated as one in 445! In one of Warton's beautiful poems, which begun with these lines:—

Within what mountain's craggy cell
Delights the goddess, Health, to dwell?

After asking where the abode of this coy goddess was to be found, whether on 'the tufted rocks' and 'fringed declivities' of Mortlock, near the springs of Bath or Buxton, among woods and streams, or on the sea-shore, it certainly would have been an extraordinary solution of the perplexity of the poet, if, when he inquired—

In what dim and dark retreat
The coy nymph fixed a fav'rite seat?

it had been answered, that it was the cotton-mills of Messrs. Finlay and Co., at Glasgow; yet such was the evidence respecting this mill, that its salubrity appeared six times as great as that of the most healthy part of the kingdom. This was the sort of evidence which had been brought to disprove the evidence of disinterested persons, of medical men, and even of persons who had an interest opposed to the measure before the House." Mr. Hindley.

"But," says the honourable member for Leeds, "factory children are so well fed, and so well clothed"—an assertion which I cannot help contrasting with another, made in a letter to Mr. Rickards by Mr. Harrison, a surgeon of Preston, whose authority must be conclusive, as he states he had measured and weighed upwards of 1200 factory children.

"If factory children were as well fed and clad as other children, and if their abodes were as cleanly and as well ventilated as those children employed in other branches of labour, I believe that few employments would be found equally healthy."

But, sir, I would appeal to higher considerations than those which have reference to these children, as mere animals, brought into the world for the sole purpose of spinning thread and weaving calico. Surely we ought to regard them as moral agents, endued with powers of mind, in the improvement of which the whole of our social system is materially interested. To what is it owing, that our machinery is brought to such a high state of comparative perfection? To the effects of mind. And if you give the individuals connected with its daily operation time to combine with the exercise of physical power the workings of mental ingenuity, who can tell what new triumphs of genius, what new developments of the resources of our country, may not be manifested?

Under the present system, you crush all the exertions of men of science and benevolence. It is in vain, that to young people working twelve hours a day, who, in addition to the hours consumed in eating and sleeping, have only one spare hour, we offer the advantage of mechanics' institutions or village libraries. They are worn out with fatigue, and cannot read; and if, in consequence, they form habits of intemperance, who can wonder, and upon whose head will fall the responsibility? Ought it not to rest, in some degree at least, upon this House, if it sanction a proposition by which the children shall be deprived of the opportunities of education? And is it at this time of unexampled prosperity that we are to be called upon to make this sacrifice of the intellect and the physical energies of our youthful population? I had intended, sir, to have considered the argument respecting foreign competition; but seeing the evident desire of the House to come to a division, I shall not enter upon it further than to say, that since the first factory bill was passed in 1819, the import of cotton has more than trebled, and though the mill belonging to the honourable member for Manchester, which spins for the Russian market, has been reduced by law in its hours of working from seventy-seven to sixty-nine hours per week, the export of yarn from Russia has increased from 4,500,000*l.* to upwards of 18,000,000*l.* per annum, notwithstanding the reduction of hours, and the protective duty of 6*d.* per pound, which the Russian government has imposed upon all yarn imported into that country.

Mr. BOLLING.—We, who advocate a moderate system of legislation, are pointed at as persons who are acting contrary to the dictates of humanity; but this I deny; and I know it will be believed in the district to which I belong, that there is nothing I say here, which I shall not be prepared to assert, maintain, and answer for, there. The honourable gentleman who has just sat down has adverted to the gentlemen who came to London as a deputation, charging them with a want of sufficient boldness; but I say that such an accusation comes but with a bad grace from him as a manufacturer, and well acquainted with those gentlemen, who certainly are not liable to the Mr. Bolling.

Mr. Bolling. same imputation that he himself is, who, reading the law, and knowing it, yet acted in violation of it,—as is shewn by the evidence on the table, which I take it for granted is true, a circumstance to which I certainly would not have alluded, but for the charge of insincerity which he has thought proper to make against his Majesty's government: (an accusation which, if just, cannot apply to myself; because, not approving of it, I was not among the advocates of the measure of 1833, and I stated my opinion upon the subject to that effect, when it was before the House.) Since that time, an improving spirit has sprung up in the consideration of this measure; and I rejoice to see the calm and dispassionate manner in which it is now discussed, in the temper and spirit of which I, also, shall refrain from charging any man with want of humanity, because he does not pursue his object in the manner I do myself.

I will not refer the House to the evidence of the medical men: the doctors differ on this subject, as they do on most others; but I will refer to the sequel of the evidence laid upon the table of the House three years ago, and widely circulated amongst the honourable members of this House. I consider this a document against which mere theory cannot stand; it comes from the most unquestionable source, and it shews the condition of the population of the borough I represent, which may be taken as a fair specimen of the state of other manufacturing districts. In legislating on behalf of the children, you have gone beyond your own intention. It was admitted in debate that they had a right to be raised to an equality with the other residents in the community,—and have you done so? You have raised their condition 100 per cent. above their fellows, and—all the while led away by false humanity—those who do it believe they are conferring the greatest benefit upon them in so doing; instead of which, they have merely driven them from one trade to another, which is worse for their health, and worse for them in every other respect than that from which they have been driven. This being the case, I call upon the House to make a stand; if you do so you will attend to the wants of those for whom you are legislating. All the operatives want is some restriction; they are satisfied that the factory labour does not injure a child of ten years; the document I allude to proves it beyond disputation. It is for the House and the country to consider whether they will continue a restriction which has not been acted upon hitherto, and which, if it is, will throw 35,000 children out of employment. I cannot vouch for the correctness of this number upon the whole; but it is a pretty good proof of it, when I state that the concern to which I belong, employing 1400 hands, will be obliged to discharge at least 250 of them if this act is enforced; and where will those persons go? Is it for their own benefit that they will be deprived of employment? Not for the benefit of their health, certainly, for no trade can be more healthy than the cotton trade is; and with respect to education, if the bill be acted upon, I cannot say what may be done under the eye of the inspectors, but I will undertake to say, that if the honourable member who spoke last, will erect a Sunday-school in his district on the same plan as in my own, he will soon prove the truth of the statement contained in the documents to which I refer; and if he do not, it must be because he does not wish to confirm them. The honourable member, alluding to a meeting at Bolton, asked, where are the clergymen of the district to be found? Not here, said he, because there are no preferments to be obtained; but I will tell him that the clergyman of my parish is always on the spot; he goes every morning to the infant school, in which there are no less than 150 children; he regularly attends and delivers a lecture at the workhouse once in every week; he attends a parochial meeting every Saturday at the library, for the purpose of seeing that proper books are distributed; he attends the savings' bank on the same day, so that he is pretty well qualified to judge of the state of the parish with respect to education; and I appeal to the honourable gentleman below me, whether the children in the Sunday-school, perhaps receiving no other education, are not as well qualified to read their bible, the only true source of all true education, as any others? All education beyond that is a matter of opinion; and some, I cannot help saying, appear to me to be disposed to fill the minds of the working classes with a species of knowledge which is like giving people drams instead of good sound home-brewed. I am disposed to support the measure of the right honourable president of the Board of Trade, because I think it is the better of the two. We must have a relaxation of the present bill—the inspectors say the same. The only question then is, how shall we retreat? Why, if we have taken a false step, let us acknowledge it fairly,—let us repeal the clause, and the best results may follow. It is better that the children should work under the care of the parents; and if you prevent this, you do an injustice both to the parent and to the child. If you look the matter in the face, I have no doubt that you will soon be able to bring it to a fair and satisfactory settlement. Mr. Bolling.

MR. JOHN FIELDEN.—I cannot suffer this debate to close without offering a few observations to the House; for if there be any subject discussed in this assembly, which I am acquainted with, it is this one, as I am so extensively engaged in manufactures myself, that I cannot but thoroughly understand the necessities and condition of the working people employed in them. The honourable member for Bolton has boasted of the superior condition of the factory hands of Bolton over other classes living in that borough; but before I listen to the comparison, I must know what is the condition of the people with whom the honourable member compares the factory hands. Now, I happen to recollect that a committee of this House sat for two years to examine into the condition of the hand-loom weavers, and as a member of that committee I can assert that it was proved that one-half of the inhabitants of Bolton were hand-loom weavers, and, moreover, that they and their families lived throughout the year upon an average sum of $2\frac{3}{4}d.$ a head per day. Mr. John Fielden.

Several HONOURABLE MEMBERS.—Oh! Oh!

MR. JOHN FIELDEN.—Why—I know this to be the fact; it was proved to the satisfaction of the committee. Let honourable members contradict it if they can. But if this be the case, what becomes of the comparison? It is good for nothing; for it is only comparing the factory hands of Bolton with a large body of the poorest part of the community. This says nothing, therefore, in favour of factory labour at Bolton; and I say that so futile a comparison ought not to sway the House for a moment. Now I put it to honourable members whether they have not heard enough this evening to determine them to resist the bill of the right honourable president of the Board of Trade. That right honourable gentleman has said that 35,000 children will be thrown out of work if this bill should not pass. I am convinced that not 35 would be thrown out of work by throwing out the bill. I feel satisfied of it as a manufacturer. Again, the House is told that the manufacturers would suffer by yielding to the noble lord's amendment. This is the worst appeal that could be made to the House; for I am sure that if there is a spark of humanity in it, the House will never set private interests against the life and happiness of these poor little over-worked children. At any rate, I, as a manufacturer, and a large one, too, will say that I would throw manufactures to the winds rather than hesitate upon such a point for a moment. But the House will do its duty; it will not drive back these 35,000 little children to labour incessantly for twelve hours a day, in

Mr. John Fielden.

addition to the time they often have to spend in walking to and from their homes. It surely will not do this.

Allusion has been made, in the course of this debate, to the statement of Mr. M'Williams as to the number of miles that a little child has to walk in a factory. It was stated at fifteen miles. Now, I recollect that, at a meeting of members of parliament and operatives, at Manchester, last December, similar calculations were brought under the attention of myself, and the members of parliament present, by the operatives. One made a statement shewing that a child in one mill walked twenty-four miles in the day, merely walking after the machine. I was surprised at this statement, and I observed that few could believe it possible. I, however, was not satisfied until I had tried its correctness; and therefore, when I went home, I went into my own factory, and, with the clock before me, I watched a child at her work, and having watched her for some time, I then calculated the distance she had to go in a day, and, to my surprise, I found it to be nothing short of twenty miles. Talk to me of "lightness of factory work" after this! It is monstrous. And yet it is this system that the honourable member for Bolton wants to perpetuate. I, on the contrary, call on the government to enforce the present act, and not attempt to repeal it; to send more inspectors down into the cotton districts, and to have it put in force rigidly. All the work-people work harder than they ought to do; but the children are unmercifully treated. The inspectors have given their opinion as to what quantity of work a child can bear, and they have referred also to medical men in the districts where they are employed. But I should like to know what is the value of such evidence as this, collected by the inspectors, when compared with that of such men as Sir Anthony Carlisle, Dr. Farre, Mr. Green, Dr. Blundell, and the other eminent men who have pronounced our factory system to be nothing short of infanticide? There has never yet been an efficient act of parliament on this subject, and if I had to bring in one, it should not allow more than ten hours' labour for any age in factories. We are reminded of foreign competition. I believe it to be the greatest humbug in the world. But this is "political economy." Now, we have been warned by one of the able physicians examined by the committee, that we have no right to trench on "vital economy" to support "political economy;" and I say, that if this House should pass this bill, and make these poor little children go back into slavery, then it will want another reforming. But what do the political economists say? When the noble marquis the member for Buckinghamshire, brought on his motion about agricultural distress, they said it would not signify if England did not grow a bushel of wheat or barley, so prosperous were manufactures, and so completely were we independent of the land; and yet it seems that we cannot go on, in manufactures, without working these poor little children for twelve hours a day. That is our prosperity! My fear is, that this is only a beginning of a total repeal of Lord Althorp's act, and that the right honourable gentleman will bring in a bill next year to repeal that act. I know that the present act is inconvenient, but it does, in some measure, save the little children. It is inconvenient, but it is not impracticable, as it is represented to be. I know it—for I observe every clause of it myself, schooling clause and all. But I always knew that the "relays"—as they are called by those who speak of the working people as they speak of cattle—could not be had. My opinion, however, as a manufacturer, and one possessed, I trust, of some feelings of humanity, is, that Lord Althorp's act ought to be maintained and enforced; and if it be repealed by this night's vote, I will go on working for a ten hours' bill, and will never cease to advocate such a measure while I have life. We have always been told that shortening the hours of labour of these poor children would ruin us.

When the hours were ninety hours a week, eighty-nine hours, seventy-two hours, sixty-eight hours—at every time when we wanted them reduced —our opponents have said the lessening the time would ruin us; and yet we have shortened them: and we are told now that we are more prosperous than ever! Then I ask the House, whether such predictions are not false? and I implore it to adopt the amendment of the noble lord. Mr. John Fielden.

Sir ROBERT PEEL.—I wish to separate the appeal which the honourable member for Oldham has made to the reason and deliberate judgment of the House, from that which he has made to its passions—an appeal, I must say, however, which, standing as the honourable gentleman does, free from the imputation of all interested motives, came from him with peculiar grace. There is no speech that has been delivered this night, however creditable it may have been to the honourable gentleman who pronounced it, that has had so great an effect in convincing me of the impropriety of acceding to the amendment, as the speech of the honourable member for Oldham, unless, indeed, it was that of the honourable member for Ashton. What is the result of the argument of the honourable member for Ashton? A bill passed into a law in 1833, which provided that children under the age of thirteen should not be employed more than eight hours a day in cotton or other factories. That bill assumed that there could be two relays of children, and that, therefore, there might be working in the factory during sixteen hours a day. The honourable member for Ashton, however, has informed the House (and I understood the honourable member for Oldham to concur in the truth of that statement) that the system of relays is impracticable, and therefore the law does prohibit the working in factories for more than eight hours a day. I came down to the House perfectly unfettered as to the course I might pursue with respect to this question. I have seen many parties, but have uniformly refused to give any pledge with respect to my vote; but the speeches of the two honourable members to whom I have alluded have done more to convince me of the impolicy of allowing the existing law to remain unamended, than all the arguments advanced by the right honourable gentleman who introduced the bill before the House. I believe that legislative interference, with respect to this subject, is necessary. I think that, in this case, we cannot trust to the natural affections of parents. The object which we ought to aim at is this—to regulate the hours of labour in factories so as, on the one hand, to prevent the undue working of children, and, on the other, to avoid imposing restrictions which would operate as a check upon that important branch of the national industry. Sir R. Peel.

I think the right honourable president of the Board of Trade ought not to have confined his bill to the single point to which it refers, because the law has been proved to be inoperative in other respects. By voting for the bill, however, I reserve to myself the right of proposing other amendments in committee. If, on the other hand, I vote for rejecting the bill altogether, the inference to be drawn from that act would be, that I was content with the law as it stands at present. Should I, after all, be acting as a friend of the children by refusing to alter the law? The House has been informed that, in the course of last year, 3000 false certificates were granted as to the age of children employed in factories, by which means the law was evaded, and children who were under the prescribed age were actually employed. The honourable member for Oldham wishes for a ten hours' bill, and I understood him to argue in favour of the existing law; not because it was good in itself, but because it was so restrictive and inconvenient to the masters, that he thinks they will at last be induced to consent to a ten hours' bill. Now, I ask, is this the way to make the law respected? This is the ground upon which the honourable and gallant member for Hull is prepared to legislate; he thinks the present law absurd, and yet he means to vote for its continuance.

Sir R. Peel. The honourable and gallant member alluded to the little difference which existed between *tweedledum* and *tweedledee;* but perhaps he will excuse me for saying that, in my opinion, he would have done wisely if he had acted the part of *tweedle "dumb"* upon this occasion. If the existing law is bad, here is a proposition for its amendment; for the title of the bill is, "An Act to amend the Acts 3 and 4 of the present reign, for regulating the labour of young persons employed in factories." That title will admit of any amendment of the law. It will admit of an amendment with respect to the granting of certificates. If that part of the law which requires the production of certificates as to age be worth anything, it ought to be enforced. At present no penalty can be inflicted upon a person who may grant a false certificate, unless an information be laid within fourteen days from the commission of the offence. The inspectors report, that under this restriction it is almost impossible to visit the offence with punishment. After reading the reports of the commissioners, and seeing the manner in which the law is violated, it is impossible, on the score of humanity, to leave the law in its present state. I must say that the inspectors, judging from their reports, appear to be disinterested witnesses, and to give their evidence free from any undue bias. One of these gentlemen, Mr. Horner, states, and he is confirmed by the testimony of Mr. Howell, that, through the abuse of certificates, the provisions of the existing act were constantly violated by the employment of children in factories under the age limited by the statute; and that new provisions might be devised for obviating these infractions of the act.

Now I appeal to my noble friend (Lord Ashley), for whose intentions I entertain the utmost respect, and I ask him whether he is content to leave this law in its present state, when Mr. Horner says that an amendment of it would prevent its being evaded. I know that it is necessary to take some precaution against the cupidity of parents, and I think that that might be done by a more simple law than the present. I think it would be preferable to the existing law to declare that no children below a certain age should be allowed to work in factories, (making provision at the same time against the granting of false certificates,) and then to prescribe the number of hours during which all persons above that age should work. By sanctioning the system of relays of children, we hold out an inducement to adults to over-tax their strength, for they will then work sixteen hours a day. Then, again, how would you extend the system of relays to remote districts. There are a great number of mills in England not in the vicinity of large towns, and which do not work by steam, but avail themselves of water-power. Suppose one of these with a hundred hands, and forty children, how could you, I ask, extend the relay system to it? Your restrictions here give you no security, for the law cannot be acted on. Why, then, preserve a clause which, if you fail of carrying it into effect, may, in consequence of its frequent violation, bring into disrespect and contempt those which might otherwise prove operative? Supposing the system of having relays of children limited to working eight hours each were adopted, would not a premium be held out to adults on whom no such restriction was imposed, to tax their powers to the utmost, in order that, by working double, they might be able to procure that quantity of employment which their wants might require? Whilst, therefore, you profess to support the children by compelling the masters to adopt the system of relays, you will do an act of great injustice to the adult labourers. Then with regard to schools, as they would be affected by the relay system. Is the school to be connected with the factory? If not, it must be but a short distance from it. Is the school to be opened during the entire time of labour? If not, you have made no corresponding provisions for the hours of work and the hours of education. As I understand the proposition of the right honourable gentleman, it is this: that children under twelve years of age shall continue subject to the present law, (or, in other words, that their period of labour shall be eight hours,) but that children above twelve years may be called on to work for sixty-nine hours a week, or twelve hours for five days, and nine for the sixth, being Saturday. I think that even with this amendment the law would be in an unsatisfactory state; but feeling that it is quite impossible to please all parties, and that the cure is not quite as bad as the disease; feeling, also, that the existing law is not consistent with humanity, I am willing to entertain the right honourable gentleman's proposal. Sir R. Peel.

Then, one word as to the danger of foreign competition said to be likely to result from the existing law, and urged as an argument against it. The danger from competition is a perfectly good and tenable ground for the reduction of the duty on cotton wool; but it does not form a good or valid reason for endangering the health of factory children, and no dread of foreign competition would induce me to vote for any measure which would have that result. I am certainly opposed to the adoption of any severe restrictive laws on labour, as they are, in my opinion, calculated to undermine the commercial energies of the country, and thereby to strike a blow at the happiness and comfort of the people; but when I am asked whether I will resist any attempt at amendment, or whether I shall support the present law, my answer is, that I am in this instance so convinced of the necessity of amendment, as to feel it my bounden duty to support the bill of the right honourable gentleman.

Lord F. Egerton.

Lord FRANCIS EGERTON.—I was not a member of this House when the law which it is now proposed to amend was passed; and I came down to the House with my mind perfectly unprepossessed upon the subject. The law which it is intended to amend was the result of a compromise between the manufacturer and the operatives, and until the experiment had been tried, it would not have been fair to interrupt it after the long and laboured discussion which the subject underwent in this House. The experiment, however, has been tried, and the law has proved a failure. I certainly feel a difficulty in the vote which I am about to give; but on a question which has been so much canvassed, it is impossible not to come to some conclusion. I should be sorry to join in the censure of those who feel inclined to throw discredit on the course pursued by his Majesty's government, because I think the right honourable gentleman opposite (Mr. Poulett Thomson) must, from his position with relation to the government, be possessed of sources of peculiar information. The question is one beset with difficulties, which meet us whichsoever way we turn. I have listened with the utmost attention to the speech of my honourable friend, the member for Bolton (Mr. Bolling); but his argument appears to be levelled as much against the bill generally as against the particular clause. He admitted that there could be little doubt but that the last bill was inoperative; yet he added, that the clauses of this measure would be found equally so. It has been used as an argument during this discussion, that the present measure, if passed into a law, will affect the position of a great number of persons, which number has been variously stated at 20,000 and at 35,000. Why, it must necessarily be so with any similar act of legislation, upon any subject whatsoever. I should be glad to find in the measure some substitute for a domestic education, for I should look upon such a provision as highly valuable; and I entertain—and I must express—my hope, that some mode will be adopted, as well by means of establishments based upon the principle of infant schools, as by procuring an education for persons more advanced, of accomplishing that object. I shall not, at this late hour, enter upon that part of the question, but I would strongly press upon government the necessity of devising some plan for carrying this object into

Lord F. Egerton.

effect. I shall not trespass longer on the time of the House, except to make one observation with respect to what has been said as regards competition. I must confess that, in this point, I am not without my fears, when I perceive that the growth of the cotton manufacture in America has proceeded with a rapidity almost equal to our own, and that in France, although the progress has not been so great, it is still but little short, and quite sufficient to make me deeply apprehensive. I therefore think that in any legislation by which our manufactures are likely to be effected, we should keep the increased production of foreign countries closely in view.

Mr. Finch.

MR. FINCH.—I have listened with great attention to the arguments of the right honourable baronet, the member for Tamworth; and although I concur in many of his observations, I differ with him as to the conclusion to which he comes as respects children employed in factories. The arguments which I have heard, so far from making in favour of the measure, have, on the contrary, induced me to negative the motion. What does the proposed bill affirm? Why that children of the tender age of twelve years shall be liable to be laboured through five days of the week for a period of twelve hours a day. This would be to affirm a monstrous doctrine, which the House, in the just performance of its duty, is rather bound to negative than assent to. At the same time, however, in setting this bill aside, it should be clearly understood, that some more efficient measure will be brought forward at as early a period as possible. If we suffer it to go abroad, that this House is careless of, or has no sympathy with, the children employed in factories,—the tocsin of alarm will be sounded, meetings of delegates will take place, and a cry will be raised throughout the entire country; but by negativing this bill, and exhibiting a disposition to bring forward a more efficient measure, this evil will be avoided.

Mr. Goulburn.

MR. GOULBURN.—I concur, in some degree, with the arguments used by my right honourable friend the member for Tamworth, but not to the extent which would induce me to vote with him on this occasion. I must confess that, in voting in opposition to my right honourable friend, I have a great distrust of my own judgment; but after having heard the discussion which has taken place on this subject, I feel that I have no alternative left but to give my vote for the amendment. I shall briefly state the reason why I shall vote so. I concur in thinking, with the right honourable baronet, that the bill of 1833 requires great and considerable amendment; and if the bill now before the House were calculated to effect that object, I should willingly give it all the support and assistance in my power; but when I am called upon to amend that bill of 1833, by the introduction of another which has, coupled with it, a clause subjecting to twelve hours' labour children under thirteen years of age, the question to me takes a totally different shape. I cannot assist in carrying forward the measure; for, in my mind, the clause which the present measure is intended to repeal in the bill of 1833 is not the one which requires amendment. I look upon it that the amendments are required in other parts. I do not think it advisable, under any circumstances whatever, to admit children from twelve to thirteen years of age to undergo that degree of labour which is sanctioned by the measure of the right honourable gentleman. After what has been stated by gentlemen connected with factories, and the medical opinions which have been quoted to the House, I find my own opinions more strongly fortified. We have heard that many frauds have been committed, notwithstanding the clause in the existing act. Admit that there have been; still this bill would only open the way more widely for subjecting children to a degree of labour beyond their capability of endurance. It has been also said, that certificates have been forged to make children of twelve years of age pass for sixteen or eighteen. Surely such fraudulent certificates as these are easy of discovery; but let the clause be repealed, and I ask whether it is not as easy to certify—indeed much more easy than the other, and with less liability to detection—that a child considerably under twelve years is above that age. I have stated the grounds of my vote. I am prepared to consider and amend any defects in the factory act, but I cannot consent to twelve hours' labour for children. Mr. Goulburn.

Several HONOURABLE MEMBERS.—Question! Question!

Mr. WAKLEY.—I feel an intense and deep interest in this subject, and I trust the House will indulge me whilst I say a few words. If I am not heard, I shall certainly move that the debate be adjourned. I am always reluctant to press myself upon an unwilling House. The measure has been considered medically, as it affects humanity, and as it affects our commercial interests; and I think after it has been discussed in those various lights I might be indulged for a few minutes. I shall just call the attention of the House to two or three sentences in the report of one of the inspectors, and I think I can shew that there has been a conspiracy to defeat this act; of which conspiracy we now, in the bill before the House, perceive the climax. I find Inspector Saunders stating that, in many instances, he found the act had been violated. What does he do? Does he visit the violation with punishment? Does he sue for the penalty? No; he contents himself with saying, that he would not abstain long from doing so, if he found the offence repeated. This warning had, of course, but little effect; for he says that, after repeated warnings—mind, repeated—there were two mill-owners who were constantly in the habit of working the children more than sixty-nine hours in the week; and then he says he had no other resource but to lay informations. From this we must presume, that there were several instances in which he did not feel himself compelled to inform the superintendent. It was said, in reply to a question as to whether or not the efficiency of the factory act had been tried by endeavouring to enforce its enactments, that indeed the provisions were of such a nature as to render it impossible to enforce them. The whole of the inspectors stated that the bill would not be carried into effect; but where is the proof that it would not? I have listened to the debate with great attention, but I did not find any such proof; and must we now, at a time that these poor factory children are beginning to hope for some relief, and that the act would be made operative,—must we throw them back into the position in which they were in 1833? This is just as I predicted it would be. The inspectors neglected to do their duty, and the blame of their negligence is thrown upon the act. From beginning to end, the whole is a conspiracy to defeat the act; and I will never be a party to it by consenting to adopt this bill. Mr. Wakley.

Mr. POULETT THOMSON.—Certainly the right honourable member for the university of Cambridge has a right to oppose the bill if he objects to this clause; for this is its main feature, which alters the old act. I cannot consent to abandon this clause; but when the law is changed, I shall do all in my power to carry its provisions into effect. It is most repugnant to the feelings of government to know that there is a law in existence which, either through supineness or inability, proves inoperative. The committee recommended to us to take one age—namely, ten years—for all the children; but we objected to this, because we thought it would not be advisable, after all the discussion which the subject has undergone, to open up the whole question again. Now, the great inconvenience which followed from these restrictions appears to have been to the children themselves. The letter of Mr. Ashton, Mr. P. Thomson.

Mr. P. Thomson. which has been read in the course of the debate, shews that by the operation of this clause, 25,000 children have been thrown out of employment. This has not been denied by any person; on the contrary, it has been admitted and argued upon. Now, Sir, it has been said by an honourable member, that I introduce this bill more as the honourable member for Manchester than in concurrence with my official colleagues; but I will tell him that it is this which renders the subject so painful to me; and I think before he made a charge of that kind, he might have considered whether, putting all other feelings aside, it is an agreeable thing to any honourable member to introduce a bill in opposition to a petition signed by so large a number of his own constituents as the one which has been presented against this bill. If the bill go into committee, and the clause so much objected to (but to which I still adhere) be carried, I shall endeavour to introduce such provisions as will make the other measure operative, but I cannot recommend a re-opening of the whole subject.

The House then divided, when there appeared—

For the second reading . .	178
For the amendment . . .	176
Majority	2

The bill was then read a second time, and committed.

AN ACT

TO

REGULATE THE LABOUR

OF

CHILDREN AND YOUNG PERSONS

IN THE

MILLS AND FACTORIES

OF

THE UNITED KINGDOM.

ANNO TERTIO ET QUARTO GULIELMI IV. REGIS.—CAP. CIII.

[29*th August*, 1833.]

WHEREAS it is necessary that the hours of Labour of Children and young Persons employed in Mills and Factories should be regulated, inasmuch as there are great Numbers of Children and young Persons now employed in Mills and Factories, and their Hours of Labour are longer than is desirable, due Regard being had to their Health and Means of Education: Be it therefore enacted, by the King's most Excellent Majesty, by and with the Advice and Consent of the Lords Spiritual and Temporal, and Commons, in this present Parliament assembled, and by the Authority of the same, That from and after the First Day of *January* One thousand eight hundred and thirty-four, no Person under Eighteen Years of Age shall be allowed to work in the Night, (that is to say,) between the Hours of Half past Eight o'Clock in the Evening and Half past Five o'Clock in the Morning, except as herein-after provided, in or about any Cotton, Woollen, Worsted, Hemp, Flax, Tow, Linen, or Silk Mill, or Factory, wherein Steam or Water, or any other mechanical Power is or shall be used to propel or work the Machinery in such Mill or Factory, either in scutching, carding, roving, spinning, piecing, twisting, winding, throwing, doubling, netting, making Thread, dressing or weaving of Cotton, Wool, Worsted, Hemp, Flax, Tow, or Silk, either separately or mixed, in any such Mill or Factory, situate in any Part of the United Kingdom of *Great Britain* and *Ireland:* Provided always, that nothing in this Act shall apply or extend to the working of any Steam or other Engine, Water-wheel or other Power in or belonging to any Mill, or Building, or Machinery, when used in that Part of the Process or Work commonly called fulling, roughing, or boiling of Woollens, nor to any Apprentices, or other Persons employed therein, nor to the Labour of young Persons above the age of Thirteen Years, when employed in packing Goods in any warehouse or Place attached to any Mill, and not used for any Manufacturing Process: Provided also, that nothing in this Act shall apply or extend to any Mill or Factory used solely for the Manufacture of Lace.

Persons under 18 Years of Age not allowed to work at Night in the Mills or Factories herein described.

Persons under 18 not to work more than 12 Hours a Day.

II. And be it further enacted, That no Person under the Age of Eighteen Years shall be employed in any such Mill or Factory, in such Description of Work, as aforesaid, more than Twelve Hours in any one Day, nor more than Sixty-nine Hours in any one Week, except as herein-after provided.

Extension of Hours of working in certain Cases.

III. Provided always, and be it further enacted, That if at any Time in any such Mill, Manufactory, or Buildings, situated upon any Stream of Water, Time shall be lost in consequence of the Want of a due Supply or of an Excess of Water, or by reason of its being impounded in higher reservoirs, then and in every such Case, and so often as the same shall happen, it shall be lawful for the Occupier of any such Mill, Manufactory, or Building, to extend the Time of Labour in this Act prescribed at the rate of Three Hours *per* Week until such lost Time shall have been made good, but no longer, such time to be worked between the hours of Five of the Clock in the Morning and Nine of the Clock in the Evening: Provided also, that no Time shall be recoverable after it has been lost Six Calendar Months.

Providing for unavoidable Time lost in Cases of Accident.

IV. And be it further enacted, That when any extraordinary Accident shall happen to the Steam Engine, Water-wheel, Weirs, or Watercourses, Main Shafting, Main Gearing, or Gas Apparatus of any such Mill, Manufactory, or Buildings, by which not less than Three Hours' Labour, at any one Time, shall be lost, then and in every such Case, such Time may be worked up at the Rate of One Hour a Day, in addition to the aforesaid and herein-after restricted Hours of Labour, for the twelve following working Days, but not after.

Loss of Time from the Want or Excess of Water in the Day time provided for.

V. And whereas, during Periods of Drought and of Floods the Power of Water-wheels on some Streams is wholly interrupted, or so far diminished that the Machinery or Part or Parts of the Machinery dependent upon such Power cannot be regularly worked at one and the same Time, and in consequence thereof a certain Portion of the Time of such Persons as are employed in the Working of such Machinery may be lost in each Day during such Period of Drought or Floods; be it therefore enacted, That it shall be lawful for the Occupier of any Mill, Manufactory, or Building, when Time is so lost, then and in every such Case, and so often as the same shall happen, to extend the Hours between which Persons under Eighteen Years of Age are hereinbefore allowed to work, (*videlicit*, from Five of the Clock in the Morning till Nine in the Evening,) as herein-before limited, to such Period as may in such Case be necessary to prevent the Loss of Time, and no longer: Provided always, that no Child or young Person within the respective Ages prescribed by this Act shall be actually employed a greater number of hours, within the Twenty-four Hours of any one Day than this Act declares to be lawful; and provided also, that no Child under Thirteen Years of Age shall be employed after the Hour of Nine of the Clock in the Evening nor before the hour of Five in the Morning.

Time for Meals.

VI. And be it further enacted, That there shall be allowed in the Course of every Day, not less than One and a Half Hours for Meals to every such Person restricted as herein-before provided to the Performance of Twelve Hours' Work daily.

Employment of Children under Nine Years prohibited.

VII. And be it enacted, That from and after the First Day of *January*, One thousand eight hundred and thirty-four, it shall not be lawful for any Person whatsoever to employ in any Factory or Mill as aforesaid, except in Mills for the Manufacture of Silk, any Child who shall not have completed his or her Ninth Year of Age.

The Employment of Children under 11, 12, and 13 Years of Age for more than Eight Hours a Day prohibited.

VIII. And be it further enacted, That from and after the Expiration of Six Months after the passing of this Act it shall not be lawful for any Person whatsoever to employ, keep, or allow to remain in any Factory or Mill as aforesaid for a longer Time than Forty-eight hours in any one Week, nor for a longer Time than Nine Hours in any one Day, except as herein provided, any Child who shall not have completed his or her Eleventh Year of Age, or after the Expiration of Eighteen Months from the passing of this Act any Child who shall not have completed his or her Twelfth Year of Age, or after the Expiration of Thirty Months from the passing of this Act any Child who shall not have completed his or her Thirteenth Year of Age: Provided nevertheless, that in Mills for the Manufacture of Silk, Children under the Age of Thirteen Years shall be allowed to work Ten Hours in any one Day.

IX. And be it further enacted, That all Children and young Persons, whose Hours of Work are regulated and limited by this Act, shall be entitled to the following Holidays; *videlicet*, on *Christmas Day* and *Good Friday* the entire Day, and not fewer than Eight Half Days besides in every Year, such Half Days to be at such Period or Periods, together or separately, as may be most desirable and convenient, and as shall be determined on by the Master of such Children and young Persons: Provided nevertheless, that in *Scotland* any other Days may be substituted for *Christmas Day* and for *Good Friday*, both or either, as such Master may determine.

Holidays to be allowed.

X. And be it further enacted, That if any Child within the Age herein-before restricted to Nine Hours of Day Labour shall have been employed in any one Day for less than Nine Hours in One Factory or Mill, it shall be lawful for any Person to employ such Child in any other Factory or Mill on the same Day for the Residue of such Nine Hours; provided that such Employment in such other Mill or Factory shall not increase the labour of such Child to more than Nine Hours in any one Day, or to more than Forty-eight Hours in any one Week.

Children employed in any one Mill less than 9 Hours not to be employed in any other Mill more than the Residue of 9 Hours.

XI. And be it further enacted, That from and after the Expiration of Six Months after the passing of this Act it shall not be lawful for any Person to employ, keep, or allow to remain in any Factory or Mill any Child who shall not have completed his or her Eleventh Year of Age without such Certificate as is herein-after mentioned, certifying such Child to be of the ordinary Strength and Appearance of a Child of the Age of Nine Years, nor from and after the Expiration of Eighteen Months after the passing of this Act any Child who shall not have completed his or her Twelfth Year of Age, without a Certificate of the same Form, nor from and after the Expiration of Thirty Months after the passing of this Act any Child who shall not have completed his or her Thirteenth Year of Age, without a Certificate of the same Form, which Certificate shall be taken to be sufficient Evidence of the Ages respectively certified therein.

Children not to be employed without a Certificate from a Surgeon as to Strength and Appearance.

XII. And be it further enacted, That for the Purpose of obtaining the Certificate herein-before required in the Case of Children under the Age of Eleven, Twelve, or Thirteen Years respectively, the Child shall personally appear before some Surgeon or Physician of the place or Neighbourhood of its Residence, and shall submit itself to his Examination; and unless the Surgeon or Physician before whom the Child has so appeared shall certify his having had a personal Examination or Inspection of such Child, and also that such Child is of the ordinary Strength and Appearance of Children of or exceeding the Age of Nine Years, and unless also such Certificate shall within Three Months of its Date be countersigned by some Inspector or Justice, or in that part of the United Kingdom called *Scotland*, by some Inspector or Justice or Burgh Magistrate, such Child shall not be employed in any Factory or Mill.

Certificates to be made by a Surgeon or Physician.

XIII. And be it further enacted, That the Certificates herein-before required in the Case of Children under the Age of Eleven, Twelve, or Thirteen Years respectively shall be in the Form following:

Form of Certificate of Surgeon or Physician.

"I [*Name and Place of Residence*] Surgeon [*or* Physician] do hereby certify, That *A.B.* the Son [*or* Daughter] of [*Name and Residence of Parents, or if no Parents, then the Residence of the Child*] has appeared before me, and submitted to my Examination; and that the said [*Name*] is of the ordinary Strength and Appearance [*according to the Fact*] of a Child of at least Nine Years of Age [*or if apparently above Nine, say* exceeding.]"

Children between 11 and 18 not to be employed in Factories more than Nine Hours a Day, or at Night, without a Certificate of Age.

XIV. And be it further enacted, That from and after the Commencement of the several Periods herein-before appointed for restricting the Employment of Children under the Ages of Eleven, Twelve, and Thirteen Years respectively, it shall not be lawful to employ, keep, or allow to remain in any Factory or Mill any Person between the said Ages respectively and the Age of Eighteen for more than Nine Hours in any Day, nor between the Hours of Nine o'Clock in the Evening and Five o'Clock in the Morning, without first requiring and receiving from such Person a Certificate in proof that such Person is above the Age of Eleven, Twelve, and Thirteen respectively, which Certificate, if a new Certificate shall be required. shall be in such Form as may be ordered by any Inspector.

Penalties against Persons not having Certificates not to be levied if it shall appear that the Person employed was above the Age required.

XV. Provided nevertheless, and be it enacted, That the Penalties and Punishments herein-after provided against any Person not requiring or not receiving such Certificate shall not be levied, if upon the Complaint or Proceeding for the Enforcement of such Penalties it shall appear to the Satisfaction of the Inspector or Justice, or in that Part of the United Kingdom called *Scotland*, to the Satisfaction of the Inspector or Justice or Burgh Magistrate by or before whom such Proceeding shall be had, that the Person so employed more than Nine Hours in the Day, or between the Hours of Nine o'Clock in the Evening and Half past Five o'Clock in the Morning, without such Certificate, was at the Time of the alleged Offence above the Age of Eleven, Twelve, or Thirteen respectively.

Provision in case Magistrates refuse to countersign Certificate.

XVI. And be it further enacted, That in case any Inspector or Justice or Burgh Magistrate shall refuse to countersign any such Certificate, he shall state in Writing his Reasons for such Refusal, and the Parents of such Child may thereupon take the Certificate to the Justices of the Peace at Petty Sessions for the Place or District of the Child's Residence, who are hereby empowered and required to decide upon the Validity of such Refusal; and every such Act of any such Petty Sessions shall be free of all Charge, Cost, or Expense whatsoever.

42 G. 3. c. 73.

XVII. And whereas by an Act, intituled *An Act for the Preservation of the Health and Morals of Apprentices and others employed in Cotton and other Mills and Cotton and other Factories*, passed in the Forty-second Year of the Reign of His late Majesty *George* the Third, it was amongst other things provided, that the Justices of the Peace for every County or Place in which such Mill was situated should appoint yearly Two Persons not interested in or in any way connected with such Mills or Factories in such County to be Visitors of such Mills or Factories, which Visitors so appointed were empowered and required by the aforesaid Act to enter such Factories at any Time they might think fit, and examine and report in Writing whether the same were conducted according to the Laws of the Realm, and also to direct the Adoption of such sanitory Regulations as they might, on Advice, think proper: And whereas it appears that the Provisions of the said Act with relation to the Appointment of Inspectors were not duly carried into execution, and that the Laws for the Regulation of the Labour of Children in Factories have been evaded, partly in consequence of the Want of the Appointment of proper Visitors or Officers whose special Duty it was to enforce their Execution;

Inspectors to be appointed.

be it therefore enacted, That upon the passing of this Act it shall be lawful for His Majesty by Warrant under His Sign Manual to appoint during His Majesty's Pleasure Four Persons to be

Inspectors of Factories and Places where the Labour of Children and young Persons under Eighteen Years of Age is employed, and in the Case of the Death or Dismissal of any of them to appoint another in the Place of such deceased Inspector, which said several Inspectors shall carry into effect the Powers, Authorities, and Provisions of the present Act; and such Inspectors or any of them are hereby empowered to enter any Factory or Mill, and any School attached or belonging thereto, at all Times and Seasons, by Day or by Night, when such Mills or Factories are at work, and having so entered to examine therein the Children and any other Person or Persons employed therein, and to make Inquiry respecting their Condition, Employment, and Education; and such Inspectors or any of them are hereby empowered to take or call to their Aid in such Examination and Inquiry such Persons as they may choose, and to summon and require any Person upon the Spot or elsewhere to give Evidence upon such Examinations and Inquiry, and to administer to such Person an Oath.

Powers and Duties of Inspectors for the Enforcement of this Act.

XVIII. And be it further enacted, That the said Inspectors or any of them shall have Power and are hereby required to make all such Rules, Regulations, and Orders as may be necessary for the due Execution of this Act, which Rules, Regulations, and Orders shall be binding on all persons subject to the Provisions of this Act; and such Inspectors are also hereby authorized and required to enforce the Attendance at School of Children employed in Factories according to the Provisions of this Act, and to order Tickets or such other Means as they may think fit for Vouchers of Attendance at such Schools; and such Inspectors are also hereby required to regulate the Custody of such Tickets or Vouchers, and such Inspectors may require a Register of them to be kept in every School and Factory; and such Inspectors are also hereby authorized and required to order a Register of the Children employed in any Factory, and of their Sex and Hours of Attendance, and of their Absence on account of Sickness, to be kept in such Factory; and all Registers, Books, Entries, Accounts, and Papers kept in pursuance of this Act shall at all Times be open to such Inspectors, and such Inspectors may take or cause to be taken for their own Use such Copy as they may think proper; and such Inspectors shall also make such Regulations as may be proper to continue in force any Certificates, Tickets, or Vouchers required by this Act, and such Certificates, Tickets, or Vouchers so continued in force shall have the same Operation and Effect as new Certificates, Tickets, or Vouchers; and such Inspector shall order and is hereby authorized to order the Occupier of any Factory or Mill to register or cause to be registered any Information with relation to the Performance of any Labour in such Mill or Factory, if such Inspector deem such Information necessary to facilitate the due Enforcement of any of the Provisions of this Act or of any of the Regulations which he may make under the Authority of this Act; and such Inspector is hereby authorized to order such Occupier of any Mill or Factory to transmit, in such Manner as may be directed in such Order, any Information with relation to the Persons employed or the Labour performed in such Mill or Factory that such Inspector may deem requisite to facilitate the Performance of his Duties or any Inquiry made under the Authority of this Act.

One of the Secretaries of State may appoint Persons to superintend, under the Inspector, the Execution of this Act.

XIX. And be it further enacted, That it shall be lawful for One of His Majesty's Principal Secretaries of State, if he shall see fit, upon the Application of any Inspector, to appoint any One or more Persons to superintend, under the Direction of any Inspector, the Execution of the Provisions of this Act, and of all Rules, Regulations, and Orders made under the Authority thereof; and such Person shall be paid by such Salary as may be determined by One of His Majesty's Principal Secretaries of State; and such Person so appointed shall have Authority to enter any School-room,

Counting-house, or any Part of any Factory or Mill, excepting such Part or Parts as may be used for manufacturing Processes; and if any Constable or Peace Officer shall be required by any Inspector to perform any continuous Service, it shall be lawful for such Inspector to allow a special Recompence to such Constable or Peace Officer for such Service : Provided nevertheless, that any such Orders may be altered or diallowed by One of His Majesty's Principal Secretaries of State, on Complaint made to him by Memorial from any Party interested.

Children in Factories to attend a School.

XX. And be it further enacted, That from and after the Expiration of Six Months from the passing of this Act every Child herein-before restricted to the Performance of Forty-eight Hours of Labour in any one Week shall, so long as such Child shall be within the said restricted Age, attend some School to be chosen by the Parents or Guardians of such Child, or such School as may be appointed by any Inspector, in case the Parents or Guardians of such Child shall omit to appoint any School, or in case such Child shall be without Parents or Guardians; and it shall and may be lawful, in such last-mentioned Case, for any Inspector to order the Employer of any such Child to make a Deduction from the weekly Wages of such Child as the same shall become due, not exceeding the Rate of One Penny in every Shilling, to pay for the Schooling of such Child; and such Employer is hereby required to pay the Sum so deducted according to the Order and Direction of such Inspector.

Schoolmaster's Voucher required.

XXI. And be it further enacted, That after the Expiration of Six Months from the passing of this Act it shall not be lawful to employ or continue to employ in any Factory or Mill any Child restricted by this Act to the Performance of Forty-eight Hours of Labour in any One Week, unless such Child shall, on *Monday* in every Week next after the Commencement of such Employment, and during every succeeding *Monday* or other Day appointed for that Purpose by an Inspector, give to the Factory Master or Proprietor, or to his Agent, a Schoolmaster's Ticket or Voucher, certifying that such Child has for Two Hours at least for Six out of Seven Days of the Week next preceding attended his School, excepting in Cases of Sickness, to be certified in such Manner as such Inspector may appoint, and in case of any Holiday, and in case of Absence from any other Cause allowed by such Inspector, or by any Justice of the Peace in the Absence of the Inspector; and the said last-mentioned Ticket shall be in such Form as may be settled by any Inspector.

Means of providing additional Schools.

XXII. And be it further enacted, That wherever it shall appear to any Inspector that a new or additional School is necessary or desirable to enable the Children employed in any Factory to obtain the Education required by this Act, such Inspector is hereby authorized to establish or procure the Establishment of such School.

Inspector may disallow Order for Salary, if Schoolmaster or Schoolmistress incompetent.

XXIII. And be it further enacted, That if upon any Examination or Inquiry any Inspector shall be of opinion that any Schoolmaster or Schoolmistress is incompetent or in any way unfit for the Performance of the Duties of that Office, it shall and may be lawful for such Inspector to disallow and withhold the Order for any Payment or any Salary to such Schoolmaster or Schoolmistress as herein-before provided.

Mill Owner liable to Penalty for Child remaining on the Premises more than Nine Hours.

XXIV. And be it further enacted, That if any Child within the several Ages herein-before restricted to the Performance of Nine Hours of Day Labour shall be kept or allowed to remain in any Room or Place whatsoever where any Machinery is used, or shall be kept or allowed to remain on any Premises within the outer Walls of any Factory or Mill, for any longer Time than Nine Hours during any one Day, or for any longer Time than the Residue of such Nine Hours in the Case of any Child which has been previously employed for any shorter Time during the same Day in any other Factory or Mill, the Occupier of such Factory or Mill shall, without any Evidence of the Employment of such Child, be liable to the same Penalty and Punishment as for employing such Child for such longer Period: Provided nevertheless, that no Place, Yard, or Play-ground open to the public View shall be considered Part of the Premises on which Children shall not be allowed to remain beyond the Hours herein-before stated: And be it further provided, that the Children may be allowed to remain in any School-room attached to such Factory or Mill, or in any other Waiting-room or Parts of the Premises where no Machinery is used, and which shall at all Times be open to the Inspection of any Mill Warden or Peace Officer duly appointed under the Provisions of this Act.

Proviso as to Play-grounds and Schools.

XXV. And be it further enacted, That Notice of any General Order or Regulation applying to more than One Mill or Factory, made by any Inspector, if published for Two successive Weeks in One or more Newspapers published in the Town, Place, or County where any such Mill or Factory is situate, shall in all Cases, at the End of Seven Days after the Second Publication thereof, have the same Effect in attaching a Responsibility upon any Offender against such Order or Regulation as a Notice personally served upon such Offender: Provided nevertheless, that such Notice shall not be to the Exclusion of any other special Notice which any Inspector may deem expedient or proper.

Notices by Inspectors.

XXVI. And be it further enacted, That the interior Walls, except such parts as are painted, of every Mill or Factory or Building where the Process of manufacturing is carried on, shall be limewashed, and the Ceilings of all rooms which have rooms or lofts above them, and all Ceilings which are plastered, shall be whitewashed once every year, unless Permission to the contrary, in writing, be granted by any Inspector.

Interior Walls of every Mill, &c. to be limewashed.

XXVII. And be it further enacted, that a Copy or Copies of such Abstract of this Act, and also such Copy or Copies of any Regulation or Regulations made in pursuance of this Act, as any Inspector shall direct, shall be hung up and affixed in a conspicuous Part or in the several Departments of every Mill or Factory; and such Copy or Copies of such Abstract and of such Rules or Regulations, so hung up and affixed, shall be signed by the Master or Manager or Overseer of such Mill or Factory; and such Copy or Copies shall be renewed by such Master, Manager, or Overseer so often as any Inspector may direct.

An Abstract of this Act, and such Rules and Regulations as any Inspector may determine, shall be hung up in Mills.

XXVIII. And be it further enacted, That if any Person shall give, sign, countersign, endorse, or in any Manner give currency to any false Certificate, knowing the same to be untrue, or if any Person shall forge any Certificate, or shall forge any Signature or Endorsement on any Certificate, or shall knowingly and wilfully give false Testimony upon any point material to any Certificate of any Inspector or Schoolmaster, such Person shall be deemed guilty of a Misdemeanor, and shall, on Conviction thereof before any Inspector or Justice, be liable to be imprisoned for any Period not exceeding Two Months in the House of Correction in the County, Town, or Place where such Offence was committed.

Punishment for Forgery of Certificates.

XXIX. And be it further enacted, That in case of the Employment of any child contrary to the Provisions of this Act, or for a longer Time than is herein-before limited and allowed, or without a due Compliance with the Provisions of this Act touching the Education of Children, or the Certificates of Surgeons or Magistrates, the Parent or Parents of such Child, or any Person having any Benefit from the Wages of such Child, shall be liable to a Penalty of Twenty Shillings, unless it shall appear to the Satisfaction of the Justice or Inspector that such unlawful Employment has been without the wilful Default of such Parent or Person so benefitted as aforesaid.

Parents liable to Penalty of 20s. for the Employment of Children beyond the legal Hours, &c.

Agents and Servants of Factory Owners to be personally liable.

XXX. And be it further enacted, That if any Offence shall be committed against this Act, for which the Master of any Factory or Mill is legally responsible, and it shall appear to the Satisfaction of any Justice or Inspector that the same has been committed without the personal Consent, Concurrence, or Knowledge of such Master, by or under the Authority of some Agent or Servant or Workman of such Master, it shall be lawful for such Inspector or Justice to summon such Agent or Servant or Workman before him to answer for such Offence, and such Agent or Servant or Workman shall be liable to the Penalties and Punishment for such offence herein provided, and such Inspector or Justice shall convict such Agent or Servant or Workman in lieu of such Master.

Penalties for Offences against this Act.

XXXI. And be it further enacted, That if any employer of Children in any Factory or Mill shall, by himself or by his servants or Workmen, offend against any of the Provisions of this Act, or any Order or Regulation of any Inspector made in pursuance hereof, such Offender shall for such Offence (except in the Case of any Offence for which some other Penalty or Punishment is specially provided) forfeit and pay any Sum not exceeding Twenty Pounds, nor less than One Pound, at the Discretion of the Inspector or Justice before whom such Offender shall be convicted: Provided nevertheless, that if it shall appear to such Inspector or Justice that such Offence was not wilful nor grossly negligent, such Inspector or Justice may mitigate such Penalty below the said Sum of One Pound, or discharge the Person charged with such Offence.

Penalty for obstructing Inspectors.

XXXII. And be it further enacted, That if any Person shall knowingly and wilfully obstruct any Inspector in the Execution of any of the Powers entrusted to him by this Act, such Person shall for every such Offence forfeit and pay a Sum not exceeding Ten Pounds.

Inspectors to have same Powers over Constables as Justices.

XXXIII. And be it further enacted, That such Inspector shall have the same Powers, Authority, and Jurisdiction over Constables and Peace Officers, as regards the Execution of the Provisions of this Act, as may by Law be exercised by His Majesty's Justices of the Peace over such Constables and Peace Officers.

Proceedings under this Act may be had before any One Inspector or any One Magistrate.

XXXIV. And be it further enacted, That all Proceedings for the Enforcement of any Penalty or Punishment imposed by or under the Authority of this Act may be had before any Inspector or Justice of the Peace acting in or for the Town, Place, County, or Division where the Offence shall be committed; and the Inspector or Justice before whom any Person shall be summarily convicted and adjudged to pay any Sum of Money for any Offence against this Act may adjudge that such Person shall pay the same either immediately or within such Period as the said Inspector or Justice shall think fit; and in case such Sum of Money shall not be paid immediately or at the time so appointed, the same shall be levied by Distress and Sale of the Goods and Chattels of the Offender, together with the reasonable Charges of such Distress; and for Want of sufficient Distress such Offender shall be imprisoned in the Common Gaol for any Term not exceeding One Calendar Month where the Sum to be paid shall not exceed Five Pounds, or for any Term not exceeding Two Calendar Months in any one Case, the Imprisonment to cease in each of the Cases aforesaid upon Payment of the Sum due.

Complaints to be preferred at or before the Visit of the Inspector; and previous Notice given.

XXXV. And be it further enacted, That all Complaints for Offences against this Act shall be preferred at or before the Time of the Visit, duly notified, of any Inspector next after the Commission of such Offence; and written Notice of the Intention to prefer the Complaint for such Offence shall by the Complainant be given within Fourteen Days after the Commission of such Offence to the Party or Parties complained against: Provided always, that no more than One Penalty for a Repetition of the same Offence shall be recoverable, except after the Service of the written Notice as aforesaid.

Proviso as to Penalties.

XXXVI. And be it further enacted, That it shall not be deemed necessary in any Summons or Warrant issued in pursuance of this Act to set forth the Name or other Designation of each and every the Partners in any such Mill or Factory, but that it shall be lawful to insert in such Summons or Warrant the Name of the ostensible Occupier or Title of the Firm by which the Occupier or Occupiers employing the Workpeople of every such Mill or Factory are usually designated and known.

In case of Partnerships, One Name sufficient for Summons, &c.

XXXVII. And be it further enacted, That the Service of such Summons or Warrant on any Occupier, principal Manager, Conductor, or Agent of any such Mill or Factory shall be good and lawful Service.

Service of Summons.

XXXVIII. And be it further enacted, That it shall be lawful for the Inspectors or any of them, or for any Justice of the Peace, upon any Complaint, or upon any Investigation under this Act without any Complaint, to administer an Oath to any Witness, and to summon any Witness forthwith to appear and give Evidence before him or them, or at a Time and Place appointed for hearing such Complaint or making such Investigation, or to order such Witness to be brought before him by any Constable or Peace Officer; and if such Witness shall not appear according to such Summons, Proof upon Oath having been given of the due Service of such Summons, or shall resist such Constable or Peace Officer, or shall not submit to be examined as a Witness, it shall be lawful for such Inspectors and Justices by Warrant under their Hand and Seal to commit such Person for such Non-appearance, Resistance, or Non-submission to the County Prison, or Prison of the Place where such Offence was committed, there to remain without Bail or Mainprize for any Time not exceeding Two Calendar Months.

Inspectors and Justices may summon Witnesses to appear and give Evidence, and on Neglect may commit to Prison;

XXXIX. Provided nevertheless, and be it enacted, That, except in the Case of Resistance to any Constable or Peace Officer, it shall be lawful for the Inspector or Justice by whom such Person shall have been committed to discharge such Person from Prison, if such Person shall shew any Cause to such Inspector or Justice which shall be deemed satisfactory as an Excuse for such Non-appearance, or if such Person shall afterwards submit himself to be examined to the Satisfaction of such Inspector or Justice, and the Order of such Inspector or Justice for such Discharge shall be a sufficient Warrant to any Gaoler or Prison-keeper.

and afterwards discharge them on sufficient Excuse or Compliance.

XL. And be it further enacted, That every Conviction under this Act before any Inspector or Justice may be made according to the Form in the Schedule to this Act annexed; which Conviction shall be certified to the next General Quarter Sessions, there to be filed amongst the Records of the County, Riding, or Division, and shall have the Force of an Act of Record, whether the same shall be by an Inspector or by a Justice of the Peace for such County, Riding, or Division; and no Conviction or other Proceeding of any Inspector or Justice under this Act shall be deemed illegal for any mere Informality.

Convictions to be filed amongst Records of County.

XLI. And be it further enacted, That if any Person who shall have been sentenced or adjudged to pay any Penalty or Forfeiture under this Act shall neglect or refuse to pay the same, it shall be lawful for the Inspector or Magistrates before whom such Person shall have been convicted to issue his Warrant to distrain the Goods and Chattels of such Person; and if no sufficient Distress shall be found, it shall be lawful for the said Inspector or Magistrates, upon such Fact being certified by the Constable having the Execution of such Distress Warrant, to commit such Person to the House of Correction or Common Gaol of the Town, County, or Place where such Offence was committed for any Time not exceeding Two Months; and the

Inspector or Justice may commit to Prison for Two Months in case Payment of Penalty is refused or Distress is insufficient.

said Warrant of Distress, Commitment, and Certificate of the Constable may be in the Forms contained in the Schedule to this Act annexed.

As to Appeal.

XLII. And be it further enacted, That no Appeal against any Conviction under this Act shall be allowed, except in the Case of a Conviction for the Forgery of any Certificates, Vouchers, or other Documents required by this Act, or by any Inspector under the Authority of this Act, neither shall any Conviction, except in the Case herein last excepted, be removable by Certiorari or Bill of Advocation into any Court whatever.

As to Convictions.

Application of Penalties.

XLIII. And be it further enacted, That any Justice or Inspector by whom any Complaint under this Act is determined shall, if he so thinks fit, give to the Complainant or Prosecutor One Half of any Penalty imposed for any Offence against any of the Provisions of this Act, together with all Costs of Prosecution and Conviction, and the Remainder of the Penalty, or the Whole if he shall think fit, shall be applied as such Justice or Inspector may direct for the Benefit of any School wherein Children employed in Mills or Factories are educated in such Township or Place where such Offence shall be committed: Provided always, that only One Penalty shall be recoverable for any One Description of Offence from any One Person for any One Day; and that it shall not be deemed necessary for the Complainant or Prosecutor to name in any Summons the particular Township in which such Offence shall have been committed, but it shall be lawful to set forth in such Summons the Name of the Parish where such Offence may have been committed: Provided always, that such Summons shall be issued upon Complaint being made upon Oath.

Restriction as to Penalties.

Summons.

Inspector may order Constable to provide a convenient Place for holding Sittings.

XLIV. And be it further enacted, That every Inspector shall be and is hereby authorized to order any Constable or Peace Officer to provide for such Inspector a convenient Place for holding any Sitting; and the Expense of providing such Place shall be defrayed in the Manner and Proportions and by the Person or Persons herein appointed for the Payment of any special Remuneration to any Constable or Peace Officer.

Inspectors to make Annual Reports.

XLV. And be it further enacted, That every Inspector shall keep full Minutes of all his Visits and Proceedings, and shall report the same to One of His Majesty's Principal Secretaries of State twice in every Year, and oftener if required, and shall also report the State and Condition of the Factories or Mills and of the Children employed therein, and whether such Factories or Mills are or are not conducted according to the Directions of this Act and of the Laws of the Realm: And whereas it is expedient that the Proceedings, Rules, Orders, and Regulations of the several Inspectors appointed under this Act should be as nearly alike as is practicable under all Circumstances, therefore such Inspectors are hereby required, within Three Months next after they shall have commenced the Execution of their several Duties and Powers under this Act, and twice at least in every Year afterwards, to meet and confer together respecting their several Proceedings, Rules, Orders, Regulations, Duties, and Powers under this Act, and at such Meeting to make their Proceedings, Rules, Orders, and Regulations as uniform as is expedient and practicable; and such Inspectors are hereby required to make and keep full Minutes of such Meetings, and to report the same to such Secretary of State when they make the Report herein-before required.

Proceedings of Inspectors required to be uniform.

Burgh Magistrates in Scotland to exercise same Powers as Justices of Peace in England.

XLVI. And be it further enacted, and it is hereby declared, That in all Cases in which any Justices or Justice of the Peace are or is required to act or do any thing in any Manner under this Act, or are or is named therein, and whenever the Subject Matter of any one of the Enactments or Provisions of this Act shall arise within that Part of the United Kingdom called *Scotland*, the Burgh Magistrates shall be held to have and shall have within the Limits of their own Jurisdiction the same Powers, Duties, and

Authorities, and which they are hereby required to exercise, as are by this Act conferred upon the said Justices of the Peace, and are required to be exercised by them.

XLVII. Provided always, and be it enacted, That nothing in this Act contained shall apply to Mechanics, Artisans, or Labourers under the prescribed ages working only in repairing the Machinery or Premises. **Act not to extend to Persons on Repairs.**

XLVIII. And be it further enacted, That from and after the First Day of *January* One thousand eight hundred and thirty-four the Act passed in the First and Second Years of the Reign of His present Majesty, intituled *An Act to amend the Laws relating to Apprentices and other Persons employed in Cotton Mills, and to make further Provisions in lieu thereof*, shall be repealed, and the same is hereby repealed, except as to any Act or Acts repealed by the same. **The Act 1 & 2 W. 4. c. 39. repealed, except as it repeals any other Acts.**

XLIX. And be it further enacted, That any Words in this Act denoting the Masculine Gender shall be construed to extend to Persons of either Sex, and any Words denoting the singular Number shall be construed to extend to any Number of Persons or Things, if the Subject Matter or Context shall admit of such an Interpretation, unless such Construction shall be in express Opposition to any other Enactment. **Construction of Terms.**

L. And be it further enacted, That this Act shall be deemed and taken to be a Public Act, and shall be judicially taken notice of as such by all Judges, Justices, and others. **Public Act.**

The SCHEDULE to which this Act refers.

Form of Conviction.

County of [Town of *as the Fact may be*] to wit. } BE it remembered, That on the Day of in the Year *A. B.* [*describe the Offender*] was, upon the Complaint of *C. D.* [*or* upon the View of *C. D.*, one of His Majesty's Inspectors of Factories], convicted before *E. F.*, one of His Majesty's Inspectors of Factories, *or* Justices of the Peace of and for, *&c.* [*as the Case may be*], in pursuance of an Act passed in the Fourth Year of the Reign of His Majesty King *William* the Fourth, for [*describe the Offence*]. Given under my Hand and Seal the Day and Year above mentioned.

Warrant to distrain for Forfeiture.

To the Constable, &c.

County of to wit. } WHEREAS *A. B.* of in the said County, is this Day convicted before me *C. D.*, one of His Majesty's Inspectors of Factories, [*or* Justices of the Peace in and for the said County,] upon the Oath of a credible Witness, [*or* upon my own View, *as the Case may be*], for that he the said *A. B.* hath [*here set forth the Offence, describing it particularly in the Words of the Statute or Rule, as near as can be*], contrary to the Statute [*or* Rule, *if the Offence is against some Rule or Regulation or Order of an Inspector*,] in that Case made and provided, by reason whereof the said *A. B.* is adjudged to have forfeited the Sum of £ , to be distributed as herein-after mentioned: These are therefore in His Majesty's Name to command you to levy the said Sum of £ by Distress of the Goods and Chattels of him the said *A. B*; and if within the Space of Four Days next after such Distress by you taken, the said Sum of £ together with the reasonable Charges of taking and keeping the same, shall not be paid, that then you do sell the said Goods and Chattels by you so distrained, and out of the Money arising by such Sale that you do pay [*according to the Award of the Justice*], returning the Overplus, on Demand, to him the said *A. B.*, the reasonable Charges of taking, keeping, and selling the said Distress being first deducted; and if sufficient Distress cannot be found of the Goods and Chattels of the said *A. B.* whereon to levy the said Sum of £ that then you certify the same to me, together with this Warrant. Given under my Hand and Seal the Day of

C. D.

Return of Constable upon Warrant of Distress where no Effects.

I *A. B.*, Constable of in the County of , do hereby certify and make Oath, That by virtue of this Warrant I have made diligent Search for the Goods of the within-named , and that I can find no sufficient Goods whereon to levy the same. As witness my Hand the Day of

A. B.

Sworn before me the Day and Year

C. D.

Commitment for Want of Distress.

County of } To the Constable of in the County to wit. } of , and to the Keeper of the Common Gaol [*or* House of Correction] at in the said County.

WHEREAS *A. B.* of in the said County, was, on the Day of convicted before me *C. D.* Esquire, one of His Majesty's Justices of the Peace in and for the said County, [*or* Inspector of Factories, *as the Fact may be*], upon the Oath of a credible Witness, [*or* upon my own View, *as the Case may be,*] for that he [*here set forth the Offence*] contrary to the Statute made in the Year of the Reign of His Majesty King *William* IV. for [*according to the Title of the Act, or* contrary to a certain Rule or Order or Regulation of His Majesty's Inspectors of Factories,] and the said *A. B.* by reason thereof hath been adjudged to forfeit and pay the Sum of : And whereas on the Day of in the Year aforesaid, I did issue my Warrant to the Constable of to levy the said Sum of by Distress and Sale of the Goods and Chattels of him the said *A. B.*, and to distribute the same as in my said Warrant was mentioned: And whereas it duly appears to me, upon the Oath of the said Constable, that he hath used his best Endeavours to levy the said Sum on the Goods and Chattels of the said *A. B.*, but that no sufficient Distress can be had whereon to levy the same: These are therefore to command you the said Constable of aforesaid to apprehend the said *A. B.*, and him safely to convey to the Common Gaol [*or* House of Correction] at in the said County, and there deliver him to the Keeper thereof, together with this Precept; and I do also command you the said Keeper to receive and keep in your Custody the said *A. B.* for the Space of , unless the said Sum shall be sooner paid, pursuant to the said Conviction and Warrant; and for so doing this shall be your sufficient Warrant. Given under my hand and Seal the Day of

C. D.

FACTORY STATISTICS.

THE OFFICIAL TABLES

APPENDED TO THE

REPORT OF THE SELECT COMMITTEE

ON THE

TEN-HOUR FACTORY BILL,

VINDICATED IN A SERIES OF LETTERS,

ADDRESSED TO JOHN ELLIOT DRINKWATER, ESQ.,

ONE OF THE FACTORY COMMISSIONERS,

BY THE LATE

MICHAEL THOMAS SADLER, Esq., F.R.S.,

CHAIRMAN OF THE COMMITTEE.

☞ Before the Reader turns to the Letters, it may be desirable to prefix a very brief explanation of what is meant by the "Tables of Mortality," and the "Notes" thereto appended; for this purpose some extracts from the Appendix to the Report of the Committee, of which Mr. Sadler was the Chairman, will here be inserted. The Title of the Volume in which the Documents first appeared is as follows: "Report from the Committee on the 'Bill to regulate the Labour of Children in the Mills and Factories of the United Kingdom,' with the Minutes of Evidence, Appendix, and Index. Ordered by the House of Commons to be printed, 8th August, 1832."

APPENDIX.

COMPARATIVE TABLE OF THE DURATION OF LIFE.

Number of Persons Buried, and at what Ages Buried, during Eighteen Years (1813 to 1830) in certain Counties and Places (sixteen are here enumerated, for which see next page); shewing the Number Buried under five years of age—from five to ten—from ten to fifteen—from fifteen to twenty—from twenty to thirty, and so on for each Decennary period to the end of Life: with Decimal Results annexed, for the purpose of Comparison.

Factory Statistics.

(The following are the several headings prefixed to the Tables of the Sixteen Counties and Places; two towns are afterwards selected to shew the plan on which the Calculations in the Appendix were given :—)

RUTLAND.

(Healthy County.)

METROPOLIS.

NORWICH.

(Old and closely built, and Manufacturing, but with few or no Factories.)

CARLISLE.
1818—1830.

(Partly Manufacturing, and partly Spinning.)

MACCLESFIELD.

(Silk Throwing and Spinning.)

PRESTON.

(Cotton Spinning, &c.)

STOCKPORT.

(Cotton Spinning, &c.)

LEEDS TOWN.

(Manufacturing and Woollen, Flax and Silk Spinning.)

ESSEX.

(Marshy County.)

CHESTER.

(Old and closely built, but non-manufacturing.)

CARLISLE.
1779—1787,

(inclusive, according to Dr. Heysham.)

BRADFORD (Yorkshire.)

(Worsted Spinning.)

WIGAN.

(Manufacturing and Spinning.)

BURY (Lancashire.)

(Cotton Spinning, &c.)

BOLTON-LE-MOOR.

(Cotton Spinning, &c.)

HOLBECK.

(Flax - Spinning.)

(The following are here reprinted as Specimens of the Sixteen Tables, the headings of which have been given in the preceding page.) Factory Statistics.

CHESTER.

(Old and closely built, but non-manufacturing.)

AGE.	Living.	Died.	Decimal.	
			Living.	Died.
Under Five Years	8,693	3,107	10,000	3,574
5 — 9	5,586	341	6,426	392
10 — 14	5,245	197	6,034	227
15 — 19	5,048	299	5,807	344
20 — 29	4,749	670	5,463	771
30 — 39	4,079	658	4,692	757
40 — 49	3,421	661	3,935	760
50 — 59	2,760	630	3,175	725
60 — 69	2,130	749	2,450	862
70 — 79	1,381	837	1,588	962
80 — 89	544	508	626	585
90 — 99	36	29	41	33
100 and upwards	7	7	8	8

PRESTON.

(Cotton Spinning, &c.)

AGE.	Living.	Died.	Decimal.	
			Living.	Died.
Under Five Years	9,498	4,699	10,000	4,947
5 — 9	4,799	498	5,053	524
10 — 14	4,301	274	4,529	288
15 — 19	4,027	308	4,241	324
20 — 29	3,719	694	3,917	731
30 — 39	3,025	615	3,186	648
40 — 49	2,410	525	2,538	553
50 — 59	1,885	533	1,985	561
60 — 69	1,352	525	1,424	553
70 — 79	827	505	871	532
80 — 89	322	283	339	298
90 — 99	39	36	41	38
100 and upwards	3	3	3	3

Factory Statistics.

The Appendix thus proceeds :—The COMPARATIVE RESULTS of the foregoing TABLES are as follows:—

In every 10,000 *of the Persons buried, there died*

	Under 20 Years old.	Under 40 Years old.	Lived to 40 and upwards.
In the Healthy County - -	3,756	5,031	4,969
In the Marshy County - -	4,279	5,805	4,105
In the Metropolis - - -	4,580	6,111	3,889
In the City of Chester - - -	4,538	6,066	3,934
In the City of Norwich - -	4,962	6,049	3,951
In the City of Carlisle (former state)	5,319	6,325	3,674
In the City of Carlisle (present state)	5,668	6,927	3,071
In the Town of Bradford (Worsted Spinning) - - - -	5,896	7,061	2,939
In the Town of Macclesfield (Silk Spinning and Throwing) -	5,889	7,300	2,700
In the Town of Wigan (Cotton Spinning, &c.) - - -	5,911	7,117	2,883
In the Town of Preston (ditto) -	6,083	7,462	2,538
In the Town of Bury (ditto) -	6,017	7,319	2,681
In the Town of Stockport (ditto) -	6,005	7,367	2,633
In the Town of Bolton (ditto) -	6,113	7,459	2,541
In the Town of Leeds (Woollen, Flax, and Silk Spinning, &c.)	6,213	7,441	2,559
Holbeck (Flax Spinning) - -	6,133	7,337	2,663

(Then come the following " Notes," so often since alluded to.)

SO THAT ABOUT AS MANY HAVE DIED BEFORE THEIR TWENTIETH YEAR WHERE THE FACTORY SYSTEM EXTENSIVELY PREVAILS, AS BEFORE THEIR FORTIETH YEAR ELSEWHERE.

The effect of the increase of the Population of a place by migration would be to diminish its general mortality.—*See* Milne on Annuities, § 713.

☞ The Official Returns of the BURIALS of MANCHESTER are so incomplete as to render them useless for this comparison; the Ages of a small portion only of the Dead being given in such Returns; namely, those only who have been interred in the Burial-places of the Established Church.

LETTER I.

The "Tables of Mortality" attacked by Mr. Drinkwater and his coadjutors.—The result of the "Tables," if true, admitted to be "Frightful."—Secret Examinations.—A series of gross blunders exposed.—Mortuary Registers of Manchester confessedly defective.—Official note respecting them.—Town of Leeds, and the Town and Liberty.—The distinction not observed by Mr. D.—Unfairness of his objections.

Factory Statistics.

Sir,—The first report of the factory commissioners contains, I perceive, a laboured attack, under your signature, upon the authenticity and validity of the tables of mortality appended to the report of the select committee of the House of Commons on the factory bill. That attack, evincing the importance which you, in common with most others, attach to those tables, admitting their correctness, seems to demand from me, as chairman of the committee, some consideration. The rest of your report, as far as I have hitherto examined it, I do not deem worthy of particular notice; otherwise, notwithstanding you have framed it entirely as you pleased, it is still sufficiently open to reprehension in many important particulars, such as in the attempts it makes to invalidate the credit of evidence given before the late factory committee (and indeed preceding ones), at the very time your own selected witnesses were confirming the main features of the factory children's case. The remarks regarding "leading questions," which, had they been consistent with the acknowledged duty and practice of parliamentary inquiries, are strange as put forth by you, who had purposely screened yourself from all such imputations by resolutely refusing to allow your own questions to be publicly heard, much less recorded; the continued attempt to exonerate, as far as practicable, the rich and powerful culprits from the load of cruelty and oppression which it were now in vain to deny has long been inflicted by the factory system, by casting it upon their humble instruments, not to say victims; and that which, however, I hardly think worth alluding to—the personal insinuations with which you have saturated your public document,—these, together with matters of a more amusing nature which are presented throughout, afford me a tempting opportunity for suitable comments, which, however, I am not inclined to avail myself of at present; still less am I disposed to notice the last printed letters which you and your coadjutor, Mr. Power, have addressed to me: their perusal afforded me at the moment some amusement and gratification; and as you say, if I recollect rightly, that you have all along regarded the controversy between us as merely personal, I am bound to confess that you have paid more attention to me as an individual than I intended rendering to you in pursuing it a single moment. The only points of the slightest interest between us, were, whether I acted in the behalf, and by the appointment, of the operatives' committee, and whether, in refusing to co-operate with you in the inquiry excepting on the condition that the proceedings should be *public*, or, at all events, *fairly* and *fully recorded*, I acted under their express authority, as well as agreeably to the plainest dictates of justice, liberality, and common sense. Now, since it appears by your printed report, that these only material points were certified to you "upon oath," by one of your own witnesses and attendants, I as little envy the position you have taken in regard to what you choose to consider as our personal dispute as I do your public demeanour, in being the avowed advocate of secret and uncontrolled proceedings in matters deeply affecting the interests and feelings of the public. While, however, the course you have pursued may have materially assisted you in certain of your purposes, it will be seen how little it has conduced to your credit in other respects; especially in permitting you to put forth, in your official character, a series of the grossest blunders (wilful errors I should be sorry to impute to you) probably ever published; and the public will naturally judge of the worth of that part of your report, which has been so framed as to elude all verification, when it is made acquainted with the manner in which you have ventured to deal with facts open to general examination.

In proceeding to the proposed examination of your report, it seems proper to premise, that during the whole course of the struggle that has been made in behalf of the factory children, it has been invariably asserted by those who have interested themselves in their unhappy condition, that the early and excessive labour to which they have been doomed, has not only been injurious to their morals and health, but, in multitudes of cases, destructive of life itself, at an age when, of all others, the human frame is the most tenacious of existence, and when to destroy it by any other means than direct violence involves a degree of long-suffering and sorrow which it is distressing to contemplate, even in imagination. To these affecting statements many distinguished individuals of the profession best qualifying them to judge on scientific principles concerning this important subject, among whom there have not been a few whose personal knowledge and experience have rendered their opinions still more decisive, have constantly borne witness in terms the most strong and unequivocal; while the sufferers themselves have invariably asserted such to be the effects of the system, that few of those engaged in it survive a comparatively early period of life.

Fully convinced of the reality of these distressing representations, but at the same time quite aware that the whole of them had not only been denied, but actually reversed by the interested advocates of the system and their meaner parasites, and knowing that in the present improved state of our national statistics it was quite possible to put these conflicting arguments and assertions to the infallible test of facts, I made a series of computations with that view, the results of which were fully confirmatory of my worst apprehensions on the subject. Hence originated the official tables, against which you inveigh so strongly; with the framing of which, however, I beg to inform you I had nothing whatever to do, however near the results were to those which I had myself previously computed.

Admitting the truth and validity of these tables, it is impossible to express in stronger language than you and your coadjutors have employed, the baneful effects of the system as at present pursued. Under that supposition it is termed by yourself "*dreadful;*" it is described as "frightful," as "awful," as "a system more hostile to human life than the most depopulating ravages of war or pestilence, or the most sanguinary forms of superstition;" and as existing "at a period when philosophy is proclaiming its triumphs, and the medical art is appropriating to itself the distinction of having contributed to prolong human life." "How frightful must be the immolations to this inexorable Moloch!" These, and a variety of similar phrases, are employed to express the fatal results which it is acknowledged would be fixed upon the system were the tables correct in themselves and properly applied. But that idea you treat with the utmost derision. You deny at once their authenticity, their correctness, and their application. I proceed to defend them in all these respects, under the equal certainty that I shall succeed in demonstrating their validity, and that having done so, the advocates of the system will desert their own conclusions, and seek in some other subterfuge an apology for the certain effects which they only described thus fearfully, while they professed to believe them unreal.

You speak of these tables as being "introduced it does not appear by whom," and "without being authenticated in any manner whatever." I will inform you on these points.

First, then, the tables were introduced under a sanction and formality which documents inserted in reports of committees do not generally possess: they were regularly moved for by the chairman of the committee, and were ordered by the House (indeed by both Houses) to be printed; and were consequently placed in the hands, not only of every member of the committee, but of every member of the legislature. They were again and again recognised and referred to by the committee in their sittings, as you might have seen by their minutes, which were in your hands at the time you made your charge.

Secondly: these tables were "authenticated" by the signature of "Jno. Rickman," in whose office the decimal results annexed for the purpose of comparison were also computed, without any instruction on the part of the committee at large, or any individual of it. I am not aware that that gentleman took any interest in the inquiry; and I hope, therefore, you will not retract your eulogy upon him, under an apprehension that he might have favoured the ten hours' bill.

You are equally wide of the fact respecting the notes. On that regarding Manchester, couched in three or four lines, you make remarks quite apart from the real question, which I shall not now notice further than just to observe, that the unhappy bias of your mind has plunged you into the same mistake respecting it, as that under which you labour regarding the whole table. I may just notice here, that the person or persons to whom the order of the House was sent, understanding, probably, that the returns called for were for the purposes of comparison, inserted, very properly, the required information regarding a few additional places (the required computations concerning which had perhaps been already made) in order to render the table more complete. But the town of Manchester not being included among the number, I moved, as before, for a copy of the mortuary registers of that important place, which were in like manner obtained and printed, with an accompanying note, however, of precisely the same import as the one in question; but still the document was not arranged as in the other instances, nor any comparative results given, which I was assured could neither be made nor recommended, owing to the accounts delivered in being so greatly defective. You are ignorant that this circumstance at which you cavil is matter of record; but it seems strange that you should have overlooked it in a public document, which, it appears, you consulted just so far as you thought it answered your particular purpose: I mean the Census of 1831. Appended to the abstract of the Manchester registers is the following note:—"The population of Manchester, second only to that of the metropolis, will appear eminently to demand a table of mortality, which, however, is *purposely omitted*, because the numerous classes of society burying at Chorlton Row, and other burial-grounds, where no registry of ages is kept—such a table would be *fallacious*. The number and sex of those irregularly buried has now been obtained and included, averaging at 3500 per annum; but in 1821 such register was not obtained, from which defect the number of marriages (1811—1820) actually exceeds that of the registered burials; indeed, the marriage registers include all the marriages in the parish, added to those of the inhabitants of the town and suburbs; which parish contains about 90,000 inhabitants additional. No calculation can be made or recommended other than jointly with the marriages of the Hundred of Salford, including the town of Manchester. The same observation is applicable in some degree to Liverpool, as situate in the West Derby Hundred."

You perceive now, I presume, how inapplicable and misdirected are your remarks on the note in the table regarding the intentional omission of the town of Manchester; and in your comment you are entirely wide of the real point in dispute. It is not the exact annual mortality of certain places calculated on the existing population which the table professes to shew, but the

Factory Statistics.

comparative mortality at certain periods of life; which, it is obvious, may be computed with sufficient accuracy for the purpose of such comparison, if the ages of a considerable proportion of the dead of the *average* condition of society be obtained; in both which respects the registers of Manchester are pronounced deficient by those to whom the duty of arranging the mortuary registers of the kingdom is confided by parliament.

While on the subject of authenticity, I may here mention another instance of the strange ignorance you manifest on the subject you have undertaken to discuss. You say, "Since this report," meaning your own, "was written, I have been enabled to consult a complete copy of the population returns of 1831, and have found, with some surprise, that although the Rutland table of the appendix is copied from Mr. Rickman's table for both sexes in the population returns, the Leeds table is not a copy of any table there, but appears to be probably an imperfect copy of Mr. Rickman's table for males only." You subjoin a copy of both, to enable the reader to share in your "surprise," but it is unnecessary for me to give either; a very few words will suffice to "surprise" yourself at your own inadvertency. The table given in the appendix is no "imperfect copy" of Mr. Rickman's "males;" it is, as far as it goes, as exact a transcript as that of Rutland; it occurs in the parliamentary paper already quoted (No. 638, session 1832), and is just what it purports to be, namely, the table of the *town* of Leeds, whereas the one you give, under the supposition of its being the correct one, is no such thing; it is not even meant as that of the town of Leeds, but the town including the *liberty* of Leeds, which liberty extends, I think, over a district of upwards of thirty square miles, comprising the population of many distinct townships, several of which are non-manufacturing, and unquestionably among the most healthy of the kingdom.* Your comment, then, upon these tables, that they are neither "useful for the purpose they have been made to serve," nor "authentic," is a suitable comment upon your own blunder, and is ridiculous enough without an additional word on my part.

So much for the authenticity of what you have evidently supposed to be my "tables." I will now proceed to examine the "authenticity" of what you have made *yours*, by adopting the data they furnished you with as facts, and founding upon them a series of deductions, precisely in character with them; but I cannot help remarking, in passing, that your ideas upon this subject also are somewhat singular; scrupulous to the highest degree regarding the conduct of the committee—most exact as to the signature and verification of the papers put forth under their authority; you, nevertheless, claim to yourself the privilege of inserting in your own report, without the slightest check or control, just what you please, and from whomsoever you please, without the least effort on your part to detect errors the most gross and palpable ever perhaps put forth on any like occasion. Regarding myself, indeed, you express a wish that you might have made use (doubtless in your own private way) of documents which, you thought it was probable, I had in my possession; but you begged, it seems, that it might be made known to me that you would not "promise to make use of any *results*, unless you were at the same time permitted to have free access to the *elements* from which these results professed to be deduced." Commendable suspicion! honourable scrupulosity! I not only approve of the idea—I suggested the very course, when I publicly advised you to apply to the sexton, rather than the mercenaries of the system: you would then have had "free access" enough, and the elements which, it appears, you wished so much to examine, would have suggested, I think, very different deductions to those which you have adopted.

* Population of the township of Leeds in 1831 was 71,602, of the parish or liberty 51,791; of the town and liberty together, 123,393.

But you were otherwise occupied, or it perhaps required your whole time and exertion to pass the poor witnesses on the factory bill through the crucible of your secret cross-examination, when, without the needful presence of a single friend, by an agonizing process of contradictions, warnings, and threats, continued for many hours together, and pursued to the extremity of endurance, you might be enabled, and wrongfully, after all, to fix the stigma of prevarication on some point immaterial to the main inquiry;—upon, for instance, a poor youth already ruined in health, deformed in his limbs, and weakened in mind, by the system against which he had dared to bear witness; and whose very appearance would be evidence sufficient, I will dare avouch, to determine every disinterested mind, as to the merciless wrongs he had endured. This ill-fated boy had been examined, and cross-examined, before the select committee, among whom, I will venture to assure you, without any disparagement to yourself, there were persons as competent to elicit the truth,—nay, fully as competent in reference to the subject under examination, on which they entirely differed from me, as yourself; a fact which, in speaking of the select committee, you always find it convenient to forget; but his evidence given before that committee, as it was open, and fully recorded both questions and answers, is, of course, in your estimation, "worthless;" it is only that given before you, as minuted, commented upon, and published on your single authority, which is to be regarded as "authentic." But enough, I think, has been said about authenticity, at least of form, authority, signature, and of oaths; in all of which respects, excepting in the matter of swearing, the tables you strive to impugn are infinitely superior to your own: but there is, however, a species of authenticity infinitely beyond all these, essential to all just and righteous evidence, but of which I proceed to show you that your most material documents are utterly destitute—the authenticity of TRUTH.

LETTER II.

Mr. Drinkwater's Authorities; Mr. Thorpe, Mr. Baker, Dr. Williamson, and Q.—Burials in Leeds Township.—Mr. Thorpe wrong in the principle of his calculations respecting them.—Table (No. I.) of Burials for Leeds now, and fifty years ago.

SIR,—The main authorities to which you appeal are, Mr. Thorpe, a flax-spinner, near Knaresborough; Mr. Baker, surgeon of a flax-spinning establishment at Leeds; Dr. Williamson; and an anonymous writer in the *Leeds Mercury*, under the signature of Q, who probably had the merit of suggesting your line of argument against the tables in question.

As I mean to exhibit the documents adduced by your authorities, and adopted by you in a sort of climax of errors and absurdities, I will commence with Mr. Thorpe, whose statements are, perhaps, on the whole, the least distant from truth, though sufficiently far from it to reverse every deduction which he and you have drawn from them.

Regarding Mr. Thorpe you intimate, properly enough, though in gentler phrase than you employ when you allude to the "framer" of the tables, that he is not familiarly acquainted with the subject, and talk of the necessity of "modifying his conclusions before they can be safely adopted;" but as to his facts, these, you assert, with an air of triumph over such evidence as is adduced by the committee, were "given upon oath." Mr. Thorpe refers to two periods; you make him state, with the utmost confidence, and with professions of the great accuracy of himself and his two clerks, as to the latter, that the burials in the township of Leeds in the years 1828, 1829, and 1830,

Factory Statistics.

amounted to 5,158, the population of the township being in 1831, 71,602; making, according to his method of computation, the annual average of mortality one in $41\frac{1}{2}$: a pretty high rate, by-the-bye, as contrasted with that of the kingdom at large. Now I assert, not indeed upon oath, or under your "authentication," but on the authority of the official returns to which yourself might have had as free access as myself, and which it were absurd to suppose can err in *excess*, that the burials in the township of Leeds in the years 1828, 1829, and 1830, amounted to 5,473 persons, viz., 2,776 males, and 2,697 females; making the average mortality, even as calculated by your authority, one in $39\frac{3}{4}$, instead of one in $41\frac{1}{2}$. But this is by no means the amount of the error. Should not you have pointed out the palpable deception in the very principle of the calculation, instead of adopting it? The population during the decennary 1821—1830, increased upwards of 47 per centum; but you and your authority calculate the annual mortality, not by dividing the mean population of the period by the mean number of the annual deaths occurring in it, but by dividing the population of 1831 by the deaths, which (on the average) occurred two years previously. It might have been hoped that an error so materially affecting the computation would not have been made, while you were advancing such high pretensions to accuracy in your facts and deductions. Sir, the mortality of Leeds at that period was at least one in 36, instead of one in $41\frac{1}{2}$, as stated in your report. It will avail you nothing to say that the calculation exhibits the *comparative* mortality only at two periods alluded to by Mr. Thorpe, for it is against comparative results that you are arguing. But the comparative results, even were the data on which they are founded correct, are, according to your own reasoning, "worthless;" except you were to show that they are computed upon a population in the same state of increase. After this sample of Mr. Thorpe's accuracy, it will hardly be expected that I should take the trouble of examining his facts regarding the mortality of Leeds in the years 1798, 1799, and 1800, or follow him to the other places to which he refers. I have shewn the quality of his pretended facts, and you have admitted his incompetency to the subject. But still you talk of his returns justifying the assertion, that the mortality of Leeds has diminished since 1801, at which time, say you, there were scarcely any factories established there; an assertion which I mean to dispute. But I will first challenge you both to a better and more indisputable proof, namely, the *comparative* duration of human life in that place fifty years ago, when unquestionably there were *none*. Now, without making any deduction for the ravages of the small-pox at the former period, I will hazard an assertion, that such is the deleterious nature of the present system of employment and labour there, and the consequent habits and condition of the people, that the mean duration of human life, calculated upon the mortuary registers of the last three years, viz., 1830, 1831, and 1832, is less than that of the corresponding years half a century ago. I have thought it worth while to examine the ages, at the former period, of those interred in the burial-ground of the parish-church, till quite lately the principal burial-place of the township, and then almost the only one, the registers of which record the ages of those buried with the greatest apparent accuracy for a long time past, and I will contrast the results with the table furnished by order of the House of Commons, founded on the interments, from 1813 to 1830 inclusive. The following are the results:—Of those buried during the three former years, the ages of 1,431 were registered; the ages at which these died, and the decimal results, for the purposes of comparison, are given in the same manner as those furnished by Mr. Rickman, which I have annexed

TABLE I.

LEEDS TOWN, 1780, 1781, 1782.						LEEDS TOWN, 1813—1830.					
AGES.	Living.	Died.	Decimal. Living	Decimal. Died.	Per Cent.	AGES.	Living.	Died.	Decimal. Living	Decimal. Died.	Per Cent.
0 — 5	1431	638	10,000	4,458	44	0 — 5	18,366	9708	10,000	5,286	53
5 — 9	793	71	5,542	497	9	5 — 9	8,658	764	4,714	416	9
10 — 14	722	31	5,045	217	4	10 — 14	7,894	420	4,298	229	5
15 — 19	691	49	4,828	342	7	15 — 19	7,474	517	4,069	282	7
20 — 29	642	72	4,486	503	11	20 — 29	6,957	1172	3,787	638	17
30 — 39	570	94	3,983	657	16	30 — 39	6,785	1083	3,149	590	19
40 — 49	476	99	3,326	692	21	40 — 49	4,702	1100	2,559	599	23
50 — 59	377	102	2,634	713	27	50 — 59	3,602	1101	1,960	599	31
60 — 69	275	130	1,921	908	47	60 — 69	2,501	1089	1,361	593	44
70 — 79	145	99	1,013	691	68	70 — 79	1,412	941	768	512	67
80 — 89	46	41	322	287	89	80 — 89	471	414	256	225	88
90 — 99	5	4	35	28	80	90 — 99	57	54	31	29	94
100&upwds.	1	1	7	7	100	100&upwds.	3	3	2	2	100

Such, then, are the results which a due examination of the real facts presents, sufficiently striking even without taking into consideration the important circumstance that the small-pox was almost totally eradicated in the latter term, though in the former one very nearly one-fourth of the deaths under five, and one-fifth between five and ten, were occasioned by that then dreadful scourge. I am aware that you pronounce, for the first time I believe, all computations founded on mortuary registers worthless, which makes your lengthy and confident deductions from Mr. Thorpe's researches in that direction somewhat curious; but I shall notice more particularly your objections, and also those of your coadjutors, Dr. Williamson and Q., on this subject, hereafter. I proceed, meanwhile, to examine, in my next letter, the pretended facts of another of your sworn authorities, Mr. Baker, who states how long he has been engaged in collecting them; "taking," as he says, "an interest in the question;" and who informs Dr. Loudon that his "attention has been a great deal directed to medical statistics."

LETTER III.

Another Table (No. 2) of Burials in Leeds Township.—Mortuary Registers.—Mr. D. pronounces them worthless, but avails himself of Mr. Thorpe's deductions from similar Sources.—Mr. Baker's Comparative Statement of Bastardy confuted.

Sir,—The object of Mr. Baker being to shew not merely an equal, but a superior, state of health, morality, and longevity, on the part of the factory labourer as compared with others, he commences, very properly, by adverting to the mortality of the township of Leeds, during the last decennary (1821 to 1830), which he puts down at 12,025; whereas it amounted to 16,085. It would be an affectation of candour not to say (using your own term) that every deduction drawn from such pretended facts as his is utterly "worthless." The blunder is so large, that it must appear almost

Factory Statistics.

incredible; I will therefore contrast his statements with the real truth, as given in officially, and which cannot be supposed to err in excess:—

TABLE II.

BURIALS IN THE TOWNSHIP OF LEEDS.		
Years.	Mr. Baker's Statement.	The Truth.
1821	868	1185
1822	930	1325
1823	1209	1684
1824	937	1358
1825	1319	1727
1826	1613	2041
1827	941	1292
1828	1634	2087
1829	1255	1647
1830	1319	1739
Totals....	12,025	16,085

I may appeal to you, whether it is quite necessary or useful to proceed with our examination of this "authority" any further; otherwise the opportunities are abundant. Thus he gives the baptisms in the township, during the same period, as amounting to 22,252, whereas the number officially returned was 25,212. He is wrong also in the number of the marriages. Exhibiting, then, these gross errors, in every instance where he can be checked by known and recorded facts, what reliance, I would ask, can, in reason or justice, be placed upon the statements that rest merely on his individual authority? It appears to me that the greatest errors occur both in the number and the ages of those employed in factories. For instance, it is rather a rare case, if the ages of the piecers and the spinners in the silk mills here average upwards of twenty-one years; at all events, it is difficult to credit that the hands employed in the mills in Holbeck, amount altogether to 1500 only. But all this falls far short of his moral statistics, if I may use the term. All that was ever heard of as to profligacy is exceeded by his comparative statement of bastardy in an agricultural district, which he brings forward for the purpose of comparing it with a manufacturing one. He says, that in a certain village, about fifteen miles from Leeds, of 500 inhabitants, which is purely agricultural, the average number of natural children born annually is nine; while in Leeds, in a population of 71,602, there were affiliated, in 1830, only 58. I cannot suppose that he means to make a quibbling distinction between the illegitimates who are affiliated, and those who are not; or that he is ignorant that numbers of illegitimate children, in this and other large towns, are not affiliated; whereas very few omissions of that sort occur in country places. If he gives the number in question as the gross amount of illegitimate children in Leeds, I will inform him that the statement errs most grossly in deficiency; a fact with which it is rather surprising his practice as a surgeon has not made him acquainted. As to the former statement, I beg of Mr. Baker to give us the name of the place to which he alludes, that his statement may be examined into; when, I am morally certain, it will turn out to be precisely of a similar character to those already noticed. Nine illegitimate children annually, on an average,

Factory Statistics.

in a population of five hundred!! Why, as he professes to have given so much attention to statistics, he ought to know that the average proportion of births to the population, in the strictly agricultural districts, seldom exceeds one in thirty-five; indeed, the ratio calculated by Mr. Rickman as that of the whole kingdom, falls short of that proportion. What, then, does Mr. Baker's assertion imply? Why, that out of about fourteen annual births, or, to give the utmost latitude to the computation, fifteen, nine are illegitimate! Thus, for every legitimate child in this strictly agricultural village, there are nearly two bastards on the average! The statement sets, not decency merely, but moral possibility, at defiance. Let Mr. Baker name the place, in England or in the civilized world, where such a state of things exists, and I will give up the question; but perhaps, it is also the prerogative of your commission to keep places, as well as persons, in this inquiry secret; a necessary rule, I am sure, in many instances.

But I will now turn to his statement about Leeds, when it will be seen, in an instant, that this gentleman has reversed the glass in which he had been contemplating the immorality of the poor agriculturists, and presents us, as regards this manufacturing population, results, if possible, as preposterous as those we have been examining—with the same intent indeed, but results of an opposite kind.

There were only fifty-eight illegitimates born in Leeds in 1830, is what he means to intimate, the population being 71,602; instead of above 1289, which, he argues, is the proportion in his strictly agricultural place, fifteen miles distant! What a happy thing is the establishment of these factories, not only as to their moral effects, on which he dilates at large, but in their physical consequences, on which he thus pronounces:

NOTE—"Puberty, in female factory operatives, is generally retarded two years!"

It is scarcely possible to treat these statements seriously; I think it quite unnecessary to explain the source of this monstrous error, any more than that concerning the births and deaths of the place. I will only just state that, on reading this account, I had the curiosity to inspect the book of my own medical attendant, in which the applications of females desirous of being wet-nurses are usually entered. I could easily shew that a limited proportion only of those so circumstanced apply; as to females in a very wretched state, or factory females, (whom their very employers, Mr. Baker perhaps knows, would seldom condescend to take as domestic servants,) they very rarely apply to be so entered. Still, I found in that one book alone, entered as *unmarried*, above the number Mr. Baker mentions as the whole amount of illegitimate children during the corresponding period. There are —— surgeons in Leeds.

Once more: Mr. Baker states he knows of many men who have been working in factories all their lives, who have reached to the age of sixty and seventy. Who, and where, are they, I ask? They cannot be in Hull or Newcastle, where, he informs you, he lived three years, I think. Since then, he says, he has been upwards of eight years in Leeds, where he speaks of his practice. Now, Sir, you assert, to answer a particular purpose, that there were scarcely any factories in Leeds before 1801. Mr. Baker says, he knows of many men that have worked in factories all their lives, who have reached sixty or seventy years of age; these must consequently have commenced their employment fifty and sixty years ago! Which is to be believed, you or your witness? Your assertions cannot stand together. For my part, I venture, most confidently, to contradict them both.

LETTER IV.

Mr. Baker's alleged Facts.—Mr. D. remonstrated with for adopting them so hastily.—The late Mr. Thackrah's Views of the Factory System misrepresented.—Mr. T.'s Distinction in favour of certain Cotton Woollen Processes approved.—Mr. T. friendly to the Ten Hour Bill.—Mr. Thackrah's Conviction recorded in his Work on Trades, that undue Mortality prevails in Leeds.—Mr. D. charged with adopting Mr. Baker's most egregious blunders.

Factory Statistics.

Sir,—As to Mr. Baker's other opinions, especially on medical points, upon which he has ventured in his anxiety to contradict Mr. Smith, whose evidence on the factory question has most deservedly excited the highest admiration, I shall only say, that he contradicts, point blank, the highest authorities of the profession in existence, as well as others now no more, such as Dr. Perceval, Dr. Aikin, Dr, Fruin, amd many others intimately acquainted with factory labour and its necessary effects, from personal experience. But Mr. Baker's *opinions* are of little importance; it is with the alleged *facts* which he has put forth, that I have had to do. Review these, Sir, or indeed only one of them—the strange statement, for instance, of the amount of the deaths in the town of Leeds, on which his whole argument rests—and I think you will feel ashamed that you have been made the instrument of putting forth such documents; more especially as you might yourself have examined the "*elements*" of his calculation in a few minutes, when the delusion must have been instantly apparent.

Allow me now to make a few observations upon the *deductions* you have drawn from certain of Mr. Baker's "facts," which are precisely worthy of them. Your object being to prove that factory labour is not so injurious to the population as the tables imply, an idea which you treat with the utmost contempt, you first, on the supposed authority of the late Mr. Thackrah, exclude workers in the staple manufacture of the district from the computation entirely; and then take Mr. Baker's authority as to the remainder. I will here first remark, that it is not the mere opinion of any man which is to decide, when the subject in dispute can be determined by an appeal to matters of fact; but, I believe, you are the first individual that has ever represented Mr. Thackrah as an authority in favour of the factory system in any branch whatever, as at present pursued. Your own witness, Mr. Thorpe, very properly recognised him as an antagonist of the factory system, and collected his "facts" to confront his arguments and statements against it. The town and neighbourhood of Leeds well know the conspicuous part he acted there in behalf of the ten hours' bill; and though he made a distinction in favour of certain processes in the woollen manufacture, as being naturally less deleterious than others, as I have constantly done, still he was a strenuous advocate, to the last, for the remission of the undue labour and confinement imposed upon children and young persons in them all. As to a few passing observations on certain pursuits connected with them, generally of a comparative nature, and I think unfairly presented, I shall not remark, but shew you at once what little authority you have from Mr. Thackrah for excluding the woollen branches from your mortuary computations. I will give you his deliberate opinion, not delivered in the select committee—there, according to your notions, it would be dictated by my "leading questions"—nor from, probably, the last observations he made upon the subject, which he gave me in MS.—those could not be authenticated to your satisfaction—but from the last edition of his Work, which you quote at the very time you make the exception referred to: one of the first arguments of which is, to contrast, in a tabular view, the extreme unhealthiness of the West Riding, the seat of the woollen manufacture, with the East and North Ridings, in great measure exempt from it; and, for the same reason, Leeds, with Ripon. But to give his own words:—"We see wool converted into cloth, in establishments so numerous and extensive as almost to supply the civilized world; we see the light-blue flowered product of the field (flax) formed into thread which passes through the eye of the needle, and into the canvass which bears our ships to every region of the globe; we see rough and massy materials drawn from the bowels of the earth, converted, on the one hand, into instruments which surpass in power the united strength of the largest animals, and, on the other, formed into the finest and most delicate pieces of mechanism. These, and works like these, are assuredly wonderful; but while we admire, let us examine. What are the effects of these surprising works—effects, I mean physical and moral? I say nothing of the wealth they produce or have produced; for wealth is good or evil according to its application: I refer to the health of the millions who spend their lives in manufactories, or live by trade, civil arts, or professions. I ask if these millions enjoy that vigour of body which is ever a direct good, and without which all other advantages are comparatively worthless." As to the very point at issue, the mortality of Leeds, and its causes, he says:—"We may fairly say, that at least 450 persons die annually in the borough of Leeds from the injurious effects of manufactures, the crowded state of population, and the consequent bad habits of life! We may say that every day of the year is carried to the grave the corpse of an individual whom nature would long have preserved in health and vigour; every day we see sacrificed to the artificial state of society one, and sometimes two victims, whom the destinies of nature would have spared. The destruction of 458 persons, year by year, in the borough of Leeds, cannot be considered by any benevolent mind as an insignificant affair. Still less can the impaired health, the lingering ailments, the premature decay, mental and corporeal, of nine-tenths of the survivors, be a subject of indifference."—Nor did he, agreeably to the horrible morality of certain apologists and worshippers of Mammon, extenuate these evils because he supposed there were other pursuits equally or even still more pernicious. On the contrary, he comes to this conclusion:—"Assuredly an examination of our civil states and employments has long been demanded, alike by humanity and by science." You, on the contrary, aided by Mr. Baker's medical statistics, treat the proposition, that the present system can have conduced to the degree of mortality the tables indicate, as "a proposition almost too absurd to be seriously refuted." On whose side soever the absurdity lies (which remains to be seen), affecting, as does the point at issue, a question of vital moment to the health and lives of the factory children, I shall take the trouble of seriously refuting your counter proposition.

First, you get rid of the woollen mills altogether, in the manner we have just noticed, mentioning only three mills in Holbeck, of at least nine separate concerns (how many you admit in Leeds you do not state); then, partly on the authority of Mr. Baker, and partly by a sort of guess (the documents relating to the very point being all the while in possession of your Central Board, and might have been consulted by you quite as readily as the "complete census of 1831"), you limit the number of children and young persons so employed, in both Leeds and Holbeck, to 4000. Upon this, you immediately break out thus—"We are therefore required to believe that the general mortality of a population of 83,000 persons is affected to the dreadful extent supposed by an increased mortality in one-twentieth part of the population. Even if we refrain from entering into the considerations above detailed, and take the whole 10,674 persons to be all under twenty,

Factory Statistics.

and all engaged in the most deleterious process that Factory labour can afford, the proposition is almost too absurd to be seriously refuted."

Let us now see whose is the "absurdity." Turning again to your own authority, Mr. Baker, I see that he has furnished you, for another purpose, a table ("absurdly" applied, I grant you), that makes the time during which the workers in several branches specified had been employed, to amount, on the average, to nearly five years each. Indeed, the great majority of the male children employed in mills, who, it is hoped, leave early enough to learn how to earn their bread in other pursuits when grown up, and the early marriages of factory girls, render it highly probable that this term is sufficiently correct. Otherwise, if you mean to argue that the survivors remain in such employments. Mr. Baker's own table of ages, as well as other censuses of factories, which you doubtless have in your possession, will "require you to believe that *their* general mortality is affected" to a far more "dreadful extent" than evidenced by the tables; those above twenty years of age so employed (including, of course, overlookers and others, necessarily adults), being, according to your own observation, only a fourth part of the whole. Now, turning to Mr. Baker's table, No. 2, I find that the average age of those under twenty-one, is a little upwards of fifteen.

If, then, you will turn to "The Summary of the Ages of the Persons in England," in the Population Abstract of 1821, you may find, that in every 20,000 of the inhabitants, there were, between the ages of ten and twenty, about 4313; giving the proportion of $431\frac{3}{10}$ as the number of the mean age. Turning to the table of the ages of Leeds in the same record, you will see that the proportion would be about $436\frac{2}{10}$. The mean population of Leeds in 1821—1831 is 58,992; which would give, as the proportion living at fifteen, $1286\frac{6}{10}$:—or, taking the population of 1831, for both Holbeck and Leeds, on which you choose to found your computation at "83,000" (it amounted to 82,792), the proportion of persons of the age of fifteen would be $1810\frac{2}{10}$. Now, if we multiply this number by five, or add to it any two years in the decennary above and under that age and equidistant from it, the result will be about the same; and we shall have the whole number of children of every class of society, that could be engaged for five years, between the ages of ten and twenty, in factory labour. That number will be, then, in a population of 83,000, about 9051. But, you tell us, that, if we take the whole 10,674 to be under twenty, and all employed in the most deleterious process, &c., it were too absurd to suppose that the effects in question could be produced! I presume you see, ere this, with whom the absurdity rests, and that no possible rectification of your own data, already sufficiently managed, can avail to screen such a conclusion from the fate it merits. I suppose you are well aware that the advocates of the children's cause have always alleged it as one of the peculiar hardships of the entire system, that the great majority of them (as your own documents indeed fully show you) are only employed by it, while children and young persons, a vast majority of them (of the survivors, at least,) having to seek other occupations as they grow up. Such, you know, to be almost entirely the case in certain branches in which great numbers of hands are employed. The advocates of the ten hours' bill assert, that during those few years of hardship and cruelty, at so critical a period of human existence, multitudes of them are grievously injured as to their future health, many crippled, and numbers of them destroyed. How do you rebut this charge? Do you attempt to do it by shewing how small a proportion of the number of children at certain ages, and of a particular class, are employed in factories, compared with the whole number of children of the same ages, and of the same class, or of all classes which the population comprises? No such thing; but by founding your comparison upon the entire amount of the population, including all classes, and every age, from infants of a span to the hoary remnants of past generations; and then inferring, that the proportion is too small seriously to affect the general health and mortality. Just as well might you argue, that it would be absurd to suppose that a five years' education to the whole, or even to half the children of the country, somewhere between the ages of nine and twenty, would have any striking effect upon the morals and intelligence of the community at large. Only think how consoling your mode of computation would have been to Napoleon in the very heat of his contests: "Your Majesty may rest assured, that it is too absurd to suppose that a constant conscription, to the amount of one-twentieth of your population, can have any important effect on the numbers of your people!" But, to exemplify your error by a more close illustration: in the American Census of 1821, there is a section that gives the number of males between sixteen and eighteen, which amounted to 182,205, or to about a forty-third of the whole population; and to about one in twenty-two of the males. Supposing that these were constantly, and successively, engaged in excessive labour, in "the most deleterious process," so as to cripple and deform the limbs, to injure the future health, and, in many cases, to destroy life—do you contend, that because they were only one forty-third part of the whole population, or, rather, one twenty-second of the males, the whole of the males beyond that age would not be necessarily affected? But no illustration can make the mistake appear more absurd than it is at first view. Sir, if you will take the trouble of again thinking upon the subject, and, distrusting as much as you please my calculations, if you will construct a scheme of your own, adhering to your exceptions, and retaining your own proportion, you will find, that, to make up your one-"twentieth part" of the population from the youth of the country, for the average term fixed upon by your sworn authority, there would pass through the crucible of factory labour "of the most deleterious processes" almost exactly ONE-HALF of the entire inhabitants, entailing upon them all the effects of the system, physical and moral, be they what they may!

Factory Statistics.

LETTER V.

Injurious effects of Factory Labour on the Second Generation treated as a "new Assertion" by Mr. Drinkwater.—The "Assertion" shewn to be the reverse of new, and borne out by all the Medical Witnesses.—The Factory System established in Leeds so far back as 1795.—Dr. Aikin's History of Manchester, &c. appealed to for the fact.—Female Factory Workers, parents early.—Puberty, when established, according to the Commissioners.

SIR,—Your argument next presents another of the most egregious mistakes, or rather, mis-statements, ever hazarded. In attempting to prove that the early and excessive mortality exhibited by the tables does not result from the factory system, you say that those "who are bent on finding a result in these tables will say, that the effects are to be traced to the second generation, and account thus for the increase of infant mortality," which, you add, is "a new assertion." Permit me to assure you, that I have never been more painfully affected in pursuing any discussion, than at the utter (I am compelled to call it) recklessness of assertion that you manifest in the present controversy: a habit, inexcusable, I think, under any circumstances, but doubly reprehensible in those in which you have been placed, when you ought not to have availed

Factory Statistics.

yourself of the opportunity afforded you for very different purposes, to put forth, under the cloke of official authority, statements not only void of foundation, but the very reverse of correct. The pernicious effects of the factory system upon the offspring of those doomed to its tyranny, a "new assertion!" Why, sir, it has constantly formed one of the principal objections to the system, and has been put forth by the most active and intelligent friends of the factory children, ever since the question was originally mooted: with regard to myself, it has been one of the topics which I have felt it my duty to press, when advocating their cause, as well as I have been able, and constituted one of those "leading questions," as you term them, of which you complain so much, and as such was put, I believe, to almost all those eminent medical authorities who appeared as witnesses before the committee, to which they returned, in effect, an unanimous and confident opinion—and still you say that it is a "new assertion," and, as yet, "unsupported by proof!" To argue with such assertions as these is a waste of words; a candid mind turns from the task with utter loathing. But I tell you what is really a "new assertion," and equally unfortunate with the former one, namely, when you state that you are required to believe that the extent of the mortality is affected by an increased mortality in one-twentieth part of the population; or whatever be the proportion actually in such employment. Is it really necessary to inform you again, that, from the very first, the advocates of the factory children have asserted, that suffering and mortality were not merely the immediate, but, in perhaps a still greater degree, the ultimate effects of the system; the operatives themselves, also, constantly asserting that premature decay* and early death are the common lot of those who escape an earlier fate? I refer you to Mr. Thackrah's, to Sir Anthony Carlisle's, and indeed to the whole mass of medical evidence, for proof that you are required to believe no such thing as you have stated.

But supposing our "new" assertion, as you call it, be admitted, still,—such is the exuberance of your facts and arguments,—you are not at all at fault. Hence, if the pernicious effects of the system are hereditary, still you assert that that system scarcely existed in Leeds before the commencement of the present century; and, as the period over which the returns extend from which it is attempted to deduce these conclusions, runs back for twenty-one years (1813—1830) "from the present time," "a large proportion of those who thus appear to have died in Leeds, under the age of five years, could not possibly be the children of those whose early and long-continued labour in factories might be asserted to have fitted them to become the parents of a weak and prematurely disappearing offspring." Here, again, I must deny your facts, and correct your deductions. First, it is quite incorrect that the factory system scarcely had existence in Leeds before the commencement of the present century. Besides several other concerns (some of which, if I mistake not, have been discontinued since that period), two of the largest works of the kind in the kingdom had then been long established and in full operation. Two out of the three concerns you mention in Holbeck were then established, and even the third was a split-off from one of them; the population of Leeds and Holbeck at that period was probably short of a third of what it amounts to at present; the hands employed, therefore, must have been much fewer, and still Leeds was, in proportion to its population, justly denominated a manufacturing place.

* How thoroughly Mr. Commisioner Mackintosh sets this question at rest by the following statement in his report (made in June 1833)!—"Although prepared by seeing children occupied in such a manner, it is very difficult to believe the ages of the men advanced in years as given by themselves, SO COMPLETE IS THEIR PREMATURE OLD AGE."—First Report of the Factory Commissioners, A. 2, page 96.

Factory Statistics.

I could give a list of the mills then in existence, but I prefer taking a representation from a well-known work, which contains brief notices of the principal towns in this part of the kingdom. I mean "Dr. Aikin's Description of the Country round Manchester," published in 1795, and compiled probably somewhat before that date. Describing the various branches of industry pursued in Leeds, the account goes on to state: "There are also in the town two carpet manufactories, and a large work has lately been erected for spinning flax by machinery. On the river Air, and the streams that fall into it, (besides numerous mills for grinding corn, dyers' wood, rape-seed, &c.) are mills for fulling cloth, and for turning machinery to spin and card wool. Several cotton mills have also been lately erected, but these are worked chiefly by means of steam-engines."

But even admitting that the factory era commenced with this century, how could you possibly suppose that the children of few of the parents who worked in the factories then, and from that period, could be found in the mortuary registers of the place, commencing with the year 1813? You will probably admit, that the factory workers married as early at that time as at present, and as early here as elsewhere; and also, that they had as many children in as short a space of time as in other factory towns. Then, on the authority of one of the witnesses under your own commission, and who, I will admit, is fully as competent to decide on the subject as any person that by possibility could have been selected, we find that "the women belonging to the factory classes are delivered of their first child—sometimes, but seldom, at sixteen, *very often* at eighteen, *generally before* twenty." Suppose, then, we take the average age of the female factory workers at the commencement of the present century, one year less than it is at present, and fix it at fourteen years; and take the very highest age mentioned by your own witness as that at which they generally become mothers, must not these, on the average, have become mothers six or eight years, and many of them have probably borne the whole number of their children before the first year of the tables to which you so unreasonably object? To these would succeed other factory mothers, increasing, of course, with the increasing population, and, at all events, swelling with their full quota the surge of mortality among infants under five from the very commencement of the table. I can hardly conceive what can have betrayed you into this mistake; you were perhaps thinking of the grandchildren of the factory workers of 1801; or, more probably, confounding the average age of the children in the factory with the period of their birth; which reminds me of a similar and common error, long made by the apologists of the system, in answer to the assertion of the factory operatives, that few of them survive the age of forty; namely, that it was impossible there could be such survivors, the system not having been (then) longer established—forgetting that the infant just dropped from the womb is not equal to its fated task; its bones must acquire some consistency, and its sinews some strength, before it can be delivered over to the tormentors. This, however, is, according to your commissioners, a brief period, as puberty (though, according to Mr. Baker, retarded in factory females two years) is, nevertheless, established at thirteen; at which venerable age the little labourers are, in the proper sense of the word, "free agents!"

LETTER VI.

Dr. Williamson and Q.; their ignorance of the proper uses of Mortuary Registers.—The quarrel not with the "Tables of Mortality," but with the illustrious Halley, Simpson, &c. &c.—Dr. Price appealed to in support of the principle on which such Tables are constructed.—The principle of the "Tables of Mortality" daily recognised in the common transactions of life.

Factory Statistics.

Sir,—I now proceed to notice your coadjutor, Dr. Williamson's argument, if such it may be called, who I understand has, on sundry occasions, as well as in his evidence before you, demonstrated, after a fashion, the "worthlessness" of the tables in question. As, however, a writer under the signature of Q. in the *Leeds Mercury*, who evidently thinks himself acute, pursues the very same line of reasoning, and is also more particular and explicit as an anonymous writer than perhaps he is under his own signature, I shall advert more especially to him, whom, in justice to my argument, I shall be obliged to treat with equal plainness, sincerely hoping that, in so doing, I shall not be exposing the doctor, who, notwithstanding many assurances to that effect, I can hardly bring myself to believe can have been the author of such a series of ignorant blunders. I shall also refer to this last objection more at large, because the opponents of the tables share it in common;—yourself, in particular, found upon it almost the entire argument against their validity.

This, then, is the sum of the objection. The mortuary registers of any place recording the ages as well as the numbers of the dead, will not shew the mean duration or the expectation of life precisely, unless the number of the living, at all ages, during the same term, be also known, and the population remain stationary.

To this it would be sufficient to answer, that the exact expectation or mean duration of life is not professed to be given by, or deduced from, these tables; but, as their very title imports, the comparative mortality at the different ages specified, distinguishing, for the purposes of the inquiry, manufacturing and non-manufacturing towns.

But, when you speak of the worthlessness of the tables, even had they been applied as you seem to suppose, you add, that "it is very well known that a mere register of burials alone" (including, of course, the ages of the dead), "even supposing it to be perfectly accurate, can never be sufficient to give the rate of mortality at different ages, unless in a stationary population." And you proceed to say, that "this is so obvious, that I shall scarcely think it worth while to detain the Central Board by the remark, were it not that I observe that persons who have acquired for themselves a certain sort of reputation for examining these subjects, have not scrupled to make the most absurd statements in connexion with this question;" which, you go on to intimate, are even thought "too ridiculous to be contradicted."

Now, sir, though it is quite obvious to whom you advert in these contemptuous terms, your shafts are misdirected, and fall upon those who can well sustain them. It is not I that have made use of such tables as those in the appendix for the purposes you intimate, but such men as the illustrious Halley, Simpson, De Moivre, Maseres, Kerseboom, Susmilch, and many others: indeed, Dr. Price said, that he knew of no observations extant which furnish the means of forming tables of the probability of the duration of hnman life at every age, other than on the principle of which you speak thus contemptuously, excepting those published by the late M. Wargentin.* Nevertheless, I need not inform you that calculations of the kind in question have been founded upon the mortuary registers of many places both at home

* Dr. Price, Observations on Reversionary Payments, &c., vol. 2, pp. 7, 8.

and abroad, and the computations so formed, though not deemed absolutely exact, have never been rejected; on the contrary, when used for the investment of money, for the creating of annuities, for securing reversions, in short, for the purposes which wealth and riches contemplate to realize by such calculations, they have been relied upon; but the moment similar computations are presented, even in the way of comparison, for the purpose of demonstrating the necessity of serving the cause of poverty and distress, of preserving innocent infancy and childhood from a system of oppression, which, notwithstanding the objections you raise, I shall shew presently, even by your own method of computation, terminates in a frightful degree of early mortality, then, forsooth, the tables, by whomsoever constructed, are utterly "worthless," and all who advert to them for any such purpose, are thus visited with your official contempt and indignation.

I admit, however, that tables, formed from mortuary registers alone, are not, in a fluctuating population, critically exact. The important question then arises, whether the incorrectness, whatever be its extent, in the instances offered in the tables, and in all similar cases, be of such a nature as to falsify, weaken, or render uncertain, the general conclusion deduced from them. It is difficult to suppose you so entirely ignorant of the subject as not to be aware that the circumstances which render those tables otherwise than critically exact, have a direct and necessary tendency to represent the rate of mortality in the manufacturing, as compared with the other places, too low, and not too high; though, if your objections have the slightest validity, they imply directly the reverse.

On this subject, however, you "fight shy." Your assistant, Q., in the *Leeds Mercury*, is more valorous, but less discreet, and, consequently, takes violent umbrage at a note appended to the tables, which you had more discreetly allowed to pass without observation, disposed, as you are, to criticise those few lines, with the utmost severity of which you are capable. The note (which Q. says, is "artfully introduced") is this:—"The effect of the increase of the population of a place by migration, would be to diminish its general mortality." The authority of Milne is given, to whom Q. says he has referred; and, he adds, he finds that Mr. Milne speaks opinionatively, and not from facts, in reference to the city of Carlisle.* All which is untrue:—that able writer speaks authoritatively, and from facts, which he gives, and on which he founds his calculations. But was this point ever disputed? Dr. Price speaks at length upon the subject, and says, that the mortuary registers of a place increased by settlers, will represent the probabilities of life too great before the age at which settlers resort to the place.† Dr. Percival, in his observations on the mortality of Manchester, says, "The proportion of deaths is now considerably less than it was in 1757; but this is chiefly to be ascribed to the large accession of new such settlers from the country. For, as these recruits come in the prime of life, they must raise the proportion of inhabitants to the deaths * * * higher than it otherwise would be."‡ Baily, whom, if at all familiar with the subject, you will allow to be one of the highest authorities extant upon the doctrine of life annuities, explains (though, I think, in nearly Dr. Price's terms) the whole subject in thus expressing himself: "There are two sorts of data for forming tables of the probability of the duration of human life; one is furnished by the registers or *bills of mortality*, which shew the numbers dying at all ages; the other, by the proportion of deaths at all ages to the numbers living at those ages, as discovered by surveys or *enumerations*. Those tables which are deduced from the former of these data are correct

* Milne, on Annuities. 8vo. 1815. pp. 415, 416.
† Price, on Reversionary Payments. 8vo. 1803. Vol. 2, pp. 79—83.
‡ Philosophical Transactions. Vol. 64, p. 58.

only when there is no considerable fluctuation among the inhabitants of a place, and when the births and burials are equal; for when there are more removals from, than to, a place, and the births exceed the burials (as is almost always the case in country parishes and villages), tables so formed give the probability of living too low; and when the contrary happens (as is generally the case in cities and large towns), they give the probabilities of living TOO HIGH."* But as these writers are perhaps among those who have, in your estimation, only "gained a sort of reputation," by considering these subjects, I will give you one, whom, if I mistake not, you prefer to most others—I mean, yourself. You believe in, and account for, so low an annual rate of mortality as one in eighty-six, in a place where your witness, Mr. Thorpe, has, I suppose, his works, because, say you, the increase in the population arose from "the immigration of workers in the mills, as the greater number of such are found between the ages of nine and twenty-one. This at once explains the apparent incongruity, as a large proportion of the population in the most healthy period of life were thus suddenly added to it, without the corresponding number of infants to carry up the general mortality to its ordinary level." Your supposition, however, that the immigration of children generally takes place without that of their parents, is incorrect. Q. states the case better, where he says, that these parents, in search of employment (generally, therefore, in the youthful period of their life), come with their ordinary "baggage"—to use his contemptuous and contemptible phraseology—"of young children" with them. But, as the expectation of life at the average age at which parents commonly remove at least equals that of a child at its birth in the places to which they resort, and as that of the children who accompany them greatly exceeds it, it follows that, on the average, the general mortality is lessened by such accessions. But I feel no disposition to make the most of the argument, being inclined to think that the relative proportion of the juvenile to the prolific classes of inhabitants is more nearly preserved, in all cases, than is generally supposed, and that when that proportion is materially disturbed or varied, it is from different causes to those assigned either by yourself, or by Dr. Williamson and Q.; always excepting the metropolis, in which, for obvious reasons, that proportion is always materially disturbed, and on the mortuary registers of which I would, in this argument, lay no stress.

LETTER VII.

The Author proceeds to argue with Mr. D. on his own principle, viz. that of taking the living from the *Census*, instead of the *Mortuary Registers*.—Still more "frightful" results disclosed.—The Census of 1821, and that of 1831.—The first short Note, appended to the "Tables of Mortality."—Mr. D. reminded that the Note gives comparative, and not absolute, results.—Two new Tables (Nos. 3 and 4) furnished, one of them precisely on Mr. D.'s own principle.—Supposed tendency of Immigration and Increase to account for the appearance of undue Mortality, inquired into.—Mr. D. and his coadjutors in error on that head.—A Table (No. 5) illustrative of the argument.

SIR,—I think I have, by this time, fully succeeded in vindicating the authenticity and validity of the appendix to the factory committee's report, and completely "turned the tables upon you." I will now proceed to show that, on your own principles, you ought to have formed computations which would have presented the results which those tables exhibit, in a still stronger and more striking point of view. But probably, when you laid down so rigidly the exact method which ought to have been adopted, you were not quite

* Baily, on Life Annuities. 8vo. 1810. p. 11.

Factory Statistics.

aware that there were the means of enabling us to pursue it. They, however, are at hand. You quote, indeed, my remarks, which, it appears, are "forcible and true" when they suit you, "absurd" and "ridiculous" when they make against you. Thus, quoting my expressions, used some years since, in reference to the state of national statistics, you add, that notwithstanding the credit due to Mr. Rickman's assiduity, and the addition made to our knowledge by the population returns of 1831, they have left the present question nearly in the state in which it was when I wrote on the subject. Sir, as one of the parliamentary committee to whom the legislature confided the directions relative to the census of 1831, and more particularly in reference to the great credit due to Mr. Rickman, I totally deny your assertion: on the contrary, it is probable few countries in the world have ever possessed data of extent and correctness equal to those of which we are now in possession. The last census gives us the ages of the burials for eighteen years, viz., from 1813 to 1830 inclusive, to the vast amount of four millions, particularizing also the counties, hundreds, and places, in which they occurred. Then, sir, we have in the census of 1821, admitting that census to have been taken in the middle of the year, the mean population of that term divided into sections corresponding with the mortuary registers, and also distinguishing as before, the counties, hundreds, and places, in which they were so classed and enumerated. Thus, sir, you perceive at once, we have the very materials you demand, and on so vast a scale as to ensure every practicable and necessary degree of precision. It is true, that in the latter classification of the United Kingdom, there was a deficiency amounting to the extent of nearly one-thirteenth part of the whole; and that the similar classification of the dead was not quite complete, though still less deficient. The surprise is, that they are so nearly complete; while it must be obvious to any one capable of comprehending the subject, that, had the deficiency been greater, so long as the numbers so enumerated and classed in any place, whether of the living or dying, were sufficiently large, and taken indiscriminately, they would indicate the relative proportions, and these again—the totals in both cases being accurately known—would give the absolute numbers in every section, with all necessary or attainable accuracy; otherwise, no tables that could be contrived in any considerable country could long be of the slightest practical utility; much less could computations founded upon a given number of the same individuals, however exact, such computations being always formed upon a select class, and, therefore, never indicating general results with any tolerable degree of accuracy. But these observations are altogether so obvious, and have been so often repeated by those who have discussed this branch of science, that it was superfluous to make them.

With these public and official documents, then, in our possession,—which I think you will by this time see are rather of a different character to those of Mr. Thorpe or Mr. Baker, which you have so unhesitatingly adopted,—we can put these cavils at the tables to a final test.

You, and Dr. Williamson, or Q., vehemently object to the first short note appended to those tables, which is thus expressed:—"So that about as many have died before their twentieth year, where the factory system extensively prevails, as before their fortieth year, elsewhere;" and you object, forsooth, because the proportion in question was calculated upon the ages of the dead, and not upon the numbers of the living at those periods of life respectively. The question is, whether this note be true and applicable, or otherwise. I proceed to show you that it is; and equally so under any description of proofs to which the subject can be submitted.

You will pardon me for again reminding you that this note has reference only to comparative results, and no more than the tables, from which it is deduced, does it profess to give absolute ones; a distinction, vital to the

Factory Statistics.

whole argument, which you sedulously omit making throughout your whole tirade. When the tables were delivered, I observed to the very able individual who had the principal share in the computation, that they would not accurately express the absolute duration or the expectation of life, in the places they comprised. His answer was, "Certainly not; their title imports to the contrary; but they are good for the purpose required—that of comparison." Has any other use been made of them? You assert that they are worthless for that purpose. I shall show you, even by your own precise rules, that they are conclusive. Calculated on the principle of the "Comparative Tables," the following table will exhibit the relative number of the buried who have lived to an age under twenty, compared with those who have lived to any age under forty, in all those places and towns comprised in the tables, concerning which the facts that you appear to demand are in my possession, or, I believe, at present accessible to me; by including one or two of which, I think I have given you an unnecessary advantage. I have distinguished the factory from the agricultural and the manufacturing but *not* factory places. Ten thousand is assumed as the proportionate number under twenty, and made the radix of the calculation, for the purpose of easy comparison; I need not say the proportions would be the same were the absolute numbers given.

TABLE III.

	Proportion of the Living.		Proportion in the "Tables," Dying.	
	Under 20.	Under 40.	Under 20.	Under 40.
Rutland	10,000	13,666	3,756	5,031
Essex	10,000	13,613	4,279	5,805
Chester	10,000	13,579	4,538	6,066
Norwich	10,000	13,596	4,962	6,049
Carlisle, Old......	10,000	13,455	5,319	6,325
	50,000	67,909	22,854	29,276
Carlisle, New ...	10,000	13,253	5,668	6,927
Bradford (*) ...	10,000	13,074	5,897	7,061
Macclesfield......	10,000	12,966	5,889	7,300
Wigan	10,000	13,076	5,911	7,117
Leeds	10,000	13,005	6,213	7,441
	50,000	65,374	29,578	35,846

Now, I beg to ask you, in what respect the note, stating, that "about as many have died before their twentieth year, where the factory system extensively prevails, as before their fortieth elsewhere," is erroneous? You deem this deduction worthless and absurd, because the number of the living at those ages is not made the basis of the computation. I have calculated these

* I obtained the census of Bradford (in which the ages and sexes of the inhabitants in 1821 were discriminated) some time ago, and for another purpose.

from the official documents, and, arranged as the preceding table, the results are as follow :—

TABLE IV.

	Proportionate Number of the Living calculated on the Population of the Mean Term, 1821.		Proportionate Number of Deaths from 1813 to 1830, Census 1831.	
	Under 20.	Under 40.	Under 20.	Under 40.
Rutland	10,000	15,312	3,756	5,031
Essex	10,000	15,426	4,279	5,805
Chester............	10,000	16,695	4,538	6,066
Norwich	10,000	16,985	4,962	6,049
Carlisle, Old ...	10,000	16,627	5,319	6,325
	50,000	81,045	22,854	29,276
Carlisle, New ...	10,000	16,127	5,668	6,927
Bradford	10,000	15,741	5,897	7,061
Macclesfield ...	10,000	15,565	5,889	7,300
Wigan	10,000	15,065	5,911	7,117
Leeds	10,000	15,139	6,213	7,441
Factory Places...	50,000	77,637	29,578	35,846

Now, Sir, is not the "note" just as applicable and true in this method of computation as by the other; and is the comparative mortality of the factory, compared with the other districts, in the slightest degree varied in either? In both tables, not merely about as many, but more, die in the factory districts under twenty than elsewhere under forty; while the difference between the proportions under forty, in the two methods, comparing the factory with the other districts, only varies half a unit per centum ! !

To shew Dr. Williamson, or Q., and yourself, how erroneous are your assumptions, that "immigration" and "increase" materially affect the proportion between the prolific and the juvenile classes of inhabitants of any place,—an idea that lies at the foundation of all these frivolous objections,—I will take a certain and equal number of towns and counties, including among the counties four Scotch and one Welsh county, and among the cities one Scotch, to render the comparison as satisfactory and general as possible, classing those together first where there has been the least increase, or none at all; and secondly, those where there has been the greatest increase; a method which, I think, must prevent me from making any partial selection so as to serve the argument; and I will thus shew you that the proportions concerning which you raise this dispute vary but little, and that that variation, for reasons which ought to have prepared him for the fact, lies in an opposite direction to what Q. supposes.

Taking, then, twenty places, five counties and five towns in which the increase had been the smallest, or even nothing, during the decennary before the classification of 1821, and five counties and five towns in which that increase during the same term had been the greatest, the following will be the results required :—

Factory Statistics.

TABLE V.(*)

Exhibiting the Proportion of the Inhabitants in 1821, *between* 20 *and* 40, *to every* 10,000 *under* 20, *in the Counties and Towns in which the Increase had been* smallest, *and also in those in which the Increase had been* greatest, *during the preceding Decennary.*

COUNTIES.	Increase per cent. 1811 to 1821.	Population between 20 and 40, to every 10,000 between birth & 20.	TOWNS.	Increase per cent. 1811 to 1821.	Population between 20 and 40, to every 10,000 under 20.
SMALLEST INCREASE.					
Peebles...........	0	5414	Woolwich ...	0	5609
Sutherland	0	5654	Yarmouth ...	0	5289
Salop	5	5271	Shrewsbury ...	5	5445
Hereford	8	5591	Hull	6	5872
Rutland	11	5314	Plymouth	9	5972
Mean Proportions	$4\frac{8}{10}$	5449	Mean Proportions	4	5637
GREATEST INCREASE.					
Lanarkshire	27	5820	Bradford	68	5741
Caithness	27	5789	Glasgow	46	6182
Lancashire	27	5190	Durham	45	5888
Surrey	22	6433	Liverpool	44	5845
Anglesea	20	5184	Macclesfield ...	44	5565
Mean Proportions	$24\frac{6}{10}$	5683	Mean Proportions	$49\frac{4}{10}$	5844

Sir, I now conceive that I have shewn that the fact pointed out to the British parliament and the public in the note in question, which you, and Q., and Dr. Williamson, concur in acknowledging presents a most frightful state of things, is TRUE; that any examination of the data on which it is founded would present, in a still stronger light, the appalling fact, that "about as many die under the age of twenty, where the factory system extensively prevails, as under forty elsewhere!" And, whatever confusion of intellect Q. might labour under, I am sure that you, from the first, were well aware that the comparison instituted extends no further, and that it was not in the contemplation of the framer of those tables, or any of those who have made use of them for the purpose of shewing the urgent necessity of passing the ten-hour factory bill, to give the actual rate of mortality at those two periods, much less at all ages;—you must have known that none of those parties had referred to them as tables shewing the expectation or mean duration of life, or the mortality at all ages; consequently, your criticism, which is false as applied to the periods of life in question, is ridiculous as applied to any other.

* The four great divisions of the United Kingdom give these proportions:—England, 5639; Wales, 5327; Scotland, 5871; Ireland, 5462.

LETTER VIII.

Further Challenge with respect to Leeds.—Mortality of Leeds, in particular, calculated on Mr. Drinkwater's own principle.—Errors of Dr. Williamson and Q.—Table (No. 6) shewing the rate of Increase in six places.—Supposed tendency of Immigration and Increase to account for undue Mortality still further inquired into.—Table (No. 7) of Leeds and Wigan.—The Author takes Leeds, in particular, because Mr. Drinkwater, Q., and Dr. Williamson, have so done.—Table (No. 8) of living and dead in six different places. —Table (No. 9) giving the proportions involved in the preceding Table.

Sir,—I might here, properly enough, terminate my argument, but that I fear lest some advantage should hereafter be taken of my not pressing it yet further. I shall therefore, lastly, calculate, upon your own principle, the mortality of this place [Leeds] in those several periods into which the population is usually divided for that purpose, and contrast it with that of those other districts which Q. has particularized, and touching which he has attempted certain computations, which he of course thinks indicate the fallacy of the tables in question, and the deduction from them expressed in the before-mentioned note. In doing this, I am compelled to expose his utter incompetency to the task he has ventured upon, or, indeed, to any other of a similar nature.

I have already observed, that this Q., and Dr. Williamson, if not yourself, have argued upon the mistaken supposition, that there is a difference in the relative number of the inhabitants of the factory districts of the ages before specified, sufficient to account for the frightful excess of mortality that appears to prevail in them. Thoroughly mistaken in the first place, as to the fact, Q. attempts to account for it in a manner equally erroneous; attributing it to the greater degree of increase which takes place in such towns and districts. The anonymous gentleman takes the trouble of calculating this increase in a given number of instances, in every one of which he makes an egregious blunder; indeed, I am very imperfectly acquainted with anything which figures and facts have to decide, if such a series of mistakes, and of so gross a character, were ever before made and published. Thus, he says, the increase during twenty years in the population of Rutland, was 9 per cent.; of the Metropolis, — per cent.; of Chester, 16 per cent.; of Leeds, 44 per cent; of Liverpool, 37½ per cent.; and of Nottingham, 31 per cent.

I must confess, this string of assertions has put me to some little trouble. Seeing the monstrous errors it involved, I conceived that, perhaps, the printer might have done the computist some injustice, by printing twenty instead of ten years; but I soon saw that any such correction, dated from whatever period since the first census, would leave the error as striking as before; and also, that whether the twenty years were dated from 1801, or 1811 (terminating, therefore, in the first case, in 1821, and in the latter, in 1831), the errors are equally incorrigible. I will again copy Q.'s statement, and contrast it with the truth, calculated from both periods.

TABLE VI.

PLACES.	INCREASE PER CENTUM IN TWENTY YEARS.		
	According to Q.	According to Truth, 1801—1821.	According to Truth, 1811—1831.
Rutland	9	13	18
Metropolis	—	—	—
Chester	16	32	32
Leeds	44	57	97
Liverpool	37½	53	100
Nottingham	31	40	48

Factory Statistics.

Really, one's mind is almost too much filled with pity or disgust at such evidence, to be able to proceed with the subject, and the reader will probably think it unnecessary. I cannot, however, refrain from remarking, that every one of his computations is full of errors. Thus, professing to proportion the population of Rutland, as divided into ages in 1821, to a radix of 20,000 (not to mention differences not exceeding a unit, a latitude which may be allowed in calculations of this nature, where it is desirable to avoid the introduction of decimal fractions), I find his second, third, fifth, ninth, and tenth sums are wrong; some to the amount of 100 or 200 each, the total falling short of that intended by some hundreds. His calculation, on a similar plan, for the West Riding, is wrong in the third, seventh, eighth, and tenth sums, the total, in this instance, exceeding what it ought to be. His computation regarding Leeds is still more incorrect, though his inaccuracies are so numerous and so jumbled together, that the total happens to be nearer the amount at which he aims than in the former cases; but his first, second, third, fourth, fifth, sixth, seventh, ninth, and tenth sums—in short, all but one, are again wrong! And still the anonymous writer can talk about correct conclusions, and I know not what beside.

But I will give this calculator's argument in a correct form, especially as it is yours also, and apply it finally, as you both appear to wish, but which neither of you has, after all, attempted to do.

The ages of 18,449 persons (or nearly every inhabitant) in the county of Rutland, in 1821, were given in the census of that year; the mean number of the population, 1821—1831, is 18,930; the number of deaths in the county during the decennary (including the unentered burials) was 3698.* The ages of 686,208 persons in the West Riding were returned at the same time; the mean number of the inhabitants in the decennary is 883,432. The deaths during that term, including the unentered, amount to 181,913. The ages of 22,822 inhabitants of Leeds were also then returned,—an amount falling far short of the full number, but quite large enough to indicate, with sufficient precision, the true proportions of the whole; nor can Q. have the face to make objections against conclusions of the nature required, drawn from the ages of nearly three-and-twenty thousand of the inhabitants, taken with no possible bias to this or any similar question, since he has built certain of his own conclusions, absurdly enough it is true, upon 120 families, selected according to his own pleasure. Indeed, he has attempted to make a similar calculation himself. In order first, however, to be thoroughly satisfied that the classification was relatively exact, and therefore as strictly correct for our present purpose as though it had been complete, I have calculated the proportions it gives on a radix of 20,000, and compared them with those of other towns similarly circumstanced, one of which I shall instance, not because the results are the most exactly coincident with those of Leeds throughout—rather the contrary—but because the computation may, in this case, be examined; the place being one of those the population of which was so classed in the printed census of 1821. I allude to Wigan, which had been increasing for twenty years before that period, about as rapidly, and indeed a trifle more rapidly, than Leeds. I have only to add, that the classification of the population of Wigan was complete. I conclude the table, as Q. does, at the age of seventy.

* It is for the purpose of including the unentered burials, that I found the ensuing calculations on the decennary, instead of the eighteen years; such unentered burials not being specified for the latter term.

TABLE VII.

AGES.	CENSUS OF 1821.	
	Leeds.	Wigan.
0 — 5	3522	3492
5 — 9	2579	2678
10 — 14	2326	2331
15 — 19	2105	2031
20 — 29	3001	3123
30 — 39	2335	2290
40 — 49	1797	1800
50 — 59	1242	1156
60 — 69	739	711
70 & upwds.	356	388
	20,000	20,000

I think it were a waste of words further to prove the correctness of the relative proportions of the population of Leeds, as given in the official Census of 1821. Indeed, I should not have offered a word upon the subject, but have taken some other factory town where the classification was complete, only that you, as well as Q. and Dr. Williamson, fix upon that town exclusively as the one on which to ground your whole argument. Well, then, the mean population of Leeds during the last decennary was 58,992; the number of the funerals, including the unentered ones, was 16,085. These sums will respectively be divided according to the proportions of the population in 1821, and according to the deaths in that town, at different ages, during eighteen years, as classed in the returns to Parliament for that town. I have also done the same regarding York, observing, however, in fairness, that in this case about one-fourth of the population is found without the walls, and in the ainsty or county of the city. The mean population of the whole, during the before-mentioned period, was 32,909, and the deaths 6665. I have also made a similar computation regarding Norwich, as I understand that city contains a greater degree of that poverty, which the apologists of the present factory system present to us as the alternative of its absence or entire reformation, than most other places of considerable magnitude in England; and, as indigence is unquestionably unfavourable to human existence, the mortality of Norwich ought to be much higher, according to the doctrine of our modern economists, than that of so prosperous a place as Leeds. The deaths in Norwich during the decennary were 13,160, and the mean population 55,438. In the three last instances I have also distinguished the sexes, for the purpose of meeting an observation of yours, or Mr. Baker's, or both, that the system is favourable to female life. Lastly, I have given the same results for Carlisle from 1779 to 1787—when trade and manufactures had established themselves in that city (which had been increasing for many years before with great rapidity), but when the spinning and present manufacturing processes had not been introduced.

TABLE VIII.

In which the Mean Number of the Inhabitants in certain places (1821—1831) is divided into different Ages, agreeably to the Census of 1821; and in which the Number of the Annual Deaths (including "unentered" ones), during the Decennary, is similarly divided, according to their classification in the Mortuary Returns, from 1813 to 1830 inclusive.

AGES.	RUTLAND. Agricultural.		WEST RIDING. Manufacturing, Factory System, and Agricultural.		YORK. Non-Manufacturing.		NORWICH. Manufacturing, but not Spinning.		LEEDS. Factory System.		CARLISLE (Old). Manufacturing, but not Spinning.	
	Living.	Dying.	Living.	Dying.	Living.	Dying.	Living.	Dying.	Living.	Dying.	Living.	Dying.
Under 5	2,634	106·	144,925	7,162·5	4,144	233·8	7,321	555·2	10,299	850·1	1,097	90·2
5— 9	2,387	11·7	122,255	806·5	4,004	30·	6,203	45·2	7,899	66·9	967	9·9
10—14	2,050	9·7	104,676	545·8	3,754	18·9	5,273	22·3	6,874	36·8	761	3·8
15—19	1,965	11·5	92,041	686·5	3,391	23·2	5,130	30·1	5,992	45·3	719	4·9
20—29	2,816	26·3	133,805	1,392·9	5,159	47·6	9,720	72·6	9,210	102·6	1,415	10·7
30—39	1,986	20·9	100,143	1,154·5	4,034	42·7	6,992	70·4	6,754	94·9	934	9 9
40—49	1,809	19·9	75,186	1,049·5	3,504	44·4	5,511	75 3	5,309	96·3	914	13·1
50—59	1,423	28·1	53,202	1,137·5	2,381	48·6	4,250	80·4	3,410	96·4	626	11·4
60—69	1,140	44·	36,013	1,518·9	1,585	61·9	2,930	115·3	2,096	95·4	466	19·2
70—79	566	52·8	16,691	1,662·3	717	65·9	1,662	144·8	938	82·4	204	16·9
80—89	141	34·7	4,162	950·5	218	43·8	419	91·6	196	36·3	62	10·9
90—99	13	4·2	329	119·8	18	5·5	27	12·3	10	4·8	11	3·1
100 & upw.	—	—	4	4·1	—	·2	—	·5	5	·3	2	·4
	18,930	369·8	883,432	18,191·3	32,909	666·5	55,438	1316·	58,992	1608·5	8,178	204·4

TABLE IX.

The foregoing Table gives the following proportions; viz., one death annually in the number and at the ages specified.

Ages.	Rutland.	West Riding.	York.	Norwich.	Leeds.	Carlisle, Old State.
Under 5	25·5	20·2	17·7	13·2	12·1	12·1
5 to 9	208·2	151·6	133·5	137·2	118·2	97·7
10 to 14	215·4	191·8	198·6	236·5	186·8	201·3
15 to 19	174·	134·1	146·2	170·4	132·3	147·
20 to 29	109·1	96·1	108·4	133·9	89·7	132·7
30 to 39	96·8	86·7	94·5	99·3	70·1	94·3
40 to 49	92·6	71·6	78·9	73·2	55·	69.8
50 to 59	51·6	46·8	48·9	52·8	35·3	52·2
60 to 69	26·4	23·7	25·6	25·4	21·9	24·2
70 to 79	10·9	10·	10·9	11·4	10·4	12·
80 to 89	4·1	4·4	4·9	4·5	5·4	5·7
90 to 99	1·2	3·	3·3	2·1	2·1	3.5
100&upwds	....	1·	....	....	1·6	·5

I have only a single "note" to append to this table. The last column, exhibiting the mortality of Carlisle, about half a century ago, includes the ravages of the small-pox, of which there died, in the nine years included in Dr. Haygarth's tables, 238 under ten years of age; deduct these, and the annual mortality of Carlisle at that period of life would be one in $27\frac{8}{10}$, of Leeds at present being one in $19\frac{8}{10}$!

LETTER IX.

Mr. Baker's Observations on the Comparative Longevity of Female Factory Workers, refuted.—Table (No. 10) shewing the Mortality of the Sexes respectively, at different Ages specified, in York, Norwich, and Leeds.—Dr. Kay, of Manchester.—His account of Deaths among certain Weavers, chiefly Females, investigated.—Dr. Kay shewn to be mistaken in his Inferences.—Advice to Dr. Kay.—Conclusion.

Sir,—There are only two or three minor topics remaining now to be noticed. In answer to the observations of Mr. Baker, as to the comparative longevity of female factory labourers (an assertion, by-the-bye, which all the advocates of the system share in common, fully sanctioned by your reports, and which it was necessary to adopt to bolster up the system you defend, in face of the contrary opinion of all the medical men who had been directly questioned on that subject), I have calculated, as before, the mortality of the sexes distinctly in the three following terms.

TABLE X.

Shewing the Mortality of the Sexes respectively, at different Ages specified, in York, Norwich, and Leeds.

AGES.	YORK.		NORWICH.		LEEDS.	
	Males.	Females.	Males.	Females.	Males.	Females.
Under 5	16·	19·6	11·7	14·2	11·2	12·6
5 to 9	116·1	155·5	128·1	147·8	114·4	122·1
10 to 14	166·2	242·6	242·1	230·8	170·8	206·3
15 to 19	142·	149·7	170·8	171·2	136·5	128·5
20 to 29	103·6	114·2	128·7	138·2	92·2	87·6
30 to 39	87·4	101·9	101·3	97·9	72·6	69·8
40 to 49	79·1	79·	68·3	78·1	53·3	57·2
50 to 59	43·2	55·7	49·6	55·9	33·	38·2
60 to 69	22·2	29·8	22·6	28·1	21·2	22·7
70 to 79	8·9	12·7	10·7	12·5	12·4	10·5
80 to 89	4·2	5·5	4·1	5·1	5·4	5·3
90 to 99	2·3	3·5	1·1	2·6	2·4	1·8
100 & upwds.	—	—	—	—	—	—

Comment on this table were surely unnecessary. While, however, upon the subject of this excess of female mortality as attendant on the factory system, I will just advert to a fact, which Dr. Kay, of Manchester, has stated in favour of some mills of Mr. Ashton, of Hyde. It is a circumstance often adverted to by witnesses in favour of the factory bill, that records of disease and mortality in mills are rarely to be depended upon; since, however correct (as far as intention goes) they may be, it rarely happens that the invalided operatives remain in such works, or those whom the system has destroyed, continue in the mills up to the period of their death. But Dr. Kay informs us, that the weavers in the above-mentioned works (chiefly young women) have a funeral club, in which, he says, forty deaths occurred in thirteen years; assuming that thirteen of these deaths occurred in 200 females, during the first six years; and twenty-seven in 400, during the last seven; or in the former period, one death in 92; in the latter, one in 103. He adds these words: "These facts indicate, that the present hours of labour" (fourteen, I think, including meals, as is proposed by your late bill to be perpetuated) "do not injure the health of a population *otherwise favourably situated;* but that when evil results ensue, they must chiefly be ascribed to the combination of this with *other causes of moral and physical depression.*" Dr. Kay will pardon me for observing, that his facts "indicate" directly the contrary. If the other advantages of the operatives in these works be as striking as they are stated to be, it is not "other causes," but the hours of labour alone, that produce, what I will shortly shew him to be, this frightful degree of mortality. Supposing these young weavers average eighteen, which, I understand, is over the usual age, then, turning to the Carlisle tables, as calculated by Mr. Milne, founded on a rate of mortality in that city fifty years ago (a rate it may be hoped hardly applicable to the present improved state of health in the country at large), he will find that the average rate of

Factory Statistics.

mortality ought to have been only one in 143, instead of one in 92 and 103. But this is not the whole. No fact connected with this subject is more clearly ascertained, than that the rate of mortality at corresponding ages is much smaller in the female than in the male sex. If, therefore, we refer to the table No. IV., in the same work (Milne, vol. ii. p. 566), we find that in Sweden (where the general mortality is much higher than in England), the annual mortality of females, at the age of eighteen, is less than one in 157. Hence it follows, that the proportion of deaths among the females in this factory was greater by above 60 per cent. than even in Sweden; in other words, that, to every three persons which nature doomed to expire, the factory system puts out of being, two additional. I trust that when Dr. Kay publishes another edition of his valuable tract, he will expunge the ill-deserved eulogium upon works where one young female in ninety-three annually expires; and that he will not be the apologist of the unmerciful and destructive degree of labour to which they are doomed, by whomsoever exacted. I will only offer, most respectfully, another piece of advice; namely, that if he really wishes for the thorough reformation of the factory system, he will not mix up the sufferings of the poor children with political notions, nor imagine that corn-laws, competition, taxation, and I know not what besides, have been the real causes of the monstrous cruelties so long inflicted. A small matter of research will enable him to see, that the enormities to which they have, from the first, been subjected, existed, and, if possible, to a still greater extent, when English spinners had no competitors in the world; when England had not operative corn-laws; when taxation was comparatively light:—they exist now even in that boasted land of liberty, the United States:—in a word, oppression of the helpless ever did, and ever will, exist, without the strong arm of the law interpose, to shield the weak and the defenceless from the domination of power, and the merciless cupidity of selfish oppressors!

Leeds. M. T. S.

CORRESPONDENCE

BETWEEN

MR. WILSON,

SECRETARY TO THE CENTRAL BOARD OF FACTORY COMMISSIONERS,

AND

MR. STUART,

ONE OF THE COMMISSIONERS.

[This Correspondence is of a most extraordinary character. It shews that the three Commissioners in London have taken upon themselves the fearful responsibility of recommending the *relay system*, in open and direct contradiction to the opinion and advice of several of the District Commissioners. Further, that they have paid no attention to the most essential parts of the recommendation of the District Commissioners, relative to the *wet-spinning of flax*, and the *web-dressing* in power-loom-weaving factories, —employments which Mr. Stuart, himself a commissioner, states to be attended with "imminent danger to the health of young people." And lastly, the London Commissioners are charged, in the most unequivocal terms, with having "omitted all notice of the mass of evidence, both in Scotland and Lancashire, unfavourable to their views."]

LETTER I.

From Mr. Wilson to Mr. Stuart, dated Factory Commission, July 24th, 1833.

Sir,—I am directed by the Central Board to inform you of their intention to print an Appendix, chiefly consisting of statistical and other details, to the first and second report already published. I am further directed to request that you will furnish the Board with any document suitable to such Appendix which may still remain in your hands; and that you will also have the kindness to offer any recommendations which may occur to you with regard to the order in which they should be printed.

I have the honour to be, &c.

LETTER II.

From Mr. Stuart to Mr. Wilson, dated 1, *Park-row, Knightsbridge, July* 29*th.*

Sir,—I have been favoured with your letter of the 24th instant, informing me that the Central Board of Factory Commissioners had directed you to apply to me to furnish them with any document suitable for the Appendix to their reports, which they are about to print, remaining in my hands, and to offer any recommendations which might occur to me as to the order of printing.

Mr. Stuart to Mr. Wilson.

I have therefore, in answer to your communication, now to acquaint you, that I am not aware of my being in possession of any document, not long ago transmitted to you, which ought still to be printed. I have looked through my papers and have not found any such document; but although I have no new documents to send to you, the terms of your present application seem to me to render it necessary that I should state, that there have been material omissions in selecting for the press the documentary evidence formerly transmitted by me, omissions which ought, as I conceive, in the very first place to be supplied, before arranging for the press any additional information now to be printed as an Appendix to the former reports.

I do not know, never until now having been at all consulted on the subject of those reports, why the answers of the mill-owners of the United Kingdom to the circular queries of the Central Board, as printed on pages 88, 89, and 90, of their first report, were not engrossed in it. A tabular view, at least, of those answers, might, I conceive, have been very easily prepared, and could not fail to have afforded very valuable information, most of all in pointing out, as it appears to me, the absolute impracticability, everywhere except in the great towns, and even there, the manifest inconvenience and risk of limiting the hours of labour for children until the commencement of their fourteenth year, to eight hours; above all, the answers to the query relative to the employment of relays of children ought to have been printed, as affording the best information which the Commissioners of the Central Board had obtained on the subject of the chief recommendation in their report; I allude, of course, to the regulation requiring the employment of relays of children under fourteen years of age. I saw enough of the answers of the mill-owners, which I, from time to time, on our journey, forwarded to you, and of the state of the population of the factories which we visited in Scotland, very many of which, especially of the flax factories, are situated on rivers and rivulets, altogether remote from towns, to be thoroughly satisfied that any legislative enactment, rendering the employment of relays of children necessary, cannot be enforced, and that its only effect will be to stop the mills altogether.

I now hope that the information derived from the mill-owners will be detailed in the Appendix, and that it may occupy the first place to which its importance entitles it.

The omission of printing in the first report the answers of Mr. Archibald Buchanan, of the Catrine works, to the queries of the Central Board, together with the appendix to his population list, has already been publicly noticed in the House of Commons. I cannot conceive how it took place, considering that the attention of the Central Board had been most particularly called to the Catrine works; 1st, by the information given to you after our departure, which led you, on the 30th of May, to write anxiously from London to the Commissioners in Glasgow, not to omit to visit those works on account of "the valuable results expected from the personal examination of them:" 2ndly, by my Report, of June 5, from Glasgow, expressing my sense of the great value attached to Mr. Buchanan's opinions, (see A. 1, pages 93, 98, and 99): and 3rdly, by the fact, that his answers respecting the impossibility of employing relays of children at almost the largest and best-regulated of the manufacturing establishments out of the great towns in Scotland are given in detail, and are directly opposed to the recommendations of the Central Board.

I am next to call your attention to another not less material omission in printing my letter of May 14th, from Dundee, to be found A. 1, pp. 21, 22. After the last paragraph, ending with the word "London," my letter, as you will at once perceive on looking at it, contains the following passage, which has been entirely left out in your printed edition of it, on the pages

Mr. Stuart to Mr. Wilson.

above-mentioned:—" I have given Messrs. Shiell and Small (the solicitors for the mill-owners at Dundee) copies of the lists sent by you on the 1st of May.

" While writing to you I was favoured with your letter of 10th May, and accompanying papers, forwarded from Dunfermline. I am surprised at your addressing to me at Dunfermline, to which there is not even a mail coach, or any mail conveyance, but by a boy on horseback. I hope your letters are now addressed according to the route I sent you.

" Since receiving your letter I have, in consequence of the importance which seems to be attached to the answers to the queries adjected to your letter, again seen Mr. William Brown, of the firm of Messrs. William and James Brown, and Mr. William Baxter, of the firm of Messrs. Baxter, Brothers, and Co., two of the chief houses here engaged in the spinning business. I saw them separately, but they entirely agreed in thinking any regulation to obtain the work of young workers for half days, and by relays, impracticable; the parents could not afford to support them if they only got half wages, which was all that the mill-owners could give, supposing them only to work for half of the working hours.

" Mr. Baxter did not think such a regulation in other respects unfavourable to the mill-owners; on the contrary, it seemed to be his opinion that, if it had been practicable, they would be placed in better circumstances, by having a superabundance of hands brought up to the business.

" Mr. Brown was rather inclined to believe that the teaching of the two classes would incommode and create waste, and that inconvenience would be felt by all concerned in shifting the hands during the hours of labour.

" I have not been able to obtain the particulars you require respecting persons receiving parish relief. There is no officer here possessed of such information, so far as I am able to learn. I am, &c. J. STUART.

" Since writing, I have received the enclosed letter from Mr. Hackney, of the company of Messrs. Chalmers and Hackney, who, you will see, are considerable flax-spinners here."

If my recollection is correct, Mr. Hackney's letter also alluded to the employment of relays of children, and it ought therefore now to be printed. The information in that part of my letter to you, to which I have now so particularly referred, was obtained and transmitted in consequence of your letter of 10th May; that part of which relative to this information contains the following paragraphs:—

" It is considered highly desirable by the Central Board that particular attention should be paid to obtaining information in reply to the following queries, as such information may be conducive to important results:—

" What would be the objection (if any) to restricting the employment of children between the ages of nine and thirteen to six or eight hours in the day, and thus to work in two sets, according as the whole day's work might extend to twelve or sixteen hours?

" (It is suggested that such change of sets might be made at the hour of dinner, when the machinery is already stopped.)"

" Would a change of hands so made entail any waste or inferiority of the work?

" If so, describe the nature of such waste or inferiority."

I need not add that I presume you will now (in the Appendix) print my letter of 14th May exactly as you received it.

I have farther to mention, that you have entirely omitted, in the report of my evidence, my letter of 4th May relative to Mr. Cowan's paper factory, near Edinburgh, which ought to have been inserted in page 3 of the evidence taken by me.

There are, I observe, numerous typographical errors in that part of the evidence sent by me, but none, so far as I have noticed, of very great consequence, excepting in my report of 10th June, from Greenock, A. 1, p. 125, in which, and in other parts of my evidence, *wet*-dressing rooms are almost always printed, instead of web-dressing rooms. In reference to that report of 10th June, which I was called on by your letter of 6th June (misdirected and missent to Greenock instead of to Paisley, according to my instructions of the 1st June to you) to make out without the delay of an hour, that it might be in Viscount Melbourne's hands before 14th June, I must take the liberty of observing, that I cannot suppose that his lordship would have required the District Commissioners to frame such reports, had he not intended that, in preparing the general report, some attention should be paid to them, and to the suggestions contained in them, the recommendations adopted where sufficient reasons were shewn, and ground stated for their rejection, where such seemed to be the fittest course. This line of proceeding has not been followed; and the general recommendation of the Central Commissioners, or, properly speaking, of the Commissioners in London, is directly opposed to the opinion and advice of several of the Commissioners as well as of myself, to whose views and observations little or no consideration seems to have been given by the Commissioners in London. Mr. Cowell told me the other day, that he, who was one of the Commissioners in Lancashire and at Manchester, was certain, that the plan of the Central Board could not be carried into effect in the district, certainly one of the most important in the kingdom, which he visited. I cannot discover from the terms of the general report that the Commissioners of the Central Board had paid any attention to what, I am persuaded, are the most essential parts of the recommendations of the District Commissioners, especially of myself, A. 1, p. 125, relative to the *wet-spinning* of flax, and the *web-dressing* in power-loom weaving factories, employments obviously attended with imminent danger to the health of young people.

Mr. Stuart to Mr. Wilson.

There was quite enough of time, after our return, for the Commissioners, who had seen the state of the factories, and of the population throughout the manufacturing districts of this country, to have had a meeting, and to have agreed on a general report, after perusing the *whole* evidence. The course which has been pursued seems to me the most unsatisfactory that could well be devised; and I fear the Appendix, however it may be arranged, will not adequately supply the omissions and obvious defects necessarily attending the plan hitherto followed. I was well aware of their existence as soon as I saw the first report; but *I of course* abstained from all communication of my sentiments, when I observed it authoritatively stated in your report, that the three gentlemen subscribing it had been, on the 20th June, after we had returned to town, and had reported our return to you, and at the Board, directed by the Secretary of State for the Home Department, "to state any opinions or recommendations formed on consideration of the whole of the evidence," all mention of the opinions or recommendations of the twelve Commissioners who had inspected the factories being entirely omitted.

If, therefore, I had not received your application of the 24th, I should not have thought myself justified in sending you this letter, however important the observations contained in it still seem to myself to be.

LETTER III.

From Mr. Wilson to Mr. Stuart, dated July 30th.

Sir,—I have to acknowledge your letter of the 29th instant. In reply to your general observations on the subject of the answer returned by the mill-

Mr. Wilson to Mr. Stuart.

owners to the queries of the Central Board, I am directed to say, that these answers will form a principal part of the Appendix to the Reports already published, but that their earlier publication has been impracticable, first from their bulk, and secondly, from the desire of the Board that the results obtained from the tabular lists which were circulated along with them should be published at the same time.

With regard to the particular errata and omissions of which you complain as having occurred in your own reports, I can only repeat the regret which I have already expressed personally, that my offer to send proofs from time to time for your correction was not accepted by you, as it probably might have obviated most of the errors complained of, and might have secured the insertion in the body of the reports of any documents of which immediate insertion might have seemed advisable. I must, however, except from this description the returns of Mr. Buchanan, which are referred to in your letter, and which cannot well appear in any other manner than in company with others of a similar nature.

I am further directed to state that the Board have been far from inattentive to the suggestions conveyed in your reports, and in those of the other District Commissioners. It, however, was impossible to shape their own recommendations so as to coincide with others, which were frequently conflicting. They have, therefore, acted simply in obedience to the directions of the Secretary of State for the Home Department, by "stating the opinions and recommendations which they had formed on consideration of the whole of the evidence, and have published the above-mentioned suggestions of the District Commissioners in a sufficiently prominent manner to direct attention to their merits. They direct me to add, that after all the information since received by them, they have seen no reason to retract the recommendation contained in their report, that the labour of children should be restricted to eight hours a day, subject to the stages suggested by them for the practical operation of the restriction; and with regard to the practicability of their plan, they deem it sufficient to have received the acquiescence of the principal manufacturers now in town, as deputies from their respective districts. I beg to refer to the evidence taken before the Central Board, and to the printed copy herewith enclosed, of resolutions entered into at a meeting of the deputies from the principal seats of the cotton manufacture, held in London on the 16th instant.

I have only further to add, with respect to the order proposed for the Appendix, that the first place will probably be occupied by some further examinations by District Commissioners, which were only placed in the hands of the Board on the 23rd instant, and that the documents which form an Appendix to your Report can follow next in order with other supplementary documents of your colleagues.

LETTER IV.

From Mr. Stuart to Mr. Wilson, dated July 31.

SIR,—I am sorry to find, on perusing your letter of yesterday, that it is incumbent on me to give you the annoyance of a reply. I might otherwise be supposed to acquiesce in some of your statements and observations, which do not appear to me to be by any means well founded.

With respect to the errata and omissions in printing the evidence sent to you by me, as pointed out in my last communication to you, I cannot admit the propriety of your remark, that probably they might not have taken place

Mr. Stuart to Mr. Wilson.

had your offer to send me the proofs of the evidence from time to time, when on our journey, been accepted. Now I pray you to notice, that my former letter, though specifying one very remarkable error of the press, relates almost entirely to palpable omissions, one of which, consisting of a page and a half of a letter, could not have happened without special directions. I have no notion, that if I had read over this letter in any of the inns, when occupied in taking the depositions of witnesses, which was our constant employment, that the omission would have occurred to me, or that in that case I should have rendered myself more responsible than I now am for the whole of my letter not having been printed. But further I beg you to recal to your recollection the precise terms in which you expressed yourself, when sending us the first and only proof on 23rd May. "I find them printed so correctly (you write), that unless you wish them regularly transmitted, as they come from the printer, *I think I could engage to have them struck off free from errors of consequence.*" In point of fact, the evidence sent by me was most distinctly written out, either by myself or by a clerk, all compared and read over by myself, so that I could not foresee errors, far less material and obvious omissions, and accordingly I wrote you on the 27th May, that we thought our time might be more profitably occupied than in rigidly comparing the proofs. I even on that occasion, however, on 27th of May, begged you to delete a passage, elicited by a question put by Sir David Barry, and which should only have appeared in a medical report, but no attention was paid to my request, and the passage remains. In like manner, when you, on our return on 19th June, gave me my evidence for a single day to correct, and I returned it at the appointed time with various corrections, none of them were adopted, although the report did not appear for ten days afterwards. You cannot, therefore, as it appears to me, shift from the shoulders of those appointed to manage the business here, any part of the responsibility obviously attaching to them, to lay it on the District Commissioners; and I submit, with perfect confidence, that no part of that responsibility, or of your duty, was so clear and plain as that which bound you that your report, which was expected to be the ground-work of parliamentary legislation, should convey full and impartial details, whatever might be the tendency of the evidence of the whole depositions and reports communicated by the District Commissioners. It is, therefore, not without astonishment that I have read that part of your present communication in which you give me to understand, that even if I had particularly, more particularly than I did, suggested its insertion, Mr. Buchanan's answers to the queries would not have appeared in former reports, nor "in any other manner than in company with others of a similar nature," that is to say, not until the recommendation of the Central Commissioners had been successful by the carrying of Lord Althorp's motion on the 5th of July, which was expected to be the result of the discussion, when the consideration of Mr. Buchanan's views must of course have become altogether useless, by a measure directly opposed to them being carried.

I have, in my last letter, as I think, mentioned sufficient reasons why Mr. Buchanan's answers ought most especially to have been printed in the first report; and I now take the liberty to add, that on reference to his deposition, to which they were appended, A. 1, p. 98, and to my letter, p. 93, it is clear that I intended them to form part of the evidence to be printed. The Central Board do, therefore, even now, it is obvious, maintain the right to withhold, until after the period had expired during which they had reason to expect it was to be of any use, what I maintain to be the most material evidence which they could procure on the subject of their own recommendation. I hardly, I must confess, expected such an avowal as this from the Central Board, an avowal which amounts to neither more nor less than this, that they, in order to shew their adherence to the merest point of form, which they themselves had established, with what views I know not, conceived themselves entitled, "because Mr. Buchanan's returns could not well appear in any other manner than in company with others of a similar nature," to suppress evidence of the "valuable results" from which they had previously declared themselves so fully sensible, that they specially directed Sir David Barry and me to travel above thirty miles on purpose to procure it, and which I, who obtained it, had shewn myself so anxious to be noticed and founded on, that I incorporated it with Mr. Buchanan's deposition, which I did in no other instance, writing at the same time to you, A. 1, p. 93. "At Catrine, I found it hardly necessary to commit any of the information I obtained, on inspecting the great manufacturing establishment of Messrs. James Finlay and Company there, to paper, Mr. Archibald Buchanan, the resident and managing partner, having, in his answers to the queries from the Commissioners, which he has printed, described the works, their origin and history, and communicated all necessary details with so great accuracy and care, that I can refer the Central Board to them, as containing an account of these interesting works far more to be relied on than I could otherwise have procured on a short visit to the spot. I therefore only took the evidence of Mr. Buchanan on two or three points in the answers which seemed to require further explanations, and which you will find in the report, to which is attached a copy of Mr. Buchanan's answers."

"I presume that the Central Board is aware that Mr. Buchanan is a gentleman of the greatest respectability, the oldest cotton-spinner in Scotland, originally educated in the spinning business with Sir Richard Arkwright, and afterwards having successively had the management of the works at Barlindalloch, Deanstown, and at Catrine. The Board will therefore judge of the value attaching to communications received from such a quarter."

The Board may rest assured, that whatever may have been their views in omitting all notice of such a document as Mr. Buchanan's answers, the general impression which that omission has created of important evidence on the subject of the inquiry having been withheld, because opposed to their recommendation, will not be removed by the mere assertion contained in your letter, that "the Board have been far from inattentive to the suggestions contained in my reports, and in those of the other District Commissioners;" for where, I would ask, am I to find evidence of the correctness of this assertion?

Is it in the omission of that part of my letter from Dundee, alluded to in my last communication to you, containing important information altogether unfavourable to the recommendation of the Central Board?

Is it in the absence of all mention of the decisive facts stated by Mr. Buchanan, which apply to almost all the country factories in Scotland? Mr. Buchanan states that "a double set of hands could not be collected, and sufficiently trained, for a series of years; that, supposing a relay of hands under twenty-one years of age could be got, the adults could not be supposed to work above fourteen hours in the day, allowing the least possible time for meals and refreshments, which would only give seven hours to those under twenty-one, and the produce in that time could not afford more than half the present wages to the younger hands, which would be found totally inadequate to afford the means of support and comforts which that class now enjoys, and that, since all his workpeople are paid by the piece, he does not see the possibility of distinguishing the produce belonging to each so as to pay for their individual exertions."

Is it in the absence of all reference to the evidence of Mr. Houston, one of the proprietors of three cotton factories, at and near the village of John-

Mr. Stuart to Mr. Wilson.

stone, in Renfrewshire, the cotton works at which are the second in point of extent in Scotland? Your attention was particularly directed to Mr. Houston's evidence by my report of the 7th of June, A. 1, p. 109—"Mr. Houston is very extensively concerned in the cotton-mills of this district. His evidence is very important, *if any plan is in agitation for preventing the employment of workers under fourteen years of age.* The children or young people at his mill, which I saw, are very well attended to, and seemed to me quite contented with their situation, but I beg to refer to Mr. Houston's statement." The whole of his deposition contains valuable matter, but with respect to a relay of hands, it bears expressly that "he is of opinion, that any such regulation could not possibly be carried into effect, so far as respects the children engaged in the throstle-spinning here, *because the requisite number of children could not be procured in this neighbourhood*, and, at all events, the children could not be procured for the half of the wages they have at present."—A. 1, p. 110.

It will not do to tell me, as you now do, that the suggestions of the district commissioners, (although not noticed in your report, were published in your report,) were published in a sufficiently prominent manner, the fact undeniably being that you have not only in the first report referred to but printed any information which you could find at all favourable to your recommendation—such as the evidence of Mr. Ashworth, and the proprietors of the Burley Mill, near Leeds, pp. 54, 58, of the report of the Central Board; while you at the same time make not the slightest allusion to the direct and positive testimony of Mr. Houston, Mr. Buchanan, Mr. Baxter, Mr. Brown, and many other most respectable well-informed individuals on the other side, both in Scotland and England.

Is it in the absence, generally, of all evidence, unfavourable to your recommendation of relays of children, contained in the reports, which you have quoted, of the District Commissions, in England? I confine my remarks, among a multitude of examples, to the two following remarkable instances:—1. Mr. Cowell reports, D. 1., p. 58, that Edward Chatterton, overlooker, employed forty-seven years in a factory; John Warmby, employed thirty years; Benjamin Wood, thirty years; James Whittaker, overlooker, thirty-two years; Thomas Cope, forty-two years, and John Entwistle, general overlooker, twenty-three years, "sensible and respectable men speaking their real sentiments, having among them nineteen children employed in factories, all sent for examination by the committee of master spinners opposed to the bill, declared on oath, that 'relays of children were not practicable.'" D. 1, p. 62.—2. Mr. Cowell reports the opinion of Samuel Hunter, and of twenty-six work people, whose names he gives, employed in the cotton-mills of Messrs. Ashworth, near Bolton, who declare, that "a bill to reduce the hours of labour for children below fourteen years of age to six or eight hours per day would have the effect of deranging the order of mills altogether, for they could not be carried on without that class of children, *and a sufficient number could not be found to form relays or changes of hands*, neither could parents maintain their children upon half wages."—D. 1, pp. 130, 131. Be it observed, that Mr. Cowell's evidence is derived altogether from Lancashire.

Finally, is it in the absence of all allusion to that part of my report, and of the reports of the other commissioners, respecting wet flax-spinning and web-dressing employments, that I am to look for evidence of the Central Board's having been "far from inattentive to the suggestions contained in our reports?" To the employments in question, though most obviously tending to produce dangerous disease, and to shorten life, you have not made the slightest reference. Here is a clear case, hitherto free from "conflicting recommendations," undoubtedly requiring a legislative remedy,

Mr. Stuart to Mr. Wilson.

which can be applied with ease and safety, nothing more being necessary, as is proved (see A. 1, p. 128), than that the manufacturer shall incur some additional expense by employing adults. The dangers of wet spinning to health can in no way be prevented by any precautions hitherto discovered, but the Central Board, although alive to the suggestions of the commissioners, do not bestow the slightest attention on them, notwithstanding the following statement in my report from Greenock, A. 1, p. 125. "I witnessed a more painful sight, again and again, in beholding the miserable, unhealthy-looking beings, in the wet-spinning and web-dressing apartments, though comparatively few in number, than in all the other parts of the many factories I have now visited."

You are pleased, in the concluding part of your letter, to inform me, that the Central Board have seen no reason to retract the recommendation contained in their report, that the labour of children should be restricted to eight hours a day. I am not surprised to receive this communication of their present opinion, because I know well how difficult it is to convince those who in the closet have adopted a theoretical notion, in accordance with their general and preconceived views, that it is at variance with facts, and with the evidence afforded by the senses. Still I have so much faith in the candour of the gentlemen composing the Central Board, as to be thoroughly satisfied that had they seen, as we did, the population of the factories, and had they thus been made aware of the lamentable effects to the younger part of it to be apprehended from giving effect to their recommendation, they would have paused in the mischievous and hazardous course they appear to me even at this hour to be pursuing.

I do not attach, I am bound to say, the slightest value to the minute of a meeting of deputies "from the principal seats of the cotton manufacture," held at Palace-yard on the 16th instant, and of which you have sent me a printed copy. I do not know how many manufacturers attended this meeting of the thousands in the United Kingdom, but I have reason to believe, that only one manufacturer from Scotland, except the gentlemen in parliament, was present, viz. Mr. Holdsworth, of Glasgow, and that gentleman had no authority to act for the manufacturers of Scotland. That gentleman was as decidedly opposed to the system of relays of children, when I was at Glasgow last month, as all the other eminent manufacturers there, or whom I saw in Scotland. But he has told me, since the meeting of the 16th of July, that he now gave his approval to the measure, because he knew it to be impracticable, as the inspectors to be appointed would report within the first year. But if you are to be influenced by the opinions of the master spinners assembled at Palace-yard, you should, I apprehend, give fair notice of your intention, so that the population of the factories, and especially the younger population, may have their representatives at this bit of a parliament, to whose wishes you are now paying so much deference. Many of the great master spinners are said to be anxious to put down the small establishments in the country. Your recommendation, if carried into effect, would, of course, be attended with this to them beneficial result. And so far they, or some of them, may give it their approbation.

I have only to add, that you are, as I sincerely think, incurring a tremendous responsibility by acting on such evidence as you describe, opposed as it is, not by opinions alone, but by facts stated on oath by the manufacturers to whom I have alluded, and by the reports of the commissioners appointed by the Government to give information respecting the factories in the most important parts of the kingdom—viz., Lancashire and Scotland.

If your views shall be carried by the force of the Government, the consequences will inevitably be to convert a considerable part of the population called into existence by the establishment of our manufactures into paupers, and to do irretrievable mischief by depriving great numbers of families of the means of support.

Had you been in the House of Commons on the last discussion relative to the subject, you could not fail to have observed, that the view in which the House generally concurred was that so ably stated and so well illustrated by Mr. Bolling, one of the members for Bolton, who contended, most successfully, as I thought, that nothing more could, with prudence, be done this session, than to extend Sir John Hobhouse's act, with a clause authorizing the appointment of inspectors to all factories, and to delay to legislate further until next session, when the reports of the inspectors might afford better information than has yet been obtained.

It is possible, however, that such a measure as you advise may be carried in a thin House towards the close of the Session. And I therefore still feel it to be my duty, from the conviction I sincerely entertain of its danger, to do all in my power, by availing myself of the opportunity afforded me by this correspondence, to advertise you of the fatal consequences to be apprehended from success.

I write on the spur of the moment, having heard that you are actually engaged in framing a bill, giving effect to your recommendation.

LETTER V.

From Mr. Wilson to Mr. Stuart, dated August 1.

Sir,—Your letter was placed in my hands about half-past five o'clock yesterday afternoon, as I was leaving the office, and not having communicated since with any member of the Board, I can only answer a small part of its contents. I hasten to do so, because, should you think any further correspondence necessary, it is better that I should *now* disentangle those matters of detail which relate to the printing of your reports from the remaining part of your discussion with the Board. I beg to say, therefore, that I am now answering merely for myself, and merely with reference to those points to which I can speak, from the share which I had in giving directions to the printer.

It was very early discussed between myself and the printer what should be done with the various documents sent to the Board, along with your minutes of evidence and those of your colleagues. The insertion of them in the body of the evidence appeared so unsightly, and so unlike the usual form of reports, that it was agreed to reserve such documents for an appendix, which should contain (with proper references to the several reports) all such matters as appeared inconvenient for insertion in the reports themselves. Even the letters which you sent to the Board from time to time, were left out in the first instance in the printing, as it was not known whether they were meant for any inspection besides that of the Board, and it was only on receipt of your letter of May 27, from Glasgow, that I proceeded to place those letters in that part of the minutes to which they belonged, in doing which I remember only general directions from the board to omit whatever was not strictly public. I perceive that the first part of the omitted passage, in your letter of the 14th of May, is of a private nature, and conclude that the omission of the following passages has arisen from mistake in extending the *dele* meant for the first passage. Had the omission been discovered or pointed out by you in time, I will take upon me to say that it should have been rectified in the second report, as I beg to remind you was done in the case of an omission

Mr. Wilson to Mr. Stuart.

equally manifest (more so indeed, as it occurred in a large section of the papers of the Central Board itself). The only excuse for such omissions was, that of the haste and distraction from one branch of the business of the office to another, which, though the occasion of it might seem trifling, was harassing to a degree, which, I believe, was felt by persons more experienced in affairs than I can pretend to be.

I have only a word to add with regard to the corrections made by you in town; so far as I remember they were very slight indeed, with the exception of one, I think, at page 84, which was attended to. My general directions to the printer, with regard to the rest, and those of some of the other district commissioners, was to correct as far as he could, without disproportionate delay and expense. Though the report was not actually published for a week or ten days afterwards, yet at the moment I am speaking of I was enjoined to hasten the completion of the printing without an hour of delay, while the composition of the report of the Board was going on at the same time, and affording, of course, additional occupation, however subordinate my part or lot in the matter.

I can only say, in conclusion, that whatever blame may be due to myself for the original notion on which I acted, in keeping documents for the appendix, I must bear it. I have no hesitation in expressing regret that the notion should have been acted upon in your case more completely than in that of some of your colleagues, though it was only deviated from on pressing requests to that effect from some of them. Had you expressed, yourself, on arriving in town, and looking over your evidence, any such sense of important omissions having been made, or of supplementary matter being requisite, as you now entertain, inconvenience and expense would have been yielded to, had you urged it.

LETTER VI.

From Mr. Stuart to Mr. Wilson, dated August 2.

SIR,—I am favoured with your letter of yesterday. I hope that you do not entertain the erroneous notion, that I meant to apply any part of the remarks contained in my two last letters to you individually, or in any other capacity than as acting with and for the Central Board. The whole of my correspondence has been with you as secretary to the Board, although in the first printed report my letters are by some strange mistake said to be addressed to the chairman of the Board, and not to you.

Conceiving that it was the duty of the Central Board to see that the evidence sent by the district commissioners was fully and fairly laid before the House of Commons, I cannot admit your right to separate my observations respecting what I conceive to have been very material omissions in printing the evidence, for which you seem to wish that you alone should be considered answerable, "from the remaining part of my discussion with the Board." The Board is obviously responsible for the whole, and for the delay still taking place in publishing the very valuable documents to be contained in the appendix, while Parliament has again, as I observe from the newspapers this morning, been called on, in the absence of the important information to be afforded by them, to legislate on the subject.

I have now, with a view to free *myself* from all responsibility, to request that copies of this and of my two last letters may be sent in the course of a day or two to the Secretary of State for the Home Department, to the president of the Board of Trade, and to the Chancellor of the Exchequer, and that they may be inserted in your appendix now printing.

I hope to receive a note from you, not later than to-morrow, to let me know that the Board has authorized you to comply with my request.

LETTER VII.

From Mr. Wilson to Mr. Stuart, dated August 2.

Sir,—I am directed, in answer to the favour of yours of the 31st ult., to state that the Central Board will always be glad to have the benefit of any remarks which you may wish to make on the conclusions which they have come to in their report, or on their method of arriving at those conclusions. And they will most readily confer with you on the subject whenever it may suit your convenience to call here for that purpose. But they cannot consent to enter into a written defence of every part of their proceedings which may appear to you to be liable to objection; for if they were to allow themselves to be drawn into a controversial correspondence by the remarks which may be made by individual commissioners on the several points embraced by their report, there would be no end of the labour which might thus become entailed upon them—a labour at least equal to that of framing the report itself. There are, however, one or two points in your communication to which they think it right to address some reply.

I have already explained, so far as lay in my power, in my letter to you yesterday, the omission of a passage in your letter of the 14th of May. Having communicated to the Board a copy of the first-named letter, I have only to add that they concur in that part of my statement which concerns them. They regret the omission referred to, as you now lay so much stress upon it, but it was most certainly undesigned on their part. Indeed, they were not aware of it at the time of its occurrence. And it may serve as a presumption of their having had no motive for suppression in the instance abovementioned, that they have given insertion, without the smallest reserve or alteration, to statements equally strong against their plan from highly respectable quarters.

With reference to your strictures on the delay which has occurred in publishing the very bulky returns made by the mill-owners to the printed queries issued from the Central Board, I beg to say, that amongst other reasons why, as the printing of all the documents in the hands of the Board could not take place simultaneously, the returns to the printed queries and tabular lists were those which it was thought might be postponed with least disadvantage; one was, that the printed queries, having been framed at the very outset of the commission, did not appear to elicit information so directly bearing on the plan afterwards contemplated, and eventually recommended, by the Board for the restriction of the labour of children under their fourteenth year. Among the printed queries, the principal question referring to sets of children includes (in the terms of the former bills) all persons under twenty-one. Accordingly, the answer to that question by Mr. Buchanan, which you quote as containing decisive facts adverse to the plan of the Board, decides, in effect, nothing at all either for or against that plan. Without disputing the correctness of the statements made by Mr. Buchanan (which, however, properly speaking, are statements not of fact but of opinion), I beg to observe, that his answer refers distinctly and expressly to *relays of hands*

Mr. Wilson to Mr. Stuart.

under twenty-one years of age. Now, as the Board recommends that all restriction of the hours of labour should cease after the thirteenth year complete, they cannot have the slightest conceivable motive for dissembling the impossibility of finding relays of hands up to the twenty-first year.

I have just received your letter of this morning, and as the Board are not now sitting, must defer reply till to-morrow.

LETTER VIII.

From Mr. Stuart to Mr. Wilson, dated August 3.

Sir,—Your letter of yesterday was delivered here this morning.

I beg leave to notice, with reference to the first paragraph of that letter, that my present correspondence with you on the part of the Central Board was not of my seeking, but has grown out of your written application to me for documents for your appendix, and for advice as to their arrangement. I called for no defence of your conduct, as you will see on referring to my first letter, but merely requested you to supply former omissions, which I pointed out.

I am very glad to close the correspondence, as soon as I point out certain mistakes contained in your letter now received; but, in the first place, allow me to mention, that although I am not aware that any beneficial object is likely to result from my having a conference with the Central Board now for the first time proposed, I can have no objection to wait on them, should they wish it, on any day after Tuesday next. I have already formed engagements for Monday and Tuesday.

You labour, I have now to observe, under a most extraordinary misapprehension in maintaining, as you do, in decided terms, "that the printed queries from the Central Board did not appear to elicit information so directly bearing on the plan recommended by the Board." I pray you, Sir, to read the 32nd question, which is in these terms:—"Has a system of relays, or change of hands in the employment of children, when extra hours are required, been tried? And if so, how has it answered? If it has not been tried, why not?"

Here is Mr. Buchanan's answer to that question:—"A change of hands for extra hours has never been practised by us, because, in our situation, we deemed it not only unprofitable, *but impracticable.*"

This answer is absolute and unqualified, applying to children of all ages. Of the same description is that part of Mr. Buchanan's answer, formerly quoted by me, in which he asserts, "that a double set of hands could not be collected and sufficiently trained for a series of years."

I make no comment on the terms of these quotations, because they are expressed in language so plain and easily understood that no one who reads them can doubt—1st, that the queries of the Central Board were calculated to elicit information whether relays of children were practicable; and, 2nd, that Mr. Buchanan's answer, not as matter of opinion, (as you, I must say, without due consideration, assert,) but as matter of fact, depending on his knowledge of the population at Catrine, a population entirely connected with his own works, as I can testify, after being on the spot, is explicit and positive, that "a change of hands in his situation is impracticable."

How is it possible, in the face of such evidence as this, to contend, as you now do, that Mr. Buchanan's answer "decides in effect nothing at all either for or against the plan of the Board" recommending a scheme rendering relays of children absolutely necessary for its execution. Mr. Buchanan also gives a very decided opinion unfavourable to this scheme, in his general answer to the last and general query of the Board,—"Have you any further observations to make with reference to the subject of the present inquiry?" "Should it be judged necessary (he answers) to regulate the hours of labour, he is of opinion that could be most conveniently done by stopping the moving power, whether water or steam, at a given time on each day, and after having wrought the established hours, under a heavy penalty to those who exceed," adding, "that he thought the twelve hours' work in the day would be found the best for all parties."

Such is the evidence, both in point of fact and of opinion, of the most experienced master cotton-spinner in Scotland, the managing partner of one of the greatest manufacturing establishments in that country, with reference to the scheme which the Central Board persist in recommending to the sanction of parliament.

It is perfectly true, as you notice in that part of your letter in which you allude to the omissions in the printing of documents, that evidence received from the district commissioners, though unfavourable to the recommendation of the Central Board, has been printed by them. Certainly this is true to a considerable extent; but I have, in my letters of the 29th and 31st of July, pointed out great, extensive, and palpable omissions in the printing of such evidence, and shewn that the Central Board have, in their general report, the only document at all likely to be read, omitted *all reference* to the evidence unfavourable to their recommendation, while they found on and quote such evidence as they could discover in any degree of an opposite tendency.

I hope you will be so good as to send me copies of the amended bill, and of your appendix, as soon as ready. I have to beg, that if the Board should wish to have a conference with me, they will send me these documents in such time that I may peruse and consider them, and be ready to refer to the evidence, before the period appointed for our meeting.

LETTER IX.

From Mr. Wilson to Mr. Stuart, dated Factory Commission, August 3, 1833.

Sir,—I am directed by the Central Board, in answer to your letter of yesterday, to state that they have no wish to shift from themselves on me, in the capacity of secretary, any responsibility which may attach for the omission which has taken place in the printing of your letter of the 14th May. They are quite willing to take upon themselves whatever blame may be fairly imputable for that omission, which, however, they must again, in the strongest manner, declare was on their part most entirely unintentional. And this being the case, they trust that they shall stand excused in the eye of the impartial part of the public from a charge of culpable negligence, when it is considered how much they had to do in the way of preparing papers for the press, as well as in drawing up the report, and the very short time within which it was all to be done. But with reference to your letter of the 14th May, the Board can distinctly state that they have a clear recollection of having perused it, and that it entered fully, along with the rest of the evidence and district reports, into their consideration when they were forming their judgment on the whole of the question before them.

The Board desire me to say that they will most readily insert in their supplementary report or appendix your letter of the 14th May, with an explanation of its accidental omission in the first report; but they cannot, unless they receive express directions from the Home Office or Board of Trade so to do,

Mr. Wilson to Mr. Stuart.

consent to insert such parts of your letter as merely consist of objections to their plan, or animadversions on their general mode of proceeding. All that was received from you before the report was made up may fairly be inserted, but the Board cannot agree to the introduction of new matter of the nature above mentioned—possibly after-thoughts—which, at all events, would entail on them the necessity of renewed discussions and counter statements of opinion.

In addition to the reply, contained in my letter to you of yesterday, to your former complaint relative to the returns made by the mill-owners, I am further directed by the Board to state that these returns are replies to their own queries, and that they do not recognise any right of complaint or direction on your part with regard to the time or mode of their publication. You appear to view the opinions of the mill-owners as "decisive" evidence. On the other hand, the Board regard the opinions of persons interested so deeply against change of any description by which they may be put to expense and inconvenience, as the last to be taken instead of the first, and then only taken when supported by collateral evidence.

The virtual restriction of adult labour to twelve hours daily, effected by Sir John Hobhouse's act, is avowedly favoured by some of the larger manufacturers, for the purpose of restricting the small mill-owners from working more actively than they themselves think it proper, or find it convenient, to work. It ought therefore to be entitled a bill for the protection of the larger mill-owners against the competition of the smaller ones. The Board, therefore, entertain no doubt that strong opinions may be obtained in favour of such a measure, especially as against a plan which is solely intended to benefit the children, while it is unavoidably attended with more or less of inconvenience to the masters, as well as the parents and adult work people.

I am farther directed to say that the Board have always been disposed to listen with respect to your opinion, but that they cannot attribute any weight to such a statement as that contained in your letter of the 31st ult., "that you do not attach the slightest value to the minute of a meeting of deputies from the principal seats of the cotton-manufacture." That minute, as I beg to remind you, was cited by the Board in corroboration of the *practicability* only of the plan proposed by them. Had a minority, instead of a large majority, of the deputies admitted the *practicability* of a measure inconvenient to them, such a sanction would, perhaps, have been deemed by the public conclusive as against any majority, especially at a time when there appeared a probability of the plan proposed by the Board being entirely superseded by a measure so agreeable to many powerful interests as a twelve-hour bill, or the act of Sir John Hobhouse.

With reference to your request that copies of your two last letters may be sent the secretary of state for the Home Department, the president of the Board of Trade, and the chancellor of the Exchequer, the Board beg to decline undertaking the office of transmitting those documents. If, however, you should yourself see proper to do so, and have not kept copies, I will furnish you with copies for the purpose on receiving notification of your desire to that effect.

Since writing thus far I have received and laid before the Board your letter of this day, in reply to which I am directed to state that the Board do not consider the printed query, No. 32, and Mr. Buchanan's answer, as having any relevancy to the subject of their recommendation. In the first place it is clear, from the context, that the relays there referred to apply to the same class of hands as were designated in No. 31. Moreover, Query 32 applies only to relays in cases where extra hands are required to be worked, and Mr. Buchanan qualifies his answer with an express reference to extra hours. Whereas the plan of the Board relates to ordinary hours. It is clear, also, that Mr. Buchanan, in his answer to No. 32, must have considered the question as referring to the same class of hands as No. 31, because, if he had not so considered it, he would naturally have defined the age to which he extended the term "children." Now as regards the question of practicability, even a single year is most important. And there seems no reason to suppose that if Mr. Buchanan had not in view the limitation of age set down in the factory bill, he would have omitted to give his own definition of it. At any rate it is obvious that without such definition his opinion cannot be brought to bear on the question of practicability.

As you intimate that you have no objection to a conference with the Board at any time they may appoint after Tuesday next, and as there are several points connected with their report and other proceedings on which you appear to them to labour under some degree of misconception, which may more easily be removed by verbal than written explanations, they propose to you to favour them with an interview on Wednesday next, at this office, at four o'clock in the afternoon.

LETTER X.

From Mr. Stuart to Mr. Wilson, dated Aug. 3.

Sir,—I have received your letter of this date, the contents of which do not at all agree with the terms of your communication of yesterday, in which you inform me that the Central Board "would most readily confer with me, but could not consent to enter into a written defence of every part of their proceedings;" now, however, when a conference is agreed to, and a day fixed by you for our meeting, you send me a laboured answer to my notices of omissions, and to my remarks on the evidence called for by your former communication, in which I observe, with regret, far too much the appearance of special pleading. It is amusing, indeed, to notice the conflicting views in which matters appear to you on different days. Three days have hardly elapsed since you confidently appealed to the evidence of the mill-owners assembled at Palace-yard. To-day you desire to throw the testimony of the mill-owners overboard, as the last to be taken, and only when supported by collateral evidence. To-day you send me an argumentative defence of your conduct, while yesterday you told me "that there would be no end to your labour," were you to be "drawn into a controversial correspondence by individual commissioners," even when you are charged with omitting in your report all mention of the evidence relative to your recommendation, which is unfavourable to your views. With more remarkable inconsistency, you, who now deprecate written discussion, even with the commissioners, have all along shewn yourself so sensitive to the remarks of the public press, that neither the labour nor the time thus lost to you in preparing the report, could deter you again and again from writing and conveying to the journals articles in defence of your proceedings, sometimes publicly authorized by you, and signed by the secretary, but more frequently anonymous, of the incorrectness of one of which, in point of fact, I had occasion, on the 3rd of June (see A. 1, p. 89), to write you from Glasgow.

The first part of your letter is occupied with the discussion of a question now not of the slightest importance—viz., whether you, as a secretary to the Board, or the Board, are chargeable with culpable negligence in suppressing an important passage in my letter of 14th May. I have said no more in my letters than this, "that it was the duty of the Central Board to see that the evidence sent by the district commissioners was fully and fairly laid before the House of Commons;" but I have not been so childish as to apply this

Mr. Stuart to Mr. Wilson.

remark more to my letter, although obviously containing information entitled to some consideration, than to the other omitted evidence. And, I beg to add, that I view the suppression of the answers to the queries from the mill-owners, probably 2000 or 3000 in number,* respecting the practicability and expediency of employing relays of children, to be by far the most important omission for which the Central Board has to answer. It is in vain for you to tell me that the suppressed or omitted evidence was considered fully by the Board when forming their judgment. The judgment of the Board, or their recommendation, or their opinion, was not to have the force of law, and they are most obviously responsible for not having accompanied their report to the House of Commons with the whole evidence on which it was founded. This was the more necessary, because the recommendation of the Central Board is not only unsupported by, but is in direct opposition to, the valuable part of the written evidence. For what intelligible purpose were a dozen of individuals despatched on different routes through the manufacturing districts of England, Scotland, and part of Ireland, if no deference nor attention were to be paid to their opinions deliberately formed, after inspecting the factories, and becoming acquainted with the state of the population in point of numbers, health, food, and other comforts, education, and morals? The report of the factory commissioners is no more the report of the twelve persons appointed to see things with their own eyes, and to report their observations on them, than of any twelve gentlemen whom one may by chance meet in St. Paul's church-yard. It is the report of three gentlemen residing in London, who, for aught that appears in the report, never visited a cotton-factory, nor a flax-factory, in their lives. The "impartial part of the public," to whom you now appeal, and to whom, as already mentioned, you have made so frequent appeals in the course of the proceedings, will now be able to judge what authority ought to attach to your report and to your statement, that you had considered all the evidence sent you by the district commissioners. I am aware that you have publicly given all the weight of *your own* authority to *your own* report, by declaring in a letter, published anonymously, which, however, it will be admitted proceeded from your office at Whitehall, and appeared in the *Spectator* newspaper of 13th July, that "your report contains a clear and faithful analysis of the evidence, and a perfectly intelligible statement of the opinions and recommendations of the Central Board founded on that evidence." I apprehend, however, that far more than enough has been said to convince every sensible person who peruses this correspondence, and most especially your own admissions contained in it, that "your opinions and recommendations are not founded on that evidence." I cannot allow myself again to detail the evidence of the gentlemen at Dundee, obtained by your express instructions, on the very point of your recommendation, but directly opposed to it; and to which it was unquestionably your duty, since you now declare that you remember to have perused and considered it, to refer in your report, nor to do more than to recal to your recollection this material fact, that you have omitted all notice of the mass of evidence, both in Scotland and in Lancashire, unfavourable to your views. My former statement, containing this averment, and some references to the striking parts of the evidence not alluded to in your report, as contained in my communication of 31st July, remains unanswered by you, and is unanswerable. But farther, I appeal to one special case, as shewing, in the clearest and most distinct manner, your total inattention to, and carelessness about, the suggestions or recommendations of the commissioners, as to which I have again and again written to you without obtaining a word in reply. I therefore once more ask you, to what part of your report I am to look for any notice of the evidence, respecting the

* These were totally withheld till many months after the legislature had settled the question, by passing the present law.

most dangerous employments, *wet flax-spinning and web-dressing*, to which children in factories are subjected, as communicated to you by Sir David Barry, Mr. Mackintosh, and myself, and also contained in the report of the committee on Mr. Sadler's bill. I maintain that if we had sent you no other information than that which describes and proves the noxious nature of those employments to the human constitution, and suggests the remedy, that the information was not acquired at too high a price by the expense of the commissioners sent to Scotland. Yet on this interesting subject you have hitherto, in your reports and correspondence, preserved inviolable silence. Depend on it, it will not be enough for you, when you appeal so magnanimously to the impartial public, and are called to account for such an omission as this, to say, as you do in your last letter, "Oh, we fully considered the whole evidence in forming our judgment on the whole of this question. And you have no right to cross-examine us, the members of the Central Board, respecting the suggestions or details of the travelling commissioners, or on what ground we thought fit to omit all notice of them."

Mr. Stuart to Mr. Wilson.

I suspect that I do not understand the meaning of your remark, that the "returns of the mill-owners are replies to their own queries." They are replies to the circular queries of the Central Board, as printed on pp. 87, 88, 89, and 90, of their report, of which you sent copies to the mill-owners of the United Kingdom. Very many of them, of which I have lists, were forwarded to you by me, and the statements contained in the answers, which I perused frequently, prevented me from taking any trouble in procuring further information. Mr. Buchanan's answer was incorporated in his evidence, and formed part of it. These documents form part of my report, which I conceive I was fully entitled, even without your application of 24th July, to have seen communicated to the House of Commons, with the rest of the evidence. I am therefore at a loss for words to express my surprise to be now told by you that "you do not recognise any right of complaint or direction on my part as to the time or mode of the publication of those returns," part of the very evidence, and of the most important part, which I sent to you. I must be permitted, allow me to observe, to judge for myself as to my right to complain of your omissions, and as to the propriety and mode of my doing so; but when you inform me that you recognise no right of direction as to the time or mode of publication of the returns, I must confidently refer (even if, as one of the commissioners, I was not perfectly well entitled to remonstrate with you) to your letter of the 24th of July, requesting me to have the kindness to offer you advice as to the order of printing the appendix, as utterly inconsistent with the expression—not altogether of the most courteous description, though I am far from complaining of it—which you have used with reference to the suggestion which I felt it to be my duty to make in compliance with your request. That suggestion referred to the returns sent to you by myself, and most especially to Mr. Buchanan's answer, the omission of which, as you now state it to have taken place advisedly, I consider to be of itself fatal to the accuracy and correctness of your report. Mr. Buchanan's answer, and all those received, apply to children from nine years old to twenty-one, and prove relays of children for the whole period from nine, when they begin to work, till twenty-one, to be impracticable. And why? For the best of all reasons, that the double sets of such children do not exist. Your critical remarks on this part of the discussion hardly appear to me, I must candidly confess, to require notice. Do you maintain, that although Mr. Buchanan has sworn that a change of hands in the employment of children at his works is *impracticable*, that he does not mean to include children employed who are from nine to thirteen years old? If this is your meaning, you are maintaining a proposition directly contrary, as it appears to me, to the express terms of Mr. Buchanan's oath, and to the mass of information sent in the returns.

This letter is already by far too long, and I must therefore conclude by

Mr. Stuart to Mr. Wilson.

once more reminding you that you seem to-day to set a very different value on the opinions of mill-owners sent from the country, from all parts of it, and contained in their returns to the queries hitherto suppressed, from that which you three days ago allowed to the sentiments expressed by confessedly only a majority of a few—comparatively a very few—manufacturers collected in Palace-yard. We have a maxim in the law of Scotland, to which I would solicit your attention, that one cannot approbate and reprobate the same deed. You cannot be heard to maintain that a deed should be approved and confirmed in one of its provisions, unless you waive all objection to support it in all it parts. In like manner I consider myself entitled to hold, that you cannot, in justice, found on the opinions of the mill-owners when favourable to your scheme, and repudiate them when they afford evidence of a contrary tendency. But the impracticability of your plan is proved not only by the great body of the mill-owners, but of the workers, and of the travelling and inspecting commissioners, especially in Scotland and Lancashire, who are aware that the population necessary to carry it into effect does not exist, and that half the present wages would be found to be totally inadequate to support the children to whom your scheme applies.

As you have declined to communicate copies of this correspondence in the official quarters to which I alluded in my former letter, as well as to publish it in the appendix to your reports, and I have no copying clerk, I cannot with justice be accused of acting disrespectfully anywhere, by forwarding this correspondence for insertion in one of the daily journals. It will thus, too, be more readily and generally read. To this course you, the members of the Central Board, who have so often officially resorted to the public press to set you or keep you right with the public, cannot reasonably object.

I shall take care that the whole letters shall be fairly and fully laid before the public.

I must beg you to have the goodness to fix the hour of meeting, on Wednesday, at *one* o'clock. I did not think of your proposing so late an hour as four for a conference which may last for some time; and I am engaged on that day to dine at perhaps an earlier hour than the members of the Central Board. I have already, however, warned you that I do not now see any good likely to arise from our meeting, more especially if you do not previously send me in time for perusal and consideration, the amended bill and appendix. Of that part of my letter relative to these documents you have taken no notice. If the appendix does not immediately appear, the important information to be contained in it will not be seen by the members of the legislature previous to the discussion of the amended bill on Friday next.

MEDICAL STATISTICS.

I give the following information, and which I am satisfied to leave to the consideration of the public, from Dr. Hawkins' Elements of Medical Statistics, and which I may justly call a very valuable work.

"In discussing the mortality of the manufacturing towns or districts, it is just to remark, that the small proportion is not always *real;* because a constant influx of *adults* is likely to render the number of deaths less considerable than that which would occur in a stationary population composed of all ages.

"The following Table of the Baptisms, Burials, and Marriages in England, during twenty years, has been formed by Mr. Rickman:—

NUMBER OF BAPTISMS, BURIALS, AND MARRIAGES.

YEAR.	BAPTISMS.			BURIALS.			MARRIAGES.
	Males.	Females.	Total.	Males.	Females.	Total.	
1801	120,521	116,508	237,029	101,352	103,082	204,434	67,228
1802	139,889	133,948	273,837	99,504	100,385	199,889	90,396
1803	150,220	143,888	294,108	102,459	101,269	203,728	94,379
1804	150,583	144,009	294,592	91,538	89,639	181,177	85,738
1805	149,333	142,868	292,201	91,086	90,154	181,240	79,586
1806	147,376	144,553	291,929	92,289	91,163	183,452	80,754
1807	153,787	146,507	300,294	97,996	97,855	195,851	83,923
1808	151,565	144,509	296,074	102,614	98,149	200,763	82,248
1809	152,812	147,177	299,989	97,894	93,577	191,471	83,369
1810	152,591	146,262	298,853	104,907	103,277	208,184	84,470
1811	155,671	149,186	304,857	94,971	93,572	188,543	86,389
1812	153,949	148,005	301,954	95,957	94,445	190,402	82,066
1813	160,685	153,747	314,432	93,726	92,751	186,477	83,860
1814	163,282	155,524	318,806	103,525	102,878	206,403	92,804
1815	176,233	168,698	344,931	99,442	97,966	197,408	99,944
1816	168,801	161,398	330,199	103,954	102,005	205,959	91,946
1817	169,337	162,246	331,583	101,040	98,229	199,269	88,234
1818	169,181	162,203	331,384	107,724	105,900	213,624	92,779
1819	171,107	162,154	333,261	106,749	106,815	213,564	95,571
1820	176,311	167,349	343,660	104,329	104,020	208,349	96,833

" On account of the acknowledged omissions in the registers of deaths in most of the parishes of Scotland, few just inferences can be drawn from them. In the parish of Cross-michael, in Kirkcudbright, the mortality at the close of the last century was published as only 1 in 98; a proportion which would imply the most unheard-of healthiness; but there can be little doubt that it was principally occasioned by defects in the registry of interments. From the returns of 99 parishes, which alone were given in the population abstracts of 1801, it appears that the average mortality was 1 in 56; and if the details were just, Scotland might at that period boast of the least considerable number of deaths ascertained to exist in any country. We have seen that in twenty years subsequent, England has attained to a still more favourable proportion; and I have not been able to ascertain how far Scotland has kept pace with her. The expectation of an infant's life in Scotland was, in the middle of last century, thirty-one years at birth, when calculated for the whole country; but in some parishes it was forty and forty-six.

" In the peculiar circumstances of Ireland, it would be very interesting to know the average mortality. But, unfortunately, no correct parochial registers have been kept; and the information, however much to be desired, is unattainable.

The following Table of the Annual Baptisms, Burials, and Marriages of the several Counties of England, has been formed by Mr. Rickman, on an average of the ten years from 1811 to 1821 :—

COUNTIES of	ANNUAL PROPORTIONS.			COUNTIES of	ANNUAL PROPORTIONS.		
	One Baptism to	One Burial to	One Marriage to		One Baptism to	One Burial to	One Marriage to
Bedford	36	62	131	Northampton . .	36	58	134
Berks	34	58	145	Northumberland .	38	58	145
Buckingham . . .	35	56	144	Nottingham . . .	33	58	133
Cambridge . . .	32	58	126	Oxford	35	61	153
Chester	36	55	136	Rutland	36	62	148
Cornwall	34	71	151	Salop (Shropshire) .	35	58	155
Cumberland . . .	34	58	154	Somerset	37	63	149
Derby	35	63	153	Southampton (Hants)	32	58	117
Devon	32	61	127	Stafford	32	56	128
Dorset	36	66	154	Suffolk	35	67	139
Durham	34	55	143	Surrey	40	52	148
Essex	35	59	150	Sussex	33	72	151
Gloucester . . .	37	64	119	Warwick	37	52	123
Hereford	38	63	170	Westmoreland . .	35	58	155
Hertford	34	58	179	Wilts	37	66	145
Huntingdon . . .	35	63	132	Worcester	34	56	143
Kent	31	50	130	York, East Riding .	33	57	127
Lancaster	32	55	126	Ditto, North Riding.	36	63	151
Leicester	36	59	133	Ditto, West Riding .	35	61	131
Lincoln	32	62	138				
Middlesex . . .	38	47	106	ENGLAND . . .	35	57	133
Monmouth . . .	47	70	154	WALES	41	69	156
Norfolk	33	61	136				
					35	58	134

To explain the rate of increase of the population of England, the female sex is chosen, as it affords a more accurate standard, from the circumstance of being less exposed to the influence of immigration and emigration.

FEMALES.

1801.	Increase per cent.	1811.	Increase per cent.	1821.
5,492,354	14 or 14.02	6,262,716	15 4-5ths. or 15.82	7,253,728

The following tables of the Number of Individuals living at various Ages are formed by Mr. Rickman, from the census of 1821, and relate to a supposed given proportion of 10,000 males and 10,000 females:—

ENGLAND (*collectively.*)

	Under 5 Years.	5 to 10	10 to 15	15 to 20	20 to 30	30 to 40	40 to 50	50 to 60	60 to 70	70 to 80	80 to 90	90 to 100	100 and upwards
Males .	1538	1343	1169	988	1470	1155	941.0	665.6	447.6	221.9	56.25	4.15	.12
Females	1444	1268	1056	995	1684	1210	932.6	653.3	458.0	228.2	64.85	5.75	.22

WALES (*collectively.*)

	Under 5 Years.	5 to 10	10 to 15	15 to 20	20 to 30	30 to 40	40 to 50	50 to 60	60 to 70	70 to 80	80 to 90	90 to 100	100 and upwards
Males .	1514	1407	1210	1009	1433	1109	871.4	646.3	474.8	243.6	74.09	7.54	.09
Females	1382	1281	1093	1003	1560	1163	911.6	672.6	535.5	281.4	104.76	10.95	.50

SCOTLAND (*collectively.*)

	Under 5 Years.	5 to 10	10 to 15	15 to 20	20 to 30	30 to 40	40 to 50	50 to 60	60 to 70	70 to 80	80 to 90	90 to 100	100 and upwards
Males .	1494	1357	1247	1032	1490	1095	895.4	649	458.1	216.3	58.22	6.71	.43
Females	1294	1177	1057	1048	1769	1204	937.9	711.6	502.2	225.5	65.18	7.42	.60

Ages of the Inhabitants of LONDON, *arranged under the same divisions.*

Fractions are omitted below 90.

	Under 5 Years.	5 to 10	10 to 15	15 to 20	20 to 30	30 to 40	40 to 50	50 to 60	60 to 70	70 to 80	80 to 90	90 to 100	100 and upwards
Males .	1397	1095	936	865	1718	1548	1203	730	353	128	22	1.69	.21
Females	1216	995	834	959	2062	1567	1092	690	388	156	34	3.93	.32

DIVISION

UPON

MR. P. THOMSON'S MOTION FOR THE SECOND READING

OF THE

FACTORIES ACT AMENDMENT BILL,

MONDAY, MAY 9, 1836.

Order for Second reading read; Motion made, and Question proposed, "That the Bill be now read a second time;"—Amendment proposed, to leave out the word "now," and at the end of the question to add the words "upon this day six months:"—Question proposed, "That the word 'now' stand part of the Question."—The House divided; Ayes, 178; Noes, 176.

MAJORITY.—AYES.

Adam, Sir C.
Ainsworth, P.
Astley, Sir J.
Baines, E.
Bannerman, A.
Barclay, D.
Baring, F. T.
Baring, W. B.
Barron, H. W.
Barry, G. S.
Beckett, Sir J.
Bellew, R. M.
Bentinck, Lord G.
Biddulph, R.
Blackburne, J.
Blamire, W.
Blunt, Sir C.
Bolling, W.
Bowring, Dr.
Brady, D. C.
Bridgeman, H.
Brocklehurst, J.
Buller, Sir J. Y.
Burrell, Sir C.
Burton, H.
Byng, G.
Byng, Rt. Hon. G. S.
Campbell, Sir J.
Campbell, W. F.
Cavendish, Hon. C.
Cavendish, Hon. G. H.
Chalmers, P.
Childers, J. W.
Clive, E. B.
Colborne, N. W. R.
Cole, Lord
Collier, J.
Cowper, Hon. W. F.
Crawford, W.
Crawley, S.
Curteis, H. B.
Curteis, E. B.
Dalmeny, Lord
Denison, J. E.
Divett, E.
Donkin, Sir R.
Duncombe, T.
Dundas, Hon. T.
Dundas, J. D.
Dunlop, J.
Ebrington, Lord
Egerton, W. T.
Egerton, Sir P.
Ellice, Right Hon. E.
Entwisle, J.
Evans, G.
Ewart, W.
Fazakerley, J. N.
Fielden, W.
Fergus, J.
Ferguson, R.
Fitzroy, Lord C.
Forster, C. S.
Gisborne, T.
Gordon, R.
Graham, Rt. Hn. Sir J.
Greene, T.
Grey, Sir G.
Hale, R. B.
Hastie, A.
Hawkins, J. H.
Hay, Sir A. L.
Heathcoat, J.
Heneage, E.
Heron, Sir R.
Hobhouse, Rt. Hn. Sir J.
Hodges, T. L.
Holland, E.
Horsman, E.
Houldsworth, T.
Howard, Hon. E.
Howard, P. H.
Hume, J.
Ingham, R.
Johnstone, Sir J.
Johnstone, J. J. H.
Kearsley, J. H.
Knight, H. G.
Labouchere, Rt. Hon. H
Lee, J. L.
Lees, J. F.
Lefevre, C. S.
Lemon, Sir C.
Lennox, Lord G.
Lennox, Lord A.
Loch, J.
Long, W.
Lowther, Lord
Lynch, A. H.
Mackenzie, S.
M'Leod, R.
M'Namara, Major
M'Taggart, J.
Maher, J.
Marjoribanks, S.
Marshall, W.
Maule, Hon. F.
Morpeth, Lord
Mullins, F. W.
Murray, Rt. Hon. J. A.
Nagle, Sir R.
O'Brien, C.
O'Connell, D.
O'Connell, J.
O'Connell, M. J.
O'Connell, M.
O'Conor Don
O'Ferrall, R. M.
O'Loghlen, M.
Oswald, J.
Parker, M.
Parker, J.
Parnell, Rt. Hn. Sir H.
Parrott, J.
Parry, Sir L. P. J.
Patten, W. J.
Pease, J.
eel, Rt. Hon Sir R.
Pelham, Hon. C. A.
Pendarves, E. W. W.
Philips, M.
Philips, G. R.
Pinney, W.
Potter, R.
Power, J.
Price, Sir R.
Pryme, G.
Pusey, P.
Reid, Sir J. R.
Rice, Rt. Hon. T. S.
Ridley, Sir M. W.
Roche, D.
Rolfe, Sir R. M.
Russell, Lord J.
Russell, Lord
Ruthven, E.
Ryle, J.
Sandon, Lord
Sanford, E. A.
Scott, Sir E. D.
Scott, J. W.
Scrope, G. P.
Seale, Colonel
Sheppard, T.
Smith, R. V.
Speirs, A.
Stanley, Lord
Strutt, E.
Stuart, Lord J.
Stuart, V.
Talbot, J.
Thomas, Colonel
Thomson, Rt. Hn. C. P.
Thornely, T.
Trelawney, Sir W.
Troubridge, Sir E. T.
Turner, W.
Villiers, C. P.
Vivian, J. H.
Walker, R.
Warburton, H.
Ward, H. G.
Wemyss, Captain
Westenra, Hon. H. R.
Winnington, H.
Wood, C.
Wortley, Hon. J. S.
Wrightson, W. B.
Young, J.

Tellers.

Steuart, R.
Stanley, E. J.

MINORITY.—NOES.

Aglionby, H. A.
Agnew, Sir A.
Alford, Lord
Alsager, Captain
Angerstein, J.
Ashley, Lord
Attwood, T.
Bagot, Hon. W.
Baillie, H. D.
Balfour, T.
Baring, T.
Barnard, E. G.
Bateson, Sir R.
Benett, J.
Bentinck, Lord W.
Bethell, R.
Bewes, T.
Blackburne, I.
Bonham, R. F.
Borthwick, P.
Bramston, T. W.
Brotherton, J.
Brownrigg, S.
Bruce, C. L. C.
Bruen, F.
Buller, C.
Buxton, T. F.
Canning, Rt. Hon. Sir S.
Cayley, E. S.
Chandos, Marquis of
Chaplin, Colonel
Chapman, A.
Chichester, A.
Churchill, Lord C.
Clive, Viscount
Clive, Hon. R. H.
Codrington, C. W.
Compton, H. C.
Corbett, T. G.
Crawford, W. S.
Darlington, Earl of
Dillwyn, L. W.
Dottin, A. R.
Dowdeswell, W.
Duffield, T.
Dunbar, G.
Duncombe, Hon. W.
Duncombe, Hon. A.
East, J. B.
Eastnor, Viscount
Eaton, R. J.
Egerton, Lord F.
Elley, Sir J.
Elwes, J. P.
Ferguson, G.
Fielden, J.
Finch, G.
Fleetwood, P. H.
Fleming, J.
Foley, E. T.
Forbes, W.
Fremantle, Sir T.
Freshfield, J. W.
Gaskell, D.
Gaskell, J. M.
Gladstone, T.
Gladstone, W. E.
Gore, O.
Goring, H. D.
Goulburn, Rt. Hon. H.
Grimston, Lord
Grimston, Hon. E. H.
Gully, J.
Halford, H.
Hamilton, Lord C.
Hardinge, Rt. Hn. Sir H.
Hardy, J.
Harland, W. C.
Harvey, D. W.
Hector, C. J.
Henniker, Lord
Hill, Sir R.
Hogg, J. W.
Hope, J.
Howard, R.
Hoy, J. B.
Hughes, H.
Jackson, Serjeant
Jervis, J.
Inglis, Sir R. H.
Johnstone, A.
Jones, W.
Jones, T.
Irton, S.
Kemp, T. R.
Kerrison, Sir E.
King, E. B.
Knatchbull, Sir E.
Knightley, Sir C.
Langton, W. G.
Lawson, A.
Lefroy, A.
Lincoln, Earl of
Lister, E. C.
Longfield, R.
Lowther, Hon. Col.
Lowther, J. H.
Lushington, Dr.
Lushington, C.
Mackinnon, W. A.
Mahon, Lord
Manners, Lord C. S.
Marsland, T.
Maunsell, T. P.
Maxwell, J.
Mordaunt, Sir J.
Musgrave, Sir R.
Neeld, J.
Neeld, J.
North, F.
O'Brien, W. S.
Owen, H. O.
Palmer, R.
Penruddocke, J. H.
Perceval, Colonel
Plumptre, J. P.
Plunket, Hon. R. E.
Polhill, F.
Pollen, Sir J. W.
Pollington, Lord
Poulter, J. S.
Praed, W. M.
Price, S. G.
Pringle, A.
Robinson, G. R.
Rundle, J.
Rushbrooke, Colonel
Scarlett, Hon. R.
Scholefield, J.
Scourfield, W. H.
Shaw Right Hon. F.
Sibthorp, Colonel
Sinclair, Sir G.
Smith, A.
Smyth, Sir H.
Strickland, Sir G.
Sturt, H. C.
Talfourd, Serjeant
Thompson, Alderman
Thompson, Colonel
Townley, R. G.
Trevor, Hon. A.
Trevor, Hon. G. R.
Tulk, C. A.
Twiss, H.
Vere, Sir C. B.
Verner, Colonel
Vesey, Hon. T.
Vivian, J. E.
Vyvyan, Sir R.
Walter, J.
Wason, R.
Welby, G. E.
Wilbraham, G.
Wilbraham, Hon. B.
Wilde, Serjeant
Williams, T. P.
Williams, W.
Williamson, Sir H.
Wilmot, Sir J. E.
Wilson, H.
Wodehouse, E.
Wynn, Rt. Hon. C. W.
Wyse, T.
Young, G. F.

Tellers.
Hindley, C.
Wakley, T.

Mr. Humphery paired off with Mr. Beilby Thompson against the second reading.